THE PHYSIOLOGY OF PLANTS UNDER STRESS

THE PHYSIOLOGY OF PLANTS UNDER STRESS

Abiotic Factors

Erik T. Nilsen
David M. Orcutt
Virginia Polytechnic Institute and State University

JOHN WILEY & SONS, INC.

New York • Chichester • Brisbane • Toronto • Singapore • Weinheim

This text is printed on acid-free paper.

Library of Congress Cataloging in Publication Data:

Nilsen, Erik T.
The physiology of plants under stress / Erik T. Nilsen, David M. Orcutt.
p. cm.
Includes bibliographical references (v. 1, p.).
Contents: v. 1. Environmental factors.
ISBN 0-471-03512-6 (cloth : alk. paper)
1. Plant physiology. 2. Plants, Effect of stress on. I. Orcutt, David M. II. Title.
QK711.2.N55 1996
581.2—dc20 96-5443

Printed in the United States of America

10 9 8 7 6 5 4 3 2

With Love

for Karen and Joel

—ETN

for Pam and Andrea

—DMO

CONTENTS

APPENDICES

PREFACE

The study of plant stress physiology is important to basic and applied botanical sciences. Morphological, anatomical, and metabolic responses to stress are some of the primary processes of microevolution by natural selection. Therefore, one of the major forces that shapes the structure and function of plants is environmental stress. The significance of adaptive response to environmental stress also is highlighted by the many cases of convergent evolution in plants. Similarities in form and function among phylogenetically unrelated plants are often a consequence of environmental driven coevolution.

The applied significance of research on plant response to stress is supported simply by observing the enormous impact of environmental stress on agricultural productivity. Major efforts to breed for traits that confer tolerance of drought, cold, heat, nutrient, and salinity stress occur each year throughout the world. An understanding of the mechanisms that regulate form and function, and the significance of those processes to plant physiology, ecology, and agriculture, must include knowledge of plant stress physiology. This is the reason why we have written this book and have taught plant stress physiology for the past 15 years.

The primary purpose of this book is to provide a basic understanding of plant stress physiology for use by those students and scholars in a diversity of botanical fields. We hope that the information contained in this volume provides a foundation that scholars in botanical fields can draw upon when they are interpreting the significance of various processes in their fields to plant form and function. A second purpose is to provide a starting point for the development of scholars of plant stress physiology.

A precursor of this book was prepared by two active scientists in the agricultural area (M. G. Hale and D. M. Orcutt, *Physiology of Plants Under Stress*, John Wiley & Sons, 1987). It was one of the first treatments that focused on many aspects of the physiology of plants under stress. The present book was conceived during the last 15 years, when the authors have been teaching plant stress physiology to students at Virginia Tech. During those years of teaching we have developed an understanding that it is very important to link the fields of agricultural plant physiology and physiological ecology of native species. We were not able to identify books that brought together these two fields for our plant stress physiology class. Thus arose our motivation for producing this work. This version is completely revised text with a large expansion of material compared to the first. We sat down at our desks and redesigned the entire book, from the ground up, based on our experiences in the classroom and in research. Emphasis is placed on physiological response to environmental stress and both ecological and agricultural interpretation of those responses. The significance of evolutionary processes in the development of plant stress responses has been greatly expanded.

THEMES AND PHILOSOPHIES

The major theme of this book is an evaluation of the various mechanisms used by plants to cope with environmental stress. We are aware that we have not covered every possible type of environmental stressor, nor have we covered biotic stressors. It was not possible to cover all aspects of this field in a comprehensive manner and still produce a reasonably sized volume. Consequently, we focused this work on above ground abiotic factors including water relations. We hope that in the near future we will be making available a second volume of this work that covers major aspects of soil and biotic stress factors. In each chapter we develop the basic physical, chemical, and biologically important aspects of the environmental component under consideration. Detrimental influences of the environmental factors on plant metabolism are considered, followed by an analysis of the mechanisms that have evolved in plants to cope with these detrimental affects. Throughout this volume we have emphasized both the interactive influences of evolutionary processes and climate on metabolism and the short-term responses of metabolism to environmental stress. Another general theme of this book is to bring out the multiple levels of adaptation that occur in plants in response to any environmental factor. Adaptation and acclimation occur through a combination of behavioral, morphological, anatomical, physiological, and biochemical processes.

Our philosophy on literature was to cite some of the important foundation articles in the particular topic and to present some of the current examples. It is impossible to present all possible research in each topic. There are many excellent pieces of work that we were not able to incorporate into the book. We hope that our peers understand that if their work was not cited, that does not mean that we think it is of less importance than the articles we have cited.

We have tried to link figures and tables carefully to text material. Figures and tables were selected to provide the best possible illustrations for the text. In many cases we have generated our own material, but in other cases we have reproduced figures or tables from other sources. We have cited all sources for the tables and figures, obtaining permission to use those that were reproduced, and we have also cited the sources that gave us ideas about the design of our own figures. In the rare possibility that a figure we have designed looks like one previously published and we have not cited that publication, this is a coincidental and inadvertent occurrence and we will rectify it as soon as possible.

ORGANIZATION OF THE TEXT

This text is organized into three parts. In the first part, Chapters 1–7, we present information that is important for understanding research in plant stress physiology. Growth regulation is covered since this aspect of plant metabolism is central to interpreting the effects of environmental stressors. Membranes are covered because these are the sensors of environmental change. Phytohormones are covered because these are the transducers of information from membranes to metabolism. This is the only treatment we know of that brings together all influences of environmental factors on phytohormones. Carbon balance is included with the first section because this metabolic process is the master integrator of plant response to environmental stress. The use of stable isotopes in studies of plant stress is considered because it is becoming an important tool in the field. This section of seven chapters should provide the students with an excellent foundation for interpreting metabolic and ecological responses to environmental stress.

The second part of the book, Chapters 8–12, covers plant responses to particular environmental conditions. In terms of water relations, the chapters on drought and flooding are coordinated with the foundation chapter on water relations. The chapter on influences of light intensity focuses on the immediate impact of light on metabolic processes. Influences of radiation on energy budget of plant tissues is relegated to the chapter on high-heat influences. The influences of temperature on plants is covered in a chapter on high heat, chilling, and freezing, and it is included in the chapter on multiple-factor interactions. There is no chapter devoted to the impact of nutrients on plant function. Our planned next volume will have extensive coverage of soil environments, the influence of nutrients on plant metabolism, and the ecophysiological adaptations of plants to nutrient variation (limitations and overabundance). This book has some material on nutrient limitations in the chapter on multiple-factor interactions. There is also no chapter on salinity, as this will also be covered in the next volume. Some consideration of salinity is available in the chapter on drought.

The third part of the text, Chapters 13–15, covers integrative topics. The coverage on multiple environmental impacts on plants integrates those influences of individual environmental factors. The chapter on biotechnology combines the metabolic influences of environmental factors with technological advances in molecular biology and genetic manipulation. The final chapter brings together general attributes of physiological responses to stress among all environmental factors.

IN-TEXT LEARNING AIDS

There are two main mechanisms of study assistance built into the book. The study-review outlines at the end of each chapter are meant to provide the student with the main topical information for the chapter. This outline summarizes and rephrases main concepts to reinforce learning and assist studying the material. Also, self-study questions are provided. Students should try to answer these questions on their own and then refer to the text to see if they have grasped the concepts presented in the text. Answers to these questions are available from the author at a nominal cost (E. T. Nilsen, Biology Department, Virginia Tech, Blacksburg, VA 24061) if students would like them or if instructors would like to make them available for their class. Supplementary reading suggestions are also listed at the end of each chapter to provide alternative sources for material concerned with the topic covered in the chapter. A glossary is provided at the end of the book, covering many of the specialized terms utilized in the text.

Although this material has been reviewed by many scientists and two classes of students, the real test of any textbook is how effective it is at helping students learn. We hope that this book will be useful for classes on topics such as plant physiology, plants and their microclimate, plant stress physiology, agronomy, and physiological ecology, as well as useful as a reference text for scientists in a wide variety of applied and basic disciplines in plant sciences. We welcome comments and criticisms from instructors and students who utilize the book. These comments will be valuable for revision of this material in the future. Please address your comments directly to us at either of the following addresses.

DR. ERIK T. NILSEN

Biology Department
Virginia Polytechnic Institute and State University
Blacksburg, VA 24061

DR. DAVID M. ORCUTT

Department of Plant Physiology, Pathology, and Weed Sciences
Virginia Polytechnic Institute and State University
Blacksburg, VA 24061

THE PHYSIOLOGY OF PLANTS UNDER STRESS

1 Introductory Comments

Outline

Objectives

1. Introduce the extent of stress conditions in natural and agricultural environments.
2. Discuss the motivations, both applied and theoretical for studying the physiology of plants under stress.
3. Outline some general concepts in plant stress physiology.
   a. Discuss the terminology of stress and strain.
   b. Discuss the general pattern of tolerance and avoidance.
   c. Discuss the significance of scale.
   d. Cover the concept of windows of sensitivity.
   e. Differentiate between chronic and acute stress.
   f. Define the concept of dose.
   g. Discuss general aspects of spatial and temporal variability.
   h. Discuss the fact that stresses interact in a diversity of ways.

I. INTRODUCTION

Throughout the world, plant species inhabit a wide array of environments with diverse combinations of abiotic conditions and biotic interactions. Plants grow where air temperatures dip to −35°C in Antarctica or soar to over 55°C at Death Valley in the United States, and they grow where irradiance may be near the maximum, or close to total darkness. Evolutionary processes have sculpted plants with elaborate combinations of morphological, anatomical, physiological, and behavioral traits that maintain populations successfully in diverse environmental conditions. However, biophysical and biochemical limitations on plant structure and function constrain the number of traits and the possible suits of traits that an individual plant can have in a particular habitat. Furthermore, there are trade-offs between specific adaptive syndromes (particularly compatible trait combinations) and growth rate, competitive capacity, defense against herbivores, or reproductive potential. Thus, long-term processes (on an evolutionary scale) have brought together particularly favorable combinations of traits due to selective pressure by environmental conditions (both abiotic and biotic). The number and nature of these trait combinations are constrained by physiochemical limitations of life and trade-offs between adaptation to climate and adaptation to biotic interactions. The short-term performance of species in response to environmental change is reflective of the longer-term evolutionary heritage of the species.

The physiology of plants under stress is an important subject area for many disciplines, including physiological ecology and plant physiology. A physiological ecologist would consider the combination of physiological responses of a plant population to a variety of physical or chemical factors as the *fundamental niche* (Figure 1.1). These multidimensional surfaces, which plot physiological responses to environmental characteristics, represent the physiological framework defined by genetic attributes of individuals in the population. In nature, or in agricultural systems, no species or population can realize its fundamental capabilities. This is because biotic factors of the environment, such as competitors, herbivores, and parasites, limit species acquisition of resources. Furthermore, not all combinations of resource availability are present in the environment, restricting the population from inhabiting all possible resource availability combinations (Figure 1.1). All species have *realized niches* which are smaller than their fundamental capabilities.

A plant physiologist would consider that plants have physiological pathways or processes that are dependent on the nature of the compounds under consideration, their molecular regulation, and the physiochemical constraints on their performance. Each species must have the same basic physiological processes to maintain a functional metabolism, but the specific nature of the compounds involved and their regulation can be different among species within the bounds of physical and chemical constraints. Therefore, a plant physiologist studying environmental physiology evaluates the structure and function of active enzymes and proteins, their molecular regulation, and their responsiveness to environmental factors.

All plant species are under environmental conditions that limit their performance (Osmond et al. 1987). In natural systems, this has resulted in adaptation, acclimation, and speciation. In agricultural systems, the domestication process has promoted specific trait combinations that maximize yield under optimum growth conditions. Combinations of traits that may increase success in natural environments with limiting resources may not result in high yield because of trade-offs between tolerating low resources and investment in reproduction (crop yield). Accordingly, combinations of traits that maximize yield in agricultural fields may not be appropriate for coping with locations or time periods with low resource availability. This relationship argues for the cooperation of plant physiologist and physiological ecologist. Physiological ecologists determine appropriate trait

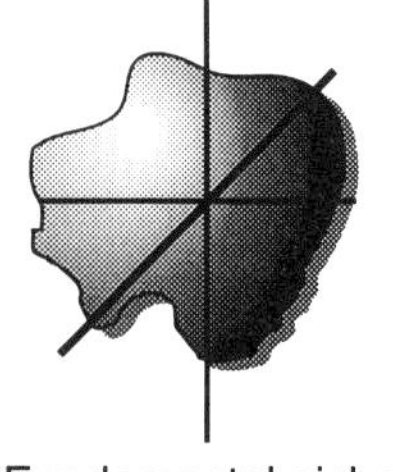

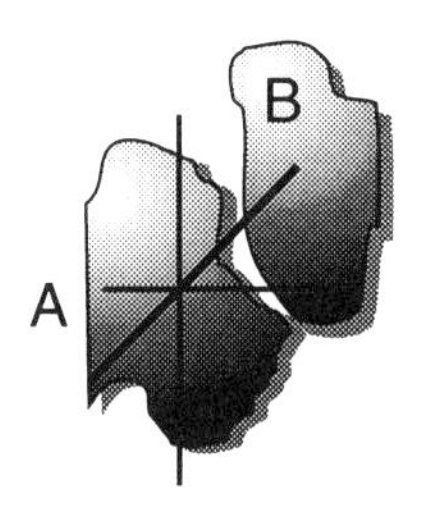

Figure 1.1 Diagrammatic representation of the fundamental and realized niches on the basis of three resource axes. The fundamental niche space includes all those resource combinations the species can utilize. The realized niche occupies a smaller area than the fundamental niche because of resource limitation in the habitat (space A) and the presence of other species (space B).

combinations that enable plants to cope with a diversity of environmental conditions by studying natural populations. Environmental crop physiologists evaluate the consequences of resource limitation or resource overabundance on crop physiology, and discover the mechanisms behind the inhibition By combining the efforts of both types of scientists, the selective pressures caused by environmental conditions will be elucidated, potential trait combinations that can cope with the environmental situation can be identified, and physiological mechanisms of those trait combinations can be studied. In a similar manner, this book addresses the subject of plant stress physiology by combining the expertise of a physiological ecologist with that of a plant physiologist to cover the impacts of environmental stress on plant physiology and the mechanisms that have evolved in plants to cope with those difficult environmental conditions. This combination of topics is also particularly germane to the use of genetic engineering for crop improvement in regions with stressful environments.

II. MOTIVATIONS FOR STUDYING PLANT STRESS PHYSIOLOGY

There are several important motivations for studying the physiology of plants under stress. In the most basic sense, the study of plant stress physiology will reveal the mechanisms by which the physiology of living organisms can be adjusted to cope with extreme environmental conditions. Once these mechanisms are understood, the information can be applied toward understanding the processes that regulate the distribution of species across the landscape. The knowledge of physiological mechanisms that plants use to cope with environmental stress is also critical for developing mechanistic models of plant function. Such models can have predictive power to forecast the potential for species migrations or extinction during periods of climate change. In a more applied sense, the process of discovering stress-induced physiological changes can result in new tools for evaluating other physiological processes. For example, the enzyme used in polymerase chain reactions (a technique used to multiply DNA by cycling between high- and low-temperature conditions) was discovered by determining the mechanism by which a bacterium could tolerate high-heat conditions.

In agriculture it is increasingly important for scientists to improve environmental stress tolerance in crops. In general, crops are limited to about 25% of their potential yield by the impacts of environmental stress (Boyer 1982). Also, as towns and cities grow, they cover some of the best arable land. Urban sprawl and poor management of agricultural land have slowly degraded the quality (environmental conditions) of land used in crop production in developed countries, and in underdeveloped countries arable land is already of low quality. To improve the ability of crops to cope with environmental stress, understanding the mechanisms by which natural plants respond to the same stresses is critical. Breeding programs cannot be designed in an educated manner unless the attributes required for coping with the environmental stress in question are known. Also, to evaluate the severity of environmental stress in a region there need to be bioassay tests. On a larger scale, linking physiological stress indices with remote sensing techniques is an important tool for evaluating landscape-level consequences of stress. Understanding the impact of environmental stress on the physiology of designated species is necessary to develop plant bioassay indices of stressful environments.

The ability to engineer the genetics of crops in order to enhance the crop's ability to cope with environmental stress has important applied and economic implications. Scientists can discover new genes that enhance stress tolerance by studying the physiology of plants under stress in a diverse array of environmental conditions and phylogenic lineages. Once stress-tolerance mechanisms are known, locating genes in native species that provide specific metabolic mechanisms for increased yield under stressful conditions is an important first step to genetic engineering. As the amount of low-quality land under cultivation increases, there is an increasing motivation for engineering stress-tolerance traits into crops.

There is a plethora of evidence that major changes in planetary climatic conditions are likely to occur in the near future. Changes such as increased atmospheric CO_2, increased temperature, changes in precipitation patterns, and changes in atmospheric pollutants are all likely to occur. Only by understanding the physiological, and evolutionary responses of plants to these changing climatic conditions can we hope to predict the consequences of global climate change to natural systems or agricultural systems.

Land-use practices around the world have often resulted in a degraded ecosystem that will not return rapidly to its original state. For this reason, research on restoring ecosystems is a common theme in ecology today. Commonly, the disturbed ecosystem has many stressors that impact plants trying to reestablish a foothold on the site. Restoration can be assisted by judicious incorporation of species that can tolerate the stresses of these disturbed habitats. Therefore, knowledge of stress adaptation in plants will facilitate a more effective protocol for restoring damaged ecosystems.

III. GENERAL CONCEPTS IN PLANT STRESS PHYSIOLOGY

The definition of physiological *stress* can be simply stated to be a set of conditions that cause an aberrant change in physiological processes, resulting eventually in injury. Under some conditions this definition can be difficult to use because physiological processes change in response to environmental characteristics without necessarily having a detrimental effect on plant performance. For example, if plants are placed in the sun and water is withheld, leaves may wilt. Wilting could be detrimental to the plant in the short term because carbon gain is inhibited. But wilting may be critical for survival in the long run because leaf temperature may be kept low enough to avoid permanent damage (less light energy is absorbed by a wilted leaf). Therefore, wilting is both a detrimental and an advantageous response to low water availability and high light. This gray distinction between adaptive mechanism and detrimental affect on physiology prompted the consideration of the additional term *strain* (Levitt 1980).

Differentiation between stress and strain comes from applying the engineering meaning of these terms to biological systems. The biophysical definition of stress is that it is the applied force divided by the area of the force (Niklas 1994), or pressure. The term *stress* in a plant physiological sense is therefore reflective of the amount of environmental pressure for change that is placed on an organism's physiology. Strain is defined as the proportional change in a substance as a consequence of stress (Niklas 1994). Strain can be characterized as physiological change that occurs in response to environmental stress that does not necessarily result in significant reductions of growth or reproduction (Levitt 1982).

Other authors define stress as change in physiology that occurs when species are exposed to extraordinarily unfavorable conditions that need not represent a threat to life but will induce an *alarm response* (Larcher 1980). Alarm responses are defensive or adaptive responses to the stimulus. In this book we take a broader view, in part because the distinction of stress and strain has not become a underpinning of research in stress physiology, and in order to cover the diversity of scales involved in plant stress physiology. A *stressor* is considered here to be an aspect of the environment that can induce an alteration in plant physiology. Stress occurs when the stressor induces enough physiological change to result in reduced growth, reduced yield, physiological acclimation, species adaptation, or a combination of these.

There are a diversity of types of stressors that may affect plant physiological processes. In general, stressors can be classified into three groups (Table 1.1). Physical attributes of the environment can affect physiological patterns in a positive or a negative manner. For example, temperature increases enzyme function within a certain range, above which strong inhibition results. Many chemical attributes of the environment can act as stressors if they are not normally in the environment (e.g., pollution, pesticides) or at relatively high concentration (salinity, pH). Biotic stressors are those concerned with the mechanism of interaction between populations, some of which (disease, herbivores) are of particular interest to agricultural systems. However, all plants will be exposed to a number of these potential stressors at one time, and interactions between them are highly likely. In this book we are concerned with physical and chemical stressors that are not a result of human intervention.

Physiological response to stressors can be divided into two possibilities. In one case, *tolerance,* plants have mechanisms that maintain high metabolic activity (similar to that in the absence of stress) under mild stress and reduced activity under severe stress (Osmond et al. 1987). In contrast, mechanisms of *avoidance* involve a reduction of metabolic activity, resulting in a dormant state, upon exposure to extreme stress. In agricultural systems, breeding and other domestication technologies have maximized the stress-avoidance mechanisms in plants. In natural systems, the diversity of potential habitats has resulted in the development of both tolerance and avoidance mechanisms. Commonly, a plant species may have several tolerance or avoidance mechanisms or a combination of both. Species that have evolved in the same habitat conditions may not have the same set of mechanisms to cope with the local combination of stressors even if they have a similar outward appearance (Nilsen et al. 1984). In words that Tom Sawyer might have spoken, there are many different ways to paint a fence.

Another important concept in plant stress physiology is scale, both spatial and temporal (Figure 1.2). Over the evolutionary history of plants, their structure and physiology, within phylogenetic lines, have changed in response to a multitude of environmental at-

TABLE 1.1 Partial List of the Sources of Environmental Stress in Plants

Physical	Chemical	Biotic
Drought	Air pollution	Competition
Temperature	Heavy metals	Allelopathy
Radiation	Pesticides	Herbivory
Flooding	Toxins	Diseases
Wind	Soil pH	Pathogenic fungi
Magnetic field	Salinity	Viruses

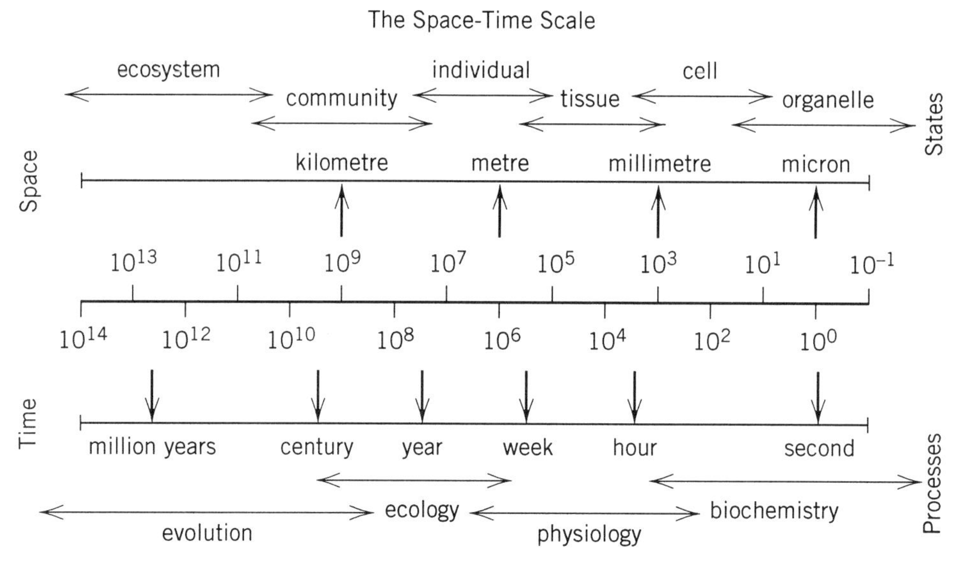

Figure 1.2 Representation of the compilation of types of research done by scientists evaluating the physiology of stress in plants. Mechanistic understanding of plant stress response at the molecular level is dependent on evolutionary processes that sculpt the basic nature of a species and its interaction with other species. (From Osmond et al. 1980).

tributes, including the diversity of biotic and abiotic stressors. The legacy of those evolutionary influences shape the basic nature of current-day responses. In studying plant physiological response to stressors, the evolutionary time scale and the significance of phylogeny should not be ignored. Many responses to stressors are also manifested at the population level in natural systems. Adaptation, or the change in genotype frequency of a population in response to a stressor, involves processes of population genetics. Clearly, agriculture has utilized selective breeding to speed up and influence the direction of adaptation in crops. Understanding the variation in response to a stressor in a population is critical to understanding the potential for population success in a changing environment. Plant stress physiology also includes processes that occur at the level of the individual plant. For example, changes in apportionment of resources to specific organs within the individual plant can be highly critical to coping with environmental stress. Of course, studies at the scale of cells over seconds of time (physiological, biochemical) are more familiar scales for the student of plant stress physiology. However, to understand the reasons that specific physiological or biochemical changes occur in response to a stressor, one cannot avoid considering the implications of evolutionary, phylogenetic, and population-level processes.

The influence of a particular stressor on the physiological processes of a particular plant species is not always equal. At different times during its life cycle a plant will be differentially sensitive to a particular stressor. A young seedling emerging near the soil surface may be very sensitive to the quantity of water in the upper soil layers. However, following growth to maturity, any sensitivity to surface soil moisture may be completely absent. Water stress that occurs late in the development of a crop will have little to no impact on yield, but the same level of water stress that occurs early in crop development may reduce yield severely. Therefore, there are *windows of sensitivity* or specific developmental stages in a plant's life cycle when plants are particularly sensitive to the influences of stressors. Sensitivity may also vary with seasonal patterns of physiology. Leaves on evergreen species of temperate environments may have a very low freezing point in the winter, but those same leaves may have a high freezing point in the summer and be very sensitive to frost damage. Thus the developmental status of plants, the age of the organs in question, and the impact of seasonal environmental patterns all influence the sensitivity of plant physiology to stressors.

The intensity of stress (pressure to change exerted by a stressor) is not easily quantified. Stress could occur at a low level, creating conditions that are marginally nonoptimal, with little effect expected. However, if this mild stress continues for a long time, becoming chronic stress, the physiology of plants is likely to be altered. In contrast, conditions could become difficult quickly, resulting in an acute condition. This shock pattern of stress is likely to induce significant changes in a short time frame.

Toxicologists, particularly those in the area of pollution studies, have developed the concept of *dose,* which has important applications to stress physiology. A dose is defined to be the magnitude of perturbation times the length of time the stress is applied. Stress can be dramatic when it is applied for a short duration and high intensity, or when it is applied for a long duration at low intensity. Plant responses to chronic stress and acute stress may be very different even though the dose is the same.

Plants experience stress in a spatially and temporally dynamic pattern. In some cases, particularly in controlled experiments, constant stress is applied. The stress is then defined by the specific level of a resource or a toxin. However, over time the physiological characteristics of the plant change. Stress decreases (even though the absolute concentration of resource or toxin has remained the same) because the plant's physiology has acclimated to a new level of performance. The ability to acclimate physiological performance is a measure of phenotypic plasticity and varies among species. Short-term acute stresses can induce an alarm or shock response. These are often short-term responses and do not reflect phenotypic plasticity but rather, defensive mechanisms.

The intensity of stress in any natural or agricultural environment generally varies both spatially and temporally. At any one time, each leaf on a plant will be receiving different intensities of light, depending on whether it is located at the top of, or inside the canopy. Specific roots in the soil may be flooded while other roots are dry. Because of the size of plants and their modular nature, there will be potential variation in stress among different parts of the plant at the same time. One of the most difficult challenges for a plant stress physiologist is to integrate all of these spatially explicit varied levels of stress. Furthermore, stress factors are not constant in time. Temperature varies on a daily and seasonal cycle. The frequency of freeze–thaw events in the winter will vary from year to year. The specific locations of sunflecks in a subcanopy environment varies dramatically over the daily cycle. The nutrients in soils are extremely patchy from one capillary pore to the next. Spatial and temporal heterogeneity of resources and stressors strongly affect the patterns of physiological response.

No plant resides in an environment where only one resource is significant to its development and survival. A multitude of abiotic and biotic factors vary spatially and temporally in a plant's milieu. Some may have interactive effects on plant physiology. Synergistic effects may be particularly deleterious. For example, increased temperature may be exacerbated by low water availability because plants cannot transpire a lot of water to cool their surfaces. High temperature can be exacerbated by high irradiance because plant surfaces absorb radiation, which increases temperature further. Antagonisms among stressors may also have an effect. For example, periods of low water availability can protect plants against the deleterious effects of freezing temperature. Plant adaptation to low nutrient availability can produce leaf structure and function that protects against water stress. In addition to synergistic and antagonistic interaction, stressors could operate entirely independently.

In plant stress physiology an important distinction must be made between ultimate (adaptation) and proximal (acclimation) plant responses. Adaptation occurs by various mechanisms at the genetic level in populations over many generations. Microevolutionary processes change gene frequencies of a population over time. In a stressful environment, it is logical to assume that specific genotypes with appropriate gene combinations (those that confer the ability to survive and reproduce) are dominant in the population. Those particularly favorable gene combinations in plants that inhabit stressful environments are called *adaptations*.

Populations that have adapted through evolutionary processes acting at the genetic level to a particular climatic regime are by no way static systems. In contrast, plants have an incredible ability to adjust physiological and structural attributes on the scale of seconds or seasons within a single genotype (acclimation). On a long-term scale, acclimation is enhanced in plants because of the modular nature of growth. Plant parts can be abscised and regrown in a new morphology or anatomy, specific organs can be enhanced by increasing their numbers or size, and biomass can be preferentially accumulated in organs with specific attributes as climatic conditions change. On a short-term basis (i.e., seconds or minutes) protein populations can ebb and wane, growth regulators can be released or activated, or transcription and translation can be regulated up or down.

Acclimation is a phenotypic response to different combinations of environmental characters. Phenotypic plasticity is an index of the amount of acclimation that is possible in a particular character suite, such as photosynthesis, to changing climates within one genotype. There is no reason to believe that there is a trade-off between phenotypic plasticity and genetic heterogeneity. Rather, phenotypic plasticity seems to be genetically regulated. This means that some genotypes, or specific genetically regulated characters, have greater phenotypic plasticity than others.

There are countless examples of morphological changes that occur in plants when they are moved to, or grown in, different climatic regimes. In fact, it is necessary to grow plants that have different physiological or morphological attributes (potentially induced by environmental differences) in a common garden under one climatic regime to determine if the character difference is genetically regulated (Figure 1.3). Another experiment that is used to determine the ability of a character to acclimate is the reciprocal transplant. In this case, individuals that have different morphology or physiology are reciprocally transplanted into the other individual's habitat (and back into their own). If the morphological or physiological character of each individual reverts to that of the other when grown in the other's climate, acclimation is indicated. The potential for phenotypic plasticity can be evaluated with a reaction norm experiment (Figure 1.3). In this experiment one genotype is grown in a number of different environmental conditions. A comparison

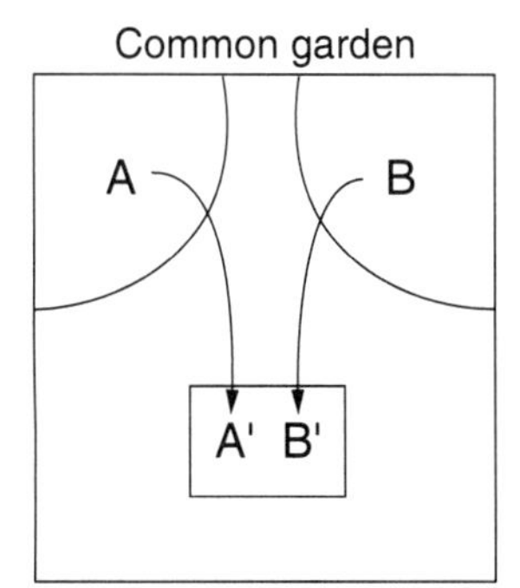

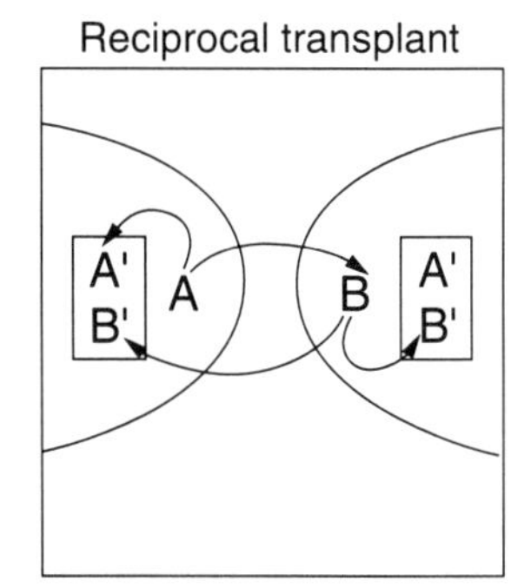

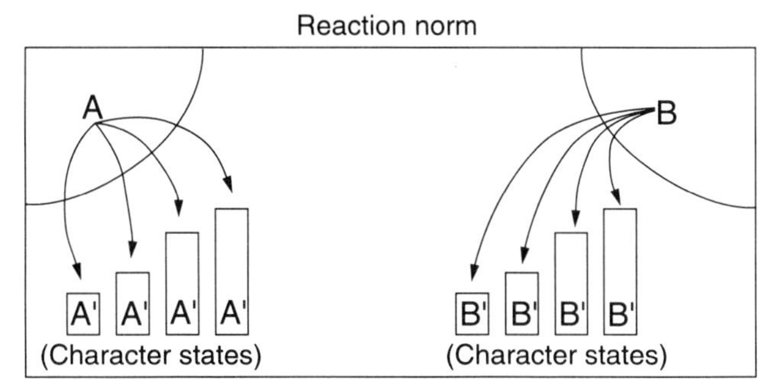

Figure 1.3 Experimental procedures used to determine the importance of genotype or phenotype in regulating morphological or physiological characteristics of plants. Common garden experiments are those in which species are taken from their respective native ranges and planted in one uniform growth condition. Reciprocal transplant experiments are those in which species are transplanted into each other's habitat. An important control in this experiment is to transplant each species into its own habitat. Reaction norm experiments are those in which one genotype is planted into a diversity of different habitats (character states) in order to determine phenotypic plasticity.

of the morphological or physiological characteristics among plants grown in these different climatic regimes indicates the magnitude of phenotypic plasticity.

IV. SUMMARY

The study of plant stress physiology is demanding and exciting. The multitude of different stressors, their spatial and temporal patterns, their variation in intensity and dose, and their potential interactions yield an abundance of scientific questions. Since plants in both agricultural and natural systems experience this complex of interacting stressors, the significance of discovering mechanisms of adaptation is great. Results from research on the physiology of plants under stress have important ramifications for other fields, such as agriculture, theoretical biology, toxicology, ecology, and developmental biology. In this book we hope to pass on this interest in the field of stress physiology in plants, and help students and scholars gain an appreciation for the significance of plant stress physiology in many other disciplines in plant sciences.

STUDY-REVIEW OUTLINE

Importance of Stress Physiology to Plants

1. In natural systems all plants experience stressful conditions.
2. Plant species do not attain their fundamental capabilities. Instead, they realize a portion of that fundamental capacity because of interference from other species, limited resources, or other stressors in their habitat.
3. In agricultural systems crops are limited to approximately 25% of their potential, due to environmental stress.
4. Environmental stress has helped to shape the physiology, anatomy, and behavior of plant species by the processes of adaptation, acclimation, and speciation.
5. Any study of the immediate responses of plants to an environmental stressor must also consider the evolutionary history of the species.
6. Responses to stress occur as suites of traits, the nature of which is constrained by physical and chemical limitations to life.

Motivations for Studying Plant Stress Physiology

1. The basic mechanisms by which plants cope with environmental perturbation will be elucidated through research on plant stress physiology.
2. Information from studies of plant response to stress at the physiological level form the basis for understanding a whole plant-integrated view of plant function.
3. Plant response to environmental parameters is used to develop predictive models that can forecast the consequences of climate change on plant performance.
4. The discovery of stress-induced physiological processes can lead to new technology in agriculture and genetic engineering.
5. If genes that regulate tolerance of stress conditions are discovered in wild plants, they can be transferred to crop species to enhance crop productivity in regions with poor agricultural conditions.
6. Breeding programs used to select for stress-tolerant genotypes cannot be effective without a basic understanding of physiological mechanisms of stress tolerance.
7. Understanding plant response to stress can lead to the development of bioassay technology for toxicology testing.

General Concepts in Plant Stress Physiology

Terminology

1. The term *stress* is difficult to define. Stress occurs when environmental conditions are such that physiological change is induced that may result in reduced growth, reduced yield, physiological acclimation, or species adaptation.
2. The term *strain* has been used in the past to represent the physiological change that occurs in plants in response to environmental stress.
3. A stressor is a characteristic of the environment (or combination of characteristics) that causes stress in plants.

Types of Stressors

1. Stressors can be divided into three basic classes: physical, chemical, and biotic.
2. Tolerance occurs when a plant is under stress, yet is able to maintain high metabolic activity, which is similar to the optimum natural conditions.
3. Avoidance occurs when plants have a severe reduction in metabolism during stress and go into a dormant state.
4. Tolerance and avoidance mechanisms are not mutually exclusive, so that both can occur in the same plant at the same time.

Importance of Scale and Sensitivity Issues

Scale

1. The study of plant stress physiology must include a wide range of temporal and spatial scales.
2. Information about a plant's evolutionary background is needed to understand why a particular stress tolerance or avoidance mechanism exists.
3. Adaptation at the population level, or acclimation at the level of the individual plant, may also depend on processes at the molecular level.

Sensitivity

1. The first time that a plant experiences a certain level of a stressor the greatest response occurs. Subsequent impacts by the same stressor at the same intensity will not have as large an impact on physiological processes. This is one view of acclimation.
2. At specific developmental stages, plants are either more or less sensitive to particular stressors. The sensitive stages of development are called windows of sensitivity.
3. The same intensity of a stressor can cause differential response among organs on the same plant.
4. Plant species can be sensitive to a certain level of a stressor during one season and insensitive to the same stress level during another season.

Intensity

1. Chronic stress occurs when the pressure placed by a stressor on a physiological process is mild but is maintained for an extended period of time.
2. An acute stress occurs when intense pressure is placed on a physiological process by sudden exposure to extreme conditions.
3. Plants respond in a different way to acute or chronic stress.
4. The acute and chronic stress definitions constitute the endpoints for a continuum of different intensity and duration of stress.
5. Dose, defined by toxicologists as the magnitude (concentration of toxin) times the time of exposure, accounts for the influence of both intensity and duration on physiological performance.
6. Stresses are not homogeneous in the environment. There is distinct spatial and temporal heterogeneity in resource availability or toxin abundance in the environment.

Interaction

1. All plant species are exposed to several stressors at the same time.
2. Stresses can interact with each other in a antagonistic, synergistic, or neutral manner.
3. One antagonistic interaction occurs when water-stressed plants gain protection against freezing stress.
4. One synergistic interaction occurs when high light intensity heats leaves and water stress interrupts transpiration, causing the leaf to heat further.

SELF-STUDY QUESTIONS

1. Describe the potentially positive interaction that can occur between the research of a physiological ecologist and that of a crop physiologist interested in stress.
2. What factors regulate a plant's sensitivity to a stress?
3. Differentiate between the terms *stress, strain,* and *stressor.*

4. What are some positive and negative interactions between stresses in relation to plant physiology?
5. Describe the major factors that can regulate the intensity of stress.

SUPPLEMENTARY READING

Dirzo, R., and J. Sarukhan. 1984. Perspectives on Plant Population Ecology. Sinauer Associates, Sudbury, MA.

Hall, D. O., J. M. O. Scurlock, H. R. Bohlhar-Nordenkampf, R. C. Leegood, and S. P. Long. 1993. Photosynthesis and Productivity in a Changing Environment: A Field and Laboratory Manual. Chapman & Hall, New York.

Levitt, J. 1980. Responses of Plants to Environmental Stress. Academic Press, New York.

Osmond, C. B., O. Bjorkman, and D. J. Anderson. 1980. Physiological Processes in Plant Ecology: Toward a Synthesis with *Atriplex*. Ecological Studies, Vol. 36. Springer-Verlag, Berlin.

Pianka, E. R. 1994. Evolutionary Ecology. HarperCollins. New York.

2 The Physiological Basis of Growth

Outline

OBJECTIVES

1. Describe the molecular mechanisms that regulate cell division with particular emphasis on start (G1–S transition) and G2–M transition.
2. Explain how plant growth regulators affect cell division.
3. Describe how plant cells expand, including the physiological factors that regulate cell expansion.
4. Describe the important attributes of plant growth as they pertain to plant stress physiology and in comparison to other organisms.

5. Describe the importance of differential organ growth to whole plant growth and survival in different environments.
6. Explain the significance of organ turnover rate in association with allocation patterns.
7. Describe various patterns of allocation and the ecological processes that are related to those patterns.
8. Differentiate between determinate and indeterminate growth.
9. Consider the physical limitations placed on the architecture of herbaceous and woody plants.
10. Identify and explain the main factors that regulate whole plant growth.
11. Discuss the various techniques that are used to measure plant growth, including direct measurement, indirect measurement, and relative indices.
12. Include a discussion of root growth measurement and the difficulties inherent in measuring the growth of the rhizosphere.

I. INTRODUCTION

Our discussion of the physiology of plants under stress begins with a consideration of the underlying mechanisms that control growth in plants, because growth is one of the best indices for evaluating plant responses to environmental or biotic stress. Scientists from various disciplines have different views about which aspects of growth are most important. Agriculturists are interested in the production of economically important organs (fruit, flower, leaf, etc.), forest scientists are interested in the production of wood products, ecophysiologists may be interested in whole plant allocation of growth, while population biologists are interested in reproductive success. Also, the growth of plants can be viewed on a number of time and spatial scales by different scientists. A physiologist frequently considers the growth of cells and tissues over seconds to hours, an agronomist is usually interested in the growth of organs over days and months, and an ecophysiologist is often interested in the growth of whole plants over seasons and years. Although scientists in all these disciplines are interested in different temporal or spatial scales of growth, all growth in plants will ultimately depend on cell growth. In this chapter we consider the underlying control of growth in cells, its impact on organ and whole plant growth, and the use of growth analysis in studies of plant stress physiology. The goal of this chapter is to develop an understanding of mechanisms that control growth from the cellular level up to the whole plant, in order to integrate the many spatial and temporal scales used to evaluate plant growth responses to stress in various disciplines.

II. COMPONENTS OF CELL GROWTH IN PLANTS

Cell growth in plants can be divided into two main components: cell division and cell expansion. Due to the presence of the cell wall, many of the properties of plant cell expansion and division are different from that of animal cells. Nevertheless, there are universal characteristics of the cell cycle in all cells. In particular, the biochemical mechanisms that control cell division and cell expansion seem to be relatively universal among organisms.

A. Cell Division

1. Regulation of Cell Division. Before the factors that regulate the cell cycle are discussed, we should review some basic aspects of this cycle. Within plants there are specific regions (meristems) which contain cells that are actively going through division. Meristems account for only a small fraction of tissues in plants, but they are strategically located to regulate growth in length and girth. Immediately following their inception by cytokinesis (the separation of two cells following division), daughter cells begin to grow. During this first growth phase (G1) the daughter cells accumulate cytoplasmic mass. Late in this phase the cell develops in one of four different ways; it can divide, arrest, differentiate, or senesce. In terms of cell division, this point is often called "start" in yeast, and R (the restriction point) in animal systems. The second phase (S) of the cell cycle is induced after the daughter cells attain a specified mass. During the S phase, genetic material is duplicated in preparation for cell division. During the third phase (G2) of the cell cycle, the cytoplasm is prepared for cell division. This phase involves a large turnover of protein populations and dissolution of the cell cytoskeleton. The mechanical aspects of cell division (M) follow the G2 phase and are commonly referred to as mitosis. Mitosis is covered extensively in all basic biology texts and will not be covered here. Rather, we concentrate on the physiological and molecular regulation of the two most important transitions between major phases of the cell cycle.

START (or R) is an important regulation component of the cell cycle. This is the point at which the cell's fate is decided. Will the cell differentiate into a specific functional type, will it become arrested in growth and development, or will cell division be initiated? There are several important conditions that cells must assess before entering into cell division. Has the last stage of mitosis been completed? Are ambient conditions appropriate for cell division (i.e., are resources available)? Has the cell enlarged enough to undergo division? Cells require molecular sensors or regulators to shepherd cells through the various stages of the cell cycle. It is these regulatory molecules that determine successful transitions among phases of the cell cycle.

The cell cycle has several control points, but two locations stand out as the most universal among cell systems: (1) the transition from the first growth phase (G1) and the DNA replication phase (S) of the cell cycle, and (2) the transition between final growth and preparation for the division phase (G2) and the mitosis phase (M) (Figure 2.1). We discuss the G2–M phase transition first because molecular regulation of this transition has the longest history of documentation. Studies on a physiological level of fission yeast, *Xenopus* eggs, and animal cell culture systems have revealed the possibility that two different mechanisms for regulation of the G2–M transition exist between different cell systems. The first possibility states that there is an activator molecule that is produced in proportion to the cytoplasmic total protein content. The activator molecule regulates cell division in growing cell systems. When the cytoplasmic protein content grows to a certain ratio with the DNA content, the mitotic process is initiated (Figure 2.2). Scientists supporting this physiological model theorize that the activator molecule may bind to sites in the nucleus. When the nuclear sites become saturated, the concentration of free activation molecule increases and the G2–M transition is stimulated.

The second physiological hypothesis concerns nongrowing cell systems, in which cell division is not dependent on the ratio of cytoplasmic protein to DNA content. This regulation mechanism requires an autonomous cytoplasmic clock that initiates the G2–M transition. This theory suggests that the autonomous clock is based on the half-life of a mito-

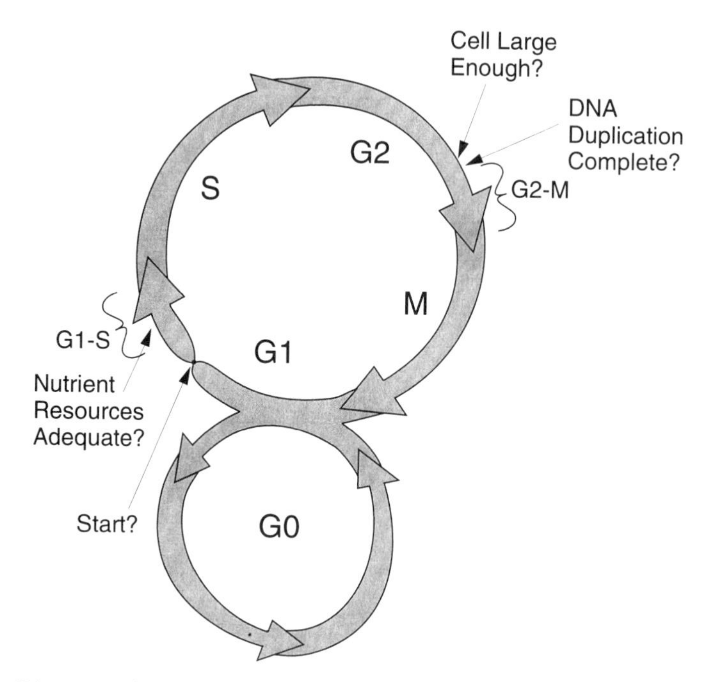

Figure 2.1 Diagrammatic representation of the cell cycle, including important parts of interphase (G1, S, G2), transitions (G1–S, G2–M), and limiting factors. If a cell does not pass through the start transition, it is said to idle in the G0 phase.

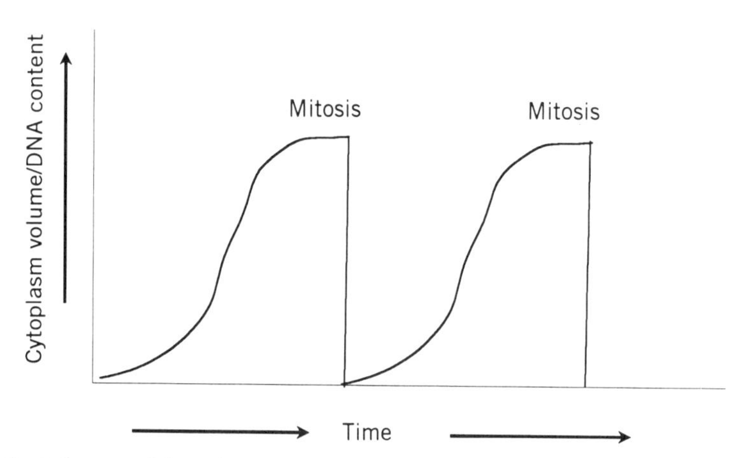

Figure 2.2 Influence of the ratio of cytoplasm protein concentration to nuclear DNA content on the initiation of mitosis. Mitosis is initiated in dividing cells when the cytoplasm volume to nuclear DNA ratio reaches a critically high level. That critical level is uniform for the cell over repetitive divisions.

sis regulation factor. As the half-life of this factor decreases (due to an increase in turnover rate), mitosis is initiated.

Scientists who use molecular biology techniques to study the cell division cycle of both cell system types have suggested that there is actually only one regulation mechanism at the G2–M transition (Jacobs 1995). The regulation system is based on many genes and gene products associated with kinase activity, but three genes have pivotal control functions. One, cdc2 (identified in fission yeast), produces a dosage-independent mitosis activator protein (P34 kinase). In many recent reviews the protein product of this gene is called CDC2 or CDK1 rather that p34 (King et al. 1994), although the latter is generally used. There is a fluctuation in p34 transcription among cell phases in animal systems, but the level of p34 in relation to total protein is constant throughout the cell cycle in all systems studied. Even though p34 abundance remains constant, there is a spike in p34 kinase activity during G2–M phase transition. It is now known that the G2–M transition is regulated by a complex of p34 and B cyclins called MPF (M-phase promoting factor). P34 by itself has little or no kinase activity. However, once associated with B cyclins, the activity, substrate specificity, timing, and cellular location of activity are determined. Since the activity of p34 is dependent on cyclin, this protein is classified as a cyclin-dependent kinase (CDK1).

A second gene, wee1, codes for a dosage-dependent mitosis inhibitor protein ($p107^{wee1}$, a protein kinase). The protein kinase coded for by wee1 inhibits the G2–M transition by phosphorylating tyrosine 15 (^{15}Tyr) of the p34–B cyclin complex (Gould et al. 1991, Lundgren et al. 1991). The third gene, cdc25, encodes a dosage-dependent activator protein ($p80^{cdc25}$, a protein tyrosine phosphatase) that dephosphorylates ^{15}Tyr, on the p34–B cyclin complex, and phosphorylates threonine 167 (Millar and Russell 1992). Therefore, the transition into the M-phase of mitosis is triggered when ^{167}Thr is phosphorylated by the protein product of cdc25 (Figure 2.3), also called the CDC2 activating ki-

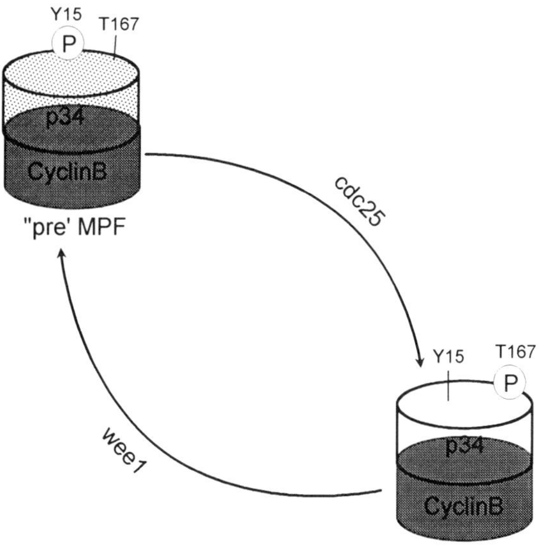

Figure 2.3 Model of molecular control over the G2–M transition in the eukaryotic cell cycle. MPF is a kinase that initiates the M phase of cell divisions by starting a cascade of protein phosphorylations. MPF is activated by the protein product of the gene CDC25 by phosphorylating tyrosine 167.

nase. Once the MPFs become active they become associated with the nuclear envelope and initiate a cascade of phosphorylations of proteins associated with mitosis. This molecular regulation model probably accounts for both of the theoretical physiological regulation mechanisms. The molecular model was developed through studies of fission yeast, *Xenopus leavis* oocyte, and several other animal cell systems, but recent research indicates that the MPF regulation of the G2–M transition applies to plant systems as well.

Homologous genes for cdc2 and other CDK coding genes have been found in a number of plant systems, including *Arabidopsis,* pea, alfalfa, rice, soybean, and maize (Jacobs 1995 and citations therein). The p34s of these systems contain several peptide sequences that are critical to the function of p34 in animal and fission yeast systems. In fact, the ATP-binding site on p34 kinase is conserved in all plant system p34s. MPF control of plant cell G2–M transition is further supported by finding mRNA transcript for cyclins that can be injected into a G2-arrested *Xenopus* primary oocyte to illicit maturation, which is the in vivo test for MPF activity (Hata et al. 1991). The G2–M transition is probably regulated by an MPF system in all eukaryotes. Plant systems may have different biochemical functions leading to and following after MPF regulation, because not all aspects of the plant cell division are identical to those of animal or fungal cells.

One indication of unique cell division processes in plants, related to the MPF regulation system, is suggested by the interaction of p34 and the preprophase band (PPB) in some plant cell division systems. The PPB is a cortical ring of microtubules that appears immediately prior to the M-phase transition. The microtubules vanish during division, but the new cell wall develops during cytokinesis in the same location as the PPB. Immunochemical studies have demonstrated that a protein containing an active component of p34 co-localizes with the PPB after the PPB has vanished (Mineyuki et al. 1991). These results indicated that p34, or a member of this class of proteins (another CDK), may establish the location of the new cell plate required for cytokinesis. These results are preliminary, and no cause-and-effect relationship has yet been found between the PPB, cell division, p34, or cyclins.

The transition between G1 and S in the cell cycle is less well understood because research on this transition began later than that of the G2–M transition. The G2–M phase transition was studied earlier because this is the final transition from cell growth to mitosis. Furthermore, the G1–S transition was confused with the G0–G1 (or "start) transition (that which determines if the cell is in an arrested growth state or in a cell division mode). It is well established that the G1–S transition involves extensive gene transcription of mRNA populations involved in the DNA replication system. Through studies with cell division in fission yeast and mammalian systems it has been found that the initiation of this transcription involves several protein and histone systems, including cyclin–p34 dimers and other CDKs, as well as other proteins, such as retinoblastoma protein (pRb) and E2F (a transcription regulation element).

In mammalian systems, pRb will bind to a number of compounds that can initiate mitosis called mitogens. For example, an important transcription factor for G1–S transition are E2F–DP1 dimers, to which pRb binds and inactivates (Sherr 1994). Phosphorylation of pRb during middle to late G1 will reverse S-phase suppression by increasing the activity of E2F, enabling E2F to activate genes required for DNA replication. In fact, injection of E2F into quiescent cells can initiate G1–S phase transition (Johnson et al. 1993). E2F probably regulates the genes that produce various cyclins. The G1–S phase transition is regulated by dimers of cyclin-D and both cdc4 and cdc2. Since cyclin-Ds are sensitive to growth regulatory hormones, this may be the site of hormonal regulation of cell division.

In a manner similar to that of G2–M phase transition, phosphorylation of these CDK complexes (in this case on threonine residues) activates them and initiates the S phase. Thus both important phases of the cell cycle are initiated by the abundance of cyclins, their dimers with cdc proteins, and their activation by regulatory kinases. Thus it can generally be stated that progression through the cell cycle is driven by a underlying biochemical cycle in which different cyclin-dependent kinases (including the cdc proteins) are sequentially activated.

Although the molecular control of the G1–S phase of the cell cycle is not well understood in plant systems, several important criteria must be met before the S-phase is initiated. First, adequate cellular resources of nitrogen, carbon, and resources for ATP production must be available. Second, the cell must be large enough to go through the division cycle. Third, the previous mitotic event must have reached completion. There is a negative regulation of S-phase initiation by the cyclins that regulate mitosis so that the S phase cannot happen twice in one cell cycle (Heichman and Roberts 1994). Fourth, the G0–G1 phase transition must have occurred. Therefore, the molecular regulation that initiates the S-phase transition in the cell cycle cannot occur if the cell does not have enough resources, particularly nitrogen, to initiate mitosis. Nitrogen resources are required for the synthesis of mRNA and proteins transcribed and translated during the G1–S transition, as well as purines and pyrimidines needed for DNA synthesis during S phase. Thus, even if the cell has initiated the G1 phase, if resource flow to the cell is disrupted by some external factor, cell division will not continue.

A unique characteristic of plants is the restricted location of cell division in meristematic zones. Rapidly dividing cells are located at shoot and root tips, as well as in radial and marginal meristems. What factors cause the rapid fixation of cells in the G0 phase upon leaving the meristematic zone? One observation is that the population of p34 decreases in differentiating cells as they become more distant from the meristem that produced them. In fact, a correlation between the levels of cdc2, mRNA transcript (as a proportion of total RNA), and the proportion of dividing cells in different plant organs has been found in maize, pea, and *Arabidopsis* (Colasanti et al. 1991, Ferreira et al. 1991). In addition, the cyclin mRNA transcript decreases rapidly (on a per cell basis) in pea root as distance from the root apical meristem increases (Jacobs 1992). Although growth regulators such as cytokinins and gibberellins have important interactions with the cell cycle, no studies have explained the link between the p34–cyclin regulation systems and plant growth regulators. Research that links studies of traditional plant growth regulators and molecular cell cycle regulation systems will be a productive technique for understanding cell division processes in plants.

2. Initiation/Retardation of Cell Division by Phytohormones. The primary growth regulators that affect plant cell division are cytokinins and gibberellins. Late in the 1940s and early 1950s, the laboratory of F. Skoog experimented with many compounds in an effort to identify cell division initiation factors, and found that herring sperm extract stimulated cell division in tobacco pith cell culture (Skoog and Armstrong 1970). In the mid-1950s, a pure compound from the herring sperm extract was isolated and called *kinetin* (Miller et al. 1955) because it initiated cytokinesis in tobacco cell culture. It was not until the late 1960s that a plant kinetin-like growth regulator, was isolated from *Zea mays* (Letham 1973) called *zeatin* (Figure 2.4). Kinetin is probably not synthesized by plants, but several other compounds, such as benzyladenine (Ernst and Schafer 1983), benzyladenine riboside (Nandi et al. 1989), and several forms of zeatin (Ambler 1992) have

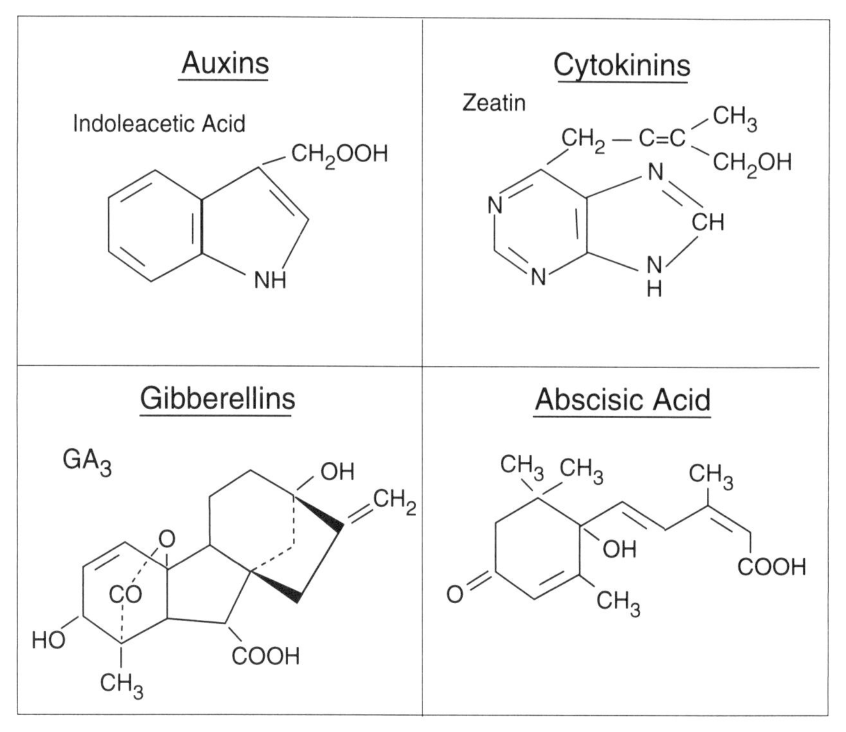

Figure 2.4 Molecular diagram of four main classes of growth regulators that have an impact on cell growth.

been isolated from plants. All of these compounds are adenine derivatives, with hydrocarbon substitutions, which initiate cytokinesis in specific cell systems (tobacco pith cells, carrot phloem, and soybean stems). The mode of action of the kinetins in promoting cell division has not been established, but kinetin-like compounds have been found as a component of tRNA with a uracil-binding codon, as well as free in the cytoplasm as the free base or nucleoside. Most of the evidence to date implies that the free base is the active form of cytokinin (Van der Krieken et al. 1990).

The mode of action for cytokinins in cell division must relate to the production of nuclear-encoded proteins. When exogenous cytokinins are applied to apical meristems, the G2–M transition is stimulated such that cells spend less time in the S phase (Houssa et al. 1990). Further studies find that the application of cytokinins causes a rapid increase in protein synthesis, associated with increased polysome levels. In addition, specific nuclear-encoded mRNA populations increase in the presence of cytokinins (Chen et al. 1987). It is not known whether the increase in mRNA is a result of transcription regulation or mRNA protection from degradation in the cytoplasm (Flores and Tobin 1987). It is possible that cytokinins may interact with cell cycle regulation genes by stabilizing their mRNA transcripts and increasing polyribosome translation of the G2–M phase regulation proteins.

Gibberellins are growth regulators with a diversity of function, including cell division, cell elongation, floral inducement, seed dormancy, and seed carbohydrate storage mobilization. Gibberellins were first discovered in Japan when extracts from the fungus *Gib-*

berella fujikuroi were shown to induce cell elongation in rice, causing the rice to become weak. An active compound was isolated in 1930 and called *gibberellin* (Thimann 1980). There are now 84 compounds that have demonstrated gibberellin activity, of which 73 occur in higher plants (Takahashi et al. 1990). Individual species may have up 20 or more gibberellins in seed tissues. All gibberellins are derived from a complex ringed structure (ent-gibberellane) and are acidic (Figure 2.4). Thus they are all designated as GA_x. The x subscript refers to the specific gibberellin and relates to substitutions and minor modifications of the ent-gibberellane skeleton. Although the mode of action of gibberellins in cell division remains obscure, initial evidence suggests that the G1–S phase transition is stimulated in apical meristems by the addition of exogenous gibberellin (Liu and Loy 1976).

3. Resource Limitation on Cell Division. All the regulation models for plant cell division described above are first controlled by the availability of resources. Although the regulation of transition between stages of the cell cycle may require the transcription of a few genes, mitosis is a complex process that requires many biochemical pathways. Furthermore, the S stage of the cell cycle involves the duplication of the entire genome. Consequently, cells remain in the G0 phase until they have the resources required by the cell division process. Conversion from the G0–G1 phase of the cell cycle is regulated by nitrogen resource availability. Many studies show that under nitrogen limitation cell division slows and mitotic cells remain arrested in the G0 phase (Jacobs 1992). However, even if nitrogen resources are available, growth may be curtailed because the requirements for cell expansion have not been met. Most plant growth is a consequence of cell expansion and differentiation of cells behind the meristematic tissues, where most cell division is occurring. Cell expansion, is also controlled by a complex set of biochemical processes associated with cell wall synthesis and carbohydrate metabolism. Thus, carbon and water resources are important limiting factors for cell expansion, due to their requirement in carbohydrates and the hydrostatic pressure needed for cell wall growth and expansion.

B. Cell Expansion

1. Cell Wall Structure. The plant cell wall structure is a complex combination of polysaccharides and proteins. The general structure could be compared to a glass fiber composite, except for the fact that the cell wall is a constantly changing dynamic structure. Most plant physiologists believe that there is a higher degree of physiological function in the cell wall than has traditionally been ascribed to this cell component.

In a manner similar to that of animal connective tissue, the cell wall is composed of a fibril skeleton embedded within a matrix phase. The fibrils are composites of cellulose molecules and are approximately 10 nm in diameter with a crystalline core of 4 nm (Figure 2.5). In the primary wall, these microfibrils of cellulose are interwoven into a complex network. Following primary wall expansion, and in the secondary wall components, the microfibrils are parallel; however, each successive layer of the primary wall has microfibrils traveling in different directions. The individual cellulose strands in each microfibril are antiparallel and connected by hydrogen bonding at a distance of 2.7 Å, making these structures rigid. The microfibrils are synthesized in a rosette of six cellulose synthetase enzymes which traverses the plasmalemma (Giddings et al. 1980). The rosettes are mobile in the plasmamembrane, and the spatial distribution of these rosettes controls the orientation of microfibrils in the wall.

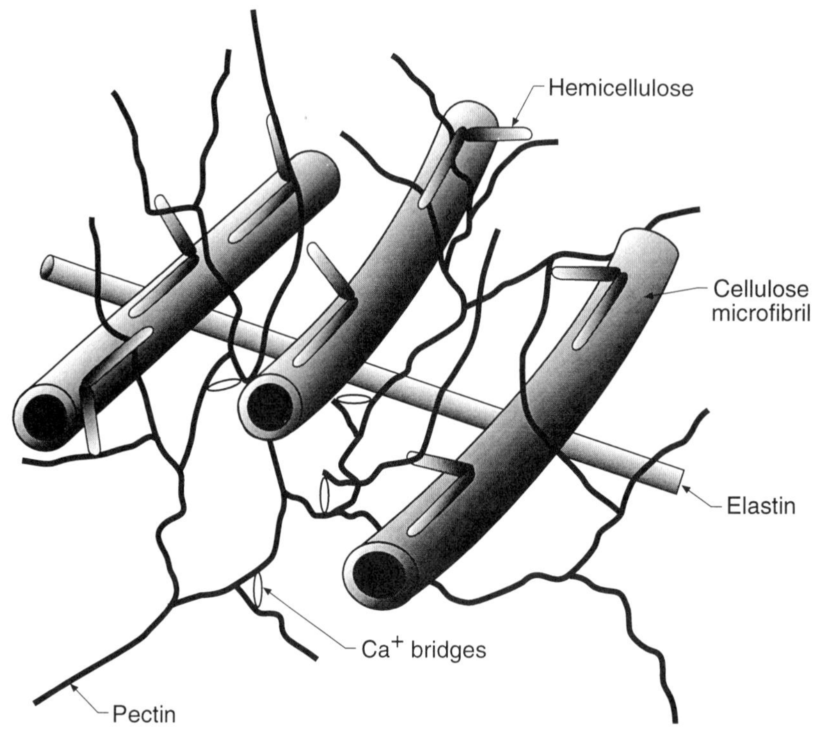

Figure 2.5 Diagrammatic representation of the main components of the cell wall and their structural interrelationship. (Modified from Taiz and Zeiger 1991.)

The matrix phase of the cell wall is composed of a diversity of polysaccharides. This group can be divided into the pectin and hemicellulose subgroups. The matrix-phase polysaccharides are synthesized in Golgi apparatus located near the plasmalemma and excreted by Golgi vesicles.

Taken as a group, the matrix-phase polysaccharides are diverse and variable at several levels. The group is polydisperase by nature (Northcote 1972). Therefore, any extraction of cell wall polysaccharides will contain a diversity of polysaccharide molecules with microdifferentiation, as a result of incomplete metabolism rather than genetic regulation. The composition of polysaccharides in the cell wall also changes with wall developmental stage, as particular components are degraded and others synthesized. Furthermore, the composition of polysaccharides varies with position in the wall and among different tissues. The specific function of the microdifferentiation among matrix polysaccharides is unknown, but specific function can be ascribed to major classes of matrix polysaccharides.

Hemicellulose is generally composed of polyxylans, such as β-1,4–linked D-xylopyranose. These are long helical molecules whose structure is maintained by a coating layer of water molecules. Hemicellulose fibers lie parallel to the cellulose microfibrils and probably form bridge connections between the other polysaccharides and the cellulose microfibrils (Figure 2.5). Gymnosperms also contain a group of glucomannans that form structures similar to that of cellulose and link to the cellulose microfibrils in a paracrystalline shape.

Pectic substances differ from the hemicellulose class by being many branched polysaccharides. Neutral (arabinogalactans) and acidic (galactonorhamans) pectic substances are present in cell walls. The structure of these compounds is readily affected by the nature of the cell wall environment. For example, in the presence of mono- or divalent cations, the physical status of the pectic substances changes dramatically. In general, the pectic substances form the gel-like nature of the cell wall in the presence of water. The pectins serve as a wall filler substance (Figure 2.5), they control the water distribution in the wall, they affect the strength of hydrogen bonding between hemicellulose and the cellulose fibrils, and they affect the texture and mechanical properties of the cell wall. Consequently, changes in cell wall expandability can be related directly to the physical status of the pectin substances.

The secondary wall in plants becomes impregnated with lignin. Lignin is a complex aromatic structure based on over 100 different types of aromatic alcohols related to *p*-hydroxycinamyl alcohol. Lignin penetrates the secondary wall from the outside (primary wall) and serves as a hydrophobic replacement for the matrix polysaccharides. Strong hydrogen bonding occurs between the lignin and the polysaccharides; thus lignin forms a hydrophobic cagelike structure embedded around the cellulose microfibrils of the secondary cell wall. Such a cage of lignin is extremely rigid and no wall expansion can occur. Thus, all cell expansion must be complete before secondary wall lignification occurs.

Of the many proteins that the primary cell wall may contain, elastin is by far the best documented. This is a hydroxyproline-rich polymer with multiple eight-unit oligosaccharides. The oligosaccharides are arabinose and galactose rich. It is thought that the glycoproteins of the cell wall are associated with the microfibril lattice, because they co-separate with the α-cellulose units of the cell wall. Although the specific function of the glycoproteins is unknown, they may be associated with the connections between microfibrils and may strengthen the wall (increase rigidity).

Water is also an important component of the cell wall structure. The polysaccharide gel is dependent on water hydration. In addition, water molecules must be available to reduce cohesion between microfibrils and adhesion between matrix polysaccharides and microfibrils. Furthermore, the cell wall must be permeable to water-soluble nutrients, as all cell nutrients must first traverse the cell wall. Therefore, water is not only an integral component of cell wall structure, but the physiological function of cell wall components is dependent on the presence of water molecules.

2. Physiological Mechanism of Cell Expansion. Expansion of plant cells is dependent on cell wall structure and water pressure (turgor pressure) in the cell. In the absence of adequate water pressure, cell expansion is limited and growth is severely reduced. However, mature cells (postdifferentiation) do not undergo significant expansion even when there is adequate turgor pressure in the cell. Thus, the cell wall structure and function must change after differentiation, which inhibits further cell expansion in response to turgor pressure.

The early work of Lockhart (1965) related the expansion of cell walls to that of fiber–polymer substances that are studied in material science labs. The extension of the wall (dV/dt) was predicted to depend on the wall extensibility (m), the wall yield threshold (Y), and the cell turgor pressure:

$$\frac{dV}{dt} = m(\Psi_p - Y) \qquad \text{cm}^3\ \text{s}^{-1} \tag{2.1}$$

Cell expansion occurs when turgor pressure (Ψ_p) is greater than the yield threshold (*Y*) and the cell wall extensibility (*m*) is greater than zero. The wall extensibility factor approaches zero when the wall is impregnated with protein (such as extensin), impregnated with lignin, or when water is removed from the wall. In cells that have random orientation of primary wall microfibrils, expansion will occur equally on all sides. During cell expansion new fibers are synthesized on the plasma membrane surface, the wall thickness remains the same, but the inner microfibrils have a longitudinal orientation. Cells that have the primary fibrils oriented in a latitudinal axis (as in a spring) cause cell expansion to occur predominantly in the long direction of the cell. Cell expansion will continue until the rigid protein extensin (or a similar protein) is embedded into the wall. At this point the extensibility of the wall approaches zero and little to no more cell expansion will occur.

Cell expansion, permanent enlargement of the wall surface area, is termed *plastic growth* and involves new cell wall synthesis and cellulose microfibril reorientation. Hydrogen bonds between hemicellulose and gel polysaccharides break during plastic growth, so that cellulose microfibrils slip past one another and can change orientation. In contrast, elastic growth of the cell wall is a temporary expansion of cell volume due to changes in turgor pressure that are below the yield threshold, or occur after the wall extensibility has approached zero.

The specific impacts of *Y* and *m* on cell expansion can be separated by plotting the change in cell volume against the specific cell pressure (Figure 2.6). During the initial phases of cell expansion (that before *Y* is reached) cell volume increases slowly in response to changing Ψ_p. When *Y* is reached, the rate of volumetric increase with increasing Ψ_p becomes higher. Cell wall extensibility (*m*) is calculated as the slope of the volume/Ψ_p relationship when $\Psi_p > Y$. The determination of *Y* and *m* by this technique is difficult because it must be done microscopically, a pressure probe should be used to measure Ψ_p, and Ψ_p changes during cell expansion events (Cleland 1986).

Cell wall elasticity can be measured by several means. However, elasticity is compared most frequently to a material science definition. *Young's modulus of elasticity* is an index of the force/area ratio required to change the relative length ($\Delta l/l$) of a material:

$$\text{Young's elastic modulus} = \frac{\text{force/area}}{\Delta l/l} \quad \text{MPa} \qquad (2.2)$$

The larger the value of Young's elastic modulus, the stiffer the material. Measurements of Young's elastic modulus on plant fibers have approached 10,000 MPa for cotton fibers,

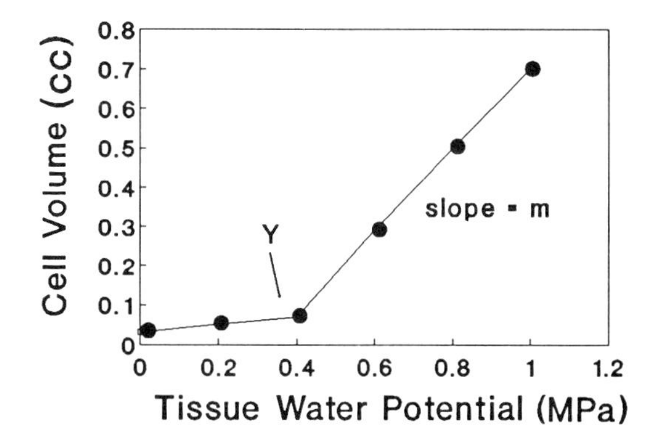

Figure 2.6 Method by which the yield threshold (*Y*) and extensibility coefficient (*m*) of cell walls in a tissue can be determined.

which is 5% of that for steel. Equation (2.2) can be modified in order to apply these concepts to living plant cells by using a volumetric, expansion index:

$$E = \frac{\Delta P}{\Delta V / V} \quad \text{MPa} \tag{2.3}$$

Thus the volumetric elastic modulus (E) of plant cells can be determined by examining the change of cell pressure (ΔP) required to cause a relative change in cell volume ($\Delta V/V$). The volumetric elastic modulus of plant tissue ranges from 1 to 50 MPa. This means that the cell volume will change by 0.2% ($E = 50$) to 10% ($E = 1$) for each 0.1-MPa change in turgor pressure.

3. Influence of Growth Regulators on Cell Expansion. Plant growth regulators could affect cell wall expansion by influencing any one or combination of the factors regulating plastic cell expansion (yield threshold, extensibility factor, or turgor pressure). The strongest influence of growth regulators on cell wall expansion is their impact on m. All four classes of growth regulators shown in Figure 2.4 have an impact on cell wall expansion among which auxins have been shown to most strongly stimulate cell expansion.

For synthetic or natural auxin to affect cell wall expansion, the wall in question must be able to respond. Thus the Ψ_p must be greater than Y, and the wall cannot be impregnated with lignin. Auxins do not affect the chemistry of the cell wall directly and do not affect Y. Rather, elevated auxin concentration stimulates the release of a wall loosening factor (WLF) from the cytoplasm into the cell wall (Cleland 1986). Calcium and hydrogen ions are two possible WLFs that have been identified. An increase in the H^+ concentration in the wall increases m, and a decrease in the Ca^{2+} increases m. Both of these WLFs relate to the quantity and strength of hydrogen bonding between the pectin and cellulose microfibrils.

III. WHOLE-PLANT GROWTH

Growth of the whole plant is dependent on the activity of meristematic regions (cell division) and the temporal and spatial differentiation (cell expansion) of the cells produced by the discrete meristems. Plant growth is unique among the kingdoms of organisms, by having "perpetually embryonic tissue" (Jacobs 1992), which continues to develop new organs throughout the life of the organism. However, the temporal and spatial patterns of organ growth and development have great variation among plant species.

A. Proportional Growth of Organs

Anyone who works with plants realizes that plants are unique because of their organ redundancy. Almost all plants have a diversity of leaves, stems, flowers, and roots. Since each of these organs has specific purposes, the proportional quantity of each organ will regulate general growth attributes of the plant. Those plants with a high proportion of leaves compared to roots may have a high ratio of carbon gain by leaves compared to carbon loss by roots; however, they also will have a limited ability to supply water and nutrients to the leaf mass. Therefore, a balance between the proportional developments of various organs and environmental conditions must be met for a plant to succeed in any specific environment. The term *allocation* is commonly used in reference to the differential allotment of materials among organs. Often it is written that a plant allocates more to one organ system than another. Although this statement sounds teleological, it is simply referring to a ratio of mass between organs that is not one to one. Ratios between the surface area or mass of two organ systems (usually the root/shoot ratio) are often considered when evaluating plant response to environmental stress.

The patterns of allocation among organs are different among species, within species under different environmental conditions, or within plants at different times. In the simplest sense, woody species allocate a higher proportion of mass to stem and wood growth than that of herbaceous species. Among herbaceous species, some allocate little to no mass to roots, while other herbaceous species are dominated by root mass development. Although species can be classified as woody or herbaceous, there is still a diversity of allocation patterns within either group (e.g., phreatophytes allocate a high proportion of total mass to roots, in comparison to that for surface-rooted trees and shrubs). In fact, there is a continuum of possibilities for differential allocation of mass and nutrients to various organs. The important controlling factors of allocation patterns in plants are environmental conditions (abiotic and biotic), evolutionary heritage, and physical constraints to plant architecture (Niklas 1994).

Every species has a certain set of developmental attributes, but allocation patterns in plants are also labile in time or among sites. A vine will develop long thin stems in comparison to a shrub, trees form central trunks, and shrubs will have multiple trunks. However, environmental conditions can influence the allocation patterns within this set of developmental attributes. Many tropical vines produce one thin and long stem that uses the trunk of other species as it grows to the canopy. When the vine reaches the upper canopy, high-light conditions stimulate branching and induce the vine to convert into a shrub growth form. Many shrub species grown under high-light conditions allocate a higher proportion of mass to stems, resulting in a short, robust stature with small numerous leaves. In contrast, under low-light conditions the same shrub allocates more mass to leaves, resulting in a tall spindly shrub with large leaves. In fact, the proportional allocation of mass or resources to leaves, stems, or roots can be an excellent indicator of environmental conditions under which a plant was grown. Since the proportional development of organs on plants must reflect the interface between environmental conditions and physiological attributes, allocation patterns can be critical to plant survival in varied habitats.

Differential allocation to various plant organs also can affect the potential for whole-plant growth. If a species allocates a large proportion of mass to leaf area, this will increase the species ability to maintain a positive carbon balance. In a sense, allocation to leaf area pays back in increased growth potential. Conversely, preferential allocation to root or stem mass may reduce future growth rate. During the lifetime of many trees, growth rate decreases in relation to overall mass because of the increasing proportion of stem and root compared to leaves. Of course, the relationship between allocation pattern and growth rate is tempered by environmental conditions. In a high-heat environment, preferential allocation to leaf area instead of root mass may induce wilting and thus could reduce rather than increase growth potential. In an environment with low soil nutrition, excessive allocation of mass to leaf area may result in diluted leaf nutrients and may reduce growth potential.

B. Determinate Versus Indeterminate Growth

There is a tremendous diversity in the way that plants grow. This is in part due to their modular nature (organ redundancy) and to the diversity of allocation patterns. However, there are some specific architectural attributes of plants that can be pointed out. Plants can have shoots that grow for a period of time, then terminate in an inflorescence. This is termed *determinate growth,* since upon developing a flower the growth of that specific shoot is terminated. In contrast, *indeterminate growth* occurs when shoots grow and the terminal apical meristem does not differentiate into a floral meristem. Inflorescences develop from axial meristems, enabling continual growth of the shoot. Most herbaceous plants have determinate growth. Some herbaceous species produce one or a few shoots that terminate in an inflorescence, while others produce abundant lateral shoots from axial buds that also terminate in inflorescences. Woody species can be either determinate or indeterminate. Both growth forms are found in a diversity of habitats, so these different growth forms have not developed because of any specific environmental attribute.

C. Considerations of Plant Architecture

Biophysical constraints to plant form are reflected in the outward appearances of both herbaceous and woody plants. The size and structural complexity of herbaceous plant canopies is severely limited by their low investment in woody tissues. Only the lack of structural tissues (secondary xylem) limits their attainable size. However, since no mass is invested in wood, a higher proportion of plant mass is photosynthetic and a higher growth rate is possible compared with many woody taxa. If herbaceous plants invest in growth potential, yet do not invest in structural support (e.g., a proliferation of collenchyma or schlerenchyma), they must have a short life span or a high organ turnover rate.

Branching patterns of plants are the result of the interaction between genetic characteristics, environmental conditions, and biophysical processes. The phylotaxy of a plant defines the spatial placement of lateral branches on the axis of shoots (shoots have a general definition here and include main stems, branches, and inflorescences). Genetic factors regulate the phylotaxy of shoots in a species. The distance between lateral shoots and the rate at which they are produced are influenced by environmental characteristics, but the placement of those branches on the shoot (spiral, alternate, opposite) is determined genetically. The angle of branch growth out from the main axis is determined by biophysical relationships. Shear forces that are developed by wood mass of branches and the forces of gravity can result in structural instability unless branches are supported with reaction wood and are held at specific angles. Computer programs have been developed with a capability to simulate plant structure based on biophysical constraints. Since the growth potential of plants is dependent on the number of growing and expanding meristems, and the architecture of plants will regulate the number of potential meristems, plant growth rate is also influenced by plant architecture.

D. Genetic Regulation of Growth Rate

The growth rates of individual plants in a population are not all exactly the same. Some individuals have high growth rates, and others seem to be runts. The differences among the growth rates of individuals is under genetic control. In an experimental program with a fast-growing shrub from Mediterranean habitats (*Spartium junceum*) it was found that relative growth rate was heritable. Clones made from genotypes that had a fast growth rate had higher growth rates than clones made from a genotype with slower growth rate (E. T. Nilsen, unpublished manuscript). Growth rate is a quantitative trait such that there is a statistical distribution of different growth rates among individuals in any population. Since there are probably many genes that regulate growth rate, determining the important regulatory genes has been difficult. In fact, scientists have not determined the genetic factors that regulate growth rate in any plant species. Yet breeding programs have been implemented successfully to select for crop genotypes that have the highest potential growth rate.

IV. MEASURING PLANT GROWTH

Research on any process of plant stress physiology will involve some index of plant growth or development. The spatial scale of the index may vary from the biochemical construction of molecules, to cell division, to organ development, to development of the whole plant, to growth of a stand of individuals. Whatever scale is chosen for the growth index, it should always be based on an increment caused by the accumulation of external abiotic resources. In plant growth processes, organs are continually being produced while others senesce; thus the determination of new living material produced by plants is confounded by the continual loss and recycling of living material. The accumulation of external abiotic resources is also not easy to measure in plant growth, because extensive cell expansion can occur without new protoplasm, and resources are recycled from senescing tissues for incorporation into new tissue. The dynamic and recycling potential of plant growth processes must be firmly understood before one can interpret the impacts of a stressor on plant performance.

The measurement of whole plant growth can be accomplished on several temporal scales. Instantaneous resource accumulation can be determined by measuring net assimilation of CO_2 (photosynthesis). Daily carbon gain can be determined by integrating photosynthesis measurements over leaf area and the day cycle. Seasonal whole plant growth can be measured by determining net primary production. Dendrochronological techniques (the study of trunk ring width in woody species) can be used to measure long-term growth in woody plants. The particular temporal or spatial scale utilized to measure growth must be chosen appropriately for the particular biological function under consideration. In the following sections we outline a diversity of mechanisms utilized for measuring growth in plants and discuss the limitations and appropriate applications of each.

A. Productivity and Yield

Productivity is defined as the net gain in material (carbon, mass, energy, etc.) divided by a time increment (Whittaker 1975). The time increment varies depending on the life span (ephemeral, annual, biennial, or perennial) or seasonality (deciduous, or evergreen) of the species in question. Productivity can be defined as the total amount of material gained by the organism [*gross primary production* (GPP)] or the amount of material gained after respiratory losses of CO_2 has been subtracted [*net primary production* (NPP)]. *Plant productivity* normally refers to the gain of materials by the individual plants, while *stand* (or *crop*) *productivity* refers to growth on a ground area basis. Frequently, agriculturalists will use plant productivity in early stages of the crop development and crop productivity after the canopy has closed.

Measurements of NPP must take into account several continuous losses of tissues from plants during the growing season. For example, large losses of tissue can occur by leaf, twig, flower, fruit, and root abscission during the growing season. Some of the accumulated resources can be lost by leakage out of the roots into the rhizosphere, as well as leakage from aboveground tissues during rainfall. In addition, herbivores are always present on most plant organs, and they remove accumulated resources by a number of methods. All of these losses should be accounted for in a measurement of NPP, but they are often ignored. In the case of perennial plants, a true increment (TI) can be determined as the net increase in standing live material following a complete growth cycle.

Productivity of any growth compartment (wood, leaves, flowers, fruit, roots, etc.) may be used as an index of stress effects on plants. This is frequently the method of choice in agricultural studies, in which the scientists are interested in the impacts of the stressor on yield. Yield is normally reserved for describing a harvestable unit of the particular growth compartment. Thus seed production of wheat may increase under high precipitation, but seed yield may decrease due to increased lodging (plants fall and become densely packed on the ground), which causes decreased harvestability. Similarly, environmental conditions may increase the growth of wood in trees, but that excessive wood growth may be of low quality and the yield of harvestable high-quality wood may decrease. Yield will always be a fraction of the TI of any growth compartment, because harvest efficiency is not 100%.

B. Techniques for Measuring Productivity

We will classify growth analysis techniques on the basis of functional approach. Direct sampling techniques (sequential harvest) have been employed for over a century in growth studies. Indirect measures (dimension analysis, quantitative phenology) were developed for studies where harvesting was not practical. Relativized measures were developed to compare organisms of varying size and growth efficiency. Modeling techniques have been developed to incorporate predictive capability into growth analysis.

1. Direct Sampling Techniques. The *sequential harvest technique* is most frequently employed in agricultural settings for measuring plant productivity or yield. The researchers harvest the desired plant component (whole plant, fruit, inflorescence, etc.) at appropriate time intervals during the growing season. The total of the increments between harvest times is the NPP for that component (Figure 2.7). Problems arise with this technique when there is loss of tissue (component abscission) between harvest intervals. Under this scenario, the measured total NPP will underestimate the actual NPP. This can be avoided by collecting abscised tissues (litter traps) or shortening the harvest interval. This technique is also unable to account for losses to herbivores; therefore, sequential harvesting is an appropriate measure of true increment or yield but not NPP. If the time interval between harvests is short enough, a summation of all increments may partly compensate for tissue losses by abscission or herbivory. However, concurrent quantitative phenology sampling must be used to describe the dynamic aspects of growth, to utilize sequential harvest for accurate NPP determination.

The sequential harvest technique is most applicable in experiments that employ large sample sizes. Each harvest must be representative of the entire population. If the harvestable population is to small, several small individuals may be harvested in one time period, and several large individuals may be harvested in the next time period, which would inflate growth. Therefore, uniformity of growth is also preferred for sequential harvest studies.

Productivity measurement by sequential harvest is applied primarily to smaller (herbaceous perennials or annuals) species with continuous growth and little tissue ab-

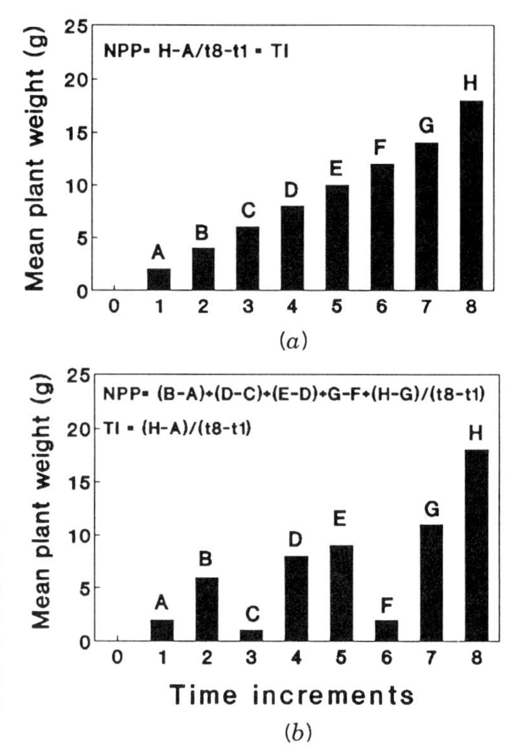

Figure 2.7 Plot of the increments in biomass of a stand used in a sequential harvest analysis of growth. (A) Stand with uniform growth. (B) Stand with intermittent growth pulses and tissue losses. NPP, net primary production; TI, true increment; t_x, a timed sample. Letters represent individual sample weights.

scission during the growth period. Crops are commonly studied by this technique because a large uniform sample size is available. Measurements of NPP in woody species by sequential harvest is restricted to very large forestry experiments. Similarly, NPP measurement in natural populations cannot normally utilize sequential harvest because of the smaller, less uniform sample size and high tissue turnover rates. However, this technique is frequently used in grassland systems (Vickery 1976, Sala et al. 1981, Birch and Cooley 1982).

2. Relativized Measures of Growth. Comparing the absolute magnitude of growth among species can be complicated by the large differences in plant size. It is also difficult to compare growth among different stages of a plant life cycle because of changing plant size. Therefore, relative measures (based on an index of plant size) have been developed for comparative studies of growth. The *relative growth rate* (RGR) is the most common index of growth and one of the first indices applied to growth analysis (Blackman 1919). The RGR is an index of the instantaneous growth increment divided by the size of the plant at the beginning of the increment (growth efficiency) and can be defined by the differential equation

$$\text{RGR} = \frac{dW}{dt}\frac{1}{W} \qquad \text{g kg}^{-1}\,\text{d}^{-1} \tag{2.4}$$

where t is the time increment and W is plant weight.

This index assumes that larger individuals will have proportionally higher growth rates. However, under many circumstances larger individuals may have a larger mass of respiring, supportive tissue compared to growth points, and RGR decreases with plant size. RGR will remain constant with time when growth rate is exponential. If the growth rate increases in a sigmoidal or linear fashion, RGR will decrease with time.

Growth has also been related to photosynthetic surface. This relationship predicts that the larger the leaf surface, the greater the growth. *Net assimilation ratio* (NAR) is defined as the change in plant weight per time increment divided by the total assimilation area and is described by the differential equation

$$\text{NAR} = \frac{dW}{dt}\frac{1}{\text{LA}} \qquad \text{g m}^{-2}\,\text{d}^{-1} \tag{2.5}$$

where LA is total leaf area at the beginning of the increment.

In traditional growth analysis, an *interval* approach is used. This works well for RGR, because the differential equation can be explicitly integrated:

$$\text{RGR} = \frac{\ln W_2 - \ln W_1}{t_2 - t_1} \qquad \text{g d}^{-1} \tag{2.6}$$

NAR cannot be integrated explicitly because the resulting equation is based on knowing the relationship between weight (W) and assimilation surface area (LA). Several possible relations between W and LA are possible (linear, exponential, quadratic, etc.), which would result in different integrated equations (Radford 1967). If there is a linear relationship between LA and W, and growth is exponential, RGR will remain constant among sequential determinations and NAR can be a useful index of growth efficiency. A similar set of potential interval equations can be formulated for other indices, such as the *leaf area ratio* (LAR), the ratio of RGR and NAR. Therefore, before one uses these relative growth analyses the relationships between LA, W, and t should be known so that the appropriate equations can be utilized.

Several other possible relative indices are used in growth analysis. Techniques that use an integrated approach to growth measurement are common (e.g., LAD or BMD indices). The excellent reviews by Evans (1972), Hunt (1978), and Chiarello et al. (1989) provide further information about the diversity of indices and their respective limitations.

3. Indirect Sampling Techniques. The *dimensional analysis technique* depends on a consistent allometric relationships of plant growth among all individuals of a species (Whittaker and Woodwell 1968, Whittaker and Marks 1975). There are biophysical constraints on the amount of supporting material required to maintain a specific amount of growth potential. Thus, one can formulate a relationship between a specific dimension (trunk diameter, stem diameter, branch length, etc.) and biomass or productivity of the organs supported by that dimension. Dimension analysis has been used primarily in the study of biomass and productivity of woody species, because these species have the most consistent allometric relationships (Niklas 1994).

Regressions of plant growth compartments have been formulated against many dimension indices (Telfer 1969, Gholz et al. 1979). Dimensions can be used directly (untransformed), mathematically transformed (i.e., the natural log of the stem diameter), or combined (height × basal diameter). Normally, researchers will harvest approximately 50 to

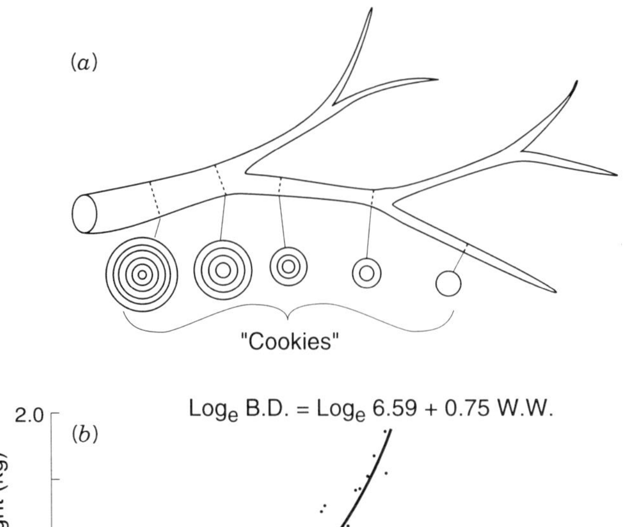

Figure 2.8 (A) Diagrammatic representation of a branch system used in dimension analysis of growth. Cookies are slices through the stem, at various stem diameters, used for determining average wood growth rate. (B) Plot of wood weight against branch basal diameter for 60 branches. The equation is a power regression model for wood weight versus branch basal diameter.

100 growth units (branches or trees) to measure a series of dimensions (length, width, basal diameter, etc.) as well as measuring the biomass of all growth components and the increments of wood and new twigs. The wood increment is calculated by taking slices of the wood (cookies) at various distances along the length of the growth unit. The width of annual rings is measured, and wood volume increment for the whole branch can be calculated geometrically for each segment of the growth unit, then summed (Figure 2.8). Multiple regressions are formulated using any combination of dimensions in any possible equation format until the best relation is determined. Dimension analyses are a primary measure of merchantable tissue in forestry applications (Gholz et al. 1979). These regressions usually involve a combination of transformed dimensions regressed against merchantable timber (board feet). The power regression format (Baskerville 1972, Sprugel 1983) is frequently the best regression between a unit of basal diameter and a unit of biomass or productivity:

$$\ln y = \ln a + b \ln x \tag{2.7}$$

where x is the allometric dimension and y is the weight of the growth compartment.

The regressions vary among studies due to the number of techniques used to formulate them, the diversity of growth forms among plants, and the multiple goals of the research projects (ecology, forestry, range science, horticulture, etc.). However, if only one regression formulation is used to determine the relation between one dimension and a specific growth component, the regression tends to be similar within growth form. For example, when branch basal diameter is compared to branch wood weight in a power regression model, the a and b values are similar within growth forms (Figure 2.9). In addition, ecologists are aware that the dimension analysis regressions of whole trees are similar within each major growth form (deciduous, evergreen broadleaf, evergreen conifer, etc.) and different among growth forms.

A problem with dimension analysis is the static nature of the measurement. Regressions are formulated from harvests of the measured unit at one time, but plant growth is dynamic. Furthermore, the timing of leaf, fruit, twig, and wood production is not uniform within any species. Therefore, a family of independent regressions must be formulated at the time of maximum biomass of each growth component. In addition, the only way that dimension analysis can compensate for tissue loss before maximum biomass is by coupling this technique with a detailed measurement of quantitative phenology (Nilsen et al. 1991) and/or litter fall.

Quantitative phenology involves a nondestructive measure of growth by intensive monitoring of growth units. A reasonable sample size (200) of growth units (usually, small branches) is permanently labeled. Bird bands serve excellently for this purpose. The presence and number of clipping production units (leaves, flowers, fruits, new twigs) are recorded for each node individually. In addition, the developmental stage of each growth

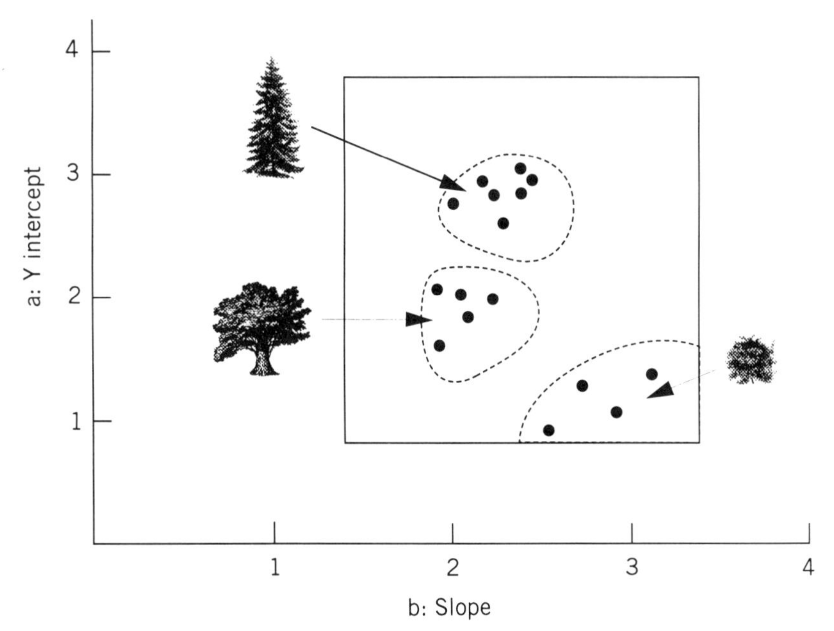

Figure 2.9 Plot of the a and b coefficients from power regressions of branch weight versus branch basal diameter for shrubs, conifers, and hardwoods. Data for this figure were collected from a large number of sources.

unit is recorded. The census is repeated on a schedule that is selected based on the rate of growth. This technique will allow the determination of clipping production seasonality as well as losses to abscission and herbivory (Nilsen et al. 1987a). To convert the numbers to mass, the average dry weight per production unit is measured with samples collected from other similar branches.

Quantitative phenological assay is excellent for determining the dynamic nature of clipping production. However, measurements of basal diameter and length of the small growth unit do not adequately describe the magnitude or seasonality of wood growth. Changes in small branch diameter due to changes in branch hydration or temperature are frequently greater than the growth increment. Therefore, a combination of quantitative phenology (to determine production flux) and dimension analysis (to determine wood production) may be an appropriate measure of plant productivity (Nilsen et al. 1987a).

Once the regressions between plant dimensions and growth compartments have been formulated, plant productivity can be determined by measuring individuals. Normally, basal diameters of all major branches are measured and the trunk is measured separately. The regressions are applied to the branch measurements to calculate clipping productivity (total of leaf, flower, fruit, and new twig) and branch wood production. Trunk wood production is determined independently and added to branch wood production (Sharifi et al. 1982). The total aboveground production of the plant is then related to a dimension of the plant. Commonly, canopy volume, canopy area, or leaf area index (LAI) are used as the whole plant dimension (Murray and Jacobson 1982, Sharifi et al. 1982). Stand productivity can then be estimated by determining the canopy relationships of the stand (area, volume, LAI) by community sampling techniques (Whittaker 1975) and applying the individual plant productivity relationship (i.e., NPP/canopy volume) to stand-level measurements.

Clearly, dimension analysis is limited to woody plants that have discrete production events with little abscission of tissue between events. The advantage to dimension analysis is that only a limited sampling is required to formulate the regressions, and most measurements in this technique are nondestructive. The technique is most useful for forest and rangeland applications in which extensive harvest is impractical. The technique is also normally limited to aboveground tissues because the root system cannot be effectively removed to formulate dimensional regressions. Attempts have been made to measure belowground tissues with this technique by explosive removal of the root system (Whittaker and Woodwell 1968), washing out the soil from the root system (Kummerow et al. 1977) or by estimated taper (Sharifi et al. 1982). None of these techniques adequately account for small root biomass. The complexity of measuring root biomass and production justifies a separate section in this chapter.

4. Plastochron Index. The *plastochron index* (PI) is a developmental index of growth. The index was developed to facilitate the selection of chronologically uniform tissues for physiological analysis. Askenasy (1880) first defined the plastochron as the time between the initiation of successive leaf primordia. Erickson and Michelini (1957) increased the scope of the concept by developing a plastochron index to define the developmental age of a growth unit that contains leaves (Figure 2.10). The plastochron index can be defined as

$$\mathrm{PI} = n + \frac{AB}{AC} \tag{2.8}$$

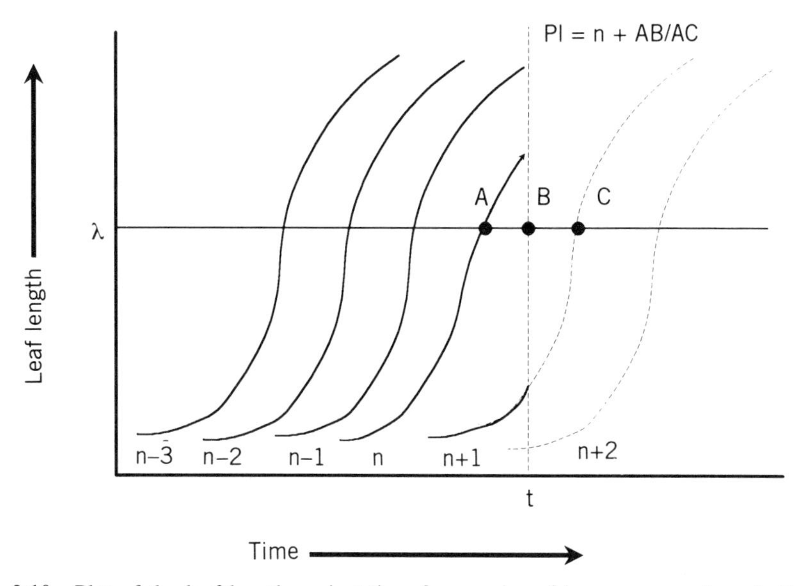

Figure 2.10 Plot of the leaf length against time for a series of leaves on one branch. Each line represents the growth of one particular leaf on a branch axis. The marked points (A,B,C) on the plot are used in the calculation of the plastochron index.

where n is the serial number of the smallest leaf greater than the reference size (λ), A the time that leaf n reached a size equal to (λ), B the time when λ will be reached for leaf $n - 1$, and C the time that leaf $n + 1$ was size λ.

The measurement of the plastochron index can be based on a specific time (t) and the reference length λ by using a geometric formulation

$$\mathrm{PI} = \frac{\ln L_n(t) - \ln\lambda}{\ln L_n(t) - \ln L_{n+1}(t)} \tag{2.9}$$

where $L_n(t)$ is the length of leaf n at time t and $L_{n+1}(t)$ the length of leaf $n + 1$ at time t. The geometric PI refers to the developmental status of a shoot, but the formulation can be adjusted to represent the developmental stage of a particular leaf. The leaf plastochron index can be defined as LPI = PI − i, where i is the serial number of the leaf in question relative to leaf n.

The plastochron index can also be used to measure a developmental index of growth rate by calculating the average number of days that it takes to complete one plastochron (Quinby et al. 1973, Snyder and Bunce 1983), the plastochron rate:

$$\mathrm{PR} = \frac{\mathrm{PI}_{\mathrm{final}} - \mathrm{PI}_{\mathrm{initial}}}{\text{time interval}} \quad \mathrm{d}^{-1} \tag{2.10}$$

Two important assumptions must be met before using PR to measure growth rate. First, growth must be indeterminate over the interval in study. This means that shoot apex meristems do not differentiate into flowering stalks and stop growing. Second, the growth rate must be exponential. Thus, growth must be at a steady state in which the growth rate of each leaf is similar and successive plastochrons are the same. A rough measure of the validity of these assumptions for any species with indeterminate growth is to measure the internodal length. If internodal length is the same between all leaves, a steady-state growth is likely.

The plastochron system of measuring growth is most applicable to the determination of developmental age. The index is best applied to rapidly growing plants with high leaf/stem ratio. Although there is a potentially wide set of possible applications for this index (Lamoreaux et al. 1978), its use has been restricted to studies requiring careful determination of chronological tissue age.

*5. **Physiological Measurement.*** We consider empirically based photosynthesis measurements as an indirect measure of growth, because the assay is based on CO_2 accumulation rather than total weight gain. Simple measurements of net CO_2 accumulation (P_n) are insufficient to predict growth because changes in the LA/respiring volume, or changes in the environment, will cause changes in the relationship between P_n and growth. Thus the P_n measurement must be scaled up to the canopy level in order to predict growth.

Canopy-integrated P_n can be measured on small individuals by measuring CO_2 accumulation rates in closed, environmentally controlled chambers, which enclose the entire canopy. If the canopy is too large to enclose, P_n measurements must be taken on leaf selections from the canopy over a diel (24-h) cycle (Figure 2.11). The leaf selection must include all developmental leaf stages and all microhabitats in a proportion equal to that of the whole canopy. This careful selection of leaf sample is rarely done (often measurements are taken of outer sunlit young leaves only). Therefore, converting diurnal measurements of P_n to whole canopy carbon gain is a labor-intensive project.

Response surface analysis of gas-exchange parameters can be utilized to estimate

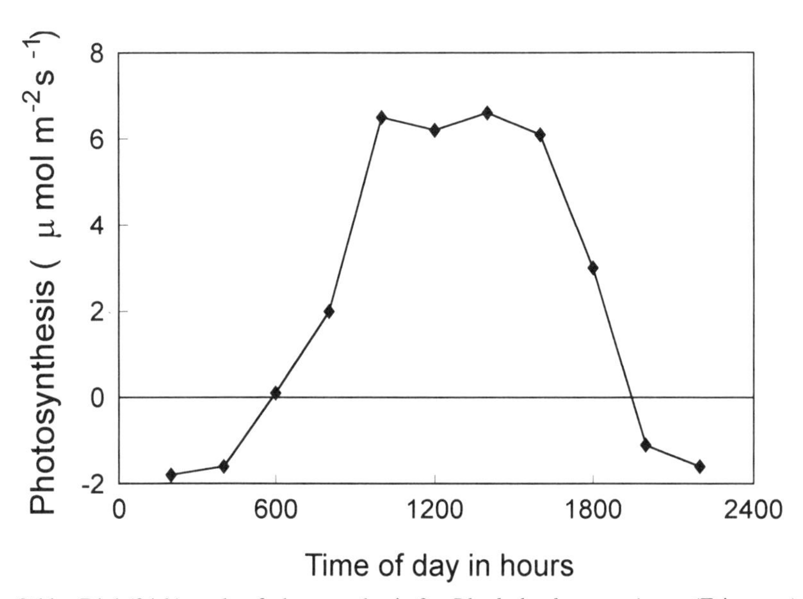

Figure 2.11 Diel (24-h) cycle of photosynthesis for *Rhododendron maximum* (Ericaceae), a large evergreen shrub that grows in the subcanopy of the southern Appalachian mountains. Each point is a mean of five leaves in the upper canopy.

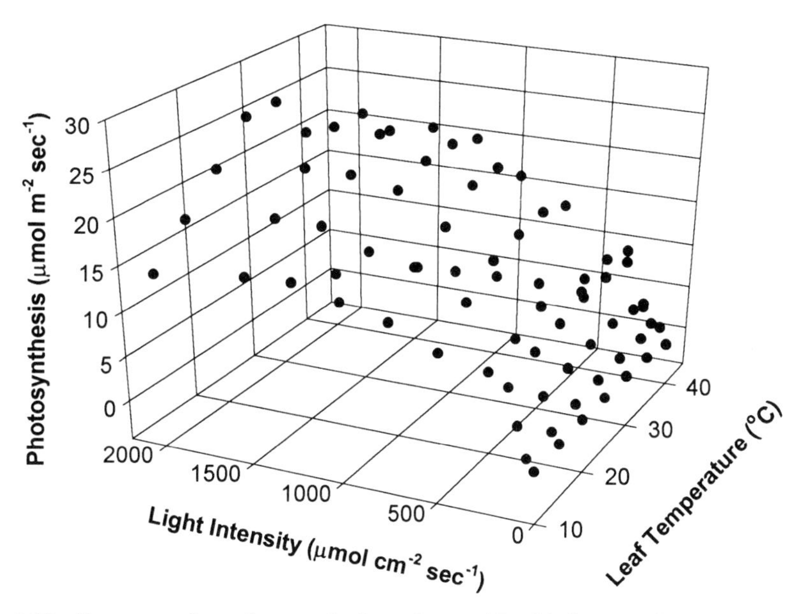

Figure 2.12 Response of net photosynthesis to the combined influence of temperature and light for an average C_3 species of plant.

canopy carbon gain. Leaves that represent the developmental classes in the canopy are used in repetitive measures of P_n to formulate relationships with environmental parameters. Often, three-dimensional plots of P_n versus two environmental factors for various age classes are developed (Figure 2.12). The response surfaces are used to predict daily carbon gain by application to microclimatic measurements of temperature, light, and vapor pressure in the canopy. Just as in direct measurement of diurnal cycles, the results have to be weighted by the leaf developmental and microclimatic heterogeneity of the canopy. Therefore, in both methods a quantitative phenological and microclimatic analysis of the canopy must be done concurrently to scale leaf measurements up to the canopy level. Due to the necessity of determining response surfaces for all leaf developmental stages and microhabitats, the empirically based photosynthetic measurement of growth is used infrequently. Most frequently, biochemically based assimilation models are used to predict whole plant carbon accumulation. However, physiological measurement of canopy carbon gain may be a better index of whole plant growth compared to other indirect measures, because it includes the carbon gain for root production.

6. Remote Sensing. Due to the current interest in global ecological phenomena, there has been a renewed effort to utilize remote-sensing techniques for predictions of large-scale productivity (Sellers et al. 1989, Asrar 1990). Remote sensing utilizes a nondestructive optical technique to measure canopy processes such as transpiration or productivity. The optical instrument could be hand carried, borne on an aircraft, or mounted on a satellite. Most of these optical procedures concern correlations between spectrographic reflectance signatures of a plant canopy with the physiological activity of that canopy (Sellers 1987). Reflectance patterns of plant canopies are distinctly different from those of soil, wood, or any other nonleafy surface. Thus attributes of canopy structure and function affect canopy spectral reflectance. Initially, it was postulated that an infrared analysis of canopy reflectance could be correlated with the transpirational activity of the leaves. This was postulated because the near-infrared reflectance signature represents the canopy heat balance relative to the surroundings, and canopy heat balance is in part dependent on transpiration. Since evapotranspiration has been correlated with productivity across a wide group of ecosystems (Rosenzweig 1968, Leith 1973), the near-infrared reflectance index of canopies should reflect transpiration as well as productivity.

Some of the earliest remote-sensing applications involved hand-held instruments that determined the ratio of near-infrared and red reflectance from canopies (NIR/R). Red-light reflectance tends to decrease with an increase in vegetation because chlorophyll absorbs red light. In contrast, NIR reflectance tends to increase with increasing vegetation because NIR light is reflected by mesophyll cells. Thus, as vegetation per ground layer increases, the ratio or NIR/R also increases. These ratios have been used to measure biomass for grassland systems in which such a sensor can be held over the canopy (Pearson et al. 1976, Hardisky et al. 1984, Prince 1991).

The ratio of NIR/R light reflectance is termed the *simple ratio*, while the *normalized difference vegetation index* (NDVI) is calculated as

$$\mathrm{NDVI} = \frac{R - \mathrm{NIR}}{R + \mathrm{NIR}} \tag{2.11}$$

The NDVI is affected by the leaf area index of a community, which is reflective of the community biomass and productivity. The relationship is not always consistent because the NDVI is also affected by the physiological status of the canopy. Stressful conditions that affect leaf function will change the NDVI. This is precisely why remote-sensing technology is being used to develop indices of large-scale stress in plants. Many studies have been performed with hand-held or supported instruments over monoculture canopies that are monitored for physiological status. The spectral properties of the canopy are then correlated with mathematical models of canopy functions such as photosynthesis.

Although this simple direct relationship was an attractive way to predict large-scale productivity, the initial associations between spectral properties of canopies and photosynthesis suffered from using simplified photosynthesis models. Predictions of photosynthesis at the canopy level were limited because they were based on simple empirical functions, had no linkage between stomatal function and carbon assimilation, and leaf properties were assumed to be uniform throughout the canopy. During recent years the correlations between a diversity of spectrographic signatures (including various ratios among numerous wavelengths) and sophisticated photosynthesis models are being investigated (Asrar 1990).

At this time, the relationships between spectral vegetation indices and canopy photosynthesis (carbon gain) have been developed for monoculture stands (Hall et al. 1990) and are not robust enough to be applied to multispecies native stands. In the future, scientists hope to utilize robust spectrographic signatures to predict productivity over large regions through the use of satellite images such as those from the Landsat multispectral scanner, the Landsat thermatic mapper, the French Satellite SPOT, and NOAA satellites.

7. *Modeling Techniques*

Curve-Fitting Procedures. Generally, any equation that can be fitted to a change in mass can be considered a growth curve. In some individual organ systems, such as roots, a linear relationship can exist between change in length and time ($dx/dt = a$) which results in the linear equation $x = x_0 + at$, where x is a growth characteristic (weight, length, etc.), t the time, and a is a growth constant. However, whole plant systems will have a type of compound interest growth capacity. One possible compound growth equation is a differential equation ($dx/dt = rx$), which results in an exponential relationship between growth and time ($x = x_0 e^{rt}$), where x_0 represents size at time zero, e is the natural logarithm, and r (a rate exponent) corresponds to the RGR [equation (2.5)]. Other possible growth equations are logistic, power, Gompertz, or Bertalanffy (Table 2.1). In a practical sense, one must first collect the data on growth for the specific species or organ. Then a *curve-fitting procedure* can be used to select the appropriate equation. The result of the curve-fitting procedure may be an arbitrary equation (i.e., not based on biochemical factors), but it may effectively predict growth in future experiments.

Curve-fitting techniques are frequently rendered inaccurate because of the error carried by repetitive size estimations. Each time a measurement of growth is taken there will be an error in that measurement. The errors are additive in the growth curve-fitting process (Erickson 1976); thus numerical smoothing of the data may be required to remove the erratic nature of replicate measurements.

Curve-fitting growth analyses techniques are excellent for systems that have a regular pattern of growth (see review by Hunt 1978). However, many plant systems have punctuated or extremely variable growth. The cause for the variability may be changes in environmental conditions or changes in internal control over growth. For example, if a plant is water limited, and water application is erratic (such as precipitation), growth will be erratic and unable to fit a growth equation. Or changes in environmental cues may cause changes in physiological receptors (such as phytochrome), which cause changes in growth. Therefore, the application of traditional curve-fitting procedures to analysis of growth in natural populations of plants has limited utility.

Demographic Analysis. Growth of whole plants can be described by the sum of the activity of small growth units (apical meristems, shoot apices, etc.), often termed *modules.* One can treat the population of modules on the plant similar to the way that populations of plants are studied on a landscape. Thus the plant canopy is considered a *metapopulation* of growing units (White 1979). Growth of the organism can then be studied by applying a matrix transition model based on module size (Lefkovitch 1965), age (Leslie 1945), or both (McGraw and Antonovics 1983) to the population of growth modules on a

TABLE 2.1. Various Common Equations Utilized to Model Plant Growth. Integrated Equations Are the More Commonly Used Forms in which the Size (S) Is a Function of Time (t). The Parameter t_0 Is a Constant of Integration. The Differential Equations Define Specific Growth Rates (G) Based on the Integrated Equations. S_∞ Is the Asymptote of the Equations in which It Is Used. Modified from Kaufmann 1981.

Equation	Integrated Version	Differential Version
Exponential	$S = \exp[b(t + t_0)$	$G = b$
Power	$S = \{ab(t + t_0)]^{1/a}$	$\text{InG} = -a \ln S + \ln b$
Gompertz	$S = S_\infty \exp[-\exp - a(t + t_0)]$	$G = -a \ln S + b$
Logistic	$S = S_\infty[1 + \exp - b(t + t_0)]^{-1}$	$G = aS + b$
Bertalanffy	$S = S_\infty[1 - \exp - b(t + t_0)]$	$G = a\ 1/S - b$

plant. In plant population studies the growth or success of any individual is assumed to be independent from all other individuals. However, the modules of a plant canopy are not independent; thus the *demographic analysis* of plant growth may violate the assumption of independence inherent in population-based models.

Demographic modeling of plant growth processes has been used to describe leaf populations (Nilsen et al. 1987b), tillers of a clone (Huiskes and Harper 1979, Fetcher and Shaver 1982), or shoot modules on shrubs and trees (Maillette 1982, McGraw and Antonovics 1983). In each case the module is defined differently, but a matrix model is the basis of growth description. Demographic techniques are similar to that of quantitative phenology because a population of shoots are labeled and monitored for growth components. When demographic analysis is used to predict whole plant annual growth, the visitation period can be limited to once a year.

The summated fate of a population of growth modules can be described by determining the probabilities of any module class entering any other class. Classification of modules in canopies is normally based on size (leaf area on the module) or occasionally on both age and size. The general equation for projecting the population size transition from time t to time $t + 1$ may be written

$$\begin{bmatrix} a_{11} & a_{12} & a_{13} & \cdots & a_{1j} \\ a_{21} & a_{22} & a_{23} & \cdots & a_{2j} \\ \vdots & \vdots & \vdots & & \vdots \\ a_{i1} & a_{i2} & a_{i3} & \cdots & a_{ij} \end{bmatrix} \times \begin{bmatrix} M_1(t) \\ M_2(t) \\ \vdots \\ M_j(t) \end{bmatrix} = \begin{bmatrix} M_1(t+1) \\ M_2(t+1) \\ \vdots \\ M_j(t+1) \end{bmatrix} \tag{2.12}$$

where a_{ij} represents the probability that module i will be derived from a module of size j during the time interval between t and $t+1$; the set of values for M is the state vector and $M_j(t)$ is the number of shoots of size j at time t (Lefkovitch 1965). The equation can be formulated after the second data collection period for plants with an annual growth cycle. Plants that have repeated flushes of growth would require more frequent sampling. Once the sampling frequency has been established, the equation can be iterated until a stable relationship is derived. The stable transition relationship can be used to predict the population structure of shoots over a longer period.

Plant modules can often take a number of possible developmental paths (Figure 2.13). For example, a growing shoot may change into a flowering shoot, or a growing shoot may branch to form multiple shoots. Therefore, a consideration of the population-level changes in reproductive versus growing modules must be considered in the demographic model (McGraw and Antonovics 1983). One can then determine the probability that a shoot of any size category will complete a given size transition in all module pathways (growing, flowering, branching):

$$a_{ij} = g_{ij} + r_{ij}(1 - \alpha_j)(\text{MB}_j)(b_{ij}) + r_j(\alpha_j)(\text{MF}_j)(f_{ij}) \tag{2.13}$$

where g_{ij} is the proportion of shoots that grew from size j to size i, r_j the proportion of shoots of size j that reproduced between time t and $t + 1$, α_j the fraction of reproducing shoots of size j that flowered, $1 - \alpha_j$ the fraction of reproductive shoots that branched, MB_j the mean number of branches for each branching shoot of size j, b_{ij} the proportion of branching shoots of size j that grew to size i, MF_j the mean number of shoots derived from flowering shoots of size j, and f_{ij} the proportion of flowering shoots of size j that grow to size i.

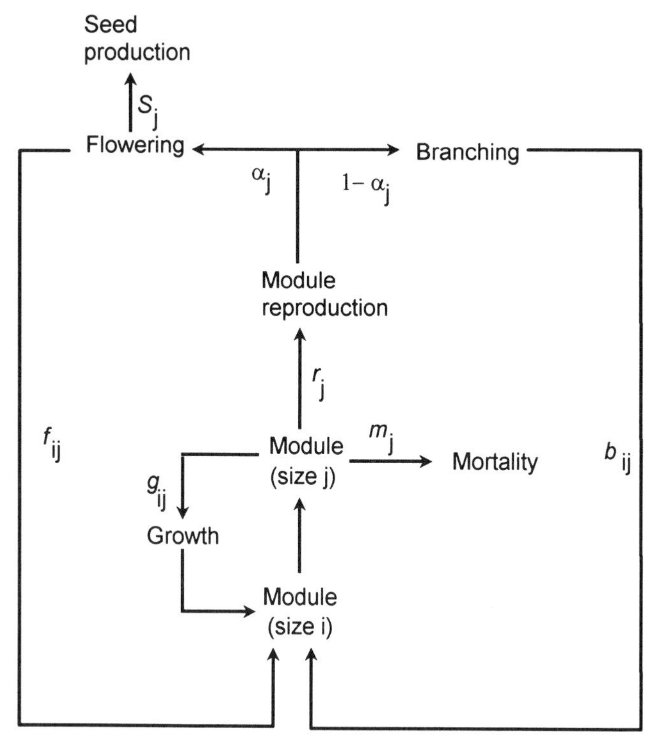

Figure 2.13 Various developmental paths for a growth module (small shoots) from a population of shoots in a canopy. (From McGraw and Antonovics 1983.)

A matrix model based on the a_{ij} values described in equation (2.13) can be iterated several times until a stable distribution is attained. At the stable distribution, the rate of increase of the shoot population can be determined (γ) which has been defined as the long-term prediction of whole plant growth (McGraw and Antonovics 1983) and is analogous to in matrix models for populations of individuals.

Important assumptions of the matrix models may limit the utility of this technique. The model assumes that the transition matrix is consistent among time intervals (commonly years). Therefore, if variability of climatic or biotic pressures on the plant change from year to year, the model will not predict long-term growth accurately. Furthermore, the model assumes that each size class is made up of one size or a stable distribution of sizes. Small increments can be used for size classes to minimize the error from this assumption, but this increases the number of shoots to sample. Shoot populations can form a continuum of sizes that would make the establishment of size categories arbitrary. Classes based on age and size may be a better classification scheme. In addition, the model assumes that there is no interaction between the size classes. If a canopy becomes denser with time, there would be an increase in shoot interaction (shading) that would change the transition probabilities. Shoots of the same size, in different canopy positions (leader, lateral, subcanopy) can have different transition probabilities. If all these problems with the assumptions behind demographic descriptions of shoot populations can be addressed adequately, such techniques can be a powerful method to forecast future growth.

Physiological Modeling. During the 1980s there were important advances in biochemically based photosynthesis models because of careful analyses of the relationships between CO_2 assimilation and water loss (Farquhar et al. 1980, Farquhar 1989). A biochemical model of leaf photosynthesis, based on enzyme kinetics and electron transport properties of chloroplasts, can realistically predict photosynthesis. In addition, semiempirical models of stomatal function can predict stomatal conductance responses to atmospheric carbon dioxide concentrations, temperature, light, or vapor pressure (Ball 1988, Collatz et al. 1991). However, to use any of these realistic models of leaf gas exchange to predict growth of a plant or stand, the model must be scaled from the leaf to the whole canopy level (Ehleringer and Field 1993).

The main inherent problem in scaling from a leaf to canopy is the diversity of leaf developmental stages and microclimates within the canopy. The relative synchrony of leaf developmental stages may vary greatly between species. Short flush species may have highly uniform leaf cohorts (Nilsen et al. 1987b), while other species may have a wide variety of developmental stages on each branch. Some taxa may have a large proportion of leaves in the outer sunlit canopy area, while others may have a deep canopy with leaves in a diversity of microclimatic conditions.

The architecture of a canopy can also interfere with the direct application of a biochemically based model to the canopy level. Canopy density and geometric architecture will affect the gradients of resources from the upper turbulent air to the inner canopy. The orientation (angle, azimuth) and leaf position (whorled, tufted, regularly spaced) will affect the fluid dynamics of the atmosphere around the canopy. Temporal changes in leaf orientation (heliotropism, wind movement) will affect the relation between environmental factors and leaf physiology.

The diversity in canopy architecture and the great number of factors that need to be accounted for when scaling from the leaf to the canopy have stimulated the search for a simple scaling mechanism. For many years (Wallace 1920) a multitude of possible relationships have been investigated by preparing a hierarchy of simple models describing each level of the scaling process (Thornley 1976). These plant–environment models are based on the interaction between biochemically based models and initial boundary layer conditions (Norman and Arkerbauer 1991). Although a great number of models have been developed by these techniques, that model produced by Wallace in 1920 predicts corn yield better than any other model (Johnson 1976). Various mechanisms of scaling and the assumptions contained therein are reviewed by Norman (1993).

Physiological modeling of growth can also be approached by evaluating the concepts of optimal resource distribution and growth efficiency indices. Resources such as nitrogen may be distributed within the canopy in an optimum manner to ensure the greatest possible growth per nutrient investment (Field 1983). Once one establishes a relationship between canopy nutrient concentration and photosynthesis, this relationship can be scaled to the canopy level based on the distribution of nutrient resources in the canopy.

*8. **Root Productivity Measurement.*** Root distribution and abundance has been quantified frequently by labor-intensive methods (excavation, soil removal, trenching) or by multiple soil coring. However, these methods have yet to determine the growth or production of roots accurately. Quantification of root production is difficult because of the abundance and rapid turnover of small rootlets. Any technique for extracting roots from the growth medium will not retrieve all fine roots (50 to 250 μm in diameter). Bearing this limitation in mind, several techniques have been employed to measure root productivity.

A sequential harvest is frequently used in field measurements. After the depth and spatial profile of roots for the particular species has been determined, multiple soil cores are taken at various time intervals in the rooting zone. The roots are floated or separated out of the cores and scored by diameter class and whether the root tissue is live or dead. Hand root separation by liquid or dry sieving (only recommended for separating roots of >2 mm diameter) is highly variable and yields only limited percentages of fine roots. Although an elutriation device (Smucker et al. 1985) can reduce variability in rootlet separation by floating, some initial separation is still required due to removing soil and organic matter, and roots must still be identified as live or dead.

The determination of living roots is critical to any root production assay. Living roots have been separated from nonliving by buoyancy differences (Caldwell and Fernandez 1975), radioactive labels such as ^{14}C or ^{32}P (Singh and Coleman 1983, Svoboda and Bliss 1974), or living tissue staining with tetrazolium (Joslin and Henderson 1984) or fluorescent dyes (McGowan et al. 1983).

Root sampling and separation from soil or growth medium also can be contaminated by the medium. Most types of rooting medium will adhere to the fine roots and frequently be inseparable from the roots. Soil contamination can be accounted for by ashing the root samples and subtracting the ash from the root mass measurement. Root mass is then reported as ash-free weight. Contamination by highly organic rooting media may be removed by evaluating the carbon stable isotope signatures ($\delta^{13}C$) of the roots and root medium. Stable isotopes are discussed in Chapter 5. If the $\delta^{13}C$ is significantly different between the root and the medium, the proportion of contamination in the sample can be determined by measuring the $\delta^{13}C$ of the sample. Stable isotope ratios of other atoms, such as N, H, or O, may be used for this purpose, as they are fractionated during decomposition processes.

In field studies, the soil core may contain the roots of a number of species. It is infrequent that the morphology of roots from the test species is unique among roots in the sample. Therefore, other techniques are used for separating the roots from various species in the sample. Some separation may be possible with stable isotope ratios if these differ among species (Caldwell and Fernandez 1975). The isotope separation mechanism may work best when comparing species with different photosynthetic pathways (C_4 versus C_3; Svejcar and Boutton 1985) or when comparing dinitrogen fixers to nonfixers. Other species separation techniques may be possible by using near-infrared spectroscopy (Rumbaugh et al. 1988). It is intriguing to postulate that molecular techniques, such as RAPD fingerprinting, may be useful for separating roots of individual plants in root productivity studies.

Root mass production is a limited index of the interaction between plants and the soil nutrient environment, because nutrient accumulation is dependent on the surface area of fine roots. Various techniques have been devised to estimate root surface area in samples without tedious root measurement. One technique uses a grid upon which the roots of a sample are spread. The number of contacts in the grid can be used to determine total root length per sample. Image analysis techniques are commonly employed to determine fine root surface area.

All these sampling techniques require an estimation of total root system biomass per volume of soil. Roots are not evenly distributed vertically or horizontally in the soil profile. Thus the heterogeneity of root mass among replicate soil cores may outweigh the temporal changes in live biomass. Due to this potentially large sampling error, techniques for determining root growth without large, spatially replicated sampling have been developed.

New root growth can be assayed by determining the dilution of a temporally applied root system tag (Caldwell and Camp 1974, Milchunas et al. 1985). Normally, ^{14}C is applied to leaves for a specific duration so that some label will be incorporated into root mass. The root mass is then sampled over several time periods to quantify the dilution of label. Although the ^{14}C dilution technique may have complicating factors (Milchunas et al. 1985), such as temporal variation in ^{14}C flow into the root system, this technique may have fewer and partially compensatory errors compared to sequential harvest techniques (Caldwell and Eissenstat 1987).

New root growth can also be assayed by ingrowth into new growth medium. Root growth in potted plants can be measured by transplanting into new media and measuring roots that grow into the new media over a specified time. Field root growth has been measured by burying soil bags, or screens, at specific depths. The bags or screens are removed after a specified time period and assayed for root ingrowth. The ingrowth techniques suffer from the initial disturbance of the root systems (cut roots produce many new rootlets). Furthermore, root growth into screens or mesh bags may be enhanced or inhibited compared to root growth in the root medium.

Among all plant organs, productivity of roots is the hardest to quantify. The large spatial heterogeneity, contamination by the media, resistance of soil for harvesting, difficulty in separating among species, and difficulty harvesting fine roots are all complicating factors. Roots are an important component of productivity because a large proportion of productivity is allocated to roots, and the root system controls soil nutrient acquisition. Therefore, improvements in the techniques for measuring root productivity should take a high priority among studies of whole-plant growth.

V. SUMMARY

The overall objective of this chapter was to cover the regulation of plant growth from cellular to whole-plant processes. During that consideration, several factors of central importance were considered. One of those factors is that plant growth is very plastic. Growth rate of cells, organs, and organisms, developmental trajectory of particular cells and tissues, and allocation patterns among organs are labile processes. They vary in time and space depending on the genetic history of the individual and environmental attributes. Another central characteristic of plant growth is its modular nature. The modular nature allows for temporal plasticity in growth form, tissue turnover rate, and whole-plant growth rate. Measurements of growth rate in plants must account for this modular growth pattern. A third central characteristic is that growth processes in plants are inextricably linked to the environmental conditions of their surroundings. Therefore, it is not surprising that variation in growth patterns is both an adaptive mechanism to and a symptom of environmental stress. The interrelationships of environmental stress and plant growth is of central importance in subsequent chapters.

STUDY-REVIEW OUTLINE

Physiological Basis of Growth: General Issues

1. Growth is studied by scientists on a diversity of temporal and spatial scales.
2. Any spatial or temporal analysis of growth will be ultimately based on cell growth.

Components of Cell Growth in Plants

Cell Division

1. There are many stages in the cell cycle that can regulate cell division, but the G2–M and G1–S phase transitions are best known.
2. G2–M phase transition is regulated by an MPF protein complex made up of p34 and cyclin B.
3. Regulation of MPF is accomplished by phosphorylating and dephosphorylating ^{15}Tyr on p34.
4. Many of the proteins and genes for G2–M phase transition regulation in animal systems have been found in plants, but some aspects of plant cell division must be unique, such as that controlling the pre-prophase band.
5. The G1–S transition is probably regulated by a protein complex system similar to that of the G2–M phase transition.
6. The G1–S transition is also regulated by the availability of important resources, such as nitrogen.
7. Cytokinins stimulate the G2–M phase transition in plants, possibly by stabilizing mitosis-regulating mRNA populations.
8. Gibberellins stimulate the G1–S phase transition, although the mechanism is unknown.

Cell Expansion

1. Cell growth in plants is mostly a result of cell expansion.
2. The cell wall is composed of cellulose microfibrils embedded in a polysaccharide gel.
3. The rigidity of the primary cell wall is dependent on hydrogen bonding between hemicellulose and the gel polysaccharides as well as that between cellulose and cell wall proteins.
4. Water in the cell wall is very important to the maintenance of flexibility, chemical activity, and structure of the wall.
5. Cell wall plastic expansion occurs when turgor pressure exceeds the yield threshold and the wall extensibility is greater than zero (i.e., the Lockhart equation).
6. Cell wall elasticity refers to cell wall stretching below the yield threshold. No hydrogen bonds are broken, so the wall returns to its original shape after the pressure is removed.
7. Volumetric modulus of elasticity describes the elasticity of plant cells and ranges from 1 to 50 MPa.
8. Four growth regulators influence cell wall expansion, but the mechanisms by which auxin affects cell expansion is the best understood.
9. Auxin stimulates the release of a wall loosening factor into the wall, which reduces m and increases the rate of cell expansion.

Whole-Plant Growth

Proportional Growth of Organs

1. Plants invest a variable proportion of their resources among organs in relationship to environmental conditions.
2. The patterns of resource allocation among organs of a plant differ among species due to genetic, evolutionary, and environmental factors.
3. Allocation to organs also varies spatially and temporally within one species or one individual.
4. Differential allocation patterns can influence the potential for whole plant growth.

Determinate Versus Indeterminate Growth

1. Plants have modular growth, which means they produce multiples of each organ.
2. In plants with deterministic growth, shoots elongate and terminate with an inflorescence each growth period.
3. In plants with indeterminate growth, shoots do not terminate in reproductive structures and continue their growth through several growth periods.

Considerations of Plant Architecture

1. Genetic factors that regulate plant architecture are constrained by biophysical imitations.
2. The phylotaxy of a species defines the spatial arrangement of organs on the plant axis, and environmental conditions adjust the distance between organs and the rate of organ production.

Genetic Regulation of Growth Rate

1. The growth rates among individuals of a population are not all the same even when grown under identical environmental conditions.
2. Growth rate is heritable and regulated by many genes.

Measuring Plant Growth

Productivity and Yield

1. Gross primary productivity is the total abiotic resources accumulated by living material over the growth period.
2. Net primary productivity is the GPP minus respiratory losses.
3. The true increment is the NPP minus other losses, such as to herbivory, abscission, or leakage.
4. Yield is TI minus the losses during harvest.

Techniques for Measuring Growth

Direct Sampling Techniques

1. Sequential harvesting, the most frequent sampling technique, is limited to studies with a large uniform sample base of relatively small species.
2. Sequential harvesting does not account for tissue turnover unless very short sampling periods are used.
3. Sequential sampling is appropriate for many agricultural systems, but the variability in growth of natural systems limits the effectiveness of this technique.

Relativized Measures of Growth

1. Relativized indices of growth compare instantaneous growth to the total mass or leaf area.
2. Relative growth rate (RGR) relates growth to total plant weight. This value is constant for plants with exponential growth.
3. Net assimilation ratio (NAR) measures the efficiency of growth by relating instantaneous growth to total leaf area.
4. RGR can be integrated to form an interval description of growth, but the integration of NAR results in a number of possible equations, depending on the relationship between increments of weight and leaf area.

Indirect Sampling Techniques

Dimension Analysis

1. Plants have consistent allometric relationships between dimensions and growth.
2. Regressions between dimensions, such as basal diameter, and plant productivity, biomass, or allocation can be formulated.
3. The regressions require initial harvest of plants or plant parts, but following the regression formulation, productivity measurements are nondestructive.
4. These techniques are appropriate for natural stands of larger species with small and variable sample size.
5. Quantitative phenology, a measure of the tissue turnover rates, can help provide the information to account for tissue abscission and herbivory when productivity is estimated by dimension analysis.

Plastochron Index

1. The plastochron index (PI) was developed to identify chronologically uniform leaves.
2. PI can be modified to measure the developmental stage of a stem containing many leaves.
3. Plastochron days (PD), the number of days to complete one plastochron, can be used as a measure of growth rate.

Remote Sensing

1. Relationships between spectrographic signatures of canopies and canopy productivity can predict growth over large regions.
2. The preliminary relationships between infrared reflectance from a canopy and productivity suffer from using inappropriate models of photosynthesis.
3. Relationships between spectral signatures (combinations and mathematical manipulations of several wavelength registers) and photosynthesis of monospecies canopies have been developed, but carbon gain by multispecies canopies has yet to be effectively associated with spectral signatures.

Modeling Plant Growth

Curve-Fitting Techniques

1. Plant growth can be fitted to a number of possible equations, the most common of which are the logarithmic, power, Gompertz, and Bertalanffy relationships.
2. Mathematical curve-smoothing techniques can develop a tight fit between growth data and a predictive equation, but the resulting equation is an arbitrary prediction without biochemical basis.
3. Curve-fitting techniques are inadequate for the prediction of growth in systems with punctuated or delayed growth.

Demographic Techniques

1. Population demography models, particularly matrix models of population structure and growth, can be applied to growing shoots or leaves of a canopy.
2. The canopy is defined as a population of growth modules based on module size, age, or both.
3. The population of modules is monitored for transitions between size classes and developmental categories (branching, flowering, growing).
4. The Leslie or Lefkovitch matrix models are then used to predict the stable probabilities of transitions between module classes.
5. The probabilities of transition describe a stable shoot population and can be used to predict growth rate.
6. The model assumes that the matrix is constant, and year-to-year climatic variables or developmental changes with plant age do not affect long-term growth.

Physiological Techniques

1. Several biochemical based models of photosynthesis and stomatal function can accurately predict photosynthesis in a variety of situations.
2. Leaf photosynthesis must be scaled to the canopy level in order to predict whole plant growth.
3. Leaf to canopy scaling is difficult because of the leaf developmental and microclimatic heterogeneity in the canopy.
4. Models of canopy carbon gain, such as CUPID, are based on a hierarchy of simple relationships that each account for an important aspect of canopy architecture, leaf display, or within-canopy microclimate.
5. Simple scaling mechanisms centered around light-penetration models in the canopy have been employed to avoid the model complexity that comes from trying to account for the within-canopy heterogeneity.

Root Production

1. Roots are the hardest plant organs to include in productivity sampling because of removal difficulty and growth-medium contamination of root samples.
2. Sequential harvests of roots cannot account for the rapid turnover of fine roots.
3. Root productivity can be estimated by a dilution of radioactive label over time.
4. Root productivity can be measured by ingrowth into new medium, mesh bags, or mesh screens. However, these techniques disturb the root system and lead to over- and underestimates.

SELF-STUDY QUESTIONS

1. Describe the current knowledge about the molecular regulation over the G2–M and G1–S phase transitions.
2. What special characteristics of the plant cell cycle will involve molecular control different from that of animal cell cycles?
3. How do growth regulators affect growth of the whole plant?
4. What factors regulate the capacity of a cell to expand in volume?
5. What characteristics of the cell wall affect elastic and plastic properties of the wall?
6. Define the appropriate techniques that one should use to study net productivity in (a) tropical forest; (b) potted plants; (c) cornfield; (d) shrub rangeland.
7. In what ways are demographic modeling and quantitative phenology similar but also in opposition?
8. Design the optimum combination of techniques for measuring the growth patterns of a field-grown woody plant.
9. Compare and contrast the appropriate applications and limitations for sequential harvest and dimension analysis techniques.
10. What is the value of the plastochron index, and in what systems should it be applied?
11. Compare and contrast the RGR and NAR indices of growth. What is the basis of their determination, and how can they be applied to interval measurements of growth?

SUPPLEMENTARY READING

Abrahamson, W. G., and H. Caswell. 1982. On the comparative allocation of biomass, energy, and nutrients in plants. Ecology 63:982–991.

Biscoe, P. V., and K. W. Jaggard. 1985. Measuring plant growth and structure. Pp. 215–228 in: B. Marshall and F. I. Woodward (eds.), Instrumentation for Environmental Physiology. Cambridge University Press, Cambridge.

Drew, M. C., and L. R. Saker. 1980. Assessment of a rapid method, using soil cores, for estimating the amount and distribution of crop roots in the field. Plant and Soil 55:297–305.

Hardwick, R. C. 1984. Some recent developments in plant growth analysis—a review. Annals of Botany 54:807–812.

Nicholls, A. O. and D. M. Calder. 1973. Comments on the use of regression analysis for the study of plant growth. New Phytologist 69:32–36.

Niklas, K. J. 1994. Plant Allometry, the Scaling of Form and Process. University of Chicago Press, Chicago.

Poorter, J. R. and C. Lewis. 1986. Testing differences in relative growth rate: a method avoiding curve fitting and pairing. Physiologia Plantarum 67:223–226.

Sabins, F. F. 1987. Remote Sensing: Principles and Interpretation, 2nd ed. Freeman, New York.

Tilman, D. 1988. Plant Strategies and the Dynamics and Structure of Plant Communities. Princeton University Press, Princeton, NJ.

Williams, R. F. 1946. The physiology of plant growth with special reference to the concept of net assimilation ratio. Annals of Botany 10:41–62.

3 Plant Membranes as Environmental Sensors

Outline

Objectives

1. Discuss the general ways that membrane structure and composition interact with environmental stress.
2. Describe important structural attributes, and molecular composition, of membranes as they relate to environmental stress.
3. Define the various states of membrane structure, and focus on the importance of fluidity.
4. Present intrinsic factors that regulate membrane fluidity, including the structure of fatty acyl groups, composition of hydrophilic heads of phospholipids, sterol composition, and integral proteins.
5. Discuss the extrinsic factors that influence membrane fluidity including temperature, light quality and intensity, degree of hydration, ionic interactions, pH, and phytohormones.

I. INTRODUCTION

Unlike animals, most plants are unable to move from adverse environments and thus must rely on physiological and biochemical strategies that allow them to exist within a range of environmental extremes that may in essence be defined by the plants growing there. Neither do plants possess the complex sensory mechanisms (sight, touch, smell, taste, hearing) found in animals that enables these organisms to move to less hostile environments. Nonetheless, plants do have the ability to respond to their environment and in many instances "sense" changes through perturbations in membrane structure and composition.

Both biophysical and biochemical changes in plant cell membranes can be altered by any number of biotic or abiotic factors. Whether a plant can survive a specific environmental insult depends on its ability to maintain metabolic homeostasis, and this in turn may be controlled in part by the structure and composition of plant cell membranes. Therefore, to appreciate plant membranes fully as sensors of environmental change it is necessary to have knowledge of the structure and function of the plant and membranes and how they respond to the environment.

II. MEMBRANE STRUCTURE AND COMPOSITION

Membranes are integral parts of cellular organelles, such as plastids (chloroplasts, chromoplasts, leucoplasts, amyloplasts, and etioplasts) vacuoles, mitochondria, endoplasmic reticula, nuclei, Golgi bodies, peroxysomes, glyoxysomes, fat bodies, lysosomes, and the plasmalemma. Each organelle has a unique function which in part can be attributed to the unique nature of its membrane. This, in turn, provides a type of compartmentation of function that enhances regulation of biochemical and physiological processes within the plant cell.

Cellular membranes are comprised of lipids, sterols, carbohydrates, and proteins. The primary components are proteins and lipids, the former comprising one-half to two-thirds of the membrane dry weight. The relative proportion of membrane components and their organization depends on the organelle, plant structure, species, and environmental conditions. Some organelles, such as the chloroplast and the mitochondrion, have two separate bilayer membranes, which provides a type of compartmentation within a single organelle. Membranes are generally arranged in a bilayer configuration, as depicted in Figure 3.1. Under some environmental conditions (dehydration, high temperature, or high ionic concentrations), and depending on the membrane composition, entire membranes or domains in the membrane can develop monolayer configurations, as depicted in Figure 3.2. The extent of such changes determines whether or not such alterations are detrimental to the cell. In some instances, monolayer formation within domains of the membrane may be of common occurrence and function to provide channels for ion transport.

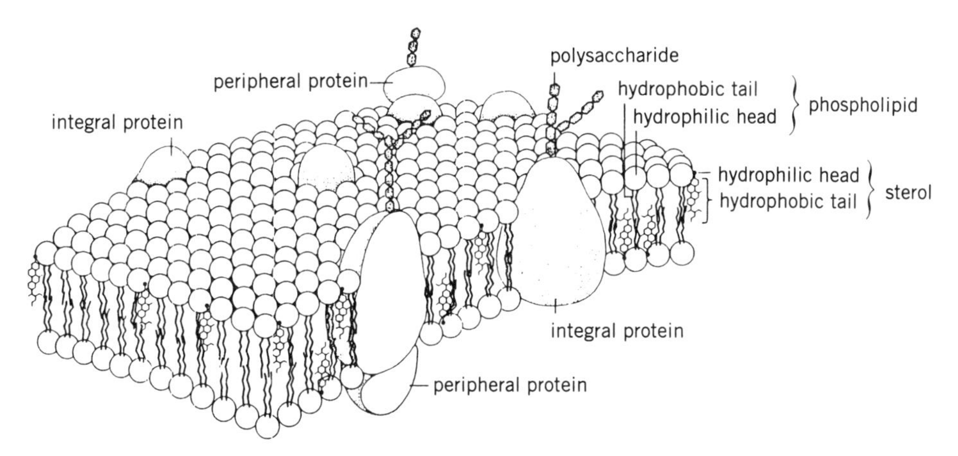

Figure 3.1 Typical bilayer configuration of biological membranes, illustrating proposed arrangement of lipids, proteins, carbohydrates, and sterols. (From Salisbury and Ross 1992.)

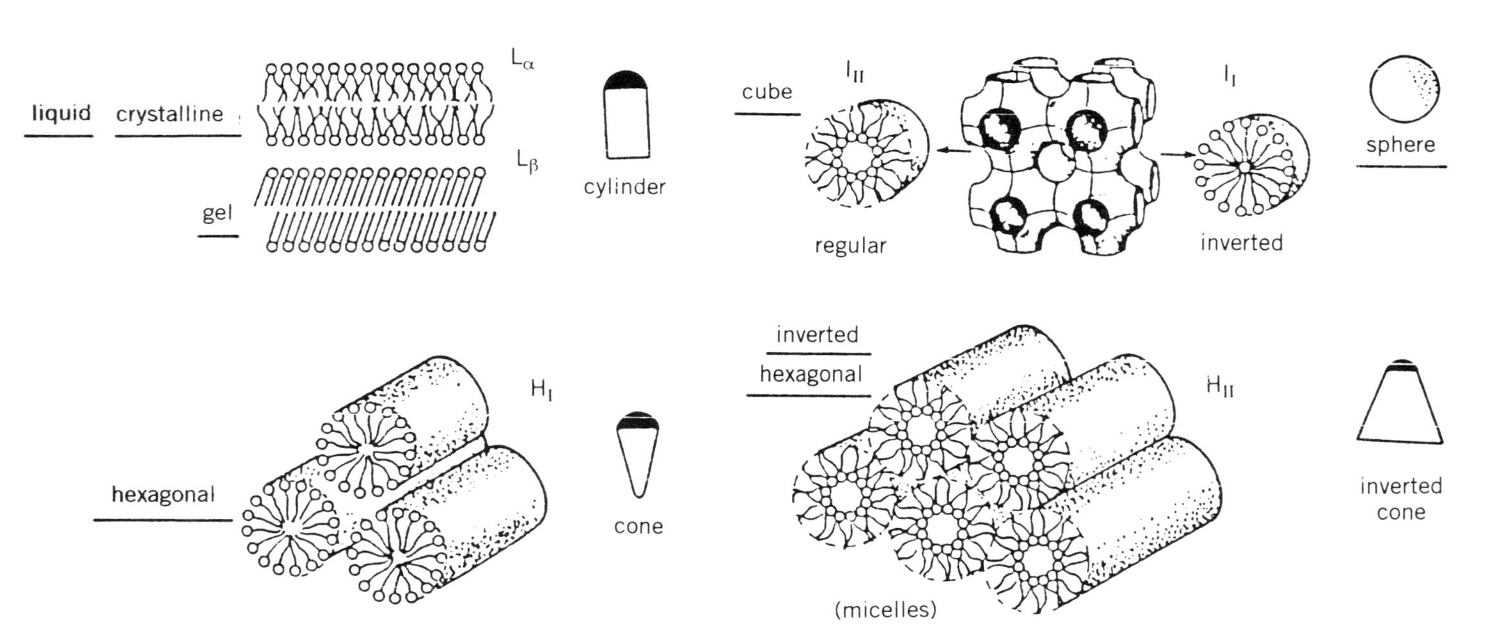

Figure 3.2 Membrane phase configurations that can occur in biological membranes. (From Leshem 1992.)

The major lipid components of cell membranes are phospholipids, which include phosphatidylcholine (PC), phosphatidylethanol (PE), phosphatidylethanolamine (PEA), phosphatidylserine (PS), phosphatidylglycerol (PG), phosphatidylinositides (PI, PI2, PI3), and cardiolipin (CL). The most abundant of the phospholipids in plant membranes is PC. All of the foregoing phospholipids, with the exception of CL, have the same basic chemical structure, which includes a three-carbon glycerol backbone to which fatty acyl groups of varying chain lengths and degree of unsaturation are attached at the 1 and 2 positions. A phosphate group is situated at carbon 3, to which may be attached a carbohydrate, amine group, or an amino acid. Cardiolipin differs from the above in that it possesses three glycerol molecules connected by two phosphate groups, and the two end glycerol molecules each have two fatty acyl groups (Figure 3.3). One other type of phospholipid is the lysophospholipid, which is characterized by having one fatty acyl group missing, which can be exhibited by any of the phospholipids mentioned above. These appear to be prevalent under stress or senesing conditions and may even accelerate membrane degradation under stress.

Three other lipid classes found in plants that do not contain phosphate groups are sphingolipids (cerebrosides), galactolipids, and sulfolipids. All have the glycerol backbone in common and two fatty acyl tails (in cerebrosides one is the 18-C alcohol sphingosine). All three lipid classes have either glucose or galactose attached to carbon 1 of glycerol. The sulfolipid sulfoquinovosyldiacylglycerol also has a sulfo group attached to a galactose sugar molecule. The sphingolipids represent the second-most-abundant lipid in the tonoplast membrane and plasmalemma.

The lipids described above exhibit what is termed *amphipathy,* the ability of a molecule to possess both hydrophilic and hydrophobic characteristics. The fatty acyl groups associated with the lipids described above give the molecules hydrophobic characteristics, while the phosphate, amine, amino, and carbohydrates provide the hydrophilic behavior. This is a very important characteristic of lipids necessary for membranes to exhibit bilayer behavior. A membrane is essentially lipid sandwiched between two water layers. The inner water layer is the cytosol, the outer layer the extracellular water. In such an environment the lipids thermodynamically orient their polar ends (hydrophilic ends) toward the aqueous phases and the nonpolar ends (hydrophobic ends) orient away from water, creating a stable bilayer.

Cell membranes also exhibit asymmetry; that is, one side of a membrane may differ in lipid and protein composition compared to the other side (Figure 3.4). Such sidedness is important relative to transport mechanisms and enzyme function which occurs at the membrane interface. Generally speaking, neutral phospholipids such as PC, PE, and PEA are associated with the extracellular surface of membranes, while negatively charged phospholipids seem to be more abundant on the cytosolic side of the membrane. Organelles possessing double membranes such as the chloroplast and the mitochondrion have also been shown to exhibit differences in membrane composition between the outer and inner membranes, as well as possessing lipids that are unique to those organelles. For example, very little phospholipid is found in chloroplast membranes, but MGDG and DGDG is abundant in the chloroplast envelope while SQDG is associated with lamellar membranes of the stroma. Similarly, in the mitochondrion, PI and CL are components of the inner membrane.

Fatty acyl groups (hydrophobic components of glycerolipids) of plant membrane lipids differ in carbon number, degree of unsaturation (number of double bonds), position of double bonds, and configuration. Plant membrane fatty acids range in carbon length from

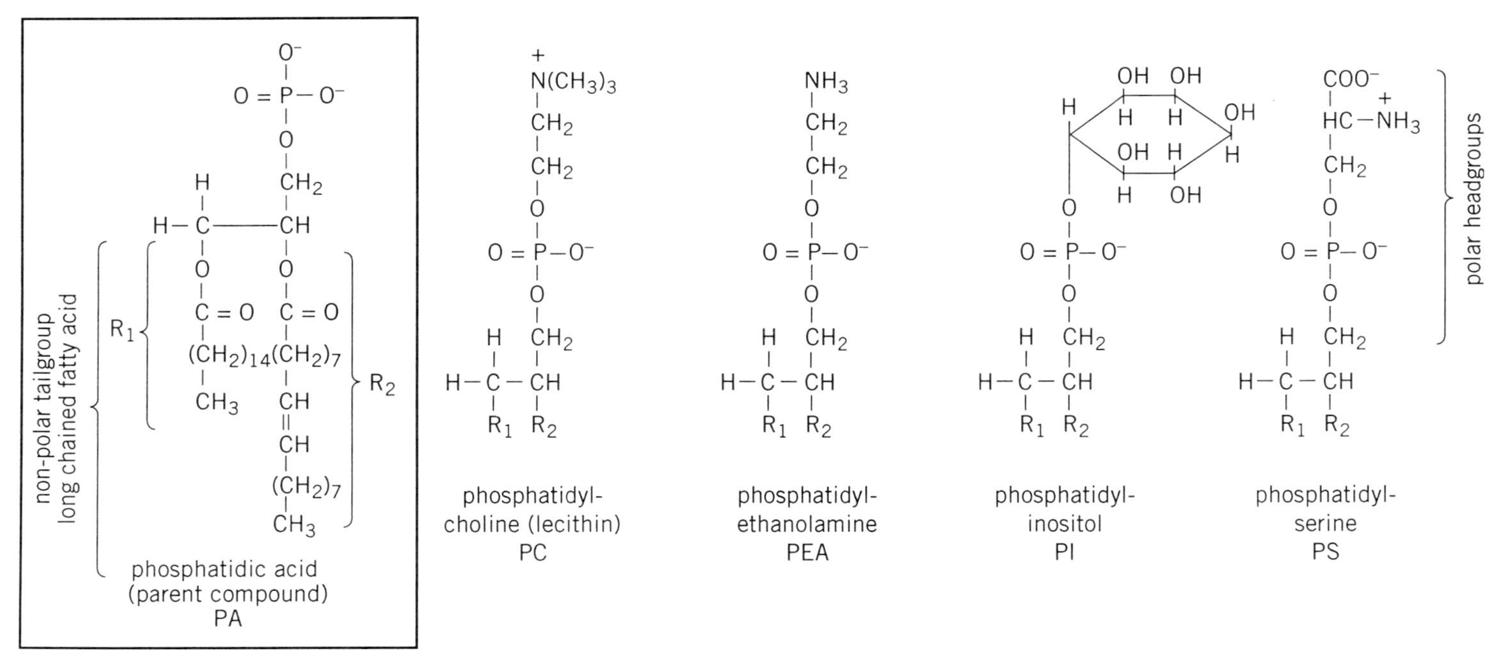

Figure 3.3 Typical membrane phospholipid structures. (From Leshem 1992.)

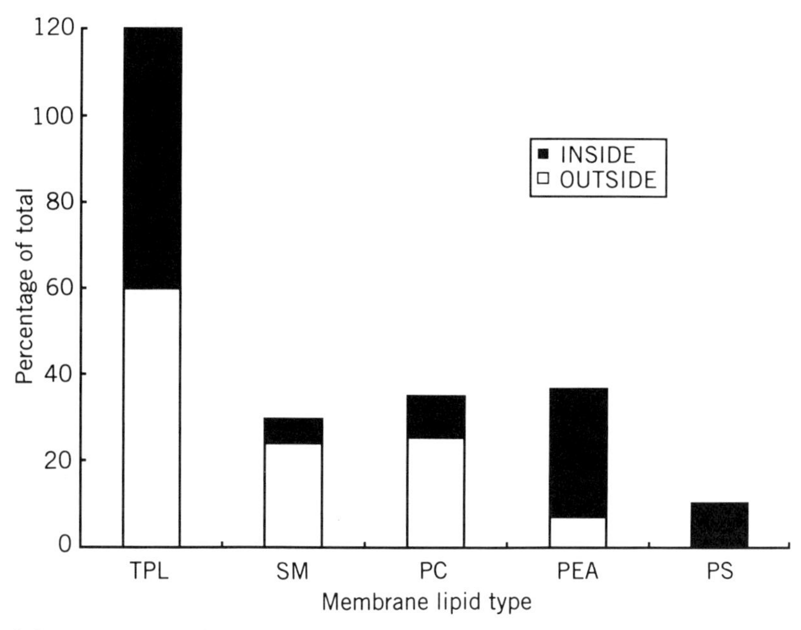

Figure 3.4 Membrane lipid composition of human blood cells, illustrating sidedness. (From Leshem 1992.)

16 to 24 carbons (even carbon numbers only), with degrees of unsaturation from no double bonds to four. Some diatoms are known to possess as many as six double bonds in long-chain fats. The double bond is usually oriented in a *cis* configuration, resulting in bending or kinking of the carbon chain, as opposed to the straight chain exhibited by saturated fatty acids. Polyunsaturated fats are twisted or kinked even more. Up to a point (two unsaturations) unsaturation will result in a greater space occupied by the fat in the membrane bilayer. The types of fatty acyl groups associated with any particular glycerolipid is variable and can change as a function of the environment. Usually, a saturated fat is associated with carbon 1 of the glycerol backbone, while an unsaturated fatty acid is linked to the second carbon. The differences described relative to chain length, configuration, unsaturation, and specificity of fatty acyl attachment to glycerol all have significant implications relative to membrane function. Also of importance are the nature and structure of the hydrophilic portion of glycerolipids. As indicated earlier, this portion of the lipid molecule can have associated with it phosphate groups, sulfo groups, amino groups, amines, and carbohydrates. Depending on how these are configured, different charge characteristics of the hydrophilic headgroups can occur. In some cases the lipids may be neutral as a result of no charge or charge cancellation, while other lipids may have a net charge ranging from -1 to -5 (Table 3.1). Such charge characteristics at the surface of membranes is very important relative to protein interactions, ion balances, hydration, and membrane structure.

Sterols are important in membranes as stabilizers of glycerolipids. Essentially all eucaryotes require them for membrane functioning. The principal sterols that occur in many higher plants are cholesterol, campesterol, sitosterol, and stigmasterol (Figure 3.5). The latter two usually occur in the greatest abundance followed by campesterol and then cho-

TABLE 3.1 Characterization of Membrane Lipids Relative to Net Charge

Lipid Species	Net Charge	Remarks
Electroneutral		
GlCer	0	Completely electroneutral since headgroups lack electric charge
MGDG	0	
DGDG	0	
PC	0	Zwitterionic in nature, having one + and one − charge, which cancel
PE	0	
PEA	0	Also a zwitterion, but + charge is on a primary NH_3, therefore tends to be more positive
Anionic (negative)		
PG	−1	Have two −1 and one + charge
PA	−1	
PS	−1	
PAF	−1	
SQDG	−1	Negative charge from the SO_3^- in the headgroup
PI	−1	
CL	−2	Has two phosphate neck sections

Source: Modified from Leshem 1992.

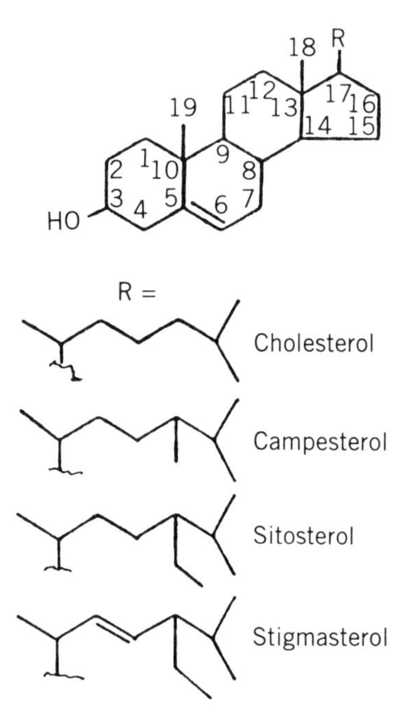

Figure 3.5 Free sterols of vascular plants. (From Grunwald 1975.)

lesterol. Cholesterol usually occurs in small amounts in plants, but relative high levels have been reported in tobacco and wheat leaves and shoot meristems of several monocots and dicots. Red algae and some diatoms have cholesterol as the predominant sterol (Howard and Orcutt 1976, Orcutt and Patterson 1975). Sterols other than the ones mentioned above have also been identified from higher plants. The algae and fungi exhibit considerable variation in sterol composition to the extent that species of the algae genus *Chlorella* have unique sterol compositions compared to other species of the same genus (Holden and Patterson, 1982). Thus the diversity of sterols is much greater in plants than it is in animals since the primary sterol found in higher animals is cholesterol. All sterols are characterized by a four-ringed hydrocarbon called cyclopentanoperhydrophenantrene. The four rings are designated as rings A to D. Attached to ring D is a hydrocarbon tail ranging in carbon number from 7 to 9. At carbon 24 a methyl or an ethyl group may be attached in an R or S configuration. A single double bond may or may not be inserted in one of several positions in the side chain; frequently, the C-22 position is unsaturated. Double bonds can also occupy several positions within the ring system. In higher plants, one or two double bonds are most commonly associated with ring B. Double bonds can also be associated with rings C and D of precursor sterols. Methyl groups are linked to carbons 10 and 13 and in precursor sterols are frequently found at carbons 4 and 14. Sterols can occur in both free and conjugated forms. Free forms are characterized by having a hydroxyl group associated at carbon 3. Conjugated sterols include steryl esters, steryl glycosides, and acylsterylglycosides (Figure 3.6). Free sterols can only be incorporated in cell membranes since the free hydroxyl group appears to be required (Grunwald 1971). The functions of the conjugated sterols are not known, but a function suggested for sterylglycosides has been that they are translocatable forms of sterol. Conjugated forms tend to accumulate in plants under a variety of stresses, while free sterols decline. Like glycerolipids, free sterols are amphipathic molecules, but not to the extent of the former. The planar ring structure and the aliphatic side chain provide the basis for the hydrophobic nature of the molecule, and the hydroxyl group represents the hydrophilic component of the amphiphile. It is thought that sterols are interspersed between glycerolipids, with the aliphatic side chain and sterol nucleus adjacent to the fatty acyl groups of the glycerolipids and the hydroxyl group associated with the polar lipid headgroups of phospholipids (Figure 3.1). It appears that certain sterols are more readily incorporated than others into membranes. Sterols without methyl or ethyl groups in the side chain seem to be incorporated more readily (cholesterol and campesterol) than sitosterol or stigmasterol, which possess ethyl groups at C-24 (Grunwald 1974). This may reflect increased bulkiness of the side chain as a result of methylation and greater difficulty of being inserted in the membrane. Thus plants have a great diversity of sterols that exhibit a variety of structural differences, which may relate to differing abilities of plants to stabilize membranes under a variety of environmental conditions.

Structurally, membrane proteins are classified as peripheral or integral depending on their position in the membrane (Figure 3.1). *Integral proteins* traverse part or all of the membrane, while *peripheral proteins* interact at the membrane surface. Functionally, membrane proteins are classified as catalytic proteins, proteinaceous carriers, solute channel proteins, and receptor proteins. Integral proteins may traverse a membrane only once or several times, creating a large protein consisting of subunits. The ability of a protein to traverse a membrane and occupy part of the inner and outer membrane domain is dependent on the charge characteristics of the amino acids, the resulting amphipathic nature of the polypeptides that comprise the protein, and the interaction of the protein with the

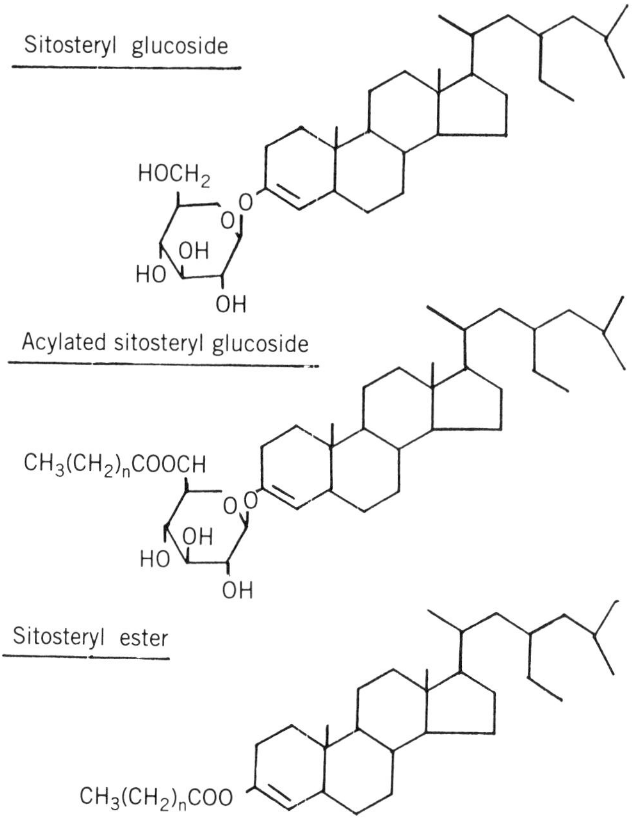

Figure 3.6 Conjugated sterols found in vascular plants.

glycerolipids comprising the membrane. Peripheral proteins interact with charged glycerolipid headgroups relative to their positioning in the membrane. Such protein associations are electrostatic in nature and easily displaced. The biophysical and chemical relationship of membrane glycerolipids and membrane proteins is very important relative to membrane stability and proper functioning.

III. DYNAMIC NATURE OF CELL MEMBRANES: MEMBRANE FLUIDITY

A. Intrinsic Factors Affecting Fluidity

Membranes are dynamic systems and are constantly in a state of motion. This motion is due largely to the kinetic movement of the lipids and proteins within the membrane. The degree of movement depends on the composition of the membrane as well as its physical and chemical environment (Table 3.2). Lipids in the membrane can exhibit rotational and translational (lateral) movement. It has been estimated that phospholipids can trade places with adjacent lipids 107 times per second but movement of proteins is 10 to 10,000 times slower. Such movement is important in membranes so as to facilitate maximum contact with enzyme substrates and other molecules, such as hormones and ions. As indicated above, membrane movement is related to composition, which determines membrane viscosity, which in turn is related to membrane fluidity and permeability. The greater the viscosity of a membrane, the lower the fluidity and semipermeability. Generally, membranes

TABLE 3.2 Factors Affecting Lipid Fluidity in Membranes

Intrinsic	Indirect
Lipid composition	Hormones
Chain length	Metabolic processes
Unsaturation	Various cellular processes
Polar head	Neoplasia
Cholesterol	Physical
Ubiquinone	Temperature
Lipid-soluble vitamins	Pressure
Integral proteins	Membrane potential
Peripheral proteins	Extrinsic
Cytoskeleton	Detergents
Hydration	Anesthetics
Ions	Synthetic polymers
pH	
Lysophospholipids	
Fatty acids	

Source: Modified from Lenaz and Castelli 1985.

exist in one of three states depending on the external environment: (1) a liquid crystalline phase, which represents a fluidity range that allows the membrane and its components to function; (2) a solid-gel phase, which represents a membrane that retains the bilayer configuration but is rigid and essentially nonfunctional; (3) and the hexagonal and cubic phases, which may reflect membrane disruption related to environmental extremes (Figure 3.7). The ability of a plant to cope with its changing environment depends in large

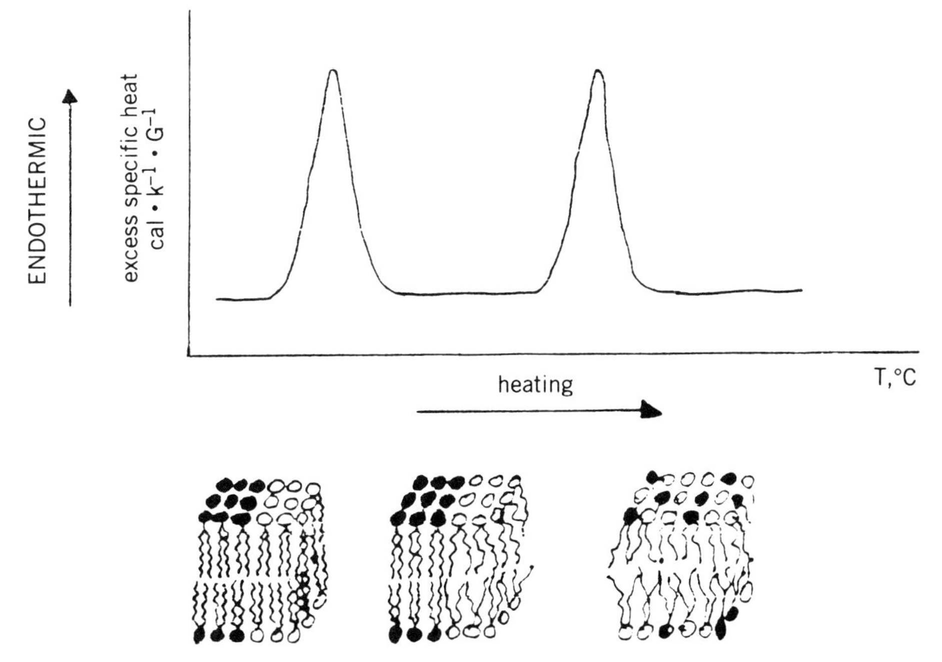

Figure 3.7 Effects of temperature on membrane phase transition. (From Lenaz and Castelli 1985.)

part on its ability to maintain membrane fluidity. This depends on the genetically determined lipid and protein composition of the plant membrane systems, the ability of the plant to alter membrane composition to modulate fluidity, and the magnitude of the external environmental stress.

In the following discussion we summarize some of the intrinsic factors that influence membrane fluidity. The composition and structure of fatty acyl groups of glycerolipids play a very important role in modulating membrane fluidity. For example, fluidity is increased by shorter chain lengths (short-chain fats exhibit low-temperature melting points, while long-chain fats exhibit high-temperature melting points) increased unsaturation (saturation increases melting-point temperature, unsaturation decreases melting-point temperatures), the *cis*-double-bond configuration is more effective than the trans configuration, and the double bonds located near the center of the fatty acyl chain are more effective than those near the end. The *cis* configuration and the central location of the double bond in the chain increase the space occupied in the membrane bilayer, thus increasing rotational movement and the kinetic motion of adjacent molecules and consequently, fluidity. Some of these relationships can be illustrated by determining the transition temperature (T_c, the lowest temperature at which the initial signs of a phase change from a liquid-crystalline to a solid-gel phase occurs) of lipids known to comprise biological membranes (Table 3.3).

Membrane fluidity can also be influenced by the nature of the hydrophilic headgroups of the glycerolipids. As indicated earlier, headgroups are comprised of phosphate and sulfo groups, amines, amino acids, and carbohydrates with differing charge characteristics, ranging from neutral to −5. Thus anionic headgroups can interact with cations, such as Ca^{2+} and Mg^{2+}, which could cause electrostatic bridges between a cation and

TABLE 3.3 Transition Temperatures for Phosphatidylcholines and Phosphatidylethanolamines

Phospholipid	[a]T_c (°C)
PC	
12:0	0
14:0	23.6
16:0	41.4
18:0	55.4
22:0	75
16:1 (c,9)	−36
18:1 (c,9)	−19
18:1 (t,9)	9.5
18:0/18:1 (c,6)	30
18:0/18:1 (c,12)	12
18:0/18:1 (c,16)	43
PE	
12:0	29.8
14:0	48.3
16:0	61.6
16:1 (c,9)	−33.5
18:1 (c,9)	−16
18:1 (t,9)	35

Source: Modified from Lee 1983.

two adjacent negatively charged lipids, resulting in condensing of the lipids in the membrane, thus reducing its fluidity. Ca^{2+} is more important than Mg^{2+} in this respect. Small amounts of Ca^{2+} are essential for membrane stability, but excessive amounts may be detrimental. Charge characteristics of glycerolipid heads are also important relative to the positioning of proteins in cell membranes. High levels of cations and the degree of hydration and pH could influence lipid–lipid and protein–lipid interactions by altering the charge characteristics of phospholipid headgroups in membranes, resulting in changes in the packing of membrane components, which in turn could affect fluidity and functioning of proteins. Comparison of the T_c values of PC and PE that have similar fatty acyl groups (Table 3.3) suggests that the choline moiety may have a more fluidizing effect on membranes than the ethanol moiety, even though both types of lipid are neutral in charge.

Sterols are essential components of cell membranes in that they tend to prevent membranes from becoming too fluid or not fluid enough. The considerable structural diversity [120 different sterols have been identified from the marine environment, and as many as 1778 have been estimated to occur naturally (Nes 1984 and references therein)] among plants with respect to sterol composition may suggest an important role in maintaining physiologically functioning membranes in diverse membrane types associated with different organelles, found in an array of plant structures associated with numerous plant species that occupy variety of environmental niches. The ability of a sterol to influence fluidity is related to how it interacts with the fatty acyl groups of the membrane bilayer. Insertion of sterols into a membrane above the T_c value tends to reduce the random motion and straighten out the fatty acyl groups, causing a rigidifying effect. Insertion below the T_c value tends to fluidize the membrane. The amphipathic nature of these molecules may explain the differences observed relative to their effects on membrane fluidity. However, such effects could possibly depend on the sterol structure. As indicated earlier, sterols that have bulky methyl or ethyl groups (stigmasterol or sitosterol) are not as readily incorporated into some membrane systems as are sterols without these structures (campesterol, cholesterol). On the other hand, it would seem that sterols with methyl or ethyl groups may tend to fluidize membranes more, if incorporated, than those without these substituents since they may occupy more space and create more randomness among the fatty acyl groups. Similarly, sterols with unsaturations in the side chain (stigmasterol) may occupy more space because of "kinking" of the side chain as observed for fatty acyl groups. It is noteworthy that ratios of sitosterol to stigmasterol are frequently reported to change as a result of numerous environmental stresses (Douglas and Walker 1983, Grunwald 1978). Sitosterol has been reported to be more readily incorporated than stigmasterol into soybean vesicles. The ability or lack of ability for a sterol to be incorporated into a membrane may be as much a function of the membrane composition and structure at any given time as it is the size or structure of the sterol molecule. It is commonly known that free sterols tend to decline and conjugated forms increase in plants subjected to a variety of environmental stresses. Metabolically, it has been demonstrated in plants that the turnover rate of conjugated sterols is very rapid (hours), while free sterols are much slower (days) (Mudd et al. 1984). This suggests that interconversions between free and conjugated sterols occurs very rapidly. Do interconversions between these forms represent possible control mechanisms that could regulate the movement of free sterols in and out of membranes under varying environmental conditions? How, when, and where different sterols are incorporated into membranes is not understood. However, interconversions between free and conjugated sterols as well as subtle differences in sterol structure (side chain size, configuration, double bonds) appear to be important in the ability of an organism to adjust to its environment.

Proteins may also influence the fluidity of membranes. However, such an effect may be site specific in addition to whether the protein is intrinsic or peripheral. Proteins embedded in the membrane may tend to rigidify adjacent membrane lipids, depending on the type of lipid–protein interaction. Peripheral proteins may have less of an effect since they are less tightly bound to the membrane.

B. Extrinsic Factors Affecting Fluidity

Having dealt with some of the intrinsic factors that influence membrane fluidity, in the following section we deal with the extrinsic environment and how plants adjust membrane composition and structure to cope with changing external conditions. This process has been referred to as *homeoviscous adaptation*. Perhaps the term *homeoviscous acclimation* would be more appropriate since adaptation has connotations of long-term evolutionary consequences, whereas acclimation implies short-term adjustments.

Some of the major extrinsic factors that can influence membrane fluidity include temperature, light, air pollution, degree of hydration, ions (heavy metals and salinity), pH, and phytohormones.

1. Temperature and Membrane Fluidity. The ability of a plant to tolerate extremes in temperature may reflect the genetic constitution of the plant as it relates to membrane lipid composition, structure, and/or its ability to acclimate by changing the same. Plants that have adapted to the various temperature climes can be shown to have differences in membrane composition, structure, and ability to adjust membrane components within the range of temperatures experienced (Smolenska and Kuiper 1977, Pearcy 1978). Several mechanisms exist in plants for acclimation of membranes to changes in temperature. These include the regulation of glycerolipid fatty acyl double bond position and number, fatty acyl chain length and branching (bacteria), attachment position of saturated and unsaturated fatty acyl groups to glycerol (sn-1 or sn-2), the insertion/removal of sterols and the removal/insertion of different glycerolipids.

High temperature increases membrane fluidity, while low temperatures decrease fluidity. This can be illustrated using some familiar analogies. Some common fats (lipids) are cooking oils (polyunsaturated plant fats), margarine (partially unsaturated fats), and lard (highly saturated animal fat). These fats have various melting points, as do the fats in cell membranes. At room temperature margarine and lard are solid and cooking oils are liquid. However, as you raise the temperature even the fats that are normally solid at room temperature will become liquid. The liquid nature of cooking oils at room temperature has to do with the higher degree of unsaturation compared to either margarine or lard. This analogy can be applied to explaining the changes that occur in cell membrane fluidity with changing temperature. As temperature increases, membranes become more fluid as a result of the greater fluidization of low-temperature melting lipids. The opposite is true as temperatures decline. However, since plant membranes are complex mixtures of saturated and unsaturated lipids, interactions of the membrane components can influence the overall melting point (T_c) of membrane lipids. The challenge plants have is to maintain membrane fluidity in a physiological range that allows the plant to carry on normal physiological and biochemical functions (homeoviscous acclimation). One mechanism that has received considerable attention in plants is the saturation and unsaturation of fatty acyl groups of glycerolipids. It has frequently been observed that as temperature increases, the degree of unsaturation is reduced, while as temperature decreases, unsaturation increases. Removal of double bonds would tend to prevent the membrane from becoming too fluid under high temperatures, whereas insertion of double bonds would tend to render the membrane more fluid under reduced temperature. Insertion and removal of sterol from membranes may also be part of the mechanism for membrane fluidity adjustment as temperatures change.

Whether such changes in membrane composition (saturation versus unsaturation of glycerolipids) represent a mechanism for changing tolerance or susceptibility to extremes in temperature remains a point of controversy since all plants do not respond in the manner predicted and some techniques utilized in measuring membrane fluidity (Arrhenius plots and fluorescent membrane probes) have been questioned.

Research in this area has evolved from that of analyzing bulk lipids in plant tissues to lipids in organelles, purified membranes, and individual lipid classes. Information from such studies suggests that membranes may not respond to environmental perturbations in a general way as thought previously (i.e., influence overall membrane fluidity), but rather, may influence specific organelle membrane lipids and/or affect isolated domains in those membranes, resulting in localized changes in fluidity or perhaps even inducing pores as a result of lateral phase transitions resulting from homologous lipid aggregations.

Recently, chilling sensitivity has been correlated with the degree of unsaturation of fatty acids in phosphatidylglycerol of chloroplast membranes (Murata et al. 1992). Plants that have high concentrations of cis-unsaturated fatty acids in the glycerolipid class (i.e., spinach and *Arabidopsis thaliana*) are chilling resistant compared to plants with low concentrations, such as squash, which is sensitive. The chloroplast enzyme glycero-3-phosphate acyltransferase is important in determining the unsaturation of fatty acids associated with PG. Tobacco, in which the gene for glycero-3-phosphate acyltransferase was introduced from chilling-resistant *Arabidopsis,* was found to be more resistant to chilling than transgenic tobacco produced from the comparable gene derived from squash. Analysis of the PG fatty acids extracted from tobacco leaves of the two types of transgenic tobacco plants revealed that the fatty acids were more unsaturated in the plant containing the gene from *Arabidopsis* than in plants containing the gene from squash. In addition, when subjected to chilling temperatures, the photosynthetic capacity of the latter was greatly reduced compared to that of the former. In a similar study, the gene coding for glycero-3-phosphate acyltransferase in *Escherichia coli* was introduced into *Arabidopsis thaliana,* resulting in increased saturation of fatty acyl groups of PG, particularly at the sn-1 position of PG (Wolter et al. 1992). A reduction of unsaturated fatty acids was observed. Associated with these changes was an increase in chilling sensitivity by the transformed *Arabidopsis*. These data strongly suggest that specific lipid classes, degree of unsaturated fatty acids associated with the lipid class, and the position of the lipid can influence chilling sensitivity in plants.

Freezing stress (0°C and below) differs from chilling stress (0 to 15°C) in that ice formation can occur in the latter, resulting in lysis of the cell, depending on whether ice formation is intra- or extracellular. If the former occurs, death of the cell is imminent upon thawing. However, if the latter occurs, the cell will usually survive; in fact; removal of cellular water is an important strategy for plants to survive freezing temperatures.

Many of the same responses occur with respect to lipid changes in plants under freezing stress as occur under chilling stress. Rate of removal of water from the cell depends on the lipid structure and composition of the plasmalemma. Thus plants that are adapted to freezing climates and those that have the ability to acclimate to freezing temperatures must be able to adjust membrane fluidity so as to accommodate the required water removal from the cell.

Heat stress not only affects membrane lipid components but also has a greater potential for denaturing proteins in membranes as compared to chilling or freezing stress. Evidence suggests that chloroplasts are quite sensitive to high temperatures and that heat stress may influence photosystem II (PSII) by causing a disassociation of the light-harvesting complex (LHC) from the PSII reaction centers. This in turn causes an increase in the reactions of photosystem I (PSI). This response appears to be a result of the restructuring of membrane lipids associated with photosystem II rather than a denaturing of the proteins associated with the LHC or PSII. This was illustrated utilizing the technique of homogeneous catalytic hydrogenation of biomembranes (Vigh et al. 1989). This technique was used to selectively saturate *cis* double bonds of lipid alkyl chains within intact thylakoids, which resulted in marked increase in the threshold temperatures at which both the thermal damage of PSII–mediated electron transport and formation of nonbilayer lipid phase occurred. Heat-induced stimulation of PSI was also controlled by the level of unsaturated lipids in the thylakoid membranes.

Studies involving theromotolerant blue-green algae have shown that species tolerant to low temperatures adjust fatty acyl groups by desaturation of 18-C fatty acids at reduced temperature and that desaturase genes can be transposed from low–temperature-tolerant species to susceptible species and increase their tolerance to reduced temperature (Wada et al. 1990).

Although several other studies suggest that saturation/desaturation of fatty acyl groups may not be important in acclimation of plants to high or low temperature, the use of transgenic techniques to isolate specific genes for regulation of this process lends strong support for this process in the temperature-acclimation process.

2. Hydration and Membrane Fluidity. Degree of hydration is very important in affecting the fluidity of plant membranes. In general, if cell water percentage falls to a level of 20% or less of the cell dry weight, this is considered to be a critical level relative to maintaining homeostatic viscosity of the membrane and may even affect the thermodynamic stability of the membrane. Lateral phase transitions tend to occur in which homologous lipids tend to aggregate into different regions or domains of a membrane. This has the potential of forming inverted micelles in the membrane, which could facilitate nutrient or ion leakage. In addition, such restructuring could influence the configuration or positioning of proteins (enzymes) within the membrane altering their functions. In some instances proteins may even be ejected from the membranes. Table 3.4 illustrates how varying lipid components in membranes exhibit differing degrees of hydration or affinity for water. Interestingly, MGDG and DGDG have the highest affinity for water. As will be recalled, these lipids are associated primarily with chloroplast membrane lipids and may reflect functional relations with respect to degree of hydration and photolysis of water in the light reactions of photosynthesis.

It is obvious that glycolipids have the highest affinity for water, which reflects the carbohydrate component of these lipids. Carbohydrates, particularly nonreducing disaccharides such as trehalose and sucrose, appear to be important in organisms capable of survival after complete dehydration (i.e., resurrection plant, plant seeds, fungal spores, nematodes, yeast, bacteria, and brine shrimp). It appears that many such organisms produce high levels of these carbohydrates when exposed to dehydration stress, and it is hy-

TABLE 3.4 Degree of Hydration of Various Plant Membrane Lipid Constituents

Membrane Lipid Component	Moles H_2O/Mole Lipid
Glucosylceramide	1
Sterolglycosides	1
PC (Lβ-gel phase)	8
PEA	<15
PC (Lα-liquid phase)	15
MGDG	17
DGDG	34

Source: Modified from Leshem 1992.

pothesized that the carbohydrates interact with cellular membranes to increase the stability of the lipid bilayers. The protective mechanism is uncertain, but one hypothesis suggests that under stress water molecules normally associated with the phospholipid headgroups are replaced with sugars, which prevent lateral phase transition and the formation of lipid domains, which could lead to inverted micelles and increased membrane leakage. In addition, high levels of these carbohydrates may be important osmotically in retaining cellular water under dehydration stress.

It is clear that drought stress has a significant influence on the qualitative and quantitative lipid composition of cellular organelles and membranes. However, there is considerable variability in such reports that probably reflect species, organ, and drought treatment effects. Generally, under extreme dehydration stress, membranes undergo a degradative process, leading to a buildup of conjugated lipids such as triglycerides, sterol esters, and sterol glucosides (Parks et al. 1984). Depending on the severity of dehydration, these may be readily reincorporated into the restructuring of membranes once the stress subsides. Phospholipid levels usually decline and frequently correspond to an increase in visible oil droplets (triglycerides) in cells. The ratios of the various phospholipids and glycolipids are known to change depending on the type of membrane.

Fatty acyl groups of the various glycerolipids tend to become more saturated, but in some cases little change occurs. Free sterols and the ratio of sterol to phospholipid usually increases, which probably reflects a less fluid membrane. Changes in the qualitative and relative composition of sterols have also been reported, with ratios of sitosterol to stigmasterol frequently being shown to change. Changes in sterol ratios have been suggested as being indicative of retailoring of membrane lipids with different sterols that have varying abilities to fluidize or rigidify membranes. Table 3.5 reflects recent data obtained from plasmalemma membranes in drought-stressed and nonstressed sunflower leaves (Navari-Izzo et al. 1993). Such changes may reflect degradative and/or retailoring responses of plants to dehydration stress.

3. Ionic Interactions and Membrane Fluidity. Divalent and monovalent cations have a considerable influence on the fluid nature of biomembranes. Divalent cations are generally more effective than monovalent cations. Ca^{2+} and Mg^{2+} are two divalent cations that have been studied extensively using both artificial and biomembrane systems. How these ions interact with membranes depends on the type and abundance of membrane components present, as well as how they are arranged in the bilayer. The phospholipids appear

TABLE 3.5 Lipid Content (mg^{-1} PM Protein) and Composition of Leaf Plasma Membranes of Sunflower Seedlings Subjected to Water-Deficit Conditions

Lipid Class	Control	Water-Stressed
Diacylglycerols	40	11
Free fatty acids	6.6	5.4
Free sterols	125	151
Glycolipids	72	49
Phospholipids	540	373
Triacylglycerols	13	16
Total	797	605
Free sterol/phospholipid (molar ratio)	0.4	0.7
Lipid/protein (mass ratio)	0.8	0.6

Source: Modified from Navari-Izzo et al. 1993.

to be the most drastically affected components of the membrane; however, interactions with proteins can also occur (Duzgunes and Papahadjopoulos 1983).

Ca^{2+} is the most effective divalent cation relative to changing membrane fluidity. It reacts primarily with negatively charged or acidic phospholipid headgroups, causing the T_c of the membrane to shift toward higher temperatures. Thus Ca^{2+} has a condensing effect on the phospholipids in the bilayer, making them more rigid or gel-like. Ca^{2+} not only has an effect on the headgroups of the phospholipids but in acidic phospholipids such as phosphatidylserine, the T_c is not affected as drastically when the fatty acyl groups are unsaturated as when they are saturated. Neutral phospholipids such as phosphatidylcholine also bind to Ca^{2+} and Mg^{2+}; however, neither element is as effective as when bound to acidic phospholipids. Ca^{2+} can also serve to position proteins in the membrane by linking proteins to other phospholipids or by linking proteins to the cytoskeleton of the cell. Figure 3.8 depicts how Ca^{2+} might interact with membrane lipids and proteins.

Mixtures of neutral and acidic phospholipids often result in lateral phase transition of lipid components, resulting in domains of neutral and acidic phospholipids in the membrane. Although acid phospholipids may be a minority in membranes, the formation of such domains around enzymes may have considerable effect on their activity. Also, changing the ratio of acidic/neutral phospholipids in specific areas of the membrane could affect the fluidity and permeability characteristics locally in the membrane. Mg^{2+} does not induce lateral phase transition. However, it does reduce the effectiveness of Ca^{2+} when applied in combination. The rate and extent to which Ca^{2+} induces domain formation in membranes is related to Ca^{2+} concentration, time of exposure, and the ratio of neutral to acidic phospholipids. In addition, Ca^{2+} replaces the water of hydration associated with phospholipid heads, and at high enough concentration, water can be displaced completely. This in turn may have a direct effect on enzyme configuration and function, thus contributing to further alteration in membrane fluidity. Mg^{2+} is not as effective as Ca^{2+} in this regard.

Exactly how Ca^{2+} and Mg^{2+} interact with the phospholipids of the membrane is not completely clear. However, it is known that reduction in negative surface charge density

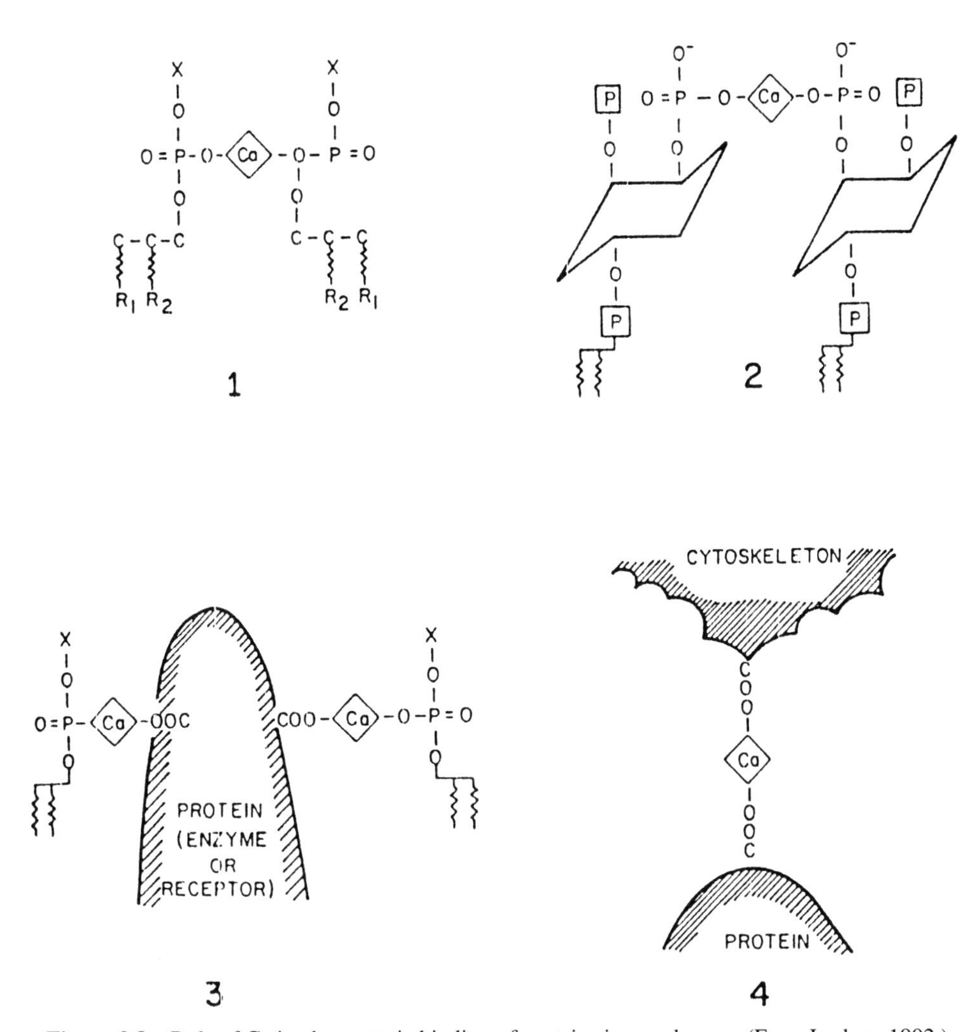

Figure 3.8 Role of Ca in electrostatic binding of proteins in membranes. (From Leshem 1992.)

occurs, which could affect electrical potential differences across membranes. Divalent cations also affect hydrogen bonding, water of hydration, and packing of polar headgroups. In addition, it has been shown that Ca^{2+} can bind with the hydroxyl group of cholesterol so that perhaps a phospholipid–Ca^{2+}–sterol interaction is part of the mechanism for "tighter" packing of phospholipid polar headgroups.

It is of interest that cytosolic levels of Ca^{2+} are generally lower than Ca^{2+} levels outside the cell. Slight increases in the cytosolic concentrations of Ca^{2+} level can result in numerous physiological and morphological changes in plants. These changes are mediated through the activation of a protein called calmodulin, which is known to activate several membrane-bound enzymes (Ca^{2+} ATPases, NAD-kinases, and phosphorylation enzymes). What the relationship is between the function of Ca^{2+} as an activator of calmodulin and its role as a membrane stabilizer is unclear. However, fluidity changes in membranes could influence the fluxes of Ca^{2+} ions across membranes, thus altering the critical balance required to activate calmodulin.

pH changes in the vicinity of the membrane surface also affect the fluidity of the membrane. Hydrogen bonding between molecules of the membranes is important in maintaining membrane integrity. High pH results in complete extraction of protons, resulting in drop in the T_c. The reverse is true when pH is lowered. For a given phospholipid molecule, three states of the bilayer membrane can be recognized: fully protonated, partially protonated, and deprotonated. Fully protonated and deprotonated systems differ in T_c by about 6°C for a single-proton system such as phosphatidylglycerol and about 12°C for a two-proton system such as phosphatidic acid. T_c is higher for the fully protonated state than for the deprotonated state. The partially deprotonated state represents a T_c temperature far above the other states of protonation. This difference may be explained by greater stabilization of hydrogen bonding. Protonation can also replace Ca^{2+} and can by itself induce lateral phase transition, which can be reversed by raising the pH in a buffered salt solution containing monovalent cations. Monovalent cations seldom bind negatively charged phospholipids but greatly affect proton binding. This can influence the phase transition of negatively charged phospholipids by interacting with partially protonated systems, causing increased dissociation of the protons and a decrease in the T_c. Monovalent cations can interact in the deprotonated state at low salt concentrations to slightly increase T_c or at high salt concentrations to strongly increase T_c (Eibl 1983).

Potassium is a very important monovalent cation in plants that helps to maintain and regulate cell turgor under stress conditions. It is also known to stimulate ATPase activity in membranes that may be involved in active transport of other ions. Several *potassium pump systems* are known to exist in plants where K^+ exchanges for other ions to maintain osmotic or ionic balance. Such functions of K^+ in plants are not inconsistent with the effects of monovalent cations in membranes in general.

4. Light and Membrane Fluidity. One of the most important effects of light seems to be associated with its influence on the development and function of chloroplast lipids. It is difficult at times to differentiate between light effects relative to controlling molecular events per se and its control of developmental responses (i.e., light-induced chloroplast development).

Chloroplast membranes are comprised predominately of MGDG, DGDG, PG, and SQDG. The fatty acyl groups of these lipids are highly unsaturated, and changes in the composition and degree of lipid unsaturation are known to influence membrane fluidity and stability of light-harvesting complexes. There is evidence that light stimulates acetyl-CoA carboxylase and de novo synthesis of fatty acids and PG. Also, the use of far-red light, which does not promote normal chloroplast development, results in changes in chloroplast lipid composition. For example, barley leaves exposed to white light incorporated three times the amount of [^{14}C]acetate into the diacylgalactosylglycerol fraction than did leaves exposed to far-red light (Table 3.6) (Harwood 1983), while only traces of phosphatidylethanolamine were detected in the leaves of plants grown in white light. Fatty acyl groups became more unsaturated in the white light treatment compared to the far-red light treatment. Such responses to far-red and white (red) wavelengths of light are suggestive of phytochrome-mediated control of lipid metabolism. This raises some interesting questions concerning the differences in the lipid composition of chloroplasts of shade versus sun plants. Those plants that normally grow in shade environments, such as the lower story of a forest canopy, would probably have a lipid composition that reflects far-red irradiation, since the upper canopy would filter out most red wavelengths of light and allow far-red wavelengths to penetrate the lower story. Could there also be a relation-

TABLE 3.6 Effect of Far-Red Light on the Incorporation of Radioactivity from [^{14}C]Acetate into Lipids of Greening Barley Leaves

	Acyl Lipid Labeling (% of total [^{14}C]acyl groups)					
	DGG	DDG	SQDG	PC	PG	PE
Far-red light	14	7	9	40	9	14
White light	45	4	5	41	5	tr
	Fatty Acid Labeling (% of total ^{14}C-labeled fatty acid)					
	$C_{16:0}$	$C_{16:1}$	$C_{18:0}$	$C_{18:1}$	$C_{18:2}$	$C_{18:3}$
Far-red light	31	4	23	18	11	13
White light	15	3	4	15	45	18

Source: Modified from Harwood 1983.

ship between lipid composition and light quality/quantity that controls the inability of shade plants to readily acclimate to high-light environments compared to the greater ability of sun plants to acclimate to low light levels?

Although the exact role and location of phytochrome relative to membrane function is not known, it is frequently suggested that phytochrome is in some way intimately associated with membranes either superficially or as an intrinsic component of membranes. When influenced by either red or far-red wavelengths of light, the position or association of the pigment–protein with membranes changes and initiates a cascade of events resulting in physiological or developmental changes. Some phytochrome-mediated events are rapid (minutes), while others are much slower (hours). The initial events of either could be membrane associated, although the former is generally associated with changes in membrane permeability or fluidity.

The lipid composition depicted in barley leaves (Table 3.6) is consistent with that found in most chloroplast membranes. It is very likely that changes in quality or quantity of light could influence rates of photosynthesis since there is strong evidence that some acylglycerols (phosphatidylglycerol) may stabilize light-harvesting complexes. *Chondrus crispus,* a red alga, has been shown to accumulate [^{14}C]acetate into MGDG, DGDG, and SQDG to a greater extent in the light compared to the dark (Table 3.7) (Harwood and Jones 1989). PG also increased while PC declined. Unsaturated fatty acid synthesis also increased, while saturated fatty acids declined in the light. In the brown alga *Fucus serratus,* of several phospholipids analyzed, only PC increased to any significant extent. However, like the red alga, fatty acid unsaturation increased in the light (Table 3.8). Similarly, unsaturated fatty acids were higher in photoautotrophically grown suspension cultures of *Euphorbia characias* compared to photomixotrophic or heterotrophically grown cells (Table 3.9) (Carriere et al. 1992). Although possibly reflecting ontogenetic changes rather than response to light intensity, etiolated barley leaves transferred to light exhibited considerable changes in the fatty acid composition of PC, MGDG, and PG (Table 3.10) (Harwood 1983). After a 24-h greening period, 18:3 increased in all three lipid classes, while 18:2 declined in all classes except PG. In the PG lipid class, 16:1 increased considerably but remained relatively unchanged in the other classes. There is a strong possibility that 16:1 associated with PG may be associated with the oligomeric organization of the light-harvesting chlorophyll–protein complex.

TABLE 3.7 Effect of Radiolabeling from [^{32}P]Orthophosphate or [^{14}C]Acetate in the Acyl Lipids of *Fucus serratus* and *Chondrus crispus*, Respectively

	F. serratus		*C. crispus*	
Lipid	Dark	Light	Dark	Light
MGDG			10.3	14.3
DGDG			1.9	6.9
SQDG			9.3	13.0
PE	54.3	53.1		
PG	10.3	11.6	5.4	18.5
PC	4.9	19.3	27.0	17.6
DPG	9.8	10.8		
PA	1.4	1.5		
PI	19.3	18.9		

Source: Modified from Harwood and Jones 1989.

The use of fatty acid mutants developed in the cyanobacterium *Synechocystis* suggests that the degree of unsaturation of thylakoid membrane lipids can influence the response of the mutated organism to photoinhibition at different temperatures and light intensities (Gombos et al. 1992). The wild-type strain possessed mono-, di-, and triunsaturated fatty acids. One mutant designated as Fad6 possessed only mono- and di-unsaturated fatty acids, and a second mutant (Fad6/desA::Kmr) possessed only monounsaturated fatty acids (Table 3.11). The later mutant was more susceptible to photoinhibition at 10, 20, and 30°C than either the wild-type or the Fad6 mutant. The later two exhibited similar responses to temperature and increased light intensity (Figure 3.9). This suggests that polyunsaturated fatty acids are important in protection against photoinhibition.

High light intensities, above those required to saturate photosynthesis, can cause photoinhibition, which results from damage of membrane components and/or the light-harvesting complexes of the thylakoid membranes. Thylakoid membranes are rich in unsaturated fatty acids, which are highly susceptible to oxidation. Under high light intensities the oxygen evolved as a result of the Hill reaction could initiate free-radical reactions and

TABLE 3.8 Effect of Light on the Distribution of Radioactivity from [^{14}C]Acetate Among Acyl Chains in *Fucus serratus* and *Chondrus crispus*

	Distribution of Label (% of ^{14}C-labeled fatty acids)						
	14:0	16:0	16:1	18:0	18:1	18:2	others
F. serratus							
Dark	3.4	47.7	—	12.9	29.6	0.9	5.5
Light	3.2	31.5	—	9.9	42.6	9.4	3.5
C. crispus							
Dark	6	31	2	7	40	3	11
Light	1	19	3	tr	64	10	3

Source: Modified from Harwood and Jones 1989.

TABLE 3.9 Constituent Fatty Acids of Neutral Lipids from Seeds, Leaves, Calli, and Cell Suspension Cultures of *Euphorbia characias*[a]

	Fatty Acid Chain Length:Number of Double Bonds								Ratio
Sample	14:0	16:0	16:1	18:0	18:1	18:2	18:3	20:0	18:3/18:2
Leaves	+	21.8	0.9	2.3	6.1	18.8	42.9	4.5	2.30
Cell suspensions									
PA (E)	2.6	20.3	3.6	5.6	4.5	24.3	38.8	—	1.60
(S)	2.8	29.4	1.9	1.6	3.8	20.3	40.1	—	1.97
PM (E)	1.4	18.8	1.5	2.9	18.9	33.1	23.4	—	0.70
(S)	+	27.7	1.2	4.6	4.9	31.3	30.1	—	0.96
H (E)	3.7	26.1	3.7	8.5	18.9	31.4	7.6	—	0.24
(S)	2.8	21.4	2.1	2.6	28.5	34.4	8.0	—	0.23
Seeds	+	6.5	—	13.3	24.4	55.4	0.4	—	0.007
Calli									
PA	2.0	37.0	4.2	3.0	7.8	20.5	25.5	—	1.24
PM	2.0	31.3	—	1.7	6.6	36.5	21.9	—	0.60
H	2.6	18.5	1.6	1.8	31.0	36.9	7.6	—	0.20

Source: Modified from Carriere et al. 1992.

[a]PA, photoautotrophy; PM, photomixotrophy; H, heterotrophy; E, exponential phase of growth; S, stationary phase of growth.

TABLE 3.10 Changes in Lipid Composition of Etiolated Barley Leaves as They Are Exposed to White Light

Lipid	Amount (μg/mg fresh weight)	Acyl Lipids (%)	Fatty Acid Composition (% of total) $C_{16:0}$	$C_{16:1}$	$C_{18:0}$	$C_{18:1}$	$C_{18:2}$	$C_{18:3}$
Etiolated tissue								
PC	0.9	15	35	trace	3	5	40	13
DGDG	2.5	42	15	1	2	2	14	67
PG	0.3	5	50	1	4	1	7	30
After 4 h greening								
PC	1.5	26	37	n.d.	4	1	35	17
DGDG	1.6	27	25	n.d.	1	4	12	53
PG	0.6	9	47	3	4	2	14	25
After 24 h greening								
PC	1.6	13	30	2	3	5	25	32
DGDG	4.8	45	9	2	1	tr	2	85
PG	1.0	8	18	28	3	2	11	38

Source: Modified from Harwood 1983.

TABLE 3.11 Major Fatty Acids from Total Lipids of Wild-Type *Synechocystis* PCC6803, Fad6 Mutant, and Fad6/desA : : Kmr Transformant, Grown at 34°C

	Fatty Acid (mol %)			
Strain	16:0	18:1(9)	18:2 (9,12)	18:3 (6,9,12)
Wild type	58	7	12	17
Fad6	59	11	25	0
Fad6/desA : : Kmr	57	41	0	0

Source: Modified from Gombos et al. 1992.

could be involved in peroxidation of polyunsaturated fatty acyl residues of thylakoid membrane lipids. Lipid peroxidation can be assayed by measuring malondialdehyde (MDA) concentrations in chloroplasts. Malondialdehyde is a decomposition product of oxidized unsaturated fatty acids. Thus increases of MDA concentrations in the chloroplasts of wheat leaves correlated with increases in light intensity suggest a relationship between lipid peroxidation and photoinhibition (Figure 3.10) (Mishra and Singhal 1992). These findings also support studies with *Synechocystis* mutants which indicate the importance of polyunsaturated fatty acids in increasing tolerance to photoinhibition (Gombos et al. 1992).

*5. **Phytohormones and Membrane Permeability.*** All the classical plant hormones (auxin, gibberellins, cytokinins, abscisic acid, and ethylene) have been shown to influence membrane permeability. Generally, the effect is to increase permeability. The mechanism of how this occurs is unclear, but hypotheses include (1) attachment of hormones to membrane proteins (receptors), which may directly alter the protein–lipid interaction and hence alter permeability; (2) the hormone–protein interaction may activate second messengers that control ion channels or membrane hydrolytic enzymes; and (3) hormones may interact directly with specific lipid components or domains of the membrane bilayer. Several approaches have been taken to study the effects of plant hormones on membranes. These include studies using vesicles (liposomes) or thin films (mono -and bilayers) comprised of lipids extracted from plant tissues or developed from pure phospholipids. Ions or other solutes can then be placed within the vesicles and leakage or swelling of the vesicle monitored. Using liposomes derived from egg PC, cytokinins, auxins, abscisic acid, and gibberellic acid all increased the permeability of the liposomes to erythritol with increasing concentrations (Figure 3.11) (Hester and Stillwell 1984). Cytokinins were the most effective at lower concentrations, while gibberellic acid was the least effective. Interestingly, in similar experiments in which *cis* and *trans* isomers of zeatin were used, it was found that the *trans* isomer of zeatin had no effect on changing the permeability of PC liposomes to erythritol (Figure 3.12) (Stillwell et al. 1985). Similarly, studies using different isomers of ABA (*cis–trans* and *trans–trans*) indicated that the *trans–trans* isomer was less effective than the *cis–trans* isomer in increasing the permeability of membrane bilayers comprised of 80% dimyristoylphosphatidylcholine and 20% dilaurolyphosphatidylcholine (Table 3.12). Although both isomers of zeatin are biologically active in plants, each appears to promote different responses in plants. In the case of ABA isomers the *trans–trans* isomer has no biological activity. The fact that isomers of these hormones affect membrane permeability and biological responses differently suggests that they may influence biological responses through changes in membrane characteristics.

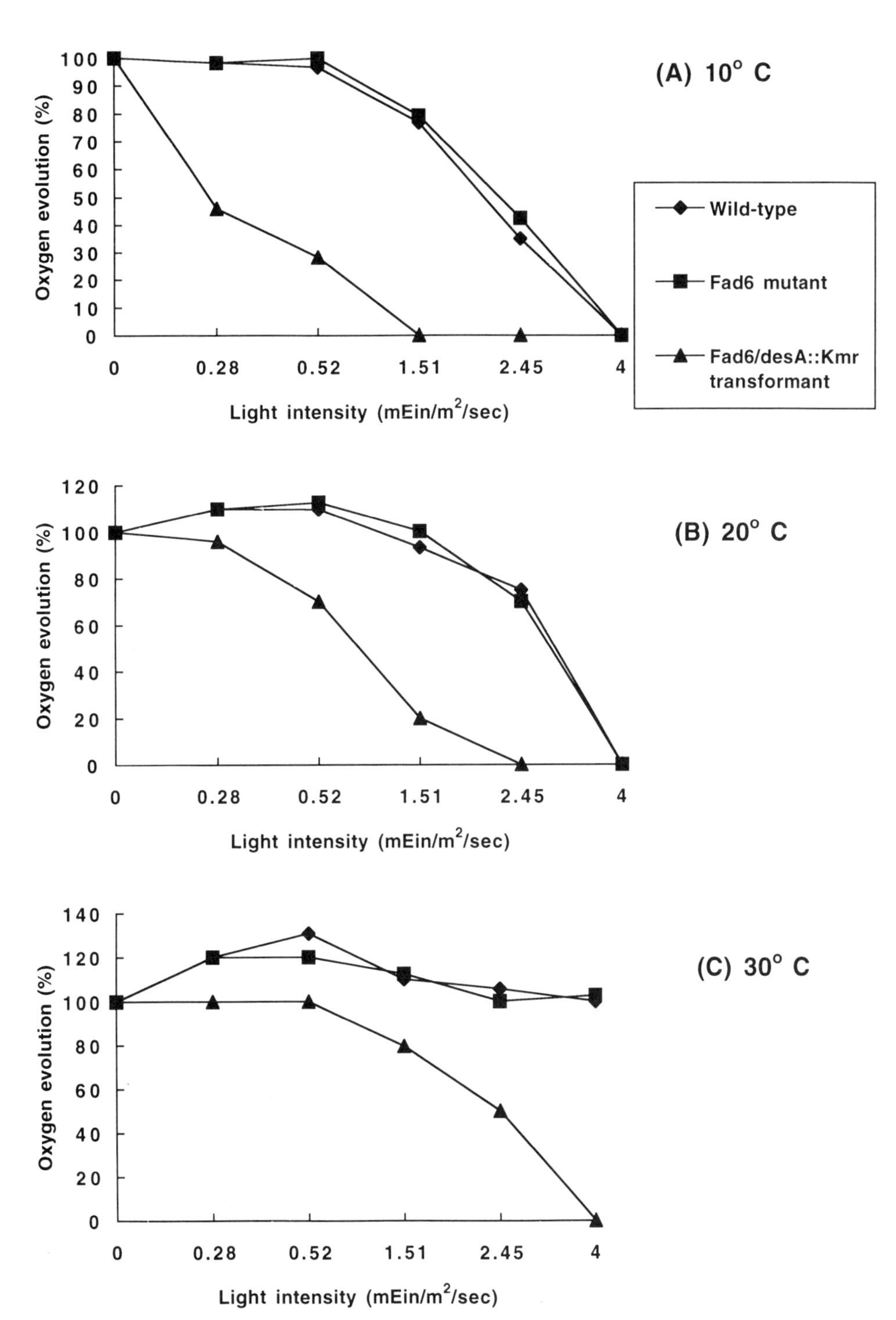

Figure 3.9 Effects of light and temperature on wild-type, mutant Fad6, and Fad6/desA::Kmr transformants of *Synechocystis*. (Modified from Gombos et al. 1992.)

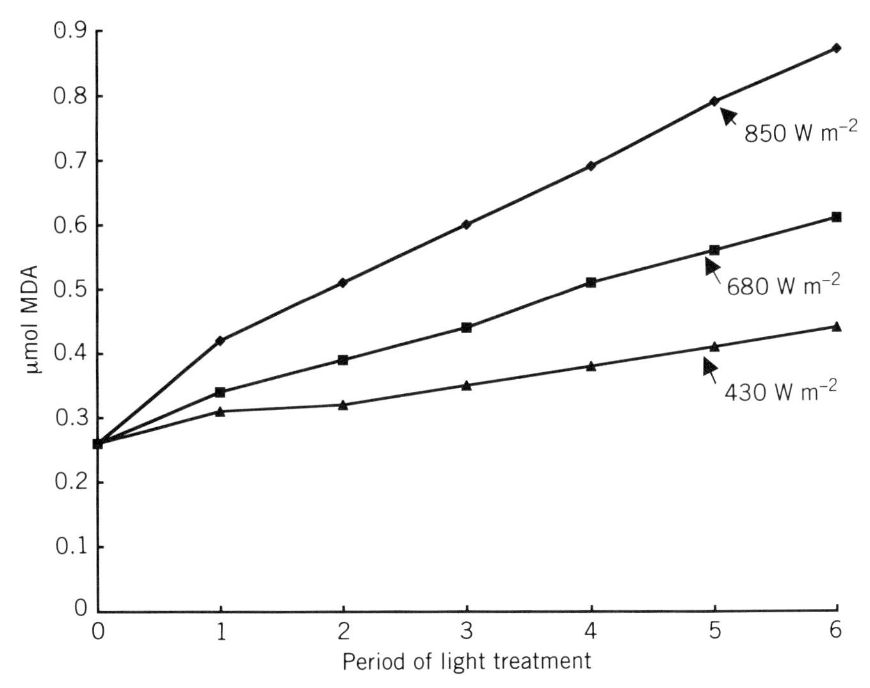

Figure 3.10 Peroxidation of thylakoid lipids on photoinhibition of wheat at three light intensities. (Modified from Mishra and Singhal 1992.)

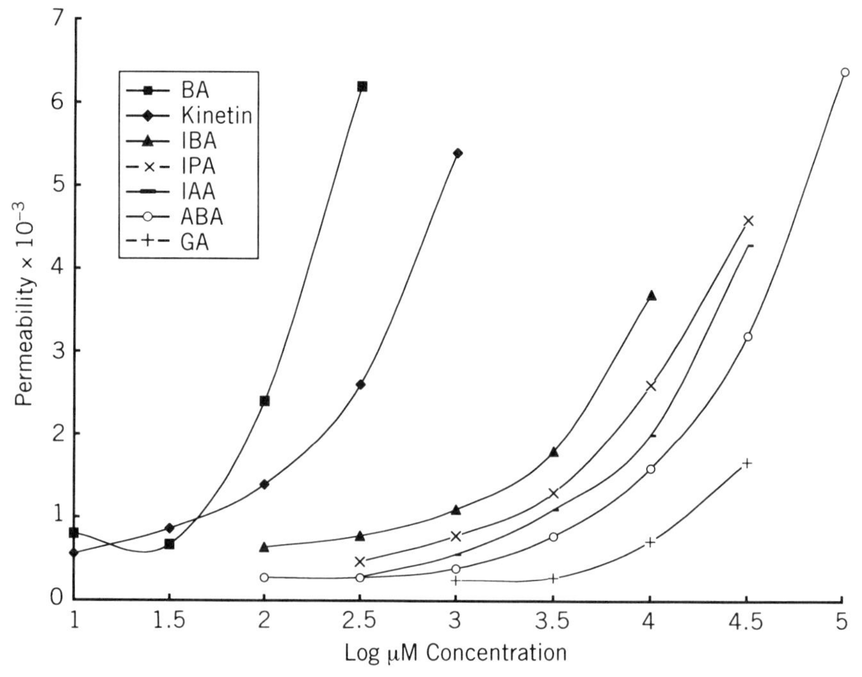

Figure 3.11 Effects of phytohormones on the permeability of erythritol to egg phoshatidylcholine liposomes. (Modified from Hester and Stillwell, 1984.)

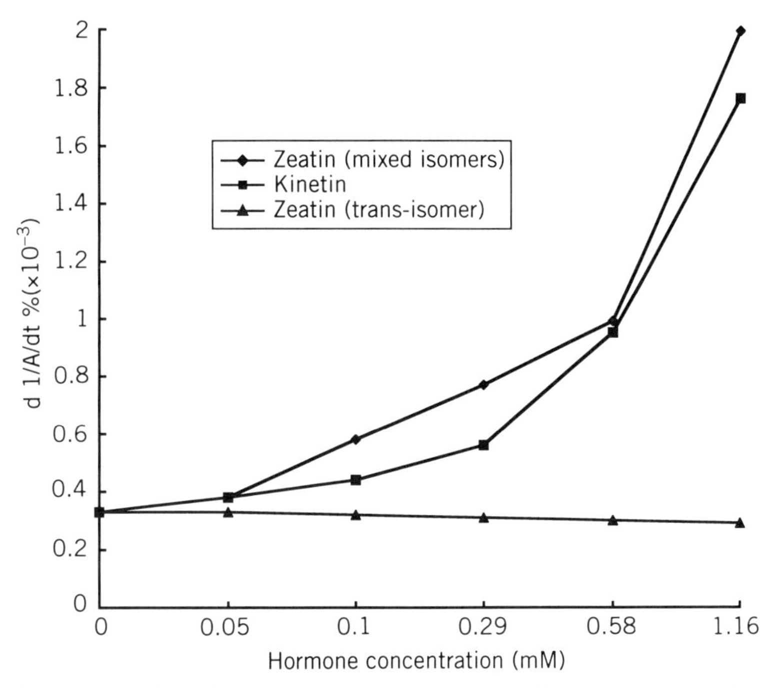

Figure 3.12 Effects of cytokinins on enhancing the permeability rate of erythritol for bilayers composed of egg phosphatidylcholine. (Modified from Stillwell et al. 1985.)

Chemical probes can also be introduced into artificial membranes, and movement of the probe within the membrane can be monitored by techniques such as fluorescence polarization, electron spin resonance, and differential scanning calorimetry. Using fluorescence polarization and aqueous interface probes, it has been determined that IAA and ABA interact differently with phospholipid bilayers (Stillwell and Wassall 1993). IAA

TABLE 3.12 Enhancement of Erythritol Permeability (Expressed as Liposme Swelling to Bilayers Composed of 80% Dimyristoylphosphatidylcholine and 20% Dilauroylphosphatidylethanolamine by Different ABA Isomers

Abscisic Acid (m*M*)	Liposome Swelling[a] (*dl*/A/*dt*%) A	B
0	0.35	0.35
0.32	0.47	0.51
0.63	0.56	0.67
1.26	0.59	0.83
2.52	0.84	2.19

Source: Modified from Stillwell and Hester 1984.

[a]A, 24% *cis–trans* ABA, 76% *trans–trans* ABA; B, 100% *cis–trans* ABA.

appears to perturb the membrane surface, while ABA acts in regions of the membrane where packing defects may occur, such as the interface between the gel state and the liquid-crystalline state of a membrane. In addition, IAA can increase the permeability of single-component lipid bilayers, while ABA requires a mixed-component bilayer to be effective. At pH 7, dissociated ABA is totally ineffective in increasing permeability, while IAA, which has almost an identical pK_a value, enhances permeability. ABA also has the ability to increase aggregation and fusion of lipid vesicles derived from mixed phospholipids, while IAA cannot do this with either mixed or single components.

Direct observation using plant plasma membranes have also been done using electron microscopy following hormone treatments. Membranes isolated from etiolated soybean hypocotyls treated with IAA or $CaCl_2$ resulted in thinning or thickening of the plasma membranes, respectively. The response is reversible, specific, and dependent on the hormone concentration and last treatment (Figure 3.13) (Morré and Bracker 1976). Such changes in membrane thickness have frequently been associated with modifications in microviscosity of membranes (i.e., thin membranes reflect greater fluidity, thick membranes greater rigidity).

Other studies utilize plant tissues such as beet roots or flower petals that have been treated with hormones and pigment or ion leakage is a measure of change in membrane permeability. As an example of the later, ion leakage and ethylene production were monitored in senescing rose petals over a period of 8 days. Ethylene production preceded ion leakage, which was followed by rapid senescence of the rose petals (Figure 3.14). Using beet segments treated with ethephon, a synthetic ethylene-producing compound, pigment leakage was found to increase as a result of treatment but could be reduced by addition of spermine or spermidine, two polyamines that have multicationic sites that may act to condense membrane lipids by binding to negatively charged phospholipid headgroups (Table 3.13) (Parups 1984).

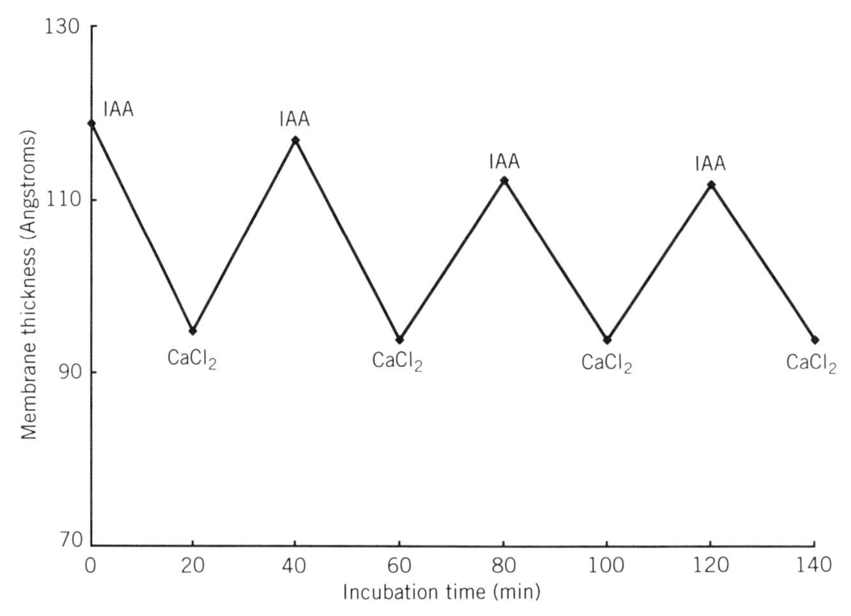

Figure 3.13 Effect of IAA and $CaCl_2$ on membrane thickness in soybean hypocotyls. (Modified from Morré and Bracker 1976.)

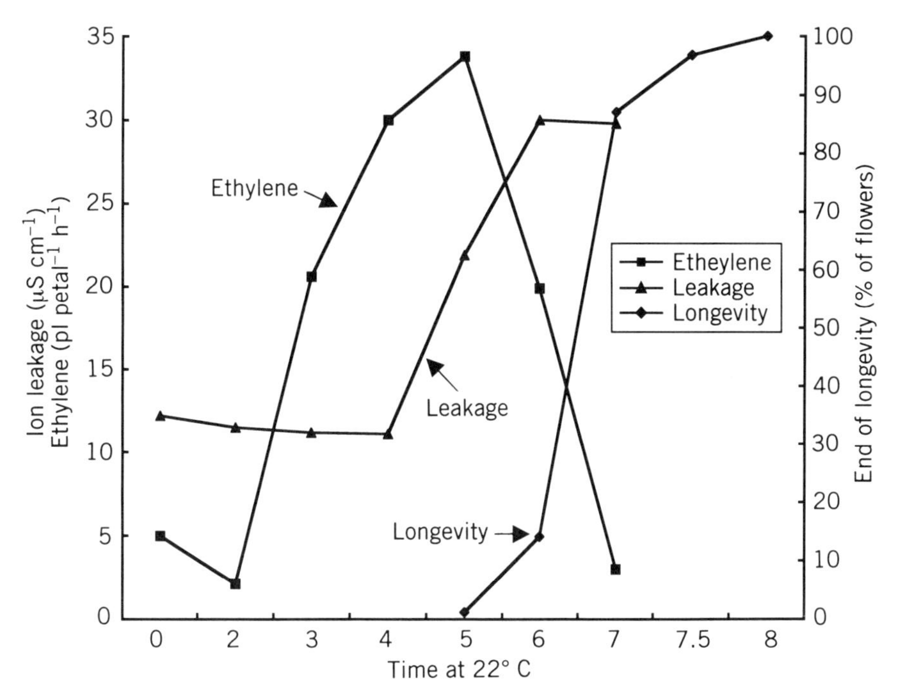

Figure 3.14 Temporal pattern of senescence of rose petals at 22°C. (Modified from Faragher and Mayak 1984.)

TABLE 3.13 Transmittance (%) of Pigments Leaked from Beet Root and Rose Petal Tissues Treated with Polyamines, Ethephon, and Ammonium Sulfate

Polyamine Treatments	Beet Root Discs			Rose Petals
	Ethephon	$(NH_4)_2SO_4$		Ethephon
	1×10^{-3} *M*	3×10^{-3} *M*	0	1×10^{-2} *M*
Spermine				
1×10^{-3} *M*	87	96	97	19
1×10^{-2} *M*	92	—	—	90
Spermidine				
1×10^{-3} *M*	67	87	95	—
1×10^{-2} *M*	98	—	—	—
Putrescine				
1×10^{-3} *M*	10	67	88	13
1×10^{-2} *M*	12	—	—	12
Cadaverine				
1×10^{-3} *M*	9	71	85	24
1×10^{-2} *M*	14	—	—	3
Control (buffer)	5	54	77	10

Source: Modified from Parups 1984.

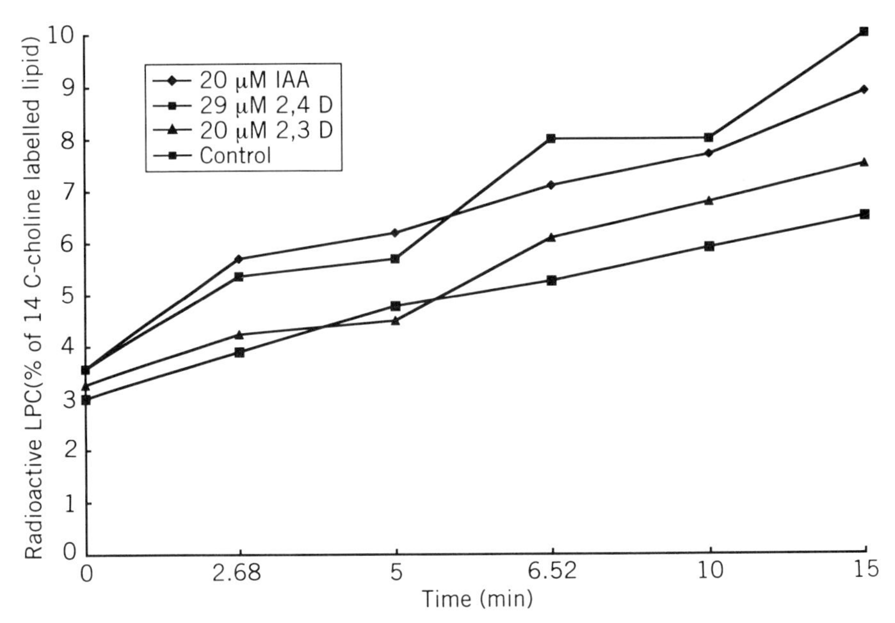

Figure 3.15 Auxin stimulation of PLA2 in microsomes of cell cultures of soybean. (Modified from Scherer and André 1993.)

Auxin has been shown to activate phospholipase A2 in microsomes of soybean suspension cells (Figure 3.15) (Scherer and Andre 1993). This may be significant for two reasons: (1) lysophospholipids formed as a result of this enzyme have been implicated as second messengers in the activation of H^+-ATPase or membrane-associated protein kinase or both; and (2) the formation of lysophospholipids could have a direct effect on membrane fluidity and permeability. In the latter case, the production of the detergent-like lysophospholipids could result in the formation of nonbilayer hexagonal configurations of membrane lipids. Also, unsaturated fatty acids produced can facilitate peroxidation reactions, which can potentially be lethal to cells. However, controlled activity of phospholipase A2 may reflect important regulatory functions relative to the restructuring of membrane phospholipids through the insertion and/or removal of fatty acids with different chain lengths and degree of unsaturation. This is particularly important with respect to the ability of plants to maintain homeostasis with changing environments. Lysophospholipid formation has also been implicated in membrane fusion and the formation of a membrane component which has a special receptor function relative to senescence induction by abscisic acid.

IV. SUMMARY

Plant membranes serve as important environmental sensors in plants. They respond to virtually every aspect of the changing environment. Such responses may either be direct or indirect, through hormone-mediated effects. Membranes occur at several different levels of organization within the cell and are very important in maintaining compartmentalization of function. Membrane structure is probably very similar among organelles, consisting of a bilayer configuration and comprised of protein, lipid, sterols, and carbohydrates. However, what is unique about membranes is that these four primary components vary greatly quantitatively and qualitatively among organelles, plant tissues, species, and cultivars. These membrane components change with the environment in an attempt to maintain cellular homeostasis, which will allow the cell/tissue/plant to survive and function under the conditions to which the plant has adapted, as well as the prevailing conditions at any given time. One of the key parameters of survival relates to maintaining a degree of membrane fluidity that permits membrane transport and enzyme function not only at optimal environmental conditions but also at the upper and lower ends of the environmental spectrum to which the plant has adapted. This is accomplished through inherent differences in the membrane composition of plants and/or the ability to alter membrane components such as fatty acids, sterols, and carbohydrates both qualitative and quantitatively. As a consequence of the sessile nature of plants, it is not surprising that plants have evolved an array of different and unique chemical strategies for dealing with changing environments.

STUDY-REVIEW OUTLINE

Membrane Structure and Composition

1. Plants "sense" changes in the environment through perturbations in membrane structure and composition.
2. Membranes are integral parts of cellular organelles and are comprised of lipids, sterols, carbohydrates, and proteins.
3. Membranes are generally arranged in a bilayer configuration but can be altered through environmental stress.
4. The lipid classes that are found in plants include phospholipids, sphingolipids, galactolipids, and sulfolipids.
5. The lipid classes differ relative to the types of fatty acyl groups, the presence of phosphate or sulfur, and types of carbohydrates associated with their structures.
6. Different organelles have characteristic lipid compositions and exhibit asymmetry relative to membrane composition.
7. Some organelles, such as chloroplasts and mitochondria, possess double membranes. Such features are extremely important with respect to proton gradients required in these organelles for the production of ATP.
8. Fatty acyl groups of lipids vary relative to carbon number, degree of unsaturation, position of double bonds, and branching. Fatty acyl groups are extremely important relative to affecting membrane fluidity.
9. Sterols are components of cellular membranes and have a function to stabilize membranes and possibly alter fluidity characteristics.
10. Many hundred and possibly thousands of different sterol structures exist in biological organisms.
11. Sterols occur as conjugated and free forms. The latter can only be incorporated into membranes. However, conjugated sterols may be important in regulating free forms by interconversion.
12. Plants have a greater diversity of sterols than do animals. Sterol structure and the ability to be incorporated into membranes may be an important regulatory function relative to membrane fluidity.
13. Membrane proteins are classified as catalytic, proteinaceous carriers, solute channel proteins, and receptors.
14. Integral proteins traverse part or all of the membrane, while peripheral proteins interact at the membrane surface.

15. Lipids and proteins can interact to affect fluidity. Such interaction can, in turn, affect protein function.

Dynamic Nature of Cell Membranes: Membrane Fluidity

1. Membranes are dynamic structures and are constantly in a state of motion exhibiting both lateral and rotational movement.
2. Movement of membrane components is important relative to facilitating maximum contact with enzyme substrates and other molecules, such as hormones and ions.
3. Membranes can exist in three phases: the liquid-crystalline phase, the solid-gel phase, and the hexagonal and cubic phases as determined by environmental conditions.

Temperature and Membrane Fluidity

1. Intrinsic factors that control membrane fluidity are related to the membrane composition and structure. This includes the composition of fatty acids, proteins, carbohydrates, and sterols.
2. T_c is the lowest temperature at which the initial signs of a phase change from liquid-crystalline state to a solid-gel phase occurs. T_c is determined by the membrane composition and may be altered simply by changing the position and/or numbers of double bonds in fatty acyl groups or by modifying types of phospholipids or other lipids in the membrane.
3. Ca and Mg cations can interact with phospholipid headgroups to either increase or decrease membrane fluidity, depending on concentration and ratios.
4. Insertion of sterols into membranes above the T_c point tends to rigidify membranes; insertion below the T_c tends to fluidize the membrane.
5. Sterol size, structure of side chain, and degree of unsaturation may influence their ability to be incorporated into membranes.
6. The extrinsic environment, such as temperature, light, salinity, pH, water availability, ion concentrations, and ratios, can all affect membrane fluidity.
7. Degree of saturation/unsaturation of glycerolipids is affected by high and low temperature. High temperature favors saturation and low temperature favor unsaturation. This is an important control mechanism for regulating membrane fluidity.
8. Changes in membrane fluidity may be controlled by alterations in specific lipid components that comprise susceptible organelles (i.e., chloroplast or mitochondria) in plants exposed to temperature extremes.
9. Extremes in temperature can also affect membrane fluidity by altering the degree of membrane hydration or the configuration of membrane proteins.

Hydration and Membrane Fluidity

1. If cellular water percentage falls below 20%, the thermodynamic stability of the membrane structure can be affected.
2. Degree of hydration can cause lipid domain formation, which could result in inverted micelles and increased leakiness.
3. Plants subjected to dehydration stress produce increased levels of carbohydrates that may help to stabilize membrane structure.
4. Plants subjected to drought stress usually have increased levels of triglycerides and other conjugated forms of lipid indicative of membrane deterioration.

Ionic Interactions and Membrane Fluidity

1. Ca is important in plants in that it interacts with phospholipid headgroups that possess a negative charge and reduces membrane fluidity. It may also be important in the positioning and linking of proteins to specific lipids within the membrane.
2. At high concentrations Ca can replace membrane-associated water which, could affect enzyme placement and function.

3. Membrane fluidity and Ca transport may be important relative to regulation of calmodulin activity, which is now know to control several physiological events in plants.
4. pH changes through protonation and deprotonation of phospholipid headgroups can affect cation interactions and degree of hydration of the membrane surface, with subsequent changes in membrane fluidity.

Light and Membrane Fluidity

1. Light quality and quantity can affect the lipid composition and degree of unsaturation of chloroplast lipids, resulting in changes in membrane fluidity and stability of light-harvesting complexes.
2. Red light–and far-red light–initiated physiological events could be related to phytochrome-membrane associated changes in membrane fluidity affecting transport or enzyme function.

Phytohormones and Membrane Permeability

1. All of the classical phytohormones have been demonstrated to affect membrane permeability. Generally, the response is to increase permeability.
2. Hormone mechanisms of action may include attachment to and activation of protein receptors, the activation of second messengers, or direct impact on membranes by interacting with specific lipids or lipid domains.
3. Isomers of ABA and CKs have differing effects on membrane permeability.
4. The insertion of chemical probes in isolated or synthetic membranes in association with fluorescence polarization, electron spin resonance, and differential scanning calorimetry have been used to study the effects of hormones on changes in membrane fluidity and permeability.

SELF-STUDY QUESTIONS

1. How do plants sense changes in their environment?

2. Describe the structure and composition of cellular membranes.

3. Outline the different lipid classes that occur in plants and indicate their chemical differences.

4. Give examples of two organelles that exhibit differences in membrane composition and summarize the differences.

5. What functions do membranes have in plants?

6. What functions do double membranes have in plants, and what is the importance of asymmetry to membrane function?

7. Describe the structural differences that exist among fatty acyl groups in plants.

8. Describe the structural differences that exist among sterols in plants.

9. What functions do proteins have in plants?

10. Describe the dynamic nature of plant membranes.

11. Describe the three phases that membranes can exist in and indicate the significance of each.

12. What is the membrane transition temperature (T_c), and what is its significance?

13. What is membrane fluidity, and how does it relate to permeability?

14. What is the significance of the concept of membrane fluidity?

15. How do cations affect membrane fluidity, and why?

16. In what ways can sterols modify membrane fluidity?

17. What extrinsic factors can modify fluidity?

18. Outline several ways that temperature extremes can alter membranes to affect various physiological responses in plants.

19. Explain how the degree of membrane hydration may affect membrane fluidity and function.

20. Explain how pH can affect membrane surface charge characteristics and how this interaction can affect membrane permeability.

21. Indicate how light quality and intensity can affect plant lipid composition and its impact on membrane function.

22. How may phytochrome be implicated with membrane function and transduction sequences leading to specific physiological responses?

23. Indicate which phytohormones affect membrane permeability and potential mechanisms by which hormones affect permeability.

SUPPLEMENTARY READING

Chapman, D. 1983. Biomembrane fluidity: The concept and its development. Pp. 5–42 in: R. C. Aloia (ed.), Membrane Fluidity in Biology, Vol. 2. Academic Press, New York.

Cossins, A. R. (ed.). 1994. Temperature Adaptation of Biological Membranes (Proceedings of the Society for Experimental Biology). Portland Press, Chapel Hill, NC.

Gibbons, G. F., K. A. Mitropoulos, and N. B. Myant. 1982. Pp. 303–342 in: Biochemistry of Cholesterol. Elsevier Biomedical, New York.

Leopold, A. C. (ed.). 1986. Membranes, Metabolism and Dry Organisms. Comstock Publishing, Ithaca, NY.

Pringle, M. J., and D. Chapman. 1981. Biomembrane structure and effects of temperature. Pp. 21–37 in: G. J. Morris and J. A. Clarke (eds.), Effects of Low Temperatures on Biological Membranes. Academic Press, New York.

Robertson, R. N. 1983. The Lively Membranes. Cambridge University Press, New York.

4 Phytohormones and Plant Responses to Stress

Outline

OBJECTIVES

1. Review the types of phytohormones (auxins, cytokinins, gibberellins, abscisic acid, and ethylene) as well as their synthesis, transport, occurrence, modes/mechanisms of action, and developmental functions.
2. Establish how environmental stresses (light, nutrition, salinity, flooding, drought, and temperature) influence endogenous phytohormone levels, balances, and plant sensitivities and how such changes influence growth and developmental responses in plants.
3. Develop a selected literature base that exemplifies the interactions between abiotic stresses, phytohormone responses, and plant development.
4. Develop a selected literature base that exemplifies the interactions between biotic stress vectors (fungi, bacteria, and viruses) and vector/host phytohormones relative to their influence on disease development and host physiology.

I. INTRODUCTION

The classical phytohormones are indoleacetic acid, (IAA), gibberellic acid, (GA), cytokinin (CK), abscisic acid (ABA), and ethylene (C=C). Structural examples of these important growth substances are illustrated in Figure 2.4. Phytohormones (plant growth regulators) occur in plants in very small amounts (picogram–nanogram). The first three are considered to be growth stimulators, while the latter two are generally considered to be growth inhibitors.

The precise location of synthesis of phytohormones is uncertain but they are generally high in apical meristems and tissues undergoing active growth. Frequently, reference is made to CKs and GAs being synthesized in the roots and transported via the xylem to other parts of the plant. Actively growing leaves, fruits, and developing seeds are active sites of synthesis of phytohormones. However, it appears that all tissues have the potential to produce any of the phytohormones; thus, defining tissues or organs that specifically produce them is difficult.

Phytohormones are transported via the xylem or phloem and by diffusion, such as in the case of C=C, which is a diffusible gas. IAA is unique compared to the other phytohormones in that it exhibits a type of polar transport that maintains a directional flow and gradient (high in shoots, low in roots) of IAA in a basipetal direction throughout the plant. Although directional transport of IAA does not appear to be associated specifically with the phloem, it may be maintained in parenchyma cells adjacent to the phloem. Why IAA exhibits this type of polar gradient is a matter of conjecture, but it may function as a type of background hormone in plants against which other hormones function interactively.

Phytohormones occur in plants as free and conjugated forms. The latter are usually conjugates of sugars, amino acids, and possibly peptides. The free forms are generally considered to be biologically active, while the conjugates are viewed as functioning in (1) controlling levels of the more active free forms, (2) transport, and (3) storage.

Generally, the mechanisms of action of phytohormones are poorly understood. However, several mechanisms or combinations may be operative. Control of genetic expression has been demonstrated for the phytohormones at both the transcriptional and translational levels. Also, hormone receptors and binding proteins have been identified on membrane surfaces that are specific for some phytohormones. The type and abundance of these proteins appear to be important in determining the sensitivity of the tissues to differing concentrations and types of phytohormones during development and changing environmental conditions. In addition, phytohormones can interact directly with cellular membranes to change fluidity characteristics (Chapter 3).

Much of the information relative to the effects of phytohormones on plant development has been obtained through experiments using exogenous applications of hormones to whole plants or excised plant parts. The limitations of such experiments include (1) application of physiologically inappropriate concentrations of chemical, (2) isolation of plant parts from the intact system, (3) uncertainty of amounts of chemical entering plants, and (4) sequestering of unknown quantities of chemical within the plant tissues. Despite these complications, much information has been derived concerning the modes of action of phytohormones in plants, and interestingly, more current analytical and molecular biological approaches are confirming many of the initial findings based on less sophisticated methods.

Table 4.1 summarizes many of the morphological and developmental responses in plants that are attributable to phytohormones. It is apparent from this list that a relative small number (five) of chemicals have a very impressive effect on plants. In addition, it is obvious that some of the phytohormones are specific in their effects, but some also exhibit overlapping functions with other hormones. The response that plants may have to exogenous phytohormone application may depend on species, stage of development, and the external environmental conditions. Such responses imply that normal internal levels may be regulated by the same factors.

The remainder of this chapter is focused on the impact of the biotic and abiotic environment on phytohormones and how they affect plant development. The emphasis will be primarily on the impact of stress response in plants, but since phytohormones respond to environmental changes even when the plant is not perceived to be under stress, it may be unclear at times whether or not the change in hormone balance is truly a stress response. For example, many perennials and biennials may require low-temperature stress to complete cyclical growth patterns (Seeley 1990). It is very clear that all aspects of the environment can influence hormone concentrations and/or sensitivities of plants to hormones and that such changes influence the acclimation and perhaps the long-term adaptations of plant to their respective habitats.

II. LIGHT STRESS AND PHYTOHORMONES

Light quality (wavelength) and quantity (intensity/duration) are very important to photomorphogenesis in plants. Virtually every aspect of plant growth, development, differentiation, and reproduction can be influenced by light. Plants have various photoreceptors that detect different wavelengths of light, two of which include phytochrome (red [R]/far-red [FR] light–sensing receptors) and cryptochrome (blue light–sensing receptor). Table 4.2

TABLE 4.1 Morphological and Developmental Effects of Phytohormones[a]

	Phytohormone Effect				
Plant Response	IAA	GA	CK	C=C	ABA
Nucleic acid synthesis	+	+	+	−	−
Cell division	+	(0)	+	−	−
Cell enlargement	+	+	+	(−)	(−)
Cell wall growth	+	0	0	0	0
Membrane permeability	+	+	+	+	+
Callus formation	+	(+)	+	(−)	(−)
Cambial activity	+	+	+	(−)	(−)
Differentiation	+	+	+	(−)	(−)
Shoot elongation	+	+	+	−	−
Apical dominance	+	−	−	(0)	(0)
Leaf growth	+	+	+	−	−
Root growth	−	0	+	−	−
Root initiation	+	0	0	(−)	(−)
Bolting	0	+	0	0	−
Flowering	+	+	0	+	−
Floral development	+	+	(+)	(+)	(−)
Sex expression	+	+	0	+	0
Parthenocarpy	+	+	0	0	0
Fruit development	+	+	+	(0)	(0)
Ripening	0	(−)	(−)	+	(+)
Dormancy	0	−	0	0	+
α-Amylase production	0	+	0	0	−
Germination	(+)	+	+	+	−
Senescence	−	−	−	+	+
Abscission	−	−	−	+	+
Exhibits polar transport	+	0	0	0	0
Nutrient distribution	+	0	+	0	0
Geotropism	+	0	0	+	+
Phototropism	+	0	0	0	0
Epinasty	+	0	0	+	0

[a]+, promotes; −, inhibits; 0, no effect; ( ), probable effect but questionable.

summarizes some phytochrome- and cryptochrome-mediated responses in plants. The structure of cryptochrome is not known for certain, but blue light action spectra suggest that the receptor pigment is similar to β-carotene or possibly riboflavin. It has a maximum absorbance value of 340 to 520 nm. The mechanism by which cryptochrome initiates physiological responses in plants is not known. It appears that cryptochrome and phytochrome may interact, the former at times regulating the sensitivity of some phytochrome-mediated responses (Mohr and Schopfer 1995).

Phytochrome is a bluish chromoprotein that exists in two forms. One form absorbs maximumly at 667 nm (P_r form), while the other absorbs at 730 nm (P_{fr} form). The latter form is considered to be the biologically active form. Figure 4.1 illustrates the proposed structure of the two forms of phytochrome. The P_r form of phytochrome is converted to P_{fr} during the day while P_{fr} is either, destroyed, converted back to P_r, or additional P_r is synthesized during the dark period (Figure 4.2). Phytochrome-mediated responses are reversible in plants if a R light treatment is immediately followed by FR light. Extending the

TABLE 4.2 Examples of Phytochrome- and Crytochrome-Mediated Responses

Phytochrome	Cryptochrome
Floral induction	Solar tracking
Nyctinastic leaf movements	Phototropism
Phototropic sensitivity	Chloroplast movements
Seed germination	Leaf rolling
Stem elongation	Carotenoid biosynthesis
Plumular hook opening	Anthocyanin synthesis
Leaf and cotyledon expansion	Stomatal opening
	Nyctinastic leaf movements
Chloroplast development	
Enzyme activation	
Protein synthesis	
mRNA transcription	
Chloroplast movements	
Surface potential in roots	
Transmembrane potential	

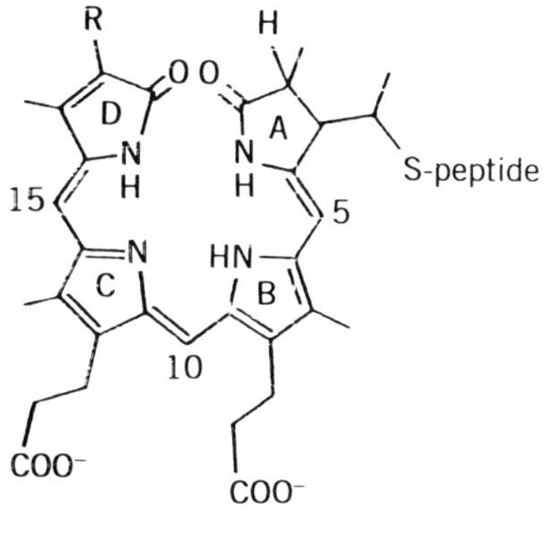

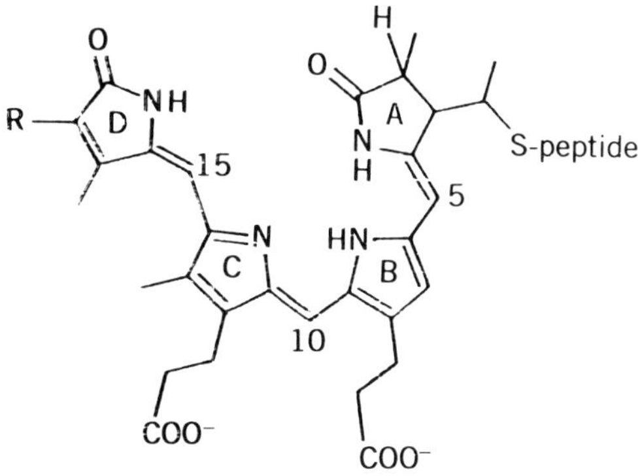

Figure 4.1 Structure of red and far-red absorbing forms of phytochrome. (Modified from Salisbury and Ross 1992.)

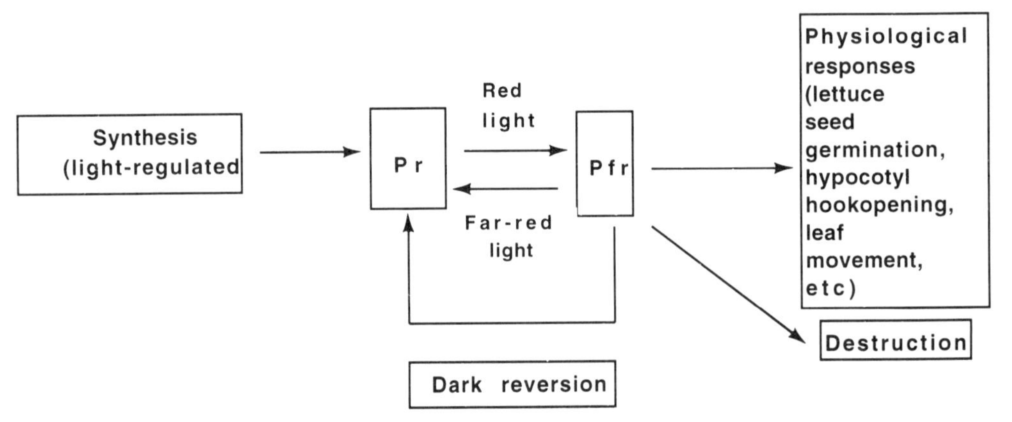

Figure 4.2 Model of phytochrome-regulated physiological responses in plants. (Modified from Taiz and Ziegler 1991.)

interval between treatments usually results in the inability to reverse the effects of R light responses in plants. The rapid interconversion of the two forms of phytochrome has been used by scientists to determine whether certain responses in plants are controlled by phytochrome. Recently, it was discovered that other species of phytochrome exist in plants (PhyI and PhyII), the synthesis of which is controlled by several different gene families (Furuya 1993). The differences in these phytochromes appears to be in the amino acid composition of the proteins that comprise these molecules. Type I phytochrome responses are generally related to deetiolation processes and accumulates in the dark and is light labile, while the type II responses are associated with green light–grown plants and is light stable. Exactly how P_{fr} initiates changes leading to morphological events is not clear, but the transduction process may include control of synthesis, degradation, sensitivity, and compartmentalization of phytohormones, which in turn may have direct or indirect effects on the metabolic processes of plants. At one time the ratio of P_r/P_{fr} was thought to be an important factor relative to the initiation of a biological event, and in some cases, but not in all, a biological response could be correlated with changing ratios. Spectral ratios of FR and R wavelengths of light, which influence phytochrome ratios, can have important morphological and biochemical effects on plants by altering the ratios of phytochrome species. For example, high ratios of FR/R light applied at the beginning of the night period to tobacco and soybean changed partitioning of photosynthate toward shoot growth, resulting in taller plants (Kasperbauer 1994). Leaves were longer, narrower, thinner, and exhibited higher ratios of chlorophyll *a*/*b*. More light-harvesting complexes (LHCII) were observed in the leaves, along with greater photosynthetic efficiency. Clearly, the ratio of FR/R wavelengths of light influenced photosynthate distribution and developmental patterns in these plants. Thus natural changes in the ratios of FR/R wavelengths to which plants are exposed can have important ramifications relative to plant growth and differentiation. The ability of plants to respond to changing ratios through photoreceptors may reflect an important mechanism for monitoring the changing environment and in avoiding stressful situations that could affect growth and development. Next, we summarize how wavelengths of light can be altered in the natural ecosystem.

Natural changes in light quantity and quality occur as a result of seasonal, diurnal, and meteorological events. For example, normal daylight consists of direct sunlight and diffuse skylight, the latter of which is enriched with blue wavelengths of light because of preferential scattering of shorter wavelengths by moisture droplets and other particulates in the atmosphere. At twilight (10°above horizon) scattering and refraction of the suns rays as they enter the earth's atmosphere enriches the light with longer R and FR wavelengths. Cloud cover and atmospheric pollutants can also cause an enrichment with blue wavelengths. However, some pollutants can absorb different wavelengths, depending on their composition (Hopkins 1995). In addition, plant canopy structure and soil texture and color can affect spectral distribution and intensity of light that can affect plants growing in adjacent areas (Kasperbauer 1994). As an example, light penetrating a forest canopy is not only reduced in intensity, but wavelengths of light reaching plants that occupy the lower stories are subjected to enriched levels of FR light. Soil color and texture is important relative to the wavelengths of light reflected. For example, plants grown in red soils compared to light-colored soils received higher ratios of FR/R light since red soils reflected FR light (Kasperbauer, 1994). Soils also absorb R wavelengths of light, allowing higher percentages of FR wavelengths to penetrate. Table 4.3 illustrates ratios of R/FR wavelengths of light that occur in various natural environments.

The following material outlines how light can influence the levels, transport, synthesis, and degradation of hormones in plants. In natural ecosystems plants have adapted to the light environments in which they are growing and in many instances do not experience light stress unless some sudden modification of the environment occurs that limits or causes an excess of incident irradiation. Although much of the information presented below reflects the impact of light on phytomorphology of plants and not necessarily light stress per se, nonetheless, to fully appreciate what transpires when plants are exposed to light stress, it is important to have a grasp of how light may "normally" affect hormone balances.

A. Light and Indoleacetic Acid Levels

IAA synthesis and/or transport may be influenced by both light quality and quantity. As one may expect, many aspects of plant development may be affected by light; therefore, only a few examples will be given that may relate directly to stress-induced responses by plants.

Dark and low-light treatments increase abscission of flowers, flower buds, leaves, and fruits. Such responses have been correlated with a reduction in IAA in such plant organs and are phytochrome (R/FR light) controlled. Using *Coleus* leaves, Mao et al. (1989)

TABLE 4.3 Light Quantity and Quality Associated with Various Environmental Conditions

	Photon Flux Density (μmol m^{-2} s^{-1})	R/FR
Daylight	1900	1.19
Sunset	26.5	0.96
Moonlight	0.005	0.94
Ivy canopy	17.7	0.13
Lakes at a depth of 1 m		
Black Loch	680	17.2
Loch Leven	300	3.1
Loch Borralie	1200	1.2
Soil at a depth of 5 mm	8.6	0.88

Source: From Smith 1982.

showed that red light (R) caused a threefold increase in diffusible IAA levels compared to dark controls and that even less IAA was recovered from leaves treated with FR light (Table 4.4). Higher IAA levels in leaves were correlated with increased break strength of leaf petioles (reduced abscission) compared to plants treated either in the dark or with FR for 4 days. Triiodobenzoic acid (TIBA), a IAA transport inhibitor, was found to increase leaf abscission even when plants were subjected to R treatment. This suggested that R inhibits transport of IAA to the leaf petiole, which normally prevents the formation of the abscission zone and retention of the leaf. The authors also demonstrated that the effect was IAA specific (CK, ABA, and GA were not effective), and diffusates collected from R-treated plants reduced leafless petiole abscission in dark-treated plants, while diffusates from FR- and dark-treated plants resulted in increased abscission that was less than or comparable to nontreated controls in the dark. The conclusion reached was that low light or darkness causes either a reduction in transport or concentration of IAA, which leads to the formation of an abscission zone in the leaf petiole, resulting in abscission. In another experiment (Wien and Turner 1989), abscission of reproductive structures in three cultivars of pepper plants increased by 38% when 80% shade was imposed for 6 days. The abscission response was correlated with increased C=C production in the floral structures, but abscission could be reversed by the application of exogenous auxin (naphthalene acetic acid). It appears that several hormones are involved in the abscission process. Auxin prevents abscission, whereas ethylene and abscisic acid increase it.

Growth of maize mesocotyls is also affected by light and appears to be a phytochrome-mediated response. Mesocotyls exposed to R irradiation exhibited a reduction in growth that was correlated with a decrease in IAA and auxin binding protein 1 (Figure 4.3) (Jones et al. 1991). Only the initial decline in growth correlated with the decrease in IAA. After IAA levels declined to 50% of the dark controls, growth continued to decline. Changes in IAA levels were attributed to reduced transport of auxin to the epidermal cells of the mesocotyl. In another study using maize mesocotyls, similar effects of R irradiation were observed along with a decrease in auxin-binding activity associated with the endoplasmic reticula of mesocotyl cells (Walton and Ray 1981).

B. Light and Cytokinin Levels

As with IAA, light quality and quantity can influence CK levels in plants. CKs are known to control many physiological responses in plants, including reduced senescence, a per-

TABLE 4.4 Effect of Light on Auxin Content in Leaf Diffusate[a]

Diffusate Source	Auxin Content (ng/sample)
Red light	84.8a
Dark	27.5b
Far-red light	17.2b

Source: Modified from Mao et al. 1989.

[a]Numbers followed by the same letter are not significantly different at $p \leq 0.05$. Diffusate collected from 20 cut leaf blades for 48 h, beginning at time zero in each light treatment.

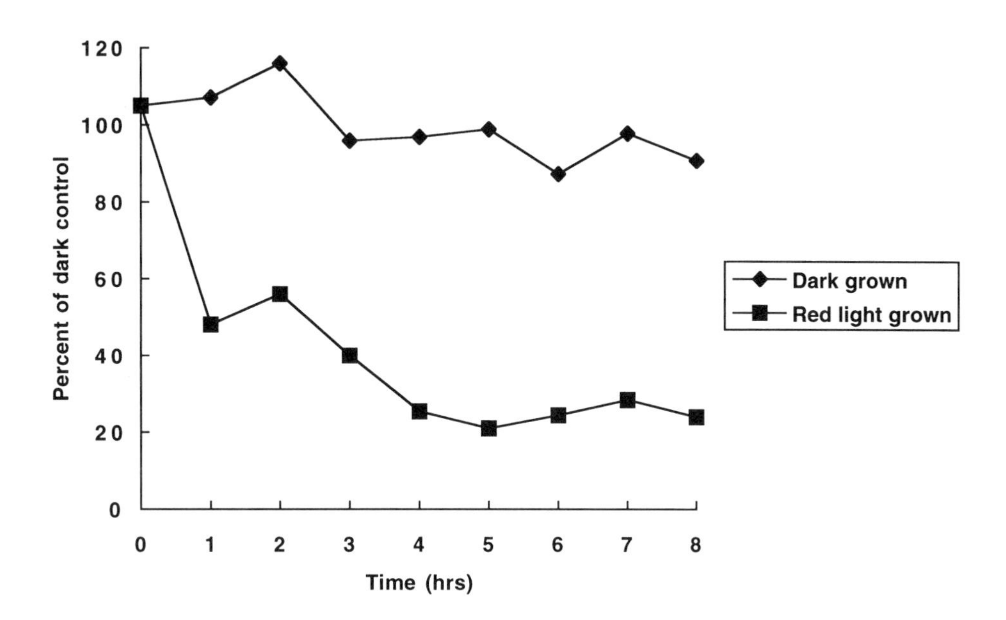

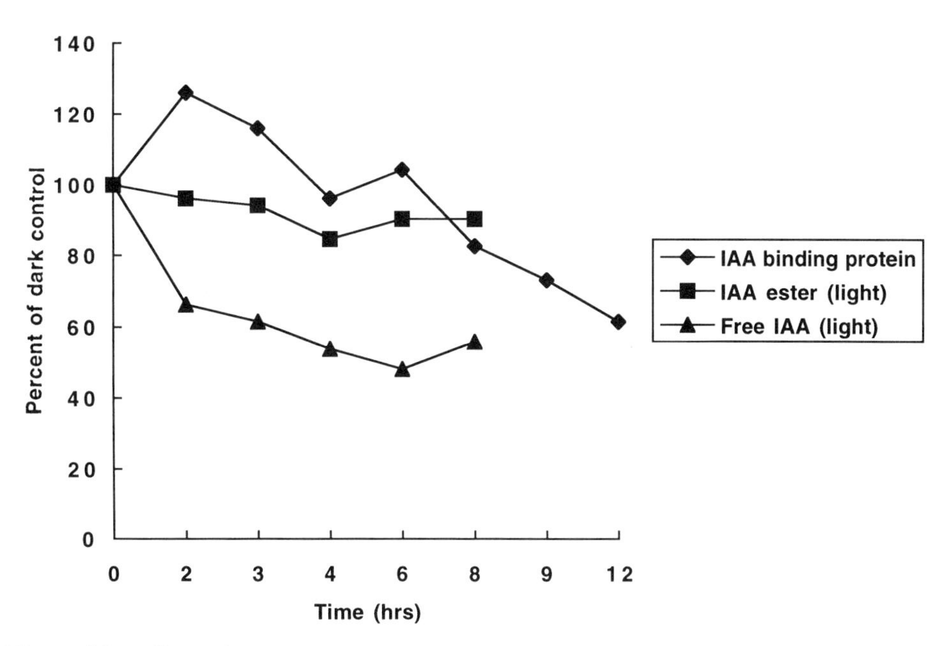

Figure 4.3 Effects of red light on growth, IAA, IAA ester, and IAA binding protein in maize mesocotyl tissues. (Modified from Jones et al. 1991.)

missive role in seed germination, breaking of apical dominance, induction of cell division, mobilization of nutrients, and tissue differentiation, just to name a few. The following examples will give the reader an idea of how light intensity and quality can affect the level of CKs in plant tissues.

Hewett and Wareing (1973) found that red light induced at least transient increases in CK levels in attached and detached leaves of poplar (*Populus robustus*). This response was usually initiated rapidly after treatment and subsided to a lower level following continuous illumination. CK riboside specifically increased after R light treatment, and diurnal changes in leaf CK in field grown poplar was prominent immediately following daybreak (Figure 4.4). As indicated in this figure, on hot sunny days, reductions in CK levels occurred earlier in the day than on warm overcast days. This probably reflects water stress due to the higher temperature. CKs are known to decline in leaves during drought stress.

Qamaruddin and Tillberg (1989) noted that R light caused a transient increase in N6 (Δ^2-isopentenyl) adenosine (iPA), a CK, in imbibed Scots pine (*Pinus sylvestris*) seeds. The effect was reversible with FR light, indicating phytochrome regulation of CK in seed germination (Figure 4.5).

Ambler et al. (1992) studied two cultivars of sorgham (*Sorghum bicolor*) that differed in senescence response. One cultivar (Tx7000) was considered to be senescent; the other (Tx2817) was nonsenescent. In field-grown plants, CK levels in the xylem sap of TX2817 was 1.5 times higher than in Tx7000. Much evidence in the literature supports the hypothesis that CKs are antisenescence factors in plants. Thus the authors suggested that differences in CK levels in the two cultivars may explain the different senescence patterns. In addition, the authors demonstrated that under optimal light and nitrogen nutri-

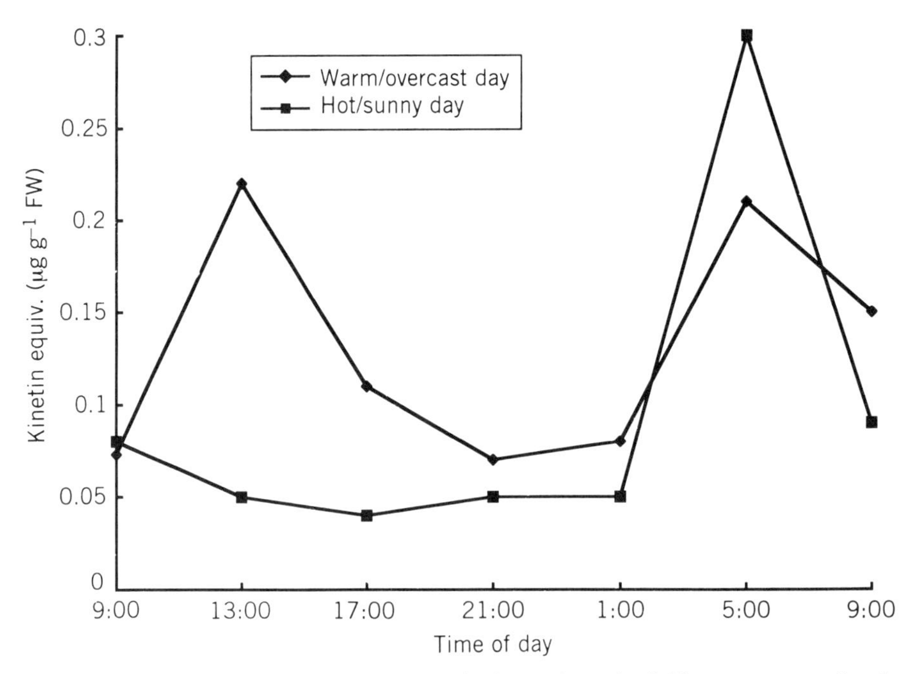

Figure 4.4 Diurnal changes in CK levels in poplar leaves from the field on two contrasting days. (Modified from Hewett and Wareing 1973.)

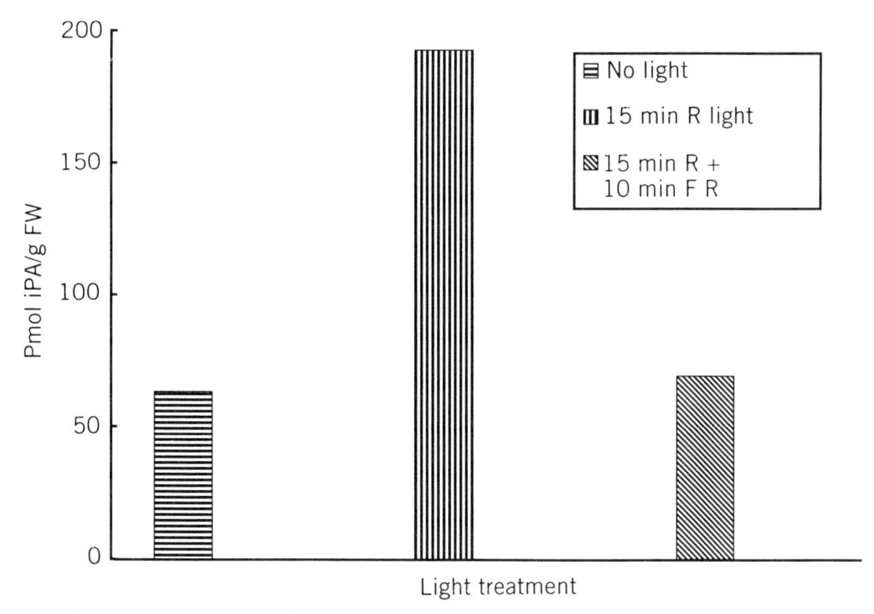

Figure 4.5 Effects of light on iPA levels in Scots pine seeds. (Modified from Qamaruddin and Tillberg 1989.)

tion, the two cultivars exhibited equal levels of zeatin and zeatin ribioside in the xylem sap of hydroponically grown plants. Tx2817 was less susceptible to reduction in CK levels under reduced nitrogen levels and actually exhibited increased concentrations of CKs in reduced light (50% reduction) compared to Tx7000 (Table 4.5).

Light intensity was also shown to increase CK levels in transgenic tobacco. Exogenous application of CKs have been shown to mimic symptoms of salt stress in plants. In an attempt to determine possible relationships between salt stress and CKs, Thomas et al. (1995) used transgenic tobacco that had been transformed to overexpress CKs. When these plants were grown at high light intensities (300 μmol m^{-2} s^{-1}), a 10-fold increase in CK was observed compared to low light (30 μmol m^{-2} s^{-1})–grown transformed plants or nontransformed control plants. High light–grown plants exhibited phenotypic symptoms similar to those of salt-stressed plants and produced high levels of proline and osmotin as in salt-stressed plants. This suggested that CK and NaCl must have common or

TABLE 4.5 Xylem Sap CK Levels (10^{-12}m in sap g^{-1} fresh wt) in Senescent (Tx7000) and Nonsenescent (Tx2817) Sorghum Cultivars[a]

	Tx7000		Tx2817	
Treatment	Z	ZR	Z	ZR
100% N, 100% light	259	246	238	252
10% N	59	73	160	92
50% light	120	196	320	381

Source: Modified from Ambler et al. 1992.

[a]Means of two replications from experiment 1.

overlapping signaling pathways that are initiated independently, which initiates gene expression that is normally initiated by developmental processes such as flowering or environmental stress. Light intensity may also play a significant role in the magnitude of the response.

C. Light and Gibberellin Levels

Gibberellins are considered to be growth promoters in plants and are known to be involved in numerous developmental functions, such as stem elongation, seed germination, flowering in long-day plants, sexual expression, and parthenocarpic fruit development, just to name a few. The first three functions mentioned are probably the most studied with respect to the effects of light on GAs. However, GAs pose a particularly unique problem compared to other hormones. Many different GAs have been detected in plants (80+, including fungi), and whether GAs have different functions or simply serve as precursors to others that have specific biological activity is only recently starting to be analyzed. It now appears that GA_1 is emerging as being the GA responsible for stem elongation in a number of plants (Taiz and Zeiger 1991). However, other GAs also appear to have biological activity, but this may be a result of further metabolism to active forms. Although little is known concerning how light intensity per se affects GAs in plants, much research has been conducted on light/dark and R/FR light effects, which has led to the conclusion that GA metabolism/sensitivity is under phytochrome control. To narrow the scope of information on the effects of light on GA levels and photomorphogenic effects in plants, the following discussion deals with light effects on GA levels that relate to shoot elongation.

As alluded to above, little information is available on the effect of light intensity on GA levels in plants (Potter and Rood 1993). However, these authors found that $GA_{1,3,8,19,20}$ increased as light intensities decreased from 500 μmol m^{-2} s^{-1} to 25 μmol m^{-2} s^{-1} PAR in canola (*Brassica napus*). Associated with this change in GA levels was increased shoot elongation, which implied a direct association of GA levels and shoot growth.

Talon and Zeevaart (1990) demonstrated that photoperiod (light/dark ratio) also cause changes in the levels of GAs. In the long-day (LD) plant, *Silene armeria* (sweet william catchfly), transfer of plants from short-day (SD) to long days resulted in a decrease in GA_{53} and increases in $GA_{19,20,1}$ in both whole shoots and mature leaves. The reverse occurred when plants were transferred from LD to SD. These shifts in GA levels were attributed to LD activation of GA_{53} oxidase, which appears to be a light-limiting step in the synthesis of the other GAs identified. The accumulation of GA_1 in shoot apices was attributed to the increased shoot growth under LD and/or the possible LD induction to increased sensitivity to GA_1.

Using spinach (*Spinacia oleracea*), Talon et al. (1991) demonstrated that light increased $GA_{53,44,19,20,1}$. Under dark conditions the same GAs declined. Evidence for light regulation of $GA_{53,20}$ oxidase activity by light was implied, and that during high light the flow through the pathway of GA metabolism was much enhanced, as was GA turnover. Supplemental LD treatments caused increased $GA_{20,1}$, but the relationship between the changes observed and shoot growth was uncertain.

Sponsel (1986), using dwarf (D) and tall (T) peas, demonstrated that R light increased $GA_{1,20,29}$ levels in both phenotypes compared to dark-grown plants. The (D) phenotype grew the same as the (T) phenotype in the dark, and (D) plants had GA_1 in both R light and dark, suggesting that GA_1 was not responsible for the growth of (D) plants in the dark.

Ross et al. (1992), using (D) and (T) phenotypes of sweet pea (*Lathyrus odoratus*), demonstrated that light generally resulted in increases in $GA_{1,8,20,29,81}$ levels in both phenotypes, although the levels of GAs were generally lower in the (D) phenotype. Basal and apical regions of the shoots were analyzed, with the highest levels of GA being associated with the apical region in both plant types. Two experiments were presented with some slight differences in results. Darkness was viewed as not enhancing sweet pea growth via GA_1.

Weller et al. (1994) using phytochrome deficient and GA_1 overproducing pea mutants concluded that phytochrome does not alter shoot elongation through GA_1 levels, and that etiolation and phytochrome increase responsiveness of wildtypes to GA_1. Levels of GA_{19} declined in shoot apices in the light, while $GA_{20,29,1,8}$ increased. Reports by Bown et al. (1975) and Neskovic and Sjaus (1974) indicated that GA-like activity in *Phaseolus coccineus* and pea declined in white and R light, respectively.

Toyomasu et al. (1992) demonstrated that whole lettuce seedlings grown for 4 days in the dark followed by 1 day in the light contained lower levels of GA_1 and higher levels of GA_{19} and GA_{20} than those of plants kept in the dark for 5 days. The same trend was also found when the upper and lower hypocotyls were compared. However, cotlyedons had lower levels of GA_{19} and GA_1 under the light treatment and elevated levels of GA_{20}. It was concluded that GA_1 controlled hypocotyl elongation in lettuce, that the levels of GA_1 are controlled by light, and that light controls the elongation of hypocotyls through regulation of GA_1.

Table 4.6 summarizes the literature relative to the effects of light quality and quantity on GA levels and shoot elongation in several plant species. The table is constructed showing the impact on GA levels of R light, white light, long days, or high light. It is clear that differences exist among the studies relative to the light treatments. This could reflect species, treatment, and analytical differences. However, in the majority of cases, light appears to cause an increase in GA levels, the exceptions being the work of Potter and Rood (1993), Neskovic and Sjaus (1974), and Bown et al. (1975). In the latter two studies,

TABLE 4.6 Summary of the Effects of Light on Gibberellins[a]

Plant	Light Conditions	Growth Response	GA Response									
			1	3	8	19	20	29	44	53	81	GA-L
Pea	D,(R)	Shoot(−)										−
Bean	D,(L)	Shoot(−)										−
Pea	D,(R)		+				+	+				
Catchfly	SD,(LD)	Shoot(+)	+			+	+			−		
Spinach	D,(L)	(?)	+			+	+		+	+		
Sweet pea	D,(L)											
Tall		Shoot(+)	+		−			+	+			+
Dwarf		Shoot(+)	+			0	0	+	+			+
Lettuce	D,(L)	Hypocotyl	−			+	+					
Canola	High light	Shoot length	−	−	−	−	−					
Pea	D,(L)	Shoot apex	+		+	−	+	+				

[a]−, inhibits; +, stimulates; 0, no effect; D, dark; (L), treatment in light; (R), treatment in red light; SD,LD, short and long day, respectively; GA-L, GA-like. Light condition in parens represents treatment condition. Plants initially grown in the dark or in SD.

bioassays were employed to detect GAs, while all other studies utilized immunoassays or gas chromatography–mass spectrometry techniques. Since some of the GAs listed in the table represent precursors to other GAs, it is not surprising that some may increase and some decrease within an experiment as a result of differences in turnover rates. In view of the current thinking that GA_1 is perhaps important in controlling shoot elongation it is rather surprising that in so many instances GA_1 increases in the light or under LD environments.

D. Light and Abscisic Acid Levels

Light intensity, duration, and quality can all affect the levels of ABA in plants. Consequently, the role of ABA in dormancy induction (seed and apical meristem), stomatal closing, inhibition of flowering, general growth inhibition, abscission, and senescence can be controlled and influenced through light quality (possibly phytochrome mediated) and light quantity (intensity and duration). Unfortunately, the data are conflicting and as mentioned before, may reflect species, tissue, experimental, and/or analytical differences. However, the following will give the reader some perspective as to the effects of light on ABA production and its potential impact on developmental and growth responses.

Prolonged cloud cover can lead to reduced yields in crops, which led Mohandass and Natarajaratnam (1988) to determine the impact of light intensity on hormone levels in light-tolerant and susceptible cultivars of rice grown in India. They determined that ABA increased more in panicles of both cultivars (35% higher in tolerant varieties compared to susceptible) under reduced irradiation levels (50% shade) than in full sunlight. ABA decreased or did not change in leaves and roots of shaded plants. High ABA levels in panicles were suggested to be associated with increased photosynthate accumulation, based on supporting data from experiments cited.

Using tomato plants, Basiouny et al. (1993) demonstrated that increasing light intensity caused a reduction in ABA in the leaves of plants subjected to high soil moisture stress (dried to 0.07 MPa and rehydrated to saturation), low soil moisture stress (dried to 0.015 MPa and rehydrated to saturation), and apparent moisture saturation (Figure 4.6). The levels of ABA were highest in saturated soils, followed by high soil moisture stress and low soil moisture stress, respectively. Contrary to the findings of Basiouny et al. (1993), Henson (1983) found ABA to increase in drought-stressed pearl millet (*Pennisetum americanum*) plants subjected to increasing photon flux densities of 0, 300, 450, and 600 $\mu mol\ m^{-2}\ s^{-1}$. Under dark conditions ABA levels declined. They concluded from this study that light strongly affects the capacity of leaves to accumulate ABA in response to water stress and that light may affect ABA content in the absence of stress (Table 4.7).

In studying shoot dormancy and the roll of ABA, Phillips et al. (1980) found that the length of the dark period had no effect on the levels of free or bound ABA in the leaves or roots of sycamore maple (*Acer pseudoplatanus*). However, light enriched with FR wavelengths increased bound ABA and induced the appearance of both free and bound phaseic acid, a degradation product of ABA. Light had no significant effect on free ABA.

In studying the effects of light on growth of maize (*Zea mays*) roots, Saugy et al. (1989) found that root growth was inhibited by white light and that an increase in free ABA levels was associated with this response. The authors also show that the induction of ABA is transient, in that after about 4 h, ABA levels drop off and root growth begins to increase. Based on root growth kinetics and the onset of ABA production, it was concluded that other factors in addition to ABA may be responsible for reduced growth.

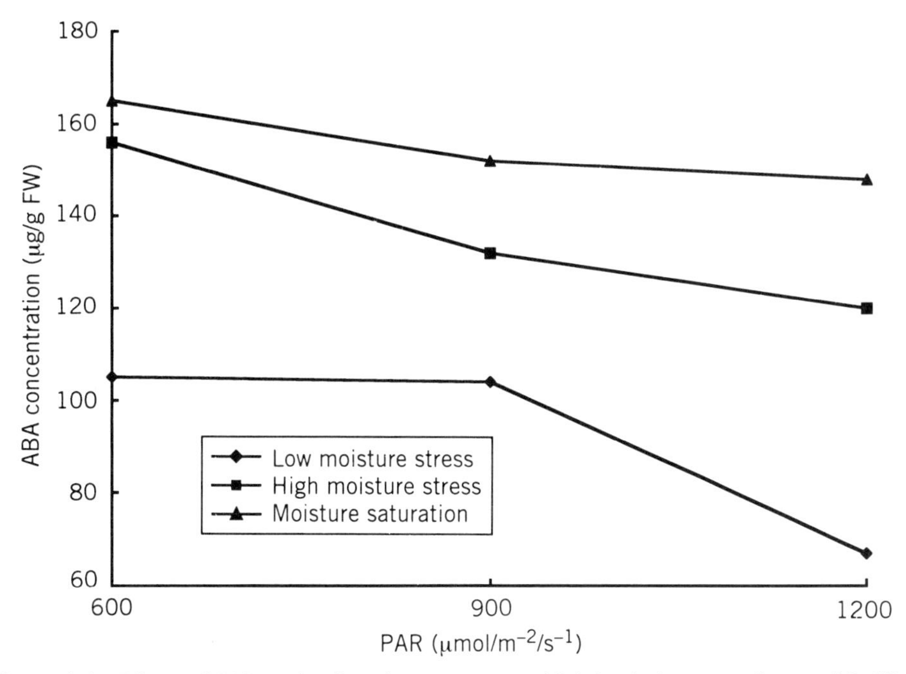

Figure 4.6 Effects of PAR and soil moisture stress on ABA levels in tomato leaves. (Modified from Basiouny, et al. 1993.)

A number of studies address the impact of light quality on ABA levels in plants. Tucker and Mansfield (1972) and Tucker (1977), studying the impact of light quality (R and FR) on apical dominance in tomato (*Lycopersicum esculentum*) and cocklebur (*Xanthium strumarium*), determined that FR light inhibited lateral bud out growth in both plants and that R light stimulated growth. This response was correlated with increased ABA levels in axillary buds of FR-treated plants and a reduction in ABA in R-treated plants (Table 4.8). In the case of decapitated tomato plants (removal of the source of IAA, which is thought to inhibit lateral growth), axillary buds of plants treated with FR light

TABLE 4.7 ABA Content (ng g^{-1} initial FW) of Water-Stressed and Unstressed Shoots from Seedlings Grown in Continuous Darkness or with Light (12-h Photoperiods)

Expt.	Days after Sowing	Growth Conditions	Turgid	Stressed
1	4	Light	25.5	153.2
		Dark	5.7	33.5
	5	Light	43.1	110.4
		Dark	6.7	12.5
		Dark (4 days) → light(1 day)	50.5	82.0
2	4	Light	19.3	79.2
		Light (3 days) → dark (1 day)	9.4	38.6

Source: Modified from Henson 1983.

TABLE 4.8 ABA Concentrations (μg/kg DW) in 7-Week-Old Plants of *Xanthium strumarium*

	Light Treatment	
Tissue	+ Far Red	− Far Red
Apical bud and three young leaves	247	<242
Axillary buds	12,220	193
Mature leaves	<57	<66

Source: Modified from Tucker and Mansfield 1972.

exhibited no growth of axillary shoots, whereas plants not treated with FR had abundant growth. It was implied that high IAA levels, also related to FR treatment in tomato, may be responsible for maintaining elevated ABA concentrations.

Loveys (1979), studying the impact of spectral quality on ABA levels in tomato, found no effect of R or FR light on this hormone in petioles or leaves. However, FR light caused a reduced level of ABA in stems (40 to 90% higher concentrations in R-light treatments), and this change was not restricted to any particular tissue (peripheral versus internal) within the stem which might affect differential growth. Changes in ABA levels appeared to be a result of a modified pathway of ABA degradation, resulting in a novel metabolite that yielded free ABA on base hydrolysis.

The fact that R and FR light seem to affect levels of ABA in plants suggests that the control of ABA is phytochrome mediated. In studies using a phytochrome-deficient and ABA-overproducing mutant (pew1) of *Nicotiana plumbaginifolia* and a double mutant produced from pew1 and aba1, an ABA-deficient mutant, Kraepiel et al. (1994) determined that the phytochrome-mediated light signal enhances ABA degradation rather than inhibiting its biosynthesis.

Williams et al. (1994), using duck week (*Lemna gibba*), found that dark causes elevated ABA levels. ABA was found to up-regulate two genes NPR1 and NPR2 while phytochrome down-regulated the same genes. The results suggested a relationship between phytochrome and ABA in regulating these genes.

E. Light and Ethylene Levels

Ethylene is generally thought of as being an inhibitor of growth responses in plants. It has been shown to be involved in numerous responses in plants, including abscission (leaf, flower, bud), flowering (bromeliads), gravitropic responses, ripening, senescence, and stress responses (flooding, wounding, pathogens, physical movements). Both quality and quantity of light affect the levels and/or sensitivity of plant tissues to ethylene.

Many C=C responses appear to be phytochrome mediated, and generally the effect of R light is to inhibit C=C levels in plants. FR light generally has the effect of increasing levels of C=C in plants. Similarly, white light (day light) tends to inhibit C=C levels and darkness promotes C=C accumulation. Table 4.9 (Michaliczuk and Rudnicki 1993) illustrates the effects of R light and darkness on C=C levels in a variety of plants differing in age. In most instances, R light inhibited the formation of C=C, and the addition of ACC, the immediate C=C precursor in its synthesis, also caused a reduction in

TABLE 4.9 Percent Change in C = C Production with and without ACC Addition in Excised Green Leaves of Various Plant Species When Grown in the Dark Compared to Red Light

Plant	Plant Age (weeks)	Without ACC	With ACC
Alstroemeria hybrida	75	7(−)	35(−)
Chrysanthemum morifolium	60	58(−)	16(−)
Codiaeum variegatum	90	60(−)	12(−)
Dahlia sp.	4	0(−)	31(+)
Dianthus caryophyllus	50	40(−)	23(−)
Dizygotheca elegantissima	120	17(−)	17(−)
Helianthus annuus	4	86(+)	76(+)
Impatiens balsamina	4	59(−)	53(−)
Lathyrus odoratus	4	33(−)	13(−)
Avena sativa	3	44(−)	34(+)
Phaseolus multiflorus	4	78(+)	20(−)
Rosa hybrida	30	60(−)	57(−)
Tagetes sp.	4	50(−)	24(−)
Zinna sp.	4	28(−)	14(−)

Source: Modified from Michaliczuk and Rudnicki 1993.

C=C but to a lesser extent. Table 4.10 (Rudnicki et al. 1993) illustrates the impact of both light intensity and wavelength on C=C levels in *Begonia hiemalsi* leaf discs. At the highest light intensity (20 μmol m^{-2} s^{-1} PPFD) all wavelengths of light (white, red, green, and blue) decreased C=C levels significantly except FR, which nearly doubled the concentration. As light intensity declined to 1 μmol m^{-2} s^{-1} PPFD, except for FR light, none of the wavelengths of light caused a significant change in C=C levels compared to the dark control, although as light intensity declined, so did the level of C=C produced under each of the light treatments except for FR, which appeared to decline and then increase again.

Problems associated with experimental procedures may have influenced how we interpret the results of light effects on C=C levels in plants since many studies have been

TABLE 4.10 Effects of Light Quality and Quantity on C = C Production in *Begonia* Leaf Discs (% change)

	Light Intensity (μmol m^{-2} s^{-1} PPFD)			
Light quality	20	9	5	1
Dark control	100	100	100	100
White	31	31	67	93
Red	36	45	93	90
Green	32	67	74	102
Blue	31	76	102	96
Far-red	191	158	164	183

Source: Modified from Rudnicki et al. 1993.

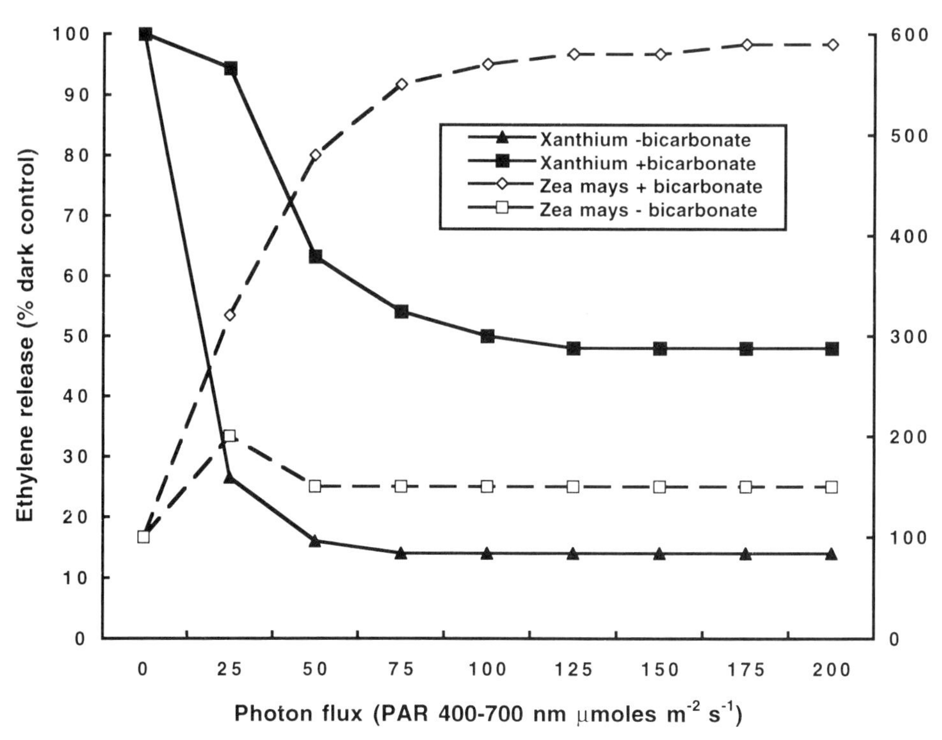

Figure 4.7 Effects of light intensity and bicarbonate on C=C production in cocklebur and corn. (Modified from Grodzenski and Boesel 1982.)

conducted on excised plant parts or leaf discs that have been enclosed in sealed containers in order to trap C=C gas. Experiments in closed containers can have varying levels of CO_2, due to differing rates of photosynthesis and respiration in the light and dark. Some studies have indicated that providing CO_2 to plant segments exposed to R light or daylight in closed containers actually exhibit higher levels of C=C than plants in the dark without the addition of CO_2 (Grodzinski et al. 1983; Kao and Yang 1982). Figure 4.7 illustrates how light intensity and bicarbonate (CO_2 source) affects the levels of C=C in a C_3 (cocklebur) and a C_4 (corn) plant, respectively. In both plants bicarbonate resulted in higher levels of C=C. However, increasing light intensity caused a reduction in C=C levels in *Xanthium strumarium* and an increase in *Zea mays*. These differences may reflect the lower CO_2 compensation points as well as the higher internal CO_2 levels exhibited by C_4 plants compared to C_3 plants. Although the precise mechanism by which light and CO_2 interact to control C=C levels is not known, it is thought that CO_2 may activate the C=C-forming enzyme, which converts ACC to C=C.

Interactions between light, ozone, and water stress have also been observed relative to their effects on C=C levels in plants. Rodecap and Tingey (1983) observed that ozone-induced stress C=C was 2.6- and 1.9-fold greater from dark-incubated than from light-incubated soybean and tomato plants, respectively. They also reported that the conversion of ACC to C=C was inhibited in the light but that both ACC and C=C accumulated in the dark but C=C declined after 6 h. ACC also declined in the dark but remained higher than expected. The authors conclude that the conversion of ACC to C=C is limited in the light.

TABLE 4.11 Effect of Light on C = C Production and Nodulation

Root Environment	C = C Production (pmol · g fresh wt^{-1} h^{-1})	Nodules/Plant on:	
		Primary Root	Lateral Roots
Dark	36.0	25.9	330
Dim light	69.8	5.1	224
Bright light	87.5	4.3	161

Source: Modified from Lee and LaRue 1992.

Lee and LaRue (1992) observed that light inhibited nodulation and induced increased C=C production by pea (*Pisum sativum*) roots (Table 4.11). Silver, an inhibitor of C=C action, increased nodule number on roots exposed to dim light. These results suggested that the inhibitory effect of light on nodulation may occur via increased C=C.

Wright (1981) reported increased C=C production in drought-stressed excised wheat (*Triticum aestivum*) leaves in the dark over that of nonstressed leaves in the dark. The levels of C=C in the former were also higher than those of wilted and nonstressed plants grown in the light (Figure 4.8). Wilted plants in the light exhibited only slightly higher levels of C=C than those of nonstressed plants in the light. No difference was observed between nonstressed plants in the light or the dark. Leaf age was also shown to be a factor relative to the effects of light on C=C production in wilted leaves (Figure 4.9). Older leaves produced more C=C in the dark, while younger leaves produced more C=C in the light. All leaves in the dark produced more C=C than did all leaves kept in the light. Exogenous application of the CK, benzyladenine, was found to in-

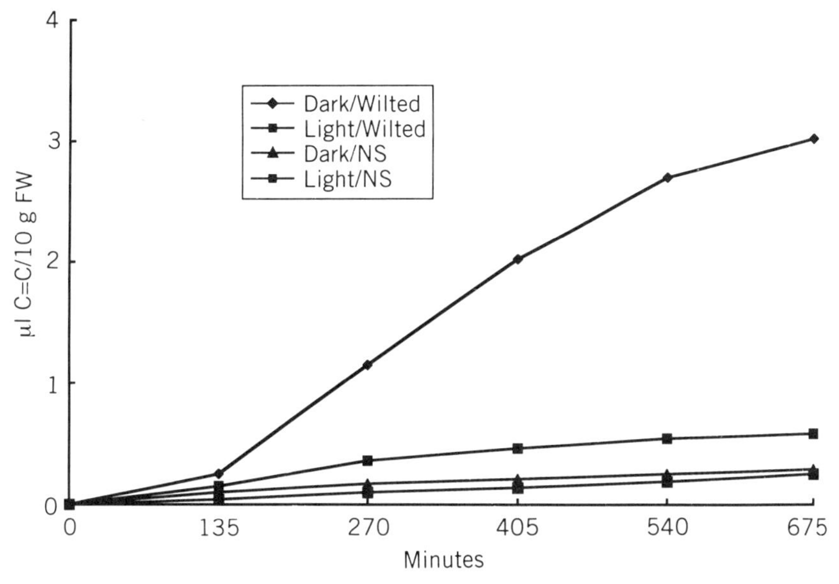

Figure 4.8 Inhibition of C=C by light in wilted and nonstressed wheat leaves. (Modified from Wright 1981.)

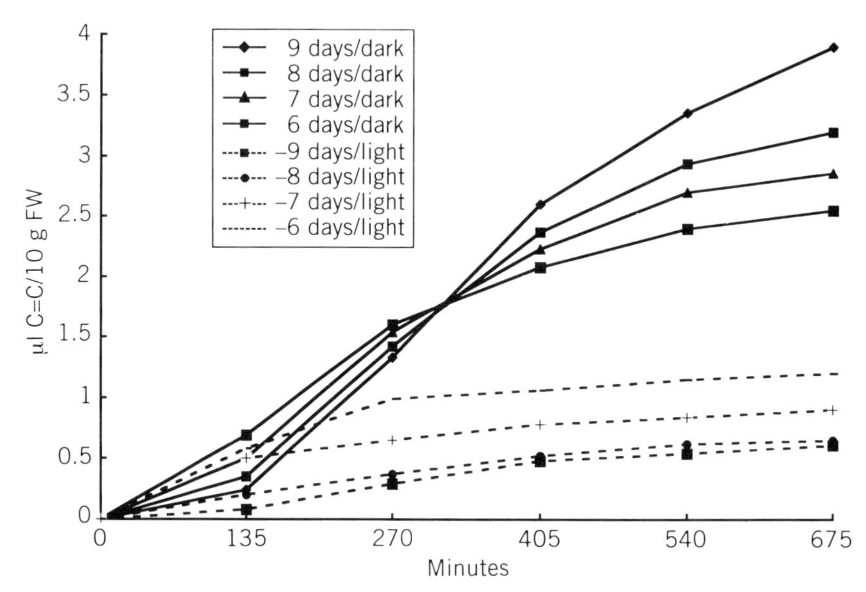

Figure 4.9 Effect of leaf age on C=C production in wilted wheat leaves. (Modified from Wright 1981.)

crease C=C levels dramatically in wilted plants and to a lesser extent in nonstressed plants (Figure 4.10).

Several studies have demonstrated that light may influence abscission of leaves, flowers, and buds through light-mediated control of C=C. Wien and Turner (1989), using three different cultivars of pepper (*Capsicum annuum*), correlated C=C production in shade-grown plants (80% shade for 6 days) with increased flower bud abscission (Table 4.12). The three cultivars of pepper exhibited varying degrees of susceptibility (Shamrock > Lady Bell > Ace) to abscission under shaded conditions and the percent abscission correlated with the levels of C=C and ACC produced in the flower buds. Exogenous applications of napthaleneacetic acid, a synthetic auxin, was found to overcome abscission in shaded buds.

Ethylene production under dark and short-day conditions was also reported to increase abscission in 2- to 3.5-cm-long flower buds in *Lilium* (van Meeteren and de Proft 1982). That abscission in shaded plants could be prevented by silver thiosulfhate (inhibits C=C action) injection into the flower buds or pretreatment of the bulbs provided further evidence for dark-mediated abscission through C=C action.

Contrary to the results above, Decoteau and Craker (1987) found little evidence to link FR-induced C=C synthesis to leaf abscission in mung bean (*Vigna radiata*). Application of AVG (aminoethoxyvinylglycine), an C=C biosynthesis inhibitor, did not alter the effects of FR/R light treatments on leaf abscission, suggesting that C=C was not responsible.

Red light was found to enhance germination of lettuce (*Lactuca sativa*) seeds at increasing temperatures (Saini et al. 1989). The ability to germinate at higher temperatures was related to increasing exposures to R light. Germination was correlated with, and was preceded by, increased C=C production by the seed (Figure 4.11). Use of the C=C biosynthesis inhibitor AVG reversed the effects of R light on seed germination, and appli-

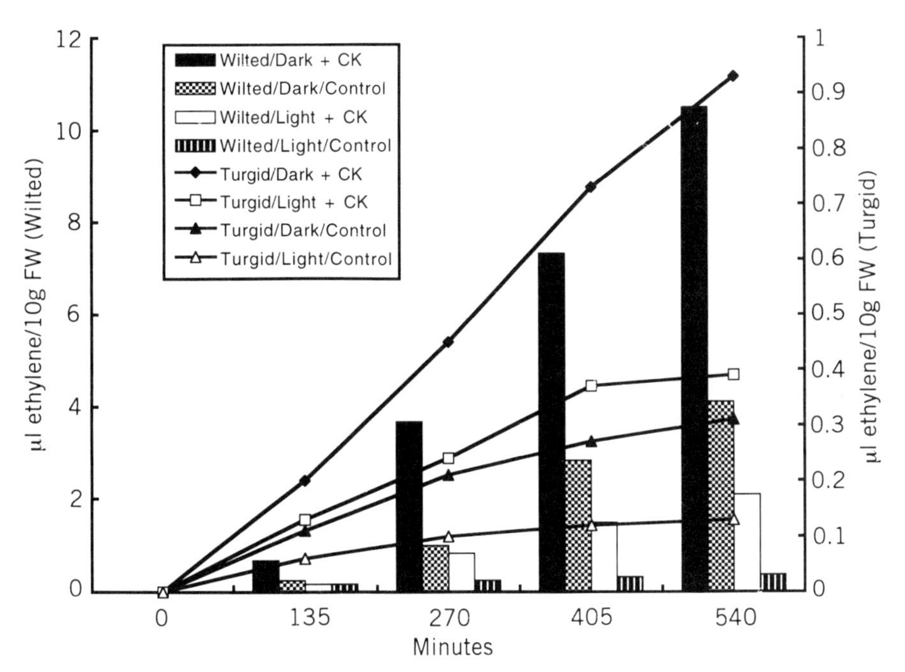

Figure 4.10 Effects of light, CK, and drought on C=C production in wheat leaves. (Modified from Wright 1981.)

cation of C=C could reverse the effect of AVG. The evidence suggested that high temperature shifted the ratios of phytochrome in favor of P_r, which inhibited germination. Repeated exposure of the seed to R light promoted the production of P_{fr}, which stimulated C=C production and germination. This is an interesting response since in this instance R light promotes C=C production in lettuce seed, while in most of the previous examples it has inhibited C=C production in vegetative tissues.

The foregoing examples of the effects of light on C=C production in plants, as well as the light/C=C interactions with other environmental factors (CO_2 levels, water stress, age, ozone, temperature), illustrates the complex nature of understanding how C=C controls responses in plants, such as germination, abscission, senescence, and nodulation.

TABLE 4.12 Effect of Shade on Abscission, C = C, and ACC Production in Pepper Flowers

	Abscission (%)		C = C ($nL\ g^{-1}\ h^{-1}$)		ACC ($nmol\ g^{-1}$)	
Cultivar	Open	Shade	Open	Shade	Open	Shade
Shamrock	45	70	0.289	1.046	0.192	0.880
Lady Bell	24	36	0.134	0.415	0.184	0.342
Ace	27	28	0.091	0.083	0.232	0.169

Source: Modified from Wien and Turner 1989.

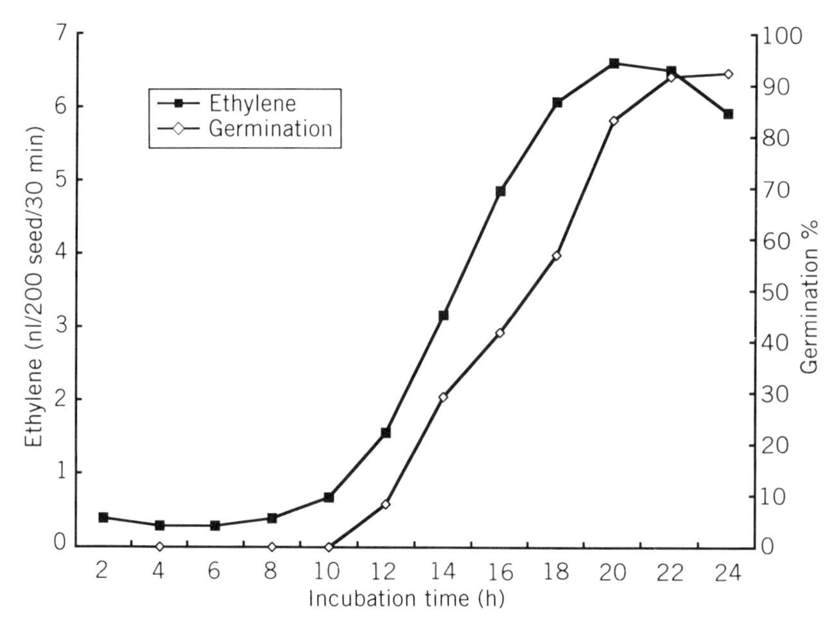

Figure 4.11 Relationship of C=C production and germination in lettuce seed incubated 24 h at 32°C in red light. (Modified from Saini et al. 1989.)

III. NUTRIENT STRESS AND PHYTOHORMONES

Hormones and nutrients can interact in several ways. Deficient and toxic levels of nutrients can affect the concentrations of specific hormones, and in turn hormones such as CKs and IAA have the capacity to direct the translocation and accumulation of nutrients in plants. The question has been raised as to whether deficiencies or elevated levels of nutrients affect growth directly or whether nutrients may affect hormones, which in turn could affect growth and developmental responses in plants (Kuiper 1988, Kuiper et al. 1989). Also related to this issue are studies demonstrating that inorganic cations, especially Ca and K, alter plant tissue responsiveness to CKs and possibly other hormones (Green 1983 and references therein). Considering the complex interactions of plant hormones and the multiplicity of plant functions they control, the impact of nutrients on hormones is an important issue. The foregoing represent examples of how nutrients can affect hormone balances in plants.

A. Nutrients and Indoleacetic Acid Levels

Nitrogen deficiency can affect IAA production in plants since nitrogen is an elemental component of its structure. Since the amino acid tryptophan is thought to be a precursor of IAA, any deficiency in nitrogen will result in a reduction in IAA levels.

Recently, studies using *Arabidopsis thalinana* auxin-resistant mutants indicate that NH_4 may disrupt synthesis, transport, or signaling of hormones (IAA and CK), which may result in the accumulation of these hormones in the roots, causing root growth inhibition (Cao et al. 1993). Auxin-resistant mutants were more resistant than wild-type plants to NH_4^+, and application of auxin or combinations of auxin and CK mimicked the symptoms induced by NH_4^+.

In another study, foliar application of urea to several mango varieties caused a 10 to

TABLE 4.13 Urea and Growth Regulator Effects on Auxin-like Activity in Three Mango Varieties

	Auxin Levels (μg^{-1} FW)		
Treatment	Nellum	Bangalora	Mulgoa
Control	7.8	7.6	8.3
Urea 0.5%	8.7	8.6	8.5
Urea + sucrose 0.5%	8.4	8.0	8.6
Urea + sucrose + CCC 6000 ppm	6.0	6.1	5.2
Urea + sucrose + TIBA 500 ppm	3.5	3.2	4.7
Urea + sucrose + Ethrel® 2000 ppm	4.7	5.0	5.7
Urea + sucrose + Alar® 2500 ppm	5.9	5.7	5.0

Source: Modified from Sivagami et al. 1988.

12% increase in auxin-like substances in two varieties but had little effect on a third (Sivagami et al. 1988). Combination treatments using urea and several growth regulators (CCC, TIBA, Ethrel®, Alar®) caused substantial reductions in concentrations of auxin in leaves (Table 4.13).

Boron deficiency also influences IAA levels in plants. However, the literature is somewhat contradictory as to the effects. Some reports indicate that IAA levels decline with B deficiency, while others indicate that supraoptimal levels accumulate, which promotes the symptoms associated with B deficiency. Other studies have focused on the impact of B deficiency on IAA oxidase activity and in one such study was found to decrease in corn and sunflower plants subjected to several days of B deficiency (Shkol'nik et al. 1964).

In squash root tips, IAA oxidase activity increased in B-deficient plants after a reduction of root growth was observed (Figure 4.12). Application of exogenous IAA to plants

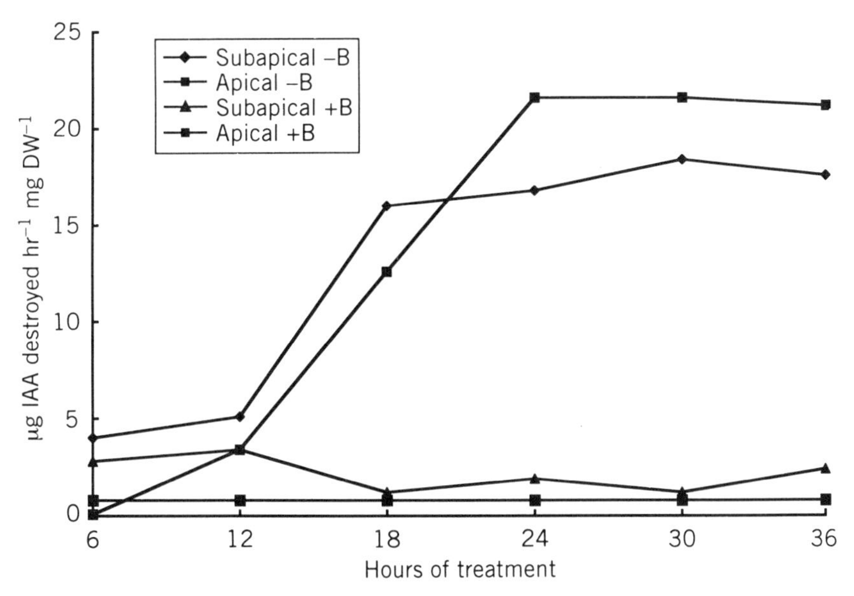

Figure 4.12 Effects of B on IAA oxidase activity in apical and subapical sections of squash roots. (Modified from Bohnsack and Albert 1977.)

resulted in the same symptoms as B deficiency, including elevated IAA oxidase activity (Bohnsack and Albert 1977). This suggested to the authors that B deficiency results in the accumulation of supraoptimal levels of IAA in the root tips, followed by reduced growth, and finally, enhanced IAA oxidase activity.

Ultrastructural studies in sunflower roots indicate that gross morphological effects such as inhibition of root elongation and a change in the direction of cell expansion from longitudinal to radial, is the same for auxin-treated and B-deficient plants (Hirsch and Torrey 1980). However, cell wall thickness increased and cell membrane integrity declined in B-deficient plants but not in auxin-treated plants. It was suggested that B may have a function in maintaining membrane function rather than influencing auxin levels. In support of this concept, B has been shown to be essential for IAA transport and has been hypothesized to affect the transport system by both influencing membrane permeability and having some effect on IAA transport proteins (Fuente et al. 1985).

Manganese deficiency has been associated with increased IAA oxidase activity in young leaves of cotton plants, and the symptoms of reduced growth and leaf abscission were consistent with those of auxin deficiency in the plants (Morgan et al. 1976). A direct correlation between IAA oxidase activity, auxin levels, and symptomology was implied (Figure 4.13).

In maize roots, increasing levels of $MnCl_2$ were shown to stimulate IAA oxidase activity in vitro (Beffa et al. 1990). The naturally occurring phenolic *p*-coumaric acid was also found to stimulate IAA oxidase activity, particularly at a concentration of 50 μm (Figure 4.14).

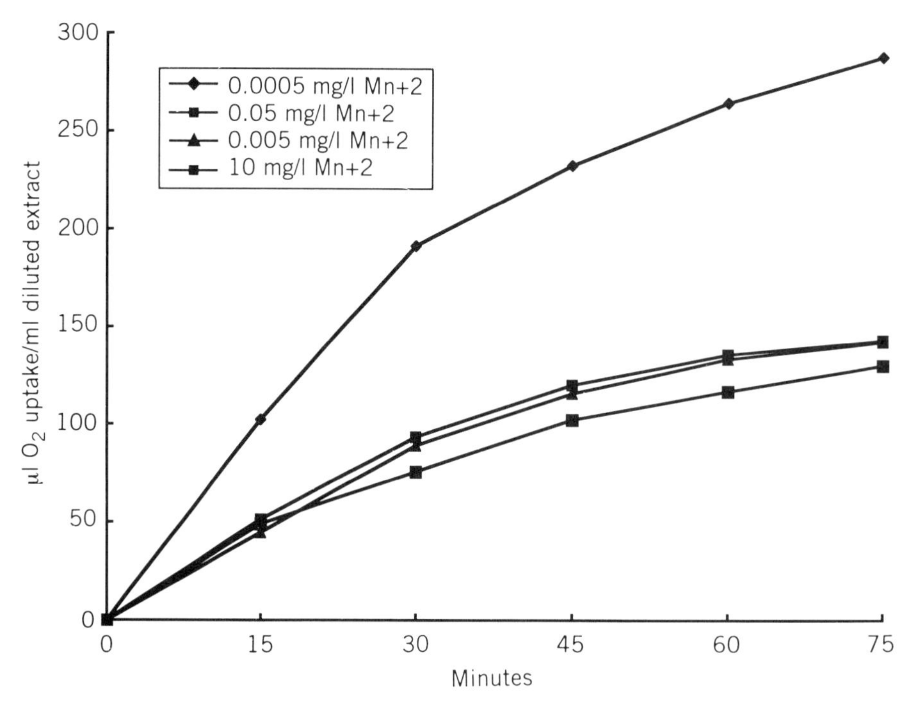

Figure 4.13 Mn effects on IAA oxidase activity in cotton leaves. (Modified from Morgan et. al. 1976.)

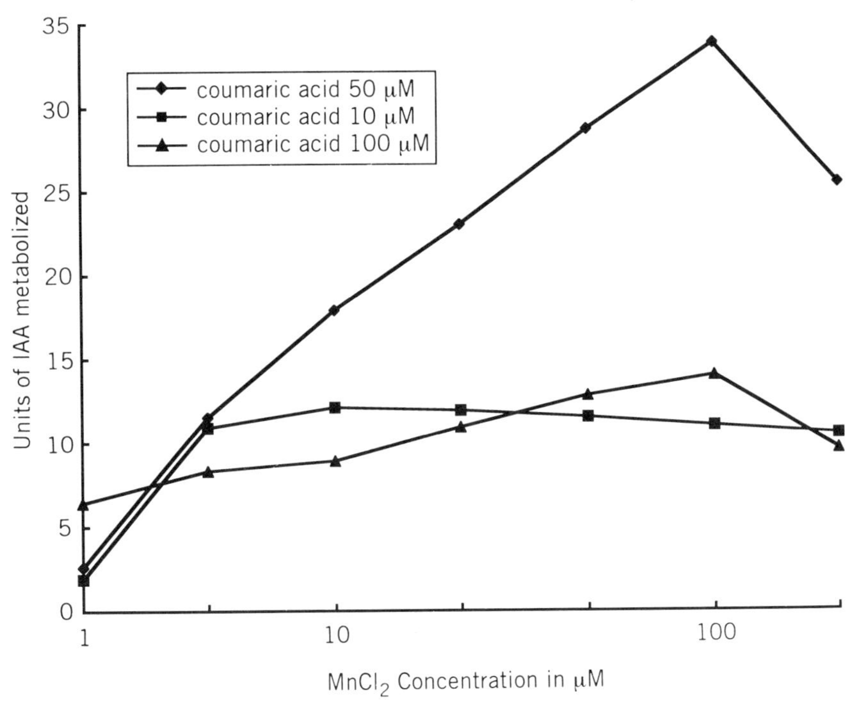

Figure 4.14 Effects of *p*-coumaric acid and $MnCl_2$ on IAA oxidase activity. (Modified from Beffa et al. 1990.)

Calcium has also been shown to influence auxin activity in plants by affecting the polar auxin transport mechanism (Allan and Rubery 1991). Calcium deficiency resulted in inhibition of polar auxin transport, and it was suggested that the lesion responsible for the diminished activity was through the auxin efflux carriers of the cells comprising the auxin transport system. However, the mechanism was unknown (Figure 4.15).

Finally, Zn is also required for the synthesis of IAA. Physiological disorders in plants such as "little leaf disease" in fruit trees is attributable to Zn and IAA deficiencies. Cakmak et al. (1989) observed a 50% reduction in IAA synthesis in Zn-deficient bean plants. Tryptophan, the immediate precursor of IAA increased, under Zn deficiency, indicating that this amino acid was not the limiting factor in IAA production. ABA levels were also reported to decline (Figure 4.16). Cytokinin levels were not affected. Resupply of Zn caused gradual increases in IAA and ABA over a 96-h period.

B. Nutrients and Cytokinin Levels

Most of the studies conducted with CKs involve the effects of N levels and sources on the endogenous concentrations of CKs. In most of these studies, limiting N results in lowering of CK levels in plant tissues and reduces the amounts translocated from the roots to the leaves. In addition to N, low levels of K, P, and Ca have been shown to affect CK production in plants.

El-D et al. (1979) reported that low N resulted in rapid decreases in CK levels in leaves, buds, roots, and root exudates of sunflower (*Helianthus annuus* L.). In addition,

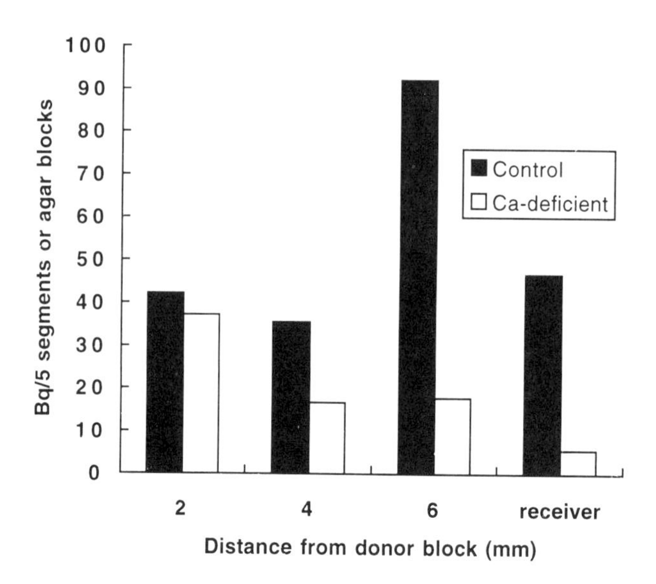

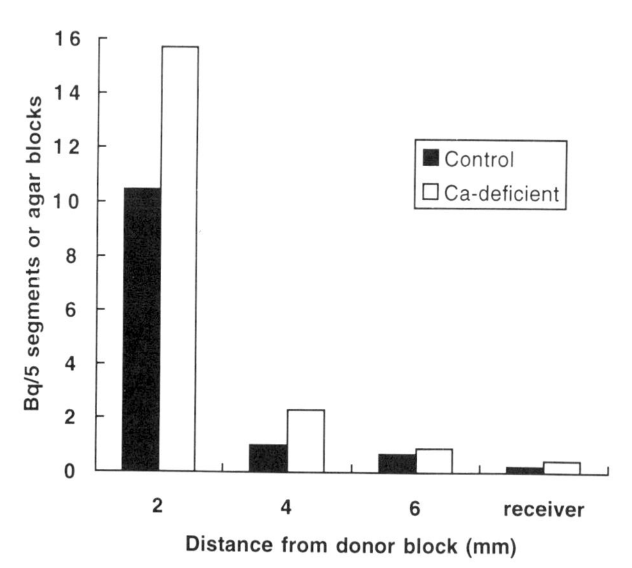

Figure 4.15 Effects of Ca on IAA transport in zucchini hypocotyls. (Modified from Allan and Rubery 1991.)

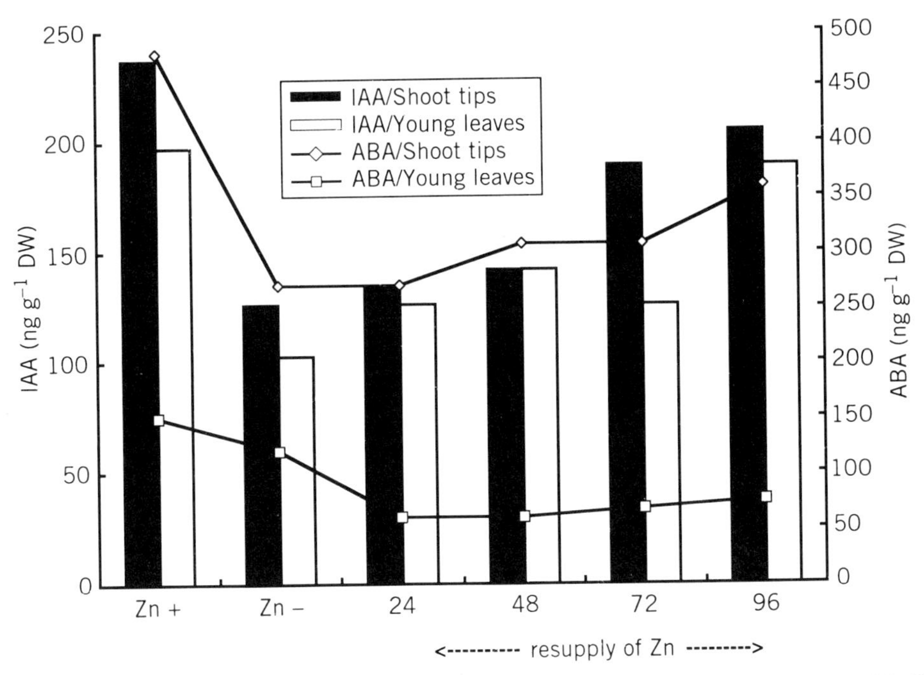

Figure 4.16 Effects of Zn deficiency and resupply on IAA and ABA levels in bean plants. (Modified from Cakmak et al. 1989.)

CK levels were greater in plants supplied with NO_3^- N rather than either NH_4SO_4 or NH_4NO_3 (Table 4.14). Growth of plants at reduced P and K levels also resulted in reduced levels of CK in leaves and roots of sunflower, but not to the extent of low N (Table 4.15).

Kuiper et al. (1989) using *Plantago major* ssp. *pleiosperma,* demonstrated that low levels of N, P, and Ca resulted in reduced concentrations of zeatin and zeatin riboside in shoots and roots. The greatest reductions were observed with reduced N and Ca levels, both exhibiting about the same reduction (Table 4.16). Low K appeared to have little effect on CK levels in this study. Reductions in growth were thought to be initiated by the impact of nutrients on CK levels rather than nutrients being the primary initiator of the growth reduction. Reduced N levels were viewed as being the primary effector of reduced growth.

TABLE 4.14 Effect of N Source on CK Content in Sunflower Roots, Leaves, and Root Exudates

Treatment	Roots (μg kinetin equiv. kg^{-1} FW)	Leaves (μg kinetin equiv. kg^{-1} FW)	Root Exudates (μg kinetin equiv. dm^{-3})
NO_3 type	2.64	4.20	3.52
NH_4 type	1.70	1.84	1.94
NH_4NO_3 type	1.28	1.16	0.95

Source: Modified from El-D et al. 1979.

TABLE 4.15 Effect of N, P, and K Levels on CK Content of Sunflower Roots and Leaves

	Kinetin Equivalents ($\mu g\ kg^{-1}$ FW)	
Treatment	Roots	Leaves
Control	2.38	3.36
1/10 nitrogen	0.94	1.06
1/10 phosphorus	1.06	1.28
1/10 potassium	1.06	2.02

Source: Modified from El-D et al. 1979.

Using duckweed (*Lemna gibba*), Thorsteinsson and Eliasson (1990) also demonstrated that low N and P resulted in reduced levels of CK (iPA and zeatin riboside) in this plant and that relative growth rates declined after reductions in CK levels (Figure 4.17). This implied that growth rate may be controlled by CK levels. Nitrogen depletion was accomplished by placing plants grown in complete nutrient solutions into solutions devoid of N or P or by acclimatizing plants to differing low levels of nutrients and measuring CK concentrations in fronds. In the latter case no deficiency symptoms were observed. However, in both instances the extent of CK depletion reflected the level of growth response. Phosphorus deficiency resulted in slower depletions of CK than did N deficiency.

Comparing senescent and nonsenescent sorghum cultivars for CK levels, Ambler et al. (1992) demonstrated that nonsenescent cultivars had higher levels of CKs (zeatin and zeatin riboside) than did the senescent cultivar. The nonsenescent cultivar also exhibited

TABLE 4.16 Cytokinin Content of Roots and Shoots of *Plantago major* ssp. *pleiosperma* Subjected to Reduced Levels of Essential Elements

Days after Transfer	Nutrient Treatment	Roots		Shoots	
		Z	ZR	Z	ZR
0	100%	78	19	52	13
1	100%	84	23	55	15
	100% → 2% N	59	17	43	12
	100% → 2% K	82	23	57	14
	100% → 2% P	85	21	51	15
	100% → 2% Ca	81	25	58	16
3	100%	83	24	65	18
	100% → 2% N	38	9	21	8
	100% → 2% K	75	21	60	16
	100% → 2% P	84	25	55	17
	100% → 2% Ca	48	13	31	8
5	100%	75	27	60	20
	100% → 2% N	29	7	17	6
	100% → 2% K	73	30	62	19
	100% → 2% P	61	20	50	13
	100% → 2% Ca	17	9	12	2

Source: Modified from Kuiper et al. 1989.

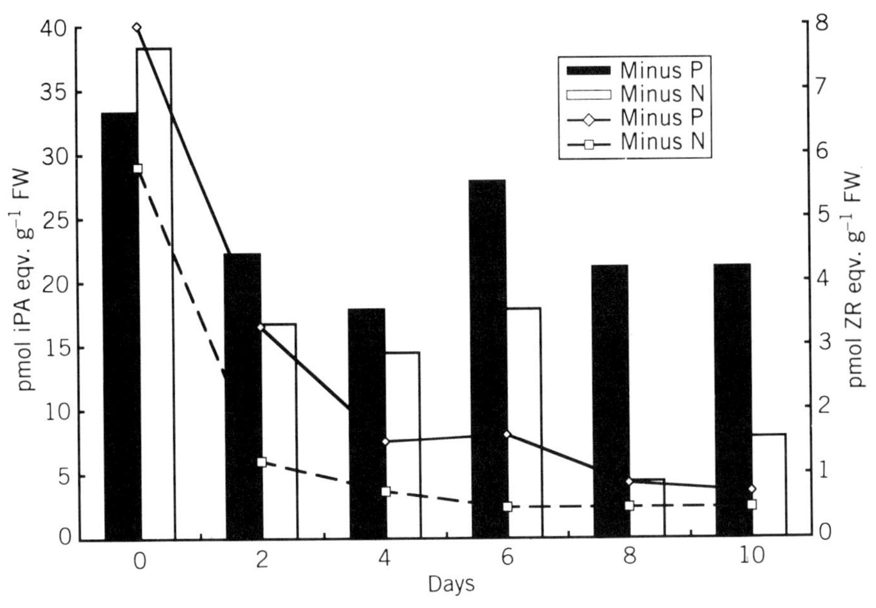

Figure 4.17 Effects of N and P deficiency on iPA and ZR in *Lemna gibba*. (Modified from Thorsteinsson and Eliasson 1990.)

greater translocatable CK from the roots when subjected to N deficiency than did the senescent cultivar. Nonsenescent cultivars also translocated more CK under reduced light.

In stinging nettle (*Urtica dioica* L.), reduced N effects were manifested in the roots and not the shoots (Wagner and Beck 1993). Cytokinin levels did not change in the shoots but were reduced by 25% in the roots and root-pressure fluid. Since only levels of *trans*-zeatin CKs in the roots (17 different CKs were analyzed) showed a linear correlation with shoot/root ratios, it was concluded that CKs play an important roll in biomass partitioning in this plant.

Differences in species response to limitations of N and P were demonstrated in birch (*Betula pendula*) and sycamore leaves (*Acer pseudoplatanus*) (Horgan and Wareing 1980). In the case of birch, N and P deficiencies resulted in continued but slow growth even though CK levels declined under both treatments. Sycamore, on the other hand, exhibited no significant differences in CK concentrations compared to controls, yet the apical buds became dormant and growth stopped. These results were contrary to the expectations relative to the action of CKs in controlling growth and in breaking bud dormancy.

C. Nutrients and Gibberellin Levels

Surprisingly, there is little information concerning the impact of nutrient stress on GAs in plants. Goodwin (1978) cites one reference in which N deficiency resulted in a reduction in the levels of extractable GA in tomato and a second reference where K deficiency reduced extractable GA in *Solanum sisymbrifolium*.

Wilkinson (1994), using different cultivars of sorghum varying in resistance to acid soils, reported that high Mn levels inhibit several enzymes involved in isoprenoid biosynthesis. This was viewed as important since GA and ABA are synthesized in this pathway. Interestingly, four different patterns (out of five cultivars tested) of incorporation of

^{14}C-labeled isopentenyl pyrophosphate into various intermediates of the pathway suggested that different enzymes were being affected in the different cultivars.

Finally, the fungus *Gibberella fujikuroi* is a major source of GA for commercial purposes. Thus growth conditions that maximize the production of GA are critical to the industry. Also, manipulation of the growth environment of this organism may also give clues as to how GA metabolism may be affected by similar changes in plants. Kumar and Lonsane (1990) found that *G. fujikuroi* cultured in urea at concentrations ranging from 10 to 70 mg % produced significant increases in the levels of GA_3. The addition of $MgSO_4$ at concentrations of 2 to 7 mg % also caused elevated production of GA_3. However, Munoz and Agosin (1993) found that addition of 10 m*M* NH_4 or glutamine caused inhibition of GA_3 in *G. fujikuroi.* Results from these studies suggested that glutamine may be the metabolite effector in N repression of GA_3 synthesis.

D. Nutrients and Abscisic Acid Levels

In a field study using kale (*Brassica oleracea*), Whenham et al. (1989) studied the impact of irrigation and N fertilization on the production of ABA and several conjugates and metabolites in an effort to see if stress-induced changes in these components could be correlated with the ability of different varieties to tolerate stressful environments. Utilizing NH_4NO_3 as a N source, they incorporated the N into the topsoil and subsoil at different combinations at concentrations of 0, 150, and 300 kg N/ha. Soil moisture was another variable studied. Nitrogen caused appreciable increases in free acids (ABA, phaseic acid, and dihydrophaseic acid) and conjugates of the same acids (Table 4.17). In addition, plants grown in dry soils exhibited higher levels of free and conjugated acids. It was suggested that metabolites of ABA provide little help in assessing the ability of plants to

TABLE 4.17 Effects of N Concentration, Soil Incorporation Method, and Soil Moisture on Free and Conjugated ABA and Its Metabolites

Irrigation Treatment	N Fertilizer Treatment	Concentration of Free Acids ($\mu g\ g^{-1}$ DW)			Concentration of Conjugated Acids ($\mu g\ g^{-1}$ DW)		
		ABA	PA	DPA	ABA	PA	DPA
Dry	0/0	0.33	1.24	3.86	0.25	0.02	0.44
	0/1	0.67	1.47	5.05	0.28	0.01	0.89
	1/2/1/2	1.12	2.96	9.61	0.23	0.44	2.07
	1/0	0.68	1.70	6.67	0.25	0.02	1.39
	1/1	0.84	1.80	8.42	0.27	0.02	1.01
	2/0	0.79	1.84	9.84	0.23	0.23	1.58
Wet	0/0	0.45	0.82	2.48	0.15	0.09	0.07
	0/1	0.72	1.20	4.21	0.35	0.01	0.40
	1/2/1/2	0.60	1.27	5.50	0.20	0.01	1.06
	1/0	0.31	0.98	6.56	0.13	0.18	1.05
	1/1	0.85	1.30	9.61	0.40	0.03	0.19
	2/0	0.55	0.84	8.47	0.14	0.20	0.95

Source: Modified from Whenham et al. 1989.

TABLE 4.18 Clonal Differences in ABA Concentrations (ng g^{-1} DW) of Poplar Subjected to High and Low N and Flooding/Drying Conditions

	Flooding		Control		Soil Drying	
N Rate	*Tristis*	*Eugenei*	*Tristis*	*Eugenei*	*Tristis*	*Eugenei*
Low	18.0	27.6	5.7	4.1	172.4	416.1
High	12.5	10.8	29.7	23.0	983.7	1262.9

Source: Modified from Liu and Dickmann 1992.

withstand stressful environments and that if ABA levels are used to screen for drought tolerance, N nutrition needs to be considered in these evaluations.

Liu and Dickmann (1992) studied the interaction of drought, flooding, and N levels on ABA production in two hybrid poplar clones (*Tristis* and *Eugenei*) and the impact of nitrogen on stomatal sensitivity to ABA. Nitrogen treatments consisted of application of NH_4NO_3 at rates of 200 kg/ha and no supplemental N. High N tended to cause small increases in ABA levels in control plants of both clones (Table 4.18). A five- and threefold increase in ABA was observed in *Tristis* and *Eugenei,* respectively, when grown under drying conditions and high N. Flooding resulted in only small increases in ABA, and when combined with high N a reduction in ABA was observed. *Eugenei* was more responsive than *Tristis* to drought and high N relative to ABA production. However, this clone was less sensitive to ABA relative to stomatal conductance. It was concluded that since ABA levels changed little under flooding, even though stomatal conductance and photosynthesis were decreased, ABA levels had little function with respect to these processes. However, direct correlations between ABA levels and gas exchange were probably mediated by the accumulation of ABA in the leaves.

Since ABA transport from roots to leaves is thought to be a signaling mechanism for drought and salinity stress in plants, Peuke et al. (1994) suggested that other factors, such as N nutrition, may have some influence on ABA levels and transport within plants. They suggested that this may be particularly true in poor or acidic soils, where NH_4-N may be high. Thus they studied the impact of NO_3 and NH_4 nitrogen in combination with NaCl stress on ABA levels in leaves, stems + petioles, roots, and xylem and phloem exudates of *Ricinus communis* (Figures 4.18 and 4.19). They discovered that as NO_3 increased, ABA concentrations increased slightly in the leaves but in no other plant structure. However, plants treated with NH_4 exhibited large increases in ABA, particularly in the leaves and roots. Addition of NaCl with NH_4 enhanced ABA levels above the levels of NH_4 treatments only in the leaves and stems + petioles but caused a reduction to near zero in the roots. In the case of xylem exudates, increasing levels of NO_3 resulted in declining concentrations of ABA, but the addition of NaCl caused a sixfold increase compared to the control. No differences were observed in the levels of ABA in the phloem of any NO_3 or NO_3 + NaCl treatment. NH_4 caused about a fivefold increase in xylem exudates. ABA levels were much higher in the phloem sap than in the xylem sap but were less responsive to N levels. ABA levels were twice as high in NH_4-fed plants compared to NO_3-fed plants, and NaCl caused ABA levels to be reduced to half those observed for NH_4 plants. NaCl had little effect on ABA levels in the phloem compared to controls. In conclusion, NO_3 deficiency did not stimulate long-distance transport of ABA. However, ABA transport in both xylem and phloem was increased by NH_4. Mild NaCl stress increased ABA transport in NO_3-fed plants but not in NH_4-fed plants. Leaf conductance was lowered

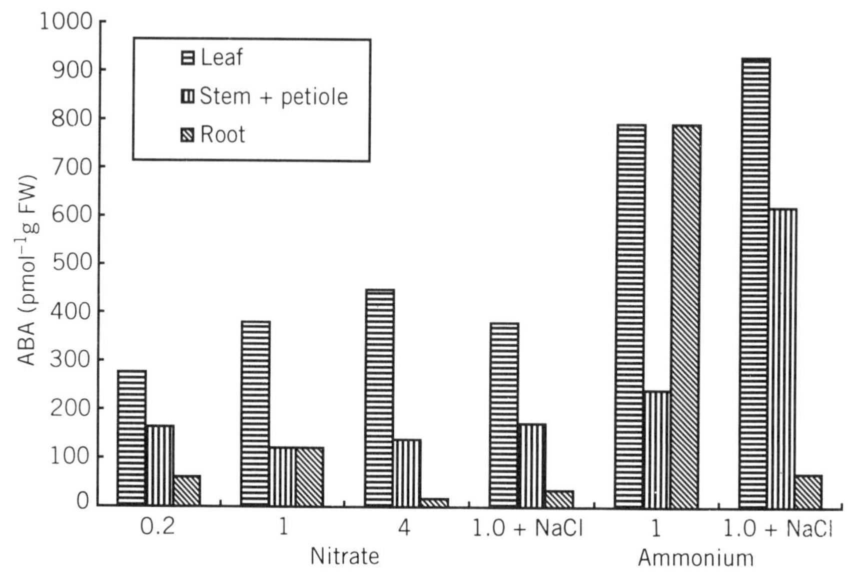

Figure 4.18 Effects of N source and NaCl on ABA levels in shoots and roots of *Ricinus communis*. (Modified from Peuke et al. 1994.)

with N sources + salt but more so with NH_4. Thus ABA does not serve as a stress signal to N deficiency.

In an attempt to shed further light on the effects of Zn on IAA synthesis, Cakmak et al. (1989) using bean plants showed that Zn-deficient plants exhibited 50% lower levels of IAA and ABA with no change in CK levels. This was observed in shoot tips and young

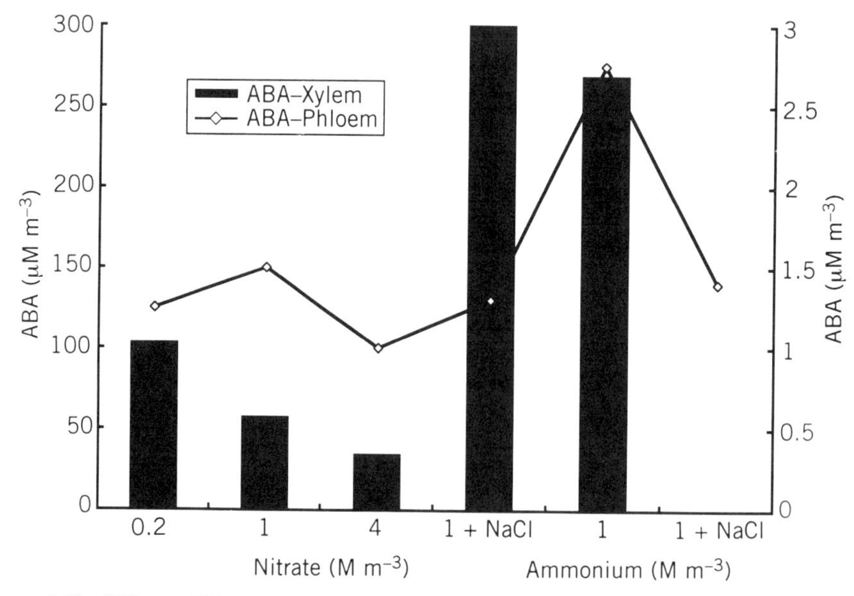

Figure 4.19 Effects of N source and NaCl on ABA levels in xylem and phloem exudates of *Ricinus communis*. (Modified from Peuke et al. 1994.)

leaves (Figure 4.16). A resupply of Zn would reverse these changes completely for IAA and partially for ABA within a 96-h period. The reduction in ABA levels was viewed as being a decrease in import rates from the phloem or a possible interaction between IAA and ABA (i.e., increasing IAA levels is often correlated with increasing levels of ABA in vegetative tissues).

Radin (1984) demonstrated an interaction between water stress, P levels, and ABA concentrations in cotton. Plants grown under low P and increasing water stress produced elevated levels of ABA sooner and at a faster rate than did plants grown under high P (Figure 4.20). Also, stomatal conductance was reduced sooner in low-P plants than in high-P plants, indicating that P may function to control the sensitivity of somates to ABA or the compartmentation of ABA in the chloroplast. Since P deficiency has already been reported to decrease CK levels and treatment of P-deficient cotton plants with exogenous kinetin increased stomatal conductance, Radin hypothesized that ABA, CK, and P interact to control stomatal opening.

Haeder and Beringer (1981) demonstrated that wheat seed grain size and ABA levels were correlated with K nutrition. Plants grown under reduced K levels exhibited small grain size and earlier production of ABA levels in the seeds than did plants grown with sufficient levels of K (Figures 4.21 and 4.22). Comparison of seeds from plants subjected to a 1-week water stress period provided further evidence that higher K levels tend to decrease the amount of ABA produced in the seed as well as extending the period before ABA is produced in the seed, thereby allowing for a longer grain filling period. It was also concluded that the flag leaf was not a source of increased ABA levels for the seeds and that the seeds probably synthesized the observed levels of ABA.

Ferritin is an iron storage protein found in both plants and animals. Iron has been shown to induce ferritin and RAB (genes responsive to ABA) mRNA in corn plantlets (Lobréaux et al. 1993). ABA also accumulates during the induction period, and exogenous ABA can induce ferritin mRNA production. The ABA-deficient mutant *vp2* exhibits

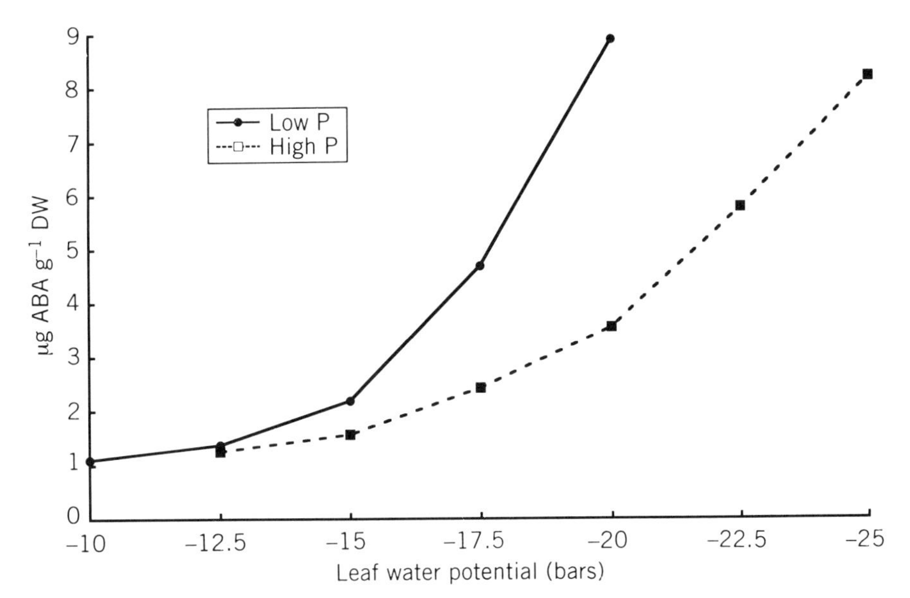

Figure 4.20 Effects of low and high P on ABA levels in cotton leaves during drying. (Modified from Radin 1984.)

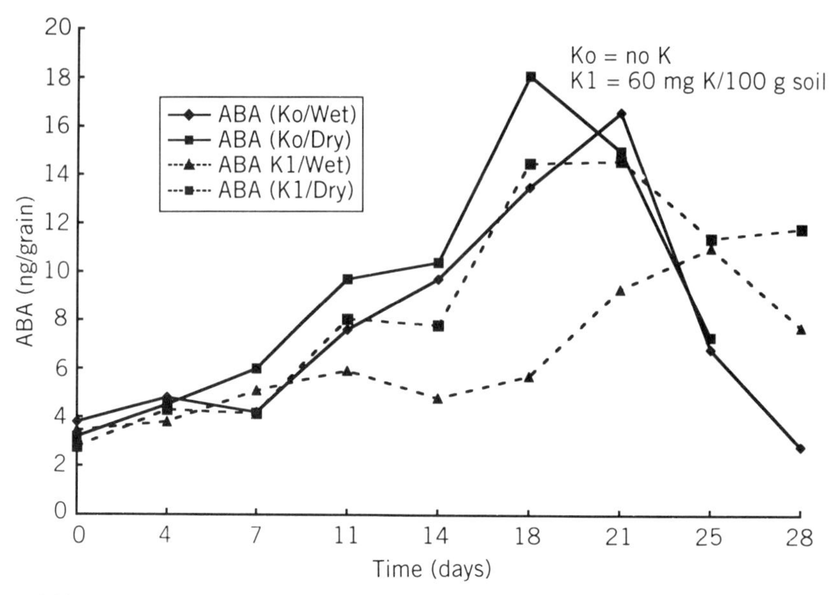

Figure 4.21 Effects of K concentration on ABA levels in wheat grown under dry and wet conditions. (Modified from Haeder and Beringer 1981.)

much reduced ferritin mRNA production when induced by Fe and can be increased by the addition of exogenous ABA. Although full expression of the ferritin genes could not be explained by an increase in ABA concentration, it was concluded that a major part of the Fe-induced biosynthesis of ferritin was achieved through a pathway involving increased levels of ABA.

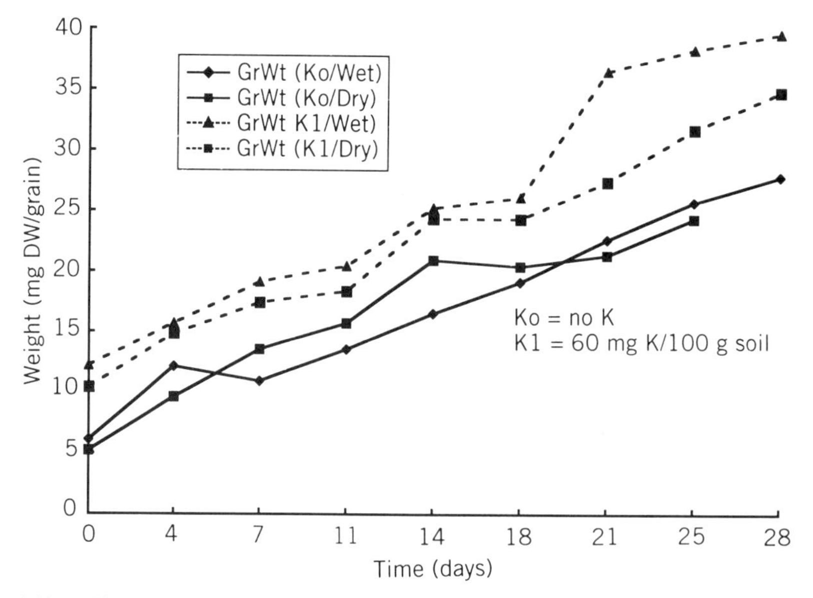

Figure 4.22 Effects of K concentration and water stress on wheat grain weight. (Modified from Haeder and Beringer 1981.)

E. Nutrients and Ethylene Levels

The impact of several essential nutrients on C=C production in plants has been studied. Barker and Corey (1988) determined that tomato plants grown in sand culture using Hoagland's solution containing either full-strength NO_3 or NH_4 as a N source, but deficient in Mg, Ca, or K, exhibited elevated levels of C=C. The effect was considerably higher in NH_4- than in NO_3-grown plants (Table 4.19). No effect of deficiencies of N, P, or S was detected on C=C levels in tomato. Plants grown using complete nutrient solutions containing NH_4 exhibited nearly a sixfold increase in C=C over plants grown on complete nutrient solutions containing NO_3. Nitrogen-deficient plants exhibited the same level of C=C production as plants grown in complete solutions containing NO_3. The use of urea as a N source resulted in C=C levels comparable to those of nitrate-grown plants (Barker and Corey 1990), while the addition of K at levels ranging from 0 to 8 m*M* to the nutrients applied reduced C=C levels below control plant levels. A similar experiment showed that 6 m*M* K could overcame the effects of NH_4 on C=C production of plants grown under acid pH conditions (Corey and Barker 1989). Ammonium-resistant varieties of tomatoes (*neg-1* and *yg-5*) produced C=C levels comparable to those of NO_3-grown NH_4-susceptible varieties. Potassium addition to plants in some instances (*neg-1*, NO_3, pH 6.7; *yg-5*, NH_4, pH 3.4) actually exhibited increased levels of C=C. Using C=C inhibitors [(aminooxy)acetic acid and silver thiosulfate] Feng and Barker (1993) demonstrated that these compounds could overcome C=C production in NH_4-sensitive plants, resulting in levels of C=C comparable to those of NO_3-grown plants. The were also able to reverse the effects of Ca- and Mg-deficient induction of C=C production in NO_3-grown plants. It was suggested that plants under stress degrade proteins to NH_4, the accumulation of which can lead to increased C=C production. Plants susceptible to high levels of NH_4 cannot detoxify the N source adequately or quickly enough, resulting in C=C accumulation and symptoms associated with this hormone.

In another study using newly emerging adventious roots of corn, Drew et al. (1989) presented evidence that N and P deficiency actually reduced C=C production in these tissues (Figure 4.23). Associated with this response was the formation of aerenchyma tissue, which is typical of C=C formation in plants exposed to flooding or hypoxic conditions. Studies of enzymes and substrates of the C=C biosynthetic pathway revealed that

TABLE 4.19 Ethylene C = C Evolution ($nL\ g^{-1}\ h^{-1}$) by Tomato Plants Grown under Full Nutrition with NH_4 or NO_3 or on Nutrient-Deficient Treatments

	Nutrient Solutions Used		
	Full Strength		
Element	NO_3-N	NH_4-N	Nutrient Deficient
Nitrogen	29	178	30
Phosphorus	29	178	18
Potassium	29	178	89
Calcium	29	178	69
Magnesium	29	178	100
Sulfur	20	290	30

Source: Modified from Barker and Corey 1988.

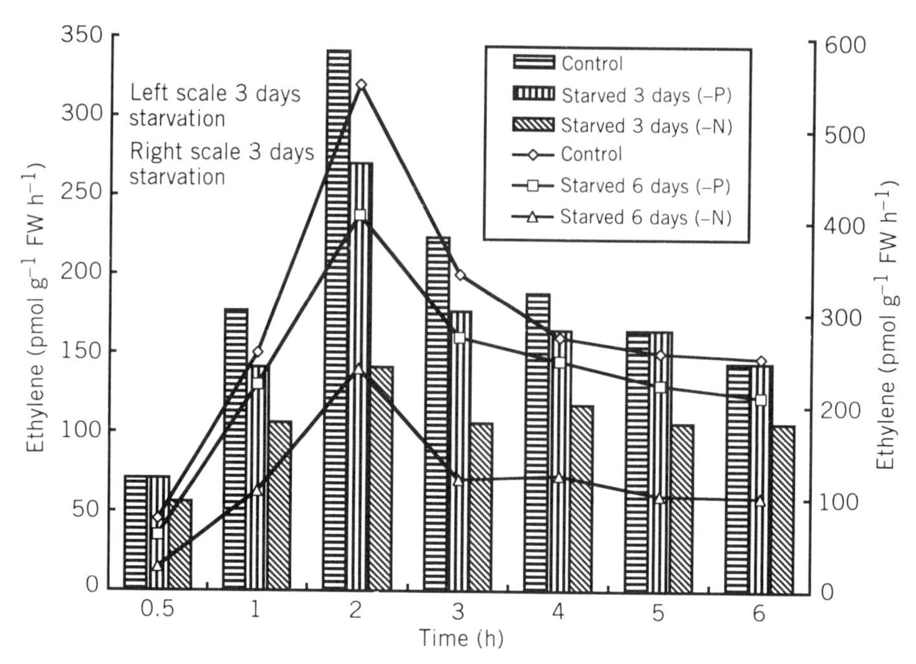

Figure 4.23 Effects of P and N deficiency on C=C production in root tips of corn. (Modified from Drew et al. 1989.)

ACC and malonyl-ACC levels paralleled the effect of N and P deficiencies on C=C production and that ACC synthase and C=C-forming enzyme was reduced. In related studies, He et al. (1994) presented evidence that anoxia, hypoxia, and N deficiency may function differently in affecting C=C synthesis in corn. As far as the impact of N deficiency is concerned, they concluded that N deficiency increases the sensitivity of the root cells of the cortex to the reduced levels of C=C observed and that there are two levels of transduction, which involve hypoxia and stimulation of ACC synthase activity and a second between C=C and the cell-degrading enzymes it induces.

Using a C=C synthesis inhibitor (AVG) and an action inhibitor (STS), Romera and Alcántara (1994) presented indirect evidence that Fe deficiency–mediated C=C induced physiological and symptomological responses in cucumber (*Cucumis sativus*) plants. Addition of C=C inhibitors to the growth medium reduced the ferric reducing capacity of the roots, which is frequently stimulated in Fe-deficient plants. Also, subapical root swelling was eliminated, which is symptomatic of Fe deficiency. The addition of ACC, a precursor of C=C synthesis, to plants with increased ferric reduction activity (Fe deficient) resulted in decreased ferric reduction activity. Taken together the data suggested that Fe deficiency induces C=C synthesis in the plant, resulting in the responses observed.

Copper at a concentration of 0.5 to 1.0 m*M* induced C=C synthesis in tobacco and the aquatic plant, *Spirodea oligorrhiza*. In both species younger leaves produced higher amounts of C=C than did older leaves. However, it was determined that different pathways of C=C synthesis existed in the two plant species. In the case of *Spirodela*, cupric ions stimulated the conversion of methionine and linoleic acid to C=C, while tobacco was more efficient in converting methionine to C=C.

The addition of Ca at a concentration of 3.5 m*M* to roses (*Rosa hybrida*) reduced the severity of postharvest *Botyrtis* blight in naturally infected flowers by 55%. Ethylene production in flowers with high Ca content was decreased by 50 to 95%. Addition of K to the growth medium negated the ability of Ca to reduce the disease, probably as a result of competition for cation uptake. This could, in turn, affect membrane stability and facilitate increased C═C production and possibly, susceptibility to pathogens.

Although not related directly to nutrient stress effects on plants, soils can produce C═C via the organisms that inhabit them. In fact, some fungi have been reported to produce very high levels of C═C (thousands of ppm versus hundreds of ppm for plants). High C═C-producing soils thus could have a large impact on plants growing in them. Several factors can affect the levels of C═C in soils, including type and abundance of organisms, type of soil, conditions in the soil (temperature, pH, water levels, aeration), and nutrient status. Arshad and Frankenberger (1991) studied the effects of pH, trace elements, and organic amendments on C═C generation by various California soil types. Amendments with glucose and methionine generally resulted in increased C═C production, as did acidic pH. Trace element effects were concentration dependent, with Ag(I), Cu(II), Fe(II), Mn(II), Ni(II), Zn(II), and Al(III) inhibiting C═C production when applied at levels of 100 mg kg^{-1} soil or greater. The most effective elements in promoting C═C production were Co(II) and As(III) at concentrations of 100 mg kg^{-1} soil. C═C generation was inhibited by Hg(II), Fe(III), and Mo(VI) at 10 mg kg^{-1} soil or above. Abiotic production of C═C was observed when Fe(II) was applied to soil at a concentration of $\geq$ 100 mg kg^{-1} soil.

IV. SALINITY STRESS AND PHYTOHORMONES

A. Salinity and Indoleacetic Acid Levels

The impact of salinity on IAA levels in plants has been studied in rice (Prakash and Prathapasenan 1990). Based on observations that GA_3 could alleviate NaCl-induced reduction in growth and yield in rice, experiments were conducted to ascertain the interactions of GA_3 with IAA and NaCl stress. The results of these experiments demonstrated that NaCl caused a statistically significant reduction in IAA concentrations in rice leaves after 5 days. Levels continued to fall up to 15 days postsalination. Application of GA_3 during the salinization period countered the effects of salinity on reducing IAA levels, but not totally. GA_3 application to the plants without salinization caused significant increases in IAA at 5 and 15 days posttreatment (Table 4.20). Little other information is available

TABLE 4.20 Effect of NaCl Salinity (12 dS m^{-1}) and GA3 (10 ppm) on IAA Content (ng g^{-1} FW) During the Growth of the Fifth Leaf of Rice

	Days after Salinization			
Treatments	0	5	10	15
Control	106	194	151	92
NaCl	106	123	95	60
NaCl + GA_3	106	161	129	85
GA_3	106	228	160	119

Source: Modified from Prakash and Prathapasenan 1990.

regarding the impact of salinity on auxin levels in plants. The known interactions of GA and IAA regarding shoot growth and the results of the study above suggests that salinity can influence hormone balances affecting growth and development. However, further studies should be conducted before generalizations are made.

B. Salinity and Cytokinin Levels

Relatively little is known regarding the effects of salinity stress on CK levels in plants. In an effort to gain a better understanding of the roll of ABA and CK in regulating acclimation responses of plants to osmotic stress, Walker and Dumbroff (1981) demonstrated that tomato plants stressed with a balanced multiple of base nutrients (−600 kPa) exhibited reduced levels of *trans*- and *cis*-zeatin over an 8-day stress period (Figure 4.24). *cis*-Zeatin decreased to a greater extent than *trans*. In addition, *cis*- and *trans*-zeatin riboside both increased after about day 2 and then declined to slightly negative values compared to controls after 8 days. Removal of the stress caused a return to normal levels of the free and conjugated CK bases along with increased growth.

Thomas et al. (1992) studied the interaction of NaCl, ABA, and CK in the conversion of ice plant (*Mesembryanthemum crystallinum*) from a C_3 form of metabolism to CAM (crassulacean acid metabolism) metabolism. Salt stress induces the accumulation of proline and an isoform of the enzyme phosphoenolpyruvate carboxylase (PEPCase) prior to the switch from C_3 to CAM. Although it was found that ABA levels increased 8- to 10-fold when this plant was stressed, the hormone was a poor substitute for NaCl for induc-

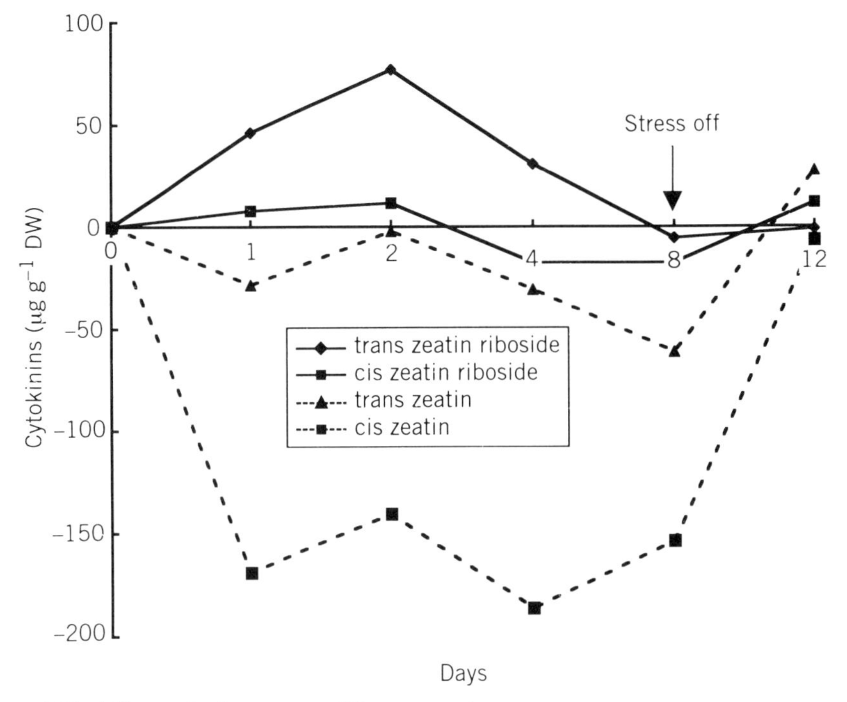

Figure 4.24 Effects of salt stress on CK content of tomato leaves. (Modified from Walker and Dumbroff 1981.)

ing proline and PEPCase. However, application of CK induced both proline and PEPCase production, imitating the effect of NaCl. However, the endogenous levels of CK in roots and leaves did not change or may have decreased slightly. This suggests independent pathways for salt and CK relative to the induction of gene expression for proline and PEPCase production. Changing levels of either ABA or CK were not attributable to the shift from C_3 to CAM metabolism.

Sorghum acclimated to grow in high-NaCl nutrient media (300 mol m^{-3}), when grown with half-strength Hoagland's solution exhibited a reduced growth rate compared to plants grown on full-strength Hoagland's solution (Amzallag et al. 1992). Addition of CK or GA, or a combination of the two, to plants grown in half-strength Hoagland's could substitute for the additional nutrients shown previously to enhance growth. This suggested to the authors that an imbalance in phytohormones was triggered by an imbalance in nutrients (i.e., salinity stress), and by the exogenous application of CK or GA, this balance could be reestablished, allowing the plants to grow better at higher NaCl levels. No internal levels of CKs or GAs were determined in this study.

C. Salinity and Gibberellin Levels

As with auxins and CKs, little information is available relative to the effects of salinity on GA levels in plants. As indicated earlier, GA can overcome high-saline conditions in sorghum to improve growth (Amzallag et al. 1992). A similar observation was made for leaf growth in rice plants subjected to NaCl stress (Prakash and Prathapasenan 1990). Again, an imbalance in hormones was suggested, particularly since treatment with GA caused an increase in IAA levels of stressed plants. It was pointed out that evidence exists for GA involvement in IAA synthesis or degradation prevention. Gibberellic acid has also been shown to improve germination of seeds under high-saline conditions (Kaber and Baltepe 1990; Begum et al. 1992).

D. Salinity and Abscisic Acid Levels

Considerable information is available concerning the effects of salinity on ABA production in plants. This probably reflects the fact that osmotic and drought stress can also induce ABA production. The isolation of specific ion stress and osmotic stress in studying salinity stress in plants is difficult. However, attempts at doing this have used comparisons of isotonic solutions of nonionic (polyethylene glycol and mannitol) and ionic substances (NaCl, $CaCl_2$ and other salts) to try and separate the two effects. The following examples will give the reader an idea of the effects that NaCl can have relative to ABA levels and the rationale for studying such effects.

Interested in understanding the adaptive value of ABA and CK relative to salinity tolerance, Walker and Dumbroff (1981) conducted long-term (8 days) salt stress (−600 kPa) studies on 10-week-old tomato seedlings. By the second day of stress, very large increases in ABA levels occurred in leaf tissues (Figure 4.25) but then began to drop off to levels below the control. Zeatin riboside increased while free zeatin declined and stayed at low levels throughout the stress period. Transfer of the plants to nonstressful conditions produced no further change in ABA levels, and zeatin and zeatin riboside returned to normal levels. Declining water loss and osmotic potential during the 8-day period, even after ABA levels started to drop, indicated to the authors that the growth response after stress was independent of any change in ABA.

As a result of osmotic (salt) stress, plants respond by producing metabolites such as

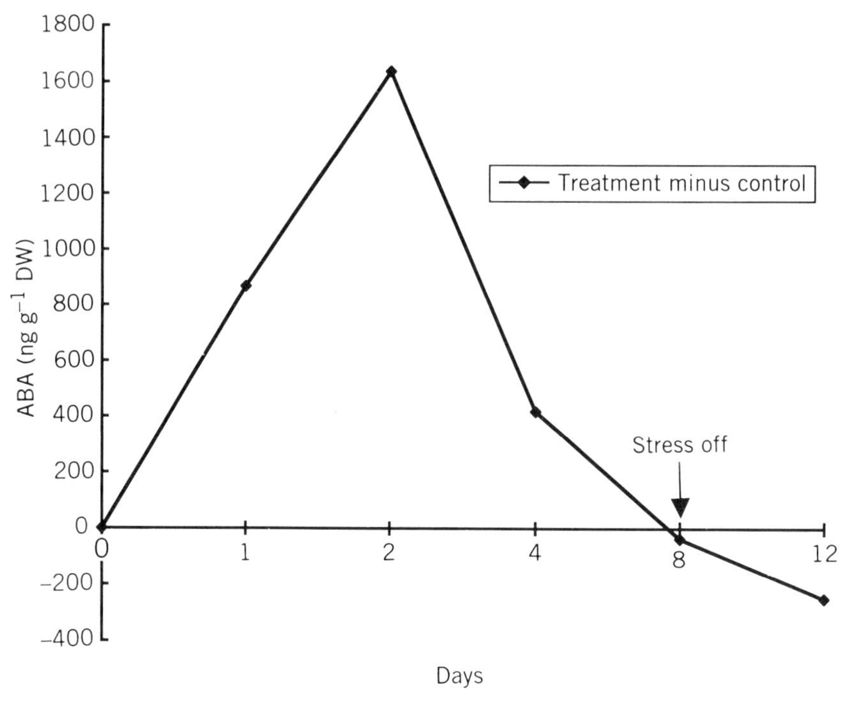

Figure 4.25 Effects of salt stress on ABA content in tomato leaves. (Modified from Walker and Dumbroff 1981.)

proline and reducing sugars which are thought to help counteract the loss of water due to the osmotic imbalance. In addition, accumulation of ions of Na, Cl, and K are frequently encountered, as is increased stomatal resistance, which also aids in control of water loss. With these responses in mind, Downton and Loveys (1981) studied the relationship of ABA to the induction of proline and an ABA metabolite, phaseic acid, by varying levels of NaCl (0, 25, 50, and 100 m*M*) over a 3-week period in grapevine leaves (*Vitis vinifera* L.). They found, as did Walker and Dumbroff (1981), that there was an initial surge of ABA and then it dropped back to lower levels. In the case of grapevines the initial burst of ABA activity occurred within 6 h and dropped back to levels slightly above the controls (Figure 4.26). The amount of ABA produced was related to the degree of the stress, with 100 m*M* NaCl causing the greatest increase. Proline did not accumulate until the day after the stress was induced, and unlike ABA, continued to increase throughout the stress period. Phaseic acid paralleled the production of ABA, suggesting the metabolic relationship between the two. Potassium preferentially accumulated in the leaves during the stress period, and the sum of K and Na ions balanced the accumulation of Cl^- ions. Reducing sugars increased, contributing equally to the osmotic potential, as did Na^+, K^+, and Cl^-. Pressure potential in the plants did not varying more than 0.1 MPa over the stress period (except for the 100 m*M* treatment, which initially lost turgor, then recovered). Abscisic acid was viewed as setting into motion the above-described events leading to osmotic adjustment and then decaying to preset levels.

In evaluating current methods for quantitating and identifying ABA in bush bean, Montero et al. (1994) found that plants grown in 25 m*M* NaCl had an 81% and 88% in-

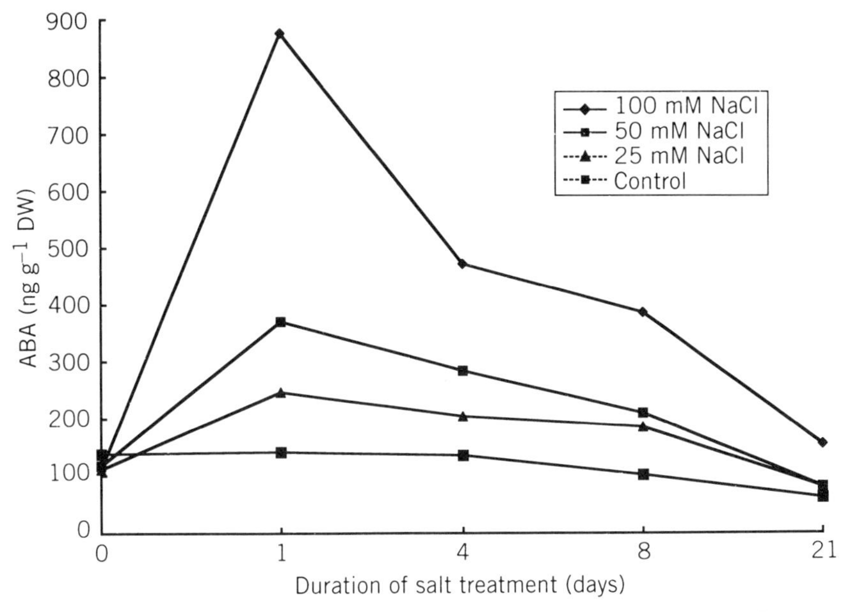

Figure 4.26 Effects of NaCl treatment on ABA levels in grape leaves. (Modified from Downton and Loveys 1981.)

crease in ABA in the leaves and xylem sap, respectively, compared to plants grown at 1 m*M* concentration. A 17% increase was observed for the roots (Table 4.21). Abscisic acid levels were highest in the leaves, followed by the roots and xylem sap.

As mentioned earlier in the section "Salinity and Cytokinin Levels," the ice plant can function as both a C_3 and a CAM plant. The latter form of metabolism can be triggered by salt stress, and associated with this is increased ABA production, with little change in CK levels (Figure 4.27) (Thomas et al. 1992). Associated with this transition is an increase in PEPCase and proline. NaCl and exogenous CK are effective in inducing these responses, but ABA is not. Since CK does not increase under salt stress in ice plant leaves or roots, endogenous CK appears not to trigger the CAM response in vivo. Thus NaCl-induced ABA, proline, and PEPCase increases were thought to be associated with direct NaCl induction.

TABLE 4.21 ABA Concentrations in Different Tissues of Bush Bean Plants Grown with 1 or 25 m*M* NaCl and determined by HPLC, GC-ECD, ELISA, and RIA

	Leaves (ng g^{-1} DW)		Xylem (μmol m^{-3})		Roots (ng g^{-1} DW)	
NaCl(m*M*)	1	25	1	25	1	25
HPLC	410	770	—	—	250	310
GC-ECD	450	830	37	78	320	370
ELISA	500	910	42	72	290	330
RIA	490	850	35	65	278	320

Source: Modified from Montero et al. 1994.

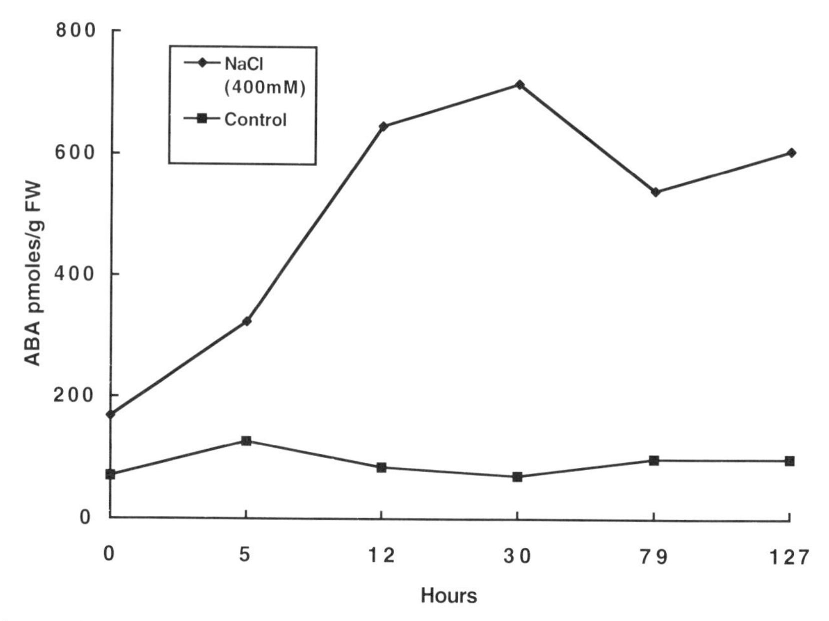

Figure 4.27 NaCl effects on ABA levels in ice plant. (Modified from Thomas et al. 1992.)

Halophytes (plants that grow well in saline environments) as opposed to *glycophytes* (plants that do not grow well in saline environments) may respond differently to reduced or elevated levels of salt relative to the function and accumulation of ABA. In the halophyte *Suaeda maritima*, ABA levels were higher when the plants were grown without salt compared to plants grown at NaCl concentrations of 200 or 400 mol m^{-3} (Clipson et al. 1988). At steady-state salt exposure the ABA levels did not change (Figure 4.28). Trans-

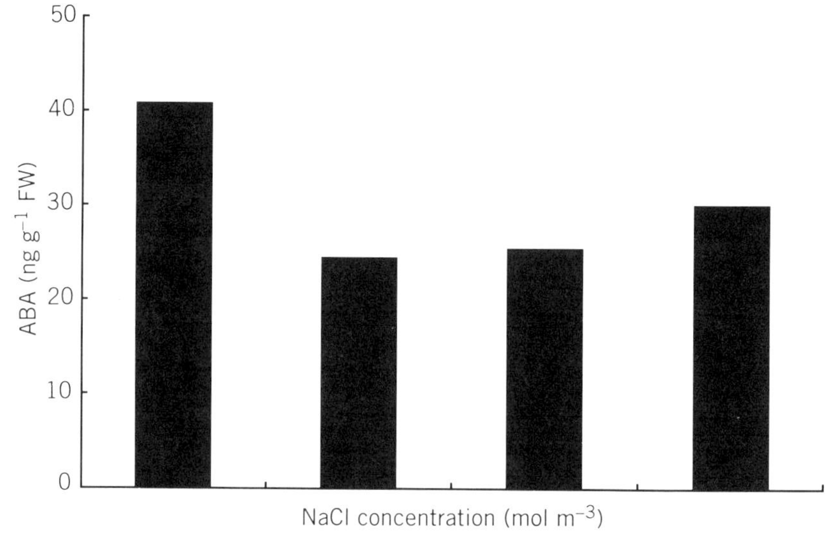

Figure 4.28 Effects of NaCl on ABA levels in *Suaeda maritima*. (Modified from Clipson et al. 1988.)

fer of plants previously grown in zero salt levels to 200 mol m^{-3} NaCl resulted in a twofold increase in ABA levels in 24 h, while plants transferred to 400 mol m^{-3} from 200 mol m^{-3} exhibited a fivefold increase in ABA in only 6 h. In both instances, ABA levels dropped off as in some previous examples. Compared to glycophytes, *S. maritima* responded more quickly to changes in salinity levels with respect to changing concentrations of ABA. It was suggested that instead of ABA being crucial for regulating stomatal closure, ABA may have a more important role in halophytes in controlling ion transport and balances under changing salinity conditions.

Another halotolerant organism, *Dunaliella salina*, a single-celled alga, has been used as a model to study ABA metabolism at the cellular level. This alga accumulates ABA under high-saline conditions just as observed with higher plants (Cowan and Rose 1991), and as in higher plants, ABA accumulates rapidly and then falls off to some constant level frequently above the level of the nonstressed controls. Increases in ABA levels in this organism have also been correlated with elevated levels of β-carotene, which upon oxygenation could give rise to postulated precursors of ABA synthesis. Thus ABA may function as a possible regulator of carotenogenesis in cells, which in turn may have a protective role relative to various environmental stresses.

The concept that ABA serves as a chemical signal between roots and shoots was addressed by Wolf et al. (1990) in studies that attempted to determine the long-distance transport patterns of ABA in NaCl-treated *Lupinus albus* plants. From these studies they concluded that net ABA synthesis occurred in the roots, basal strata of leaves, and the inflorescence. ABA degradation occurred in the stem internodes and apical leaf strata. Salt stress increased xylem ABA transport 10-fold and pholem transport fivefold. Fifty-five percent of xylem ABA originated from the roots of salt-treated plants compared to 28% in controls. The remainder of the ABA in the xylem came from fully differentiated leaves in the shoot and was translocated to the root and recirculated back to the shoot. It was proposed that both roots and shoots contribute to the synthesis of ABA and that depending on the growth conditions, one may dominate relative to the concentration being produced at any one time.

Under various stress conditions, including osmotic or salinity stress, various messenger RNAs and proteins accumulate that can be divided into (1) a set of mRNAs/proteins that are inducible by exposure to both stress and ABA, (2) a set of mRNAs/proteins that are inducible by exposure to stress, and (3) a set of mRNAs/proteins that are only inducible by ABA but not by the stress imposed (Luo et al. 1992). The Em gene encodes a hydrophilic protein of the late-embryo abundant class that is one of the most abundant proteins in embryos of dry cereals such as wheat. Levels of Em mRNA normally increase during maturation of wheat and maize embryos, but it is also expressed in osmotically stressed vegetative tissue (Bostock and Quatrano 1992). Using rice suspension cultures it was demonstrated that NaCl and ABA act synergistically to control Em mRNA production. Also, it was shown that NaCl can act through an independent pathway that changes the sensitivity of rice cells to ABA. Table 4.22 illustrates the relationship between various salinity treatments, ABA levels, and Em mRNA expression. Several treatments are represented in this table. As indicated, dehydration causes an increase in ABA and Em mRNA is induced. Increasing concentrations of NaCl also causes sequential increases in both ABA and Em mRNA, with the highest levels being produced at 0.4 *M* NaCl. Fluridone is an ABA synthesis inhibitor and when applied with the most effective concentration of NaCl, ABA and Em mRNA are reduced. Thus it appears that the ABA and the induction of Em mRNA are linked. Other evidence presented indicated that NaCl increased the sen-

TABLE 4.22 Endogenous ABA Concentrations and Relative Accumulation of Em mRNA in Rice Suspension Culture Cells After Various Treatments

Treatment mRNA	ABA (nM g^{-1} DW)	Relative Em (nM g^{-1} DW)
Control + ABA (100 μ*M*)		1.0
Control (no ABA)	0.79	0
Dehydration (12–15% initial fresh weight)	1.81	0.1
0.1 *M* NaCl	0.41	0.06
0.25 *M* NaCl	1.31	0.1
0.4 *M* NaCl	1.59	0.76
0.4 *M* NaCl + fluridone (200 μ*M*)	0.36	0.42
Fluridone (200 μ*M*)	0.12	0

Source: Modified from Bostock and Quatrano 1992.

sitivity of cells to respond to ABA. The authors proposed that NaCl can act independently to initiate a stimulus/response coupling mechanism (SRCM) or can stimulate the production of ABA, which in turn stimulates the same (Figure 4.29). Probable candidates for SRCM were invisioned as being number of receptors, ion/turgor changes, second messenger intermediates, and turnover rates of Em mRNA.

Luo et al. (1992) characterized a gene family from alfalfa (*Medicago sativa*) that encode for ABA- and environmental stress-inducible proteins. These genes could be induced by ABA, NaCl stress, wounding, cold temperature, and drought. Heat would not induce the genes. The induction of the genes was shoot specific but was not dependent on stage of development. The amount of ABA produced under each type of stress correlated with the amount of transcript produced, suggesting a relationship between ABA and the induction of the genes. However, data did not reveal whether the genes were induced by the stress or by ABA. It was hypothesized that changes in turgor initiated by the stress

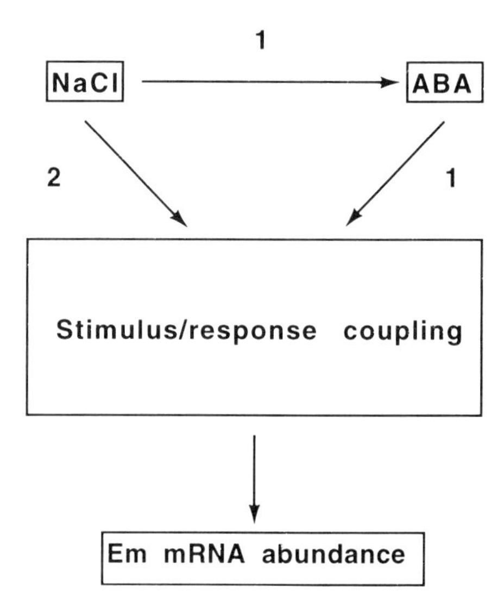

Figure 4.29 Model for ABA and NaCl induction of Em gene expression in rice suspension culture cells. (Modified from Bostock and Quatrano 1992.)

acts as a trigger to activate genes in the de novo synthesis or decompartmentalization of ABA, which in turn activates genes that are involved in stress protein synthesis.

E. Salinity and Ethylene Levels

Ethylene is generally categorized as the stress hormone in plants since it seems to be responsive to numerous types of biotic and abiotic stresses. This being the case, it is not surprising that plants respond to salinity stress usually by increasing C=C levels. However, scientists study plant responses to C=C for different reasons and plants may respond differently to salinity stress, depending on developmental stage, nutritional factors, plant part, and type of plant being studied. The following examples illustrate how C=C responds in plants to salinity stress.

Hadid et al. (1986) found that older leaves of tomato produced a twofold greater increase in C=C than younger leaves when the plants were grown in 0.4% NaCl. Lacheene et al. (1986) observed that tomato fruits from plants treated with 500 and 3000 ppm NaCl had no differences in C=C concentrations compared to controls.

Garcia and Einset (1983), using tobacco callus grown in 0, 40, 80, 120, 160, and 200 MEq L^{-1} NaCl detected no treatment effects on C=C levels. The authors were attempting to use C=C generation as an early index of stress in plants.

Chrominski et al. (1989) found that proline overcame the inhibitory effect of NaCl-induced C=C production in the halophyte *Allenrolfea occidentalis*. Sodium chloride (1.5 kmol m^{-3}) was found to cause increases in C=C production in hydrated plants taken from the field, nonhydrated plants from the field, and plants subjected to −5% dehydration. Plants dried to −10, −15, and −20% exhibited no differences in C=C concentrations compared to non-NaCl-stressed plants (Figure 4.30). Non-salt-stressed plants

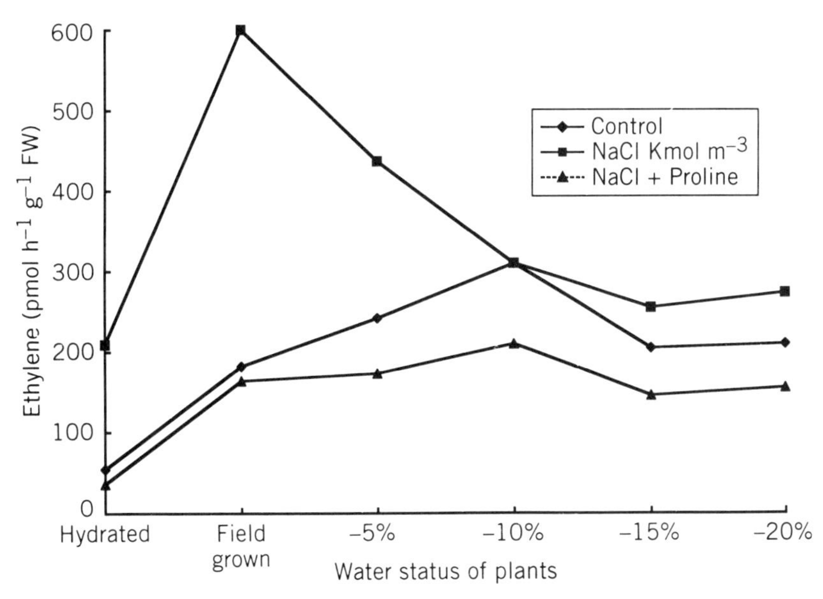

Figure 4.30 Effect of NaCl, NaCl + proline, and water status on C=C evolution by *Allenrolfea occidentalis*. (Modified from Chrominski et al. 1989.)

subjected to increased dehydration also exhibited elevated C=C levels up to −10%; then the amounts dropped off at −15 and −20%. The addition of proline completely overcame the dehydration- and NaCl-induced C=C production.

Both $CaCl_2$ and NaCl were found to increase C=C production in tomato plants. However, the response differed with variety and nitrogen source used to grow the plants (Feng and Barker 1992). Variety Heinz 1350 exhibited higher levels of C=C evolution when grown with multiple applications of 1 *M* $CaCl_2$ compared to the same concentration of NaCl. The reverse response was observed with the variety Neglecta-1 (Figure 4.31). Higher levels of C=C were produced when the plants were grown on NO_3-N compared to NH_4-N in both varieties. This is interesting in view of the fact that the tissues of NH_4-grown plants contained higher concentrations of NH_4-grown plants than NO_3-grown plants and in our earlier discussion in the section "Nutrient Stress and Ethylene Levels," high NH_4 concentrations tended to stimulate C=C synthesis (Table 4.23).

In an attempt to use C=C induction of epinasty in tomatoes as a means of determining sensitivity of tomatoes to salinity, Jones and El-Abd (1989) demonstrated a correlation between leaf angle, C=C synthesis in leaves and petioles, and NaCl concentrations (50, 150, and 250 m*M*; Table 4.24). As NaCl concentrations increased, the levels of C=C increased, as did the leaf angle measured from the adaxial surface of the petiole. The use of C=C antagonists in conjunction with salt stress inhibited the production of C=C as well as reducing the leaf angle. Co^{2+} is known to inhibit the conversion of ACC to C=C by interfering with the EFE while AOA interferes with the synthesis of ACC. The latter did not interfere with C=C production when plants were supplied with exogenous ACC, while Co^{2+} did, demonstrating that C=C production is dependent on EFE.

Seed germination can be inhibited by high saline conditions. Lettuce seed germination

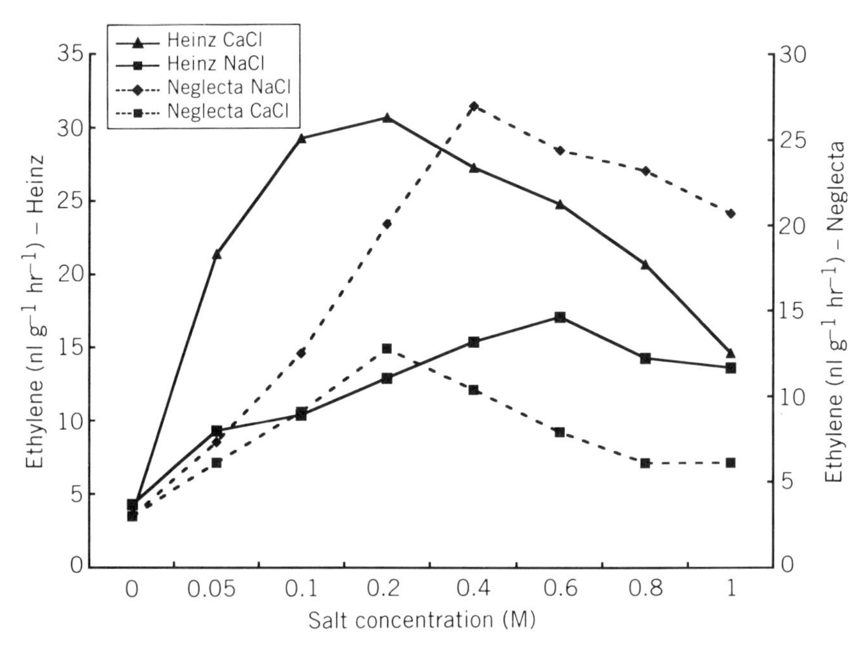

Figure 4.31 Effects of salt concentration on C=C evolution in tomato. (Modified from Feng and Barker 1992.)

TABLE 4.23 Ethylene Evolution and NH_4 Accumulation of Tomato Shoots as Affected by Salt Stress and Nitrogen Form

Treatment		Measurement	
Nitrogen	Salt	Ethylene ($nL\ g^{-1} \cdot h$)	Ammonium ($\mu g\ N\ g^{-1}$)
		Heinz 1350	
NH_4	None	9.3	189
	NaCl	10.5	200
	$CaCl_2$	13.5	236
NO_3	None	7.1	47
	NaCl	18.4	86
	$CaCl_2$	32.9	76
		Neglecta-1	
NH_4	None	12.5	203
	NaCl	14.3	219
	$CaCl_2$	17.9	217
NO_3	None	8.1	48
	NaCl	47.5	82
	$CaCl_2$	19.2	76

Source: Modified from Feng and Barker 1992.

TABLE 4.24 Leaf Epinasty and C = C Production by Leaf Petiole and Laminae Segments of Tomato Plants Irrigated with Saline Water and Treated with AOA or Cobalt Ion

Treatment	Ethylene ($nL\ g^{-1}\ h^{-1}$) Petiole	Laminae	Leaf Angle (degree change)
Control (zero NaCl)	1.45	2.68	0
50 m*M* NaCl	1.59	3.11	+5.2
150 m*M* NaCl	1.93	4.30	+17.6
250 m*M* NaCl	3.33	6.39	+27.9
1 m*M* AOA	1.30	2.52	−3.8
150 m*M* NaCl + 1 m*M* AOIO	1.63	3.62	−5.6
250 m*M* NaCl + 1 m*M* AOA	1.86	4.69	+3.4
100 μM CO^{2+}	1.18	1.36	+3.9
150 m*M* NaCl + 100 μM CO^{2+}	1.24	2.62	+8.0
250 m*M* NaCl + 100 μM CO^{2+}	2.03	4.27	+14.0

Source: Modified from Jones and El-Abd 1989.

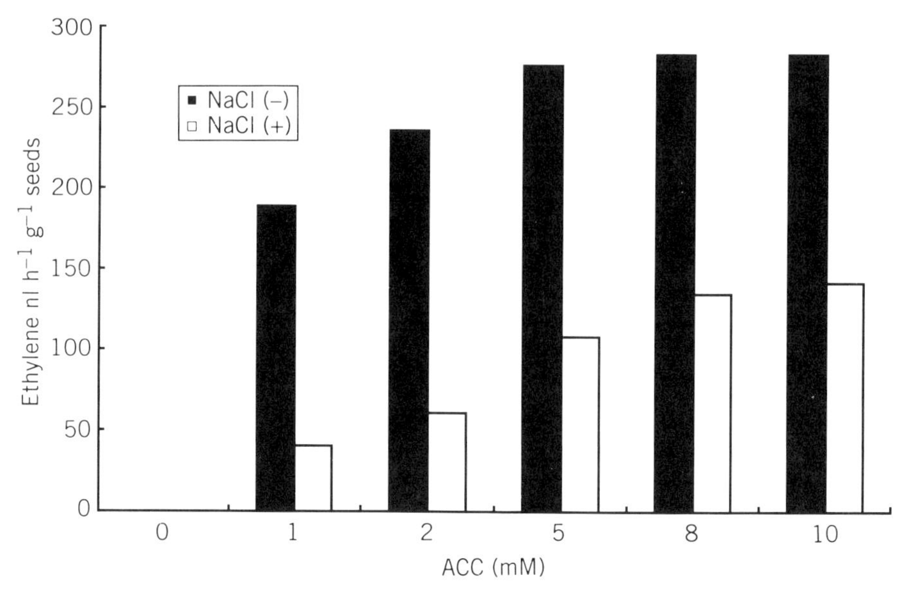

Figure 4.32 Conversion of ACC to C═C by lettuce seed soaked in 0.1 *M* NaCl or water. (Modified from Khan and Huang 1988.)

is substantially inhibited by 0.1 *M* NaCl, but the salt stress can be removed with 1 m*M* kinetin or 1 to 10 m*M* ACC (Khan and Huang 1988). The two chemicals act synergistically to enhance pregermination C═C production. Sodium chloride–treated seed produced less C═C when supplied with varying levels of ACC than did non-salt-treated seed. This implied NaCl inhibition of C═C biosynthesis (Figure 4.32) and that C═C was required for germination of seeds under salt stress. The use of C═C biosynthesis and action inhibitors allowed the deduction that kinetin alleviated NaCl stress by increasing the activity of ACC synthase, EFE, and/or increasing the action of C═C. The requirement for CK and C═C was considered obligatory only under stressful conditions such as salinity stress, hypoxia, thermoinhibition, and osmotic restraint.

V. FLOODING STRESS AND PHYTOHORMONES

A. Flooding and Indoleacetic Acid Levels

Relatively little is known regarding the effects of flooding on auxin levels in plants. The current evidence suggests that flooding causes an increase in auxin concentrations in shoots and reduced levels in roots of plants. It is well established that C═C increases in the shoots of seedlings exposed to flooded conditions and that C═C can cause a decrease in the basipetal polar transport of auxin in the shoots, which could explain the accumulation of this hormone in shoots. To support this notion (Figure 4.33) (Hall et al. 1977), it was demonstrated that C═C production precedes IAA accumulation in broad bean and that a rapid rise in hypocotyl auxin correlated with decreased basipetal transport of [^{14}C]IAA in sunflower (Wample and Reid 1979). Abscisic acid, which also can accumulate under flooding conditions, has been demonstrated to inhibit IAA transport in tree stems.

Frequently, flooding induces adventitious root development above the flood line on the

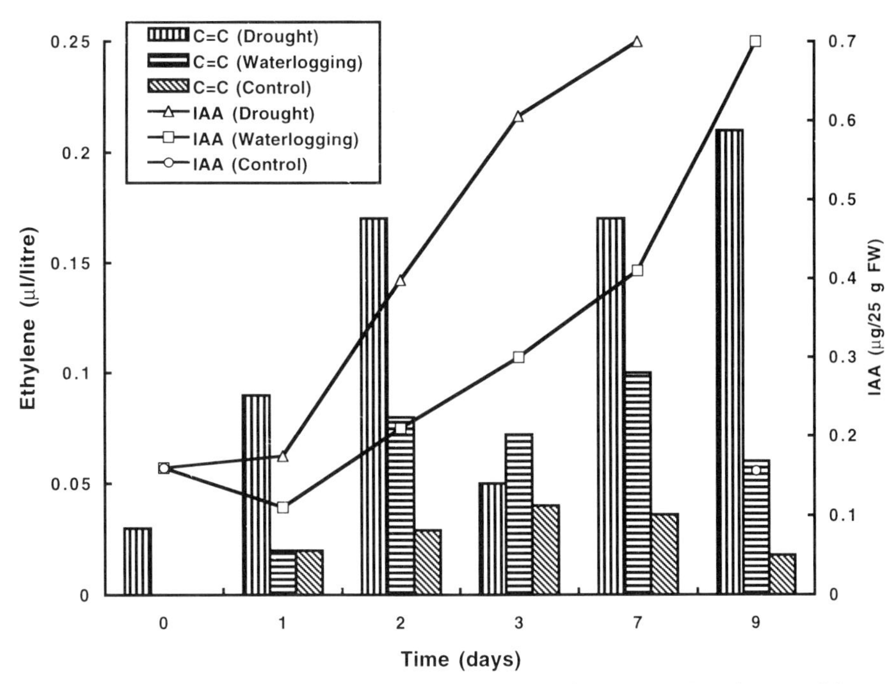

Figure 4.33 Effects of waterlogging or drought on levels of C=C and IAA in *Vicia faba* leaves. (Modified from Hall et al. 1977.)

shoots of plants. Whether this is an IAA or C=C-induced response has not been established. Recent studies (McNamara and Mitchell 1991) using IAA and C=C transport inhibitors and antagonists, respectively, concluded that the anoxic condition created at the level of the flood line causes an accumulation of IAA as a result of transport inhibition by high levels of C=C in the root zone, which in turn stimulates adventitious root development. Certainly, additional studies need to be conducted to determine how C=C, IAA, and ABA may interact under flooding conditions to produce the observed changes in plants in response to flooding.

B. Flooding and Cytokinin Levels

Various root stresses can cause changes in the levels and distribution of phytohormones in plants. Cytokinins have long been thought to be synthesized in the roots and transported to the leaves via the xylem. Thus conditions in nature leading to reduced levels of oxygen (hypoxia), such as flooding stress, may have an important impact on the synthesis and transport of CKs in plants. Plants under hypoxic stress exhibit increased stomatal resistance and rapid reductions in leaf growth, which has led some investigators to hypothesize that CKs along with other hormones may function as signals that can affect stomatal apperature as well as having a direct impact on leaf growth.

In an effort to correlate the changes in CK levels in the xylem and leaves with stomatal closure and reduced leaf growth in hybrid poplar and bean plants, Neuman et al. (1990) found that hypoxia caused a reduction in CK flux (concentration × mass flow rate of sap) in the xylem but no change in the concentration of CKs in the leaves of either species compared to nonstressed plants (Figure 4.34). Exogenous applications of high concentra-

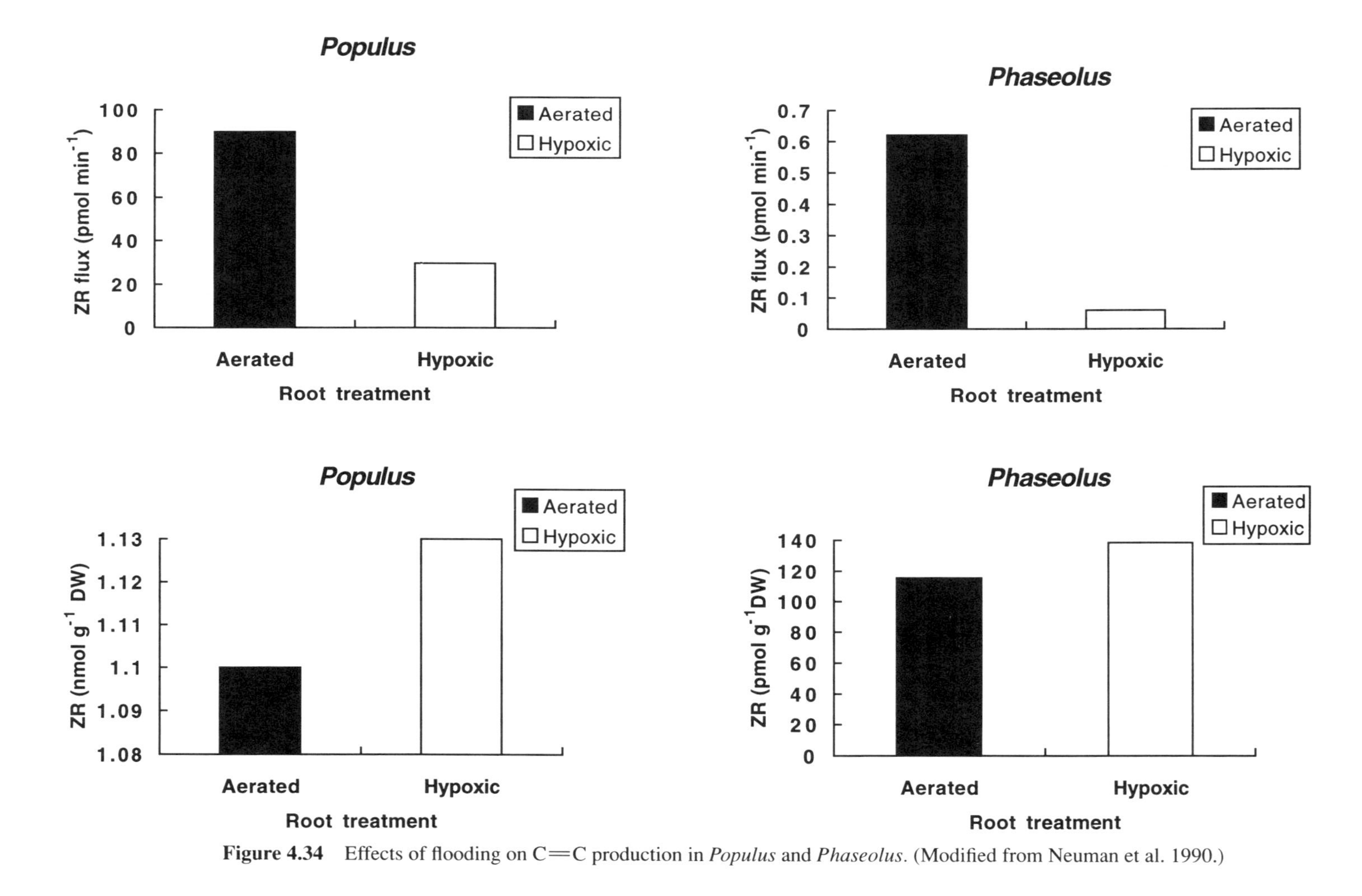

Figure 4.34 Effects of flooding on C=C production in *Populus* and *Phaseolus*. (Modified from Neuman et al. 1990.)

tions of CK could force stomatal opening and increase growth rates, but the effect was not considered to reflect normal physiological responses by the plants. Since other hormones may interact with CK to influence stomatal aperture and leaf growth, GAs were also measured in poplar. Hypoxia was found to cause a slight reduction in the levels of GA_1 and GA_3 in poplar, but the authors believed that the change was too small to have any significant physiological impact. The authors concluded that since no change in CK levels occurred in the leaves when the experimental plants were exposed to hypoxia, the total bulk concentration of CKs in the leaves was not a factor and perhaps the change in xylem flux was the signaling parameter that controlled stomatal aperture and leaf growth. Additionally, it was proposed that little is known about compartmentalization of CKs in plants and that compartmentalization may also be a factor in controlling the types and availability of CKs in the leaves. A cross reactivity of the immunoassay for different CKs was also proposed as a potential problem. The literature is sparse relative to the impact of flooding or hypoxia on CKs in plants.

C. Flooding and Gibberellin Levels

As indicated above (Neuman et al. 1990), hypoxia caused a slight but significant decrease in GA_1 and GA_3 in poplar. Contrary to these results, Hoffmann-Benning and Kende (1992) found that GA_1 and GA_{20} increased fourfold (after 3 h of submergence) and threefold (after 24 h submergence), respectively, in intercalary meristems and internode elongation zones of adult deepwater rice plants (Table 4.25). Associated with the increase in GA levels was a decline in ABA concentrations (Figure 4.35). It was previously established that C=C concentrations increased in submerged plants and that exogenously applied C=C caused a reduction in ABA levels. The ABA synthesis inhibitor fluridone enhanced coleoptile growth in rice but had little effect on wheat, barley, and oats. Since rice seedlings respond to reduced partial pressures of O_2 by increased coleoptile and internode growth, these data suggested to the authors that elevated C=C levels increased the sensitivity of coleoptiles and internodes to GA by possibly reducing the levels of ABA that antagonize GA. The fact that nonaquatic plants such as oats, wheat, and barley did not respond to fluridone treatment by enhanced growth, at least compared to the response level of rice, suggested that a unique control system exists in rice which allows rapid growth under reduced partial pressures of O_2.

TABLE 4.25 Levels of GA_1 and GA_{20} in the Intercalary Meristem and Cell Elongation Zone of Internodes of Adult Deepwater Rice Plants Grown in Air or Submerged for Various Time Periods

Duration of Treatment	GA_1 (ng g^{-1} DW)	GA_{20} (ng g^{-1} DW)
0	15.2	12.5
1	18.9	ND[a]
3	62.3	19.4
24	63.0	40.9

Source: Modified from Hoffmann-Benning and Kende 1992.

[a]ND, not determined.

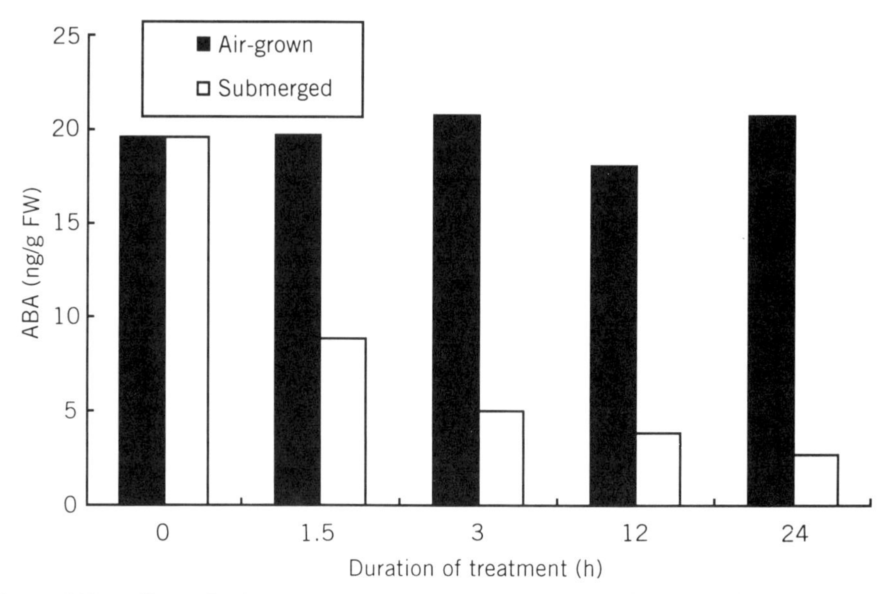

Figure 4.35 Effects of submergence on ABA levels in deepwater rice. (Modified from Hoffman-Benning and Kende 1992.)

D. Flooding and Abscisic Acid Levels

One of the purported roles of ABA in plants under stress is to control stomatal aperature. Usually, under stress, ABA in the leaves increases and it is generally thought that ABA causes stomates to close. However, whether ABA originates from the leaves through synthesis or decompartmentalization or produced in the roots and transported to the leaves has been a point of considerable debate.

Flooding of pea (Jackson et al. 1988) and alfalfa (Figure 4.36) (Castonguay et al. 1993) plants resulted in increased levels of ABA in the leaves of both plant species. However, compared to other reports of sizable increases in ABA in the roots of peas (Zhang and Davies 1987), in neither of the cases above was there any significant increase in ABA levels in the roots of alfalfa or pea. These results in part led the authors to conclude that the roots were not the source of ABA in leaves. Since low O_2 levels have been shown to result in the reduction of translocation of assimilates (including ABA) from the shoots to the roots, Jackson et al. (1988) concluded that ABA accumulates in the leaves because of reduced phloem transport.

Flooding also increased ABA concentrations (15-fold increase in 8 h) in bean (Wadman-van Schravendijk and van Andel 1986), and the increased C=C levels were proposed to interfere with ABA control of diffusive resistance. Application of the C=C generating compound, Ethrel®, along with differing levels of ABA to the leaves of bean plants, resulted in decreased diffusive resistance compared to ABA treatments only. As indicated in the preceding section, C=C was shown to reduce ABA levels in the intercalary meristem and part of the shoot elongation zone (Hoffmann-Benning and Kende 1992). Thus it appears that C=C and ABA may interact to influence stomatal opening.

The application of exogenous ABA to a variety of plants has been shown to increase tolerance to a variety of environmental stresses, including salinity, drought, cold, and

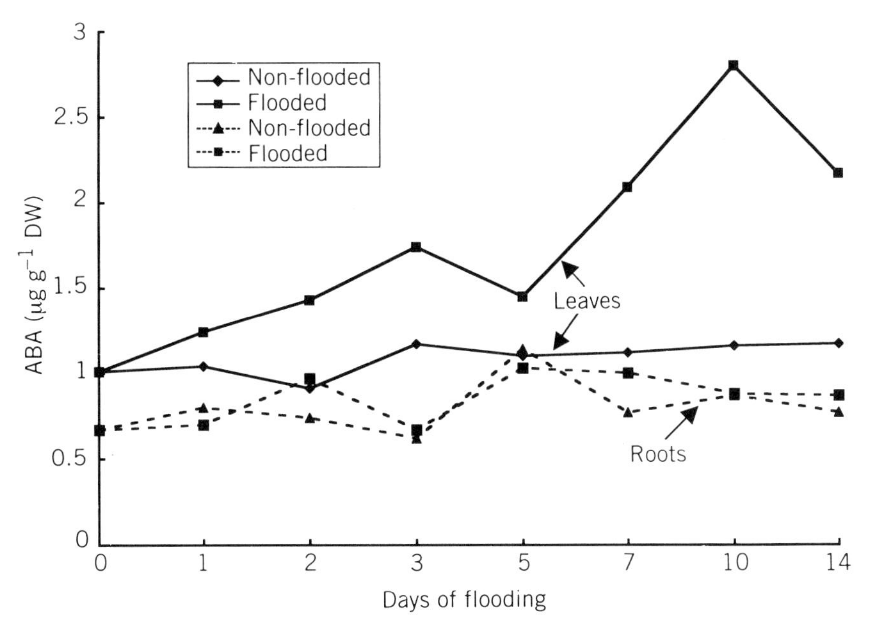

Figure 4.36 Effects of flooding on ABA levels in leaves and roots of alfalfa. (Modified from Castonquay et al. 1993.

heat. The roots of young corn plants pretreated with ABA resulted in seedlings that exhibited a 10-fold increase in tolerance to anoxia (a twofold increase compared to water controls). The survival rates of seedlings increased from 8% in nontreated controls to 87% in ABA-treated plants and 47% in water controls (Hwang and VanToai 1991). Transfer of laboratory flood–stressed ABA-treated and nontreated seedlings to the field resulted in faster recovery of seedlings treated with ABA compared to nontreated seedlings (VanToai 1993). Toward the end of the growing season, no differences were observed in growth, seed yield, and chlorophyll concentration.

E. Flooding and Ethylene Levels

Most of the research that has been conducted on flooding stress and phytohormones has been relative to the effects on C=C concentrations and the subsequent impact on the physiology and morphology of plants. Many of the plant symptoms associated with flooding stress are typical of high ethylene levels in plants. Table 4.26 summarizes some of the symptoms associated with flooding stress and increased C=C levels in plants, but exceptions have been reported. Generally, flooding causes an increase in C=C concentrations in plants. The increase can occur in submerged tissues or above-water organs (Table 4.27) (Yamamoto and Kozlowski 1987a). It appears that under anoxic conditions (situations where O_2 is totally lacking) C=C cannot be synthesized. Thus high levels of ACC accumulate in the roots and can be transported to the shoots, where available O_2 facilitates the conversion of ACC to C=C. Hypoxic conditions in the root zone will allow C=C accumulation in the roots and/or move throughout the plant. As indicated earlier in the section "Nutrient Stress and Phytohormones," deficiencies in N and P levels causes a reduction in C=C concentrations but mimics the symptoms of C=C accumulation (He

TABLE 4.26 Plant Symptoms and Responses to Flooding Stress

Leaves
Wilting
Epinasty
Stomatal closure
Reduced transpiration
Reduced photosynthesis
Reduced growth and size
Thickened leaves
Chlorosis
Nutrient deficiency
Accelerated senescence
Necrosis
Abscission
Shoots
Reduced elongation growth
Increased diameter growth
Lenticular and stem hypertrophy
Roots
Adventitious roots formed
Inhibited root hair formation
Increased membrane leakage
Reduced water and nutrient uptake
Root death (lack of O_2 and accumulation of ethanol)

et al. 1992, 1994). It was hypothesized that N and P deficiencies increased the sensitivities of corn root tissues to lower levels of C=C, thus inducing the typical C=C symptomology.

Donovan et al. (1989) studied the interaction between flooding, water temperature, and C=C production in swamp seedlings of bald cypress (*Taxodium distichum* L.) and water tupelo (*Nyssa aquatica* L.). Although statistical treatments showed few significant differences

TABLE 4.27 Effect of Flooding for 55 Days on C = C Content (nmol g^{-1} DW) of Stem Segments Taken at Various Stem Heights[a]

Treatment	2.5–3.5 cm BWL	0.5–1.5 cm BWL	0.5–1.5 cm AWL	2.5–3.5 cm AWL
10 days				
Unflooded	5.58	6.31	5.17	4.53
Flooded	17.69	13.08	13.73	8.71
20 days				
Unflooded	10.15	8.48	8.10	5.35
Flooded	8.70	7.99	12.74	9.41

Source: Modified from Yamamoto and Kowslaski 1987a.

[a]BWL, below water level; AWL, above water level.

in flooding and temperature treatment interactions relative to C=C production, trends suggested that more C=C was produced in water tupelo under ambient temperature/flooded conditions than in bald cypress under the same conditions. In addition, high-temperature/flooded conditions tended to decrease C=C production in the former to a greater extent than in the latter. The combination of high temperature and anoxic/hypoxic conditions were viewed as leading to thermal tissue injury and inability of the O_2-deprived plants to initiate C=C-related responses (i.e., lenticular hypotrophy, aerenchyma production, and adventitious root formation), which may increase the ability of the plant to survive.

Ethylene may also be formed in flood soils and could also be a source of C=C accumulation in plants subjected to flooding conditions. Hunt et al. (1981) demonstrated that intermittent or continuous flooding of tobacco plants causes an increase in C=C concentrations in the leaves of the plant as well as the soil in which they grew. Ethylene accumulated in these plants to a maximum level about 4 days after flooding but then dropped off to control levels about 5.5 days later. Although the contribution of soil C=C versus plant-generated C=C could not be determined, it is likely that soils could contribute to elevated levels in plant tissues and could affect morphological and physiological responses.

A considerable number of studies have been conducted relative to the effects of flooding on morphological and developmental changes in plants. Many of these changes have been attributed to flood-induced C=C generation in the plant or perhaps from soil microorganisms. Many such studies have focused on the relationship between C=C production and aerenchyma synthesis in roots and even in stems of plants (He et al. 1992, 1994). Arenchyma tissue forms in plants that exist in hypoxic environments and is characterized by lysis of cell walls of the root cortex, forming large intercellular spaces, which probably facilitates the diffusion of available O_2 throughout the root and shoot tissues. Another response to flooding is the development of hypertrophy in lenticels of shoot stems. This has also frequently been associated with C=C accumulation in flooded plants. It has been hypothesized that this response may facilitate gas exchange as a result of the large openings in the stem and may allow the release of toxic components as a result of flooding. This phenomenon has been demonstrated mostly in woody plants such as mango (*Mangifer indica* L.; Figure 4.37) (Larson et al. 1993) and flood-tolerant trees such as elm (*Ulmus americana*), ash (*Fraxinus pennsylvanica*), *Melaleuca quinquenervia*, and *Eucalyptus camaldulensis* (Figure 4.38) (Tang and Kozlowski 1984a & b). Adventitious root formation also develops in some plants at the waterline or on submerged tissues (Yamamoto and Kozlowski 1987b, McNamara and Mitchell 1991, Tang and Kozlowski 1984a & b). Such development is thought to provide the plant with root-absorbing capacity nearer the surface of the water, where more O_2 is available or perhaps to replace and increase root-absorbing capacity as a result of dying and decaying root tissues. The ability of plants to form aerenchyma tissue and adventitious roots and to develop lenticil hypertrophy all appear to be related to C=C accumulation. Such changes may allow plants to adjust to flooding conditions, and the propensity of different species to be able to accomplish such changes may reflect their ability to tolerate flooding stress. Anatomical changes in vascular tissues have also been noted in flooded seedlings of *Cryptomeria japonica*. Flooded seedlings increased in diameter, which was caused by increased phloem growth and intercellular space development. Xylem tracheids were larger in abovewater tissues than in submerged tissues. Flooding also increased tracheid lumen diameters, decreased tracheid wall thickness, and stimulated axial parenchyma cells in the xylem (Yamamoto and Kozlowski 1987a).

Ethylene may also have a function in growth stimulation of some species of plants ex-

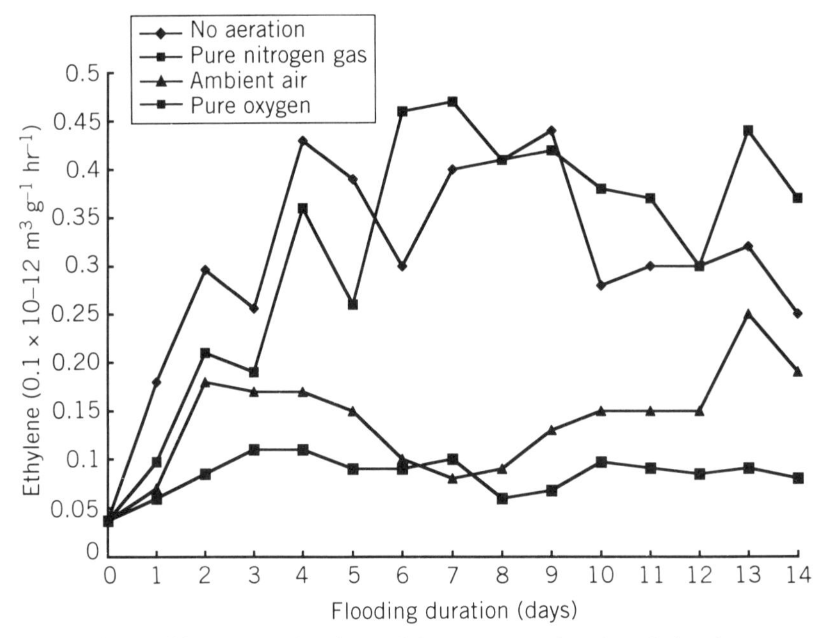

Figure 4.37 Effects of floodwater duration and O_2 content on C=C evolution from mango stem tissue. (Modified from Larson et al. 1993.)

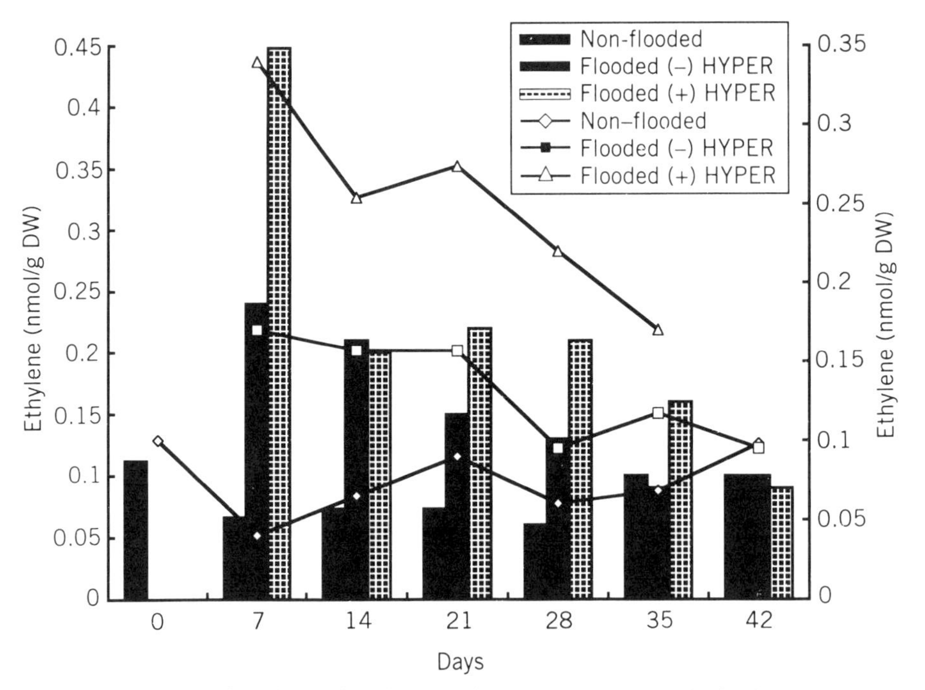

Figure 4.38 Effects of flooding on C=C production and stem hypertrophy in above (bar graphs)- and below (line graphs)-water-level tissue sections of *Fraxinus pennsylvanica*. (Modified from Tang and Kozlowski 1984a, 1984b.)

posed to flooded conditions. For example, submergence of *Rumex palustris* and *Rumex crispus* stimulates petiole growth. Petioles of nonsubmerged plants could also be stimulated by exposure to C=C–air mixtures. A third species, *Rumex acetosa,* does not exhibit increased petiole growth under flooded conditions. Compared to the former two species it does not produce as much C=C under flooded conditions. The response of these plants to C=C and flooding correlated with their natural distribution along river flood plains: *R. acetosa* normally grew along high dikes and levees, *R. palustris* occurs in low-lying areas, and *R. crispus* was found in intermediate areas (Voesenek and Blom 1989; Voesenek et al. 1990, 1993).

Other plants have been reported to respond to flooding by increased growth. These are mostly aquatic plants such as rice species. Khan et al. (1987) found significant positive correlations between several rice cultivars relative to C=C production and elongation growth (Figure 4.39). This finding was thought to be of value relative to selecting different cultivars (based on C=C evolution) for different areas of the world prone to differing degrees of flooding. As noted previously, GA also has been observed to increase in deepwater rice cultivars (Hoffmann-Benning and Kende 1992). The elongation response appears to involve GA, ABA, and C=C. According to these authors, C=C increases the sensitivity of elongating tissues to GA, perhaps by inhibiting or reducing ABA levels. This response appears to be a complex interaction among several hormones.

Waxapple (*Syzygium samarangense* Merr.) is a tropical fruit tree that is flooding tolerant, and farmers in Taiwan induce early flowering by continuously flooding orchards for 30 to 40 days in the summer (Lin and Lin 1992). In this species flooding does not result in C=C accumulation or the synthesis of C=C precursors such as ACC or MACC, as has been observed previously. Also, waxapple does not exhibit the normal symptoms frequently associated with C=C stress, such as epinasty, chlorosis, abscission, and wilting. Both flooded and nonflooded plants exhibited aerenchyma tissue development in the roots after 9 days of flooding. Flooding tolerance in this species was attributed to its ability to produce several isozymes of alcohol dehydrogenase, an important enzyme involved in the

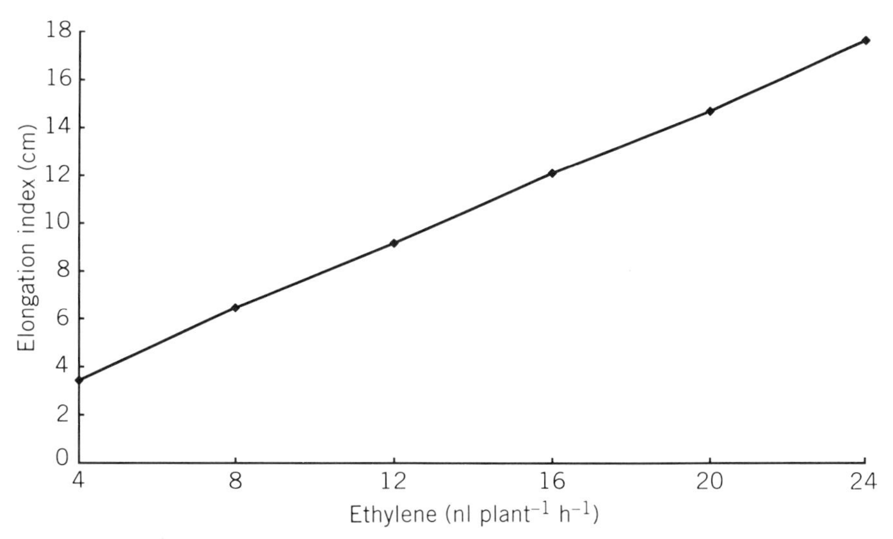

Figure 4.39 Relationship of C=C production and elongation index in 14 cultivars of rice during submergence. (Modified from Khan et al. 1987.)

degradation of toxic ethanol, which accumulates in plant roots under hypoxic conditions, low O_2 requirements under flooding conditions, and the presence of aerenchyma tissue, which facilitated the diffusion of available O_2.

VI. DROUGHT STRESS AND PHYTOHORMONES

A. Drought and Indoleacetic Acid Levels

Generally, drought stress or osmotically induced stress causes a reduction in free IAA concentrations and perhaps an increase in conjugated forms. However, there are a few reports which suggest that drought has no effect on IAA levels or that IAA levels increase under drought.

Auxin levels were observed to decline in apple and plum woody shoots during a normal cyclical drought but were found to recover after a rainy period (Hatcher 1959). In Sitka spruce, elevated internal water stress resulted in a reduction in cambial diffusible and acidic IAA concentrations, while in the leaves, acidic IAA declined (Figure 4.40) (Little and Wareing 1981). Water withheld from potted *Pinus resinosa* seedlings resulted in a reduced level of IAA-like substance extracted from the terminal shoots. This reduction was correlated with a zone of narrow-diameter latewood tracheids and decreased needle elongation (Larson 1963). Free IAA declined in field-grown and water-stressed cotton bolls and abscission zones while conjugated IAA increased. After irrigation the reverse trend was observed (Table 4.28) (Guinn and Brummett 1988).

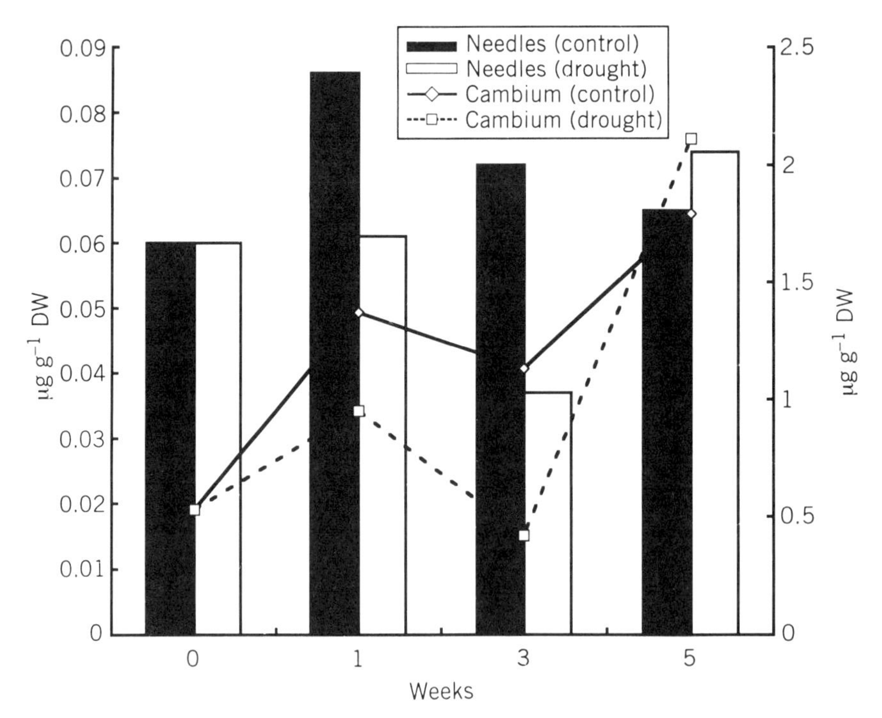

Figure 4.40 Effects of drought on cambial and foliar levels of IAA in spruce. (Modified from Little and Wareing 1981.)

TABLE 4.28 Changes in Water Potential, Boll Retention, and Free and Conjugated IAA in 3-Day-Old Bolls and Abscission Zones through Irrigation Cycles in 1985 and 1986

	1985[a]			1986[b]		
Measurement	July 8	July 12	July 15	June 16	June 26	June 30
Xylem (ψ)	−1.69	—	−2.90	−1.80	−2.81	−1.82
Boll retention (%)	74	51	3	96	27	78
		(ng g^{-1} DW)			(ng g^{-1} DW)	
Bolls						
Free IAA	105	114	61	120	93	57
IAA ester	493	808	868	1422	1625	471
Abscission zone						
Free IAA	91	64	36	72	43	61
IAA ester	17	97	241	57	221	185

Source: Modified from Guinn and Brummett 1988.
[a]Plots irrigated July 3 and July 18, 1985.
[b]Plots irrigated July 13 and July 27, 1986.

IAA oxidase is important in regulating the concentrations of free IAA in plants. High activity in this enzyme typically would cause a reduction in IAA levels. In tomato grown under drought conditions in the field (Figure 4.41) and pea plants subjected to mannitol-induced stress, levels of IAA oxidase increased (Darbyshire 1971). Although endogenous levels of IAA were not measured in this experiment, it is implied that drought may control IAA levels through IAA oxidase activity.

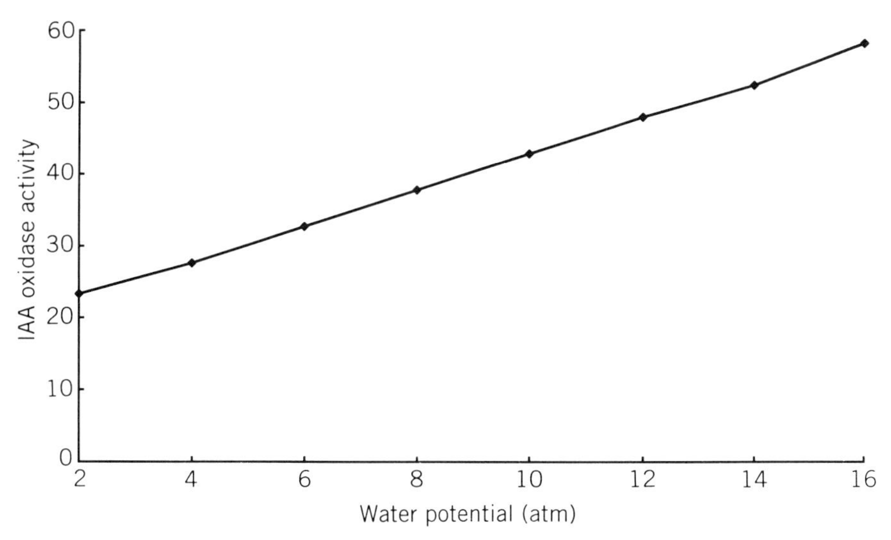

Figure 4.41 Relationship between IAA oxidase activity and water potentials of field-grown tomato plants. (Modified from Darbyshire 1971.)

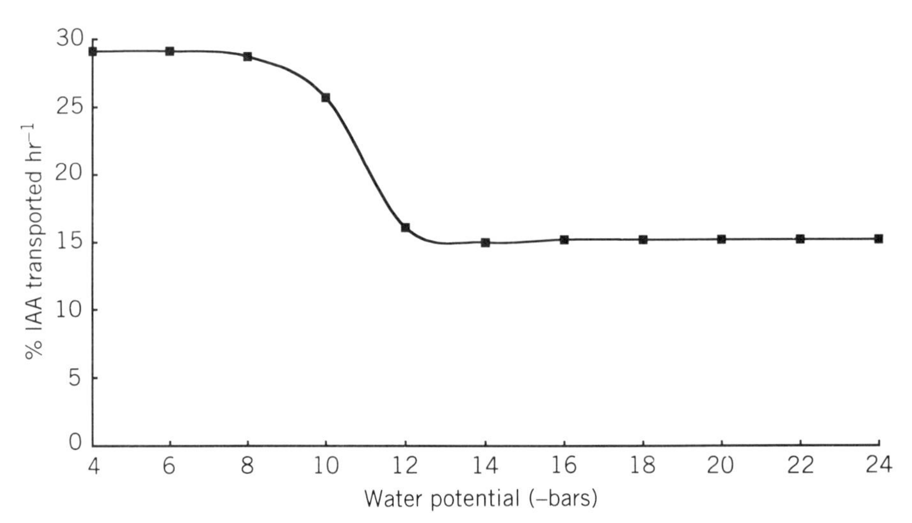

Figure 4.42 Effects of water potential on basipetal IAA transport in cotton cotyldonary petioles during stress and 24 h after rewatering. (Modified from Davenport et al. 1977.)

As indicated earlier, IAA exhibits polar transport in plants, which is important in maintaining a gradient throughout plant tissues. Leaves and apical meristems are important sources of auxin, which helps to sustain the gradient of IAA in the polar transport system which is thought to be associated with the cambial initials. Drought stress has been reported to reduce the rate of polar IAA transport in several instances. Indoleacetic acid transport decreased in cotton cotyledonary petiole sections subjected to decreasing plant water potentials. This decline was correlated with abscission of the petioles (Figure 4.42) (Davenport et al. 1977). A similar effect was observed for the abscission of cotton leaves of plants subjected to drought. Internal water stress decreased the movement of IAA through branch sections of balsam fir exposed to drought stress. The reduced transport of auxin was believed in part to be responsible for reduced cambial activity (Little 1975).

Dark-grown (etiolated) squash seedlings exhibited reduced levels of IAA in-hypocotyls of both drought-treated and control plants. However, plants stressed with polyethylene glycol exhibited a slower rate of reduction than did control plants (Sakurai et al. 1985). No change in IAA levels were observed in sorghum leaves of field-grown and drought-stressed plants (Kannangara et al. 1982, 1983). Drought caused an increase in IAA levels in broad bean leaves with increased stress. However, data for the control plants were only provided on day 9 following imposition of the stress. All other treatments include data from at least four other time periods prior to day 9, making comparisons with the control difficult (Hall et al. 1977).

As indicated from the examples above, drought does have an impact on levels of IAA in tissues as well as influencing IAA transport. The increased levels of IAA oxidase may have some influence on the amounts transported, which could influence the final levels that occur in a tissue at any one time. These changes then could have ramifications relative to cambial activity as well as abscission of fruits, flowers, and leaves.

B. Drought and Cytokinin Levels

Most of the information available regarding the relationship of CKs to water relations in plants has been relative to their effects on stomatal movement and if CKs interact with ABA in an opposing manner to affect stomatal apperature and thus transpiration. Generally, the studies that have been conducted on the effects of drought on CK levels in plants are limited, but those that have been done indicate that drought decreases CK synthesis and/or transport from the roots, where they are thought to be predominately synthesized. The reduction in levels of CKs in the shoots and the large increase in ABA that follows drought has led scientists to believe that the two hormones act antagonistically to affect stomatal apperature. CKs are thought to promote opening, while ABA promotes closing of stomates. Some studies have demonstrated that stomates can be opened under drought stress with application of exogenous CKs to leaf surfaces. These experiments have often been done with leaf epidermal preparations and have not always been successful in all plants. The following examples will give the reader some idea of recent contributions to the literature relative to the impact and significance of drought stress on CK levels in several different plant species.

Hubick et al. (1986b) found that aeroponically grown sunflower plants subjected to differing misting regimes exhibited half as much CK in shoots as did nonstressed plants, and stressed roots accumulated nearly twice as much CK as nonstressed roots. High concentrations of zeatin glucoside were reported to accumulate in stressed roots and was not present in nonstressed tissues. ABA levels were 32 times higher in the roots of stressed plants than in nonstressed roots and 6.7 times higher in stressed shoots. Minimal effects of drought were observed on GA and C=C levels. The authors suggested that bound CK accumulated in the roots as a result of the interaction with increased ABA levels and that the conjugated CK served as a storage form that could not be transported.

In another study using sunflower, Bano et al. (1994) found that xylem sap of stressed plants contained 3.8 times less zeatin/zeatin riboside than did nonstressed plants. Although older plants exhibited higher levels of zeatin compounds than younger plants, the relative amount of CKs did not change between stressed and nonstressed plants. The same trend was observed for isopentenyladenine and isopentenyladenosine, but lower levels of these CKs were observed than of zeatin and zeatin riboside in the xylem sap.

Cytokinins were also studied in the xylem sap of 1- to 4-year-old desert-grown almond (*Prunus dulcis*) trees cultured in lysimeters of different volumes, receiving differing amounts of water and subjected to an annual drying cycle (Fusseder et al. 1992). Zeatin-type CKs were always prominent and in a limited number of days showed a peak concentration in the morning and rapid decline in the afternoon. However, correlations between water status and CK concentrations based on daily variations could not be established. It was concluded that some threshold level of ABA that is attained when plants are stressed counteracts any influence CK may have on stomatal conductance but that CKs may influence stomatal conductance once ABA levels fall below that critical threshold after the stress is relieved. Thus it was viewed that CK may affect stomatal behavior on a short-term basis and ABA acts as an opposing signal, the size of which reflects long-term water deficit.

Drought-resistant (*Lycopersicon esculentum* cv. 'New Yorker') and susceptible (*Solanum pennellii* and *Lycopersicon chilense*) species of tomato were subjected to short-term osmotic stress using PEG as an osmoticum (Pillay and Beyl 1990). Zeatin riboside was found to decrease in the roots of both resistant cultivars (dramatically in *S. pennellii*

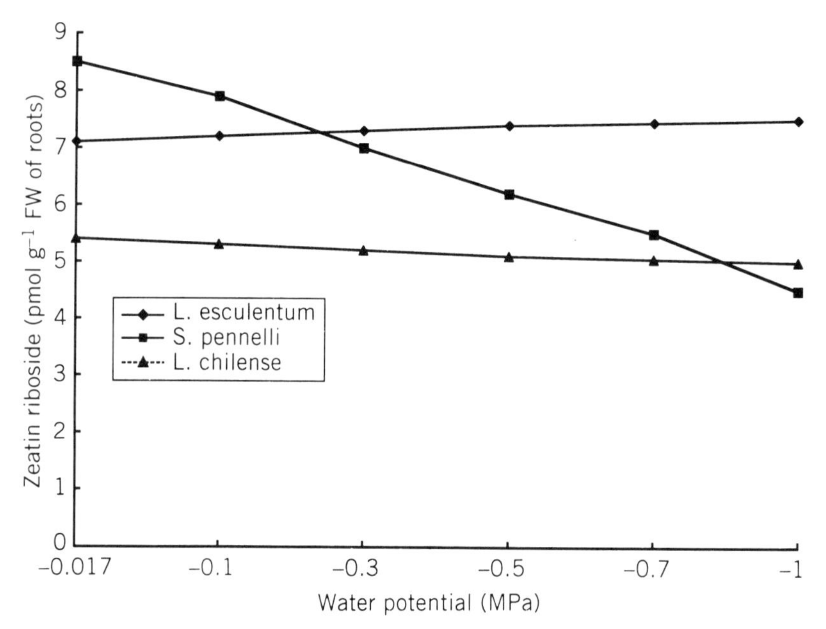

Figure 4.43 Effects of water stress on CK levels in three tomato species. (Modified from Pillay and Beyl 1990.)

and only slightly in *L. chliense*) and to increase slightly in *L. esculentum* as osmotic potential decreased (Figure 4.43). There was a highly significant correlation between leaf water potential and zeatin riboside concentrations and a significant correlation between transpiration and CK levels. The authors concluded that CK derivitization, transport, and metabolic systems could differ in the tomato species studied and each possesses a different mechanism for the early response to water deficit. There appears to be little conclusive evidence of the role of CKs in the control of stomatal apperature in intact plants, but certainly more work in this area is warranted.

C. Drought and Gibberellin Levels

Most of what is known about the effects of drought stress on GAs relates to the control of vegetative and/or reproductive growth. In coffee (*Coffea arabica*), water-deficit stress breaks dormancy and is considered mandatory for normal flower development (Schunch and Fuchigami 1992 and references therein). This treatment, followed by irrigation or exogenous GA_3 treatment, is required to stimulate flower bud development. Studies conducted by Schunch and Fuchigami (1992) determined that exogenously applied GA_3 compensated partially for insufficient water stress required to initiate flowering in coffee. This was viewed as indirect evidence that the degree of water stress had not been sufficient to induce GA production thought to be essential for flower opening.

As a general rule, in conifers (Pinaceae), flowering can be stimulated by hot and dry conditions, and an even stronger effect can be achieved if certain GAs are applied (Moritz et al. 1990 and references therein). Sitka spruce (*Picea sitchensis*) can be induced to flower by exposure to high-temperature and dry conditions (Moritz et al. 1990). Along with this induction process was a change in GA composition (Figure 4.44) . It was hy-

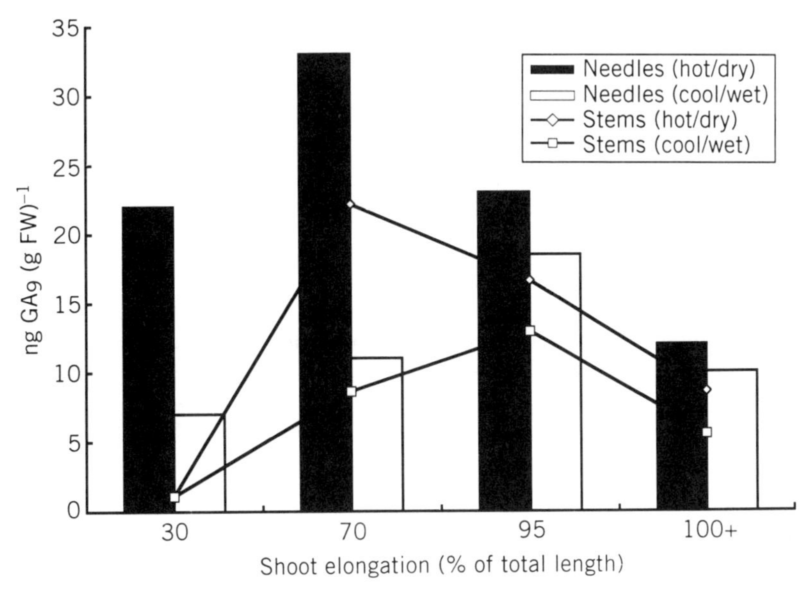

Figure 4.44 Effects of temperature and moisture conditions on GA levels in Sitka spruce at different stages of shoot elongation. (Modified from Moritz et al. 1990.)

pothesized from their results that conditions favorable for induction of reproduction (hot–dry) resulted in the synthesis of nonpolar GAs (GA_9 and GA_4), which appear to regulate flower induction. Conditions favoring vegetative growth (cold–wet conditions and/or poor flowering clones) tend to result in the formation of polar GAs (GA_1 and GA_3), which appear to be effectors of vegetative growth.

Bensen et al. (1990) provided evidence that low water potential causes a reduction in endogenous GA concentrations in the hypocotyls of dark grown soybean. This was particularly true of GA_1 and GA_{19} (Table 4.29). Saturated conditions resulted in increased

TABLE 4.29 Endogenous GA Content (pg GA mg^{-1}) of Dark-Grown Soybean Hypocotyls after Growth in Saturated (Wet) or Low ψ (Dry) Vermiculite

Sample	GA_1	GA_{20}	GA_{19}	GA_{53}
Experiment 1				
0 h	37.7	3.5	29.2	ND
24 h wet	95.5	1.8	32.1	ND
24 h dry	21.3	ND	9.8	ND
24 h dry + 2 h wet	20.8	1.1	7.4	ND
Experiment 2				
0 h	97.0	0.9	80.3	ND
24 h wet	196.7	trace	132.4	17.8
24 h dry	16.6	ND	trace	ND
60 h dry	65.2	trace	ND	ND

Source: Modified from Benson et al. 1990.

concentrations of these GAs. Trying to show some relationship between water stress–induced growth reductions in hypocotyls and GA levels, the authors concluded that changes in GA_1 and ABA levels, the latter of which increased under water-deficit stress, play some role in adjusting hypocotyl elongation rates but were not the primary effectors since the changes observed in the two hormones were not fast enough nor of sufficient magnitude to account for the changes observed in elongation rate and rapidly changing plant water status.

Expressing an interest in hormonal control of stomatal opening relative to water-deficit stress, Hubick et al. (1986b) conducted simultaneous analysis of GA, C=C, ABA, and CK in the roots and shoots of aeroponically grown sunflower plants. The results indicated no significant effects of water stress on GA and C=C levels in shoots and roots, but ABA increased considerably in shoots and roots. The levels of CKs declined in the shoots and increased in the roots. Considerably more could be accomplished relative to our understanding of how water-deficit stress affects physiological responses of plants through changes in GAs.

D. Drought and Abscisic Acid Levels

The most studied hormone with respect to drought stress is ABA. It is general knowledge that ABA increases in plant tissue when subjected to water-deficit stress or osmotic stress. There are many physiological implications of increases in ABA levels in plants, which will be the major focus of this section rather than biochemical or molecular implications. The student interested in the latter is referred to several review articles that address the subject (Bray 1991; Chandler and Robertson 1994; Taylor 1991; Thomas et al. 1991).

Some of the physiological implications of drought stress and ABA induction in plants include abscission (leaf, flower, fruit, and bud), dormancy [bud, seed, and perennation organs (i.e., tubers, corms, rhizomes)], flowering (inhibition and stimulation), stomatal closure, growth inhibition, increased root hydraulic conductance, leaf heterophylly, increased production of trichomes and spines, decreased tillering in grasses, increased root/shoot ratio, inhibition of germination, decreased pollen viability, decreased seed set, and acceleration of embryo maturation. Most of the foregoing responses can be induced either by water stress or by ABA application, and the two have been either weakly or strongly (of 25 listed responses, 4 were considered to be strongly, 13 moderately, and 8 weakly correlated) correlated relative to the induction of the response (Trewavas and Jones 1991).

One of the most studied of the physiological responses above is that of the regulation of stomatal closure by ABA. There appear to be two mechanisms by which plants may control stomatal closure when under stress. One mechanism involves the decompartmentalization of free ABA from the chloroplasts of the mesophyll cells into the apoplast and binding of the hormone to the exterior surface of the guard cell plasmalemma, resulting in membrane changes that initiate ion transport, resulting in loss of guard cell turgor and subsequent stomatal closure. This mechanism may be quite turgor sensitive in some plants and may be responsive to rapid changes in atmospheric water potentials. The second mechanism may be through ABA "signal" transduction via the root system of the plants. It is well established that plants subjected to drought stress produce high levels of ABA in the roots and can transport ABA via the xylem to the shoots. This hypothesis suggests that soil moisture levels dictate the degree of synthesis of ABA in the roots, and in turn ABA acts as a chemical signal that is transported to the leaves to initiate stomatal closure as described previously. Evidence for this hypothesis is supported by experiments conducted with plants in which split-root systems have been employed. One half of the root system is grown in pots containing soil that is fully hydrated, while the other half is grown in a separate pot containing soil that has been depleted of moisture. In this type of system the shoot and leaves do not loose turgor, but the stomates close and ABA accumulates, indicating that some signal originates in the roots and is transported to the shoots and affects stomatal closure. These experiments argue against leaf turgor as being implicated in stomatal closure, although both processes may operate, depending on species and environmental circumstances. In some cases, very large amounts of ABA can accumulate in the leaves of plants as a result of drought stress and can remain high for long periods of time even after the stress has been removed. In such plants, stomatal opening has been observed even though ABA levels remain elevated in the leaves. In other plants, there is an initial rise in ABA levels in the first few hours/days of stress and then a gradual decline of ABA to near control levels with stomates remaining closed. In addition, plants that have been previously drought stressed seem to be more sensitive to water deficit than plants not previously hardened. This is not always true, however.

Some of the best information generated relative to a direct role of ABA in stomatal closure has been demonstrated with the tomato mutant *flacca,* which is deficient in ABA synthesis. Under drought stress the stomates of *flacca* do not close, and consequently, the plant wilts. The application of exogenous ABA to the plant allows for stomatal closure and the plant does not exhibit wilting symptoms. Thus some role of ABA in stomatal closure in this plant seems apparent.

In summary, it appears that ABA is some how involved in stomatal closure in plants. This mechanism may differ with species, developmental stage, and/or environmental conditions, all of which may reflect inherent differences or changes in sensitivity to ABA. Sensitivity changes may be a function of species or initiated through regulation of the number of ABA receptors during development or exposure to differing environmental conditions. Sensitivity may also be under control of other hormones, such as gibberellins, which seem to counter the effects of ABA in many responses. However, stomatal control seems to be one of the exceptions. It is also possible that plants have a rapidly responding stomatal closure mechanism that is sensitive to turgor changes in leaves and quickly responds to adverse atmospheric changes and a second, slower mechanism that is linked to moisture conditions in the soil, which, depending on the type of soil, would lose water at a slower rate. The second mechanism appears to maintain a higher "titer"of ABA throughout the plant for a longer period of time. This may, in essence, maintain the plant in a hardened condition, which would allow for quicker responses to intermittent water-deficit stress and maintain a higher level of drought resistance.

Abscisic acid is well known for its growth inhibition properties. Most experiments have been pharmacological in nature, with little experimentation being done with growth correlation and endogenous levels of ABA. Growth inhibition of plants as a result of water-deficit stress thus may be a result of ABA accumulation, although turgor loss and/or reduced assimilation of carbon as a result of decreased stomatal conductance, or direct effects on photosynthesis could contribute. The information available on the subject of drought-induced ABA and its affects on growth is quite variable and inconclusive.

Abscisic acid levels in various cultivars of wheat subjected to drought stress correlated with drought resistance but not with plant height (Table 4.30) (Chumakovskii 1986). Using soybean hypocotyl elongation as a basis to study the effects of water stress on growth related to changes in ABA and GA levels, Bensen et al. (1990) demonstrated that GA_1

TABLE 4.30 Leaf Content of ABA in Leaves of Wheat Varieties with Different Stem Height and Drought Resistance (Booting Stage)[a]

	Height of Plants (cm) at End of Vegetation		Content of ABA (μg g^{-1} dry mass)		
Variety	60% TFM	30% TFM	60% TFM	30% TFM	Drought Resistance
Diamant	103.3	78.8	0.21	2.25	−
Saratovskaya 29	93.0	79.5	1.18	9.10	+
Grekum 25h84	77.8	59.0	2.05	7.02	−+
Red River 68	68.3	53.3	0.41	1.25	−
Siete Cerros	66.3	58.6	3.37	12.73	++

Source: Modified from Chumakovskii 1986.

[a]−, Drought unresistant; −+, weakly drought resistant; +, drought resistant; ++, strongly resistant.

and ABA levels declined and increased, respectively, in hypocotyls of soybean seedlings subjected to water-deficit stress. Rewatering caused a reversal of the stress effect and the authors concluded that the two hormones play a role in adjusting hypocotyl elongation rates, but changes in their levels do not correspond temporally with changes in growth rates of the hypocotyls.

Sakurai et al. (1985) found a significant correlation between the log concentration of endogenous ABA and the growth rate of squash seedling hypocotyls subjected to PEG osmotic stress. The accumulation of ABA was claimed to precede the reduction in growth rate (Figure 4.45). Studies comparing dwarf and tall varieties of different plant species in-

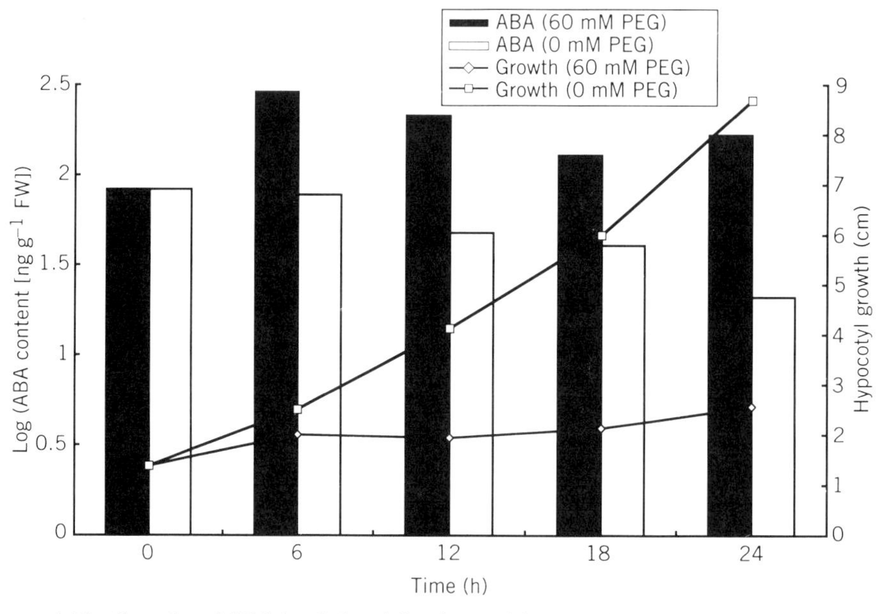

Figure 4.45 Growth and ABA levels in etiolated squash hypocotyls under water stress. (Modified from Sakurai et al. 1985.)

dicate little to no difference in ABA levels, and young actively growing seeds and leaves contain high levels of ABA, which is inconsistent with ABA growth inhibition (Bradford and Hsiao 1982 and references therein).

Furthermore, ABA has been reported to accumulate in apical meristems in some plants subjected to drought stress, suggesting to some that ABA may be involved in the induction of dormancy in meristems. However, in maize, drought stress has been reported to cause an accumulation of ABA in corn tassels, causing growth inhibition in that reproductive structure but releasing axillary inforescences for further growth and development (Aspinall 1980). This is suggestive of a role for ABA in controlling apical dominance of plants under stress.

Adventitious root formation was reported to occur in pea seedlings subjected to PEG-induced osmotic stress and was more pronounced under high light intensities (38 W m^{-2}) compared to low light (Rajagopal and Andersen 1980).

Other root responses to drought stress/ABA include increased hydraulic conductivity, ion flux in and out of roots, and increased root/shoot ratio (Aspinall 1980). The latter observation, as well as the previously mentioned effects of ABA on the stimulation of adventitious root formation in squash hypocotyls, suggest that ABA differs relative to its effects on shoot and root growth. The authors implicated elevated levels of ABA in the osmotically stressed plants to increased root development. The effect could also be induced by exogenous application of ABA.

Finally, the accumulation of ABA in the phloem–cambial region of balsam fir was hypothesized to interfere with IAA and carbohydrate transport to the cambial region, thus causing an inhibition of cambial activity in trees subjected to water stress (Little 1975).

Fruit, flower, leaf, and bud abscission are also responses of plants to drought. Abscisic acid has been implicated in this process along with IAA and C=C. As an example, Guinn and Brummett (1988) found that free IAA decreased in the abscission zones, while conjugated IAA increased in cotton fruits (bolls) and fruit abscission zones of plants subjected to moisture stress. Both free and conjugated forms of ABA increased. Associated with these changes was increased boll abscission. Rewatering caused a reversal of this response with better retention of cotton bolls. Accumulation of high levels of IAA esters in the abscission zone of cotton bolls of water-stressed plants was hypothesized as being a readily available source of free IAA that could potentially prevent abscission when the plant recovered from stress. In addition, it was proposed that increased C=C production in cotton bolls associated with drought stress could be induced by high ABA levels, resulting in decreased boll retention.

Although there is much evidence that supports a role for ABA in drought stress–induced physiological responses, some evidence supports geneotypic differences relative to the response of plants to exogenous applications of ABA. Trewavas and Jones (1991) cite several examples, including cultivar differences among soybeans relative to abscission, growth, and senescene responses (14 out of 34 responded), differences in abscission and growth of two races of *Acer rubrum,* and ranges in sensitivity (very sensitive to insensitive) of stomatal closure.

E. Drought and Ethylene Levels

Elevated levels of C=C in plants can have a significant impact on plant physiology and growth. High C=C levels have been correlated with reduced height growth, increases in stem diameter, abscission of leaves, flowers, buds and fruits, rates of fruit ripening, and accelerated senescence of tissues. Such responses have also been correlated with drought stress in plants. Thus interest in drought stress and how it affects C=C production and its relationship to the responses above has been of interest to some scientists. One of the techniques that has frequently been used to test the effects of desiccation on C=C production in plants is to remove plant parts, such as leaves or fruits, and allow them to dry for differing periods of time. The tissue or tissue part is then placed in a sealed container and over time the C=C levels are measured in the head gas, which is removed via a syringe inserted through a port sealed with a rubber septum. Usually, the result has been an increase in C=C concentration as the tissue dries. Another approach has been to stress the intact plant, then excise the plant part, place it in a sealed container, and measure the C=C levels as described above. The problem with this approach is that in sealed containers CO_2 can build up and O_2 depletion can occur, both of which can inhibit C=C production. Also, wound C=C is generated by the plant part as a result of the excission process. Some examples using these techniques follow.

Abscission of plant parts is a common response of plants to drought stress. Tudela and Primo-Millo (1992) found no significant increase in C=C levels in the leaves of Cleopatra Mandarin (*Citrus reshni* Hort. Ex Tan.) seedlings subjected to drought stress, and no leaf abscission was observed during the stress period. However, release of the stress resulted in elevated concentrations of C=C and ACC in the leaves 2 h after rehydration and was associated with the onset of leaf abscission. When roots or shoots were treated with either AOA (inhibits ACC synthase) or cobalt (inhibits EFE) prior to the stress period, leaf abscission was inhibited (exception being AOA application to shoots). The concentration of ACC was 10-fold higher in xylem sap than in controls after rehydration of stressed plants. It is hypothesized that ACC accumulated in the roots of water-stressed plants and was translocated to the shoots upon removal of the stress and oxidized to form C=C, which in turn promoted leaf abscission. Graves and Gladon (1985), using weeping fig (*Ficus benjamina*), also demonstrated that leaf abscission was induced in this plant 24 to 48 h after PEG-induced stress was started. However, most abscission occurred within the first 24 h after the stress was relieved.

Osmotic stresses corresponding to water potentials of −1, −11.7, −14.0, −16.8, and −18.3 bar all induced a surge of C=C production in fig leaves within 6 h of treatment (Figure 4.46). Concentrations declined and then proceeded to increase gradually through a 48-h time period. Correlated with the highest C=C levels after 48 h was the initiation of leaf abscission. No differences were observed among the stress treatments relative to C=C production or abscission. However, the authors did note what appeared to be a threshold osmotic stress level (225 to 265 g kg^{-1} PEG 8000), and any stress above this level produces the same amount of C=C and abscission of leaves.

Michelozzi et al. (1995) demonstrated clone differences in eucalyptus (*Eucalyptus grandis x E. camaldulensis;* clone 2814 and *E. grandis x E. robusta;* clone 2798) relative to water stress, C=C production, and leaf abscission. Clone 2814 and 2798 exhibited a six- and sevenfold increase in C=C, respectively, at leaf water potentials of −0.4 to −1.3 and −0.3 to −0.8 MPa. Ethylene levels then decreased to near below prestress levels at leaf water potentials of −1.2 to −2.5 MPa. Rewatering caused a large increase in C=C production in 2814 (100 to 1400 pL g^{-1} DW h^{-1}) which was associated with significant leaf abscission. In clone 2798, C=C production increased only slightly (200 to 270 pL g^{-1} DW h^{-1}) on rewatering without leaf drop. Clone 2814 seemed to avoid drought stress, as indicated by rapid closure of stomates, massive leaf abscission, and greater root biomass. Plantlets of clone 2798 appeared to tolerate drought by keeping

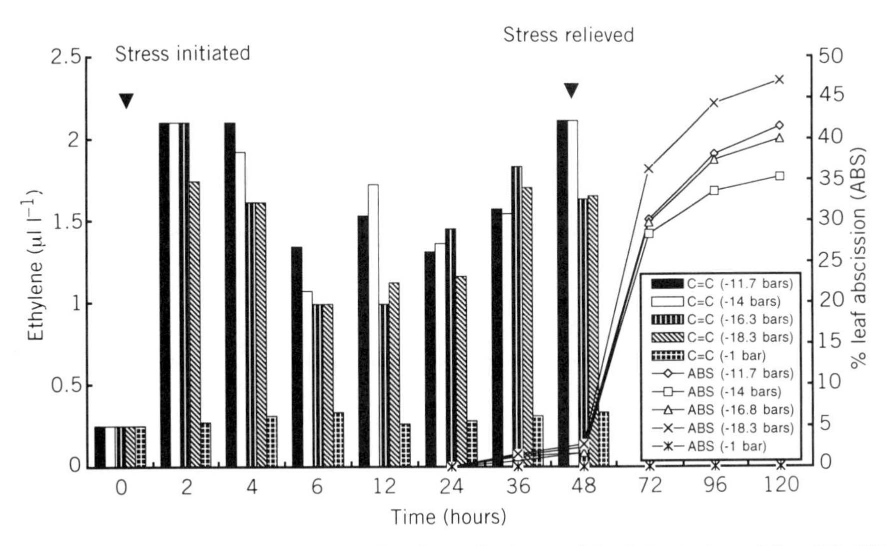

Figure 4.46 Effects of water stress on C=C production and leaf abscission of fig. (Modified from Graves and Gladon 1985.)

stomata open longer, maintaining higher chlorophyll concentration and leaf area during the stress period.

Ben-Yehoshua and Aloni (1974) demonstrated that C=C production in detached leaves and leaves of detached branches of Valencia orange *(Citrus sinensis)* maintained at 55% relative humidity exhibited significant increases within 2 to 4 h compared to leaves and branches maintained in a saturated mist chamber. Rehydration of plants previously stressed caused reductions in C=C levels back to control levels if the stress was relieved within 5 h. Rehydration after 10 h caused a reduction to intermediate levels observed between leaves kept in the mist chamber and leaves maintained at 55% RH. Attempts to rehydrate leaves after 20 h of stress resulted in continued increases in C=C concentrations in the leaves. Water stress caused the abscission of leaves from detached branches maintained under stress conditions within the first week after detachment.

Using intact cotton petioles, McMichael et al. (1972) measured C=C produced in these tissues utilizing a closed chamber positioned around the petiole and periodically removing gas samples from plants exposed to water-deficit stress. They found that as water stress increased so did C=C production by the petioles. The level of C=C produced could not be correlated with leaf abscission, but it is of interest that as in previous studies, abscission was initiated after rewatering of the stressed plants. Ethylene was measured in young (third node) and old petioles (twelfth node) of stressed plants and found to be slightly higher in young petioles in three separate experiments (2.75, 1, and 4 μL C=C kg^{-1} FW h^{-1} higher than in old petioles). The authors concluded that differing sensitivities of petioles as a result of age may be a more likely factor controlling abscission rather that absolute C=C concentration when plants are exposed to drought. They also propose that other hormones may have a regulatory role in the development of the abscission zone and when these become out of balance, the induction of the abscission layer in the petiole is initiated. It was further suggested that C=C may function in this mechanism by inhibiting the polar transport of IAA, which may initiate a series of events leading to abscission.

The examples above seem to have one thing in common. In most instances leaf abscission occurred during the rehydration phase of the experiment. The findings of Tudela and Primo-Millo (1992), where ACC was thought to be produced in high concentrations in the roots and transported to the shoots on rehydration with the effect that it increased C=C levels in the leaves with subsequent abscission, should possibly be explored further relative to a similar mechanism operating in other plant species.

Drought stress and light appear to interact to affect C=C production in plants. Rajagopal and Andersen (1980) found that high light caused an increase in ABA levels in the leaves of osmotically stressed pea leaves, whereas the levels of C=C declined under high light intensities. More severe osmotic stress resulted in higher C=C levels in the pea leaves. High light treatments caused a reduction to control levels after 6 h. Increased ABA levels and decreases in C=C concentrations were attributed to increased rooting potential of pea seedlings exposed to high light under osmotic stress.

High soil moisture stress also caused elevated levels of C=C and ABA in tomato leaves (Basiouny et al. 1993). However, in tomato, increased light intensity resulted in reduced levels of C=C and ABA. The intent of this study was to ascertain how light intensity and water-deficit stress interact to affect levels of ABA and C=C in tomato.

In broad bean (*Vicia faba*), C=C concentrations increased to a greater extent in osmotically (0 to -6 bar, varying strength of NH_4NO_3 nutrient solution) stressed leaves and stems of younger tissues than in older plant parts (Hall et al. 1977). Also, drought caused an initial increase in C=C and then a reduction to near control levels in leaves 3 h after the onset of drought. Ethylene then increased to 0.22 $\mu L\ L^{-1}$ after 9 h of stress compared to less than 0.02 $\mu L\ L^{-1}$ in the controls. Flower senescence was found to increase at a faster rate in *Freesia* inflorescenses that were cut and stored dry in the dark compared to inflorescenses that were maintained in deionized water (Spikman 1986). In both instances, C=C concentrations increased over time, but those stored dry exhibited higher levels of C=C up to about 4 days after the stress was initiated, and thereafter, concentrations declined and fell below those of nonstressed inflorescences. Stressed inflorescences exhibited a maximum in C=C production about 1 day earlier than nonstressed plants.

Water stress is used to break dormancy in coffee flower buds and initiate the reproductive cycle (Schuch and Fuchigami 1992). Application of GA_3 can also aid in this process, but C=C was shown to decrease in the flower buds of coffee under drought stress of -2.65 and -3.50 compared to -1.20 and -1.75 MPa (Figure 4.47). Relief of the stress resulted in increased levels of C=C production in plants subjected to the latter two water potentials; however, C=C production did not increase in the former two water potentials until 9 days after stress relief. Unlike the flower buds, the two highest water potentials caused an increase in C=C concentrations in leaf discs of coffee trees.

Seasonal variation in C=C concentration of sapwood and heartwood (Figure 4.48) was followed in stems of 70 to 100-year-old Scots pine (*Pinus sylvestris*) (Ingemarsson et al. 1991). Ethylene concentrations rose to 3 to 7 ppm in sap wood during the growing season and fell to 0.1 to 0.3 ppm during the winter. In response to extreme drought, C=C increased to 30 ppm and then declined following a rainy period (Figure 4.49).

Some tomato cultivars (*L. esulentum* cv. 'Hossen') are susceptible to drought stress in that they exhibit a condition referred to as *pithiness,* in which stems undergo autolysis of the stem pith (Huberman et al. 1993). Drought stress caused an increase in apoplastic polygalacturonase and cellulase. Reirrigation caused a large transient increase in the former and a decrease in the latter. Mechanical perturbation (MP) and drought stress cause large increases in C=C concentrations in the pith tissues of this tomato cultivar. However, the latter is more effective. When MP preceded drought stress, the amount of C=C produced was significantly less than drought stress alone. When MP or ethephon (an

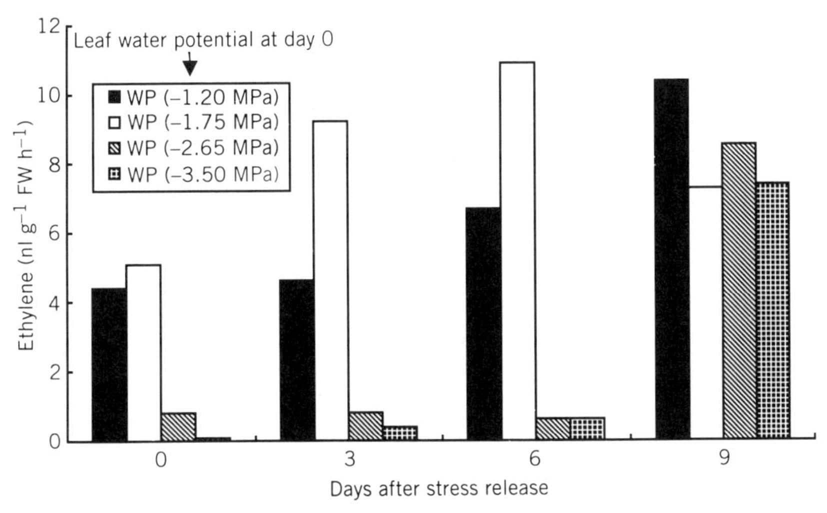

Figure 4.47 Effects of water stress on C=C evolution from flower buds 0, 3, 6, and 9 days after irrigating to release water stress. (Modified from Schunch and Fuchigami 1992.)

C=C-generating chemical) treatments preceded drought stress, the concentrations of apoplastic cellulytic enzymes declined. It was concluded that drought stress induces C=C production, which in turn stimulates apoplastic cellulytic enzymes, which results in pithiness in tomato.

It should be pointed out that the use of plant parts in closed systems to measure C=C

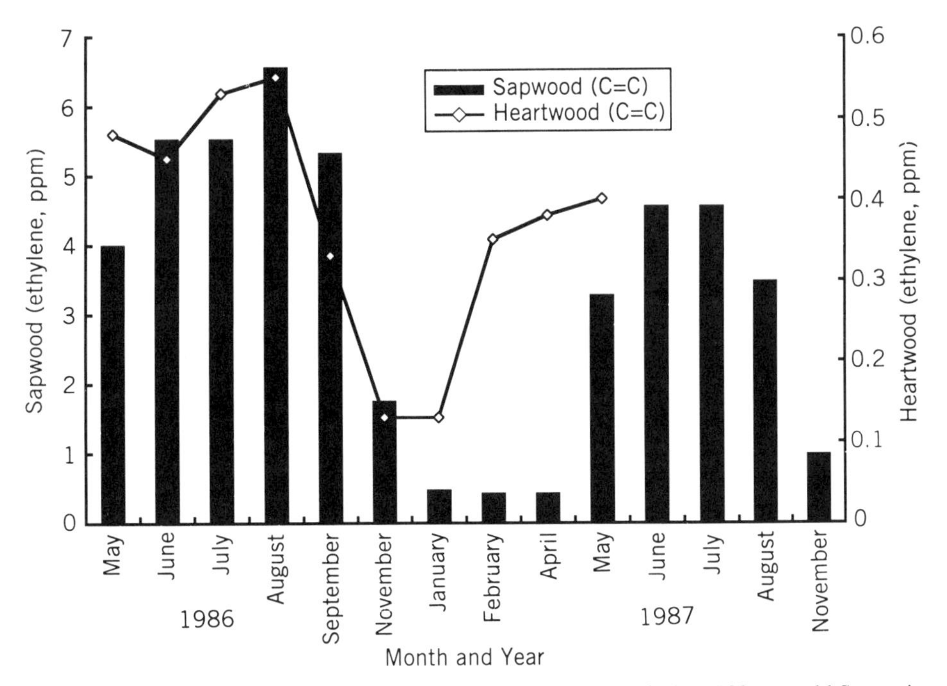

Figure 4.48 Seasonal variation in C=C concentration in boles of 70 to 100-year-old Scots pine. (Modified from Ingemarsson et al. 1991.)

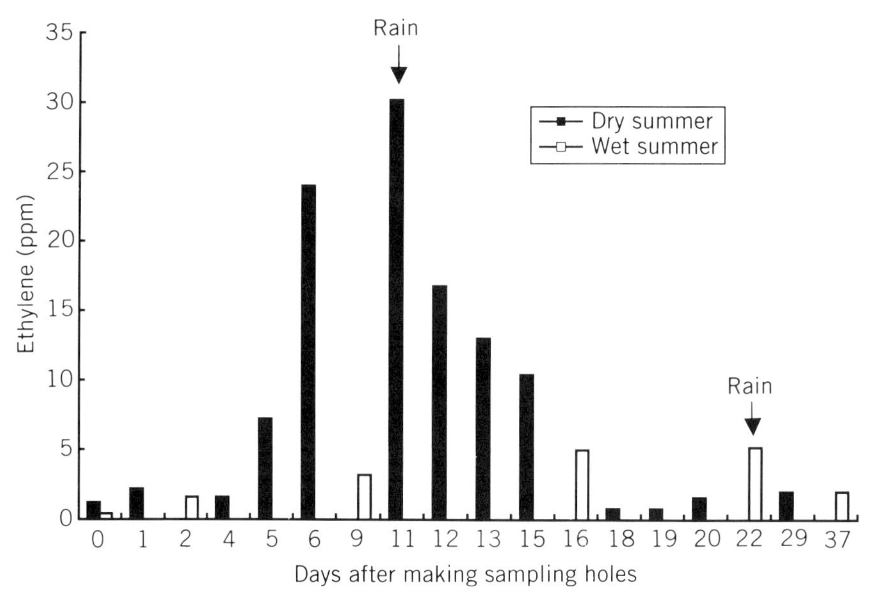

Figure 4.49 Effects of rainy period during a dry and wet season on C═C production in Scots pine. (Modified from Ingermarsson et al. 1991.)

production may result in artifact production relative to what actually occurs in intact plants. Using intact whole wheat seedlings to determine the effects of drought stress on C═C production, Narayana et al. (1991) demonstrated that in a continuous flow system, C═C production did not change significantly in stressed plants at −2.3 MPa compared to control plants. This technique was compared with previous methods where leaves were excised, dried to some predetermined water content, and then placed in sealed containers and C═C withdrawn and analyzed over time. In these experiments the authors found significant increases in C═C production in dried leaves compared to nondried leaves, with younger leaves producing much higher levels than older leaves. They also observed that excised leaves from intact plants stressed for 2, 4, and 6 days actually generated less C═C (determined as total C═C accumulated in closed vessels) the longer the duration of the stress. From these experiments the authors conclude that the utilization of excised plant tissues and measuring C═C production in closed vessels can result in artifactual data that are not representative of how drought stress may influence C═C levels in intact plants.

VII. TEMPERATURE STRESS AND PHYTOHORMONES

A. Temperature and Indoleacetic Acid Levels

Very little information is available regarding the effects of temperature on IAA levels in plants. Considering the known interactions of IAA and C═C (IAA enhances C═C production) it is surprising that more information is not known, particularly in the area of postharvest physiology, where the literature on the effects of temperature on C═C is extensive.

Indoleacetic acid and GA levels were measured in floral buds, open flowers, and young fruits of two tomato cultivars (L387 and CL-lld) subjected to 38°C for 5 h after seedlings had formed the second inflorescence (Kuo and Tsai 1984). Hormone levels

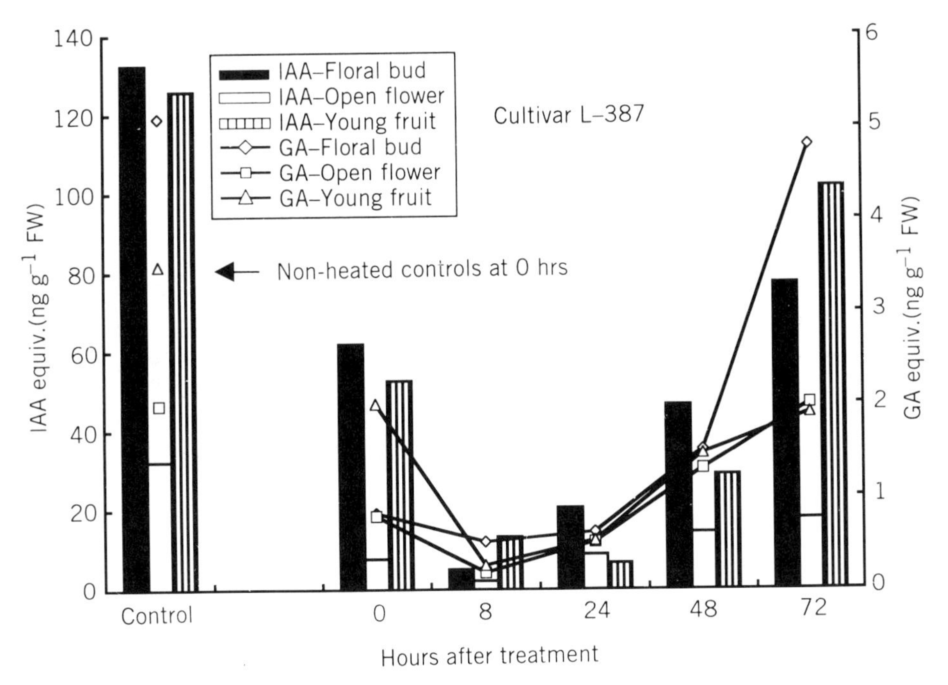

Figure 4.50 Effects of heat treatment (5 h at 38°C) on IAA and GA-like activity in tomato reproductive organs. (Modified from Kuo and Tsai 1984.)

were measured at 0, 8, 24, 48, and 72 h posttreatment. Indoleacetic acid levels were greatly reduced immediately after treatment and continued to decline after 8 h. Levels began to recover 24 h after treatment and continued to increase up to 72 h (Figure 4.50). After 72 h the levels were still lower than the controls. Both cultivars responded in a similar manner. Gibberellic acid concentrations followed a pattern similar to that of IAA concentrations. However, GA concentrations were several magnitudes lower than IAA. Recovery of GA levels in the cultivar CL-lld were not as rapid as L-387, particularly for floral buds. The importance of these determinations relates to the involvement of IAA and GA in fruit set and development. High temperatures are known to cause reductions in tomato fruit set, and it has been demonstrated that exogenous applications of IAA can improve fruit set under high-temperature stress.

Increasing temperature was also found to reduce CK and IAA levels in two morphologically distinct tissue culture lines of tobacco (Table 4.31) (Pence and Caruso 1986).

TABLE 4.31 Auxin and CK Levels in Leafy and White Tobacco Tumors Cultured at Three Temperatures

Hormone	Leafy Tissue			White Tissue		
Concentration	21°C	27°C	33°C	21°C	27°C	33°C
ZR (ng g^{-1} FW)	2818	251	16	1987	210	81
IAA (ng g^{-1} FW)	176	37	<3	135	42	15
Molar ratios	8.0	3.4	2.7	7.4	2.6	2.7

Source: Modified from Pence and Caruso 1986.

One line, designated as leafy, produced green abnormal shoots, while the other, white, produced hard spherical tissue. Manipulating tissue culture differentiation with differing ratios of CK and IAA has been used to induce shoot or root production. However, in the study above, increasing temperature caused a reduction in both IAA and CK and differences in ratio were not observed at the temperature where the distinct morphological changes are normally observed, suggesting either that very subtle differences in hormone ratios may be required or that other factors may be involved.

Indoleacetic acid oxidase is an enzyme thought to control the levels of free IAA in plants. Thus it has been assumed that environmental factors such as temperature that affect IAA oxidase would in turn affect IAA levels (high levels of the enzyme would result in low concentrations of IAA, and the reverse with low levels of the enzyme). Bolduc et al. (1970) and Omran (1980) observed that cold treatment caused an increase in IAA oxidase in wheat (2°C) and cucumber seedlings (5°C), respectively. Seed treatment of wheat with GA resulted in a higher activity of IAA oxidase in seedlings than in plants derived from non-GA-treated seeds (Figure 4.51) (Bolduc et al. 1970). Treatment of seeds with AMO-1618, a known GA synthesis inhibitor, caused a reduction in IAA oxidase activity in wheat seedlings subjected to chilling. The findings above both suggest a role for GA in regulating IAA oxidase activity and subsequent IAA levels.

B. Temperature and Cytokinin Levels

As alluded to in the preceding section, high-temperature stress can cause a reduction in CK levels as well as IAA in morphologically distinct tissue culture lines of tobacco (Pence and Caruso 1986). However, no obvious differences were observed in the ratios of the two hormones that would account for the change in morphology of the two lines.

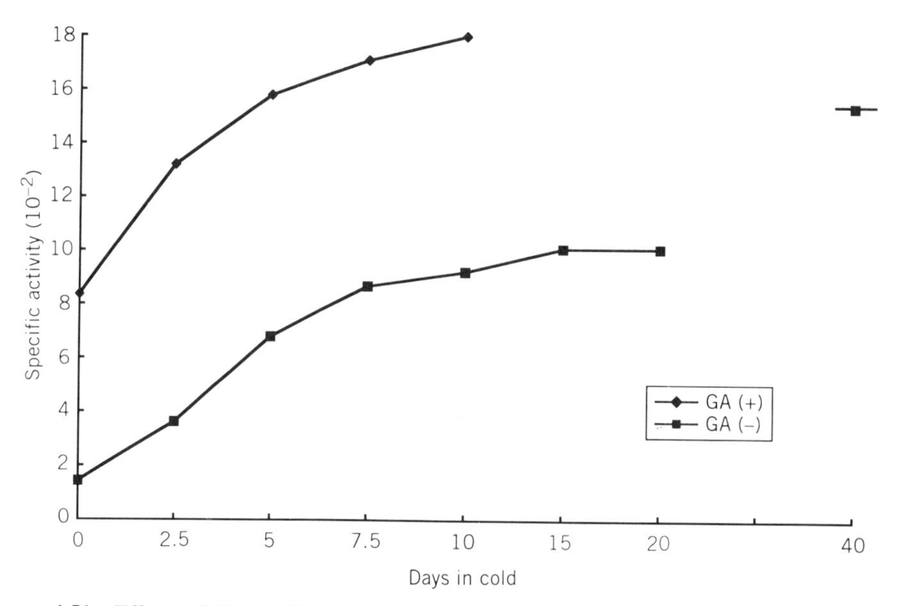

Figure 4.51 Effects of GA on IAA oxidase activity in wheat seedlings grown at room and cold temperatures. (Modified from Bolduc et al. 1970.)

TABLE 4.32 Cytokinin Levels in Tumor Lines Cultured at 27°C and 34°C

	Cytokinin Concentration (nmol zeatin equiv. kg^{-1} FW)	
Tumor Line	27°C	34°C
W38T37	25	6
W38EU6	40	8
W38B6	70	7

Source: Modified from Amasino and Miller 1983.

Amasino and Miller (1983) also observed that CK levels declined substantially when the growth temperature of crown gall teratoma tissue is raised from 27°C to 34°C (Table 4.32). Associated with the reduced CK levels was a change in tissue morphology toward an expression of increased shoot elongation and apical dominance. One line exhibited increased root formation. Exogenous application of CK to tissues grown at 34°C prevented the observed changes in morphology and suggested to the authors that high CK levels were responsible for preventing root formation and the lack of apical dominance in tumor shoots.

Heat stress (45.5°C for 2 min) applied to the roots of etiolated or green corn seedlings resulted in differing effects on leaf CK levels. Compared to nonstressed controls, etiolated plants subjected to heat stress exhibited no differences in ZR and no detectable levels of glu-ZR (Caers et al. 1985). Green seedlings, on the other hand, exhibited no detectable levels of ZR in either treated or control plants, but heat caused a sixfold decrease in glu-ZR. In etiolated plants, heat stress also inhibited photosynthesis, chlorophyll accumulation, and chloroplast development. In green plants, only photosynthesis and chlorophyll accumulation were inhibited. In both instances, exogenous applications of benzyladenine could overcome the effects of heat stress. Benzyladenine applied to the controls of both etolated and green seedlings also enhanced photosynthesis and chlorophyll production. It was proposed that etiolated plants have limiting concentrations of CK and low capacities for photosynthesis since chloroplasts are not well developed. Heat stress further accentuates the problem. In the case of green plants, heat-induced reductions in CK levels become limiting and are associated with reductions in photosynthetic activity, chlorophyll production, and chloroplast development. The ability to reverse these processes in both green and etiolated tissues suggests an important role for CKs in chloroplast morphogensis that is thermally controlled.

Heat stress has also been shown to affect corn kernel development and the response was correlated with a heat-induced reduction in CK and an increase in ABA levels (Cheikh and Jones 1994). The authors found that zeatin levels fell to zero when in vitro cultured corn kernels were subjected to a heat treatment of 35°C for 4 or 8 days. Eight-day treatments also caused no detectable levels of ZR 10 days after pollination. However, 4-day treatments resulted in some accumulation of ZR but not to the levels of control plants maintained at 25°C. Stem infusion of benzyladenine was found to reverse the heat-induced abortion and development of corn kernels in intact corn plants. This led the authors to conclude that a shift in hormone balance is an important mechanism by which heat stress disrupts corn kernel development and that CKs play a pivotal role in the establishment of kernel sink potential and thermostability of kernel development.

Apple trees (*Malus domestica*) were subjected to all combinations of 10°C and 20°C root and shoot temperatures after an initial chilling requirement of 5°C. Xylem sap concentrations were found to be higher when shoot temperatures were 10°C. In all treatments, CK levels initially increased but then began to fall over a 30-day monitoring period. Bud break was found to be sooner and to progress at a faster rate when CK levels were lower (shoots at 20°C). When both roots and shoots were subjected to 20°C, CK levels initially increased, then fell abruptly after 20 days and then began to increase again. The low point in CK levels corresponded with the highest level of bud break. Root temperature had no effect on CK levels, and bud break was viewed as being correlated with decreases in CK levels.

In an effort to understand hormonal control of cold hardening in wheat, Taylor et al. (1990) measured CK and ABA levels in field- and growth-chamber-grown winter wheat plants. They observed that hardening occurred in the field in October and November, and ABA levels increased while CK levels declined. In mid to late winter the reverse started to occur. These changes in hormone ratios were also correlated with cold resistance in the plants. Growth chamber experiments also revealed that ABA concentrations increased in cold-hardened plants and CK concentrations declined (Table 4.33). It was concluded that CKs do not play a role in the hardening of winter wheat but may be involved in the dehardening process and the resumption of growth in the spring. It is clear that temperature affects CK levels in plants and more research needs to be conducted to determine the physiological implications of these effects.

C. Temperature and Gibberellin Levels

It has long been known that seed germination and flowering in some plants have a cold as well as a light requirement. In some instances GA can substitute for either or both of these environmental factors. Thus it is reasonable to assume that temperature can affect the synthesis of GA and influence physiological responses in plants. However, the mechanisms of how GAs influence such activities is poorly understood. Hazebroek and Metzger (1990) established that cold treatment (6°C) of field pennycress (*Thlaspi arvense* L.), a winter annual with a cold requirement for stem elongation and flowering, resulted in a greater rate of metabolism of [2*H*]-ent-kaurenoic acid (KA) in shoot tips than in noninduced plants. In the latter, KA accumulated to a level 47 times greater than the thermoinduced plants, indicating a very slow turnover rate. Gibberellin A_9 was the only GA in the

TABLE 4.33 Endogenous ABA (ng $^{-1}$ g DW) and CK (ng zeatin equiv. g^{-1} DW) Levels in Crown and Leaf Tissue of Winter Wheat Seedlings Exposed to Hardening or Nonhardening Conditions in a Growth Chamber

	ABA	CKs	
Growth Conditions	Crown Tissue	Leaf Tissue	Crown Tissue
Nonhardening	13.9	41.2	22.8
Hardening	32.6	104.1	19.0

Source: Modified from Taylor et al. 1990.

shoot tip that was shown to incorporate deuterium and was only detected in induced tips. Equal incorporation of labeled GA_{20} into induced and noninduced leaves of the plant led the authors to conclude that conversion of KA to GAs is under thermoinductive control and is localized in the shoot tip, which is the site of perception of thermoinduction. The authors also conclude that cold-induced synthesis of GA_9 is responsible for shoot elongation in *T. arvense.* In a related study, Hazebroek et al. (1993) established that KA levels declined 50-fold in shoot tips 10 days after vernalization compared to nonvernalized plants. The levels of KA in the leaves were not affected. Activity of the enzyme for the conversion of KA to 7-OH-kaurenoic acid was enhanced under thermoinduction, as was the oxidation reaction (ent-kaurene $\longrightarrow$ ent-kaurenol). These studies verify that reduced temperature can directly affect the biosynthesis of specific GAs (GA_9), in localized tissues (shoot tips), for a specific function (shoot elongation and floral induction).

In tomato, the number of flowers in the first inflorescence was increased by low temperature but reduced by the application of GA_3 (Abdul and Harris 1978). The effect of GA_3 was greater at low temperature (12°C) than it was at normal growth temperatures (16°C). Removal of leaves at normal temperatures caused an increase in inflorescence flowers. This was not true if leaves were removed at low temperatures or GA_3 was applied at low temperature. Analysis of leaves grown at low and high temperature revealed elevated GA-like activity in leaves of plants grown at normal temperatures versus plants grown at reduced temperatures. These results implied that GA produced in the leaves at normal temperatures reduced flower number in tomato infloresences and that reduced temperature caused a reduction in GA production in the leaves which allowed for greater flower numbers. Although not addressed in this research, it is also possible that a specific GA is produced under reduced temperature, as was the case in the previous example with field pennycress (Hazebroek and Metzger 1990).

The concept described above may also apply to *Eucalyptus nitens,* where Moncur and Hasan (1994) demonstrated that the number of flower buds formed in this plant was negatively correlated with GA concentrations. Use of the GA inhibitor paclobutrazol was found to stimulate floral induction at a younger age when exposed to cold treatment. Low GA concentrations were viewed not to be the only factor responsible for flowering since paclobutrazol-treated plants exposed to varying low temperatures had similar amounts of GA, and only plants receiving maximum cold exposure produced flower buds. The authors' purpose in this study was to find a way to reduce the time to flowering in *E. nitens,* because domestication in this species is limited by the shortage of seed production.

Low temperature (2°C for 7 days) was found to decrease $GA_{1,4,7}$ in seeds of GA-insensitive mutants (*gai*) of *Arabidopsis thaliana.* Chilling had no effect on GA levels in the wild-type seed, but light was required for germination. Gibberellic acid levels ranged from about 7- to 10-fold higher in *gai* mutants germinated in the light than in wild-type seeds germinated in the light. However, percent germination in the light of the latter was 94%, while *gai* germination in the light was 14% even though *gai* had higher concentrations of GAs. Chilling increased germination in *gai* seeds but caused a reduction in all GAs, suggesting that the concentration of GA was not decisive relative to germination, but sensitivity to GA may be more important. It thus appears that light and chilling interact to influence sensitive to GA, which in turn may affect GA. Since no obvious difference was observed in relative ratios of the GAs identified in the various chilling and light treatments, it is doubtful that germination can be attributed to shifts in the metabolism of any single GA.

D. Temperature and Abscisic Acid Levels

Abscisic acid has been studied extensively relative to its role in temperature stress. Many studies have shown that ABA increases under extremes in temperature (high and low) and have correlated changes in ABA levels with tolerance to temperature extremes. Applications of exogenous ABA to plants has met with some degree of success in inducing heat and cold tolerance to a variety of different plants. However, such treatments and correlations do not always hold true for all plants. The following examples will give some perspective of the types of temperature studies being conducted with respect to ABA and the importance of such studies relative to understanding the role of ABA in the response of plants to temperature stress.

1. High-Temperature Stress. Five-week-old tomato seedlings exposed to diurnal temperatures, ranging from low to high (day/night = 10/5, 15/10, 25/15, 35/25, 45/35°C), exhibited the greatest increases in free and conjugated ABA at the two temperature extremes, compared to controls (25/15°C). Water potentials of the plants were maintained at a constant level to eliminate the influence of plant water status on changes in ABA levels. It was concluded that heat per se can cause an increase in plant ABA levels and that high and low temperatures are effective in changing the balance of this hormone, suggesting that the magnitude of the stress determines the level of ABA in tomato, leading to speculation that ABA is involved in increasing plant tolerance to temperature stress (Daie and Campbell 1981).

Two grape cultivars (Venus and Veeblanc) exhibited differences in heat tolerance based on electrolyte leakage (conducted at 42°C) of leaf tissues (plants grown at 38°C) and cell tissue culture (grown at 36°C, Abass and Rajashekar 1993). Leaf and tissue culture cells of Venus were more tolerant to heat stress than Veeblanc. Abscisic acid levels (free and conjugated) increased prior to either cultivar reaching maximum heat tolerance. Levels of ABA were two- to threefold higher prior to maximum heat tolerance for the respective cultivars. Heat tolerance could be increased in cultured cells of both cultivars after 24 h of exogenous ABA treatments (7.6 and 9.5 μM).

Winter wheat (Avalon) and spring wheat varieties (Highbury) were acclimated to 6°C for 60 days and then some reacclimated to 25°C for 18 h in the dark (Ward and Lawlor 1990). Photosynthesis, stomatal conductance, C_i (calculated CO_2 pressure in substomatal air space), and leaf ABA concentrations were then determined in nonacclimated (plants at 6°C) and acclimated (25°C) plants. Photosynthesis, stomatal conductance, and C_i were measured over a temperature range of 5 to 30°C at 5°C increments. Plants reacclimated to 25°C (both varieties) showed increased rates of photosynthesis, up to about 20°C, and then P_n began to fall. Net photosynthesis of nonacclimated plants was lower; however the P_n value of Avalon fell off to near zero levels above 15°C, while the P_n value of Highbury was near optimum at 30°C. Somatal conductance showed similar trends and C_i did not change significantly. Abscisic acid concentrations declined in Avalon at 25°C and were negatively correlated with increased P_n at the same temperature (Figure 4.52). In addition, in exogenous application of ABA into the transpiration stream of detached leaves of acclimated (25°C), Avalon and Highbury responded similar to nonacclimated attached leaves in that Avalon exhibited a rapid decline in P_n and stomatal conductance when treated with ABA and increasing temperature, whereas Highbury did not. The authors concluded that ABA regulated the observed temperature acclimatization of apparent photosynthesis in the leaves of cold-grown wheat plants.

Several characteristic affects of ABA on plants is to reduce growth, increase senescence, inhibit photosynthesis, and subsequently, reduce yields. These affects are not necessarily contrary to the ability of ABA to increase tolerance to extremes in temperature but may actually be part of the mechanism for doing so. However, with respect to yield increases, high levels of ABA could be detrimental to yield, as has been mentioned for corn (Cheikh and Jones 1994). Lu et al. (1989) approached the issue of heat and osmotic stress effects on yield of wheat by screening for ABA-insensitive clones derived from wheat tissue culture callus and regenerating seedlings from these clones and testing them with respect to various yield parameters under heat and osmotic stress. They ultimately determined that those clones that were ABA insensitive grew longer and produced higher yields under stress than did more sensitive clones. These results suggest that stress-induced ABA could be detrimental to some plants but may not reflect the importance of ABA in long-term or extreme stress.

Seed dormancy has frequently been associated with high ABA levels in seeds. However, Ozga and Dennis (1991) were not able consistently to correlate ABA levels with germination capacity in apple seeds during stratification or after exposure to high temperature (30°C). However, it was shown that stratification at 5°C caused a reduction in seed coat ABA compared to 20°C, and transfer of seeds to 30°C for 3 and 6 days caused a further decline in ABA levels. Increase in stratification time caused an increase in ABA levels regardless of stratification (20 or 5°C) or posttemperature incubation temperature (30°C for 3 or 6 days). Stratification temperature also caused a reduction in ABA levels in cotyledons, but no consistent effects were observed relative to time of stratification. No differences were observed in embryonic axes relative to time or temperature of stratification. As indicated in the section on CK and temperature effects, ABA increases in corn

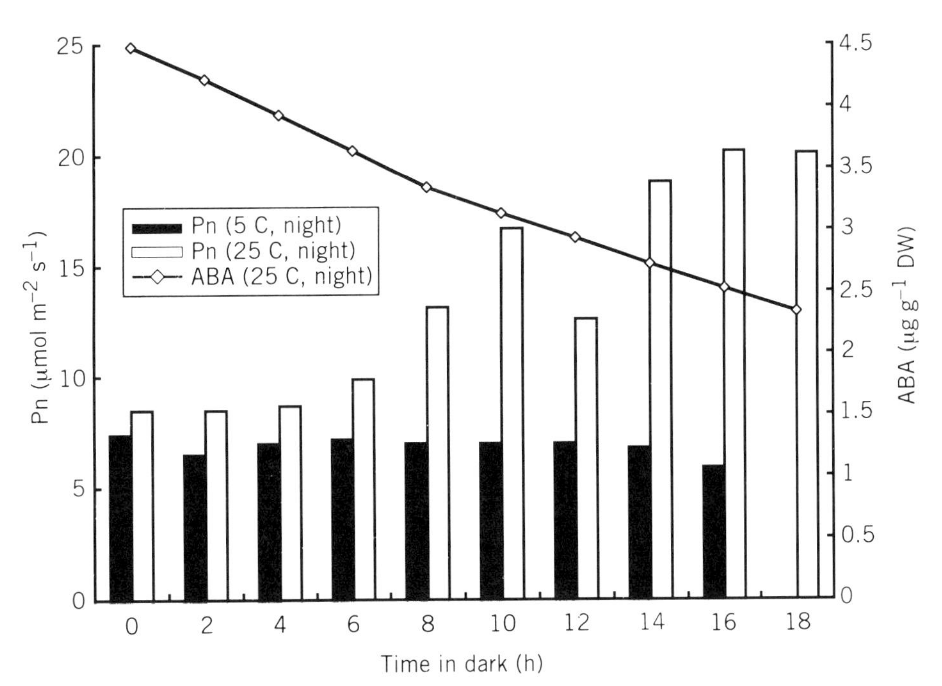

Figure 4.52 Effects of warm-night treatment (25°C night) on net photosynthetic rate and leaf ABA content in Avalon. (Modified from Ward and Lawlor 1990.)

kernels subjected to heat stress and CK levels decline (Cheikh and Jones 1994) and infusion of heat-stressed corn plants with CK increased yield. High ABA and low CK levels as a result of heat stress were proposed to cause increased seed abortion and lower yields in corn.

*2. **Low-Temperature Stress.*** It is difficult to separate the direct effects of temperature on ABA levels from the effects of temperature on plant water relations and subsequent ABA levels. Maintaining adequate soil moisture content and high relative humidities to reduce transpiration rates are usually employed in an effort to prevent water-deficit-induced ABA production during temperature experiments. Vernieri et al. (1991) were able to show that chilling stress (4°C) and drought resulted in very different responses relative to ABA production in bean leaves. It was observed that ABA levels increased more quickly in droughted plants compared to chilled plants (maintained at low relative humidity), even though chilled plants had a lower relative water content (RWC) over the time course of the experiment. Levels of ABA started increasing after 12 h in drought-stressed plants, while ABA did not start to increase until 24 h posttreatment. When plants were subjected to similar treatments in a high-relative humidity environment, no water stress was observed in chilled or nonchilled controls over the time course of the experiment. Abscisic acid levels did not increase in either chilled or nonchilled plants after 48 h. Seventy-two hours into the experiment, an increase in ABA was observed in the chilled plants but not in the nonchilled controls. This suggested to the authors that chilling stress can induce ABA production in the absence of water stress. Using bean leaf discs exposed to differing osmotic potentials at 4°C and 25°C, the authors were able to show that ABA synthesis was stimulated at the higher temperature with decreasing water potential and that at low temperature ABA levels were very low and did not change.

In another experiment (Vernieri et al. 1994) using the same system, the authors tried to determine if conjugated ABA served as a source of free ABA in chilled or drought-stressed bean plants. They concluded that very little change in the ABA–glucose ester occurred either under chilling or drought stress and that conjugated ABA played an insignificant role in ABA production during stress. A similar response was observed in ABA production in chilled plants compared to drought-stressed plants, as was observed in their previous paper (Vernieri et al. 1991) (i.e., slower development of ABA in chilling stressed plants than in drought-stressed plants). However, another difference that was not alluded to by the authors was that once ABA was produced in the chilling-stressed plants, the RWC, water potential, and pressure potential of the plant also increased, which was not the case for the drought-stressed plant.

Freezing tolerance in several cactus species was related to ABA levels and the ability of freeze-tolerant cacti to remove water quickly from the chlorenchyma tissues (Loik and Nobel 1993). Two freezing-sensitive species (*Ferocatus viridescens* and *Opuntia ficus-indica*) and a cold-hardy species (*Opuntia fragilis*) were shifted from a day/night temperature regime of 30/20°C to 10/0°C, which caused a freezing tolerance adjustment of 2°C for the sensitive species and 14.6°C for *O. fragilis*. After 14 days of the cold treatment ABA levels increased from <0.4 pmol g^{-1} FW to 84 pmol g^{-1} FW for *O. ficus-indica*, and 49 pmol g^{-1} FW for *O. fragilis*. Four days after exogenous application of ABA at 30/20°C, freezing tolerance was enhanced in *F. viridescens*, *O. ficus-indica*, and *O. fragilis* by 0.5, 4.1, and 23.4°C. The time course for acquisition of freezing tolerance after 14 days of acclimation to low temperature was similar to ABA treatments at moderate temperatures. It was concluded that water adjustments and increases in ABA may be important in low-temperature acclimation, particularly for the cold-tolerant species.

In wheat plants, the role of ABA in cold tolerance is unclear. In some cultivars ABA has been correlated with increased hardening (Taylor et al. 1990), while in others the levels of ABA actually declined during the hardening process (Dallaire et al. 1994). Taylor et al. (1990) measured the levels of ABA and CK in wheat cultivar, Norstar, grown in the field and in growth chamber experiments. In both instances they observed that ABA increased during the months of hardening while CK decreased. During the dehardening process, CK increased and ABA declined. In growth chamber experiments the same response was observed. This led the authors to conclude that ABA is involved with the hardening process and that CK initiates dehardening, leading to growth during the spring.

Dallaire et al. (1994) used a spring variety (Glenlea) and a winter variety (Fredrick) of wheat to study the relationship of ABA to freezing tolerance in these plants. They determined that exogenous application of ABA had little effect on increasing freezing tolerance in either of these varieties (3°C for both). Entry of the chemical into the plant was verified using radiolabeled ABA. Only after acclimation at reduced temperature could the maximum freezing tolerance be induced in the respective varieties. Measurement of ABA in leaves, crowns, and roots of both varieties grown under hardening conditions revealed that ABA levels actually declined in the leaves and crown of both varieties (Table 4.34). Roots also showed a tendency to decline as well but not to the extent of the other tissues. Using winter wheat calli (source not indicated) the authors could induce thermotolerance at 24°C to the same extent as cold acclimation. However, as in intact plants, ABA did not accumulate under reduced temperature acclimation. The authors conclude that in these varieties of wheat freezing tolerance by low temperature is not associated with increased levels of ABA. Further, the authors determined that a specific 32-kD protein was induced by ABA in calli that was specific to it but different from proteins induced by low temperature. This suggested to the authors that two separate mechanisms induced cold tolerance in calli.

Exposure of *Arabidopsis thaliana* to low temperature, exogenous ABA, and drought resulted in a 4 to 5°C increase in freezing tolerance. Low temperature and ABA treatments caused only a minor decrease in water potential compared to drought. Also, only minor and transient increases in ABA were detected in LT treatments compared to drought. Previously, it was shown that expression of the *rab*18 gene in *A. thaliana* in response to LT and drought was ABA mediated (Lång and Palva 1992). The *rab*18 mRNA that resulted from these stresses was six- to 10-fold lower in LT-treated wild-type plants than in ABA- or drought-treated plants. In addition, the mutants of *A. thaliana, abi*1, and

TABLE 4.34 Endogenous ABA Content (ng g^{-1} DW) in Leaf, Crown, and Root Tissues of Two Wheat Genotypes During Acclimation at Low Temperature

	Leaves		Crowns		Roots	
Treatment[a]	Glenlea	Fredrick	Glenlea	Fredrick	Glenlea	Fredrick
Control	146	182	109	213	137	191
LT, 1 day	101	102	134	99	137	72
LT, 8 days	95	81	115	59	113	81
LT, 28 days	77	65	82	61	122	108
Water stress, 4 days	1207	1006	ND	ND	ND	ND

Source: Modified from Dallaire et al. 1994.

[a]LT, low temperature (6/2°C day/night), 10-h photoperiod; control, 24/20°C day/night, 15-h photoperiod.

aba-1 are insensitive to and deficient in ABA, respectively. Both have impaired freezing tolerance, even though *abi*1 produces twice as much ABA as the wild type and maintains high levels throughout the LT period. Contrary to low-level induction by LT of the *rab*18 gene, *lti*78, a gene recently discovered to be induced in overwintering or annual plants, can be induced by LT in wild-type, *abi*1, and *aba*-1 *A. thaliana*. Thus the expression of this gene appears not to be ABA mediated. This lead the authors to conclude that partially different acclimation mechanisms to freezing temperatures coexist in *A. thaliana* and that ABA may have at least an indirect role in freezing tolerance, as implied by deficiencies in freezing tolerance of ABA mutants.

Dieback in trees is probably caused from a number of interacting factors, including diseases, acid rain, poor nutrition, and environmental extremes such as very cold winters, which could lead to root damage through dehydration or freezing stress. In maple trees (*Acer saccharum* Marsh) soil freezing and drought were studied with respect to spring sap ABA levels and the possible relationship of ABA to decline in maple (Bertrand et al. 1994). Freezing and drought were controlled by preventing precipitation (snow or rain) from reaching the forest floor. Lack of snow cover resulted in soil temperatures well below 0°C, and soil moisture was less than 10%. Abscisic acid concentrations were significantly higher in trees subjected to soil freezing than in droughted trees. Abscisic acid levels at the end of sap flow correlated with the onset of decline and reduction in leaf area observed during two growing seasons. It was suggested that at the end of sap flow ABA levels could be used as an indicator of tree stress. Leaf ABA levels were not considered to be a good measure of tree stress. It was also suggested that ABA levels may have a direct role in the initiation of dieback in that ABA could be responsible for the reduction in leaf area and reduced bud break. Also, the importance of snow cover in protecting tree roots from desiccation and freezing injury was thought to play a role in ABA production by the roots and subsequent transport to the aboveground parts of the tree. It is difficult in this experiment to separate freezing stress from drought stress since freezing of soil can result in reduced water availability to the roots.

Kubik et al. (1992) studied the effects of storage temperature (2.5 and 10°C) on tomato fruit quality, ABA levels, and ABA metabolism. It was determined that ABA levels increased in fruits under reduced temperatures and that concentrations were slightly higher at 2.5°C. However, there was no correlation between ABA levels and increased damage symptoms of fruits stored at 2.5°C. Free ABA, ABA conjugates, and dihydrophaseic acid changed depending on storage temperature and poststorage ripening conditions. Chilling damage resulted in decreases in ABA and DPA–ester conjugates, which was associated with the degree of fruit damage.

E. Temperature and Ethylene Levels

To have a better understanding of the physiological effects of C=C on plants, a general understanding of C=C biosynthesis is beneficial. Probably, more is known about the biosynthesis of and how the environment influences C=C production in plants than for any other phytohormone. This is largely a result of research conducted in the area of postharvest physiology of fruits and vegetables, where C=C production has a direct impact on ripening processes and subsequently, the storage life of fruits and vegetables. Figure 4.53 outlines the general biosynthetic scheme for C=C and indicates how various environmental and hormonal factors may interact to affect its synthesis.

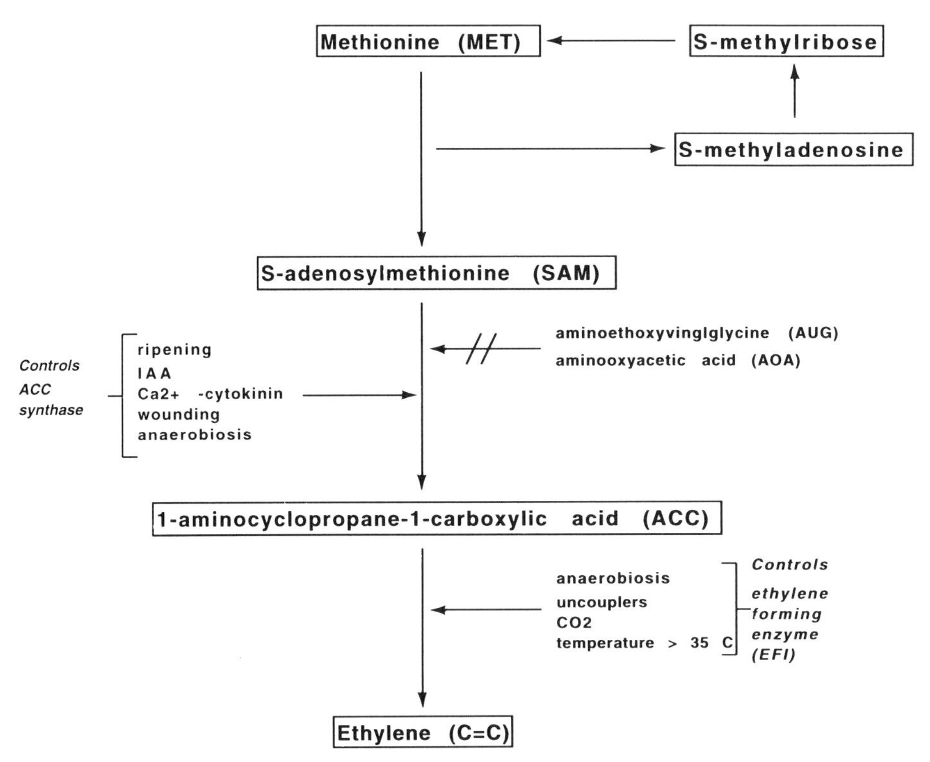

Figure 4.53 General scheme of C=C synthesis and factors controlling key enzymes. (Modified from Devlin and Witham 1983.)

*1. **High-Temperature Stress.*** Not much is known about the effects of high-temperature stress on C=C production, but the following will give the reader some idea of the physiological implications of how high temperature and C=C interact to affect growth and development.

The relationship of C=C in dormancy of Red-osier dogwood (*Cornus sericea* L.) was studied by measuring C=C evolution from excised branch segments at differing developmental stages (states of dormancy) subjected to differing temperatures ranging from 30 to 60°C at 5°C intervals (Shirazi et al. 1993). It was determined that maximum C=C production occurred around 40°C and that higher temperatures resulted in a reduction in C=C production. In addition, C=C was initially produced in early spring (March), was maximum about May, and dropped off in late summer. No C=C was produced at any temperature during the months of October through January. Thus C=C production occurred only during months of active growth and did not occur during the dormant period. Application of ACC, methionine, and IAA to dormant tissues subjected to the same temperature regimes as previously all produced significant amounts of C=C. This suggested that enzymes were not a limiting factor in C=C production in dogwood but that substrate concentration was limiting. Failure of dormant tissues to produce C=C following exposure to sublethal temperature stress was viewed as evidence that C=C may not be involved in breaking dormancy.

Thermoinhibition of seed germination can occur in some seeds if they are exposed to supraoptimal temperatures. In chickpea (*Cicer arietinum* L.), temperatures of 30 and

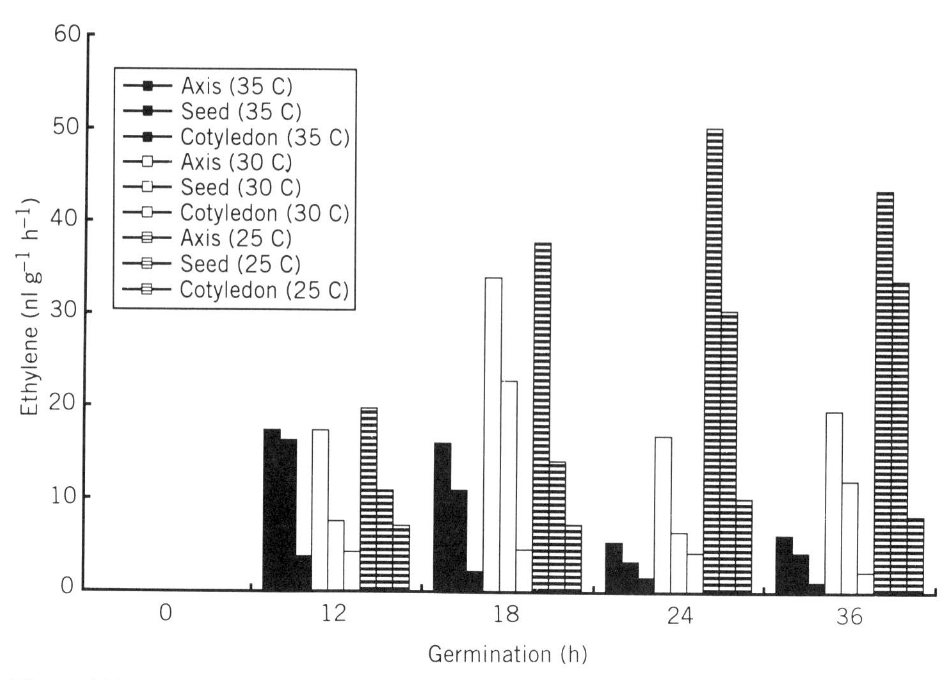

Figure 4.54 Ethylene production by chickpea seeds, embryonic axes, and cotyledons at three different temperatures during germination. (Modified from Gallardo et al. 1991.)

35°C inhibited germination and C=C production (Gallardo et al. 1991). Ethylene production was associated primarily with the embryonic axis and the whole seed, while cotyledons produced only small amounts (Figure 4.54). Germination could be restored by treatment of seed with ethephon, an C=C-generating chemical. Enzymatic and C=C precursor studies revealed that high temperature inhibited EFE, which converts ACC to C=C. 1-Aminocyclopropane-1-carboxylic acid synthase activity increased, but levels of ACC did not increase in the embryo, and ACC was probably being conjugated to form 1-(malonylamino)cyclopropane-1-carboxylic acid (MACC), which was in high concentration in the cotlyedon and declined during germination, suggesting that it may act as a source for free ACC that may be transported to the embryo during germination.

The propensity of bean plants to synthesize C=C at different rates (EER, C=C evolution rate) was demonstrated to be genetically controlled and highly variable among field-grown lines and cultivars (Sauter et al. 1990). Also, the ability of plants to heat acclimate was lower for plants that exhibited higher EERs (Table 4.35). The significance of such findings relates to the fact that high temperature can cause substantial reductions in yield, which may be related to high-temperature induction of C=C. Such increases could cause reduction in fruit set and/or abscission of flowers and possibly reductions in plant and fruit size. Thus breeding for plants with low EERs may be beneficial for production purposes in hot and dry areas.

The storage of fruits at low temperature can be detrimental to the quality of the fruit, particularly if it is subsequently transferred to a higher temperature. Such disorders as poor ripening, pitting, collapse of structural integrity, development of off flavors, and rotting can occur (Jackman et al. 1988). Tomatoes are particularly sensitive to low temperature, and if

TABLE 4.35 Leaf C=C Evolution Rate for Field-Grown Bean Cultivars/Lines and Their Heat Acclimation Potential as Determined by Heat Killing Time from Electrolyte Leakage

Cultivar/Line	Ethylene Evolution Rate ($\mu g\ g^{-1}\ h^{-1}$)	Heat Killing Time (min)
BBL47	2.52	60
Ul 111	2.45	50
Strike	1.71	95
Labrador	1.59	97
Red Cloud	1.48	—
GNUI 59	1.39	110
PI 271998	1.37	94
P. acutifolius	1.09	—
Hebei No. 1	1.08	80
G 4727 (Ancash 66)	0.96	100
85-CT-4986-3	0.62	90

Source: Modified from Sauter et al. 1990.

stored below 10°C and transferred to higher temperatures can develop chilling injury symptoms (Lurie and Klein 1991). It is well established that some environmental stresses can produce cross tolerance to others. For example, drought, cold, and osmotic stress seem to share some commonality relative to the mechanisms involved in the acclimation of plants to these stresses (i.e., drought acclimation increases tolerance to cold and osmotic stress in many plants). These observations led Lurie and Klein (1991) to consider the possibility of how acclimation of mature-green tomatoes to high temperature (3 days at 36, 38, and 40°C) might affect the quality of tomatoes subsequently stored at 2°C (3 weeks) and then transferred to 20°C. The authors discovered that heat treatment resulted in normal ripening of tomatoes, although somewhat more slowly than in freshly harvested tomatoes. Also, C=C production was higher than in nonheated plants and exhibited a normal climacteric pattern of production and lower levels of leakage, which may be indicative of greater membrane integrity. Plants that did not receive the heat treatment remained green, formed brown areas under the peel, and exhibited high rates of CO_2 evolution, which declined sharply while C=C levels were low but increased when exposed to 20°C. Prestorage heating resulted in inhibition of heat-stress C=C, protein synthesis was depressed, and heat-shock proteins appeared. The authors conclude that heat-shock proteins may be responsible for the better storage performance of the heat-treated tomatoes.

Considerable evidence exists that membrane integrity is necessary for C=C synthesis. Thus temperature or other environmental effects that might affect membrane structure and function (see Chapter 3) could potentially influence the production of C=C in plants. Although the specific site(s) of C=C synthesis is not known evidence indicates that EFE may be associated with the inner surface of tonoplast membrane. Some evidence suggests that ACC synthase is associated with the plasmalemma (Mattoo and White 1991 and references therein). Evidence for the requirement of membrane integrity for C=C synthesis was illustrated in bean leaf discs (Field 1981). Ethylene evolution was measured from bean leaf discs at 25°C after being exposed to temperatures of 25 to

47.5°C for 1 h. As temperature increased, C=C production also increased, until 47.5°C, where it was proposed that membrane integrity was lost and subsequently the ability for C=C synthesis. This contention was further supported by the large increase in leakage of electrolytes from leaf discs at the elevated temperatures (Figure 4.55).

2. Low-Temperature Stress. Low-temperature stress is usually categorized into chilling stress and freezing stress. Chilling stress usually causes injury to plants of tropical or subtropical origin and can occur at temperatures below 10 to 15°C. Freezing stress occurs in plants more tolerant to low temperature and occurs at 0°C or lower. Plants are highly variable relative to their sensitivities to low temperatures. Much of the research that has been conducted with C=C production as a function of low temperature has emphasized chilling stress. This is probably due to the interest of the agricultural industry in postharvest storage and transport of fruits and vegetables as well as the impact of C=C on ripening. Generally, chilling stress causes an increase in C=C production. In some plants this occurs during the time of chilling, but frequently a transient peak in C=C production occurs in plants when they are transferred from chilling to warmer temperatures. Various interacting factors may be responsible for this response and include the possibility that low temperature inhibits or limits the production of enzymes such as ACC synthesis or the EFE. Membrane stability factors may also play an integral part relative to C=C synthesis since evidence suggests that some enzymes (EFE and ACC synthase) may be associated with membranes and that the maintenance of membrane integrity is critical to their normal function. Thus scientists have approached research on low-temperature stress and C=C production from several perspectives, including effects on synthesis, relationship of C=C levels and chilling sensitivity, whether chilling-induced water deficits or chilling per se induce C=C production, and the influence of chilling-induced C=C on quality and storage of fruits and vegetables.

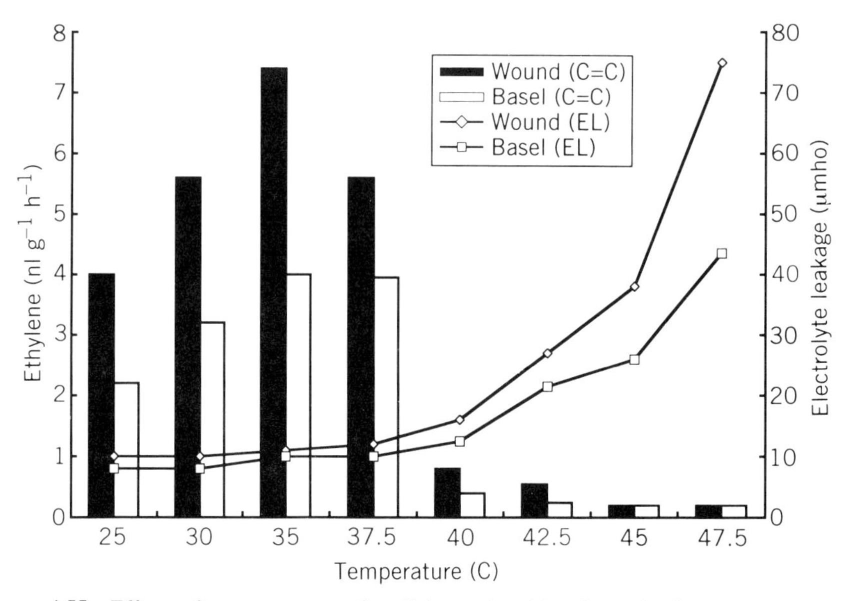

Figure 4.55 Effects of temperature on C=C (wound and basal) production and electrolyte leakage (EL) in bean leaf discs. (Modified from Field 1981.)

The transient production of C=C in plants transferred from chilling to warmer temperatures has raised questions as to how low temperature affects C=C biosynthesis. Field (1984), using dwarf bean leaf discs, found that C=C production increased 10-fold when leaf discs were transferred from 5°C to 25°C. Low C=C production and the transient overshoot production of C=C after transfer to 25°C was not associated with an accumulation of the C=C precursor, ACC. Addition of ACC to leaf discs incubated at 5°C resulted in stimulation of C=C at that temperature. Also, ACC synthesis did not occur in normally chilled discs until they were transferred to warmer temperatures. Taken together, these data suggested that reduced activity or level of the ACC synthesis enzyme may be affected under chilling stress.

Comparing chilling-sensitive and chilling-tolerant species of herbaceous and woody plants relative to C=C production and ACC production, Chen and Patterson (1985) concluded that species of *Passiflora* (as well as other plants) could be ranked relative to chilling tolerance based on the time at 0°C required for peak C=C production in leaf halves after being transferred to 20°C (Table 4.36). Also, unlike the results of Field (1984), treatment at 0°C resulted in an accumulation of ACC which persisted even after transfer to 20°C and was not reduced by severe chilling as was the case for C=C. The authors were also able to show that the rate of ACC accumulation also depended on the chilling sensitivity of the species involved (Figure 4.56).

Studying chilling-induced C=C and ACC production in chilling-sensitive bean, Tong and Yang (1987) found that whole plants differed relative to C=C and ACC production in response to chilling injury, and that whole plants rather than leaf discs should be used in such analyses when evaluating chilling sensitivity in plants. In their studies they found that with bean leaf discs ACC and C=C did not accumulate in chilled tissues until transferred from 5°C to 25°C and that at the time of transfer there was a parallel increase in both. In the case of whole plants, both ACC and C=C accumulated in chilled plants (5°C) over time and increased even further when transferred to 25°C. The reverse scenario was true for peas. When pea leaf discs were incubated at 5°C, ACC accumulated (C=C did not accumulate until transferred to 25°C as did ACC), while in whole plants, neither ACC nor C=C accumulated under chilling conditions or when transferred to 25°C. However, based on whole plant determinations it does appear that chilling-sensitive plants such as bean may be differentiated from chilling-tolerant species based on C=C and/or ACC production under chilling and warming cycles (Figure 4.57).

TABLE 4.36 Species Differences Relative to Chilling Required for C = C Production

Species	Chilling Duration Required for Ethylene Production (days)
Passiflora edulis flavicarpa	1–3
P. maliformis	3–4
P. quadrangularis	4–5
P. ligularis	5–6
P. caerulea	12–16
P. edulis	>15

Source: Modified from Chen and Patterson 1985.

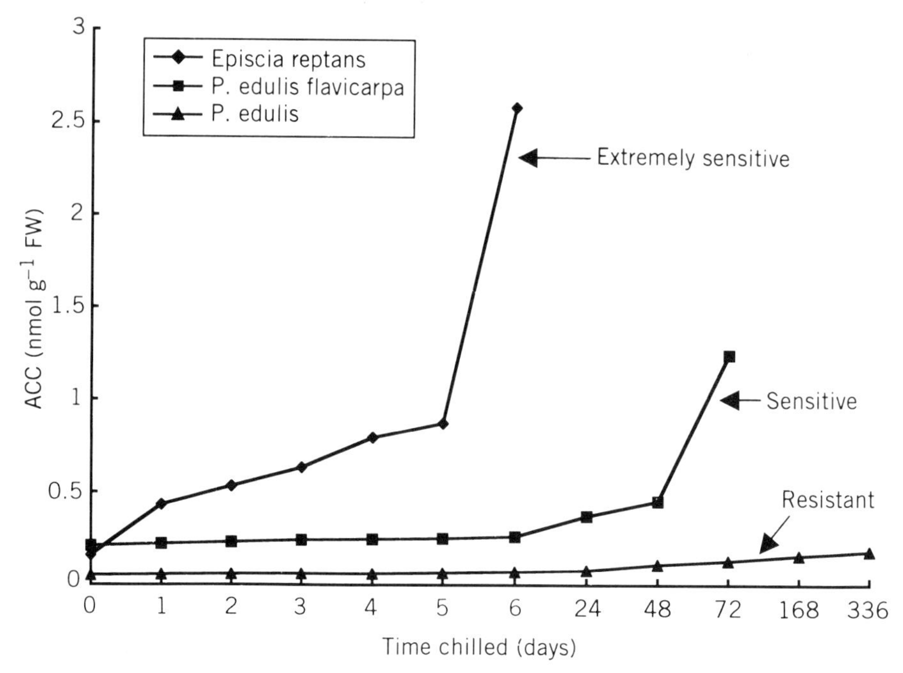

Figure 4.56 Accumulation of ACC at 0°C by three species of contrasting chilling sensitivity. (Modified from Chen and Patterson 1985.)

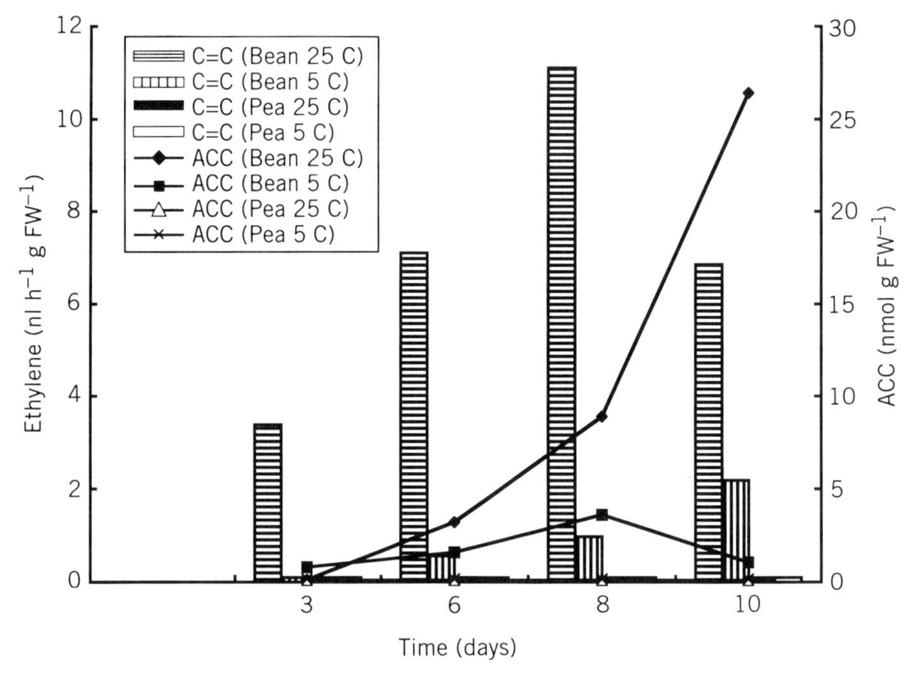

Figure 4.57 Ethylene and ACC production of pea and bean plants acclimated for 3, 6, and 10 days at 5°C and transferred to 5°C and 25°C. (Modified from Tong and Yang 1987.)

When plants are subjected to low-temperature stress, the ability of plants to absorb water and nutrients is reduced and can lead to dehydration. Low temperatures also leads to increased viscosity of soil water and results in reduced rates of movement to root surfaces. Thus water relations in plants is a factor relative to determining the impact of chilling stress on plants. Guye et al. (1987) tested several cultivars of bean differing in chilling sensitivity relative to C=C production, leaf water loss at chilling temperatures, and the effect of exogenous levels of ACC on C=C production on leaf discs subjected to chilling and nonchilling temperatures. Basically, they discovered that C=C production decreased with increasing dehydration and that plants chilled in 100% RH atmospheres produced C=C at about the same rate as did plants with a 47% reduction in fresh weight. Wright (1974), using excised bean leaves, also claimed that dehydration was primarily responsible for chilling-induced C=C production and that chilling bean leaf discs for 48 h at 5°C at 100% RH had no effect on C=C during chilling. Guye et al. (1987) also correlated chilling resistance with increased C=C production and the ability of chilling-resistant plants to retain water (Table 4.37). They also demonstrated that application of choline chloride to chilling-sensitive plants increased chilling tolerance and that associated with increased tolerance was increased water content and higher C=C levels. These data are interesting in light of other findings in plants, which correlate high ethylene levels under chilling conditions with decreased tolerance to chilling temperatures.

In light of recent studies in which the use of excised plant parts in determinations of C=C may result in artifactual data and the fact that Tong and Yang (1987) observed differences in C=C and ACC response of chilled intact versus excised parts of the same plant suggests that care must be taken relative to the interpretation of such data, particularly as they relate to whole plant physiology.

The practicality of low-temperature treatment of fruits and the ability to manipulate ripening through subjecting fruit to chilling temperatures was illustrated in pear storage (Sfakiotakis and Dilley 1974). 'Bosc' pears held at 20°C produced C=C at a low rate and resisted ripening for 12 days. Storage for 7 days at either 5 or 10°C and transfer to 20°C caused an immediate increase in C=C along with rapid uniform ripening. The

TABLE 4.37 Percentage Reduction in Leaf-Water Content and C = C Production Induced by Chilling at 5°C for Six Genotypes of *Phaseolus*

Phaseolus (+ACC)	% H_2O Loss	C = C Production: Chilled (+ACC) As % of Control
P. coccineus		
cv. Prizewinner	18.5	43
cv. Streamline	21.3	44
P. vulgaris		
cv. 222	22.0	27
cv. 251	17.3	25
cv. Tendergreen	41.8	10
P. aureus		
cv. Berken	70.0	0

Source: Modified from Guye et al. 1987.

longer the storage at 5°C up to 6 days, the faster the onset of C=C production. Longer periods lengthened the time. The results suggest that cold induces metabolic activity that increases C=C biosynthesis and subsequent ripening.

VIII. PLANT PATHOGENS AND PHYTOHORMONES

Historically, the interest in hormonal involvement in plant pathogenic reactions developed around the symptom expression in diseased plants, which in many instances is reminiscent of the effects that phytohormones have on plants. For example, some disease symptoms, such as epinasty, adventitious root formation, chlorosis, xylem hyperplasia, tylosis formation, stimulation and inhibition of shoot growth, and wilting, have either been correlated with endogenous increases in various hormones or can be induced by exogenous application.

Host–parasite interaction in plants is a very complex process involving multiple biotic and abiotic systems. Figure 4.58 outlines a general biochemical scheme of events that could occur during the infection process. As indicated in the scheme, phytohormones are key players in the initiation of other biochemical events that may render the host susceptible and/or resistant to a specific pathogen. As we have seen in previous sections of this chapter, all abiotic stresses mentioned can affect plants by changing phytohormone synthesis (increase or decrease), sensitivity, oxidative degradation (increase or decrease), compartmentalization and/or binding (and their reverse), conjugation or deconjugation, and transport form and/or mechanisms. Plant pathogens can also affect these processes and the problem is further complicated by hormonal interactions and the fact that some pathogens (bacteria and fungi) can synthesize the same hormones (i.e., IAA, ABA, C=C, CK, and GA) that plants utilize in growth and developmental processes as well as genetically control hormone synthesis in the host (i.e., the bacterial pathogen *Agrobacterium tumefaciens*). It is very difficult to separate the effects of pathogen hormone production from that of the host relative to infection and disease development in the host. The remaining portion of this section deals with how fungi, bacteria, and viruses affect host plant hormone relations and the ramifications of such interactions relative to disease susceptibility and resistance.

A. Fungal Diseases

Fungal diseases are most prevalent among plants. All the major plant hormones can be synthesized by phytopathogenic fungi, although no single fungus appears to have been studied relative to its ability to produce them all. As a general rule, all plant hormones that have been studied appear to increase in plant tissues as a result of infection of various hosts with a variety of phytopathogenic fungi.

Hyperauxiny (increased IAA levels in host tissues) is prevalent in many fungal–host interactions. Correlated with such responses has been tumorous tissue development in corn (Turian 1961) as a result of corn smut (*Ustilago zeae*) and hypocotyl elongation in safflower infected with *Puccinia carthami* (Daly and Inman 1958). Elevated IAA levels in plants may occur as a result of increased synthesis by the host and/or the pathogen, by the inhibition of IAA oxidase activity in the host by the pathogen, or by hydrolysis of conjugated IAA to form free IAA. Thus phytopathogenic fungi may alter plant host growth and development through the production of hormones or by regulation of the host hormone balance.

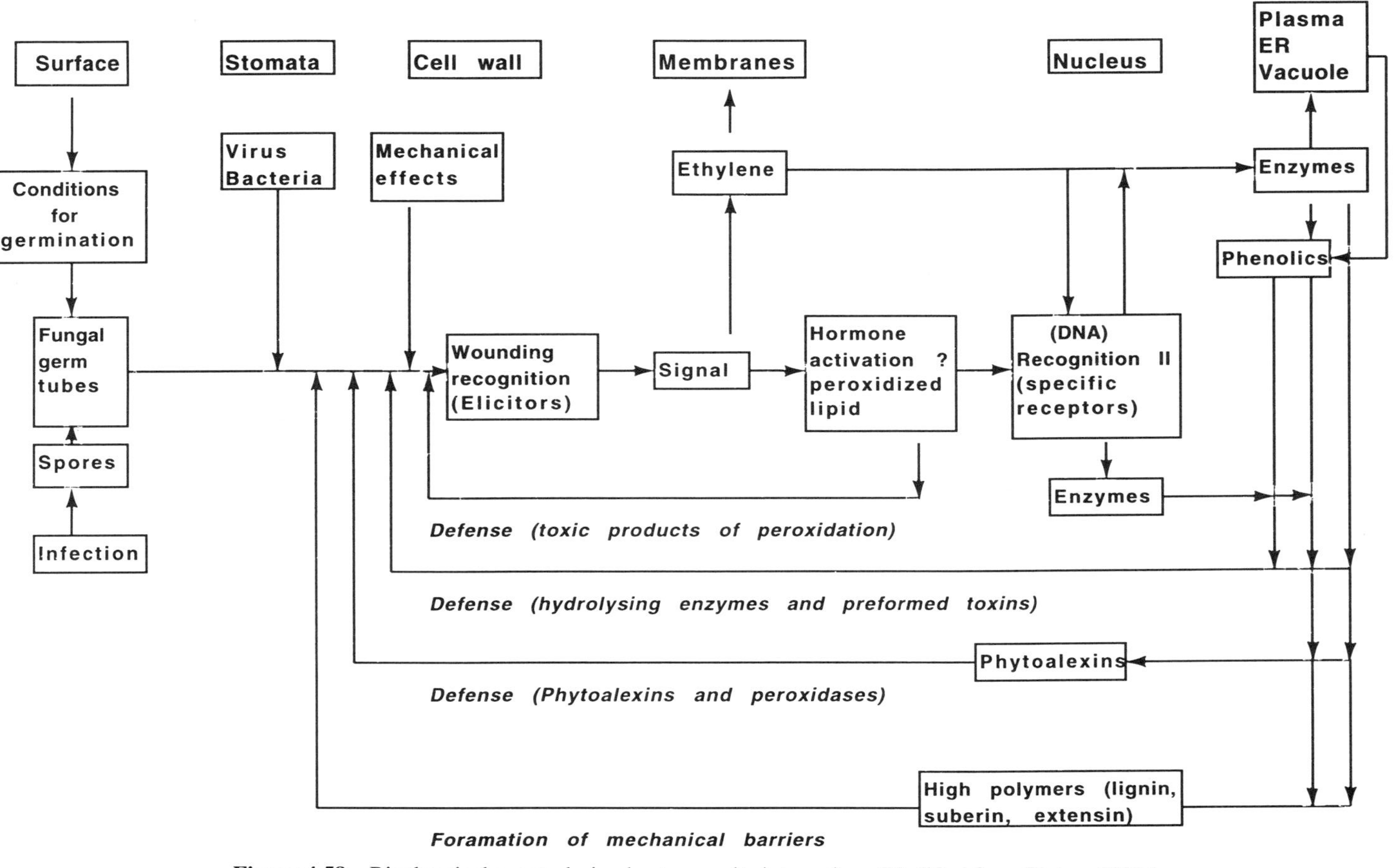

Figure 4.58 Biochemical events during host–parasite interaction. (Modified from Elstner 1983.)

That CKs are involved in disease development is again correlated with what is known about the role of CKs in normal physiological functions of the cell. For example, CKs are known to direct or control the transport of ions and nutrients to areas of the plant where high levels of CKs are produced, stimulate cell division, increase chlorophyll synthesis, and reduce senescence. Several fungal diseases caused by rusts and powdery mildews produce areas on leaves that are referred to as *green islands,* where lesions are characterized by a central green area surrounded by necrotic or chlorotic tissue. Analysis of such tissues shows that the green island areas are high in CKs, while the chlorotic/necrotic areas are low in CK. It has been hypothesized that whatever the source of the increased level of CK, nutrients are mobilized from the surrounding tissues to the area of high CK levels (green islands), resulting in senescence of the surrounding tissue. Such a response could provide nutrients to the fungus near the infection site. Cytokininlike substances secreted from germinating conidia of *Alternaria brassicae* were also reported to induce green island formation when droplets containing the secretion were placed on the leaf surface of mustard plants (Suri and Mandahar 1985). Club-root of cabbage (*Plasmodiophora brassicae*) has frequently been associated with elevated levels of CKs and IAA in tumorous tissues (Dekhuijzen 1976). Explants of tumors can be grown in tissue culture without the addition of exogenous hormones. However, over time they lose the ability to grow unless supplied with IAA and CK. This suggests that genetic transformation does not occur in the host plant as has been determined for the bacterium *Agrobacterium tumefaceins. Cronatium fusiforme* causes gall formation in pine stems (Western pine blister rust). Gall analysis revealed increased levels of CK in the infected tissues. The fungus is capable of being cultured away from the host. However, both IAA and CK are required in the culture medium. This suggests that the host may provide IAA and CK to support growth of the fungus (Dekhuijzen 1976 and references therein). Whether involved directly or indirectly, CKs are probably required in a number of aspects of disease development.

Little information appears to be available regarding the effects of phytopathogenic fungi on endogenous levels of GAs in plants. This is surprising since GA (although not chemically defined as such) was first associated with the pathogenicity of the fungus *Gibberella fujikuroi* on rice plants. The disease, called "foolish seedling disease," was first described in 1912 by the Japanese plant pathologist, Sawada (Sawada 1912). Plants infected with the pathogen exhibited abnormal shoot elongation, which resulted in plant lodging and subsequent reductions in yield. In the early 1920s, Kurosawa, a student of Sawada's, determined that fungus culture filtrates applied to rice plants caused the same symptoms as those produced by the fungal infection (Sawada and Kurosawa 1924). In 1938, Yabuta and Hayashi, in collaboration with Kurosawa, isolated crystalline gibberellin from culture filtrates of the fungus (Yabuta and Hayashi 1939). Phinney (1957) first demonstrated their presence in higher plants. Gibberellin production is probably common among fungi, but little work appears to have been done in this area. Rademacher and Graebe (1979) isolated GA_4 from a fungal pathogen (*Sphaceloma manihoticola*) of cassava, GA_3 was isolated from *Neurospora crassa* (Kawanabe et al. 1988), Pegg (1973) detected GA-like activity in five *Basidiomycete* species, GA-like activity was reported for the mycorrhizal fungus *Boletus pinicola* (Gogala 1971), and Aubé and Sachston (1965) reported GA-like substances in several *Verticilium* species. Matheussen et al. (1991) demonstrated that head smut fungus (*Sporisorium reilianum*) produced GA_1 and GA_3 and represented the first nonascomycete to produce GA. In addition, they demonstrated that noninfected sorghum plants produced 60 to 100% higher concentrations of GA in the panicles than did infected plants. The authors hypothesized that the panicles were a major source of GA in sorghum and that the fungus inhibited GA production of peduncle elongation and adjacent internode growth. Rust-infected (*Puccinia punctiformis*) creeping thistle (*Cirsium arvense*) resulted in increased levels of GA, which were correlated with a period of increased shoot growth over those in healthy control plants (Bailiss and Wilson 1967). Dwarfing syndrome of wheat caused by *Tilletia controversa* could not be correlated with decreases in GA content of the plant, and growth inhibition could not be reversed with the application of exogenous GA (Vanicková-Zemlová et al. 1981).

Crocoll et al. (1991) described 10 species of saprophytic and parasitic fungi that produce ABA. According to the authors, *Cercospora rosicola* was the first fungus described to produce ABA and ABA has been shown to be produced in eight other fungal species. Prior to the identification of ABA in saprophytes, only phytopathogenic fungi were known to produce ABA, which weakens the hypothesis of Crocoll et al. (1991) that ABA functions in pathogenesis.

Various pharmacological studies in which ABA has been applied exogenously to plants have shown that susceptibility to fungi can be increased. Abscisic acid is a well-documented inhibitor of many physiological and biochemical processes in plants and has been shown to regulate nucleic acid expression directly. In support of the pharmacological observations have been reports that ABA suppresses what are considered to be established defense responses to infection by plants. These include suppression of the production of phytoalexins, chitinase, lignin, phenols, and shikimic acid enzymes (phenylalanine ammonium lyase), which reflects the production of some of the aforementioned compounds.

Leaves of cotton plants infected with a defoliating strain of *Verticillium albo-atrum* were shown to have twofold-higher concentrations of ABA than those in healthy plants. A nondefoliating strain had no effect on ABA and no defoliation occurred, suggesting that increased ABA levels were responsible for leaf abscission by the defoliating strain (Table 4.38) (Wiese and DeVay 1970). Compatible and incompatible reactions were compared between soybean lines, showing differing resistance responses to separate races of *Phytophthora megasperma* f. sp. *glycinea* (Cahill and Ward 1989). The incompatible reaction was characterized by a 35% reduction in ABA concentration in the infection site 2 and 4 h postinoculation. Significant decreases in ABA were also noted 4 and 6 h after inoculation in tissues surrounding and remote from the infection site. No significant change in ABA levels was noted in the compatible reaction 8 h after inoculation. It was implied that ABA may be required to initiate defense responses in the incompatible reaction, such as increasing phytoalexin synthesis and defense-related enzyme production (phenylalanine

TABLE 4.38 ABA ($\mu g\ g^{-1}$ DW) in Healthy and *Verticillium Albo-atrum*-Infected Cotton Tissues

Plant Part	Assay Method	Healthy	Diseased	
			Nondefoliating	Defoliating
Leaves	GLC	4.0	3.8	8.6
	TLC	1.2	1.6	2.8
Stems	TLC	0.6	0.3	0.4

Source: Modified from Wiese and DeVay 1970.

ammonium lyase, chitinase, lignin production, phenolic production). Ryerson et al. (1993) also found that French bean inoculated with either pathogenic *Uromyces appendiculatus* var. *appendiculatus* or nonpathogenic *U. vignae* caused significant reductions in leaf ABA levels. It was concluded that the response was nonspecific and that ABA is not a determinant of successful fungal invasion. These results were said to be in contrast to those of Dunn et al. (1990), who found that increased levels of ABA in French bean was related to increased resistance to *Colletotrichum lindemuthianum.*

Ethylene is generated in plants as a wounding response. It may also be generated as a result of infection by various pathogens. Like a physical wound, infection by pathogens may trigger the same or similar responses relative to C=C production. The response of the plant to pathogen infection and wounding is similar. Enzymes and chemicals associated with defense mechanisms are induced such as chitinase, β1,3-glucanase, phytoalexins, proteinase inhibitors, polyphenols, lignin hydroxyproline–rich glycoproteins, extensins, and enzymes involved with flavenoid biosynthesis [PAL, 4-CL (4-coumarate:CoA ligase), and CHS (chalcone synthase)]. The production of these substances may protect the plant by affecting the pathogen directly (phytoalexins and hydrolase enzymes) or may bolster the defense mechanisms of the plant directly through processes that may exclude or isolate the pathogen from healthy tissues (increase lignin production, abscission, and vascular and gummosis and occulsion). In many cases, exogenous C=C can trigger many of the foregoing responses, and thus such an association has been part of the basis for the implication of pathogen-induced C=C production with disease initiation and development in plants. Evidence for and against the involvement of C=C in resistance and susceptible responses in plants appears to be related to a number of variables, including the type of plant, stage of development, sensitivity to C=C, and perhaps the ability of the pathogen to produce C=C. One might expect that during the ripening and senescence process, when C=C levels are quite high, that C=C might enhance the disease development process. This is the situation in many instances; however, in some cases the relationship is not as clear. Many fungi have been shown to synthesize C=C. In most instances exogenous methionine is required as a substrate, but several species of *Fusarium* have been isolated that produce very high levels of C=C without additional methionine. Some species utilize 2-oxoglutaric acid as a substrate to form C=C.

In considering the impact of C=C relative to its role in disease development, normal nonpathogenic soil microflora may also have to be considered when assessing the impact of C=C on plants. Arshad and Frankenberger (1988) were the first to demonstrate conclusively that microbial-generated C=C from the nonpathogenic soil fungus *Acremonium falciforme* was able to induce the "triple pea response" (height inhibition, swelling of shoots, altered geotropic response) in pea plants grown in a soil with and without methionine supplementation. Also, nutritional and other environmental interactions with C=C production in plants has to be considered. For example, Nevin et al. (1990) demonstrated that toxic levels of NH_3-NH_4 that accumulated in drought-stressed avocado trees was not a result of *Phytophthora cinnamomi* infection but was related to NH_3-NH_4-stimulated C=C production. They suggested that leaf damage was in part related to C=C and NH_3-NH_4 interaction. In a mycorrhizal fungus (*Glomus fasciculatum*) association with potato, a deficiency in P resulted in reduced activity of ACC and levels of C=C produced in the roots. This reduction was related to increased production of phenolics, which was hypothesized to inhibit the conversion of ACC to C=C. Ethylene was viewed as being linked to resistance or susceptible mechanisms that allowed mycorrhizal association with potato roots. Reduced P stimulated phenolics, which inhibited ACC conversion to C=C and subsequently allowed infection. Sufficient levels of P for plant growth reversed the process. An excellent review of the function of C=C in pathogenesis and disease resistance was developed by Boller (1991).

B. Bacterial Diseases

Bacteria and fungi both can synthesize IAA in vitro. Several biosynthetic pathways leading to the synthesis of IAA have been determined for biological organisms. Tryptophan is thought to be an important early precursor for the synthesis of IAA. In many higher plants synthesis appears to proceed via the IPA pathway (tryptophan ⟶ indole-3-pyruvic acid ⟶ indole-3-acetaldehyde ⟶ IAA). However, variants of synthesis have been observed in bacteria and fungi relative to the intermediates involved (Costacurta and Vanderleyden 1995). The IAM pathway (tryptopan ⟶ indole-3-acetamide ⟶ IAA) appears to be unique to bacteria and rare in dicotyledonous plants (Yamada 1993). Some pathogenic bacteria, such as *Pseudomonas* and *Agrobacterium,* possess, both the IPA and IAM pathways for the biosynthesis of IAA (Costacurta and Vanderleyden 1995). The ability of pathogens to synthesize IAA via different pathways may provide an advantage to the pathogen to utilize host IAA intermediates and/or control the synthesis of host IAA.

Much of our understanding of the role of IAA and CKs in plant pathogenicity as well as physiological functions has come from studies utilizing tumorigenic bacteria such as *Agrobacterium tumefaciens* and *Pseudomonas savastanoi.* The reason for this is probably related to the obvious growth abnormalities induced in plants by these bacteria, the potential relationship of observed metabolic abnormalities and cancerous growth in animals, and the ability to use *A. tumefaciens* to introduce genetic material into host plants. Both of these organisms can produce IAA and CKs, and it has long been anticipated that the production of these hormones influences the development of tumerous growths on host plants. Demonstration that bacteria-free tobacco crown gall tumor tissue could grow in the absence of exogenous IAA and CK (unlike healthy tobacco callus tissue) was the first evidence that genetic information controlling IAA and CK synthesis in the crown gall tissue was incorporated into the genome of the host. Since these early findings, specific genes have been identified that control various aspects of IAA and CK synthesis, and in the case of *A. tumefaceians* T-DNA has been shown to be transferred to the host plant and that this T-DNA segment possesses the genes responsible for control of IAA and CK synthesis in the host. *P. savastanoi* does not have the ability to transform host cells, so apparently does so by secretion of the hormones. This hypothesis is based on the observation that nonvirulent strains of *P. savastanoi* do not produce IAA, and tumors are not induced in inoculated plants. Two genes identified from *P. savastanoi* as *tms*-1 and *tms*-2 code for the synthesis of iaaM (tryptophan monooxygenase) and iaaH (indoleacetamide hydrolase) have been found to regulate new IAA synthesis in transformed cells. Another gene, *iptZ,* codes for isopentenyl transferase, which is involved with side chain synthesis of CKs. Gene 5 regulates IAA synthesis by the production of an IAA antagonist, indole-3-lactic acid. Gene 6b modulates the activity of CK by decreasing tissue sensitivity to CK. Genes 5 and 6b both affect the phenotypic expression of *P. savastanio*–induced teratomas. When gene 5 is expressed, more shoot-producing teratomas are produced, causing shifts in IAA/CK ratios in favor of CK. It is well established that changes in ratios of these hormones in tissue culture can be used to produce shoot or roots, depending on the ratios of the two hormones. An excellent review on the subject of the role of auxin and plant disease development is presented by Yamada (1993).

Plants transformed by the T-DNA of *A. rhizogenes* exhibit phenotypic alterations known as hairy-root syndrome. In transformed tobacco, the syndrome is controlled by three genes (*rol* A, *rol* B, and *rol* C). Gene *rol* A alters the morphology of the plant (wrinkled leaves, reduced length/width ratio of leaves, shortened internode length, and a delay in flowering) (Dehio et al. 1993). Analysis of the five classical phytohormones (which is rarely done—this may be the first recorded attempt to do so) by Dehio et al. (1993) determined that multiple tissue-specific alterations of phytohormone concentrations are the consequence of *rol* A gene activation. Most notable was the 40 to 60% reduction in GA_1 associated with young, fully developed, transgenic tobacco plants. An indirect relationship was claimed to exist between reduced levels of GA_1 production, symptom expression, and *rol* A gene induction. The genes *rol* B and *rol* C both possess β-glucosidase activity, and the former has been hypothesized to control the production of active free IAA by hydrolyzing conjugated forms of IAA (Estruch 1991). Activation of this gene has been shown to cause an increase in free IAA in some systems but not in others (Dehio et al. 1993). The gene *rol* C has been demonstrated to hydrolyze conjugated CKs and again, as in the case of *rol* A, the evidence is inconclusive relative to the accumulation of CK in different systems and the possible significance.

Very little information is available regarding GAs and phytopathogenic bacteria. Twenty-eight different species of endophytic pseudomonads have been reported to produce gibberellin-like substances (Pegg 1985 and references therein). Definitive identification of $GA_{1,20,4}$ was determined for free-living cultures of *Rhizobium phaseoli* using gas chromatography–mass spectroscopy. Most studies regarding bacteria–plant interactions and GAs appears to be conducted with symbiotic relationships of nitrogen-fixing bacteria and their hosts. In these studies nodule analysis has revealed, in most cases, increased levels of GA or GA-like substances in the nodules compared to surrounding tissues as well as qualitative changes (Dobert et al. 1992). Using cowpea (*Vigna unguiculata* L.) and lima bean (*Phaseolus lunatus* L.), Dobert et al. (1992) clearly demonstrated that different strains of the bacterium *Bradyrhizobium* sp. could elicit considerable changes in the relative abundance of four GAs identified in the nodules and stems of both species infected with the different strains. No qualitative differences were observed in the nodules or stems of hosts inoculated with either strain of bacteria. The contributions made by the hosts or bacterium to GA production could not be determined. However, the authors claimed that differences observed in GA_1 precursor levels observed in nodules formed by the two bacterial strains were correlated with differences observed in elongation growth in lima bean.

Little information is available concerning the production of ABA by bacteria and/or the relationship of ABA and phythopathogenic bacteria in infected plants. Fraser (1991) implies that bacteria may not produce ABA in pure culture. However, ABA has been observed to increase around the roots of maize plants inoculated with rhizosphere bacteria, and ABA-like substances have been observed to be higher in nitrogen-fixing nodules caused by *Rhizobium* than in noninfected tissues (Fraser 1991). A twofold increase in ABA concentration was observed in tobacco plants inoculated with *Pseudomonas solanacearum*. Symptoms associated with growth reduction and wilting were attributed to increased ABA concentrations in the plant. The bacterium could not produce ABA in vitro; thus it is assumed that elevated levels of ABA in tobacco were related to host production (Steadman and Sequeira 1969).

Several species of phytopathogenic bacteria are known to produce C=C in axenic culture. Examples include *Pseudomonas solanacearum, Xanthomonas campestris* pv.

citri, Pseudomonas syringae pv. *phaseolicola,* and *Erwinia rhapontici* (Ben-David et al. 1986 and references therein). However, in other species, C=C is either not produced, as in *E. caratovora* (Boller 1991) or in the cases of *X. campestris* pv. *vesicatoria, P. syringae* pv. *tomato,* and *P. syringae* pv. *lachrymans,* produced only in small amounts and only under very low (0.01 to 0.05 atm) O_2 tensions (Ben-David et al. 1986). As with other phytopathogens that produce and/or induce C=C synthesis in their host, the precise role of C=C relative to resistance, susceptibility, and/or symptom expression is not certain. Some phytobacteria can produce chemicals that can induce or inhibit C=C production in plants. For example, the nodulating bacterium *Bradyrhizobium* produces a chemical called rhizobiotoxine that inhibits the synthesis of C=C in soybean. This is of interest in view of the fact that C=C has been shown to inhibit nodulation in some plants. Rhizobiotoxine has also been demonstrated to occur in the phytobacterium *Pseudomonas andropogonis* implying that this chemical may not only be important in symbiosis but also pathogenesis (Boller 1991). On the other hand, *Pseudomonas syringae* pv. *glycinea* produces a toxin called coronatine (Ferguson and Mitchell 1985). In this case the toxin stimulated C=C synthesis and chlorosis in bean leaf discs (Table 4.39). Klee et al. (1991) isolated a soil bacterium (*Pseudomonas* sp. strain ACPC) that produced ACC deaminase, an enzyme that inhibits the conversion of ACC to C=C. The gene encoding this enzyme was cloned and introduced into tomato plants which ultimately resulted in no abnormalities in the vegetative growth of the transgenic plants but the fruits from these plants remained firmer for at least 6 weeks longer than fruits from wild type plants. The ability of phytopathogenic bacteria to produce chemicals and/or enzymes that can either stimulate or inhibit C=C synthesis illustrates the diverse capacity of bacteria to control C=C production in the host and may relate to earlier observations that increases and decreases in plant resistance to pathogens may be a function of enhanced C=C production.

Although leaf abscission is more common as a result of fungal infection some phytobacteria also induce leaf abscission in their hosts (Ben-David et al. 1986). The response is probably mediated through the production of high levels of C=C as a result of bacterial infection. The source of C=C may be either the host or the pathogen, which often is unclear. An example of a phytobacterium that causes leaf abscission is *Xanthomonas campestris* pv. *vesicatoria,* which is pathogenic on pepper (*Capsium annuum*). In an extremely well-designed experiment, the authors (Ben-David et al. 1986) were able to show that increases in C=C production in pepper plants was correlated with inoculum levels, levels of bacterium in diseased tissues, leaf abscission, susceptibility of organ to infection (young leaves more susceptible than old leaves), and disease severity index.

TABLE 4.39 Effect of Coronatine Concentration on C = C Production Collected over 24 h in Bean Leaf Discs

Coronatine (*M*)	C = C (nl g^{-1} h^{-1})
Control	2.43
5×10^{-7}	15.59
5×10^{-6}	17.54
5×10^{-5}	22.95

Source: After Ferguson and Mitchell 1985.

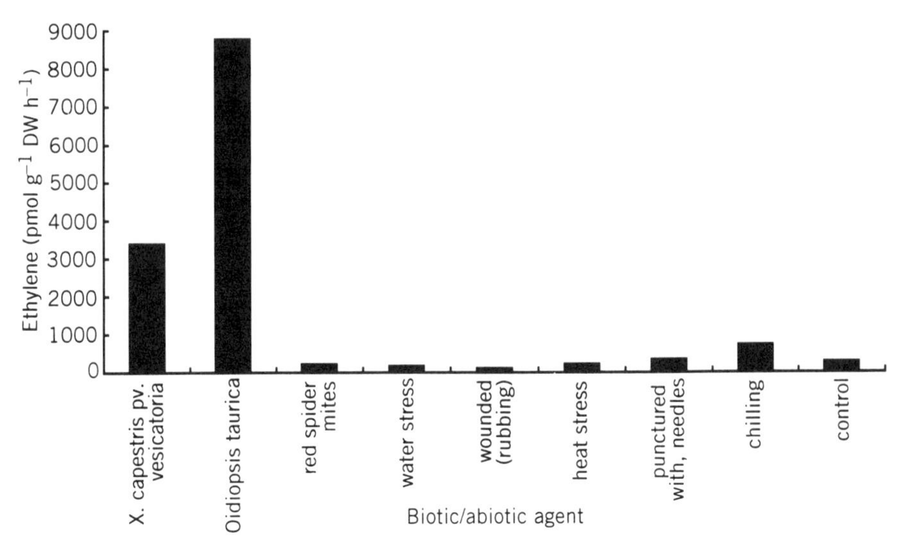

Figure 4.59 Effects of biotic and abiotic stresses on C=C production in pepper leaves. (Modified from Ben-David et al. 1986.)

Ethylene was produced mainly in the distal parts of leaf blades and around developing necrotic spots. The authors also compared the effects of fungal infection, mite infestation, water stress, wounding, and high and low temperature on C=C production in pepper. Interestingly, the highest amount of C=C produced was in plants infected with the bacterium *X. campestris* pv. *vesicatoria* or the fungus *Oidiopsis taurica* (Figure 4.59). Auxin treatment of the pepper plants at the time of inoculation, at time intervals postinoculation, or prior to inoculation all caused reduced C=C levels, leaf abscission, and disease severity in infected plants. Also, treatment of plants with aminooxyacetic acid, a known C=C inhibitor, similarly decreased C=C production and disease severity and leaf abscission. High C=C levels and declining IAA levels in leaf blades have been associated with increased leaf abscission in plants. Stall and Hall (1984), using two different cultivars of pepper exhibiting differing responses to *X. campestris* pv. *vesicatoria* (one cultivar produced yellowing around infection sites more quickly than the other), determined that C=C production and electrolyte leakage were similar for both cultivars infected with the pathogen. However, it was concluded, based on exogenous C=C and ethephon treatments, that the cultivar that initiated yellowing faster was more sensitive to C=C and that sensitivity was probably not related to resistance to the pathogen.

Using virulent and avirulent strains of *Pseudomonas* and *Xanthomonas* pathogens, Bent et al. (1992) determined that C=C is not required for active resistance against avirulent bacteria in *Arabidopsis thaliana* C=C-insensitive mutants. When *ein*2 and *ein*1 mutants (both C=C-insensitive mutants with unlinked loci) were challenged with virulent strains of the foregoing bacteria, *ein*1 mutants were susceptible, while *ein*2 mutants developed only minimal disease symptoms, and despite reduced symptoms, *P. syringae* pv. *tomato* grew extensively within *ein*2 leaves. Thus disruption of wild-type *ein*2 function conferred tolerance to the pathogen. The reason that *ein*1 plants did not respond similarly was not clear, but it was suggested that the *ein*1 gene product is active after a branch point in the C=C response pathway and is not involved in the generation of pathogen-induced symptoms.

C. Viral Diseases

Since viruses are obligate parasites and cannot be grown in vitro without host cells, it is assumed that viruses cannot synthesize plant hormones and must therefore alter the host's hormone balance directly through genetic manipulation or other indirect mechanisms. The impact of virus infection on IAA levels in plants is variable and can either be increased or decreased. However, few studies have addressed this issue in a definitive manner using more contemporary hormone analytical procedures. In some instances, viral infection causes reduced transport of IAA, which has also been correlated with elevated C=C synthesis, the latter also being implicated in the inhibition of IAA transport. In most cases, viral symptom expression, has not been associated with IAA accumulation, although at times application of auxin analogs to viral-infected plants has reduced symptom expression, while in other instances auxin enhances symptom expression. Viral infection generally increases peroxidase activity, which also correlates with increased symptom expression. Under these circumstances oxidation of IAA would probably occur, leading to lower levels, which may imply that IAA is not involved in symptom expression. Exogenous applications of auxins appear to have varying effects on viral replication as well. In some instances it is stimulated, while in others it is inhibited. This response appears to be related to leaf age, host–viral association, tissue infected, and whether the response is local or systemic.

Considering that CK is an established component of plant mRNA and the potential for CKs to influence cellular events leading to the infection and replication of viruses in plants, it is surprising that so little research has been done on the effects of viral infection on CK production in plant tissues. What information is available on the subject is inconclusive. In some viral–plant associations, CKs increase, and in others they decrease. Application studies using exogenous treatments of CKs generally indicate that treatments prior to inoculation decrease infection and symptom expression, whereas treatments after inoculation tend to work in the reverse. Timing of CK application appears to be very important since it has been shown that waiting as little as 5 min after inoculation with TMV before applying CKs to tobacco leaves can result in significantly higher virus titers than if applications are made 1 min after inoculation (Aldwinckle 1975). It has been proposed that treatments prior to infection stimulate RNA and protein synthesis in the host and that the virus is less competitive relative to similar metabolism and thus unable to replicate efficiently. Type of cytokinin and concentration have also been shown to be factors that can affect viral infection and replication. Cytokinins have also been associated with systemic resistance in tobacco. Plants that had their lower leaves inoculated with TMV or CMV exhibited systemic induced resistance. Upper leaves had slightly higher concentrations of CK and fewer lesions than did the lower leaves; however, the infectivity recovered from the upper leaves was similar to that of controls (Sziráki et al. 1980). Finally, cells within a given cell type or within an organ (epidermal cells versus mesophyll cells) appear to exhibit different susceptibilities to viral infection, which may or may not reflect differences in cellular amounts of CK at any given time relative to viral infection (Balázs et al. 1976, Kasamo and Shimomura 1977).

Where viral infection is associated with stunting, GA levels frequently are reduced in the host plant. In one instance, CMV-infected cucumber plants exhibited qualitative and quantitative changes in GA composition. This could imply that active forms involved in elongation growth are converted to inactive GAs as a result of virus infection (Ben-Tal and Marco 1980). Although application of exogenous GA to viral-stunted plants can partially overcome the stunting effect, it is doubtful that stunting viruses interfere with GA growth effects since the impact of viral infection appears to affect cell number rather than cell enlargement, the latter is generally attributable to the effects of GA. In addition, GA applied to healthy plants can also stimulate growth.

Most of the information regarding viruses and ABA has been derived from studies done with TMV on tobacco. In most cases, but not all, TMV causes an increase in ABA levels. However, in tomato, ABA levels seem not to be affected by TMV except in plants containing the Tm-22 gene for TMV resistance, where ABA levels increased (Fraser 1991). Whether the virus is systemic or produces local lesions appears to affect the level of ABA production (Table 4.40). In tobacco it appears that local lesion-producing TMV isolates (*flavum*) induce the plant to produce higher concentrations of ABA than do systemic viruses (*vulgare*). The reverse is true for the production of the ABA metabolite phaseic acid. Also, ABA production appears to correlate with symptom expression, as demonstrated with TMV isolates that multiplied to varying extents in tobacco, resulting in different degrees of symptom expression (Table 4.41) (Whenham et al. 1985). In addition, ABA concentration was correlated with symptom severity but less so with TMV replication. The relative growth rate of plants was equally correlated with virus replication and ABA concentrations. Also, chloroplastic ABA was high and apoplastic ABA low when symptom expression was mild to moderate. The reverse was true when symptom expression was severe. Interestingly, stomatal closure was not induced when apoplastic ABA increased (Whenham 1982). It appears that ABA and CK may interact in symptom development in TMV-infected plants. Typically, the infected area has light yellow or green areas separated by dark green islands located near veins. The highest accumulation of ABA is in the green islands, which are devoid of virus. Virus infection (TMV) of tobacco caused a reduction in free CK and an increase in conjugated CKs in the green island areas (Whenham and Fraser 1990). The affect on CK changes could also be simulated by treating healthy plants with ABA. Subsequent inoculation of the plant with TMV reinforced the affect.

Ethylene production was first detected in 1951 for host–virus interactions resulting in local necrotic lesions (Boller 1991 and references therein). Considerable research has been conducted since on TMV-infected tobacco plants. Generally, viruses that induce local necrotic lesions also cause increased C=C production, while systemic viruses tend not to cause elevated levels of C=C. However, chlorotic local lesion development can cause increased C=C production (Goodman et al. 1986). Whether C=C is the causal agent of symptom development or whether it simply accentuates the process has been an

TABLE 4.40 Effects of TMV Infection on the Concentrations of ABA and Phaseic Acid (PA) in Tomato and Tobacco

Host	TMV Isolate	ABA (ng g^{-1})	PA (ng g^{-1})
Tobacco	Healthy	20	745
	Vulgare	82	2436
	Flavum	280	n.d.
Tomato	Healthy	266	124
	MII-16	284	62
	Vulgare	338	54

Source: After Fraser 1991.

TABLE 4.41 Intracellular Distribution of ABA in TMV-Infected Tobacco Leaves

Virus Strain	Symptoms	ABA ($ng\ mg^{-1}$ chlorophyll) Leaf	Chloroplast	% ABA Outside Chloroplasts
Healthy		31.1	30.7	1
M11-16	Very mild green mosaic	24.8	16.3	34
V4	Mild green mosaic	50.0	29.3	41
N2	Moderate green mosaic	42.4	23.9	44
Vulgare	Severe light green/dark green mosaic	128.6	7.6	94
Flavum	Severe yellow-green mosaic	77.9	18.7	76
U5	Very severe green mosaic	86.1	10.7	88

Source: After Whenham et al. 1985.

area of considerable research but with little resolution of the issue. In many instances, C=C and C=C-forming enzymes have either coincided with or preceded necrotic lesion development (Figure 4.60). The use of C=C-generating compounds administered via a pinprick induced necrotic lesions similar to those of virus infection, lending some support to a hypothesis of C=C involvement in hypersensitive necrotic lesion induction (van Loon 1982). What appears to be clear is that increased ACC formation in host plants is the prime cause of increased C=C production in the hypersensitive viral-induced reaction.

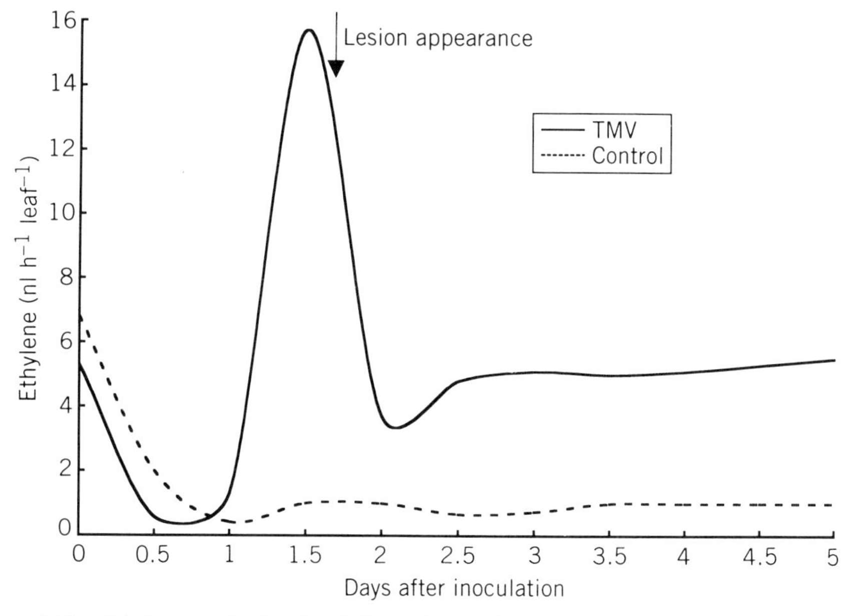

Figure 4.60 Ethylene production by TMV- and water-inoculated tobacco leaves. (Modified from van Loon 1982.)

IX. SUMMARY

As the reader can no doubt gather from the information presented in this chapter, the interactions that occur between plants and the environment are complex. The five classical phytohormones play an important role in this process, as do possibly other substances mentioned from time to time as potential hormones (i.e., brassinosteroids and polyamines). The complexity is such that it is difficult to develop a unifying theme that allows for the development of a larger concept of understanding with respect to modes and mechanisms of action of phytohormones. The concept of a complex network of hormonal interactions analogous to an integrated system of neuron and dendrites used by Trewavas (1986) as a means of illustrating aromatic amino acid synthesis in plants probably describes accurately what we have observed with phytohormones but does little to explain why.

Virtually every environmental parameter (abiotic and biotic) to which plants are exposed affects plant growth and development in some way. Whether the responses are always mediated through direct effects or even indirect effects on phytohormones may be debated. What is certain is that hormones do respond to the environment and hormones do control the biochemical, physiological, and morphological aspects of development. Several aspects of hormone metabolism are operative in the plant relative to plant response and development. Control of hormone synthesis, balance, concentration, sensitivity, degradation, conjugation, and transport are all factors that can affect plant growth response. As we have seen, all of these factors are influenced by the environment and can be species, cultivar, tissue, and developmentally specific. To complicate matters further, many plant responses that might be viewed by the "objective eye" as being stressful to a plant may in fact be part of its survival mechanism and be necessary for normal growth and development. Thus it would appear that plants have coevolved mechanisms utilizing a few simple molecules that not only control daily growth and development activities under normal or optimal conditions but utilize the same chemicals to respond to extreme or stressful environments. Thus a network of communication among the five classical hormones and their multiple interactions among other biochemical, biophysical, and physiological parameters controlling plant growth provides the flexibility required for a sessile organism to compete in a highly variable and changing environment.

Much criticism is frequently voiced in the literature regarding past studies in plant hormone research in which bioassay, isolation techniques, and the use of pharmacological approaches are viewed as less than adequate. It is true that more sophisticated techniques, such as gas chromatography–mass spectrometry analysis, immunoassays, and the development of transgenic and genetic mutant plants allows for increasingly definitive results. However, it is important to realize this has all been a part of an evolving process of technique development and careful research. In fact, much of the current literature regarding the functions of hormones in plants has confirmed findings of past carefully conceived and analyzed experiments that have used less definitive techniques and approaches.

Some of the problems with past approaches have been the inability to analyze all plant hormones simultaneously in a given experiment, the requirement for large tissue samples, limiting analysis to whole plants or large plant parts rather than specific tissues, the lack of analytical tools/approaches specific for individual hormones, the lack of attention to physiological stage of development, and the considerable time required to do the analyses. To a great extent, current techniques have eliminated such problems, but relatively few current studies have taken a multiple-hormone approach to determining the complex hormone–hormone and hormone–environmental interactions. How hormones interact under a given set of environmental conditions is the key to a better understanding of hormonal control of plant development.

STUDY-REVIEW OUTLINE

Introduction

1. Basically, there are five classical growth hormones that are genetically controlled and interact with the environment to affect numerous aspects of growth and development in vascular plants.
2. These growth substances are divided into inhibitors and stimulators of growth and include IAA, GA, CK (stimulators), C=C, and ABA (inhibitors).
3. The mechanisms of action of the plant hormones are very complex, affecting nucleic acid synthesis, membrane permeability, nutrient allocation patterns, and hormone–hormone interactions.

Light Stress and Phytohormones

1. Both quantity and quality of light can affect growth and morphological development of plants.
2. Photomorphogenic effects are thought to be mediated through pigments such as cryptochrome and phytochrome, which in turn are thought to interact with phytohormones to influence developmental responses.
3. Environmental control of red and far-red wavelengths of light can have a significant impact on phytochrome ratios in plants, which may control numerous hormone-regulated responses in plants.

Light and Indoleacetic Acid Levels

1. Red light appears to cause an increase or to maintain high levels of IAA in leaves, which in turn reduces leaf abscision, whereas darkness or shade tends to promote the abscission process. This suggests phytochrome-mediated control of IAA.
2. Conversely, red light causes a reduction in corn mesocotyl IAA levels and growth. This suggests a differential response of tissues to red light and/or IAA.

Light and Cytokinin Levels

1. Red light appears to promote CK production in plants and may be controlled in a diurnal manner. Diurnal fluctuations in ABA and CK appear to be important in regulating stomatal opening and closing.
2. Genetic differences in CK synthesis have been correlated with reduced plant senescence rates, which in turn reportedly interact with light and nutrition.
3. Germination of some seeds is controlled by phytochrome. In some cases CKs have been shown to exhibit a permissive effect relative to the action of GAs in the induction of germination. Increased CK levels have been shown to be induced in some seeds by red light.
4. High light intensity has also been shown to cause an increase in CK levels in some plants.

Light and Gibberellin Levels

1. The effect of light on GA production in plants has been studied most in relation to stem elongation, seed germination, and flowering. Most of these studies deal with phytochrome-mediated responses; few studies have focused on light-intensity effects. In one example, low light intensity caused an increase in five different GAs and an associated increase in growth. However, in other plants high light caused increases in GA levels.
2. The production of many different GAs during the developmental cycle of plants makes difficult the interpretation of the functions of the 80 or more different GAs isolated.

3. Differences that exist among plant species relative to the effects of light on GA levels reflect species, treatment, and analytical differences.
4. In general, light appears to cause an increase in GA levels.

Light and Abscisic Acid Levels

1. The effects of light intensity has been shown to both decrease and increase ABA levels in plants. Cultivar differences in crop plants may respond differently with respect to the effect of light intensity on ABA levels, which may be expressed in higher reproductive capacity.
2. Interactions between soil moisture and light intensity need to be studied further since some evidence suggests that high light intensity causes reduced ABA levels in plants simultaneously subjected to drought.
3. Red and far-red light have been shown to affect differentially the levels of ABA in plant tissues. Far-red light promoted ABA in lateral buds of cocklebur but stimulated ABA in stems of tomato. No effect of R or FR light on ABA levels in petioles or leaves of tomato has been observed. Thus light quality affects different tissues and plants relative to ABA levels.
4. Phytochrome may regulate ABA degradation rather than synthesis in some systems.

Light and Ethylene Levels

1. White light (red light) and high light intensity tends to inhibit C═C production, while FR light, reduced light intensity, and darkness tend to increase it.
2. Care should be taken in interpreting C═C data derived from excised plant parts, where C═C was determined in closed containers. Wound C═C and respired CO_2 levels can affect the results. Differences in CO_2 compensation points of C_3 versus C_4 plants could also affect C═C production by such plants in closed systems.
3. Ozone accentuates C═C production by soybeans in the dark. Light may inhibit ACC production, which in turn controls C═C.
4. Nodulation is decreased in the light and correlates with increased C═C production. Inhibitors of action/synthesis cause increased nodulation in dim light.
5. Leaf, flower, and bud abscission may be controlled by C═C accumulation in the dark or under shaded conditions.
6. Seed germination can be stimulated by C═C in some seeds. Theromodormancy can be overcome by red light treatment, which has been correlated with increased C═C production and germination in lettuce seed.

Nutrient Stress and Phytohormones

Nutrients and Indoleacetic Acid Levels

1. One question that arises relative to nutrient stress in plants is: Do changes in nutrient levels affect plant growth directly, or indirectly through alteration in hormone production?
2. Since the structure of IAA has N as a component, deficiencies in N can cause reduction in the levels of IAA.
3. Nitrogen source may also affect IAA levels. Urea reportedly increased IAA in mangos, and NH_4 may inhibit hormone synthesis, transport, or signaling (IAA or CK) in *Arabidopsis*.
4. Boron deficiency also influences IAA levels. However, conflicting results prevent absolute conclusions regarding its role. IAA accumulation has been observed in B-deficient plants, while reductions in IAA oxidase have also been observed.
5. Ultrastructural studies in B-deficient plants suggest that B may interact at the membrane level rather than on IAA levels.
6. Mn deficiency has been correlated with increased IAA oxidase activity and reductions in IAA.
7. Calcium deficiency appears to inhibit polar transport of IAA.

8. Zinc deficiency causes a reduction in IAA production but apparently does not affect tryptophan production.

Nutrients and Cytokinin Levels

1. Most studies regarding nutrient stress and CK levels involve N levels and sources. Since N is a component of CK structures, deficiencies usually cause reductions in CK levels.
2. Cytokinin levels are generally greater when supplied as NO_3^- rather than as NH_4^+.
3. Calcium and P deficiencies have also been shown to result in reduced levels of CKs.
4. Genetic propensity to produce higher levels of CKs may be beneficial relative to plant longevity under reduced N levels.

Nutrients and Gibberellin Levels

1. Little information is available regarding the effects of nutrition on GA levels.
2. Deficiencies in Mn inhibit several enzymes in isoprenoid biosynthesis that could affect GA, ABA, and CK biosynthesis.
3. A few reports indicate that N and K deficiencies can cause reductions in GA levels.
4. Gibberellin-producing fungi have been reported to produce elevated levels of GA when urea is used as a N source, but reduced levels when NH_4 is used. Magnesium sulfate also stimulated GA levels in fungi.

Nutrients and Abscisic Acid Levels

1. High levels of NH_4-N have been shown to increase ABA levels in plants compared to NO_3^--N. High nitrate levels did not stimulate long-distance transport of ABA, while ABA transport was stimulated by NH_4^+.
2. Low levels of P result in earlier increases in ABA levels in drought-stressed plants compared with well-watered plants.
3. Low K levels has been correlated with earlier production of ABA in grain seeds, which correlated with reduced yields.
4. Iron has been linked to an ABA-induced iron storage protein (ferritin). Iron deficiency could influence the expression of this protein in plants.

Nutrients and Ethylene Levels

1. Plants under stress degrade proteins to form NH_4^+. Ammonium has been found to cause elevated levels of C=C in plants. The ability to detoxify NH_4^+ may be an important part of the basis for stress resistance in some plants.
2. Deficiencies in Mg, Ca, and K have been shown to cause increases in C=C production in plants grown in either NO_3^- or NH_4^+. However, the levels were higher in the NH_4^+ medium.
3. Potassium has been observed to overcome C=C induction by NH_4^+ in some systems.
4. Deficiencies in N and P have been correlated with reduced levels of C=C in corn roots and increased aerenchyma tissue formation. Increased sensitivity to reduced C=C levels as a result of N deficiency may have resulted.
5. Iron deficiency may stimulate ferric-reducing capacity through C=C stimulation since C=C inhibitors reduced ferric-reducing capacity under Fe deficiency.
6. High Cu concentrations increased C=C production in tobacco and *Spirodea*.
7. High Ca has been shown to decrease C=C levels, which has been correlated with reduced disease severity and senescence. This response may be related to increased membrane stabilization by Ca.
8. The accumulation of heavy metals in soils can either promote or inhibit C=C synthesis by soil microflora, which could affect the growth of plants.

Salinity Stress and Phytohormones

Salinity and Indoleacetic Acid Levels

1. Little information is available regarding the effects of salinity on IAA levels in plants. One study found that NaCl decreased the levels of IAA in rice. The application of GA_3 could overcome the effects of salinity on IAA levels and growth inhibition, suggesting some interaction between the two hormones.

Salinity and Cytokinin Levels

1. As with IAA, little is known about the effects of salinity on CK levels in plants. However, in one study high salinity caused elevated levels of CKs and CK-ribosides. In another study, the conversion of ice plant from C_3 to C_4 metabolism using NaCl had no effect on CK levels but caused an increase in ABA. Neither change was linked to a role for these hormones in the metabolic conversion process.
2. Inherent ratios/balances of CK and GA may play a role in salinity tolerance in sorghum since exogenous applications of either can improve tolerance in this plant.

Salinity and Gibberellin Levels

1. Exogenously applied GA appears to increase IAA levels in salinity-stressed rice plants, which was correlated with increased tolerance.
2. Gibberellic acid has also been shown to increase germination of seeds in saline environments.
3. Gibberellic acid may be involved with IAA synthesis and/or prevention of degradation.

Salinity and Abscisic Acid Levels

1. Generally, ABA increases in plants exposed to high saline environments. However, separating specific ion effects and osmotic effects is difficult.
2. In halophytes, ABA levels increased when plants are transferred to nonsaline environments. Compared to glycophytes, ABA may function in controlling ion balances in saline environments rather than stomatal aperture.
3. Salt-induced ABA synthesis may originate from either shoots or roots but depending on growth conditions, one may dominate relative to the concentration produced at any one time.
4. NaCl may act through an independent pathway to change the sensitivity of rice cells to ABA relative to the induction of an ABA- and NaCl-inducible protein encoded by the Em gene.
5. Sodium chloride–induced ABA and related protein synthesis was shoot specific in alfalfa.

Salinity and Ethylene Levels

1. Generally, salinity stress increases C=C levels in vegetative tissues.
2. Older leaves usually generate more C=C than that generated by young leaves in saline environments.
3. Tomato fruits and tobacco callus exhibited no changes in C=C levels when grown under saline environments.
4. Proline can counter the effects of salinity stress on C=C production in the halophyte *Allenrolfea occidentalis*.
5. Tomato varieties respond differently than other crops to salt stress with respect to C=C production. Plants grown on NO_3^- produced higher C=C levels than when grown on NH_4^+. This is contrary to previous reports that NH_4^+ increased C=C production in crops.
6. Leaf epinasty was used as an index to determine tomato sensitivity to salinity. Petiole angle and C=C levels were correlated with NaCl concentrations.
7. Sodium chloride inhibits C=C production in lettuce seed, which has been found to be important in germination. Cytokinins could alleviate the stress-induced inhibition by increasing ACC synthase, EFE, and/or the action of C=C.

Flooding Stress and Phytohormones

Flooding and Indoleacetic Acid Levels

1. Concentrations of IAA increase in shoots and decrease in roots of flooded plants. This may be a result of C═C inhibition of polar transport of IAA.
2. Adventitious roots will frequently form on stems at the interface of the flood line. This may result from accumulation of IAA in this area.

Flooding and Cytokinin Levels

1. Cytokinins probably interact with ABA to influence stomatal aperture.
2. Root production of CKs may act as a signal under drought and hypoxic conditions (flooding) to influence physiological responses. Xylem fluxes of CK levels may be critical relative to the signaling process.
3. Little is known about compartmentalization of CKs in leaves compared to ABA. This may be important relative to the interaction of these two hormones.

Flooding and Gibberellin Levels

1. Plants adapted to aquatic environments, such as rice, respond to flooding by increasing GA levels, while plants that grow in more xeric environments exhibit reduced GA concentrations.
2. Under hypoxic conditions, C═C increases and may inhibit ABA levels, which antagonizes GA synthesis.
3. Higher GA levels stimulate stem elongation in the aquatic plant, allowing it to emerge above the waterline. This appears to be a different mechanism than that which occurs in terrestrial plants exposed to flooding.

Flooding and Abscisic Acid Levels

1. Abscisic acid usually increases in leaves of plants exposed to flooding stress. The origin of the ABA is uncertain but it may originate from the roots or from cellular compartments within the leaf. The result of this increase is usually stomatal closure.
2. Flooding induced C═C production may be antagonistic to ABA action relative to stomatal closure.
3. Pretreatment of plants with ABA can increase flooding tolerance in corn.

Flooding and Ethylene Levels

1. Flooding causes elevated levels of C═C in above- and below-water parts of plants. Flooding symptoms in plants mimic effects of high C═C levels.
2. Anoxic conditions inhibit C═C production in roots but not ACC synthesis. ACC can be transported to the shoot where O_2 is available, and C═C can be synthesized.
3. Plants respond to flooding and high C═C levels by forming adventitious roots, reduced root growth, wilting, stem and lenticular hypertrophy, arenchyma tissue formation, increased stem diameter, stunting, and increased phloem growth and tracheid diameters. The ability of tolerant plants to respond to flooding by altering their morphological characteristics may, in part, explain their success in growing in such environments.
4. Ethylene promotes growth of petioles of some species of *Rumex* when submerged. This may be through C═C inhibition of ABA production and increased sensitivity to GA production. This response may be beneficial to the plant relative to reaching the surface of the water and a source of O_2 and light.
5. Flood-tolerant waxapple does not accumulate C═C or precursors. Both flooded and nonflooded plants form aerenchyma tissue in the roots. Flooding tolerance is attributed to the formation of several isozymes of alcohol dehydrogenase.

Drought Stress and Phytohormones

Drought and Indoleacetic Acid Levels

1. Indoleacetic acid levels are generally reduced in drought-stressed plants. This may be a result of increased IAA oxidase activity.
2. Drought stress also reduced polar IAA transport, which may be related to reduced cambial activity in drought-stressed woody perennials.
3. Abscission of plant parts under drought stress is probably related to reduced polar transport of IAA. Maintenance of high levels of IAA in leaves and fruits appears to be necessary to prevent abscission.

Drought and Cytokinin Levels

1. Cytokinins are believed to be synthesized in the roots and transported to the leaves via the xylem. Drought generally causes a reduction in CK levels in leaves, probably as a result of reduced production in the roots.
2. Cytokinins interact with ABA to control stomatal opening. Abscisic acid closes stomates, CKs cause opening.
3. Drought-stressed sunflower exhibited reduced CK levels in xylem sap and increased CK-ribosides in roots. These may be storage forms that are not translocatable and may be induced by high levels of ABA in drought-stressed roots.
4. Diurnal fluctuations occur in CK and ABA levels. The relationship is inverse with CK increasing during the morning and declining in the afternoon. The role of CKs in plant water relations is uncertain, but a threshold level of ABA may be necessary to prevent CKs from opening stomates.
5. Drought resistance in tomato was correlated with reduced levels of CK-ribosides in osmotically stressed roots. Differences in drought resistance among cultivars may be related to cultivar variability in metabolic control of free and conjugated CK synthesis.

Drought and Gibberellin Levels

1. Hot, dry conditions are needed to induce flowering in coffee and most conifers. Addition of GA can enhance flowering in both. In spruce, increase in nonpolar GAs occurs during hot/dry conditions, which may be necessary for reproductive development. Conversely, cold wet conditions caused polar GAs to be produced and maintenance of vegetative growth.
2. Elongation growth in dark grown soybean hypocotyls was not correlated with low water potential–induced reductions of GA or increased ABA concentrations.
3. Water stress had no effect on GA levels in sunflower shoots or roots.

Drought and Abscisic Acid Levels

1. Abscisic acid increases in plant tissues exposed to osmotic or drought stress.
2. The most studied response of plants to drought-induced ABA production is stomatal closure. The mechanism of ABA increase in leaves and how it controls stomatal closure is controversial but probably involves decompartmentalization of ABA in leaf mesophyll cells as well as transport from the root to the leaves. Abscisic acid probably attaches to the exterior surface of the guard cell membrane and changes ion transport characteristics, leading to loss in guard cell turgor and stomatal closure. The response may be turgor sensitive in the leaves of some plants and osmotically sensitive in roots.
3. One system may operate on a short-term basis, sensing the atmospheric environment, while the other monitors the soil/root environment.
4. Other plant responses implicated with drought and ABA production include dormancy, abscission, apical dominance, flowering, growth, tillering, pollen viability, germination, seed set, embryo maturation, root hydraulic conductance, and leaf heterophylly.

Drought and Ethylene Levels

1. Drought stress causes increases in C=C levels in plants, which has been correlated with abscission (leaf, flower, fruit, and buds), reduced height, increased diameter, senescence, and fruit ripening.
2. Abscission has frequently been observed in plants after drought stress has been removed. This may be related to reduced translocation of ACC to leaves under dry conditions, and when drought is relieved, ACC is transported to the shoot, where it is oxidized to C=C.
3. Clonal differences have been observed in *Eucalyptus* relative to C=C production, abscission, and drought stress, suggesting metabolic differences in the control of C=C under drought conditions.
4. Older leaves appear to be more sensitive than younger leaves to C=C, and sensitivity may be more of a factor in abscission than absolute amounts of C=C.
5. Multiple hormone interactions are probably involved in the abscission process, and changes in hormone balances lead to different phases in the abscission process.
6. Light intensity interacts with drought to affect C=C and ABA levels in peas. High light and drought caused decreases in C=C and an increase in ABA. However, in tomato high light and drought caused an increase in both hormones.
7. Ethylene declined in flower buds of coffee plants subjected to drought stress. However, leaves exhibited increased levels of C=C, showing that specific tissues respond differently to drought with respect to C=C production.
8. Seasonal increases in C=C levels in sapwood and heartwood of pine were correlated with periods of drought.
9. Drought stress induces C=C and pithiness in some cultivars of tomato. Mechanical perturbation or treatment of tomato fruits prior to drought stress reduces the production of enzymes involved in the development of pithiness.

Temperature Stress and Phytohormones

Temperature and Indoleacetic Acid Levels

1. High temperatures are known to reduce fruit set in tomato. Indoleacetic acid and GA may both be involved in this response. Endogenous levels of these hormones decline under high-temperature stress in tomato. Exogenous applications of IAA are known to improve fruit set under high temperatures.
2. Elevated temperature caused reduced levels of CK and IAA in two morphologically distinct tobacco tissue lines. Morphological differences were not attributed to changes in ratios of these two hormones, as has been demonstrated with exogenous application.
3. Cold treatment has been reported to cause an increase in IAA oxidase activity in cucumber and wheat. Gibberellic acid treatment of wheat seed also increased IAA oxidase activity. Gibberellic acid may control IAA oxidase activity during vernalization, resulting in reduced levels of Indolacetic acid.

Temperature and Cytokinin Levels

1. Cytokinin levels tend to decline under high-temperature stress. Such changes have been observed in tissue culture and related to changes in root development and apical dominance, in chlorophyll retention and photosynthesis in leaves, and in abortion of corn kernels and reduction in yields.
2. Cold temperatures also cause reductions in CK levels and increases in ABA levels. Cold hardening in plants appears to be related to increased ABA and dehardening related to increased Cytokinins.

Temperature and Gibberellin Levels

1. Flowering and growth are induced by cold treatment in some plants. Application of GA has been shown to substitute for cold treatment, suggesting that endogenous GA controls these processes.
2. Reduced temperature can directly affect biosynthesis of specific GAs in localized tissues (shoot tips) and affect specific functions (shoot elongation and floral induction).
3. Low temperature caused reduced production of GAs in tomato leaves and increased flower numbers compared to tomatoes grown at normal temperatures, where GA levels were lower in the leaves and fewer flowers were produced. Flowering was negatively correlated with GA levels in *Eucalyptus,* but other factors were considered to be operative since GA levels were similar at lower temperatures, and maximum flowering occurred only at the lowest temperature.
4. Light and low temperature interact to affect GA levels and germination in GA-insensitive mutants and wild-type seeds of *Arabidopsis.* Absolute levels or specific GAs appear not to be a factor in seed germination; rather, changes in sensitivity to GA in response to light and temperature are more important.

Temperature and Abscisic Acid Levels

1. Abscisic acid can increase in plants exposed to high and low temperature. Abscisic acid application to plants has been somewhat successful in increasing temperature tolerance. This is not true for all plants.
2. Plants grown under high relative humidity and adequately watered still exhibited accumulation of ABA when exposed to both high and low temperature. Thus temperature per se appears to influence ABA levels in plants.
3. Photosynthetic capacity of winter and spring wheat varieties has been correlated with differential changes in ABA levels associated with the ability of the two varieties to acclimate to cold and deacclimate to higher temperatures.
4. Selecting for ABA-insensitive wheat clones from tissue callus culture resulted in regeneration of seedlings that grew longer, were more tolerant to heat stress, and produced higher yields.
5. Seed stratification can affect seed ABA levels. Low-temperature stratification caused a reduction in apple seed coat ABA levels, as did high temperature stratification. Stratification duration also affects ABA levels. Abscisic acid increases in heat-stressed corn kernels, causing reduction in yield.
6. ABA levels have been observed to increase more quickly in drought-stressed plants than in chilling-stressed plants, even though chilling-stressed plants had a lower relative water content than that of drought-stressed plants. Thus the mechanisms of drought and chilling stress induction of ABA seem to be very different.
7. Low-temperature acclimation of freezing-tolerant and freezing-susceptible species of cacti was correlated with the ability of tolerant species to remove water rapidly from chlorenchyma tissues and accumulate higher levels of ABA.
8. Wheat cultivars appear to vary with respect to cold acclimation and ABA accumulation. Some show a strong correlation between acclimation and ABA accumulation, while others actually exhibit a decrease in ABA levels with acclimation.
9. Abscisic acid–insensitive or abscisic acid–deficient mutants of *Arabidopsis* have impaired freezing tolerance but have higher levels of ABA than do the more tolerant wild-type. This may suggest an indirect role for ABA in freezing tolerance.
10. Dieback in maple was related to reduced soil temperatures during dormancy and ensuing elevated levels of ABA in sap, which was attributed to reduced bud break and leaf growth during the following growing season.
11. Tomato fruits stored at chilling temperatures exhibited elevated ABA levels, but increases were not correlated with increased damage symptoms.

Temperature and Ethylene Levels

1. In dogwood, C=C was found to be high in branches only during the growing season. During dormancy C=C levels were very low or not detected. Ethylene was not considered as important in breaking dormancy.
2. Thermodormancy in chickpea seed was related to decreased production of C=C when germination was initiated at high temperatures.
3. Differing cultivars of bean plants produce C=C at different rates at high temperature. Breeding for low producers was viewed as an option for reducing C=C-induced flower abscission and increased fruit set.
4. Plants subjected to one stress may develop tolerance to other stresses, a phenomenon referred to as cross tolerance. Green tomato fruits stored for short periods at high temperatures exhibited better storage capacity at low temperatures than that of tomatoes not treated at high temperatures. Ethylene increased during the heat treatments; however, membrane leakage was lower, which may indicate greater membrane integrity. Heat-shock proteins increased, which could have contributed to better tolerance to the lower temperatures.
5. Membrane integrity appears to be necessary for C=C production, suggesting that enzymes involved with C=C synthesis are membrane bound.
6. Chilling stress and C=C production have been studied extensively in the postharvest physiology of fruits and vegetables.
7. Plants are highly variable in their response to chilling stress. Some produce higher C=C levels during the chilling process, while others produce peaks of C=C when transferred from a cool environment to a warmer one. Since enzymes involved in C=C production appear to be attached to membranes, these different responses may reflect differences in membrane integrity. In other instances, evidence suggests that chilling reduces the activity of enzymes involved in C=C synthesis.
8. The rate of production of C=C produced by plants after having been chilled for varying times at 0°C has been used as a measure of chilling tolerance in some plants.
9. Intact whole plants may differ relative to C=C production under chilling temperatures compared to leaf discs, as suggested by studies with bean plants.
10. Some evidence supports the notion that chilling-induced C=C production is a result of chilling-induced dehydration.
11. Ripening rates can be enhanced in some fruits by low-temperature storage and subsequent transfer to higher temperatures.

Plant Pathogens and Phytohormones

Fungal Diseases

1. Considering how abiotic factors can influence hormone balances in plants, the interaction of plants with biotic organisms increases the complexity of these relationships considerably. This is particularly true since many microorganisms can produce the same substances that act as hormones in plants.
2. Fungal diseases are the most prevalent among plants. Fungi have the ability to produce essentially all of the five classical phytohormones. As a general rule, all hormones tend to increase in plants as a result of fungal infection. It is frequently difficult to determine if the host or the pathogen is responsible for the increased levels of phytohormones.
3. Fungal infections of plants result in symptoms suggestive of increased hormone production, such as tumorous tissue development in corn smut; which has been related to high IAA levels; green island formation on leaves caused by rusts and powdery mildews, which are indicative of increased CK levels, and club root of cabbage, which produces tumorous growths high in CK and IAA.
4. Although GAs were first associated with "foolish seedling disease" of rice, and numerous fungi have been shown to produce GAs, little research has been conducted relative to the role of GA

in disease development in plants. However, some studies have demonstrated increases in GA levels in fungal-infected plants which correlated with increased growth.

5. Abscisic acid is produced by a number of saprophytic and phytopathogenic fungi. Exogenous applications of ABA have suggested that ABA can suppress natural defense responses to infection in plants.
6. Defoliating strains of fungi have been observed to cause elevated levels of ABA in cotton plants, while nondefoliating strains do not.
7. Other studies utilizing fungi in compatible and incompatible reactions, as well as pathogenic and nonpathogenic genera, indicate that ABA may be required for initiating defense responses or has no role at all in the infection process.
8. Ethylene can be generated in plants by wounding or by fungal invasion. The plant response is similar biochemically and symptomologically. Certain chemicals, such as phytoalexins, are triggered by C=C and act to suppress pathogen development. Other chemicals act to isolate the organism to prevent its spread throughout the plant.
9. The involvement of C=C in disease development is unclear, but it is well known that fungi can produce C=C and that factors such as nutrition may interact with C=C to affect fungal associations, as has been demonstrated with P nutrition, C=C production, and the establishment of mycorrhizal associations.

Bacterial Diseases

1. Bacteria can synthesize IAA via different pathways compared to host plants, which may give them an advantage relative to disease establishment.
2. Much of our understanding of bacterial control of phytohormones in plants comes from studies utilizing *Agrobacterium tumefaciens* and *Pseudomonas savastanoi*, both of which can produce IAA and CK. The former bacterium can introduce T-DNA, which is responsible for producing bacterial IAA and CK into the host, which can subsequently control the host synthesis of these hormones. *P. savastanoi* does not have this ability but can change the host hormone balance by secretion of IAA and CK.
3. Several genes have been described and characterized that are known to interact to control free and conjugated forms of IAA, CK, and GA, as well as change sensitivity to them. Expression or repression of these genes has been shown to alter morphological expression in tumorous tissues grown in culture as well as in transgenic plants.
4. Bacteria may not produce ABA, but most infection has been shown to induce ABA or ABA-like substances in plant tissues and root nodules.
5. Bacteria can produce C=C and induce C=C production in host plants. In some cases, nodulating bacteria produce toxins that inhibit C=C production in the host, which appears to facilitate nodulation. In other bacterial species chemicals are produced that enhance C=C production by the host. Different strategies appear to have evolved relative to the manipulation of host C=C to the advantage of the pathogen.
6. Although leaf abscission is more common as a result of fungal infection, several bacteria have been shown to cause leaf abscission as a result of C=C induction in the host.
7. Some evidence indicates that C=C is not required for active resistance against avirulent bacteria and that C=C is not involved in generation of pathogen-induced symptoms.

Viral Diseases

1. Unlike other phytopathogens, viruses do not produce phytohormones.
2. Studies concerning the effects of viral infection on IAA and CK production in plants are limited and inconclusive. Virus infection can cause increases or decreases. Viral symptom expression has not been associated with IAA accumulation, although such results have also been variable.
3. Exogenous application of CKs can increase or decrease resistance to viral infection, but timing

of application relative to inoculation seems to be critical. There is also some evidence for a role for CKs in systemic resistance.

4. Where viral infection is associated with stunting, there is some evidence for changes in the quantitative and qualitative composition of host GAs.
5. Plant ABA production is generally, but not always, elevated as a result of viral infection. Viruses that produce local lesions appear to induce higher levels of ABA in the host than do systemic viruses. Also, ABA levels appear to correlate with symptom severity.
6. Apoplastic ABA levels in tobacco leaves increases with severity of viral infection. Interestingly, stomatal closure was not observed.
7. Abscisic acid may control the conversion of free CK to conjugated forms in green islands induced by viral infection.
8. As with ABA, viruses that induce local lesion tend to induce C=C production, whereas systemic viruses do not. Whether C=C is the causual agent of virus symptom development or simply accentuates the process has not been resolved. In several instances C=C production or C=C-forming enzymes coincided with or preceded necrotic lesion development.

SELF-STUDY QUESTIONS

1. Name and draw the five classical phytohormones.
2. Outline several responses in plants attributed to each phytohormone.
3. In general terms, describe the modes/mechanisms of plant hormone action.
4. What aspects of light treatment are important with respect to the effects of light on hormone control in plants?
5. Describe two pigments that are involved in photoreception, and describe how they may interact with phytohormones to affect growth and development.
6. Describe several photomorphological responses in plants that control light-perceiving pigments in plant.
7. Describe how various environmental conditions can affect ratios of red and far-red wavelengths of light that plants might receive.
8. In general terms, how would you assess the impact of high light and red light on IAA, CK, and GA levels in plants? Give physiological examples of how such changes may affect plant processes.
9. Address question 8 relative to ABA and C=C.
10. Deficiencies in five elements can affect IAA production or function in plants. Name the five elements and indicate how they can influence IAA in plants.
11. What form of nitrogen would probably maintain a higher level of CKs in plants?
12. Why is nitrogen required for the maintenance of CK levels?
13. Name three hormones produced in the isoprenoid pathway, and indicate an element that if limiting could affect their synthesis.
14. How are ABA levels affected by nitrogen source and K and P levels? What affect does the altered level in ABA have on various plant functions?

15. Which is more effective in inducing C=C production, NO_3-N or NH_4-N?

16. How does the answer to Question 15 relate to stress-induced responses in plants?

17. What is the relationship between membrane stability, Ca levels, and C=C production?

18. How might the accumulation of heavy metals in soils affect C=C production in soils and plants growing in the soil?

19. What hormone has proved effective in alleviating salinity stress in some plants?

20. Would you expect halophytes to respond the same as glycophytes relative to the function of ABA in plants? Explain.

21. How does salinity generally affect C=C production in plants, and what chemical (not a hormone) has been shown to prevent the accumulation of C=C in plants subjected to salinity stress?

22. Outline the symptoms of flooding stress in plants.

23. Associated with the symptoms outlined in Question 22, suggest one or more hormones that may be responsible for each.

24. How do aquatic vascular plants respond to flooding compared to terrestrial plants relative to the impact of flooding on GA levels and function?

25. Why does C=C not accumulate in the roots of anoxic plants but does accumulate considerably in the shoots?

26. One of the symptoms of flooding stress in plants is leaf wilting. Why would this occur with sufficient water supply?

27. Explain how ABA and CK may interact to control stomatal aperture under drought-stress conditions.

28. Abscission of leaves, fruits, flowers, and buds is commonly observed in drought-stressed plants. Indicate what happens to various hormones in a plant under drought stress, and indicate how each may be involved in the abscission process. Is this process an advantage or disadvantage to the plant? Explain.

29. How do hot, dry conditions affect flowering in conifers, and what hormone may be involved with this process? Is this a stress response?

30. In some cases, leaf abscission has not been observed in drought-stressed plants until the stress is removed. What could be controlling this response?

31. What is the potential for using C=C generated by woody perennials as a means of monitoring plant stress? What evidence suggests that this is possible?

32. What is the relationship between tomato fruit set and high temperature and IAA levels?

33. What relationship may exist between IAA oxidase activity, cold treatment, and GA levels in wheat?

34. What hormones appear to be involved in the cold hardening and dehardening process?

35. How may high-temperature stress affect kernel development in corn and why?

36. Substituting GA for cold treatments in some plants can induce flowering. What does this suggest about the relationship of GA and cold treatment in plants? Is GA concentration paramount in this consideration? What should be considered other than concentration?

37. How can the effects of temperature and drought be separated relative to the impact of each on ABA levels in plants? Can temperature per se influence ABA levels? Give evidence for or against.

38. What genetic attribute has been selected for in wheat which increased yield under heat stress? Why do you suppose this would be a effective approach?

39. Drought- and chilling stress induced–ABA production seem to be induced by different genetic controls. What evidence suggests that this true?

40. Give several lines of evidence that ABA is involved with chilling tolerance in plants.

41. What evidence indicates that ABA may be involved with dieback in maple?

42. What evidence suggests that C=C production under high temperature is under genetic control, and what practical use might be made of this information?

43. What is cross tolerance as it pertains to abiotic stress? Give an example relative to temperature.

44. Some plants release increased C=C levels during chilling, whereas others produce it after being removed from the stress. Explain why such variation may occur among plants.

45. Can rate of C=C evolved after chilling stress be used as a measure of chilling tolerance in plants? Explain and give an example.

46. Can chilling stress induce dehydration? Explain how.

47. Compared to abiotic stress, what additional variable is introduced into plant pathogen interactions that further complicates understanding of the role of phytohormones in this process?

48. In the case of pathogen infection of plants, what symptoms develop that implicate hormonal control of the response?

49. What are green islands, and what hormone/s are suggestive of their presence?

50. As a general rule, how do fungal pathogens influence hormone levels in plants?

51. "Foolish seedling disease" of rice was associated with what hormone?

52. How might C=C production in infected plants be involved with defense responses in plants?

53. What is unique about bacterial synthesis of IAA that may give these organisms a competitive advantage relative to host infection and reproduction?

54. How can bacteria such as *Agrobacterium tumefaciens* and *Pseudomonas savastanoi* alter and/or control host hormone production?

55. How can bacteria alter host C=C levels to their advantage?

56. What is unique about viruses relative to hormone production compared to other phytopathogens?

57. Discuss the impact of timing of application of CKs relative to virus inoculation and further reproduction and growth in the host plant.

58. What is unique about local lesion production versus systemic viral infections relative to C=C and ABA production?

SUPPLEMENTAL READING

Abeles, F. B., P. W. Morgan, and M. E. Saltveit, Jr. 1992. Ethylene in Plant Biology, 2nd ed. Academic Press, New York.

Davies, W. J., and H. G. Jones (eds.). 1991. Abscisic Acid: Physiology and Biochemistry. Bios Scientific Publishers, Oxford.

Goodman, R. N., Z. Király, and K. R. Wood. 1986. The Biochemistry and Physiology of Plant Disease. University of Missouri Press, Columbia, MO.

Heitefuss, R. and P. H. Williams (eds.). 1976. Physiological Plant Pathology, (Encyclopedia of Plant Physiology, Vol IV). Springer-Verlag, New York.

Kozlowski, T. T. (ed.). 1984. Flooding and Plant Growth. Academic Press, New York.

Letham, D. S., P. B. Goodwin, and T. J. V. Higgins (eds.). 1978. Phytohormones and Related Compounds: A Comprehensive Treatise, Vol. II, Phytohormones and the Development of Higher Plants. Elsevier/North-Holland Biomedical Press, New York.

Mattoo, A. K., and J. C. Suttle. 1991. The Plant Hormone Ethylene. CRC Press, Boca Raton, FL.

Morgan, P. W. 1990. Effects of abiotic stresses on plant hormone systems. Pp. 113–146 in: R. G. Alscher and J. R. Cummings (eds.), Stress Responses in Plants: Adaptation and Acclimation Mechanisms. Wiley-Liss, New York.

Pharis, R. P., and D. M. Reid (eds.). 1985. Hormonal Regulation of Development. III. Role of Environmental Factors (Encyclopedia of Plant Physiology, Vol. II). Springer-Verlag, New York.

5 Stable Isotopes and Plant Stress Physiology

Outline

Objectives

1. Explain what isotopes are and how they may help to evaluate plant stress physiology in general.
2. Describe the carbon isotope family and the mechanism by which carbon isotopes are fractionated in nature.
3. Identify the components of plant metabolism that discriminate against carbon 13.
4. Describe how ratios of carbon isotopes in plant tissues can be used to assay aspects of plant physiology.
5. Describe the isotope family of nitrogen and natural fractionation mechanisms.
6. Explain how ratios of nitrogen isotopes in tissues can be used to measure aspects of nitrogen fixation and other nitrogen metabolic processes.
7. Describe the isotope family for hydrogen and discuss natural fractionation processes.
8. Explain how ratios of hydrogen isotopes can be used to determine the source of water for plants.
9. Describe the isotope family for oxygen and its natural fractionation processes.
10. Explain how oxygen isotopes can be used to determine the source of water for plants, and uptake kinetics of phosphate or nitrate.

I. INTRODUCTION

Biochemical and physiological measurements for analyzing plant response to the environment abound. These assays are normally limited in space or time. Photosynthetic measurements can be done on chloroplasts, single cells, protoplasts, leaf discs, leaves, and shoots, but rarely are whole plants measured. Furthermore, gas exchange, enzymatic, or biochemical constituent measurements are snapshots of metabolic processes recorded over a short time period. To represent the whole plant over a season or a life cycle, the gas exchange measurements must be scaled up, by mathematical means, from the leaf to the whole plant. Other physiological measurements (enzyme activities, membrane properties, etc.), are studied in vitro as reflective of in-vivo processes. Therefore, the classical techniques for assaying physiological processes are inadequate for interpreting the significance of the process to the whole plant in a field situation over the seasonal or annual cycle.

Physiological assays are also frequently limited by sample size because the time or technical resources required for the measurement preclude multiple determinations. Recent advances in gas exchange technology does allow rapid measurement of photosynthesis (in a few seconds), which results in a larger sample size (in the range 40 to 70 samples/hour). However, this sample size may be inadequate to account for the leaf age structure and microenvironmental variability in the canopy. Pressure–volume analysis of plant water relations is limited to small sample size, due to the length of assay time (5 to 6 h/curve). Enzymatic assay is further limited by the time required to transfer the live material to the laboratory, where the assay is often performed.

The inherent problems of limited sample size, scaling up, and in vitro analysis when studying plants in their natural state have made interpretation of physiological assay for whole plant response to stressors difficult. Physiological assays have to be done frequently, over the season or year, to integrate temporal effects of stressors. To represent the whole plant, physiological assays must be done on organs of various developmental stages and interpreted in relation to the proportional abundance of organs of specific developmental stages in the whole plant. Therefore, any physiological measurement done on one developmental stage or at one time cannot be used to evaluate whole plant responses to stress. This disquieting feeling of speculating from short, infrequent, and few samples to the life cycle of whole plants has stimulated scientists to search for assays that would integrate physiological processes over larger temporal and spatial scales. The relative abundance of stable isotopes in plants has become the most important integrative assay for plant physiological processes.

All students of biological systems, particularly physiology and biochemistry, are aware of isotopes and their utility in research. Isotopes can be defined simply as atoms of a particular element with a variable atomic weight, due to different numbers of neutral particles in the kernel. Therefore, the various isotopes have a different atomic weight, but the same atomic number and no electrical charge. Each element can have a family of isotopes that often vary across many atomic weight units (e.g., $^{8}C \rightarrow ^{20}C$). Within each family of isotopes (that for one element) there is a diversity of stability among its members (Figure 5.1). Physiologists are mostly aware of the radioactive isotopes, which are unstable. The radioactive isotopes will emit energy (electromagnetic or particle) as they change from one atomic form to another. The type of emission can vary among family members such that one may emit high-energy wave (gamma emission: ^{11}C) or low particle energy (beta emission: ^{14}C). Furthermore, the stability of the isotope can vary greatly among family

Carbon isotope family

Designation	^{10}C	^{11}C	^{12}C	^{13}C	^{14}C	^{15}C
Abundance	---	---	98.9%	1.1%	---	---
Half life	19.3s	20.3m	---	---	5715y	2.45s
Standards	Pee Dee Belemnite $^{13}C/^{12}C$ = 0.0112372					

Hydrogen isotope family

Designation	^{1}H	^{2}H	^{3}H
Abundance	99.985%	0.015%	----
Half life	----	----	12.32y
Standards	Vienna standard marine ocean water $^{2}H/^{1}H$ =0.00015576		

Oxygen isotope family

Designation	^{14}O	^{15}O	^{16}O	^{17}O	^{18}O	^{19}O
Abundance	----	----	99.76%	0.04%	0.20%	----
Half life	70.6s	122.2s	----	----	----	----
Standards	Vienna standard marine ocean water $^{18}O/^{16}O$ = 0.0020671					

Nitrogen isotope family

Designation	^{12}N	^{13}N	^{14}N	^{15}N	^{16}N	^{17}N
Abundance	----	----	99.63%	0.37%	---	----
Half life	0.1s	10m	----	----	----	----
Standards	Atmosphere $^{15}N/^{14}N$ = 0.0003676					

Figure 5.1 Partial list of isotope families for carbon, hydrogen, oxygen, and nitrogen. Those isotopes without reported half lives are stable isotopes. Half-lives of radioactive isotopes are reported in: s = seconds, m = minutes, d = days, y = years. The natural abundance of each isotope is based on the total global quantity of the element.

members. Stability is measured as the half-life of the isotope. *Half-life* is the amount of time required for one-half the mass of an isotope in a specific system to change from its unstable to its stable form. An isotope's half-life may vary from a few seconds (^{10}C) to thousands of years (^{14}C) within the same family of isotopes (Figure 5.1).

Isotopes with short half-lives are frequently used as molecular identification markers in biochemical tracer studies. A phosphorus isotope ^{32}P is frequently used for labeling proteins or probing for nucleic acid structure. ^{11}C has been used to understand the processes of in vivo carbohydrate transport because, as a gamma emitter, a nondestructive assay (i.e., no tissue collection) is possible because the radioactive emission can be detected with a Geiger counter. The half-life of ^{11}C is short (Figure 5.1), so that it can only be used in research labs closely connected to a cyclotron where ^{11}C is generated. Tritium (^{3}H) or ^{14}C are used for identifying and following carbon chains or pieces of organic molecules in fairly long biochemical pathways because they have a relatively long half-life and emit radiation of relatively low energy. Isotopes with exceptionally long half-lives, such as ^{14}C and ^{235}U, are also used as geological clocks. These are only a few of the many radioactive isotopes used to identify and follow molecules through physiological processes.

Stable isotopes are not radioactive and half-life does not apply to these isotopes. Therefore, stable isotopes cannot be assayed by scintillation counting or other radioactivity tracing techniques because they do not emit any energy or particles. Stable isotopes must be identified by mass spectrometry, a technique that is able to separate elements into atoms of different weights, or by nuclear magnetic resonance. Each family of isotopes has several stable isotopes (Figure 5.1). Of course, the most common weight for the atom is stable (e.g., ^{12}C), and other atomic weights of the atom may also be stable (e.g., ^{13}C).

The natural abundance of stable isotopes is dominated by one most common form (e.g., $^{12}C = 98.9\%$ of all C). The natural abundance of an isotope is an indication of what proportion of the total global amount of the element (or other defined pool) is accounted for by that particular isotope. The relative abundance of the less frequent stable isotope varies among different compartments of the environment (organic, inorganic, etc.) or the organism (lipids, carbohydrates, protein, water, etc.). This variation, in isotopic composition, among compartments is due to fractionation of the less frequent isotopes (*isotope effects*) among compartments. Fractionation of isotopes occurs when molecular or atomic bonds are formed or broken, or when other nonreactive process occur that are affected by atomic mass (such as molecular diffusion).

The isotope effects result in fractionation of isotopes among molecules or locations. Fractionation processes are usually grouped into kinetic or thermodynamic processes (Figure 5.2). The kinetic effects are nonequilibrium processes such as the relative diffusion rates of isotopes in a specific medium or the preference of an enzyme reaction for the various isotopes. Normally, kinetic effects discriminate against the heavier isotopes and are not dependent on temperature. Thermodynamic effects are those in which two kinetic effects come to an equilibrium and result in an equilibrium isotope effect often called a *fractionation factor* (Figure 5.2). For example, the ratio of two isotopes on either side of a phase barrier (such as liquid/gas) is dependent on the thermodynamic diffusion processes regulating the movement of isotopes into or out of each phase. The resulting concentrations of each isotope in each phase will be due to the equilibrium results of two kinetic processes, controlling the diffusion of the isotopes into and out of each phase. Equilibrium isotope effects are sensitive to temperature.

Abiotic process cause fractionation of isotopes in the environment, and biotic processes cause further fractionation from the environment to the organism and among various biochemical components of the organism. Biotic fractionation of isotopes creates a ratio of isotopes in tissue that reflects the integrated result of physiological processes in the whole plant over relatively long time periods. Different physiological processes discriminate against heavier isotopes to different degrees. In addition, the factors that regulate the equilibrium isotope effects of specific plant resources (e.g., CO_2 diffusion) will affect the isotopic composition of resources available to physiological processes. It is this combined impact of abiotic isotope effects and physiological discrimination that makes stable isotope composition an integrated index of whole plant responses and has sparked the interest of physiologists in these techniques and their application to plant stress physiology. Stable isotopes have been investigated as a potential assay for whole plant, integrated physiology for over 40 years (Craig 1954). The isotope families that have received the most attention are those of C, H, O, and N (Figure 5.1). We will survey the basic nature of these isotope families and their application to physiological analyses. Study of the information that stable isotope fractionation can provide

Kinetic Isotope Effects

$$R_r \xrightarrow{\alpha_{r\to p}} R_p$$

$$\alpha_{r\to p} = \text{kinetic effect} = \alpha_{kinetic}$$

$$\alpha_{kinetic} = R_r/R_p$$

for example

$$\delta^{13}C_r = \alpha_{r\to p}\delta^{13}C_p$$

$$\alpha_{kinetic} = \frac{\delta^{13}C_p}{\delta^{13}C_r}$$

Thermodynamic isotope effect

$$R_x \underset{\alpha_{y\to x}}{\overset{\alpha_{x\to y}}{\longleftrightarrow}} R_y$$

$$\alpha_{equilib} = R_x/R_y$$

for example

$$CO_2 + H_2O \Leftrightarrow H_2CO_3$$

$$\alpha_{equilib} = \frac{\delta^{13}C(CO_2)}{\delta^{13}C(H_2CO_3)}$$

Figure 5.2 In kinetic isotope effects R_r and R_p are the $^{13}C:^{12}C$ ratios for the reactants and products, respectively. $\alpha_{r\to p}$ is the isotope effect of the reaction or process that transfers the isotope from the reactants to the products and is often referred to as $\alpha_{kinetic}$. In the case of CO_2 diffusion $\alpha_{kinetic}$ is the ratio of diffusion coefficients for $^{12}CO_2$ and $^{13}CO_2$ in air and can be defined as $\delta^{13}C_p/\delta^{13}C_s$. Thermodynamic isotope effects concern reversible reactions or balances between opposing reactions. In this case the isotope ratios of compartment x and y (R_x, R_y) are affected by two opposing isotope effects ($\alpha_{x\to y}$ and $\alpha_{y\to y}$). The equilibrium isotope effect ($\alpha_{equilib}$) for the hydration of carbon can be defined as the ratio of isotopic ratios of CO_2 and HCO_3.

for plant stress physiology is relatively young, and we expect many more possible applications in the future.

The main isotopes used in physiological analyses have different abundances and levels of natural fractionation, both of which affect the interpretation of results and the ease of analysis. Thus, an analysis of the meaning of the δ unit [see equation (5.1)] is important. Deuterium (^2H) has the smallest shift in isotopic composition per unit δ because of its small natural abundance compared with the other isotopes used in plant physiology studies (Table 5.1). However, individual fractionations are large (50 to 100‰), making them easy to measure. If the fractional difference in deuterium between reactant and products in 50‰, this would correspond to a change in fractional abundance of 7.5×10^{-6}. In contrast, ^{18}O has the greatest fractional abundance (10 times larger than ^2H), but this is counterbalanced by small natural range in δ values. If a fractionation of 50‰ were to occur for ^{18}O (an unrealistic value), this would correspond to a change in fractional abundance of 1.0×10^{-5}. Commonly, fractional differences for ^{18}O are in the range 1 to 10‰. ^{13}C has a relatively large fractional abundance in air. Although there is a small difference between isotope weights (^{13}C versus ^{12}C), and small fractionations occur by diffusion processes, biotic processes can have large discriminations [29‰ for rubisco (ribulose bisphosphate carboxylase-oxygenase)]. Tissues also have large amounts of carbon, so measurement of the isotope composition is not difficult. ^{15}N has the third largest natural abundance, and fractionation can be relatively large (up to 50‰). The commonly observed range of fractionation is lower ($-10 \rightarrow +20$‰), and equilibria established among various processes

TABLE 5.1 Natural Occurrences, Fractional Changes, and Observed δ Units for the Four Most Important Stable Isotopes in Studies of Plant Physiology and Ecology[a]

	Heavy Isotope[a]				
	^{15}N	^{2}H	^{18}O	^{13}C	^{34}S
Standard	N_2 air	SMOW[b]	SMOW[b]	PDB[c]	CD[d]
Mean fractional abundance[e]	0.00366	0.00015	0.00204	0.0111	0.0422
1 δ as fractional change of isotope composition	4×10^{-6}	1.5×10^{-7}	2×10^{-6}	1.1×10^{-5}	4.2×10^{-5}
Usually observed ranges of δ values in nature (‰)	−49 to +49	−350 to +200	−30 to +30	−40 to 0	−45 to +40
Observed range as fractional change of isotope composition	3.9×10^{-4}	8.2×10^{-5}	1.2×10^{-4}	4.4×10^{-4}	3.6×10^{-3}

Source: From Handley and Raven 1992.

[a]In each case the abundance and fractionation of the heavy isotope is compared to the most abundant form of the atom ($^{15/14}N$, $^{2/1}H$, $^{18/16}O$, $^{13/12}C$, $^{34/32}S$. The heavy isotope of hydrogen is deuterium (D); the common usage for normal abundance is δD.

[b]SMOW, standard mean ocean water.

[c]PDB, Pee Dee Belemnite (limestone); none of this remains, and secondary standards are used.

[d]CD, Canyon Diablo Meteorite.

[e]Fraction of the total element occurring as the heavy isotope.

of the nitrogen cycle result in small differences in δ units. For example, the difference in δ units between the soil and the atmosphere is commonly near 5 → +10‰.

II. CARBON STABLE ISOTOPES

The vast majority of carbon atoms are one of four different isotopes: two radioactive and two stable forms. ^{14}C is radioactive with a half-life of 5715 years. The natural abundance of this isotope is used as an assay of geological age because of its long half-life. In this way, fossilized remains of organisms can be dated to particular geological time frames based on their ^{14}C isotopic signature. In addition, the natural abundance of ^{14}C increased during intensive aboveground nuclear testing during the 1950–1955 period. Therefore, a spike in ^{14}C content in soil layers, wood rings, or recently dead plants can be used as an index of recent, historical age (Geyh and Schleichter 1990). The second radioactive isotope of carbon, ^{11}C, has a relatively short half-life (20.3 min). This isotope has been used successfully to study patterns of carbohydrate flow (Minchin and Thorpe 1982). This isotope of carbon has led to only limited use because its short half-life necessitates a cyclotron juxtaposed to the research facility.

^{12}C and ^{13}C are the stable isotopes of carbon. A majority of the world's carbon is ^{12}C (98.9%) and most physical and chemical processes discriminate against ^{13}C. The isotopic composition of organic material is constant once dried, so samples that are collected and stored in a state which prevents microbial respiration will have a constant carbon isotopic composition for many years. The ratio of ^{13}C to ^{12}C of a sample must be compared to that of a reference material. The reference for carbon is traditionally the isotopic ratio of fossil belemnite from the Pee Dee Formation in South Carolina (Craig 1954, Farquhar et al. 1989). Although the Pee Dee belemnite reference material has been exhausted, suitable substitutes (National Bureau of Standards graphite NBS-21) with a known $^{13}C/^{12}C$ ratio are available. All comparisons of sample $^{13}C/^{12}C$ ratio to that of PDB are designated as follows:

$$\delta^{13}C = \left(\frac{R_p}{R_{ref}} - 1\right) \times 1000\,‰ \tag{5.1}$$

where R_p is the $^{13}C/^{12}C$ ratio in the plant tissue, and R_{ref} is the $^{13}C/^{12}C$ ratio in the reference PDB material. The multiple of 1000 results in a parts per thousand unit (‰), which is most commonly referred to as "per mil."

The isotopic ratio of carbon can also be compared to the isotopic composition of the source (commonly the atmosphere, for carbon isotope studies). The deviation of this comparison (R_a/R_p; where R_a is the $^{13}C/^{12}C$ ratio in the air around the plant) from unity has been suggested to be an appropriate index for discrimination against $^{13}CO_2$ (Farquhar and Richards 1984) and is designated as Δ rather than δ. In fact, comparisons between the ^{13}C content of plant and the atmosphere can be used to study temporal changes in discrimination by photosynthesis in a nondestructive sampling technique. For example, the depletion of ^{12}C from the atmosphere around a plant (increase in $^{13}C/^{12}C$ ratio) can be used as an index of isotopic discrimination by the leaf without disturbing the photosynthetic organ (Evans et al. 1986, O'Leary et al. 1986).

The use of both Δ (discrimination) and δ (composition) in the literature causes some problems in interpretation of stable isotope data because Δ values are positive and δ values are negative. Classically, δ has been used because in this analysis it is not necessary to determine the isotopic composition of the source (air around each plant in the case of CO_2 fixation). Often an $\delta^{13}C$ value of $-8‰$ is assumed as an approximation for the atmospheric carbon isotope composition when calculating discrimination (O'Leary 1993). In such cases, discrimination is calculated as $-8 - \delta^{13}C_{plant}$. Comparisons among isotopic composition (δ) of plants from different habitats or locations may not reflect physiological differences among the plants if the $\delta^{13}C$ of the source (atmosphere) is different among the sites. Although the atmospheric $^{13}C/^{12}C$ composition is fairly stable, variations in atmospheric $^{13}C/^{12}C$ can result from the effects of canopy boundary layer, soil respiration (Farquhar et al. 1982a), natural annual and seasonal variation (Mook et al. 1983), or anthropogenic causes (Keeling et al. 1979). Thus, in environments where a variable atmospheric isotopic composition is suspected, a value of -8 for atmospheric $\delta^{13}C$ cannot be assumed, and the atmospheric $\delta^{13}C$ must be measured in order to study discrimination.

A. Fractionation of Carbon Stable Isotopes

The two stable isotopes of carbon have different diffusion coefficients due to their different atomic mass. The isotope effect (a) is defined as the $^{13}C/^{12}C$ ratio of the reactants (R_r) divided by the $^{13}C/^{12}C$ ratio of the products (R_p). In the case of diffusion, $R_r = {}^{13}C/^{12}C$ of the region with high CO_2 concentration. A value of a greater than 1 represents discrimination against ^{13}C. The isotope effects of CO_2 diffusion through air is 1.0044 and water is 1.0007, resulting in discriminations of 4.4‰ and 0.7‰, respectively (Craig 1954, O'Leary 1984). During diffusion of CO_2 from turbulent air into the leaf cells, CO_2 will be depleted in $^{13}CO_2$ relative to the atmosphere. Furthermore, the equilibrium dissolution of CO_2 into water has an isotopic effect of 1.0011, resulting in a discrimination of 1.1‰, which further depletes cell water of $^{13}CO_2$.

Once CO_2 is dissolved in cell water, further fractionation occurs due to the discrimination of enzymes against $^{13}CO_2$ compared with $^{12}CO_2$. Many models have been developed to describe isotopic discrimination in photosynthesis (O'Leary 1981, Farquhar et al. 1982b, Vogel 1980), all of which are based on the additive impacts of discrimination against $^{13}CO_2$ at various levels of photosynthesis. A simplified version of discrimination in C_3 plants is represented as

$$\Delta = a\frac{C_a - C_i}{C_a} + b\frac{C_i}{C_a} = a + (b - a)\frac{C_i}{C_a} \tag{5.2}$$

where C_a and C_i are ambient and intercellular partial pressures of CO_2, respectively, a is the fractionation due to diffusion in air and equilibration with water, and b is the fractionation due to carboxylation. The isotopic effect due to diffusion (a) is estimated to be 4.4‰ (a theoretical value based on mass-specific diffusion coefficients).

By examining equation (5.2) one can deduce that when C_a is close to C_i (high conductance relative to photosynthesis) Δ approaches the value of b. Conversely, when C_i is significantly lower than C_a (photosynthesis is strongly limited by stomatal conductance), the impact of a reaches its maximum value and Δ approaches the value of a (4.4‰).

The isotopic discrimination of carboxylating enzymes has been measured several times with several techniques (Farguhar et al. 1982b; Roeske and O'Leary 1984, Guy et al. 1987, Farguhar and Richards 1984). If CO_2 is taken as the sole substrate, the isotopic effect of rubisco is 1.0027. The result must be multiplied by the equilibrium discrimination between air and water (1.1‰), which results in a value of 30‰ for b. Direct calculation of rubisco discrimination of CO_2 by $^{13}CO_2$ depletion in a closed system results in a value of −29‰ for b, which is very close to the result predicted.

Discrimination against $^{13}CO_2$ by PEP-carboxylase is less than that measured for rubisco. This is in part due to the use of bicarbonate ion (HCO_3^-) by PEP-carboxylase instead of CO_2. When an atmosphere containing CO_2 is in equilibrium with a water phase, the heavier isotope of carbon concentrates in soluble HCO_3^- compared with that in the gaseous CO_2 form. Therefore, the equilibrium isotope effect of CO_2 (in air) and HCO_3^- (in water) is less than 1 (Figure 5.2). PEP carboxylase then discriminates against the $H^{13}CO_3^-$ ion with an isotopic effect of 1.0020. The combined result of the isotope effects of CO_2 diffusion into the stomata, the equilibrium of CO_2 in air and HCO_3^- in solution, and $H^{13}CO_3^-$ fixation by PEP-carboxylase is a predicted discrimination of 5.7‰.

Fractionation also occurs during the continued metabolism of reduced carbon. This results in variable isotope composition among major cytoplasmic fractions. Lipids tend to be 0.5 to 1.0‰ lighter (less ^{13}C) than amino acids and pectin (Deines 1980). In addition, cellulose and lignin tend to be lighter than sugars and pectin (Figure 5.3). Leaf tissues tend to be lighter than wood tissues (Leavitt and Long 1982). The physiological processes leading to fractionation of ^{13}C among tissues and cytoplasmic components are not well documented.

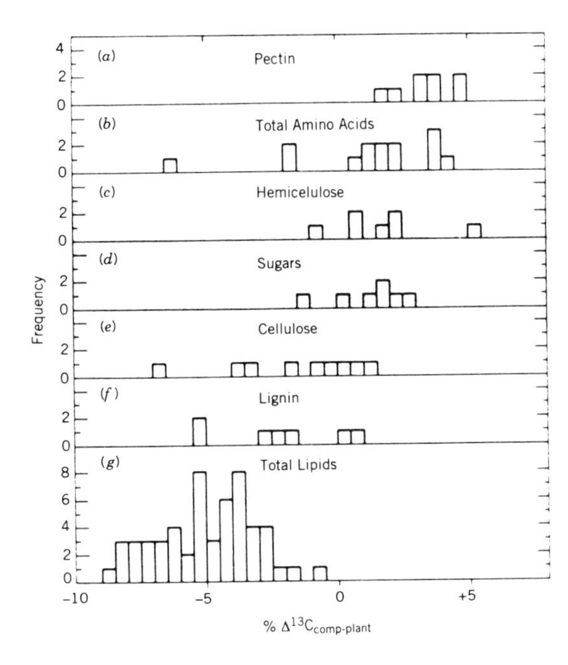

Figure 5.3 Carbon stable isotope composition of several main components of protoplasm. Notice that lipids have a distinctly lower carbon isotope ratio than that of pectin. (Used with permission from Deines 1980.)

B. Physiological Processes

Variation in isotopic fractionation among species was first employed to identity various photosynthetic patterns. C_3 plants have a $\delta^{13}C$ value ranging from -20 to $-35‰$, while the $\delta^{13}C$ of C_4 species range from -10 to $-12‰$ (Bender 1971, Smith and Epstein 1971, Troughton et al. 1974). The $\delta^{13}C$ values of C_3 plants have a larger range than that of C_4 plants, because the isotopic composition is strongly affected by diffusion of CO_2 through stomata. The $\delta^{13}C$ of C_4 plants is dominated by the equilibrium isotopic effect of $H^{13}CO_3^-$, and PEP-carboxylase discrimination against ^{13}C. When measured empirically, the $\delta^{13}C$ of C_4 plants is found to be lower than the $-5.7‰$ expected from theoretical considerations because some CO_2 leaks into the bundle sheath cells and is directly reduced by rubisco. The use of $\delta^{13}C$ to identify photosynthetic pathways in large numbers of species has resulted in a much better understanding of the physiological significance of C_4 photosynthesis and the global distribution patterns of C_4 plants.

The $\delta^{13}C$ of CAM plants is intermediate between that of C_3 and C_4 plants. Furthermore, CAM plants have an extremely variable $\delta^{13}C$ (even more so than C_3 plants). Some of the variance can be accounted for by comparing the particular CAM pathway with the $\delta^{13}C$ of plants with that pathway (Figure 5.4). PCK-type physiology has the highest $\delta^{13}C$ value, followed by NADPH-ME and NAD-ME physiology, respectively (Hattersley 1982, Farquhar et al. 1982b). Other variation in tissue $\delta^{13}C$ can be accounted for by the switching capabilities of CAM plants. Many species of CAM plants will reduce carbon with a C_3 pathway under optimal conditions. When limitations of water or high concentrations of salt occur, CAM physiology is induced instead of C_3 photosynthesis. This results in some carbon being reduced with $\delta^{13}C$ of approximately $-29‰$ (when using C_3 physiology), and some carbon will be reduced with a $\delta^{13}C$ closer to $10.0‰$ (when using CAM physiology). The result is a temporally variable isotopic composition in tissues (Figure 5.5), which reflects the recent historical photosynthetic physiology (Winter et al. 1978).

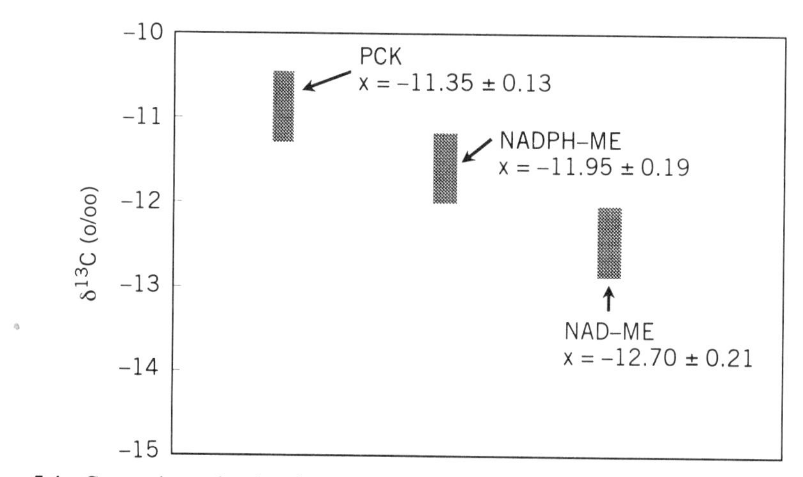

Figure 5.4 Comparison of carbon isotope composition of the three main photosynthetic pathways within C_4 grasses. Each bar refers to the range of measured $\delta^{13}C$ in leaves from plants with each photosynthetic type. (Produced from data in Hattersley et al. 1982.)

There are a number of environmental characteristics that can affect the $\delta^{13}C$ in tissues: light, water availability, salinity, vapor pressure (heat + RH together), and air pollution. Understanding the impacts of these climatic parameters on $\delta^{13}C$ and associated physiology may result in the ability to use $\delta^{13}C$ as an integrative index of physiological response to environmental stress.

Many interactions between environmental conditions and $\delta^{13}C$ are not yet well validated. For example, light intensity has been suggested to have a positive relation with $\delta^{13}C$. In some studies, the $\delta^{13}C$ of leaf tissue decreased from the outer to inner canopy leaves (Francy et al. 1985, Vogel 1978, Garten and Taylor 1992). However, interpretation of these results as representative of direct light results are inappropriate because the $\delta^{13}C$ of the air becomes lighter in the interior of a canopy or closer to the soil, due to the effects of respiration. Thus leaves at the bottom of the canopy, in low light, also experience air with a low $\delta^{13}C$ value (Schleser and Jayasekera 1985). Even though a portion of the reduced $\delta^{13}C$ of lower leaves must be due to lower air $\delta^{13}C$, low irradiance has been associated with increasing $\Delta^{13}C$ (Ehleringer et al. 1987a). The increase in $\Delta^{13}C$ under low light is probably a consequence of decreasing C_i, due to increasing stomatal limitation at low light (Zimmerman and Ehleringer 1990). Unfortunately, it is difficult to separate the influences of vapor pressure gradient and light because in field situations, low-light sites cause lower leaf temperature, and this results in a lower vapor pressure gradient even if atmospheric vapor pressure is unchanged.

Changes in tissue $\delta^{13}C$ can also be related to a change in water availability (Meinzer et al. 1990). As water availability decreases, $\delta^{13}C$ also decreases. The change in the isotopic composition of tissues during water stress can be related to an increase in hydraulic resistance (Meinzer et al. 1990), an increase in soil resistance (Masle and Farquhar 1988), or a decrease in leaf conductance (DeLucia et al. 1988, Ehleringer and Cooper 1988). In all cases, $\delta^{13}C$ is decreased because diffusional limitation to photosynthesis (stomatal) increased faster than biochemical limitation (nonstomatal) during water stress, which caused a decrease in C_i.

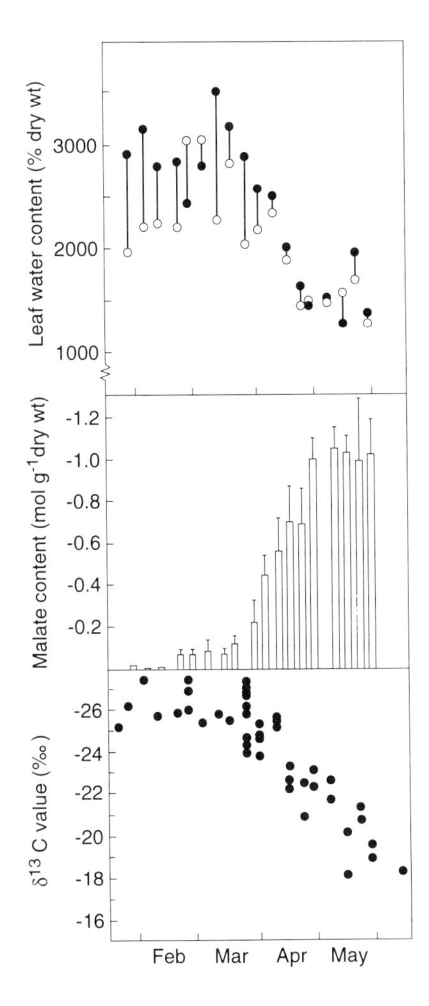

Figure 5.5 Seasonal changes in leaf water content, malate accumulation, and carbon isotope composition in a species capable of CAM photosynthesis. These data provide evidence that the photosynthetic pathway switches from C_3 to CAM as water limitation increases. Notice that as the amount of malate increases, due to water stress, the isotopic composition of the tissue increases (becomes more negative). (Modified from Winter et al. 1978.)

Salinity is also positively associated with $\delta^{13}C$. In glycophytes, leaf conductance often decreases in high-salinity situations (Greenway and Munns 1980, Seemann and Critchley 1985). The decrease in conductance due to increased salinity led to a decrease in C_i and resulted in an increase in $\delta^{13}C$. Furthermore, there is a positive relationship between salinity and $\delta^{13}C$ in halophytes (Farquhar et al. 1982a, Guy et al. 1986). In some halophytes salinity is required for optimum photosynthesis; therefore, in the absence of enough ions, photosynthesis decreases relative to conductance and C_i increases. The increase in C_i at low salinity in halophytes may explain part of the $\delta^{13}C$ association with salinity.

Other environmental characteristics also affect C_i. For example, when various forms of atmospheric pollution increase, $\delta^{13}C$ increases (Greitner and Winner 1988) because stomatal aperture often decreases. This stomatal aperture effect on $\delta^{13}C$ is due to the fact that when stomata close, C_i decreases, due to an increased resistance to CO_2 diffusion. Also, increases in the vapor pressure gradient (VPG) between the atmosphere and the leaf should lead to decreasing C_i (because of stomatal closure in response to increasing VPG) and a decrease in $\delta^{13}C$ (Farquhar and Richards 1984). Thus variations in C_i induced by changing environmental impact on stomatal conductance may be the regulator of $\delta^{13}C$ in C_3 plants (Farquhar et al. 1982b).

The most uniform observation about environmental effects on the $\delta^{13}C$ of C_3 plants is the potential association between $\delta^{13}C$ and intercellular CO_2 concentration (as discussed above). Farquhar et al. (1982b) formalized the theoretical basis of this relationship. Since $\delta^{13}C$ may represent C_i, and C_i is known to be correlated with water use efficiency (WUE), $\delta^{13}C$ may be an indicator of integrated WUE in C_3 plants.

Photosynthesis can be related to leaf conductance of water vapor as in

$$P_n = \frac{(C_a - C_i)g}{1.6} \tag{5.3}$$

where P_n is net assimilation of CO_2 from the ambient air, g is leaf conductance to water vapor, and the factor 1.6 refers to the ratio of CO_2 to water diffusivities.

Transpiration can also be related to conductance: $E = \Delta wg$, where E is transpiration, Δw is the gradient in water vapor between the leaf and air, and g is leaf conductance to water vapor. The integrated water use efficiency, P_n/E, can be expressed by the differential equation

$$\frac{P_n\,dt}{E\,dt} = \frac{(1 - \phi)(C_a - C_i)}{1.6\overline{\Delta w}} \tag{5.4}$$

where $\overline{\Delta w}$ is the mean vapor pressure gradient and ϕ is the proportion of fixed carbon lost to respiration.

Equation (5.4) can be combined with equation (5.3) in a modified expression representing integrated water use efficiency. In this expression C_i is replaced by the isotopic expression for C_i (Farquhar et al. 1982b):

$$\frac{\int P_n\,dt}{\int E\,dt} = \frac{(1 - \phi)(C_a)}{1.6\overline{\Delta w}}\,\frac{(b - \delta^{13}C_{atm} + \delta^{13}C_p)}{b - a} \tag{5.5}$$

This expression is significant because if $\delta^{13}C_{atm}$ and $\overline{\Delta w}$ are relatively constant, tissues with a higher $\delta^{13}C_p$ will represent plants that have a relatively high integrated water use efficiency.

Originally a theoretical relationship, the association between $\delta^{13}C$ and water use efficiency has been established repeatedly. Greenhouse or growth chamber environmental manipulation of water use efficiency was reasonably well associated with $\delta^{13}C$ in creosote bush (Figure 5.6), cotton, peanuts, tomato, and barley (Hubick et al. 1986a, Hubick and Farquhar 1987, 1989; Masle and Farquhar 1988; Martin and Thorstenson 1988; Sharifi and Rundel 1993). In field applications the relationship between $\delta^{13}C$ and WUE can be distorted by variation in atmospheric $\delta^{13}C$. As explained previously, $\delta^{13}C_{atm}$ can vary temporally and spatially. However, if the group of organs or species are utilizing the same atmosphere, comparisons of $\delta^{13}C$ may be a valid test for variations in WUE among organs or species. For example, photosynthetic stems have a higher instantaneous water use efficiency (Figure 5.7) than that of leaves in the same canopy, and stems have a higher $\delta^{13}C$ than leaves (Ehleringer et al. 1987b, Nilsen and Sharifi 1996). Also, hemiparasites have a

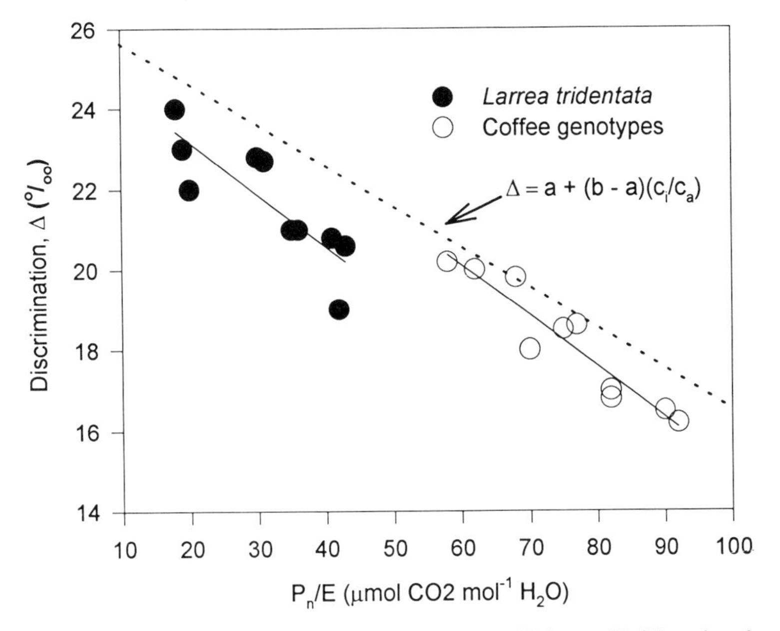

Figure 5.6 Relationship between instantaneous water use efficiency (P_n/E) and carbon isotope discrimination in coffee genotypes and creosote bush *(Larrea tridentata)*. The dotted line represents the theoretical relationship between water use efficiency and isotope discrimination at a constant intercellular CO_2 concentration of 347 μmol CO_2/mol air. (Redrawn from Meinzer et al. 1990 and Sharifi and Rundel; 1993.)

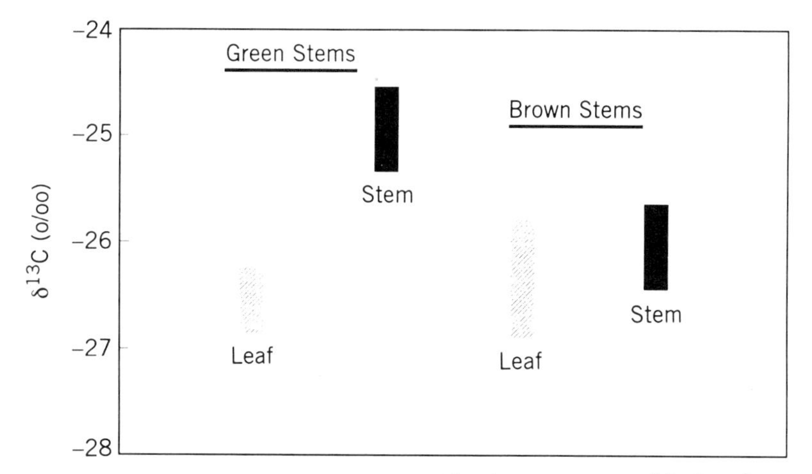

Figure 5.7 Comparisons of carbon isotope composition between stem and leaf on the same individuals. "Green stems" refers to species that have prominent green stems all year. "Brown stems" refers to species that do not have photosynthetic stems. The species tested are all desert species from Arizona and California. (Data for this figure are from Comstock and Ehleringer 1988 and Nilsen and Sharifi 1996.)

lower WUE and a lower $\delta^{13}C$ than those of their hosts (Ehleringer et al. 1985, Goldstein et al. 1988).

The relationship between $\delta^{13}C$ and WUE may be extended to a determination of canopy transpiration. If data are available on net primary production for a species, a measure of tissue $\delta^{13}C$ (and thus WUE) can be used to estimate transpiration:

$$TR = NPP \frac{1}{WUE} = NPP \times \frac{TR}{NPP} \tag{5.6}$$

where TR is transpiration of the canopy, NPP is net primary production, and WUE is the ratio of carbon gained per water lost (estimated from $\delta^{13}C$). The ability to use carbon isotope ratios to estimate total plant transpiration has not been validated.

Carbon isotope composition may serve as a screening index in plant breeding programs. The development of crop varieties with increased water use efficiency may be accomplished by separating genes, in common garden experiments, based on their leaf tissue $\delta^{13}C$. Those with high $\delta^{13}C$ will be those with greater WUE (Hubick and Farquhar 1987, Meinzer et al. 1990). Although $\delta^{13}C$ of leaf tissue is consistently well correlated with water use efficiency, its association with plant growth or seed yield is inconsistent. In some cases there is a positive relationship between $\delta^{13}C$ and growth (Condon et al. 1987), and in other cases there is a negative relationship (Hubick et al. 1988). Part of the inconsistency between $\delta^{13}C$ and growth potential (assuming uniform microclimate) is due to genetic effects.

Genetic control over photosynthesis and water relations of cells is complex because many polytropic genes and gene families are associated with these major physiological functions. However, in one case genetic differences in $\delta^{13}C$ were maintained under a number of growth conditions (Condon et al. 1987). Broad-sense heritability (the proportion of total variance that can be explained by genotype in genotype versus environment experiments) of the $\delta^{13}C$ trait ranges from 60 to 90% in a diversity of species (Cordon et al. 1987, Hubick et al. 1988, Ehleringer 1988). Also, the variance in $\delta^{13}C$ has been found to correspond to DNA markers for genome heterozygosity (Martin et al. 1991) and tends to be a dominant trait (Martin and Thorstenson 1988). Although the beginnings of an understanding of molecular and genetic control over $\delta^{13}C$ is being developed, this information will not help breeding programs until the consequences of various $\delta^{13}C$ values to growth and yield are well established.

III. NITROGEN STABLE ISOTOPES

Nitrogen has two stable isotopes (Figure 5.1), of which ^{14}N is more abundant, accounting for 99.633% of atmospheric nitrogen. There are four radioactive isotopes of nitrogen with short half-lives (10 min → 0.1 s). The short half-lives of radioactive isotopes of nitrogen make them unsuitable for most plant physiological applications. The stable isotopic composition of the atmosphere is very constant (Junk and Suek 1958, Mariotti 1983). In fact, the ratio of $^{15}N/^{14}N$ in the atmosphere is used as the standard for calculating $\delta^{15}N$ (as Pee Dee Belemnite was used as the standard for calculating $\delta^{13}C$).

Nitrogen stable isotopes are used in two ways by research scientists. In one case a source of nitrogen that is enriched in ^{15}N is used to follow events of nitrogen metabolism or nitrogen cycling in the ecosystem. This technique involves a large enrichment of ^{15}N

over the background, making the measurement of isotope effects easy because the difference between the isotopic composition of the source and plant is large. In a second type of study, the natural abundance of ^{15}N of a sample is compared to that of the atmosphere and the soil solution. Studies of ^{15}N natural abundance involve very small ^{15}N concentrations and small differences between the sample and the atmosphere. Contamination during sample preparation and measurement by the large amount of ^{15}N in the atmosphere is an important consideration when measuring the natural abundance of nitrogen stable isotopes in tissues or solutions. Consequently, samples must be isolated from the atmosphere during combustion (or digestion) and measurement of ^{15}N.

A. Fractionation of Nitrogen Stable Isotopes

Similarly to carbon stable isotopes, ^{15}N will be discriminated against compared with ^{14}N during any physical process involving mass. Diffusion of $^{15}N_2$ is slower than $^{14}N_2$ in the atmosphere. Most biological processes, with the notable exception of nitrogen fixation, utilize nitrogen in an organic form or as NO_3^- or NH_4^+. The natural abundance of ^{15}N in NO_3^- or NH_4^+ in the soil solution is the result of both kinetic and equilibrium isotope effects of the complex ecosystem nitrogen cycle that may have operated over a long time scale. Akin to that of carbon isotope discrimination, kinetic isotope effects normally result in a higher ^{15}N content in reactants than in products. Thus, $\delta^{15}N$ is normally discriminated against in a reaction process. However, the pools of nitrogen in biological materials may have been influenced by many isotope effects possibly repetitively. Hence the equilibrium ^{15}N content of nitrogen in biological systems cannot be predicted easily from one or a few discrimination processes.

Some general characteristics of nitrogen isotopic composition in biological nitrogen can be supported. Soil nitrogen is normally enriched in ^{15}N by as much as 10‰ over the ^{15}N in the atmosphere (Shearer et al. 1978, Karamanos et al. 1981). Although an occasional study reports a large variance in soil ^{15}N content (Broadbent et al. 1980), many studies report that the standard deviation of soil $\delta^{15}N$ within a particular site is less than 3‰ (Shearer et al. 1978, Bremner and Tabatabi 1973, Shearer et al. 1983). However, $\delta^{15}N$ of soil can be variable among sites, such that some sites may have soil $\delta^{15}N$ indistinguishable from that of the atmosphere (possibly a statistical quirk of variance). The $\delta^{15}N$ of various layers in the soil profile can vary as much as that among sites (Tiessen et al. 1984). In addition, $\delta^{15}N$ of total soil N may not represent that of the $\delta^{15}N$ for nitrogen available to the plant. For example, the nitrite pool mineralized from soil has a more uniform $\delta^{15}N$ than that for the soil used in the mineralization studies (Ledgard et al. 1984).

This heterogeneity of $\delta^{15}N$ in the soil profile and among different soils is due to the impact of microbial and biophysical isotope effects (Table 5.2). In regions of the soil or sites that favor denitrification, soil $\delta^{15}N$ will increase because denitrifying organisms preferentially convert $^{14}NO_3^-$ to $^{14}N_2O$ or $^{14}N_2$ (Shearer and Kohl 1988). Moreover, ammonia volatilization ($NH_4^+ + OH^- \longrightarrow NH_3 \uparrow + H_2O$), which occurs more abundantly in high-ammonia sites (fertilized fields) or high-pH soils, also discriminates gains ^{15}N (Mizutani and Wada 1988). The $\delta^{15}N$ of the soil may be decreased by the deposition of litter from N_2-fixing species. For example, the soil $\delta^{15}N$ in sites dominated by an actinorhizal N_2 fixer (chaparral) was low compared to other ecosystems (Virginia et al. 1989). Thus, although most soil N transformations tend to enrich soil ^{15}N, the rates of these processes vary between sites or regions in the soil profile, as does abiotic isotope factors such as leaching and diffusion. The summation of abiotic and biotic spatial and temporal

TABLE 5.2 Various Processes of Nitrogen Transformation: Their Reaction and Range of Measured Discrimination

Process	Reaction	Discrimination[a]
Abiotic		
Diffusion	NH_4^+, NH_3, or NO_3^- in water	1.000
Diffusion	NH_4^+, NH_3, or NO_3^- in air	1.018
Soil biotic		
N_2O reduction	$N_2O \rightarrow N_2$ or $N_2O \rightarrow NH_4^+ \rightarrow$ org. N	1.034 → 1.039
Denitrification	$NO_3^- \rightarrow N_2O$	1.028 → 1.033
Dinitrogen fixation	$N_2 \rightarrow NH_4^+ \rightarrow$ org. N	0.991 → 1.041
Nitrification	$NH_4^+ \rightarrow NO_2^-$	1.015 → 1.035
Plant		
Ammonia assimilation	$NH_4^+ \rightarrow$ org. N	1.009 → 1.020
Nitrate assimilation	$NO_3^- \rightarrow NH_4^+ \rightarrow$ org. N	1.003 → 1.030
Transamination	Glutemate + oxaloacetate → aspartate + 2-oxyglkutamate	1.001 → 1.008

Source: Data simplified from Handley and Raven 1992.

[a]A discrimination of 1 means that both ^{15}N and ^{14}N have an equal likelihood of reacting.

variability in N transformations and isotope fractionation results in site-specific and regional variations in soil $\delta^{15}N$.

The soil ^{15}N composition, being established by the fractionation processes discussed previously, will provide a pool of N to plants. Given an isotopic composition of the soil solution, fractionation occurs in the plant due to assimilation of NO_3^- or NH_4^+, translocation to the leaves, and nitrogen metabolism in cytoplasm (Table 5.2). NO_3^- uptake results in an isotopic effect between 1.003 and 1.03 or approximately 0.3 to 3‰ (Kohl and Shearer 1980, Mariotti et al. 1980). The uptake fractionation decreases with plant age (Kohl and Shearer 1980), increases with an increase in NO_3^- (Wada and Hattori 1978), and decreases with increasing light intensity. However, for actively growing mature plants, in normally fertile soil and moderate light intensity, ^{15}N discrimination during NO_3^- uptake is very small.

There can be a significant isotopic effect for nitrogen assimilated by N_2 fixation. Estimates of the isotope effect for N_2 fixation range from 0.991 to 1.041 or from −0.9 to 4.1‰ (Mariotti et al. 1980, Shearer et al. 1982). Part of the variance among measurements can be due to the selection of tissue, or potential differences in laboratory technique (Mariotti 1983). The variance in tissue $\delta^{15}N$ may be due to the isotope effect of translocation. The impact of diffusion and mass flow in the vascular system is not expected to be important to $\delta^{15}N$. However, the N transformations before translocation may have an isotope effect.

Several tests have shown that the $\delta^{15}N$ is fairly uniform (Table 5.3) among tissues of herbaceous species except for that of root nodules (Shearer et al. 1983, Shearer and Kohl 1989a). However, significant differences between $\delta^{15}N$ of tissues can develop in long-lived woody species (Shearer et al. 1983). In particular, stem and root tissues tend to be depleted compared to leaf tissues (Shearer and Kohl 1989b). One tissue with unusual $\delta^{15}N$ compared to the rest of the plant is the root nodule. Several studies on a diversity of legumes have shown ^{15}N to be strongly enriched in nodules (Shearer and Kohl 1978,

TABLE 5.3 Total N and $\delta^{15}N$ for Various Parts of Two Legume Species from the California Desert

	D. schottii		*D. mollissima*	
Tissue	% of Total N in the Plant	$\delta^{15}N$	% of Total N in the Plant	$\delta^{15}N$
Leaves	17.3	-1.5 ± 0.2	46.5	-1.6 ± 0.3
Fruits and flowers	1.4	-1.1 ± 0.7	15.8	-0.4 ± 0.5
Stems	42.1	-3.1 ± 0.1	19.1	-2.6 ± 0.3
Roots	34.5	-3.0 ± 0.6	10.5	-2.2 ± 0.2
Nodules	5.1	$+6.3 \pm 1.4$	8.4	$+2.5 \pm 0.5$
Entire plant	100.0	-2.0 ± 0.2	100.0	-1.3 ± 0.2

Source: From Shearer and Kohl 1989b.

Steele et al. 1983). ^{15}N enrichment in nodules is not always the case (Virginia et al. 1984) when nodules are inactive or have a poor bacterial symbiosis. Thus active mature nodules with extensive bacterial presence have strong enrichment tendencies. The mechanism for bacterial enrichment in root nodules is not known, but the ^{15}N enrichment is associated with the nodulating bacteria, not the host cells (Reinero et al. 1983).

B. Physiological Processes

The main physiological process investigated to date by ^{15}N isotopes is N_2 fixation. The general objective of most studies using ^{15}N isotopes is to determine the relative contribution of biological nitrogen fixation (BNF) to whole-plant N accumulation. Nitrogen isotopes can be used to assay BNF by an ^{15}N enrichment experiment or by using the natural abundance of ^{15}N. In both types of experiments the assay of BNF is based on a higher ^{15}N content of the soil solution compared to that of the atmosphere. In studies where the soil is enriched in ^{15}N, the difference between the soil and the atmosphere is very large compared to the natural abundance techniques.

1. Studies of Nitrogen Fixation by Soil Enrichment. The ^{15}N dilution technique was presented in the mid-1970s as an alternative to acetylene reduction (Fried and Broeshart 1975). Acetylene reduction techniques could not be accurately used to predict total N contribution from N_2 fixation because scaling up from the individual nodule to the whole plant or canopy introduced considerable error. The original ^{15}N dilution technique and modifications thereof (Fried and Middleboe 1977, Ledgard et al. 1985, Chalk 1985), are based on the addition of ^{15}N-enriched fertilizer to the growth medium of a nitrogen-fixing crop and to that of a nonfixing crop. Assuming that the soil N cycling process is equivalent (mineralization, denitrification, etc.) for the two crop growth mediums and that the $\delta^{15}N$ of the soil solution has been adjusted equally for fixer and nonfixer, the lower $\delta^{15}N$ of the fixer should represent the impact of N_2 fixation on whole-plant nitrogen uptake (atmospheric $\delta^{15}N$ will be much smaller than soil solution $\delta^{15}N$).

Soil ^{15}N enrichment (or dilution) techniques are used primarily in agricultural systems. In studies of annual crops, ^{15}N modification of the soil can be concurrent with seed planting; thus there will be no effect of the $\delta^{15}N$ of previous tissues on the interpretation of results. Furthermore, ^{15}N fertilizer can be applied fairly uniformly in an agricultural setting, although significant variation remains. For example, most ^{15}N fertilizer will be near the soil surface; there will be spatial microsite variability, within the soil, and the $\delta^{15}N$ of soil NO_3^- decreases with time. Since the accurate quantification of soil $\delta^{15}N$ would involve a prohibitive sampling design due to the variation described above, a nonfixing reference crop is utilized to compensate for the soil heterogeneity. The reference crop must be selected carefully and planted in an intermixed design with the test crop. Important criteria for reference crop selection include similarity of root zone to the test crop, absence of rhizosphere N_2-fixing organisms, and similarity in seasonal NO_3^- accumulation. If large areas are available for testing, a multiple-split-block design can assure accuracy of the N_2-fixation estimate in crops (Ledgard et al. 1985).

The basic ^{15}N dilution analysis depends on the mass of N available from three sources and the $\delta^{15}N$ of each source. The three nitrogen sources are those in the atmosphere (lowest $\delta^{15}N$), soil solution (moderate $\delta^{15}N$), and fertilizer (highest $\delta^{15}N$). Two classical ratios are a basis for the technique. The first is the proportion of crop nitrogen derived from fertilizer compared to total nitrogen in the mature crop (y):

$$y = \frac{n_{\text{fert}}}{n_{\text{total}}} \tag{5.7}$$

where n_{total} is the total nitrogen in the mature crop and n_{fert} represents N in the crop derived from added fertilizer.

The second proportion refers to the efficiency of fertilizer utilization by the crop and is represented by

$$\omega = \frac{n_{\text{fert}}}{N_{\text{fert}}} \tag{5.8}$$

where N_{fert} equals the total nitrogen in fertilizer added to the soil.

These two equations can easily be combined to form a relationship showing that the calculation of ω will be dependent upon estimating y, because N_{fert} and n_{total} are easily measured:

$$\omega = y\,\frac{n_{\text{total}}}{N_{\text{fert}}} \tag{5.9}$$

It is possible to estimate y from isotopic composition of the various sources of N only because all the N atoms are identical. Since the soil N pool has been strongly enriched compared to the atmosphere, y can be estimated from the ratio of the $\delta^{15}N$ in the crop compared to that in the soil. In this case, crop fertilizer use efficiency (ω) can be rewritten as

$$\omega = \frac{\delta^{15}N_{\text{crop}}\; n_{\text{total}}}{\delta^{15}N_{\text{fert}}\; N_{\text{fert}}} \tag{5.10}$$

An estimation of the total quantity of N accumulated from the atmosphere by the crop can be determined by subtracting the reference crop's available nitrogen sources from that of the test crop. Thus $N_{\text{ref}} - N_{\text{test}} = N_{\text{fix}}$. If the ω and y relationships are utilized in algebraic manipulations, a simplified equation that defines the quantity of N in the test crop derived from N_2 fixation based only upon isotopic ratio and total crop N can be derived and is represented as follows:

$$\text{amount of N derived from } N_2 \text{ fixation} = \left[1 - \frac{\delta^{15}N_{test}}{\delta^{15}N_{ref}}\right] N_{test} \tag{5.11}$$

Therefore, assuming that the constraining experimental conditions are valid, the contribution of N_2 fixation to whole-crop nitrogen accumulation can be determined by measuring the $\delta^{15}N$ of the test N_2-fixing crop, the $\delta^{15}N$ of the non-N_2-fixing reference crop, and the total crop accumulated nitrogen.

This technique has been applied to many crop systems over the past 15 years. During the application period, several sources of error have been identified. For example, the application rate of ^{15}N-enriched fertilizer must be low enough so that N_2 fixation is not inhibited by an elevated NO_3^- content in the soil. Also, denitrification will convert some of the added N fertilizer into N_2 or NO. Denitrification discriminates against ^{15}N, but the N_2 produced will have a higher ^{15}N content than that of atmospheric N_2. Thus, the soil N_2 gases may be enriched in ^{15}N compared to the atmosphere, resulting in an underestimate of N_2 fixation. The most critical source of error comes from selection of the reference crop. If the source $\delta^{15}N$ for the reference crop is different from that of the test crop, the assay of N_2 fixation by ^{15}N dilution is invalidated (Witty and Giller 1991).

2. Natural Abundance Techniques. N_2 fixation in natural communities cannot be assayed with an ^{15}N enrichment study. Any addition of fertilizer would perturb the N balance of the soil and invalidate the N_2-fixation measure. Moreover, the added ^{15}N could not be evenly distributed in the soil profile because any mechanism that homogenized the soil profile would severely alter the natural state and destroy root systems. Also, the plants are often present (woody perennials) before any fertilizer would be added. Consequently, the plants contain a large amount of N whose $\delta^{15}N$ is reflective of prefertilizer conditions. The prefertilizer N in plants will mask the effects of ^{15}N accumulated from fertilizer added to the soil. Due to these inherent problems in using the ^{15}N enrichment techniques in natural systems, and the inadequacy of using acetylene reduction techniques, scientists turned to evaluating the possibility of using the natural abundance of ^{15}N in tissues as an assay of N_2 fixation.

The natural abundance method is dependent on the fact that the soil N pool is enriched in ^{15}N compared to the atmosphere due to natural processes of the soil nitrogen cycle (discussed earlier). Furthermore, there must be a large enough difference between soil $\delta^{15}N$ and atmosphere $\delta^{15}N$ (usually about 3 to 4‰) to measure dilution effects. This technique is similar to soil enrichment experiments because a reference plant is required to account for fractionation due to metabolic processes and to avoid the tedious assay of soil spatial and temporal variation in $\delta^{15}N$.

The natural abundance of ^{15}N in biological tissues ranges from 0.0018 atom % depleted to 0.0036 atom % excess ($\delta^{15}N$ of -5 to $+10$‰). These are small differences, but modern mass spectrophotometer techniques can accurately and repetitively detect those differences. Assuming that a reference taxa (with an identical N resource) can be located, the fractional contribution of N_2 fixation to the nitrogen pools can be represented by

$$\text{FNdfa} = \frac{\delta^{15}N_o - \delta^{15}N_t}{\delta^{15}N_o - \delta^{15}N_a} \tag{5.12}$$

where FNdfa = fraction of total N accumulated by N_2 fixation, $\delta^{15}N_o$ is the $\delta^{15}N$ value for the test species using only soil N (the reference species $\delta^{15}N$ is used here), $\delta^{15}N_a$ is the $\delta^{15}N$ for the test species forced to survive only upon N_2, and $\delta^{15}N_t$ is the $\delta^{15}N$ of the test plant growing in the site where nitrogen is supplied by both N_2 and soil N. The value for $\delta^{15}N_o$ is that of the reference species; thus the appropriateness of the reference is critical. The $\delta^{15}N_a$ must be determined in the laboratory or greenhouse by growing the N_2-fixing species in a nitrogen-free medium. In a general sense, the $\delta^{15}N_a$ and $\delta^{15}N_o$ form two poles of a gradient from 100% N_2 fixation to 0% N_2 fixation (Figure 5.8). Thus the position of $\delta^{15}N_t$ between these two poles represents the relative contribution of N_2 fixation to tissue nitrogen.

This technique has been applied to natural systems for the past 10 to 15 years (Delwiche et al. 1979, Virginia et al. 1989). Several validation studies have shown that the prediction of N_2 fixation by the natural abundance technique is highly correlated with N mass balance techniques (Kohl and Shearer, 1980). Estimates of N_2 fixation have been made using the natural abundance technique in desert (Shearer et al. 1983), crops (Ledgard et al. 1985), chaparral (Virginia and Delwich, 1982), and tropical forest (Virginia et al. 1989).

The $\delta^{15}N$ natural abundance technique has some distinct disadvantages. Commonly, the maximum difference between $\delta^{15}N$ of the reference and the N_2-fixing species is only 10‰. Thus, the inherent variance among individuals, tissues, or the measurement technique may be greater than the difference of $\delta^{15}N$ between the reference and N_2-fixing species. Therefore, great care must be taken to reduce all sources of variance in the comparison.

The second disadvantage is the difficulty of selecting an appropriate reference species. Due to the heterogeneous environment in natural systems, there is a high likelihood that a non-N_2-fixing species will be using a different soil N resource compared to the N_2-fixing species. This problem can be minimized by utilizing several reference individuals. If the variance among reference individuals is small, it is more likely that the references represent the same soil nitrogen pool as that that used by the N_2-fixing species.

Most fractionation processes in plants can be accounted for by using a reference species. However, fractionation by N_2 fixation (in root nodules) must be accounted for by determining the resultant $\delta^{15}N$ ratio in tissues whose nitrogen is derived exclusively from

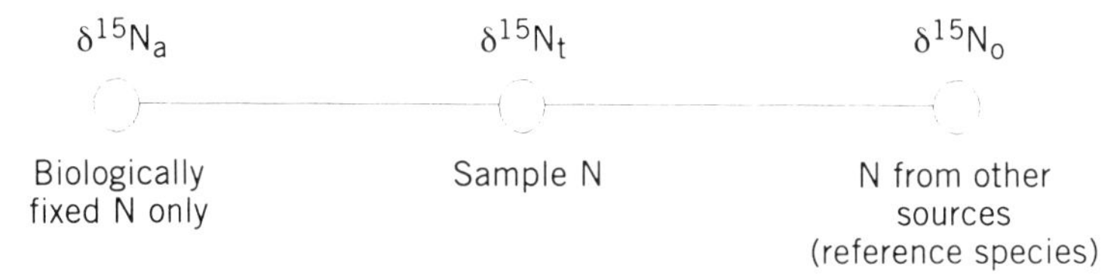

Figure 5.8 Diagrammatic representation of the natural abundance technique for determining the contribution of N_2 fixation to whole-plant nitrogen accumulation. The nitrogen isotope composition of the sample will be a value somewhere in between the lowest nitrogen isotope value (when plants are grown without dissolved nitrogen ions in the soil) and the highest value (when plants are receiving all of their nitrogen from the soil solution). (Redrawn from Shearer and Kohl 1989b).

N_2 fixation. Therefore, a greenhouse experiment is required to determine the $\delta^{15}N$ that a plant would have if all nitrogen was derived from N_2 fixation. It must be assumed that such a $\delta^{15}N$ value for a plant in the greenhouse mimics that for a plant at the field site which derives all nitrogen from N_2 fixation. Most authors believe that the $\delta^{15}N$, natural abundance technique has been adequately validated and is an appropriate quantitative index (Handley and Raven 1992).

Coupling the analysis of N isotopes with that of ^{13}C or ^{18}O may provide a more powerful signature of biological processes. For example, a successful model of oceanic denitrification processes was possible by linking $\delta^{15}N$ and $\delta^{18}O$ (Kim and Craig 1990). Moreover, seasonality of N_2 fixation in an estuary community could be defined by a linkage of $\delta^{15}N$ and $\delta^{13}C$ (Cifuentes et al. 1988). The association between $\delta^{15}N$ and $\delta^{13}C$ may be related to the carbon economy of N_2 fixation (Schulze et al. 1991), although conflicting data are present (Handley and Raven 1992). Future validation of the biotic processes that establish patterns in multiple-isotope compositions may increase the utility of $\delta^{15}N$ in evaluations of N metabolism.

Stable nitrogen isotopes may also be useful in metabolic tracer studies, a domain normally occupied by radioactive isotopes. Since radioactive isotopes of N (^{13}N) have half-lives less than 10 min, they are normally inappropriate for metabolic tracing. However, by combining gas chromatography and mass spectrometry (GC-MS) with nuclear magnetic resonance spectroscopy (NMR), the uptake and assimilation of inorganic nitrogen sources can be followed. For example, these techniques were used to verify the relative ability of species to accumulate NO_3^- or NH_4^+ (Stewart et al. 1993). Moreover, this technique was used to show that in the presence of VAM mycorrhizae, NH_4^+ assimilation occurred through the glutamate synthetase cycle in one species and the glutamate dehydrogenase assimilatory route in another species (Stewart et al. 1993). There is no doubt that the recent improvements in ^{15}N isotope detection will result in rapid development of new N-metabolism assay techniques.

IV. HYDROGEN AND OXYGEN STABLE ISOTOPES

The use of hydrogen and oxygen stable isotopes is discussed together because they are closely linked in carbohydrate metabolism and water flow in plants. There are three isotopes of hydrogen (Figure 5.1): hydrogen (^1H) and deuterium (^2H, or D) are stable; tritium (^3H) is radioactive. Tritium is commonly used in carbohydrate-labeling experiments and is often selected as the second label on a dual-labeled compound along with ^{14}C. The fractional abundance of deuterium (0.00015) in nature is particularly low compared with the stable isotopes of C, N, O, and S.

Oxygen has a moderately large set of isotopes, including several radioactive members (Figure 5.1) and three stable isotopes (^{18}O, ^{17}O, ^{16}O). However, only the $^{18}O/^{16}O$ ratio is commonly used in stable isotope studies. The fractional abundance of ^{18}O (0.00204) is in the range of that for ^{15}N (0.00366) but much lower than that of ^{13}C (0.0111) or ^{34}S (0.0422).

The reference material for both ^2H and ^{18}O (Table 5.1) is standard mean ocean water (SMOW). The original $^2H/^1H$ ratio in snow was 155.76×10^{-6} and that of $^{18}O/^{16}O$ was 2067.1×10^{-6}. The original supplies of SMOW have been exhausted and replaced by a mixture of various waters produced by the Atomic Energy Agency in Vienna (V-SMOW), which has an isotopic composition nearly identical to the original SMOW.

A. Fractionation of Hydrogen and Oxygen Stable Isotopes

The main abiotic processes that affect isotope concentrations of ^2H and ^{18}O are evaporation and photodiscrimination by ultraviolet light. In both cases the heavier isotopes ^2H and ^{18}O are discriminated against by the process. Fractionation of these isotopes that has occurred on other planets can provide some evidence about the uniqueness of Earth. Deuterium (^2H) ratios in the atmospheres of Mars and Venus are enriched compared to that of Earth. The enrichment of deuterium in the atmospheres of these planets is taken to represent a large loss of water from the atmosphere by photodissociation (Donahue et al. 1982). The atmospheric composition of deuterium is low on Earth compared to the other two planets because water has been retained in the atmosphere and on the surface of the planet. This retention of water by Earth has buffered the planet's global temperature and allowed the development of life.

Isotopic composition of carbonates in the ocean can also be used to ascertain global climatic properties. Evaporation from oceanic water discriminates against $H_2{}^{18}O$; therefore, oceanic water is enriched in $H_2{}^{18}O$ compared to rainwater. In fact, the $\delta^{18}O$ of carbonates can be used to measure the planetary ocean volume (Schlesinger 1991). During ice ages, evaporated water collected in glaciers, causing $H_2{}^{18}O$ enrichment in ocean water. Thus, the production of oceanic carbonates would be elevated in ^{18}O during periods of low ocean volume.

Fractionation processes during the formation of rainwater can be useful for determining the source of rainwater. There is a strong covariance between the relative abundance of ^2H and ^{18}O in precipitation (Gat 1980). The relationship between ^2H and ^{18}O in precipitation ($\delta^2H = 8 \times \delta^{18}O + 10‰$) is termed the meteoric waterline (MWL). In addition, the natural abundances of ^{18}O and ^2H (or D) in rainfall are highly related to the mean annual temperature (Figure 5.9A). Thus, summer precipitation is generally enriched in D and ^{18}O compared to that in winter precipitation (Dansgaard 1964). In fact, the isotopic composition of rainwater is dependent on the temperature (Figure 5.9B) at which the water condensed (Gat 1980) and the source of rainwater. Rain derived from recently formed clouds of oceanic sources will have a different isotopic signature than rain coming from clouds that have formed over land or have traveled a long distance. Groundwater has been found to have an isotopic composition similar to that of the integrated δD or $\delta^{18}O$ of annual precipitation. However, water in upper soil layers may be enriched in ^{18}O and ^2H compared with groundwater because of surface evaporation (Figure 5.10). Evaporative enrichment in soil water in upper soil layers of inland basins can be as much as 200‰ (Dansgaard 1964).

Biotic fractionation of ^2H and ^{18}O in plants occurs during metabolic use of water, transpiration, and movement of water between the apoplast and symplast. All plant H is derived from water. Therefore, the δD value of immediate products of photosynthesis, defines the isotopic effect of photosynthesis rather than the δD of metabolic water. Measurement in algal systems (where there is no fractionation of isotopes between the water source and photosynthesis) indicates a large discrimination against ^2H_2O (Estep and Hoering 1981), which can result in an isotope effect of -100 to $-170‰$ compared to water. Postphotosynthetic exchanges of H occur on carbohydrates such that as much as 50% of all H can be replaced in cellulose following synthesis of glucose by photosynthesis (Yakir and DeNiro 1990). The replacement of H on carbohydrates results in an enrichment of ^2H by a factor of about $+150‰$. The final ratio of $^2H/^1H$ is dependent on the δD value of metabolic water and the balance between photosynthetic discrimination against ^2H and postphotosynthetic ^2H enrichment. Further support that H replacement during metabolic processes affects δD comes from the finding that the δD value of starch is de-

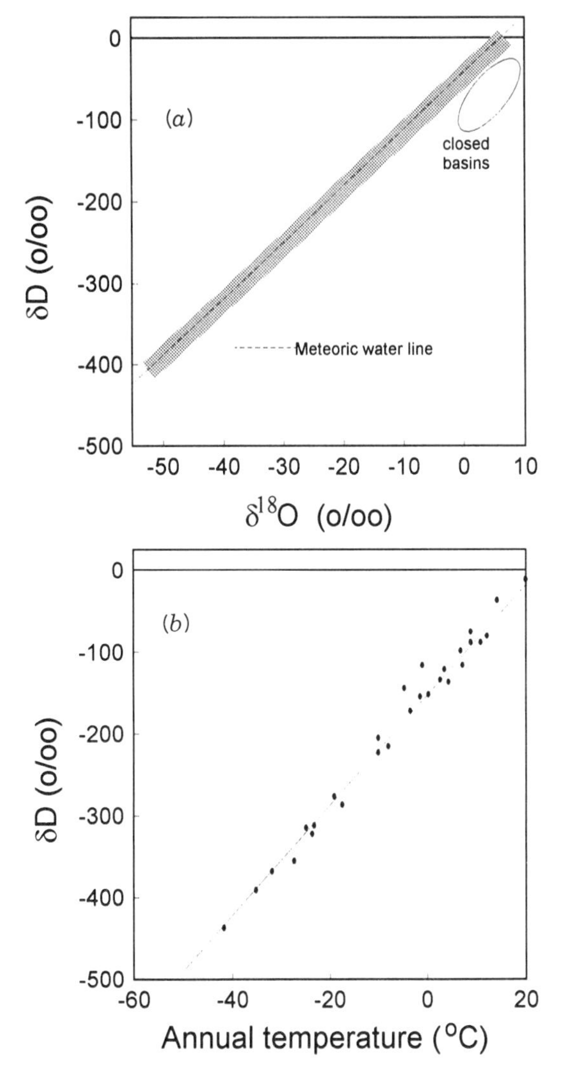

Figure 5.9 (A) Relationship between the hydrogen and oxygen isotopic composition of rainwater. The regression line is the meteoric waterline (MWL). "Closed basins" refers to precipitation collected over closed basins in which the water is internally recycled. (B) Relationship between isotopic composition of rainwater and mean annual temperature of the region where the rainwater condensed. (Revised from Fontes 1980.)

pleted compared to that of cellulose (Sternberg 1988). Also, fatty acids that receive all H from NADPH (photosynthetically derived H) have a much lower δD value than that of water (Zeigler 1989). Since all H in fatty acids is derived from the same source (tissue water) and the same mechanism, there is a high correlation between the δD value of lipids and that of tissue water (Sternberg 1988).

The isotopic content of water in the leaf is not homogeneous. When water is pushed out of the petiole of a leaf, as in a pressure–volume curve (see Chapter 7), there is a gradual increase in δD and $\delta^{18}O$ as greater amounts of water are removed (Figure 5.11). The increase in isotopic composition represents a change from apoplastic to symplastic water in the exudate (Yakir et al. 1989). Therefore, metabolic water (in the symplasm) is enriched in ^{2}H and ^{18}O compared to apoplastic water. In other words, an assay of δD or $\delta^{18}O$ of whole-leaf water will result in an under estimate of symplastic water isotopic composition because of dilution by apoplastic water that has a low δD or $\delta^{18}O$ value.

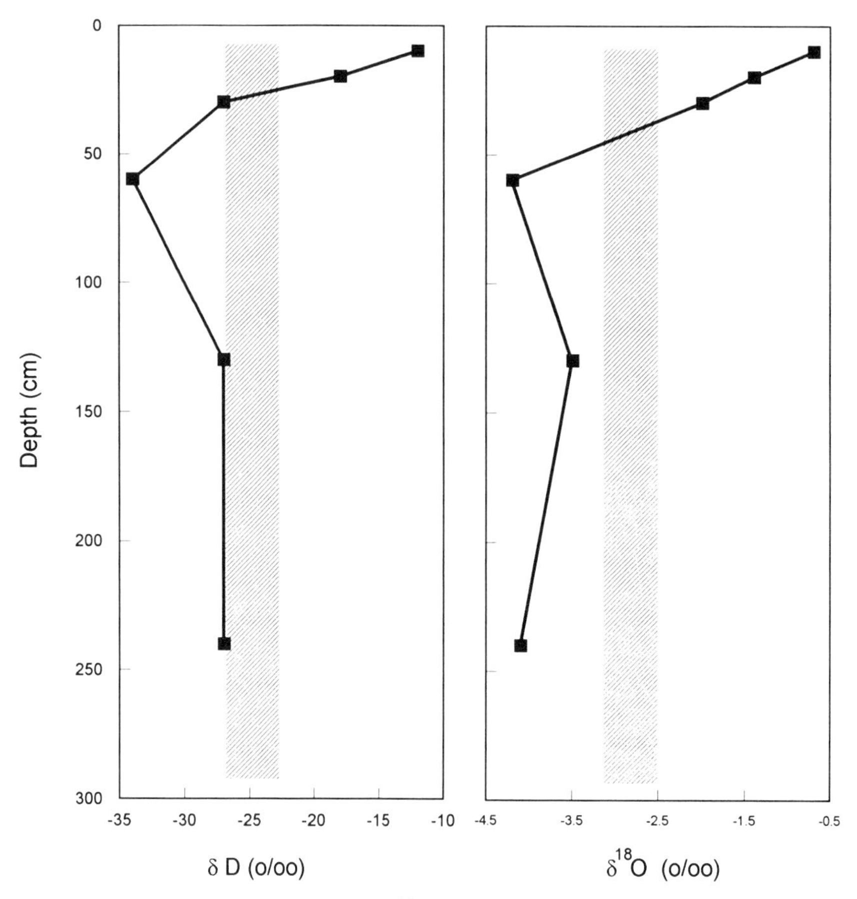

Figure 5.10 Deuterium (δD) and oxygen ($\delta^{18}O$) isotopic composition of soil water through the soil depth profile. These values are for an inland site of South Australia in July. (Used with permission from Thornburn and Walker 1993.)

The isotopic composition of bulk leaf water is dependent on the isotopic composition of the source and the impacts of transpiration. There is no evidence that discrimination of ^{2}H and ^{18}O occurs during water uptake by roots (Dawson and Ehleringer 1991) or by translocation in the xylem. The magnitude of isotopic enrichment of bulk leaf water at the mesophyll cell walls will depend upon transpiration, the humidity gradient, and the isotopic composition of water in the air (Leaney et al. 1985). The diversity of factors that affect δD in leaf water make physiological interpretations difficult. However, the isotopic composition of xylem water unambiguously reflects the isotopic composition of water in soil taken up by the plant.

B. Physiological Processes

The temporal and spatial use of water by roots is difficult to determine by conventional means. Root excavations can define the spatial root distribution for a species at one time, but this will not define the temporal changes in root distribution or the temporal activity

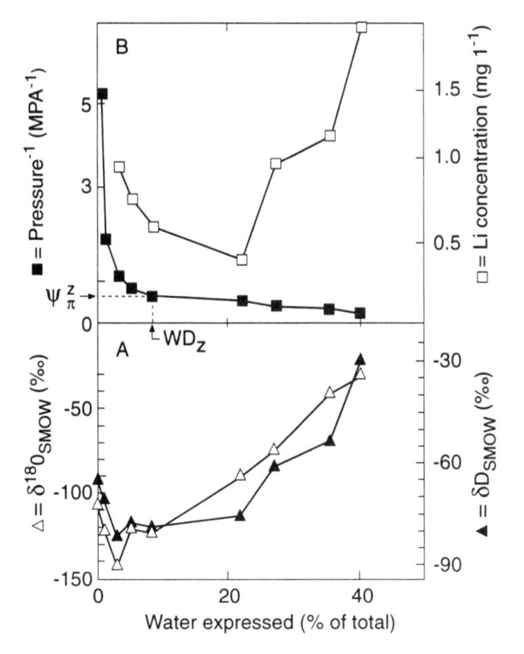

Figure 5.11 (A) Comparisons of isotopic composition of exudated water at various relative water contents (the larger the water expressed, the larger the relative water deficit). (B) Pressure–volume curve used to indicate the water potential components during drydown. Lithium concentration serves as an indicator of symplastic water. A large amount of symplastic water enters the exuded sap after the cells decrease to 20% water expressed. (Used with permission from Yakir et al. 1989.)

of various sections of the root system. Use of surface water or groundwater can be defined qualitatively by careful measurements of water potential and transpiration (Nilsen et al. 1984), but this takes extensive work over long time periods.

Due to the potential differences in isotopic composition of surface soil water (reflective of immediate past precipitation or enriched by evaporation) and deep groundwater (an integration of long-term precipitation δ^2H and $\delta^{18}O$), and the fact that no discrimination occurs during water uptake or translocation, the xylem water will have an isotopic signature identifying water resources. Several successful studies have applied this technique. For example, xylem δ^2H can be used to determine if a species utilizes summer precipitation (Figure 5.12), because summer precipitation is enriched in ^2H (White et al. 1985, Flannagan and Ehleringer 1991) compared with winter precipitation. Deuterium isotopic variation between summer and winter can be used to identify phreatophytes because groundwater has an isotopic signature close to that of winter precipitation (Thornburn et al. 1992). Oceanic water is enriched in ^{18}O and ^2H (Figure 5.13); thus the xylem isotope signature can be used to identify whether coastal species are using oceanic or fresh water (Sternberg et al. 1991). Moreover, the xylem isotope signature can be used to identify if riparian taxa are utilizing streamwater or soil water (Dawson and Ehleringer 1991).

Natural abundance of ^2H and ^{18}O in leaf water should vary on a diurnal cycle due to the enrichment effects of transpiration (Leaney et al. 1985). However, the isotopic enrichment of leaf water is dependent on the isotopic signature of soil water, differentiation of source water into apoplastic and symplastic water, as well as transpiration. Therefore, the value of δ^2H or $\delta^{18}O$ in leaf water cannot be related consistently to transpiration alone (Yakir et al. 1989). Yet the deviation between δ^2H or $\delta^{18}O$ of the leaf from that of the xylem should indicate the relative quantity of transpiration on a short-term basis. Integrated transpiration over the lifetime of the tissue cannot be determined by current technology from ^{18}O or ^2H isotopic composition of leaf water.

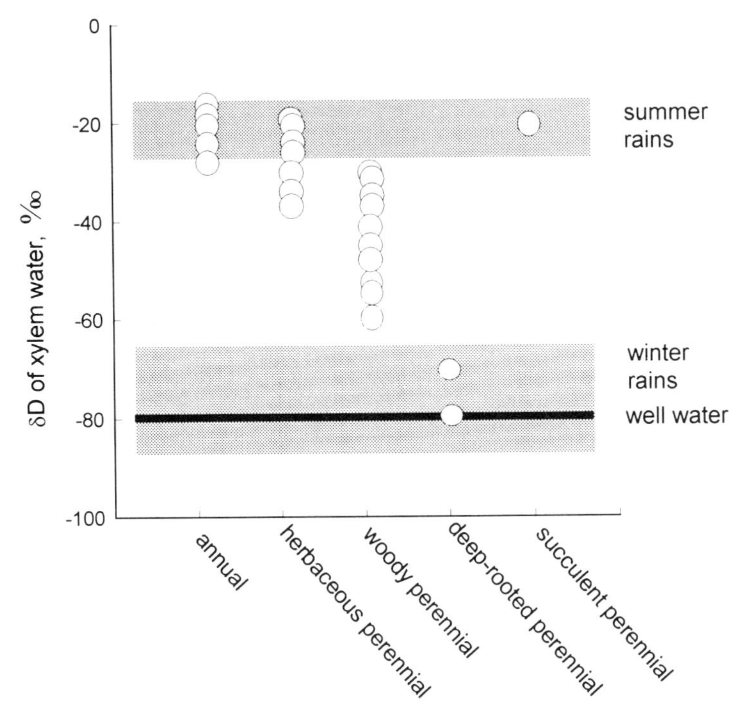

Figure 5.12 Hydrogen isotope composition (δD) of xylem water during the summer in an array of desert species in one shrub community in Utah. The shaded bands represent the δD for summer rain, winter rain, or groundwater found at this site over the year. Succulents and annuals are using summer precipitation while deep rooted perennials are using groundwater. (Used with permission from Ehleringer et al. 1991.)

Although the isotopic composition of leaf water may not be a good long-term index of transpiration, an association has been observed between δD or $\delta^{18}O$ and water use efficiency. Early tests of this relationship concerned comparisons of δD and $\delta^{18}O$ in CAM, C_3, and C_4 species. C_4 species (high WUE) tended to have a higher enrichment of deuterium than that of C_3 species (Leaney et al. 1985). In addition, plants adapted to saline environments (halophytes) have a higher WUE than those adapted to freshwater environments (glycophytes), and halophytes also have an enriched δD (Guy et al. 1986, Sternberg et al. 1991) compared with glycophytes. The increased WUE and δD enrichment probably corresponded to a greater water stress under saline conditions. If water stress induces increased water use efficiency, there should be an enrichment in ^2H corresponding to an increase in $\delta^{13}C$ under lower water potential. In fact, such a relationship has been identified in desert taxa (Ehleringer and Dawson 1992 and citations therein). Isotopic composition of CAM plant stem or leaf water may not be lower than that of C_3 or C_4 plants because CAM plants are succulent (extensive exchange of water between apoplast and symplast), CAM plants have unusual root systems (can become decoupled from the soil), and CAM plants open stomata at night (lower impact of transpiration on the isotopic composition of leaf water).

The complex linkages between transpiration, δD, and metabolic water can cause cellulose to have variable ^2H content. Variation in tree ring cellulose deuterium content has

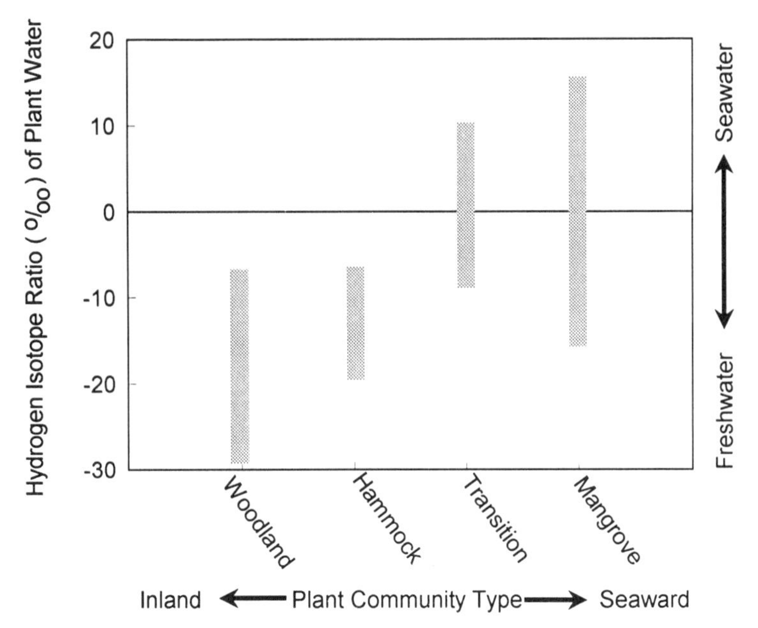

Figure 5.13 Range of hydrogen isotope ratios in plants occupying a gradient of habitats from ocean front to interior in southern Florida. Mangrove communities are on the ocean side, while woodland communities are interior. On the right-hand vertical axis, the average hydrogen isotope ratio for seawater and fresh water are represented. (Redrawn from Dawson 1993.)

been used to estimate long-term climatic conditions. Since H in cellulose has a rapid turnover, nitrocellulose is used for this assay. Adequate correlations have been found between the δD of tree ring nitrocellulose and the mean air temperature of the growth season when the growth ring was produced (Stokes and Smiley 1968, Gray and Song 1984). Moreover, a 1 : 1 relationship was found between the δD of nitrocellulose and the δD of precipitation (White et al. 1985). Tree ring δD should reflect both mean air temperature and the δD of precipitation if the tree is using groundwater that is derived from local storms. However, if the tree is utilizing surface water or is affected by stormwater from other regions, then the δD of tree ring nitrocellulose will not reflect current climatic conditions. After the relationships among the isotopic composition of nitrated cellulose, tree water use patterns, precipitation patterns, and climatic temperature conditions are further established, old tree trunks may once again become an accurate predictor of long-term climatic trends.

Nuclear magnetic resonance (H-NMR) is another technique for accurately identifying ^{2}H in plant tissues. H-NMR has an advantage over mass spectroscopy because the assay can be nondestructive and is able to identify changes in the position of ^{2}H within molecules. H-NMR assay of deuterium is difficult because the effective sensitivity of the assay with respect to ^{1}H is very small. Furthermore, the ability to discriminate chemical shifts of ^{2}H is very low (Martin et al. 1992).

Investigations of site-specific isotope fractionation (SNIF-NMR) can be used to identify molecules in a way similar to that done with radioactive labeling. For example, SNIF-NMR can unambiguously identify the sugar source (beet sugar, wine sugar, juice sugar) by evaluating δD and $\delta^{18}O$ in the ethanol products of fermentation. This technique is commonly used to identify contaminated wines and beers (Martin et al. 1991). In addition, the technique may be sensitive enough to separate among the sugars produced by a particular wine over several years. Since SNIF-NMR can discriminate annual variation in ^{2}H placement upon sugar in wine, this technique has the potential to evaluate the impact of environment on photosynthate production.

V. SUMMARY

Measurements of physiological responses and adaptations to stress in plants by classical means are snapshots of plant physiological activity. Developmental, temporal, and spatial changes in physiological characteristics are critical to understanding plant response to environmental or biotic stress. This can be addressed by replicating physiological measurements over time, development, or space and interpreting the changes in physiological activity at the level of the whole plant. In contrast, stable isotope composition can be utilized to assay integrated physiological responses to environmental or biotic factors. Both biotic and abiotic processes discriminate against heavier isotopes. For example, $^{13}CO_2$ diffuses more slowly than $^{12}CO_2$, $H_2{}^{18}O$ evaporates more slowly than $H_2{}^{16}O$, and Rubisco reduces $^{12}CO_2$ preferentially to $^{13}CO_2$. Kinetic and thermodynamic fractionation processes determine the isotopic composition of tissues and water in plants. If one can understand the relationship between tissue isotopic composition and the specific regulating kinetic or thermodynamic fractionation processes that lead to that isotopic composition, simple assay of tissue isotopic composition can be used to evaluate temporally and spatially integrated fractionation process.

Studies of plant stress physiology have shown that fractionation of C, O, H, and N isotopes can be used as indicators of integrated physiological processes. Isotopic composition of carbon in tissues can indicate photosynthetic physiology (C_3, C_4, or CAM) and can serve as an index of whole-plant integrated water use efficiency. Other factors that affect intercellular CO_2 concentration will also influence tissue $\delta^{13}C$. The natural abundance of ^{15}N in plant tissues can be used as an indicator of the contribution of nitrogen fixation to whole-plant nitrogen accumulation. Enrichment of soil N with fertilizer containing a high proportion of ^{15}N can be used to measure nitrogen use efficiency and the impact of nitrogen fixation on whole-plant N accumulation. The $\delta^{2}H$ or $\delta^{18}O$ of xylem water can indicate which layers of the soil are being used for water resources, whether plants are using seawater or fresh water, or whether plants are using stream water or groundwater. The deuterium composition of leaf water is influenced by the source water, water exchange between the symplast and the apoplast, and transpiration. The use of stable isotopes as indices of integrated physiology is relatively recent, and many more possible relationships are likely to be found in the future.

STUDY-REVIEW OUTLINE

Introduction

1. Physiological assays are limited by small sample size, inability to estimate integrated whole-plant processes, and by the snapshot nature of the assay.
2. Families of isotopes contain radioactive and stable members.
3. Radioactive isotopes are useful as molecular markers in biochemical pulse chase assay.
4. The relative abundance of stable isotopes may be useful for determining integrated physiological processes.

5. Isotope effects due to abiotic or biotic mechanisms cause differential fractionation of isotopes among environmental or biotic components.
6. Kinetic isotope effects refer to the impact of a particular process on isotope discrimination.
7. Thermodynamic isotope effects refer to the equilibrium ratio of isotopes as a result of two opposite kinetic processes.
8. The most important families of stable isotopes to plant stress physiology are C, H, O, and N.

Carbon Stable Isotopes

1. The family of isotopes for carbon contains two stable (^{12}C and ^{13}C) and many radioactive members.
2. The most useful carbon isotopes for pulse chase experiments are ^{14}C and ^{11}C, and ^{14}C can also serve as a geological clock because of its long half-life.
3. The ratio of ^{13}C to ^{12}C is often compared to the standard isotope ratio in Pee Dee Belemnite. This comparison is termed the $\delta^{13}C$ or carbon isotope composition.
4. Isotopic ratios of tissues can also be compared to the isotopic ratio of the CO_2 source (usually the atmosphere), and this is the $\Delta^{13}C$ or carbon isotope discrimination.

Fractionation of Carbon Stable Isotopes

1. The primary abiotic isotopic effect is that of diffusion, which discriminates against ^{13}C.
2. The thermodynamic isotope effect between air and water (CO_2 dissolving in water) also discriminates against $^{13}CO_2$.
3. Rubisco has a kinetic isotope effect of 1.0029, which means that it discriminates against ^{13}C by 29‰.
4. PEP-carboxylase uses HCO_3^-, which results in a discrimination of only 5.7‰.
5. Fractionation also occurs during metabolic conversions of reduced carbon such that different classes of compounds can have different isotopic composition. Lipids tend to be 5 to 10‰ depleted in ^{13}C compared to other cellular molecules.

Physiological Processes

1. The isotopic composition of tissues can be used to identify the carboxylation pathway.
2. C_3 plants have a $\delta^{13}C$ of 20 to 30‰, C_4 have a $\delta^{13}C$ of 10 to 12‰, and CAM plants have variable isotope ratios.
3. The $\delta^{13}C$ of C_4 can be used to identify whether the plant is utilizing the NADPH-ME, NAD-ME, or PCK physiology.
4. Variability in CAM stable isotope ratio can be attributed to the ability to switch between C_3 and CAM physiology.
5. Several environmental characteristics, such as temperature, light, vapor pressure, and pollution, affect the isotopic composition of tissues.
6. The underlying reason for environmentally induced changes in carbon isotope composition of C_3 tissues is the intercellular CO_2 concentration.
7. The $\delta^{13}C$ of C_3 species can be related to the integrated water use efficiency of the plant if the isotopic composition of the atmosphere and the vapor-pressure gradients are similar.
8. It may be possible to utilize the carbon isotope ratio of tissues to predict water use efficiency, then combine this information with productivity measurements to estimate water use patterns.
9. The $\delta^{13}C$ of tissues may be an important screening device for selecting and breeding crop genotypes with higher water use efficiency.

Nitrogen Stable Isotopes

1. The family of nitrogen isotopes contains two stable members (^{14}N and ^{15}N).
2. There are several radioactive isotopes of nitrogen, but these have very short half-lives and are unusable in physiological research.
3. Nitrogen stable isotopes are used in enrichment studies and in natural abundance studies.

4. Measurement of nitrogen stable isotope composition is complicated by small differences between different tissues or species and the high possibility of contamination by the atmosphere during mass spectrometer measurements.

Fractionation of Nitrogen Stable Isotopes

1. Kinetic and thermodynamic isotope effects combine, over a long time period, to create the isotopic composition of nitrogen in the soil solution.
2. Although many transformations of nitrogen occur in the soil system, soils tend to be enriched in ^{15}N (by 0 to 10‰) compared to the atmosphere.
3. The atmosphere has a very stable nitrogen isotopic composition; therefore, it is used as the standard in $\delta^{15}N$ determinations.
4. Soils may vary in $\delta^{15}N$ spatially (both horizontally and in depth) and temporally. Moreover, there is more variability between regional sites (20‰) than within each site (3‰).
5. Some fractionation of N isotopes occurs in plants after N accumulation under certain conditions. However, if the plant is growing actively in fertile soil and moderate light, little to no fractionation will occur from the root accumulation of NO_3^- to deposition of nitrogen in organic constituents.
6. A large isotopic effect is possible in root nodules, associated with the physiology of the bacteria.
7. Nitrogen isotope composition is fairly uniform among organs in herbaceous plants. In woody plants the trunk can be strongly depleted in ^{15}N compared to stems or leaves.

Physiological Processes

1. Stable nitrogen isotopes are most useful in quantification of nitrogen fixation.
2. Other nitrogen fixation determinations cannot stage up from the physiological measure (acetylene reduction of the nodule) to the whole plant or population.
3. The measurement of biological nitrogen fixation (BNF) by stable isotope abundance is dependent on a difference between the soil solution and atmospheric nitrogen isotope composition.
4. In the enrichment technique, soil is enriched in ^{15}N to enlarge the difference between soil and atmosphere.
5. It is critical to employ a reference crop (non N-fixer) planted interstitially with the test crop. This is done to negate the significance of soil nitrogen isotope heterogeneity.
6. The contribution of nitrogen fixation to whole-plant nitrogen content can be estimated from the tissue nitrogen content and the $\delta^{15}N$ values of the reference and test crops.
7. Natural populations of plants cannot be evaluated by the enrichment technique because: (a) the addition of NO_3^- or NH_4^+ would change the availability of nitrogen in the system, (b) the nitrogen could not be applied evenly, and (c) the plants already have tissues with variable nitrogen isotope composition.
8. The natural abundance of nitrogen isotopes varies by approximately -5 to $+10$‰.
9. If a (or several) suitable reference species can be located in the same region as the test species, the $\delta^{15}N$ of their tissues can serve as a proxy for the $\delta^{15}N$ of soil nitrogen in enrichment studies.
10. The $\delta^{15}N$ of plants surviving on nitrogen fixation as the only nitrogen source must be known.
11. The $\delta^{15}N$ of plants using only fixed nitrogen and the $\delta^{15}N$ of reference plants using only soil nitrogen form two poles of a scale from 100% to 0% nitrogen fixation. The $\delta^{15}N$ of the test species can be compared to this scale to determine the contribution of nitrogen fixation to whole-plant nitrogen content.
12. Although the natural abundance technique has been applied successfully in several ecosystems, there are several disadvantages to this assay: (a) the differences in $\delta^{15}N$ between the reference and test species may be very small; (b) it is difficult to find a naturally growing appropriate reference species.
13. Coupling nitrogen and oxygen isotope discrimination data have been used successfully to determine the impact of denitrification on oceanic nitrogen pools.

14. Stable nitrogen isotopes may also be useful as metabolic tags and can be assayed by H-NMR technology.

Hydrogen and Oxygen Stable Isotopes

1. There are three isotopes of hydrogen, two of which are stable (^{1}H, ^{2}H).
2. There are many isotopes of oxygen, three of which are stable (^{18}O, ^{17}O, ^{16}O).
3. The fractional abundance of deuterium is the lowest among that of stable isotopes (1.00015 or 0.015%).
4. The reference material for both hydrogen and oxygen isotopes is SMOW (standard marine ocean water).

Fractionation of Hydrogen and Oxygen Stable Isotopes

1. There is a strong covariance between hydrogen and oxygen isotopes relative abundance in precipitation. This relationship is called the meteoric water line.
2. The isotopic composition of precipitation is determined by the temperature at which the rain droplets are formed. The higher the temperature, the more enriched in deuterium and ^{18}O will be the rainwater.
3. There is no fractionation of isotopes in water during accumulation into xylem and transport to tissues. But fractionation of hydrogen and oxygen isotopes occurs during metabolism of water and transpiration.
4. Photosynthesis has a large discrimination against deuterium (about 150‰).
5. Once carbohydrates are formed, hydrogen is replaced and this process results in an enrichment of deuterium of about 150‰.
6. The final hydrogen isotope ratio is dependent on (a) the isotope ratio of the source water, (b) the isotope effects of photosynthesis, and (c) the isotope effects of hydrogen cycling.
7. The isotopic composition of water in the leaf is dependent on that of the source and the isotopic effect of transpiration.

Physiological Processes

1. The $\delta^{18}O$ or δD of xylem water can be used to determine the source of water used by plants (ground versus surface).
2. Since ocean water is enriched in deuterium and ^{18}O compared to fresh water, xylem isotopic composition can be used to determine if halophytes are using ocean or fresh water.
3. The isotopic composition of leaf water varies over the diurnal cycle as a consequence of transpiration, and over the season as a consequence of changing source isotopic composition.
4. The isotopic composition of leaf water compared to that of the xylem water can be used to compare rates of transpiration.
5. In some situations the isotopic composition of leaf water is reflective of the water use efficiency and covaries with $\delta^{13}C$.
6. Deuterium composition of wood, particularly nitrocellulose, may be a good indicator of long-term climatic conditions.
7. SNIF-HNMR technology (site-specific isotope fractionation) is able to determine the position of deuterium on specific molecules, and can identify a signature for sugars from certain organisms.

SELF-STUDY QUESTIONS

1. Compare and contrast the natural abundances of C, N, H, and O stable isotopes.

2. Why does the $\delta^{13}C$ value of C_3, C_4, and CAM plants differ when they are all growing in the same air?

3. Which stable isotopes are most useful as integrative (across time) indicators of physiological function?
4. When a CAM plant switches from the C_3 pathway to the CAM pathway, what happens to the $\delta^{13}C$ and why does it change?
5. What is the cause of the large variance in $\delta^{13}C$ of C_3 plants compared to C_4 plants?
6. Compare and contrast the advantages and disadvantages of the enrichment or natural abundance methods for measuring nitrogen fixation with $\delta^{15}N$.
7. Why is the reference plant so important to the $\delta^{15}N$ technique for measuring nitrogen fixation?
8. What regulates the oxygen and hydrogen isotope compositions of xylem water?
9. How can the isotopic composition of xylem water be used to determine the sources of water used by the plants during different seasons?
10. What physiological factors control the isotopic composition of leaf water, and how can this be used to evaluate water dynamics in plants?
11. Differentiate between a kinetic and a thermodynamic isotope effect.

SUPPLEMENTARY READING

Ehleringer, J. R., A. E. Hall, and G. D. Farquhar. 1993. Stable Isotopes and Plant Carbon–Water Relations. Academic Press, San Diego, CA.

Farquhar, G. D., J. R. Ehleringer, and K. T. Hubick. 1989. Carbon isotope discrimination and photosynthesis. Annual Review of Plant Physiology and Plant Molecular Biology 40:503–547.

Lajtha, K., and R. H. Michner. 1994 Stable Isotopes in Ecology and Environmental Science. Blackwell Scientific Publications, Cambridge, MA.

Rundel, P. W., J. R. Ehleringer, and K. Nagey. 1989. Stable Isotopes in Ecological Research. Springer-Verlag, New York.

6 Plant Carbon Balance

Outline

Objectives

1. Give a short introduction to the plant carbon cycle.
2. Discuss the processes that regulate carbon dioxide accumulation by plants. Include a discussion of the atmospheric CO_2 concentration, the CO_2 diffusion pathway, and the regulation of CO_2 flux by physiological and anatomical attributes of stomata and mesophyll.
3. Describe the functional relationship between CO_2 concentration (ambient and intercellular) with photosynthesis. Include a discussion of a technique for calculating stomatal and mesophyll limitations to photosynthesis.
4. Describe the molecular model of light harvesting and utilization in photosynthesis.

5. What are the limitations to carbon gain by the biochemical (enzymatic) processes of photosynthesis?
6. Describe the C_4 pathway in relation to its effect on the factors that limit CO_2 accumulation, and its response to intercellular CO_2.
7. Explain the impact of CAM physiology on CO_2 accumulation processes. Include limitations to the effectiveness of CAM.
8. Describe the factors that regulate assimilate storage and circulation in plants.
9. How are the immediate products of photosynthesis allocated among cellular functions?
10. Explain the mechanisms that regulate the translocation of surplus cellular photosynthate.
11. What determines the allocation patterns of photosynthate among the various organs of a plant?
12. Describe the factors that regulate carbon storage processes in plants.
13. Describe the important regulatory steps in the respiratory pathway.
14. How do photosynthesis, respiration, and photorespiration interact in cells?
15. Define whole-plant respiration and its components. Explain how this relates to whole-plant growth.
16. What are the important environmental factors that influence whole-plant respiration?
17. Describe the various types of organic carbon emissions from plants.
18. Explain volatile organic carbon pathways in plants and describe what regulates these pathways. In particular, focus on the monoterpene and isoprene pathways.

I. INTRODUCTION: BASICS OF THE CARBON CYCLE IN PLANTS

The carbon balance of plants includes a diversity of physiologically integrated carbon input and output processes, such as photosynthesis, respiration, and volatilization (Figure 6.1). Exogenous factors influence the relationships among various endogenous processes of plant carbon balance. Environmental conditions may stimulate or inhibit particular aspects of the carbon cycle, and relationships among carbon balance processes can be changed by particular combinations of environmental factors. For example, as air temperature increases (within a moderate range) photosynthesis increases, but the relationship between photosynthesis and respiration changes (respiration increases more than photosynthesis). Furthermore, plant carbon balance is affected by other species in natural communities or agricultural fields. Shading by canopy species influences the carbon balance of subcanopy species. Herbivore populations influence the net carbon gain of host plants, and host plants utilize carbon resources to defend against herbivory. The combination of endogenous regulation mechanisms (e.g., enzyme activity), external environment influences (e.g., climatic conditions), and associated species interactions (e.g., competition) makes the study of plant carbon balance complex and scientifically stimulating.

The relationship between carbon gain and carbon loss, as well as the distribution of photosynthates within plants, is critical to stress tolerance or avoidance because of their influence on growth and survival. Environmental factors that influence carbon gain or loss have profound affects on plant distribution and evolution of adaptive traits. Countless examples of these relationships can be observed in natural and agricultural systems. Growth and reproduction in all plants are tied directly to carbon balance. Those plants that invest a larger portion of carbon into constructing photosynthetic organs will have high growth rates. Pod fill and the size of crops in agricultural systems are dependent on adequate photosynthate for fruit development. Fruit growers have long known that the biomass of

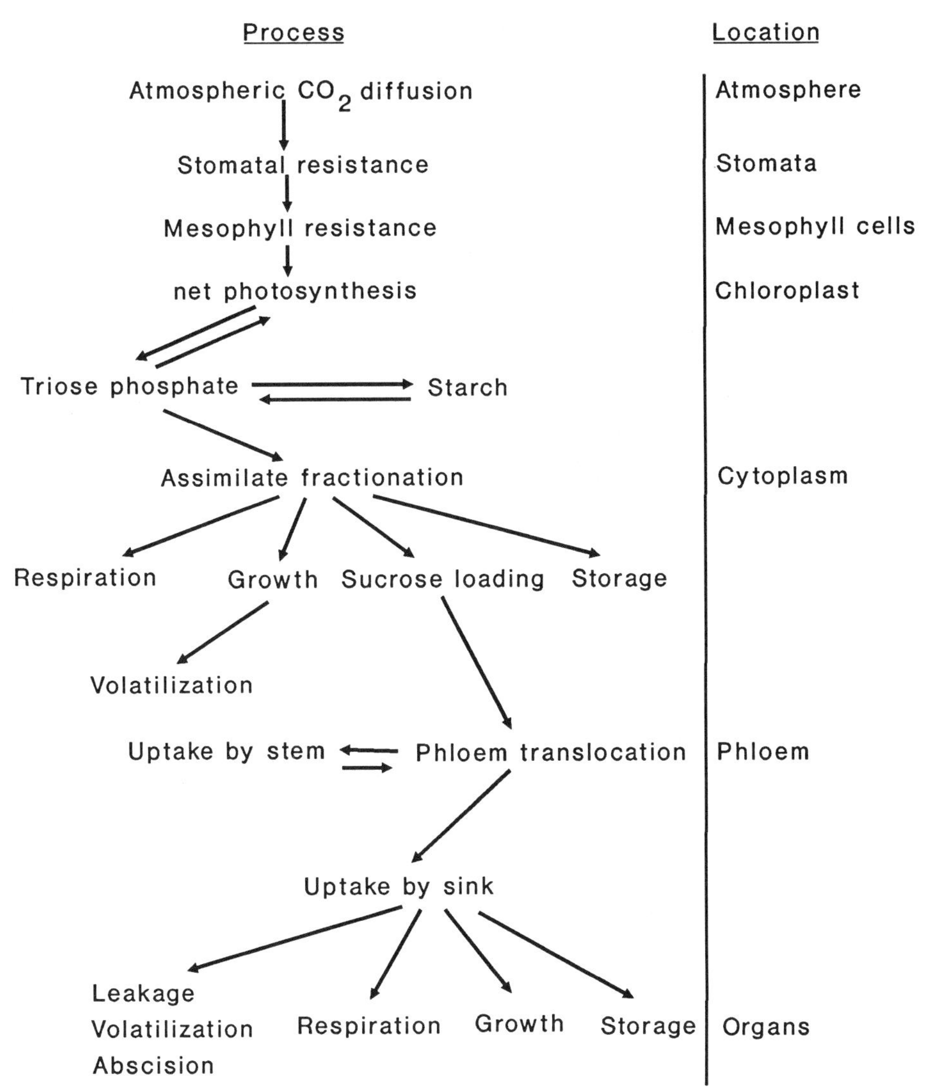

Figure 6.1 Schematic diagram of carbon flow through major regulatory processes in plants and its location in cells tissues and organs.

marketable fruit can be improved by increasing the ratio of leaf area to fruit mass. Consequently, a thorough understanding of basic plant carbon balance properties is a necessity for anyone studying plant stress physiology.

It is not possible in one chapter to cover all aspects of plant carbon balance in detail. Therefore, the objective of this chapter is to discuss several major processes of plant carbon balance and the factors that can regulate those processes. The chapter is designed to cover the pathway of carbon flow from the accumulation of CO_2 to volatilization of carbon containing compounds. The regulation of gaseous diffusion of CO_2 into the chloroplast, photochemical processes, CO_2 reduction pathways, assimilate circulation, respiration, and volatile emissions are considered.

II. CARBON ACCUMULATION PROCESSES

A. Atmospheric Carbon Dioxide Concentration

In terrestrial plants, the source of all carbon gain is CO_2 in the atmosphere. The concentration of carbon dioxide surrounding plant leaves in an unaltered atmosphere is termed the ambient condition. If the atmosphere around leaves has been altered, this is referred to as elevated or depleted conditions. Ambient CO_2 concentration is currently limiting to photosynthesis in C_3 plants. As a result, net photosynthesis (P_n) increases with an increase in elevated CO_2 atmospheric conditions. Commonly, an ambient concentration of 350 ppm CO_2 is assumed for the turbulent air around plants. However, that value is increasing every year and is predicted to double by the year 2050 (Strain and Bazzaz 1983). Furthermore, the CO_2 concentration in many ecosystems is variable in space and time. There is a gradient of decreasing CO_2 concentration with increasing height in a canopy (Figure 6.2A), and the concentration of CO_2 can be high in the morning and low in the late afternoon (Figure 6.2B). This temporal and spatial variation of atmospheric CO_2 in natural systems is a consequence of the ratio between respiration and photosynthesis for organisms in the ecosystem. The lower canopy has a greater quantity of respiring organisms and tissues (including those in the soil), while the upper portion of the canopy has a greater proportion of photosynthetic organs. Thus the concentration of CO_2 in the canopy atmosphere is lower than that near the ground surface during the day. In addition, photosynthesis of outer canopy leaves is higher than that of inner canopy leaves because of the atmospheric turbulence and high radiation in the outer canopy environment than in the subcanopy environment. Following the dark hours, atmospheric CO_2 reaches its peak concentrations because photosynthesis is inhibited. Atmospheric CO_2 reaches its minimal value in the late afternoon when net photosynthesis has had its largest effect.

Geological and anthropogenic factors can also affect the CO_2 concentration in the atmosphere. Natural geothermal vents can release geologic CO_2, or natural combustion of fossilized organic matter can create local regions of high CO_2 concentrations. Furthermore, combustion of fossil fuel by automobiles and industry can cause localized areas and/or periods of elevated CO_2. Due to the spatial and temporal variation in CO_2, local patterns of CO_2 need to be evaluated when considering the impact of ambient CO_2 concentration on plant carbon balance in natural or agricultural systems. This is necessary because diffusion of CO_2 into leaves is partly dependent on the concentration gradient of CO_2 between the atmosphere (C_a) and intercellular air spaces (C_i).

It is valuable to consider the units used to describe CO_2 in the atmosphere. At sea level, when the atmospheric pressure is 100 kPa, the molecules of CO_2 exert a pressure of 34 Pa, which is termed the partial pressure of CO_2. This is the same as saying that in 100,000 molecules of air there are 34 molecules of CO_2, giving us the traditional 340 μmol mol^{-1} (or 340 ppm) of CO_2. When atmospheric pressure is decreased to 67 kPa (by changing the altitude to 3000 m, for example), the mole fraction of CO_2 does not change but its partial pressure is proportionally reduced, to 22.5 Pa. This means that there are fewer total number of molecules per volume of space (lower "air pressure") and proportionally fewer CO_2 molecules in the same space (lower CO_2 partial pressure). In this case the concentration of CO_2 per unit volume, not per unit air, has decreased to 225 ppm. Therefore, the ppm designation describes the number of molecules in the same volume of space, not the proportional number of molecules per volume of air (partial molar fraction). This means that a room at sea level with 225 ppm CO_2 would have the same effect

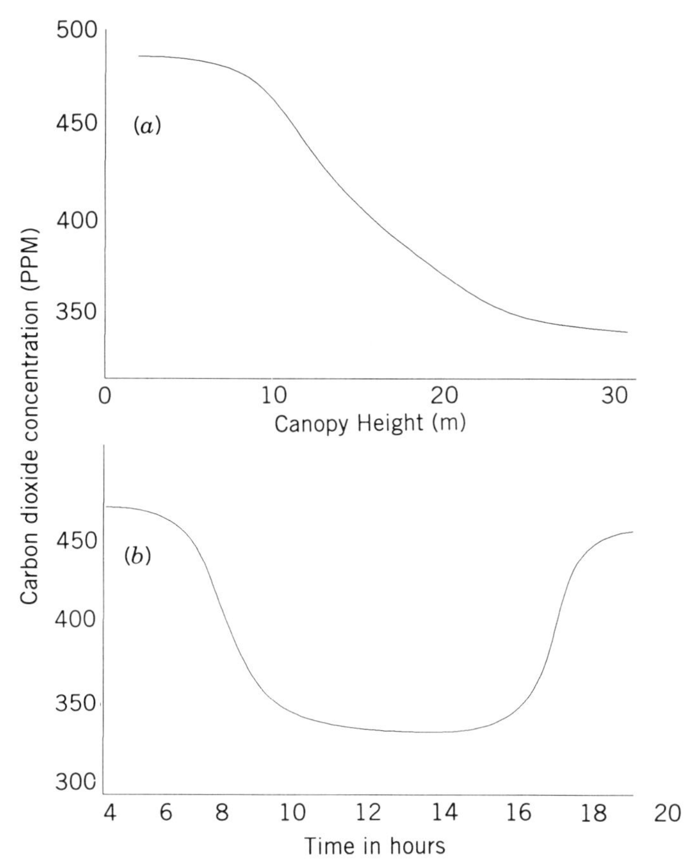

Figure 6.2 (A) CO_2 concentration in the atmosphere at different heights in a forest canopy at dawn. (B) Variation in CO_2 concentration over a diurnal cycle in a subcanopy forest environment at 1 m height.

as the air at 3000 m with 340 ppm CO_2. In this case, the mole fraction of CO_2 is altered in the experimental room instead of a change in CO_2 partial pressure that occurred at 3000 m.

B. Pathway of CO_2 Movement

Physical factors of diffusion, such as temperaure and, the partial pressure of CO_2 at the source (C_a) and sink (C_i), regulate the movement of CO_2 into the chloroplast. Along the pathway of CO_2 diffusion into plants, molecules of CO_2 meet a number of different resistances (Figure 6.3). Several of these are the same as those faced by water molecules as they move out of the intercellular spaces into the turbulent air around the plant. In fact, the pathways for CO_2 and H_2O movement are the same (but in opposite directions) from the turbulent air, through the boundary layer and the stomata, into the intercellular spaces (Figure 6.4). Once inside the intercellular spaces, CO_2 diffusion meets resistances which are different from those met by H_2O molecules (see Chapter 7 for a further discussion of diffusion). These cellular resistances to CO_2 diffusion (termed *mesophyll resistance*) in-

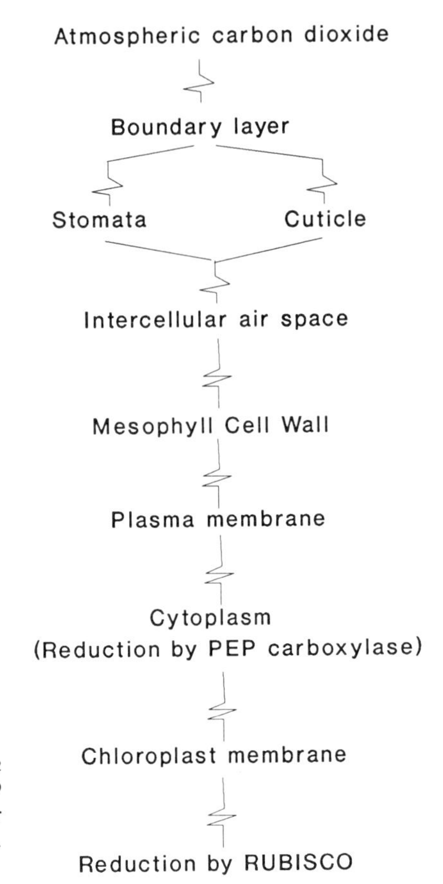

Figure 6.3 Flow diagram of the resistances to CO_2 diffusion along the pathway from the atmosphere into the chloroplast. All resistances are in series except for the stomatal and cuticular resistance, which act in parallel.

clude the pathway from the cell wall, through the plasma membrane and cytoplasm, to the chloroplast membrane, and into the chloroplast. The medium within which CO_2 diffuses is an aqueous solution for mesophyll resistance and a gaseous solution for intercellular, stomatal, and boundary layer resistances. Therefore, there is a change in the rate of diffusion due to a change in both the diffusion coefficients (in air versus in water) and the resistances.

The rate of CO_2 diffusion in the gaseous portion of its pathway into cells is regulated by external atmospheric conditions, the thickness of unstirred air near the leaf surface (boundary layer), stomatal conductance (inverse of stomatal resistance), substomatal chamber size, and the atmospheric conditions of the intercellular spaces. General leaf structural attributes that influence gaseous diffusion into or out of the intercellular spaces are covered in Chapter 8. Here we concentrate on the physiological regulation of stomatal and mesophyll conductance.

1. Regulation of Stomatal Conductance. It is generally agreed that the primary mechanism for regulation of stomatal conductance is guard cell turgor pressure (Figure 6.5). A simplified model of stomatal regulation is that as turgor pressure in the guard cells increases, the guard cells swell and the stomatal aperture increases, due to the bending moment placed on the guard cell wall. Guard cell walls are thickened on the stomatal aper-

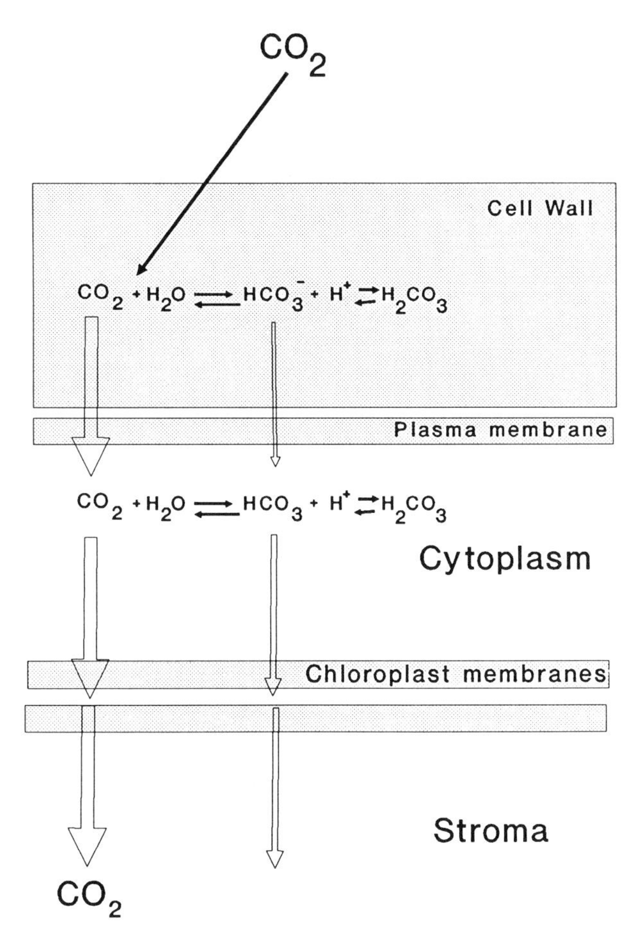

Figure 6.4 Solubility chemistry of CO_2 from the intercellular air spaces in the cell wall and cytoplasm. The size of the arrow crossing the membrane reflects the relative ease by which the substance crosses the membrane. The ion, CO_3^{2-}, is not considered in this figure because it is not normally found in the cell wall due to the pH of the cell wall water. (Redrawn from Nobel 1991.)

ture side and have thick bundles of microfibrils oriented perpendicular to the stomatal aperture. This configuration of cellulose thickening ensures that stomata will open when guard cell turgor pressure increases. This linkage between guard cell turgor and stomatal aperture is sensitive because changes in stomatal aperture in response to dry air blown over the guard cells can occur within a few minutes (Lange et al. 1971), and this rapid response of guard cells to dry air is complete within 1 min (Fanjul and Jones 1982). Our knowledge of the relationships among physiological and environmental factors that influence guard cell turgor pressure is still being refined. These relationships are important because stomatal aperature regulates an important aspect of CO_2 accumulation by plants. Therefore, when stomata close in response to water stress, CO_2 is curtailed. Many researchers are trying to develop an integrated model that links environmental and physiological regulation of stomatal aperture. Some of these relationships are discussed below.

Studies of the impact of atmospheric vapor pressure on the aperture of individual stomata has led to significant understanding of stomatal function. The direct impact of va-

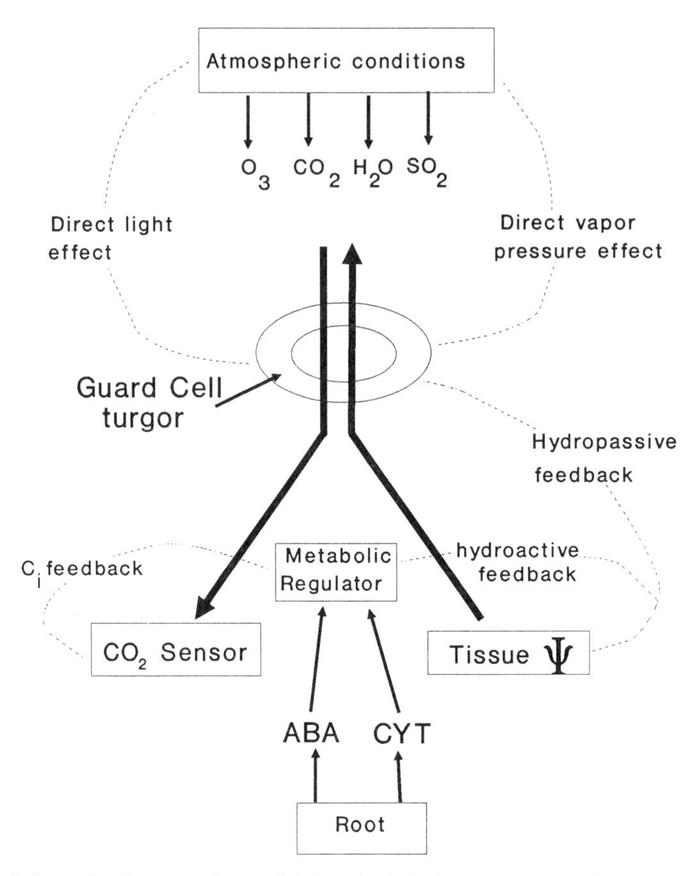

Figure 6.5 Schematic diagram of a model for physiological regulation of stomatal conductance. ABA, abscisic acid; CYT, cytokinins; C_i, CO_2 concentration of intercellular air spaces. (Redrawn and updated from Rashke 1975).

por pressure on guard cell turgor pressure can be temporarily decoupled from the bulk water potential of the leaf. Bulk water potential is the average energy of water in the leaf, in contrast to the water potential of specific cell layers such as the epidermis. As atmospheric vapor pressure decreases, the vapor pressure gradient between the leaf tissues and the air (VPG) increases and water loss from the peristomatal cells (those immediately adjacent to the guard cells) increases relative to the rest of the epidermis. Water loss from the peristomatal cells occurs through the cuticle. This localized cuticular water loss reduces the water potential of the cells around the guard cells, causing the guard cell turgor pressure to decrease (Mair-Maerker, 1979) and stomata to close even though there is no change in bulk leaf water potential. Other evidence indicates that the site of enhanced water loss is the inner walls of the guard cells (Meinder 1976) rather than the peristomatal cells. However, water loss from the inner cell walls of the guard cells will be proportional to total transpiration. This creates a dilemma for the model that invokes the importance of the inner guard cell walls to stomatal aperture. How can transpiration from the guard cell inner walls cause a reduction in water loss when transpiration of guard cell walls is low at low transpiration (Cowan 1994)? In other words, high transpiration would induce stomatal closure, but stomatal closure causes low transpiration, which would induce stomatal opening. What could reduce this potential oscillation effect? This dilemma could be resolved if one invokes a delicate balance between guard cell wall transpiration and the gain (rate of change up or down) in guard cell turgor pressure around a set point. For example, high transpiration would cause rapid stomatal closure toward the set point. As transpiration decreases toward the set point, the rate of stomatal closure decreases. The same phenomenon would happen in reverse for low transpiration. This would result in a dampening of the change in stomatal conductance (decreasing rate of change) when transpiration approaches the set point.

Other atmospheric factors besides vapor pressure can also influence stomatal conductance either directly or indirectly. As temperature increases (at constant humidity) the vapor pressure gradient between the inner leaf spaces and the turbulent air (VPG) also increases. The increase in temperature dramatically increases transpiration unless stomatal aperature decreases to moderate the increase in transpiration. This directly affects the water relations of the epidermis and turgor pressure of guard cells. Increased wind velocity decreases the thickness of the boundary layer between the leaf and turbulent air, thereby increasing transpiration (by decreasing the diffusion pathway) and causing stomatal closure. The concentration of atmospheric pollutants can also directly affect stomatal conductance. Sulfur dioxide, ozone, and other oxidizing compounds in the atmosphere can decrease the effectiveness of the leaf cuticle as a barrier to water loss (Mansfield and Freer-Smith 1984). This increases cuticular and stomatal conductance to water, causing lower epidermal water potential, which may induce stomatal closure. Atmospheric dust particles can directly occlude stomata, decrease irradiance to the leaf surface, and increase cuticular transpiration, all of which cause stomatal closure.

The discussion above points to the importance of evaporation from specific cellular locations in the epidermis for regulation of stomatal aperture. This can be considered a hydropassive regulation since there is no active metabolic regulation of guard cell turgor pressure (Figure 6.5). An additional hydroactive regulation system, in which specific metabolic processes affect stomatal aperture, has been suggested as critical to stomatal regulation. Evidence for hydroactive regulation first came from studies of the impact that light intensity has on stomatal aperture. Increasing light intensity increases stomatal aperture in most plants. In some cases, stomata open in the dark in C_3 species (Tobiessen 1982), and stomata open at night in CAM species. Thus light must have variable effects on the metabolism of different taxa, or different taxa respond differently to the stomatal changes induced by increasing light.

Light intensity is thought to induce stomatal opening by changing the osmotic potential of guard cells. Two photoreceptor systems, one absorbing blue light located in the plasma membrane, one absorbing red light and associated with the chloroplast (Schwartz and Zeiger 1984), both induce a proton efflux from the guard cells. The proton gradient then stimulates potassium-specific channels, increasing the permeability to potassium of the guard cell membrane. As a result, light induces a rapid influx of potassium from surrounding peripheral cells into guard cells, which reduces guard cell osmotic potential. The lower osmotic potential increases diffusion of water to the guard cells, which increases guard cell turgor pressure and induces stomata to open (for further information, see Taiz and Zeiger 1991).

There are two other important physiological signals affecting stomatal aperture. The first is the growth regulator abcissic acid (ABA). Very small increases in ABA concentration in relation to cytokinin concentration induces stomatal closure (Rashke 1987). The ABA signal to the stomata is induced by a stress proximal to the transpiring surface. For example, reductions in root water potential, in the absence of any changes in atmospheric demand for water, can induce a rapid increase in free ABA content of guard cells, thereby inducing stomatal closure (Lösch 1993). In contrast, many studies have been unable to find a correlation between leaf tissue ABA content and stomatal conductance. This is most likely because a bulk measure of ABA in leaf tissues is used to correlate with stomatal conductance. There is a large amount of background ABA in mesophyll tissues which masks the amount in the epidermis. In a study using split-root systems, drying of a small portion of the root system caused stomatal closure, with little change in leaf water potential (Zhang et al. 1987). Furthermore, bulk leaf ABA content remained constant but epidermal ABA increased, in agreement with reduced stomatal conductance. ABA is thought to increase the permeability to potassium in the guard cell membrane, causing potassium to flow out of the cell, thereby reducing osmotic concentration and guard cell turgor.

The response of stomata to ABA is not a simple mechanism because there are other possible signals that will interact with the ABA signal. For example, three general possible mechanisms for root signals have been proposed (Gowing et al. 1990). First, there is a negative hormonal signal from roots that counteracts the effects of ABA on growth and stomatal opening. Various researchers have suggested that a constant supply of cytokinins produced by roots may serve as this negative signal (Davies et al. 1986). Second, a positive dose-sensitive root signal could be produced during soil water shortage that increases in magnitude as root water potential decreases. ABA is thought to be the dose response signal from the roots that stimulates stomatal closure. Third, dry soil conditions will restrict flow of substances in the transpiration stream. The abundance of ions (and other substances) from the xylem stream could have a dose-responsive stimulating effect on stomatal conductance.

Stomatal conductance is also affected by the intercellular CO_2 concentration (Figure 6.4). Relationships among those factors that regulate atmospheric diffusion of CO_2 into the leaf and the demand for CO_2 by photosynthesis will be the primary regulators of intercellular CO_2 concentration (C_i). When the rate of net CO_2 reduction by photosynthesis increases compared with the limitations to gaseous diffusion of CO_2, C_i will decrease. A decrease in C_i is thought to increase stomatal conductance (within limits). Therefore, under well-watered conditions, increased photosynthesis (say as a result of increased light) often will result in increased stomatal conductance. This is not always the case, particularly when leaves are responding to sunflecks (Chapter 10).

When ambient CO_2 concentration decreases, this stimulates a lower C_i value, and stomata open. This response occurs only under relatively high light conditions. Furthermore, other factors that normally cause stomatal closure are overridden by the low ambient CO_2 effect (Kappen et al. 1994). Thus, it is thought that there is a physiological signal that is sensitive to C_i and can override the physical effects of the atmosphere on stomatal conductance.

There are many stresses that can cause an increase in C_i and therefore affect stomatal conductance. When ambient CO_2 increases, as is occurring in the atmosphere now, stomatal conductance will decrease because C_i increases. Since C_i reflects the balance of photosynthesis and stomatal conductance, it also reflects instantaneous water use efficiency (μmol of CO_2 gained/ mmol of H_2O lost), which is described in Chapter 8. The specific impacts of pollutants such as ozone may also operate through a C_i signal. Ozone and other oxidants increase the free-radical concentration in chloroplasts. This inhibits the rate of photosynthesis, which would increase C_i and cause stomatal closure (discussed further in Chapter 10).

Whatever the combination of factors that regulate stomatal aperture, transpiration and stomatal conductance are ultimately limited by the hydraulic architecture of the stem. Several authors have argued that the regulation of stomatal conductance is fine tuned to maintain the transpiration stream at a point just above that which would cause hydraulic failure by xylem cavitation (Sperry 1995, Field and Holbrook 1989) and still maximize carbon gain. In this case, maximum carbon gain is limited by the threshold for hydraulic cavitation. As a result, plants are balanced systems that adjust leaf area, stomatal conductance, and hydraulic properties to keep xylem water potential within a safety factor above that which causes xylem dysfunction by cavitation (see Chapters 7 and 8 for further discussion).

2. Regulation of Mesophyll Conductance. After a specific CO_2 molecule is located in the intercellular spaces, it has an opportunity to diffuse through the cytoplasm to the site of CO_2 reduction. This pathway is different from that through the stomata because it is an aqueous phase and the molecules must traverse membranes. The CO_2 must first dissolve in the water layer of cell walls lining the intercellular spaces of the leaf. Partitioning CO_2 between the aqueous and gaseous phases is dependent on temperature and pH. At high temperature, CO_2 has a lower solubility in water, which is a common characteristic for many gases. Dissolved inorganic carbon will take four forms; CO_2, HCO_3^-, CO_3^{2-}, and H_2CO_3 (Figure 6.4). pH affects both the amount of CO_2 that dissolves in the cell wall water and the equilibrium between HCO_3^- and CO_2. Since the H_2CO_3 in solution is only 1/400 of that for CO_2 and HCO_3^-, we will not consider the former. In addition, the amount of CO_3^{2-} is important only above pH 8 (an unusually high pH for cell wall water). Thus we focus only on CO_2 and HCO_3^- in this discussion. When the pH of the cell wall solution rises from 6 to 7, the ratio of dissolved CO_2 plus $H_2CO_3^-$ in cell wall water to gaseous CO_2 in the intercellular spaces increases by sixfold (Nobel 1991). Furthermore, the higher the pH, the smaller the ratio of HCO_3^- to CO_2. The fractional contribution of these two substances to the total CO_2 pool in the cell wall water is important when considering movement of CO_2 across the plasma membrane.

The cell wall provides little resistance to diffusion of CO_2 or HCO_3^-, but the plasma membrane has an extremely high resistance to movement of HCO_3^-. Therefore, it is likely that most carbon destined for photosynthesis enters the mesophyll cell as disolved CO_2. The actual transport process is not well defined but is adequate to support measured rates of photosynthesis. Once in the cell, the cytosolic resistance to CO_2 diffusion is small because the distance is very short. Chloroplasts are distributed around the cell near the plasma membrane (0.1 to 0.3 μm from the membrane). At the chloroplast membrane, HCO_3^- once again has a large resistance to transport. Estimates of resistances for CO_2 diffusion inside the chloroplast suggest values about three times those of the cytosol (Coleman and Espie 1985). The summation of resistances against CO_2 diffusion from the cell wall to the chloroplast lamellae is slightly smaller than that from the turbulent air to the cell wall (Nobel 1991). Therefore, changes in mesophyll conductance or leaf conductance have approximately an equal impact on C_i.

Large changes in C_i can result from changes in photosynthetic processes. When ozone enters the leaf it dissolves in cell wall water and diffuses into the chloroplast, where it severely affects photosynthesis by its oxidizing effects. The resulting lower rate of photosynthesis will cause a buildup of CO_2 in the chloroplasts, reducing the CO_2 gradient across the mesophyll cells, thus slowing mesophyll conductance. This scenario will cause an increase in C_i, and the higher C_i should stimulate stomatal closure. As another example, root-enforced water stress will induce a lower stomatal conductance (decreases stomatal aperature) by a hydroactive signal. This will lower C_i, a condition that should stimulate stomatal opening. In this case, two stomatal regulation signals (that induced by water limitation and that induced by low C_i) oppose each other, which may lead to stomatal instability or patchiness (Cowan 1994).

In general, mesophyll conductance favors the diffusion of CO_2 over HCO_3^- because of the large membrane resistances to HCO_3^- diffusion. Changes in mesophyll conductance can be physiologically modified by changes in membrane permeability to CO_2 or HCO_3^- and changes in the subcellular location of chloroplasts. Mesophyll conductance has a similar to slightly lower impact on CO_2 diffusion than that of leaf conductance. Thus, the regulation of C_i and its influence on stomatal conductance is strongly determined by both mesophyll and stomatal resistances to CO_2 diffusion.

C. Photosynthetic Response to CO_2

As mentioned above, photosynthesis is limited by ambient CO_2 concentration. Net photosynthesis will increase with short-term increases in CO_2 in a fashion similar to that of an enzyme responding to an increase in substrate concentration. This CO_2 response curve for photosynthesis has become a useful tool for determining the impact of specific stresses on photosynthetic metabolism (Figure 6.6). At any concentration of CO_2 in the ambient air (C_a), the C_i value can be calculated if one knows the net photosynthesis rate and leaf conductance to water vapor:

$$C_i = C_a - 1.56(P_n/g) \qquad \mu\text{mol } CO_2 \text{ mol air} \tag{6.1}$$

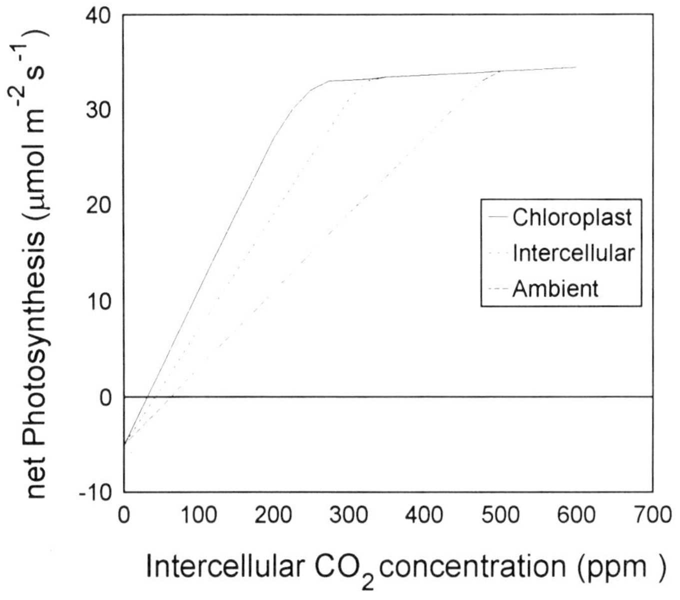

Figure 6.6 Response curves for net photosynthetic (P_n) to increasing ambient, intercellular, or chloroplastic CO_2 concentration.

where P_n is the net photosynthesis (μmol CO_2 m^{-2} s^{-1}), g is the leaf conductance (mol H_2O m^{-2} s^{-1}), and the value 1.56 (some researchers use 1.6) is the ratio of diffusion coefficients for water and CO_2 in air.

Equation (6.1) is limited in its application because it does not take into account the interference to CO_2 diffusion by water molecules diffusing out of the leaf. Thus, another expression was devised to compensate for transpiration (von Cammerrer and Farquhar 1981):

$$C_i = \frac{[(g_{tc} - E/2)C_a] - P_n}{g_{tc} - E/2} \quad \mu\text{mol } CO_2 \text{ mol air}^{-1} \tag{6.2}$$

where g_{tc} is the leaf conductance to CO_2 (mol m^{-2} s^{-1}) and E is the transpiration rate (mol H_2O m^{-2} s^{-1}). The C_i value calculated is close to that measured directly by gas exchange techniques (Sharkey 1985). Given that we can calculate C_i accurately, we can plot a relationship between net photosynthesis and C_i (Figure 6.6).

The difference between the P_n/C_a and P_n/C_i response curve is due to boundary layer effects and stomatal resistance. A plot of P_n against CO_2 concentration in the chloroplast (C_c) would define the relationship between net CO_2 fixation and the chloroplastic concentration of CO_2. Then the difference between the P_n/C_i and P_n/C_c response curves would be due to mesophyll conductance.

The relationship between net photosynthesis and C_i can be divided into several important components. The initial part of the P_n/C_i curve is linear (Figure 6.7). In this region of the curve, the primary limiting factor influencing P_n is C_i. The slope of this line is a con-

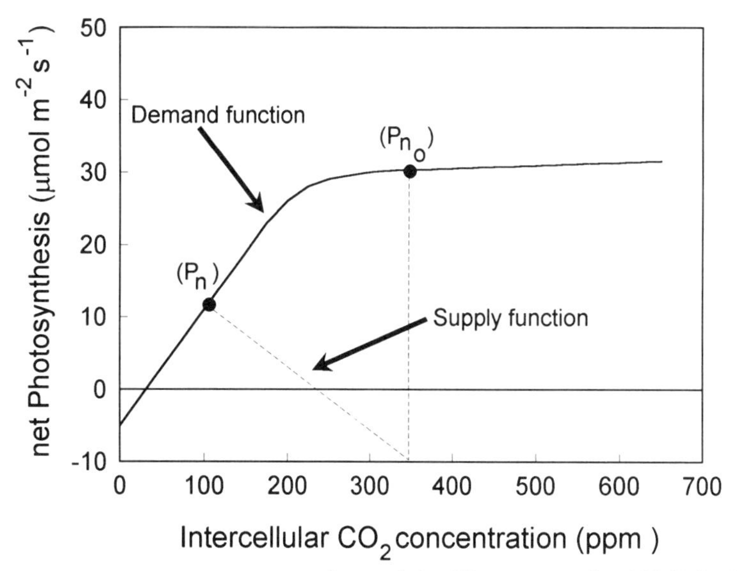

Figure 6.7 Net photosynthesis response to intercellular CO_2 concentration (C_i). Point P_{no} refers to the net photosynthetic rate under the condition where no leaf resistance exists and ambient CO_2 concentration (C_a) = C_i. Point P_n represents the net photosynthesis at the actual C_i when there is resistance to CO_2 diffusion into the leaf. The supply function represents the limitation by stomata to diffusion and is the same as that used to calculate C_i from C_a. The demand function represents mesophyll conductance and carboxylation efficiency. (Redrawn from Sharkey 1985.)

sequence of the mesophyll resistance and carboxylation efficiency. In contrast, the initial slope of the P_n/C_c (Figure 6.6) response curve is only dependent on carboxylation efficiency. When comparing among C_3 plants, the initial slope of the P_n/C_i response curve is often referred to as mesophyll conductance to CO_2 because there is an approximatly equal carboxylation efficiency (due to rubisco activation) among C_3 plants at ambient O_2 concentration. The initial slope of the P_n/C_i response curve has also been considered as a demand function for the CO_2 diffusion pathway (Sharkey 1985).

At high values of C_i, P_n becomes curvilinear and then levels out to a slight rise (Figure 6.7). This is the CO_2 saturated region of the P_n/C_i curve. At this point P_n is not limited by C_i, but is limited by biochemical capacity of photosynthesis or the availability of photon energy. Thus, under optimum light and temperature conditions, this region of the curve defines the biochemical capacity for P_n.

If there were no resistances to CO_2 diffusion from the turbulent air and from water molecules into the intercellular spaces, there would be no difference between C_a and C_i. However, the stomatal and boundary layer resistances maintain C_i lower than C_a until very high ambient CO_2 concentrations are reached (Figure 6.6). The difference between C_a and C_i can be used to define a supply function for photosynthesis (Figure 6.7). In fact, the supply function equation is the same as that used to calculate C_i from C_a.

Sharkey (1985) demonstrated that an index for the limitations of P_n by leaf resistance can be calculated from the P_n/C_i (Figure 6.8) relationship under saturating light (i.e., fully light activated Rubisco).

$$\int_{\text{leaf}} = \frac{P_{n_0} - P_n}{P_n} \tag{6.3}$$

where P_{n_0} is P_n at the C_i value equal to a specified C_a. $P_n = P_n$ at a C_i that results from applying the specified C_a to the leaf.

In addition, by comparing the P_n/C_a to P_n/C_i relationship, we can determine the limitations to gaseous diffusion due to the effects of mesophyll conductance. Sharkey (1985) defined the limitation to photosynthesis due to mesophyll conductance as follows:

$$\int_{\text{mes}} = \frac{P_n - P_{n_1}}{P_n} \tag{6.4}$$

where P_n is the photosynthetic rate at a specific C_i and P_{n_1} is the photosynthetic rate on the P_n/C_a curve at a CO_2 concentration equal to the specified C_i. Through the use of P_n/C_a and P_n/C_i relationships, it is possible to examine if a stress has affected leaf conductance limitations or mesophyll conductance limitations of photosynthesis (Figure 6.8).

Under specific conditions of light and CO_2 concentration, different limitations to photosynthesis are likely to occur (Table 6.1). Under high light conditions and low CO_2 concentration (such as ambient CO_2 concentration), photosynthesis is limited by the amount of ribulose bisphosphate carboxylase-oxygenase (rubisco) and the response of P_n to increasing atmospheric CO_2 will be strong. Under high CO_2 concentrations and high light, photosynthesis is limited by the rate at which other metabolic processes utilize triose phosphate molecules, and there will be little to no response to increasing CO_2. Under high CO_2 concentration and low light, P_n is limited by the ability of the Calvin cycle to regenerate ribulose bisphosphate (RuBP). Under these conditions, the primary limit to P_n is the light-harvesting and utilization component of photosynthesis, and the response to increasing CO_2 is moderate.

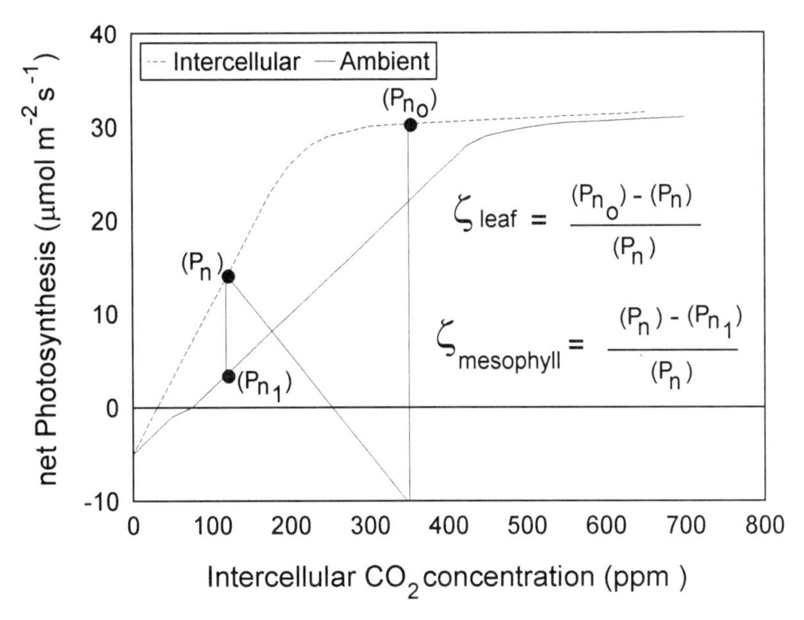

Figure 6.8 Net photosynthesis response to intercellular CO_2 concentration (C_i) and ambient CO_2 concentration (C_a). Point P_{no} refers the net photosynthetic rate under the condition where no leaf resistance exists and $C_a = C_i$. Point P_n represents the net photosynthesis at the actual C_i when there is resistance to CO_2 diffusion into the leaf. Point P_{n1} refers to the photosynthetic rate on the P_n/C_a response curve when C_a is equal to the C_i values after the effect of leaf resistance. $\int_{\text{leaf}}$, limitation to photosynthesis based on leaf resistance. $\int_{\text{mesophyll}}$, limitation to photosynthesis based on mesophyll resistance. (Redrawn from Sharkey 1985.)

D. Harvesting Photon Energy

Modifications of the photosynthetic system for harvesting and utilizing photon energy for the production of ATP and $NADPH_2$ are important mechanisms for stress tolerance in plants. Plants may reside in a diversity of different light environments that have either severe limitations, overabundance, or extreme variability of photon availability. Furthermore, many stresses on plants may affect stomatal limitation and or biochemical regulators of P_n on which the light-dependent reactions must depend. Imbalance within light-

TABLE 6.1 Atmospheric Conditions of the Intercellular Spaces under Which Certain Limitations to Photosynthesis Are Likely to Occur

Light Intensity	CO_2 Pressure	Biochemical Component	Intensity of Response	Location on the P_n/C_i Curve[a]
High	Low	Rubisco	Strong	P_n/C_i slope[b]
Low	High	RuBP regeneration	Moderate	Entire curve
High	High	Triose phosphate utilization	Weak	CO_2 saturated[c]

Source: Adapted from Sharkey 1985.

[a]P_n/C_i curve, the relationship between photosynthesis and intercellular carbon dioxide concentration.

[b]P_n/C_i slope refers to the lower light portion of the P_n/C_i curve, where the slope of the curve is linear.

[c]CO_2 saturated, the part of the P_n/C_i response curve at high C_i where increasing C_i has little influence on net photosynthesis.

dependent reactions and with associated biochemical components of photosynthesis can result in damage to photosynthetic physiology. The capacity to modify the physiology of light-dependent reactions is a critical component of a plant's ability to deal with a multitude of stresses. The mechanisms by which photon harvesting and utilization reactions are modified in variable light environments are covered in Chapter 10. The purpose of this section is to describe the molecular model of these systems as it pertains to potential developmental differences among leaves or physiological plasticity in response to stresses.

1. General Molecular Model. Photon energy capture and utilization in photosynthesis can be broken down into several important components. Light-harvesting complexes capture photon energy. This system is based on chlorophyll and associated pigments embedded in specific light-harvesting proteins. The hydrolysis of water will provide the electrons for the electron transport chain and is associated with photosystem II (PSII). The electron transport mechanism is associated with protein complexes bound in the thylakoid membranes. Photophosphorylation is associated with ATP synthetase, a protein complex in the thylakoid membranes. The production of reducing power ($NADPH_2$) is associated with photosystem I (PSI) protein complex and the chloroplast stroma. Traditionally, these systems have been linked in plant physiology texts by the Z scheme, which depicts electron energy flow. However, for a scientist interested in stress physiology it is much better to view this system by its molecular model (Figure 6.9) rather than the energetics model.

There are four main protein complexes in chloroplast thylakoid membranes. Photosys-

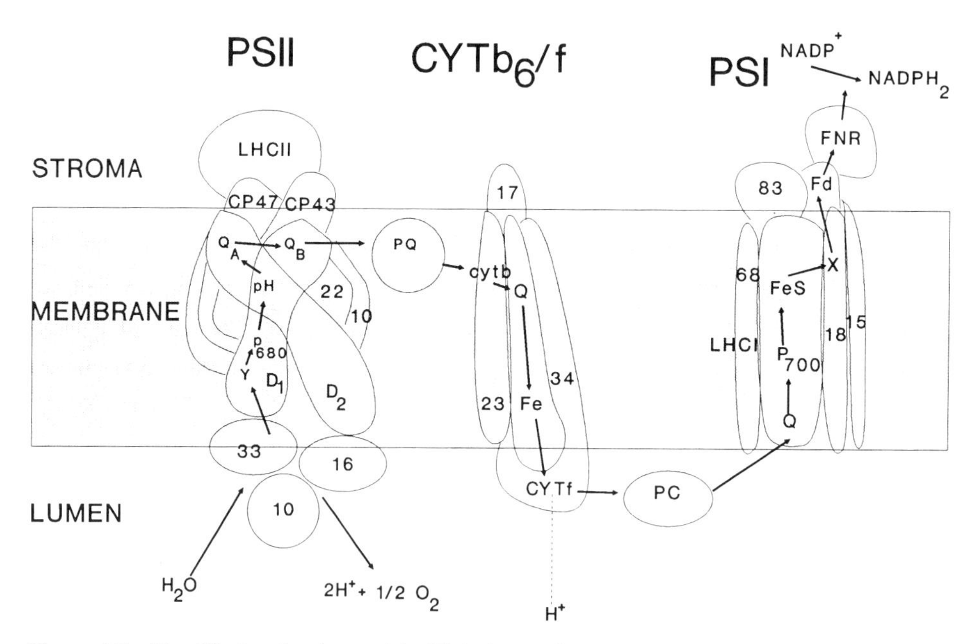

Figure 6.9 Simplified molecular model of light-harvesting complexes and photosystem core complexes found in the thylakoid membranes of chloroplasts. ATPase, another protein complex in the thylakoid membranes, is not shown in this figure. The numbers shown in proteins refer to their designated weight in kilobases (not all proteins and weights are shown). Arrows represent electron flow through the transport chain.

tem II core complex contains about 12 polypeptide chains and traverses the thylakoid membrane. Proteins attached to PSII on the inside of the thylakoid lumen are associated with water hydrolyses. The actual hydrolysis of water is associated with Mn and tyrosine 161 on proteins D1 and D2 of the core complex. The electron donated through tyrosine 161 on D1 is passed through a monomeric chlorophyll (P680) to a quinone (Q_a) of protein D2. However, it takes two excited electrons to reduce the quinone (Q_b) on protein D1 to the hydroquinone, the next step of electron transport. It is important to note that the proteins of the PSII core complex contain only 30 molecules of chlorophyll *a*. The other associated chlorophyll molecules (about 300) and accessory pigments are found on accessory proteins attached to the core complex, the light-harvesting complex for PSII (LHC-PSII). The LHC-PSII is relatively weakly attached to the PSII core antenna proteins (Figure 6.9) on the outside of thylakoid membranes.

Another protein complex in the thylakoid membrane is the cytochrome b_6/f protein complex. This complex serves to accept the electrons from reduced plastoquinone. Plastoquinone is a population of molecules in the thylakoid membrane that pick up electrons from reduced Q_b on PSII and transfer them to the cytochrome b_6/f complex. In other words, plastoquinone is reduced in association with the D1 protein in PSII complex, and oxidized by the cytochrome b_6/f complex. The cytochrome b_6/f complex is a group of polypeptides containing several cytochomes and a Fe_2S_2 reaction center. The purpose of this protein complex is to pump H^+ ions into the lumen of the thylakoid. Electrons are passed from cytochrome b_6/f complex to a protein containing plastocyanin (PC) that is soluble in the thylakoid lumen. The population of PC-containing proteins are reduced by the cytochrome b_6/f complex and oxidized by photosystem I protein complex (PSI).

Photosystem I is another protein complex in the thylakoid membrane containing two major polypeptides and several low-molecular-weight proteins (Figure 6.9). This protein complex oxidizes plastocyanin-containing proteins and eventually transfers the electrons to ferredoxin, a protein located in the chloroplast stroma matrix. The electrons are probably initially accepted by a quinone and transferred rapidly to another monomeric chlorophyll (P700). Following excitation by a photon, the electrons are passed to other electron acceptors in the PSI core complex leading to ferredoxin. The ferredoxin then reduces the matrix-soluble flavoprotein ferredoxin $NADPH_2$ reductase (FNR), which in turn reduces $NADPH^+$.

Another protein complex in the thylakoid membranes is ATP synthetase. This protein complex is composed of a group of hydrophobic proteins that are integral to the thylakoid membrane known as the CF_0 proteins. These are attached to a large complex of hydrophilic proteins, the CF_1 proteins, that extend out from the matrix (stroma) side of the thylakoid membrane. This protein complex utilizes the energy contained in the proton gradient between the lumen and matrix to phosphorylate ADP to ATP.

These four protein complex populations are not randomly organized in the thylakoid membrane (Figure 6.10). In the unstacked portions of the membrane (also called stromal membranes) protein complex pools are dominated by PSI, cytochrome b_6/f, and ATP synthetase. The PSII core complexes are most frequent in the stacked portions of the thylakoid membranes (traditionally called *grana stacks*). In fact, the presence of thylakoid stacks is a result of the abundance of PSII core complexes. Cytochrome b_6/f protein complexes are evenly distributed between stacked and unstacked membranes. The regional distribution of PSI in the stacked and unstacked membranes and PSII in the stacked membranes only requires that electrons be transported a relatively large distance by plastocyanin-containing proteins.

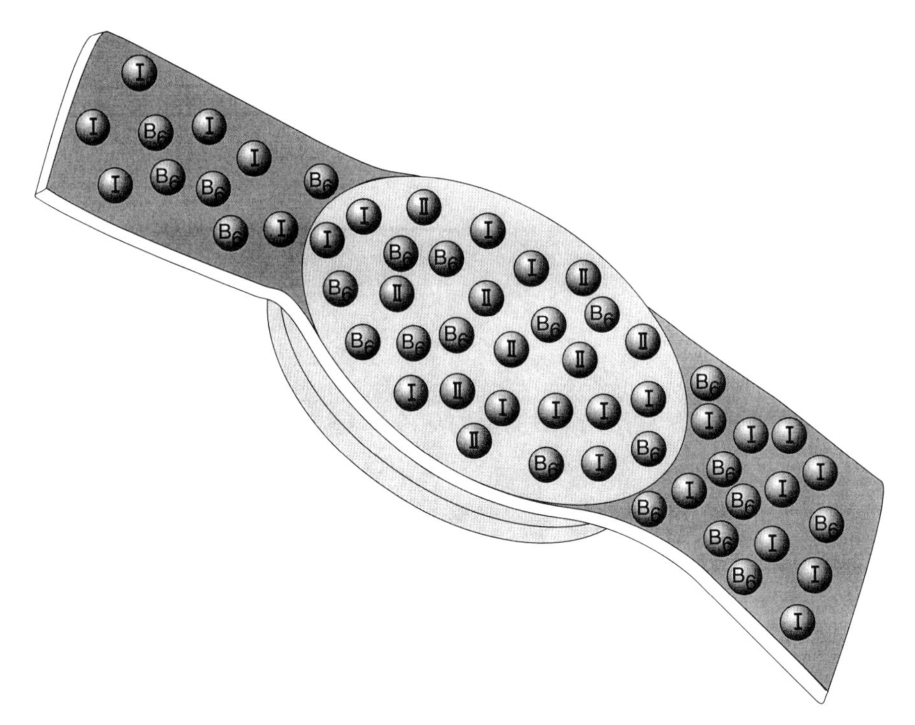

Figure 6.10 Diagrammatic representation of the spatial distribution of photosynthetic core complexes in the stacked and unstacked membranes of the chloroplast. I, Core complex of photosystem I; II, core complex of photosystem II; B_6, the cytochrome b_6/f core complex.

The purpose of describing the molecular basis for the light-dependent reactions is to begin the process of understanding the capability of this system to adjust to environmental and biotic stresses. It is important to realize that there is a population of each protein complex in the chloroplast membranes. Therefore, the relative numbers (population sizes) of the various protein complexes and their location can change during chloroplast development and in the mature chloroplast. However, there is little variation in the chemical composition of each protein complex. Furthermore, the relative numbers of LHC-PSII attached to each PSII core complex can change, which influences the relationship between photon gathering (chlorophyll antennas) and electron transport. In Chapter 10 the relative changes in protein core complex populations are discussed in regard to environmental variation and photoinhibition. For a recent review of the molecular model of light-dependent reactions, see Barber (1992).

E. Biochemical CO_2 Reduction Pathways

During most of this chapter we have been discussing aspects of photosynthesis that are common to all terrestrial plants. However, there is a diversity of CO_2 reduction physiology that affects plant response to stress conditions. The two most abundant variations of carboxylation (in comparison to C_3) are C_4 and CAM physiology. These alternative CO_2 reduction pathways are physiological adaptations to stressful environments. They have evolved several times in a number of different taxa, making these multiphylletic adaptations. The C_4 physiology is associated with species in warmer and drier sites, while the CAM physiology is commonly found in taxa of extremely dry environments or sites with extreme seasonal drought (Ehleringer and Monson 1993). However, there are many facultative CAM species that are located in other environments, including aquatic environments. Since these adaptations are found in some species from stressful environments (other C_3 species utilize other processes to compensate for the stresses), they are important to cover in this book. In this chapter we assume that the reader knows the general C_3, C_4, and CAM carbon reduction pathways. To reinforce the reader's knowledge of these pathways we suggest a recent plant physiology text, such as Taiz and Zeiger (1991). Our purpose in this chapter is to compare the three CO_2 reduction pathways with respect to environmental resources and to cover the significance of various photosynthetic pathways to environmental stress.

1. C_4 Photosynthesis. An important difference between C_3 and C_4 physiology is the enzyme used for the initial reduction of CO_2. In C_3 plants, the Calvin cycle enzyme rubisco reduces CO_2 in the chloroplast, while in C_4 plants, the first carboxylation is done by phosphoenol pyruvate carboxylase (PEPC) using HCO_3^- in the cytoplasm. Reduction of CO_2 by rubisco is inhibited by oxygen but that of PEPC is not. Therefore, at ambient CO_2 and O_2 concentrations the initial carboxylation in C_3 species is inhibited by approximately 30% (Figure 6.11). The competitive inhibition of carboxylation by oxygen is called *photorespiration*. This process utilizes photon energy, reduces O_2, and releases CO_2, which reduces net photosynthesis in C_3 plants. Three main cellular organelles (chloroplast, mitochondria, and peroxisome) are utilized in photorespiration pathways. Furthermore, the

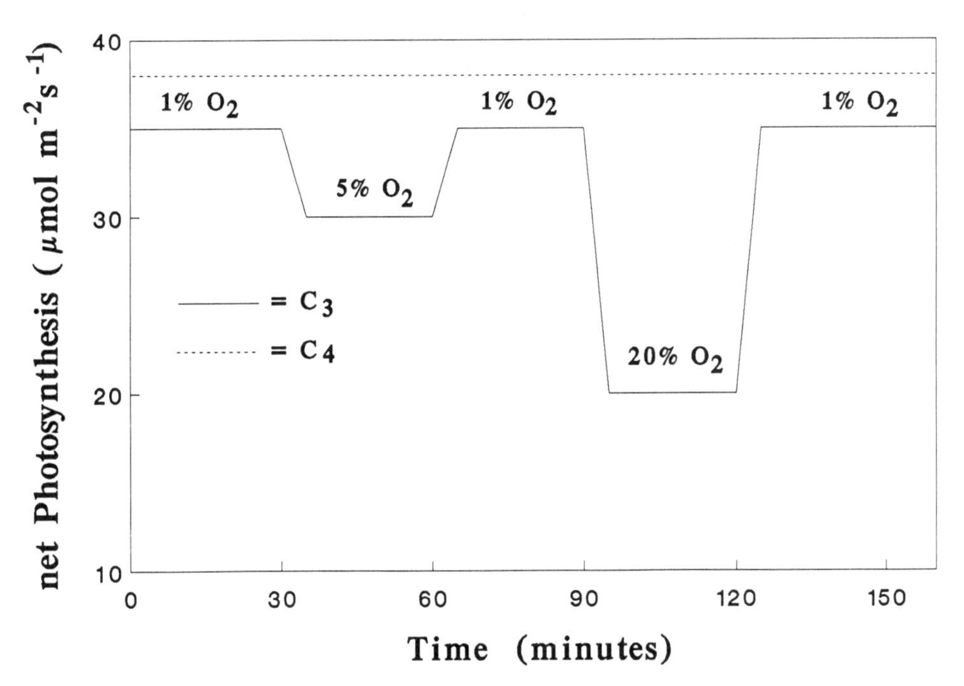

Figure 6.11 Theoretical influence of changing oxygen concentration on net photosynthesis of C_3 and C_4 species. All other atmospheric gas concentrations are kept constant.

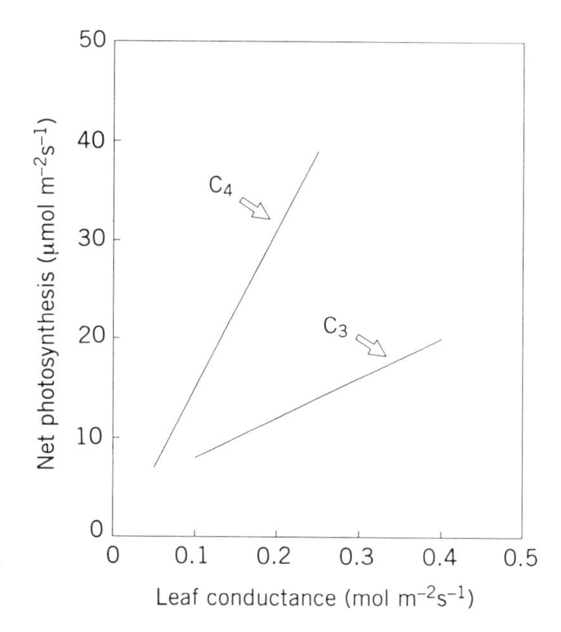

Figure 6.12 Net photosynthetic rate plotted against leaf conductance for several C_4 and C_3 species. This is a generalized response for a large diversity of species.

proportional relationship between photorespiration and C_3 photosynthesis is dependent on the ratio of CO_2 and O_2 in the chloroplast. For a more complete review of photorespiration, see Ogren (1984).

Photorespiration is absent in C_4 plants because the initial carboxylation of phosphoenolpyruvate by PEPC and the subsequent transport of malate or aspartate serves as a CO_2 pumping mechanism. The result of this CO_2 pump is a high concentration of CO_2 in the bundle sheath cells, the site of the second carboxylation (by rubisco). Thus the influence of O_2 on rubisco is minimized by the high CO_2/O_2 ratio in bundle sheath cells created by the CO_2 pumping mechanism in C_4 plants. The absence of photorespiration in C_4 plants results in a more efficient incorporation of CO_2 in photosynthesis than C_3 plants, which in turn reduces the intercellular CO_2 concentration (C_i) compared with that in C_3 leaves. Thus, in response to the lower C_i, stomata can be less open. In fact, species with C_4 photosynthesis have a higher net photosynthesis at any given stomatal conductance than C_3 (Figure 6.12). The higher net photosynthesis to stomatal conductance ratio causes C_4 species to have a higher water use efficiency (WUE) than that of C_3 species. A high WUE value of C_4 plants is an important attribute for species in hot and dry environments because it is a water conservation mechanism.

The enzyme kinetics for rubisco and PEPC also differ in an important way. The affinity of PEPC for HCO_3^- is greater than the affinity of rubisco for CO_2. In fact, the concentration of CO_2 in the intercellular spaces creates a HCO_3^- concentration in the mesophyll cells that essentially saturates PEPC. Thus there is a large difference between the P_n/C_i curve for a C_4 species and one for a C_3 species (Figure 6.13). The C_i compensation point of C_4 species is lower and the mesophyll conductance is higher than those for C_3 species. However, at high C_i the difference between the photosynthetic rate of C_4 and C_3 species is small. As a result, it is believed that as ambient CO_2 increases in the near future, the photosynthetic rate of C_3 species will increase relative to that of C_4 species.

Recall that in C_4 plants PEPC serves to increase the CO_2 concentration in the bundle

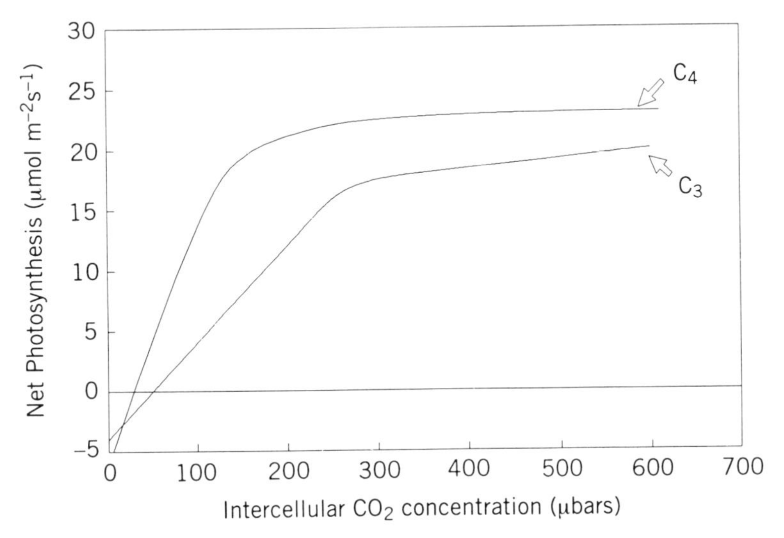

Figure 6.13 P_n/C_i response curves for a C_3 and a C_4 species under high-light conditions. This is a generalized response for a large number of species. Notice that at high carbon dioxide concentration, net photosynthesis of C_3 species approaches that of C_4 species.

sheath cells, where a second carboxylation occurs by rubisco in the Calvin cycle. This means that the CO_2 is pumped from the mesophyll cells, where it is at low concentration, to the bundle sheath cells, where there is a high concentration of CO_2. This pumping pathway requires energy in excess of that needed by the C_3 cycle. In fact, two extra ATP molecules are required to pump each CO_2 molecule, in the form of malate or aspartate, into the bundle sheath cells compared to that needed by the C_3 pathway. This added energy requirement for C_4 compared with C_3 causes C_4 plants to have a higher quantum requirement (number of absorbed quanta required to reduce a given quantity of CO_2 into triose phosphate). The high quantum requirement of C_4 plants gives them a competitive disadvantage to C_3 plants in shaded environments. A further discussion of light acclimation and adaptation is covered in Chapter 10.

The efficiency of utilizing light by C_3 and C_4 plants is further influenced by temperature. As leaf temperature increases, there is a disproportional increase in respiration compared to photosynthesis. The greater increase of photorespiration and respiration compared to the increase in photosynthesis causes a greater depression of P_n in C_3 plants than in C_4 plants at high temperature. As temperature increases, the quantum yield of C_3 plants decreases, while that for C_4 plants remains relatively unchanged (Figure 6.14). Thus at high temperatures and high light conditions, C_4 species will have a competitive advantage over C_3 plants. This physiological difference between C_4 and C_3 plants affects the distribution of physiological types among habitats. For example, the relative proportion of C_4 plants among desert winter annuals (relatively cool season) is small compared with desert summer annuals (relatively hot season). Grass species at the base of mountains have a relatively high proportion of C_4 species compared with grass species at high-elevation sites (Rundel 1980). There is a higher proportion of C_4 species in the flora of low-latitude habitats in Australia than in high-latitude floras (Hattersley 1982).

Physiological functions other than quantum requirements and the ratio of respiration to

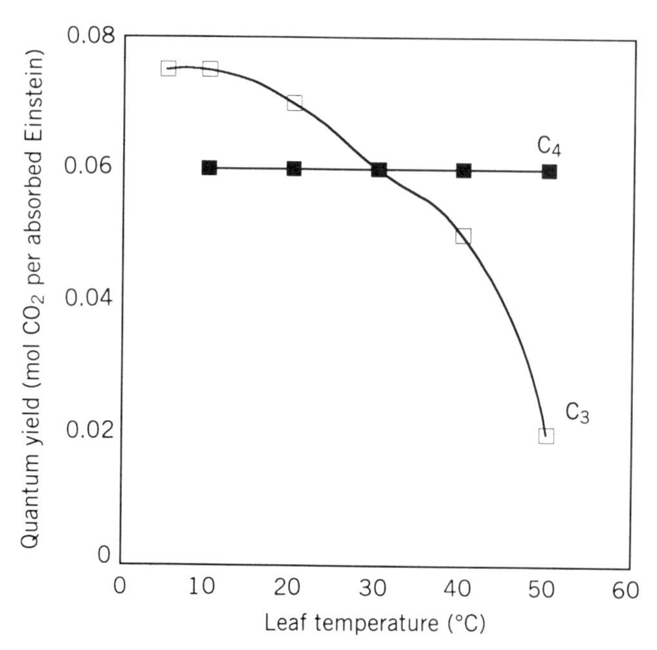

Figure 6.14 Influence of temperature on quantum yield in C_3 and C_4 species. (Redrawn from Ehleringer and Bjorkman 1977.)

photosynthesis must be involved in the mechanisms of high-temperature tolerance in C_4 plants. The greater heat tolerance of C_4 species compared to C_3 species may not be due to enzyme variants (isozymes) that have higher thermal optima or tolerances. In fact, a series of photosynthetic enzymes isolated from C_3 *Atriplex* and C_4 *Tidestromia* had inconsistent relationships among thermal responses (Osmond et al. 1980). In some cases, enzymes from C_4 species had higher thermal tolerances than that from C_3-species enzymes (RuBP kinase, NADP glyceraldehyde phosphate dehydrogenase), and these thermal differences matched the thermal tolerances of photosynthesis. In other cases, the enzymes of C_4 species had thermal attributes identical to those of the C_3 species (NADP reductase, 3-PGA kinase). Although there is variation among thermal tolerance of C_3 and C_4 enzymes, PSII-driven quantum yield always has a greater thermal tolerance in C_4 plants than in C_3. There is no such consistent difference between C_3 and C_4 PSI-driven quantum yield. Therefore, the ability of C_4 photosynthesis to tolerate higher temperatures than are tolerated by C_3 photosynthesis is the result of thermal stability of PSII electron transport (as well as several specific enzymes) at high temperatures.

*2. **CAM Photosynthesis.*** CAM photosynthesis is similar to that of C_4 photosynthesis except that the two carboxylating enzymes are separated in time in the same cells instead of being separated between two cells at the same time. The initial CO_2 reduction occurs at night, when PEP carboxylase reduces HCO_3^- in the cytosol. The initial stable product is malate or aspartate and those are stored in the central vacuole until the daylight hours. During the daylight, malate or aspartate in the vacuole is transported back to the cytoplasm, where it is decarboxylated. The resulting CO_2 is then reduced by the Calvin cycle in the chloroplast. The primary significance of this pathway is that CO_2 diffuses through open stomata at night when PEP carboxylase reduces HCO_3^-. Having stomata open at night and closed during the day results in far less water loss from the plant (compared with plants that have a reverse pattern of stomatal opening) because the lowest VPG of the diurnal cycle occurs at night. Consequently, water use efficiency of plants with CAM photosynthesis is higher than that of C_4 or C_3 plants.

There are several limitations to net CO_2 accumulation by plants with CAM photosynthesis. First, there can only be so much malate or aspartate stored in the vacuole. When this quantity is reached, net CO_2 accumulation stops. Furthermore, the phosphoenolpyruvate pool (PEP) in the cytoplasm is limited by the amount of starch produced the day before. Both of these limitations cause net CAM photosynthesis to stop several hours before dawn. A further limitation to CAM photosynthesis is the amount of energy it takes to pump the malate or aspartate into and out of the vacuole. Thus, as in C_4 plants, CAM plants have a higher quantum requirement than that of C_3 plants.

Due to these limitations placed on plants with CAM photosynthesis, these plants usually do not have as high a growth rate as C_4 or C_3 plants (although there are some notable exceptions). Commonly, CAM plants are found in habitats with extremely low or variable water availability. Their high water use efficiency makes CAM plants particularly adapted to low water sites. This does not mean that CAM plants are well adapted to low tissue water potential. Most CAM species have several mechanisms which buffer them from the low water potential of the environment. Stem or leaf succulence keeps tissue water potential fairly constant. Root systems can die back when soil moisture is low, separating the aboveground tissues from low soil water potential. In fact, several cactus species are known to recycle CO_2 internally during diurnal cycles and keep stomata closed, even at night, when water is extremely limiting.

Several species with CAM photosynthesis can switch to C_3 photosynthesis under conditions of relatively high water availability (Winter et al. 1978, Bloom and Troughton 1979). This is a mechanism for CAM plants to increase their potential productivity relative to C_3 plants. In other cases, CAM photosynthesis is magnified by increasing water availability (Hanscom and Ting 1978, Willert et al. 1985). The large diversity of species utilizing CAM photosynthesis makes it hard to generalize about the potential significance of CAM photosynthesis to competition among species. It suffices to say that CAM species are commonly located in drought-prone areas (some notable exceptions are wetland CAM species of *Isoetes* and the epiphytic bromeliad *Tillansia usneoides* or spanish moss which is commonly found in swamps), suggesting that their high water use efficiency is an important physiological attribute regulating their distribution. The low frequency of CAM plants in moist habitats could be due to competition with other species, pathogen infections, or other mechanisms. Comprehensive reviews of CAM photosynthesis and the ecology of CAM species have been published (Gibson and Nobel 1990).

3. Dark CO_2 Fixation in C_3 Plants. Although most studies consider PEP-carboxylase as an important enzyme only in C_4 and CAM plants, this enzyme is present in C_3 plants as well. The quantity of PEP-carboxylase in the cytosol of C_3 plants is greatly reduced compared to plants with C_4 or CAM photosynthesis. Just as in C_4 or CAM photosynthesis, bicarbonate ion is reduced to oxaloacetic acid in the presence of phosphenolpyruvate. Oxaloacetic acid is rapidly converted to other organic acids, particularly malic acid, which is stored in the cytosol and the vacuole. At a pH of 7.5 in the cytosol and an ambient CO_2 concentration of 350 μmol mol^{-1}, the bicarbonate ion concentration in the cytosol would be approximately 190 μmol (Brown 1985). This level of substrate concentration could lead to significant activity of PEP-carboxylase during the night in C_3 plants. In fact, night fixation of CO_2 may be a factor reducing the buildup of CO_2 (and its subsequent effect on lowering cell pH) coming from respiration. There are other carboxylases in the cytosol that may also be involved in nighttime fixation of CO_2 (Raven and Farquhar 1990). The significance of nighttime CO_2 fixation to the carbon cycle of C_3 plants has not been carefully perused. This topic may have increased importance because of the impending increase in ambient CO_2 concentration (Amthor 1994).

III. ASSIMILATE CIRCULATION AND STORAGE

After CO_2 has been reduced and converted into triose phosphate, the cell will use a fraction for biosynthesis and respiration, another fraction may be incorporated into starch, and a third fraction may be circulated to other organs of the plant (Figure 6.1). The amount of photosynthate that is circulated to different parts of the plant at different times will regulate the survival and growth of specific plant organs. There may be an optimal proportional growth among organs under each particular combination of stresses. For example, during low soil water availability, a major proportion of photosynthate is allocated to developing roots and osmotic regulation rather than to leaf production. Therefore, the factors that regulate photosynthate fractionation (among photosynthetic cell, starch, or transport), phloem loading of photosynthate, and allocation of transported photosynthate among organs are important processes to stress adaptation or acclimation.

A. Fractionation of Photosynthate Within the Cell

The immediate products of photosynthesis are converted into triose phosphates by triose phosphate dehydrogenase located in the chloroplast. The triose phosphate pool can then supply the synthesis of starch, supply the pentose phosphate cycle (regenerating RUBP), or be transported out of the chloroplast into the cytosol, where it can enter into mitochondrial respiration, the sucrose synthesis pathway, or other biosynthetic pathways leading to growth or defense (Figure 6.15). Here we are interested in the processes that regulate the proportional distribution of triose phosphates among starch synthesis, sucrose synthesis, and the pentose phosphate cycle (i.e., the Calvin–Benson cycle).

If triose phosphate remains in the chloroplast, it will be converted into fructose-6-phosphate. Fructose-6-phosphate can serve as a substrate for either starch synthesis or the pentose phosphate cycle. Otherwise, the triose phosphate will be transported out of the chloroplast into the cytosol. Regulation of whether triose phosphate remains in the chloroplast or is transported out into the cytosol is a complicated process related to feedback and feedforward regulation of both fructose-1,6-bisphosphatase and sucrose-6-phosphatase. There is some evidence that a buildup of triose phosphate in the cytosol may feed back to stimulate fructose-1,6-bisphosphatase and thereby increase starch synthesis (Stitt 1993). Therefore, one main regulator of starch biosynthesis is the utilization of triose phosphate by respiration.

Recent studies on the regulation of starch biosynthesis have involved specific mutants. For example, in *Arabidopsis*, several mutants that are deficient in specific starch biosynthesis enzymes have been isolated. The PGM mutant is deficient in chloroplastic phosphoglucomutase and the 7% ADPGPPase mutant, which has only 7% of the wild-type ADP glucose phosphorylase activity (Schulze and Schulze 1994). Both ADP glucose

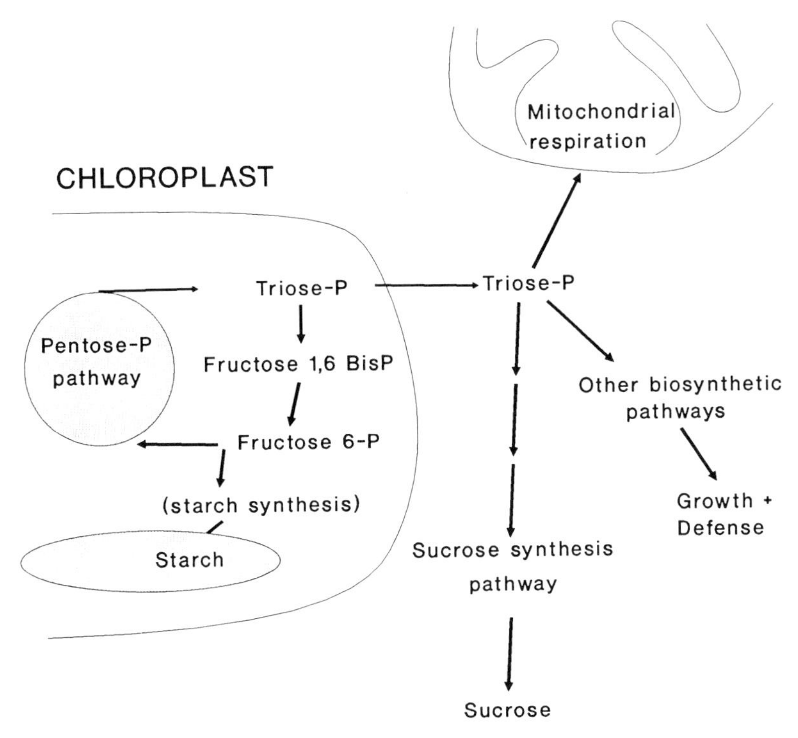

Figure 6.15 Generalized biochemical pathways for the immediate products of the Calvin–Bensen cycle of photosynthesis. Notice the two important switch points at the cytoplasmic triose phosphate pool and at the chloroplastic fructose, 6-phosphate pool.

phosphorylase and phosphoglucomutase are enzymes involved in starch biosynthesis. The cytosolic respiration rate of young leaves in PGM mutants is approximately six times that of the wild-type. This high respiration rate results in a 57% reduction in carbon supply for young leaves (Caspar et al. 1986). Even though the respiration rates of young leaves were strongly affected by the lack of starch synthesis in the PGM mutant, there was no impact on root respiration. Therefore, changes in assimilate flow out of the chloroplast have a greater impact on the immediate metabolism of the cytosol and a limited impact on sucrose synthesis. Sucrose synthesis may primarily be affected by processes that regulate phloem loading and photosynthate translocation in the phloem.

Evidence from the starchless *Arabidopsis taliana* mutants, and that from studies on other mutants, clearly indicates that starch biosynthesis affects whole plant growth. Normal leaves have a prominent diurnal cycle of starch such that the content of starch increases to a maximum late in the daylight hours. The starch is then mobilized during evening hours for growth and metabolism. Thus, starch synthesis during the day allows for a carbohydrate supply at night for growth and maintenance respiration. Furthermore, starch synthesis keeps cytosolic mitochondrial respiration low because it limits the size of the cytosolic triose phosphate pool. The latter significance of starch synthesis is particularly significant when phloem transport is limited by small demands for sucrose by the other sinks of the plant. This situation frequently arises during various plant stresses, and a buildup of starch during plant stress often occurs.

There are several important environmental factors that can influence starch metabolism. Studies with *Nicotiana tabacum* mutants, which are deficient in photosynthesis, have shown that high light can increase starch turnover rate and that this is associated directly with increased photosynthesis and growth (Fichtner et al. 1994). The photosynthesis rate of mutants increased at high light because of an increased starch synthesis rate. In addition, increasing the supply of nitrogen to plants decreases starch storage at any photosynthetic rate. At high nitrogen availability, carbohydrate that is normally stored as starch is allocated to growth and nitrogen accumulation processes. Thus, under low nitrogen availability and high light, photosynthesis and starch storage both increase, but growth does not increase. Other environmental variables also influence starch synthesis and hydrolysis. For example, during water stress the starch pool in chloroplasts decreases because photosynthesis is inhibited by water stress more than cellular respiration (discussed further in Chapter 8). Also, a reduction in starch synthesis, in response to environmental stress, is linked more closely to a proportional decrease in photosynthesis/respiration ratio than the reduction in phloem translocation of sucrose.

B. Regulation of Assimilate Transport Among Tissues

The general mechanism of phloem loading and translocation of sucrose is well known (Taiz and Zeiger 1991). The pressure-flow hypothesis states that the magnitude and direction of photosynthate translocation is determined by the relationship between the strength of the source and the sink for photosynthate. The strength of the source depends on the rate of phloem loading in the newly mature photosynthetic leaves. Phloem loading depends on the quantity of sucrose in the apoplast and symplast near the sieve tubes and the activity of the sucrose transport permease in the sieve tube cell membrane. The sucrose synthetase transport channel is a sucrose-proton *symport,* which in turn is dependent on a hydrogen gradient between the sieve tube companion cell and the apoplastic space established by an ATPase proton pump. A symport means that the transport protein moves both sucrose and hydrogen across the membrane simultaneously in the same direction. The concentrations of sucrose in the companion cells and sieve tubes are much higher than the surrounding apoplast. Sugars and other osmotically active moieties such as potassium ions cause high pressure in the sieve tubes at the source, which generates hydrostatic pressure at the source. This hydrostatic pressure enhances the diffusion of sucrose through sieve tubes from the source to other locations in the plant that have lower hydrostatic pressure.

The strength of a sink region is dependent on the magnitude of metabolic processes in the region. Therefore, both the size (number of cells) and the metabolic rate (respiration and biosynthesis) affect sink strength. Sinks can be anywhere in the plant in relation to the source. As leaves develop, they begin as a sink, convert to a source, then return to a sink once again. Therefore, recently mature leaves will provide sucrose for newly developed leaves and leaves close to senescence. Thus photosynthate must move in both directions in the phloem to meet the needs of various sinks in the whole plant.

Unloading of sucrose at the sink is a more complicated mechanism than loading sucrose into sieve elements. Among plants there is variation in whether the unloading occurs by an apoplastic route, a symplastic route, or both. In some species, an ATP-dependent sucrose transport mechanism occurs in the sieve tube plasma membrane and transports sucrose directly into the apoplast. This transport mechanism is regulated by apoplastic sucrose concentration. Sucrose then diffuses in the apoplast to the cells of the sink. In other species sucrose leaves the sieve tubes in a passive manner through plasmodesmata. Then sucrose diffuses throughout the sink through the symplasm.

There are some basic differences in phloem transport among plants. For example, some species transport other nonreducing sugars in the sieve tubes rather than sucrose. Commonly, these other transport sugars are compounds composed of sucrose bound to one, two, or three galactose molecules (raffinose, stachyose, and verbascose). In some herbaceous dicots, parenchyma cells near the sieve tube members become transfer cells. Transfer cells have folded cell walls which greatly increase the surface area of the plasma membrane for sugar transport. Recent evidence suggests that the folded cell wall structure of transfer cells is labile, appearing and disappearing on a diurnal cycle (Van Bel 1995).

Regulation of the rate of sugar translocation through the phloem is dependent primarily on metabolic properties of sources and sinks. Thus any environmental perturbation will affect phloem translocation secondarily though its influences on source and sink activity. Phloem transport is relatively insensitive to perturbations in metabolic activity between source and sink. For example, when the petiole of a source leaf is rapidly refrigerated, there is an initial decrease in phloem transport, followed by a gradual rise to pretreatment levels (Geiger and Sovonick 1975). During continued refrigeration of the source leaf petiole, phloem transport from source to sink occurs normally. In contrast, there is some evidence that reduced turgor pressure of mesophyll cells, due to water stress, can directly affect the sucrose-proton symport. Although variations in environmental conditions have limited direct long-term effects on phloem transport mechanisms, the nature of the environment does directly regulate the proportional distribution of photosynthate among various organs through its impact on source and sink strength.

C. Allocation of Photosynthate Within the Plant

Photosynthate allocation to plant organs is defined by genetics and modified by environmental characteristics. As discussed in Chapter 7, many woody species from desert habitats have extensive allocation of photosynthate to roots compared with shoots. However, root/shoot ratios can vary within such plants dependent on surface water availability. Herbaceous plants allocate a greater proportion of their total photosynthate to leaf production compared to woody species. However, within herbaceous species the relative proportion of growth directed toward leaves or roots may vary widly. Patterns of allocation may have a significant influence on growth rate. In fact, in many situations the patterns of allocation to the growth of various organs may have a more profound effect on whole plant growth than on the rate of photosynthesis (Poorter and Remkes 1990, Fichtner and Shulze 1992).

The primary factor that regulates photosynthate allocation is the proportional sizes of sinks. Thus, if root biomass is large, due to genetic factors, this will be a large sink, and small changes in root metabolism can have large changes on the relative sink strength (in comparison to other sinks). In contrast, those organs that are genetically defined to have a relatively small biomass will become a major sink for photosynthate only when metabolism increases greatly. Metabolic changes that affect sink strength and therefore photosynthate allocation can be regulated by several possible growth regulators (discussed in Chapter 4). Furthermore, sink strength can be affected by storage properties of the organ.

If the organ has an efficient mechanism of photosynthate storage, its strength as a sink may be high even though the cellular metabolism is relatively low.

D. Assimilate Storage

Photosynthates are stored in all plant organs as starch in leucoplasts or as sugars and organic acids in the central vacuole. The regulation of storage in various tissues is complex. In one sense it is convenient to speculate that regulation of starch storage is dependent on those mechanisms that regulate the fractionation of triose phosphates and photosynthate translocation. Therefore, when gain by photosynthesis or uptake exceeds demand by the cell, storage of photosynthate occurs. The term *storage* has been defined in a number of ways. It is convenient for this book to consider storage as resources that build up in the plant and can be mobilized at a later time for biosynthetic reactions. Based on this definition, not all starch in cells is storage. For example, starch grains in cells of the root cap (statoliths) serve as a gravity sensor. The starch that is not readily available to metabolism will be lost from the plant when the tissues abscise.

There are spatial and temporal aspects to storage in plants. On a short time scale, photosynthates are stored in chloroplasts of actively photosynthesizing leaves. Diurnal storage in leaves occurs only when light is relatively high, nutrient availability is relatively low, or there is mild water stress; otherwise, photosynthate is utilized in current metabolism. In many woody plants, a majority of seasonal photosynthate storage occurs in organs other than leaves. At the end of the growing season, starch is stored in parenchyma cells of roots, stems, tubers, or other nonphotosynthetic organs of many species. In many herbaceous taxa and some woody species there is a constant partitioning of photosynthate between storage and growth in nonphotosynthetic organs. In sugar beet, the ratio of photosynthate partioning between sucrose storage in the vacuole versus root growth is constant in a diversity of different environmental conditions (Watson et al. 1972). However, when sugar beets are grown at very low light intensities or very high nitrogen availability, sucrose is allocated to leaf development instead of being stored in the root (Loomis and Worker 1963). The sugar beet example demonstrates that competition between storage and growth for photosynthates is influenced by environmental conditions.

The patterns by which photosynthates are partitioned between storage and growth among the various organs has significance to stress adaptation by plants. Some long-term storage is required by plants that suffer occasional stochastic catastrophic stress. Fire, freezing, or severe herbivory are examples of such stresses. Studies with artificial clipping have shown that even after regrowth there remains a significant amount of stored photosynthate in roots that was not used for regrowth (White 1973). The unused portion of stored photosynthate may be in dead cells and therefore unavailable to regrowth, or it may not be used because other resources are limiting regrowth (Chapin et al. 1990). Many chaparral species have evolved large subterranean lignotubers that store extensive quantities of starch. Those species with lignotubers resprout vigorously after fire, whereas species without lignotubers must start from seed. Unlike the sugar beet example discussed above, those species that are adapted to environments with low resource supplies have low growth rates and higher storage than species adapted to high resource availability. The high photosynthate storage may be part of the system that allows these species to utilize short flushes of nutrients better than those species adapted to high-nutrient habitats. During a nutrient flush, stored carbohydrates are mobilized rapidly and made available for biosyntheses that permit greater utilization of the nutrient (Chapin 1980).

IV. CARBON LOSSES FROM PLANTS

A. Respiration

Reduced carbon is lost from plants as CO_2 from respiration and as organic carbon in litter fall, leaching, and volatilization. Among these processes the largest fraction of reduced carbon is lost as CO_2 from respiration. It has been estimated that 50% or greater of total photosynthate is lost as CO_2 by aerobic respiration (Amthor 1989). Respiration occurs in all live cells of the whole plant, but whole-plant respiration estimates are rare (Monteith 1972). The main reason for infrequent whole-plant respiration estimates is the difficulty of separating root respiration from microbial respiration. Respiration research on roots normally involves axenic hydroponic root media, which is far from the natural system (Farrar and Williams 1991). Although root respiration is often excluded from whole-plant estimates, respiration by aboveground tissues alone consumes a large fraction of photosynthate. When whole-plant respiration is measured and compared with growth, there is a negative relationship (Figure 6.16). Growth decreases as respiration increases in selected genotypes of one species (Wilson 1982, Kuiper 1983) or in several species (Poorter et al. 1990). In fact, the negative relationship between respiration and growth is stronger than the positive relation between photosynthesis and growth. Therefore, in any study that concerns a stress that induces decreased growth, the impact of the stress on respiration is critical to understanding the stress-altered physiology.

Cellular aerobic respiration can be subdivided into five main components that are common to all eukaryotic cells (see Amthor 1994): glycolysis, the oxidative pentose phosphate pathway, the Krebs cycle (TCA cycle), mitochondrial electron transport, and oxidative phosphorylation (Figure 6.17). Glycolysis is found in all cells and thus is thought to be the oldest component of respiration on an evolutionary scale. Aspects of respiration that involve the mitochondria are all present in eukaryotes, but there are differences in specific pathways among organisms. The biochemistry of the general respiratory pathways can be found in any plant physiology or cell physiology textbook. In this chapter we concentrate on aspects of respiration that are particularly applicable to growth regulation

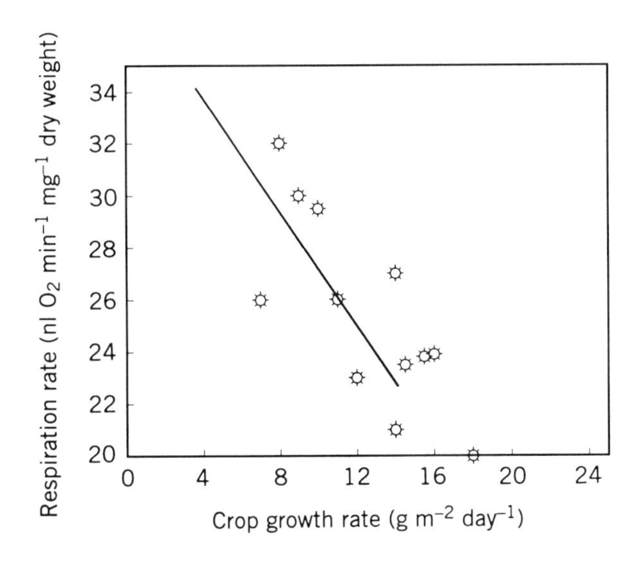

Figure 6.16 Relationship between dark respiration and crop growth rate. (Redrawn from data in Lambers 1985.)

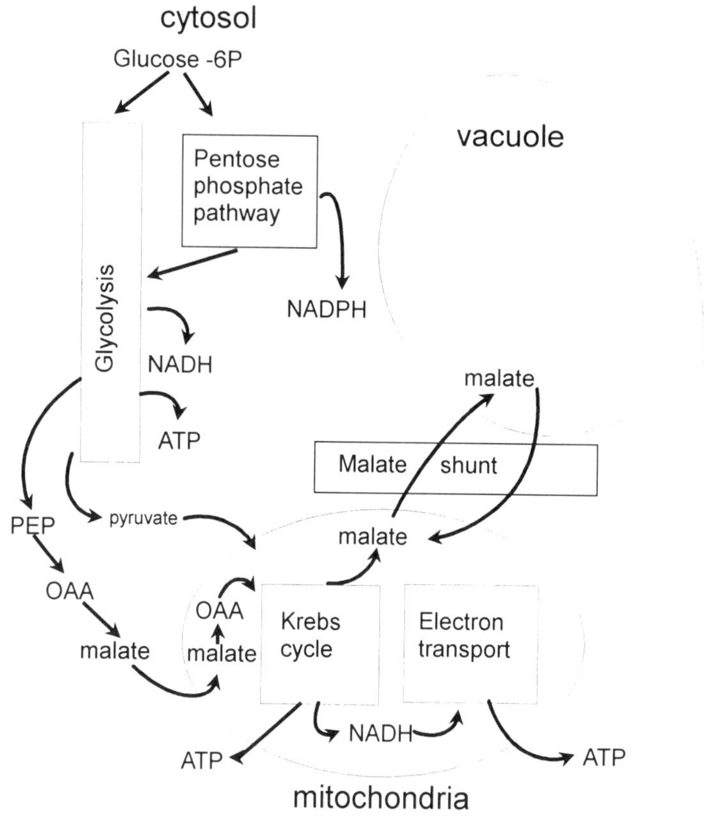

Figure 6.17 Five main portions of respiration and their linkage to energy synthesis. Each of the five components of respiration are highlighted by enclosure in a square.

and stress tolerance in plants. Since total plant growth is strongly affected by the balance of respiration and photosynthesis (Poorter et al. 1990), it is critical to understand that several aspects of the respiration pathway interact directly with photosynthesis and photorespiration (Figure 6.18).

B. Interactions Among Photosynthesis, Photorespiration, and Respiration

Mitochondria from nonphotosynthetic cells are very similar to those from photosynthetic cells, except that the former do not interact directly with photorespiration or photosynthesis. Both photosynthesis and respiration produce ATP and reductant (NADH or NADPH) that are released into the cytosol. Some of the CO_2 reduced by photosynthesis was produced by respiration, and some of the O_2 used in aerobic respiration was produced by photosynthesis. In addition, several intermediates of the Calvin cycle and glycolysis are the same. Dihydroxyacetone phosphate (DHOAcP), synthesized in the chloroplast from 3-phosphoglyceraldahyde (3-PGAL), can be transported into the cytosol, where it enters glycolysis and reduces NAD^+. Also, 3-phosphoglyceraldahyde can be used by plants in the cytosol, in a pathway that reduces $NADP^+$. Another connection between respiration and photosynthesis occurs through the sucrose synthesis pathway. Respiratory ATP is re-

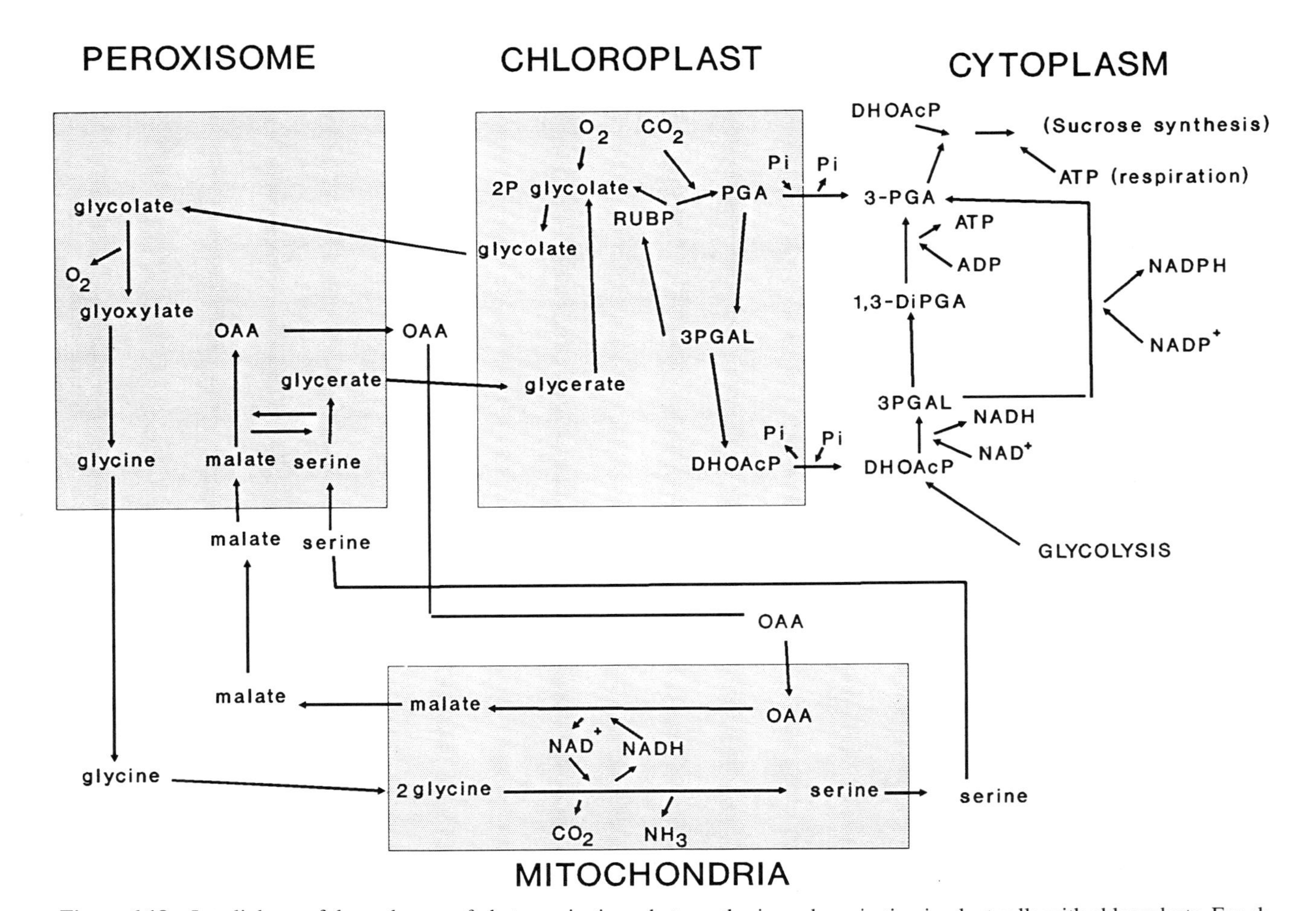

Figure 6.18 Interlinkage of the pathways of photorespiration, photosynthesis, and respiration in plant cells with chloroplasts. For abbreviation interpretation, see the list of abbreviations in Appendix A.

quired for sucrose synthesis. Thus, if respiration is slow, sucrose synthesis is inhibited and triose phosphate builds in the cytosol and inhibits photosynthesis and stimulates starch synthesis. Therefore, photosynthesis and respiration are linked not only because of similar end products, but also because several intermediate molecules are shared by both pathways, and the regulation of one pathway is dependent on the products of the other.

Photorespiration and oxidative respiration are linked in the mitochondria. Both processes utilize the mitochondria and are linked by the NAD^+ oxidation reduction process (Figure 6.19). Photorespiration requires NAD^+ to oxidize two glycine molecules and synthesize one serine. NAD^+ is provided by a reduction of oxaloacetic acid (OAA) to malate in the mitochondria. This reaction is opposite to that of the Krebs cycle, also in the mitochondria, where malate is converted to OAA. The opposing nature of the Krebs cycle and photorespiration reactions have lead researchers to suggest that the two reactions are separated in space between membrane compartments of the mitochondria. Malate is transported back into the cytoplasm and shunted back into the peroxisome to enable continued photorespiration (Figure 6.18). Malate is converted back into OAA in a coupled reaction (through NAD) with the conversion of serine to glycerate in the peroxisome. Therefore, the malate-to-OAA reaction sequence occurs in two locations (mitochondria and peroxisome). Since photorespiration is most evident in C_3 species and repressed in C_4 and CAM plants, the linkage between respiration and photorespiration is strongest in C_3 species. However, in C_4 and CAM plants, respiration interacts with photosynthesis in a unique way compared to C_3 plants.

Recall that there are three classes of photosynthetic physiology in C_4 species (NAD-malic enzyme, NADP-malic enzyme, and PEP-carboxykinase). CAM plants are often grouped into two classes: the NAD and NADP-malic enzyme (ME type), and the PEP-CK types. Each one of these classes is defined by the mechanisms of decarboxylation, which produces the CO_2 for the Calvin cycle. Decarboxylation occurs in the bundle sheath cells of C_4 plants and the mesophyll cells (during the daylight hours) of CAM plants (Figure 6.19). Hence the carbon-fixation pathways of C_4 and CAM plants are linked to respiration through their decarboxylation pathway and the malate shunt of respiration (deposition of malate into or withdrawal out of the central vacuole).

For both CAM and C_4 plants in the PEP-CK class, the primary decarboxylase (PEP-carboxykinase) is located in the cytoplasm. This enzyme requires one ATP per decarboxylation and the release of one CO_2. Therefore, ATP must be supplied to PEP-carboxykinase at an approximately equivalent ratio to the rate of CO_2 assimilation by the Calvin cycle. The ATP could be provided by chloroplasts to the cytosol in both C_4 and CAM plants by a chloroplast transport mechanism (triose-P/PGA symport) that shuttles ATP out of the chloroplast into the cytoplasm. Or the ATP could be provided by mitochondrial respiration. Thus, it is likely that high levels of mitochondrial respiration occurs in the light in PEP-CK species in order to supply part of the ATP required by PEP-carboxykinase.

In NADP-malic enzyme types (ME type), the NADP-malic enzyme is located in the bundle sheath chloroplasts of C_4 species and the cytoplasm of ME-type CAM plants. Mitochondria are not involved in NADP-malic type C_4 plants since this enzyme is in the chloroplast (Gardeström and Edwards 1983). In contrast, the cytoplasmic location of NADP-malic enzyme in ME-type CAM plants will generate NADPH in the cytoplasm. It is possible that the NADPH reduced by NADP-malic enzyme will be oxidized in the mitochondria for ATP production since plant mitochondria have an NADPH-specific uptake mechanism. Therefore, mitochondria can compete with chloroplasts for NADPH reduced in the cytosol by NADP-malic enzyme.

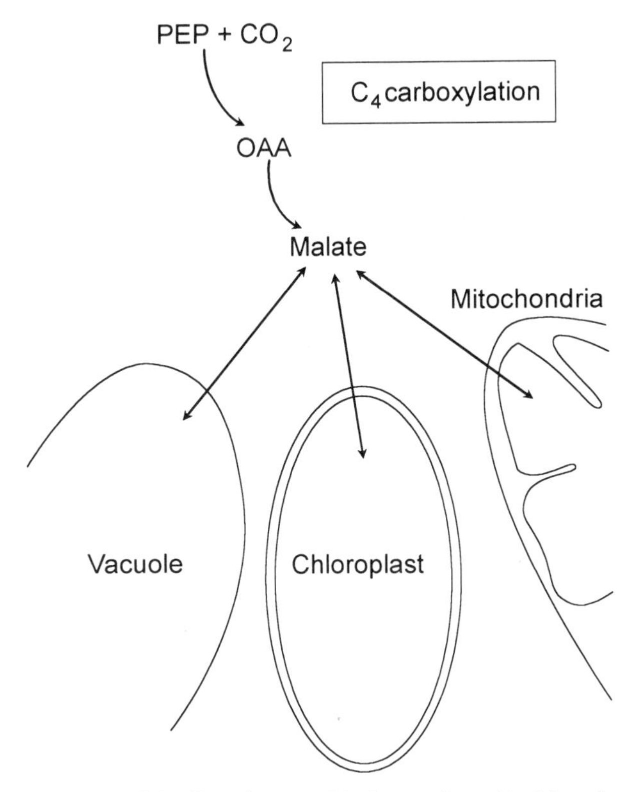

Figure 6.19 Carboxylation of the C_4 pathway and its interaction with chloroplasts, mitochondria, and the central vacuole. Note the central significance of malate in these interconnections among organelles.

All plant mitochondria (in photosynthetic or nonphotosynthetic cells) have NAD-malic enzyme. This is the basis for malate oxidation in the mitochondria. However, in C_4 and CAM species that utilize this enzyme for their primary decarboxylation in photosynthesis, the levels of NAD-malic enzyme is 40-fold that in C_3 species. In C_4 species bundle sheath cells have a much higher activity of NAD-malic enzyme than mesophyll cells. It is critical to understand that the NAD-malic decarboxylation converts either malate or aspartate into pyruvate and releases CO_2. The CO_2 then diffuses from the mitochondria into the cytoplasm and into the chloroplast for carboxylation by the Calvin cycle. However, the fate of the pyruvate is critical to C_4 and CAM photosynthesis. Pyruvate could be utilized directly by the mitochondria in the TCA (Krebs) cycle. Therefore, the mitochondria is competing for pyruvate with either the pep-carboxylase in the mesophyll cells of C_4 plants or glucogenesis (incorporation of pyruvate into starch) in CAM plants. Energetically, it is more efficient for CAM plants to convert pyruvate into starch directly by glucogenesis (requiring 9.7 ATP and 3 NADPH) instead of metabolizing pyruvate in the TCA pathway to CO_2, then reducing it to glucose by the Calvin–Bensen cycle (requiring a total of 12.7 ATP and 8 NADPH). Consequently, it is theorized that most pyruvate is converted directly into starch (Gardeström and Edwards 1983).

The interactions among respiration, photosynthesis, and photorespiration in C_3, C_4, and CAM plants exemplifies the flexibility of respiratory pathways in plant mitochondria.

In fact, many additional aspects of plant respiration are unique from that of cells in other kingdoms. For example, plant mitochondria have enzyme systems that can reduce both NAD^+ and $NADP^+$. The rate of mitochondrial O_2 uptake per unit of protein can be higher in plants than animal cells. Plants can also perform rapid detoxification through cyanide- and rotenone-resistant respiration.

Cyanide-resistant respiration involves an alternative electron transport chain that utilizes NADH, and transfers electrons to oxygen, but does not synthesize ATP. The purpose of this alternative pathway of energy use in plant mitochondria is still controversial. However, since this alternative pathway for electrons is basically constitutive in all plant species and can be up or down regulated by cytoplasmic physiology, it must have an importance to energy maintenance in plant cells. The alternative pathway has been suggested to serve as an "energy overflow" protection mechanism. When there is an excess of sugars (i.e., more than that required for ATP production, carbon skeletons for growth, osmoregulation, and storage), the alternative pathway utilizes the extra sugar without affecting the ATP pool size (Lambers 1982).

Also, plant mitochondria are unique because they can actively accumulate NAD^+ from an external medium (cytoplasm or an in-vitro experimental system) by a specific NAD^+ transfer mechanism. Plants contain the largest, most complex mitochondrial genome of all eukaryotes (Quetier et al. 1985). These are a few of the unique properties of plant respiration that result in much more flexibility than respiration in organisms from other kingdoms.

Although the flexibility and efficiency of plant respiration is impressive compared to other kingdoms, it is by no means optimized. Respiration evolved in plants not by following theoretically optimal developmental pathways but by utilizing the physiology that was available during the process of evolution. In fact, there is a certain amount of idling in respiration where ATP is produced and used by simple hydrolysis with no outcome, thereby reducing the efficiency of respiration. However, a majority of the reduced carbon used by plant respiration fuels both maintenance and growth processes of cells.

C. Whole-Plant Respiration

The utilization of total reduced carbon in plants can be divided into three main functions: (1) maintenance respiration uses reduced carbon to maintain cell function and releases all carbon that the process used as CO_2, (2) growth respiration uses a portion of total reduced carbon for construction of structure and long-term storage and another part of reduced carbon for oxidative respiration, and (3) a portion of ATP synthesized by respiration is used for ion uptake and transport. The use of respiratory ATP for ion accumulation was originally verified by the fact that when ions are added to a solution containing root tissue, the root tissue respiration rate increases (Lundgarth 1960). Others have proposed that the increased respiration in the presence of ions is due to an increase in the availability of respirable substrates (such as malate or other organic acids) coming out of the vacuole (Lambers 1980). In most whole-tissue respiration studies, total reduced carbon use can be described by

$$C = C_m + (C_t + C_r) \qquad \text{mol CO}_2\ \text{h}^{-1} \tag{6.5}$$

where C is the total reduced carbon use by the tissue; C_m the reduced carbon lost as CO_2 by oxidative respiration for maintenance of cell metabolism (including ion uptake and transport), C_t the reduced carbon invested into growth (long-term storage and structure), and C_r the reduced carbon used by oxidative respiration for growth and released as CO_2. Thus, the traditional separation of carbon use by respiration is not classified by metabolic pathway but rather by the process that uses the carbon.

Under this classification scheme, total respiration (R) is the sum of respiration used for growth and maintenance:

$$R = C_m + C_r \tag{6.6}$$

This relationship suggests that growth will be partly regulated by the proportion of respiration used for maintenance. Those species that are able to produce tissues that have a low maintenance respiration cost will have a greater proportion of respiration available for growth.

Estimations of the carbon cost for growth have been made by several scientists, the most notable being those proposed by Penning de Vries et al. (1974). A biochemical pathway analysis was used to determine the construction cost (in glucose equivalents) for the main classes of biochemicals. This technique allowed estimates of growth construction cost, in respiration units, for specific plant tissues that could be combined proportionally into a growth cost for the whole plant. Later research has shown that estimates of construction cost by biochemical pathway analysis are realistic when applied to specific organs (roots, stem, leaves) but not when they are combined into a whole-plant value. Alternative methods have been developed to ascertain the cost of growth without using extensive biochemical analyses (Williams et al. 1989). There are some tissues in which the estimated value for growth respiration are not accurate. For example, the theoretical growth respiration requirement to synthesize roots for two *Senecio* species severely underestimated experimental growth respiration (Lambers and Steingröver 1978). This was because root respiration had a large contribution by the alternative pathway (cyanide-resistant respiration), which did not produce ATP for growth. Estimates of the cost for construction of plant tissues are currently being examined as an important factor regulating the ability of species to exist in various habitats (Chapin 1989, Sobrado 1991).

Maintenance respiration is normally experimentally determined on mature tissues that are not growing, or from estimates based on major plant biochemical fractions (Merino et al. 1984). During organ growth, maintenance respiration is a small portion of total respiration. However, there are several situations under which maintanence respiration can become a large portion of total plant respiration. Following the growth phase, maintenance respiration dominates total respiration. This can occur once during the development of an organ (leaves) or on an annual basis (perennial stems and roots). Furthermore, large differences in maintenance respiration would be expected when comparing organs that are rapidly accumulating and transporting ions to organs that are not involved in ion accumulation. Separation of growth and maintenance respiration in many tissues is difficult because the two fractions of respiration co-occur. It is important to understand the relative strengths of these two sinks for photosynthate in order to evaluate mechanisms by which plants tolerate various stresses.

1. Environmental Effects on Whole-Tissue Respiration. The most dominant environmental factor influencing respiration is temperature. For many organisms respiration increases by a factor of 2 for every 10°C increase in temperature. Thus, respiration has a Q_{10} value of approximately 2. Not all plant tissues have a Q_{10} value of 2.0, and the Q_{10}

can change in one tissue with temperature or the supply of carbohydrates. Breeze and Elston (1978) determined that the Q_{10} of leaf respiration was 2.1 at 20°C and high carbohydrate content. However, at low carbohydrate content the Q_{10} decreased to 1.5 below 20°C and 1.7 above 20°C. In natural environments there are a number of internal and external factors that can affect the Q_{10} value of tissues. In general, any internal factor that increases the rate of respiration will also increase the Q_{10}. Thus an increase in carbohydrate, an increase in temperature, an increase in tissue nitrogen metabolism, and an increase in ion accumulation will increase the Q_{10} of respiration. It is critical to realize the labile nature of Q_{10} when trying to use this relationship to predict respiration rates.

There are also transient effects of temperature on respiration. Smakman and Hofstra (1982) found that cytochrome-mediated electron transport decreased, and the alternative electron transport (cyanide-resistant respiration) increased in *Plantago lanceolata* after decreasing root temperature. However, cytochrome-mediated electron transport took only 2 h to return to its original level, while the alternative pathway took 24 h to resume its original rate. Plant tissues that have been pretreated with a chilling temperature have a burst of respiration when temperature is brought back to normal (discussed further in Chapter 12). Sunflecks below a forest canopy or clouds in an open environment cause rapid changes in the light environment of leaves. If the leaf is relatively uncoupled from air temperature (large difference between leaf and air temperature), rapid changes in leaf temperature will occur. Very little is known about the impact of these "transient heat flecks" on the physiology of leaves (discussed further in Chapter 11). These and other examples highlight the importance of understanding the transient effects of temperature changes on various components of respiration in order to understand the impact of such changes on growth.

Respiration rate is also affected by light. Of course, light has a secondary influence on respiration through its effects on temperature, photosynthesis, and photorespiration in tissues with chloroplasts. Also, an increase in light will increase the flow of carbohydrates and secondarily increase respiration, because a primary regulator of respiration is carbohydrate abundance. In addition, light has been shown to have some direct effects on respiration, possibly through a blue light–sensitive regulation system. The direct effects of light on the biochemistry of respiration are as yet undefined, as is the significance of these effects. Respiration rates are also affected by salt concentration, water stress, nutrient availability, and atmospheric pollutants. In fact, any type of environmental factor that influences growth will have an effect on respiration.

V. ORGANIC CARBON LOSSES FROM PLANTS

The previous discussion demonstrated that respiration is the major process by which inorganic carbon is lost by plants. An additional fraction of reduced carbon will be lost by plants in an organic carbon form. Such losses occur from all plant organs. Tissues are directly lost by abscission or mortality of leaves, roots, and stems. In fact, many scientists utilize the term *throwaway construction* when referring to plants that invest a minimal amount of energy into the construction of an organ that has a short lifespan. The ability to lose and replace vital organs easily is a unique characteristic of plants compared to other organisms. Any study of plant carbon balance must take into account all those tissues that were lost over the lifetime of the plant. Also, the rate of tissue turnover can be an important adaptation to environmental stresses.

Organic carbon is also lost from plants by processes other than tissue death and abscission. Roots exude considerable amounts of organic carbon into the rhizosphere. This is a primary carbon source for a large and diverse community of microorganisms. Organic carbon can be leached from leaves, roots, and stems (particularly bark) during rainfall. Both throughfall and stemflow can have a considerable diversity of organic compounds derived from plant tissues (Rice 1979). Leaching of organic compounds such as phenolic acids from leaves or juglone from roots has been linked to allelopathic effects (Rice 1984).

The amount and type of organic molecules leached from plant organs depends on the nature of the cuticular covering over the organ and the intensity and quality of rainfall. Acidification of rainfall will affect the organic constituents of throughfall and stemflow because cuticles become thinner when exposed to chronic acid precipitation. However, the most ubiquitous type of organic carbon loss from plants may be the emission of volatile organic compounds rather than leachates.

A. Volatile Emissions of Carbon Compounds

Plants emit a diverse array of volatile organic carbon (VOC) molecules. These include monoterpenes, isoprene, sesquiterpenes, esters, ketones, alcohols, organic acids, alkanes, aldehydes, and other aromatic hydrocarbons (Winer et al. 1992). Volatiles are emitted from several sources, including leaves, stems, and roots. The particular suite of VOC emitted by any particular plant depends on the evolutionary lineage of the species, the environmental conditions, and the developmental stage of the tissues in question. Some classes of VOCs, such as monoterpenes, are highly concentrated in mature tissues. This results in a constant emission of the VOC independent from the rate of compound synthesis because the emission rate is only a small portion of the pool in the tissue. In contrast, some VOCs are normally at very low concentrations in tissues, such as isoprene, and there is no buffer between the biosynthetic pathway and the emission of the VOC. The magnitude of VOC emission can constitute a major output of carbon from ecosystems into the atmosphere (global estimates range from 500 to 1000 Tg y^{-1}: 1 Tg = 10^{15} g), and evergreen species have relatively high VOC emission rates compared to deciduous species. For this reason, ecosystems with the greatest emission of VOCs into the troposphere are boreal and tropical forests (Lamb et al. 1993). In contrast, VOC emissions are equal to only 1 to 2% of whole-plant carbon assimilation.

Emissions of VOCs by plants are important for plant metabolism in relation to environmental stress, for several basic reasons. First, VOCs are a major process by which organic carbon is lost from mature plants. The amount of organic carbon lost as VOCs is small compared to that lost through tissue death and abscission, but the amount of VOC is large compared to organic carbon lost through leaching of live plant tissues. Therefore, as stressors affect carbon cycling processes in plants during the growing season and before leaf abscission, emission of VOCs may be the most important process of organic carbon loss. Second, VOCs can have important interactions in tropospheric chemistry, particularly in relation to the content and reactivity of oxidizing agents such as $OH^{\cdot}$ (the hydroxyl radical), and O_3. In the presence of VOCs, ozone and other oxidizing agents can be reduced, yielding methane or carbon dioxide. Therefore, interactions between VOCs and oxidizing species will affect atmospheric chemistry and modify the concentration of pollutants in the atmosphere. Third, those VOCs that have large buffers between synthesis and emission (e.g., monoterpenes) have been shown to serve as antiherbivory agents.

Fourth, there are some alternative significances of VOCs not related to carbon balance. For example, several members of VOCs have been related to allelopathic functions (Muller and Muller 1964), and isoprene emissions may be related to high-temperature tolerance (Sharkey and Singsass 1995).

Regulation of the rate of VOC emission depends on metabolic pathways, tissue developmental stage, and environmental conditions. Biochemical pathways for VOCs are components of other important metabolic pathways. For example, the pathway for isoprene synthesis is a short branch from the mevalonic acid pathway leading to carotenoid synthesis (Monson et al. 1991b). Monoterpene synthesis is a component of the lignin synthesis pathways. Consequently, biosynthetic pathways leading to VOCs are considered a part of secondary metabolism. Most of these biosynthetic pathways leading to VOCs are constitutive, and regulation of the pathway is often determined post-transcriptionally by substrate concentration and end product inhibition rather than at the level of translation or transcription.

The emission of VOCs can be characterized by a basal emission rate and an instantaneous emission rate. The basal emission rate is determined by the species-inherent capacity to emit VOC, and is measured at standard environmental conditions (Monson et al. 1991a). In contrast, the instantaneous emission rate is that rate determined by the influence of prevailing environmental conditions on the basal emission rate. Most research on VOC has concentrated on factors regulating the instantaneous rate rather than the basal rate. If biochemical mechanistic models are to be developed which predict VOC emission rates, the factors that regulate basal emission rates are critical for determining the relationships between environment conditions and instantaneous emission rates. The factors that regulate the emission of VOCs that have a large buffer versus those that have no constitutive buffer are discussed below.

Monoterpenes are a good example of VOCs that have a large constitutive concentration in tissues of some plants. Species that concentrate monoterpenes are thought to do so as an antiherbivory defense mechanism. Although other functions for the high monoterpene concentration are possible, none have the experimental support as that for the antiherbivory model. Monoterpenes act to deter herbivores when they are ingested by insects or when some insects detect vaporized monoterpenes. In particular, several monoterpenes have been shown to inhibit mitochondrial function. To have a significant impact on herbivores, there must be a high concentration of monoterpenes in plant tissues. The same mechanism of action (mitochondrial inhibition) has been proposed as an allelopathy mechanism for several *Salvia* species against competing grass species (Muller and Muller 1964).

Monoterpenes are emitted from leaves at a basal rate that depends on the concentration of monoterpenes in the tissue (Schindler and Kotziab 1989). The basal emission rate is modified by temperature primarily to produce the instantaneous emission rate. Tissue temperature influences emission rates by affecting the vapor pressure of monoterpenes in their storage structures (Tingley et al. 1991). In some cases, monoterpene concentrations are high enough to saturate the intercellular chamber, while in other species the monoterpene storage structures (leaf trichomes, glandular hairs, oil glands) may not be saturated because of the large storage structure and low diffusion conductance of VOCs from the site of synthesis into the storage structure. In storage regions without saturated monoterpene vapor pressure, changes in temperature will have a smaller effect on instantaneous emission than they will in tissues with saturated monoterpene concentrations.

In pine needles monoterpenes are located in resin ducts and are emitted through stomata. In leaves where monoterpene concentration is saturated, emission rates should be correlated with changes in stomatal conductance. However, in tissues where monoterpene concentrations are less than saturated, the effect of decreasing conductance on emission rate should be compensated by the increase in intercellular monoterpene vapor pressure. Published evidence supports the latter case for pine needles because there was little change in monoterpene emission rates when stomata closed in response to several different stimuli (Tingley et al. 1980, Guenther et al. 1991).

If monoterpene emission rates are simply dependent on tissue concentration of monoterpenes and tissue temperature, it will be easy to predict instantaneous emission rates. A more interesting question is at what cost to the plant is the synthesis of large quantities of monoterpenes. There is a trade-off in the carbon cost between allocating carbon to growth or to herbivore defense mechanisms (Herms and Mattson 1992). In fact, several studies have shown that monoterpene synthesis is directly proportional to the balance of photosynthesis and the use of photosynthate (Loomis and Croteau 1973). When photosynthesis is relatively high compared to metabolic use of photosynthates (for any number of reasons), monoterpene synthesis increases. Therefore, under conditions of mild stress (e.g., low water availability or minimal N availability) photosynthate use decreases relative to net photosynthesis, and monoterpene synthesis will increase.

The use of monoterpenes as a defensive mechanism may also involve a nitrogen cost to the plant. The nitrogen cost of synthesizing monoterpenes for defense is not intuitively obvious because nitrogen is not a major component of these compounds. Nitrogen is an important component of other defensive secondary compounds, such as alkaloids, and nitrogen availability is a significant constraint to their deployment. However, in the case of monoterpenes, a significant amount of nitrogen is required for the metabolic machinery that synthesizes monoterpenes. Monoterpenes are required at high concentration, and many enzymes in the monoterpene synthesis pathways are not constitutive. Therefore, plants must invest a significant amount of N into secondary metabolic pathways in order to synthesize enough monoterpenes to provide defense against herbivores (Lerdau 1991).

Herbivory could have direct impacts on the induction of monoterpene synthesis or indirect effects through evolutionary processes. Under what types of herbivory pressures are plants stimulated to synthesize high amounts of defensive compound? There is evidence that wounding bark tissue can induce monoterpene synthesis, and there is some recent evidence for monoterpene induction by wounding leaf tissues. Although some monoterpene induction occurs, the answer to the question posed above requires information on an evolutionary scale, during which herbivory pressure selected for genotypes that allocate larger proportions of carbon to defense at a cost to growth.

In direct contrast to those factors regulating monoterpene emissions, isoprene emission is linked closely to its biosynthetic pathway. The most likely pathway for isoprene synthesis is that from starch to carotenoids through mevalonic acid (Figure 6.20). Evidence suggests a link between photosynthesis and isoprene synthesis. In fact, the light response curve for photosynthesis is very similar to the light response curve for isoprene emissions (Loreth and Sharkey 1990). This is supported by evidence that the carbon emitted as isoprene from leaves is derived from recently reduced carbon from photosynthesis (Monson et al. 1991b). The linkage between photosynthesis and isoprene synthesis is probably through PGA or 1,3-bis-PGA (Loreth and Sharkey 1990). If CO_2 is removed from the atmosphere, isoprene synthesis is not inhibited. Thus, isoprene synthesis rate is not related directly to metabolic pools that respond to changes in CO_2 partial pressure (Badger et al. 1984). Even though isoprene biosynthesis utilizes starch or immediate products of photosynthesis as the primary substrate, isoprene synthesis is probably not regulated by the end

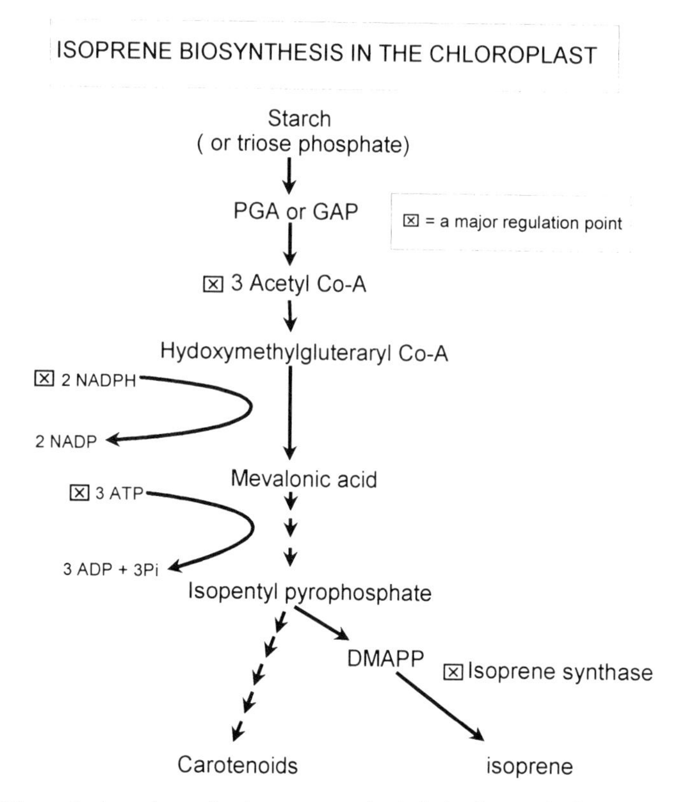

Figure 6.20 Biosynthetic pathway for isoprene synthesis in leaf mesophyll cells. Notice the close connection between isoprene synthesis and photosynthesis. Important regulation points of this pathway are energy sources from photosynthesis and isoprene synthesis. (Redrawn and revised from Monson et al. 1996.)

products of the Calvin cycle. The component of photosynthesis best correlated with isoprene synthesis is the rate of photosynthetic electron transport (Monson et al. 1991b). This linkage is probably through chloroplastic ATP and NADPH pools.

Isoprene synthesis is also temperature dependent, such that a temperature response curve of isoprene synthesis has the form of an enzyme thermal response. This is because the main regulator of isoprene synthesis is the enzyme isoprene synthase (Silver and Fall 1991), which converts DMAPP to isoprene (Figure 6.20).

It is interesting to pose some potential significances to isoprene emissions in plants. One possibility is that isoprene synthesis is a heat-stress adaptation (Sharkey and Singsaas 1995). In the case of kudzu (a fast-growing leguminous vine), isoprene emissions are thought to penetrate into chloroplast membrane and provide increased thermal stability. Isoprene synthesis may be another mechanism of dissipating excess energy from photochemical reactions. Even in the absence of CO_2, electron transport continues to feed isoprene synthesis. Since isoprene synthesis utilized ATP and NADPH, oxidized versions of these molecules are regenerated by isoprene synthesis, inhibiting the buildup of reduced forms and promoting electron flow through the photochemical reactions. This is very similar to the potential significance of photorespiration except that isoprene is the final end product (rather that CO_2 and serine), and the mitochondria and peroxisome are not involved in the isoprene synthesis pathway.

Another important reduced carbon emission from leaves is methanol. Methanol is emitted by most if not all plant species, which is in contrast to isoprene or monoterpene emissions, that occur is a small subset of species. The amount of methanol emitted by leaves is intermediate between the emissions of monoterpenes and isoprene. The emission of methanol is dependent on stomatal aperture. The higher the conductance, the higher the methanol emissions (Nemecek-Marshall et al. 1995). In addition, methanol emission is dependent on leaf age, with maximum emissions occurring in recently matured leaves. Information about the regulation of methanol emissions is just being gathered, but since all plants emit this hydrocarbon, it is likely that methanol emission is an important organic carbon flux from plants.

In summary, the basic emission rates of VOCs will be determined by the evolutionary history of the particular species. Has the species evolved significant antiherbivory mechanisms by increasing the concentration of volatile organic compounds in its tissues? Has the species evolved VOC emission pathways as a mechanism to tolerate high heat or excessive radiation? The instantaneous emission rate of VOCs will be dependent on the basic emission rates as modified by temperature and other environmental factors that affect the balance between photosynthesis, photorespiration, and photosynthate use. All plants emit various amounts of VOCs of various composition. The instantaneous magnitude and quality of VOCs emitted will be dependent on the basal emission rate and the physiological and behavioral mechanisms by which plants respond to changes in the biotic and abiotic environment. Much research is required to determine the mechanisms regulating the rates of VOC emissions, but it suffices to say that VOC emissions are an important aspect of carbon balance of plants during stress conditions.

VI. SUMMARY

The purpose of this chapter was to introduce the reader to basic aspects of the physiology determining carbon cycling in plants. Factors that regulate carbon accumulation are the CO_2 gradient between the air and the chloroplast, light absorption properties of leaves, electron transport capacity, biochemical properties of the Calvin cycle, triose phosphate use. All of these factors can be modified by plants in their responses to stress. It was clear that photosynthesis and photorespiration interact directly with respiration (the primary output of CO_2). That interaction occurs at various levels of biochemical organization and in all organelles associated with these processes. Even in plants with alternative biochemical pathways for photosynthesis, respiration is closely linked to photosynthesis. There are several ways in which organic carbon can be lost from plants, including tissue abscission, organic carbon leakage, and volatilization of reduced carbon. Growth is basically defined by the ratio of carbon gain to carbon loss, but it is the denominator of this equation that may most closely correlate with growth rate. Therefore, it is critical to understand the impact of a stressor on all aspects of carbon flow dynamics in order to unravel the mechanism by which a particular plant stress affects whole-plant carbon balance.

STUDY-REVIEW OUTLINE

Carbon Accumulation Processes

1. The relationships between carbon gain, translocation of assimilates, and carbon loss is critical for understanding plant stress physiology.
2. The relationships among carbon gain and loss processes are affected by exogenous and endogenous factors.

Atmospheric Carbon Dioxide Concentration

1. The gradient of CO_2 partial pressure between the atmosphere and intercellular spaces in leaves allows for diffusion of CO_2 into leaves.
2. Atmospheric CO_2 concentration is not stable and varies with height in canopies and times of day.
3. Other anthropogenic and natural factors can increase atmospheric CO_2 concentrations significantly.

Pathway of CO_2 Movement

1. The diffusion pathway for CO_2 in air into the intercellular spaces of leaves is the same as that for water vapor diffusing out of leaves, and these two molecules interact along that pathway.
2. Mesophyll conductance concerns the diffusion of CO_2 in water from cell walls that line intercellular spaces to chloroplasts.

Regulation of Stomatal Conductance

1. In general, stomata open when guard cells gain turgor pressure.
2. Evaporation from cells surrounding guard cells may affect guard cell turgor and cause stomatal closure. This may be the main mechanism by which stomata respond to atmospheric vapor pressure.
3. Increasing air temperature will increase the gradient of vapor pressure between leaf and air, cause increased transpiration, lower leaf water potential, and be followed by reduced stomatal conductance.
4. A hydroactive modulation of stomatal aperture is regulated by photoreceptors (red and blue light) in guard cells.
5. Root zone water limitation can cause reductions in stomata without affecting mesophyll tissue water potential. This response is regulated by a root signal, probably abscissic acid (ABA).
6. The balance of ABA and cytokinins in leaf tissue may have an important role in regulating stomatal aperture.
7. Stomatal conductance is also sensitive to intercellular CO_2 concentration (C_i). When C_i decreases, stomata are stimulated to open.
8. Stomatal aperture may respond to various endogenous and exogenous factors that regulate transpiration in order to keep tension in the xylem low enough to preclude xylem conduit cavitation.

Regulation of Mesophyll Conductance

1. Both increasing temperature and pH increase CO_2 solubility in cell wall water.
2. The cell membrane has an extremely high resistance to the diffusion of HCO_3^- compared to that for CO_2. Thus most carbon destined for photosynthesis enters the mesophyll cell as CO_2.
3. The magnitude of mesophyll resistance is similar to the magnitude of resistance to the movement of gaseous CO_2 into the leaf.
4. Both mesophyll and stomatal conductance have a strong impact on intercellular CO_2 concentration.

Photosynthetic Response to CO_2

1. As the atmospheric CO_2 concentration increases, photosynthesis will increase until C_i is high enough to saturate photosynthesis.
2. The photosynthetic response to increasing CO_2 (P_n/C_i response curve) can be used to calculate the limitations placed on photosynthesis by stomatal and mesophyll resistances.
3. The initial linear rise in photosynthesis as C_i increases is representative of mesophyll conductance effects in C_3 plants.

4. At high C_i, photosynthesis is limited by the biochemical process of photosynthesis and photon energy flow.

Harvesting Photon Energy

1. Photon energy is harvested and funneled into ATP by several protein complexes in the chloroplast membranes.
2. The photosystem II core complex is composed of about 12 proteins that traverse the thylakoid membrane. Attached loosely to the PSII core complex are light-harvesting complexes.
3. A population of plastoquinone molecules transport electrons from the PSII core complex to the cytochrome b_6/f complex. The cytochrome b_6/f complex serves to pump protons across the thylakoid membrane.
4. A population of plastocyanin molecules transport electrons from cytochrome b_6/f complexes to photosystem I core complexes.
5. Photosystem I core complexes are associated tightly to light-harvesting complexes that collect photons. PSI core complexes reenergize the electrons and transfer them to ferredoxin $NADPH_2$ reductases, which in turn reduces $NADPH^+$.
6. The various protein complexes are not evenly distributed on thylakoid membranes. PSII core complexes are more abundant on stacked membranes, and PSI core complexes are more abundant on unstacked membranes.
7. A sound understanding of the molecular model of light harvesting and electron transport is necessary to evaluate possible adaptations to light intensity.

Biochemical CO_2 Reduction Pathways

1. Variations in biochemical reduction pathways among plants relate specifically to adaptations in response to environmental stress.

C_4 Photosynthesis

1. Photorespiration is inhibited in C_4 plants because of the CO_2 pumping mechanisms founded on the enzyme kinetics of PEP carboxylase and transport kinetics of malate or aspartate from mesophyll cells to bundle sheath cells.
2. C_4 plants have higher photosynthesis at any transpiration rate compared to C_3 plants, which results in a higher water use efficiency in C_4 plants.
3. C_4 plants have a higher quantum requirement than C_3 plants because ATP energy is required to transport malate or aspartate into bundle sheath cells. The higher quantum requirement gives C_4 plants a competitive disadvantage over C_3 plants in shaded environments.
4. In high-temperature environments, quantum yield of C_3 plants decreases relative to C_4 plants, so C_4 plants are favored in high-temperature sites.

CAM Photosynthesis

1. In CAM plants the two carboxylations are separated in time rather than in space as in C_4 plants.
2. CO_2 is accumulated from the atmosphere at night when atmospheric vapor pressure is relatively high. This minimizes transpiration and maximizes water use efficiency.
3. CAM photosynthesis is limited by the energy cost of transporting malate into and out of the vacuole, the dependency of PEP upon starch stored the night before, and by the limited amount of malate stored in the vacuole.
4. Some CAM plants can increase their productivity by switching to C_3 photosynthesis during relatively cool and moist times of the year.
5. Species with CAM photosynthesis are normally restricted to regions with few competitive species.

Assimilate Circulation and Storage

Fractionation of Photosynthate Within the Cell

1. Triose phosphates from photosynthesis can be directed toward starch synthesis, sucrose synthesis, and to the Calvin–Benson cycle.
2. Fractionation of triose phosphate is dependent on the activity of fructose-6-phosphate, the switch point between starch synthesis and the Calvin–Benson cycle.
3. The rate at which triose phosphate enters the cytosol is probably dependent on the existing pool of triose phosphate in the cytosol and the activity of sucrose-6-phosphatase.
4. Mutants deficient in starch have high cytosolic respiration rates because the triose phosphates are stimulating respiration but not phloem loading or sucrose translocation.
5. Starch metabolism is highly linked to environmental processes but may not indicate changes in growth.

Regulation of Assimilate Transport Among Tissues

1. The rate of photosynthate transport among organs depends on the strengths of sinks and sources.
2. Sink strength is dependent on the mass of the sink and its respiration rate.
3. The sucrose-proton symport regulates sucrose loading into the phloem at the source by proton gradients across the cell membrane.

Allocation of Photosynthate Within the Plant

1. Proportional allocation of photosynthate among plant organs is defined by genetics but modified by environmental conditions.
2. Allocation is modified by the proportional sizes of sinks. Sink size is determined by organ mass, metabolic rate, and storage properties.

Assimilate Storage

1. Photosynthates are usually stored as starch in chloroplasts or leucoplasts.
2. The size of a storage pool is defined as stored material that can be retrieved for renewed metabolic activity. Not all starch is available for remobilization into sugars.
3. On a short time scale, starch is stored in leaves. On a longer time scale (seasonal), starch is synthesized in storage tissues such as root, stem, or tuber parenchyma cells.
4. Species that experience stochastic disturbance, such as fire, often have large subterranean starch storage structures. Some of the subterranean starch can be mobilized after a fire to fuel rapid regrowth.

Carbon Losses from Plants

1. Carbon lost from plants occurs mostly by tissue abscission and respiration.
2. However, significant losses of carbon by volatile emissions from leaves and stems or leakage from roots of organic carbon can occur.

Respiration

1. Respiration of a plant releases 50% or greater of total photosynthate.
2. The relationship between growth rate and respiration may be better than the relation between photosynthesis and growth rate.
3. The respiratory pathway can be broken down into five main processes: (a) Glycolysis, (b) oxidative pentose phosphate pathway, (c) Krebs cycle, (d) mitochondrial electron transport, and (e) oxidative phosphorylation.

Respiration, Photosynthesis, and Photorespiration

1. Plant cells are unique because of the potential interactions between respiration, photosynthesis, and photorespiration.
2. Oxygen produced by respiration is used in photosynthesis. Both photosynthesis and respiration produce ATP and reductant that is released into the cytosol.
3. Intermediates of photosynthesis and respiration are similar.
4. Respiration and photorespiration are linked in the mitochondrion through the OAA ↔ malate transformations.
5. The first carboxylation in C_4 and CAM plants produces malate or aspartate; this interacts with the malate shunt in respiration.
6. In PEP-CK plants, respiration can provide ATP for PEP-carboxykinase.
7. In ME-type CAM plants NADPH is generated in the cytoplasm and can be used as reductant by the mitochondria.
8. An alternative electron transport chain in the mitochondria oxidises NADH and transfers electrons to oxygen but does not produce ATP. This pathway has been termed an energy overflow protection mechanism for plant cells.
9. The respiratory pathway in plants has a high degree of flexibility because of the multiple interactions between photosynthesis, respiration, and photorespiration.

Whole-Plant Respiration

1. Whole-plant respiration is separated into components based on the use of energy rather than on biochemical pathways.
2. Whole-plant respiration is used for maintenance of biochemical processes, energy invested in growth, and respiration used for ion accumulation.
3. Tissue construction cost depends on the biochemical makeup of the tissue. Those tissues with high concentrations of proteins and fats have high construction cost.
4. Maintenance respiration is usually measured empirically on mature (nongrowing) tissues.

Environmental Effects on Whole-Tissue Respiration

1. The Q_{10} value of respiration is usually between 2.1 and 1.7, which means that the respiration rate approximately doubles for every 10°C rise in temperature.
2. The Q_{10} value is not stable and can change with background temperature or with the magnitude of cyanide resistant respiration.
3. Light intensity affects respiration primarily through its influence on photosynthesis and photorespiration. Secondarily, light affects respiration rate through its influence on tissue temperature.

Organic Carbon Losses from Plants

1. Much carbon can be lost by abscission of tissues. The ability to lose and replace vital organs (leaves and roots) repetitively is a unique property of plants in comparison to animals.
2. Plant tissues are also leaky, so that precipitation can wash out some forms of organic molecules from leaves. In addition, roots may exude or leak organic into the rhizosphere.

Volatile Emissions of Carbon Compounds

1. Plants emit a diverse array of volatile organic compounds, including terpenes, sesquiterpenes, isoprene, alkanes, organic acids, aldahydes, and others.
2. VOC emission is regulated by a basal rate and an instantaneous emission rate. The basal rate is dependent on tissue maturation status and genetics, while the instantaneous rate is dependent on environmental conditions.

3. Monoterpenes have a large background concentration, and the emission rate is not dependent on monoterepene synthesis rate.
4. Isoprene has a very low concentration in tissues, and its emission rate is dependent upon its biosynthetic pathway.
5. VOCs that have a large buffer between synthesis and emission (monoterpenes) may serve as antiherbivory agents.
6. Emission rates of monoterpenes are basically dependent on tissue temperature, as this influences the internal vapor pressure of the monoterpene.
7. The advent of stress increases monoterpene production rates.
8. The evolutionary interaction between tissue-damaging herbivores and the cost of producing monoterpenes may be the primary factor regulating the production of monoterpenes in plants.
9. Isoprene released from plants is related directly to photosynthesis.
10. Isoprene emission is most influenced by the functions of light absorption and the electron transport chain.
11. It is probable that photorespiration and isoprene synthesis compete for ATP and NADPH when CO_2 is low in the leaf.
12. Isoprene synthesis may also be involved in heat stress tolerance or dissipation of photochemical energy.
13. Methanol is emitted by all leaves and may constitute a major volatile organic compound loss from plants.

SELF-STUDY QUESTIONS

1. Compare and contrast the factors that regulate stomatal conductance and mesophyll conductance.

2. Diagram a photosynthesis versus intercellular CO_2 concentration response curve and label important parts of the curve.

3. In what way can one measure the limitations of photosynthesis by stomatal or mesophyll conductance?

4. What factors regulate the atmospheric CO_2 concentration?

5. Define the passive and active mechanisms that regulate stomatal aperture.

6. If you were to design a light-harvesting apparatus and electron transport chain for a plant in an extremely low light environment, how would you arrange the protein complexes and light-harvesting complexes in the thylakoid membranes?

7. What are the advantages and disadvantages of the C_4 photosynthetic pathway for plants in relation to environmental factors?

8. Why is it that CAM plants have limited productivity, and what do some CAM plants do to maximize productivity but retain their high water use efficiency during drought?

9. How is photosynthate partitioned in cells, and what regulates that fractionation?

10. Why is it that the starchless mutant of *Arabidopsis* have particularly high respiration rates?

11. Rank the relative impact of sucrose loading processes and phloem transport processes to the overall translocation of photosynthate in plants.

12. What are the primary factors that regulate the proportional allocation of photosynthate to the various organs?

13. What factors regulate diurnal and long-term storage of photosynthate?

14. In what ways do respiration and photosynthesis interact in plant cells?

15. If stomata are closed and intercellular CO_2 decreases during the day, what is likely to happen to photosynthesis, photorespiration, respiration, and isoprene synthesis?

16. How is CAM photosynthesis linked to the malate shunt of respiration?

17. In what way do photorespiration and respiration interact in the mitochondria?

18. In what way does respiration contribute to photosynthesis in PEP-CK forms of C_4 photosynthesis?

19. Whole-plant respiration is divided into what three principal uses?

20. How can the alternative electron transport chain interfere with the measurement of maintenance respiration?

21. In what way do changes in temperature and light affect respiration?

22. Contrast the factors that regulate emissions of monoterpenes and isoprene from plants.

23. How will an increase in temperature affect the emissions of monoterpenes and isoprene?

SUPPLEMENTARY READING

Baker, N. R. and S. P. Long (eds.). 1986. Photosynthesis in Contrasting Environments (Topics in Photosynthesis, Vol. 7). Elsevier, New York.

Baker, D. A. and J. A. Milburn. 1989. Transport of Photoassimilates (Monographs and Surveys in the Biosciences). Longman Scientific and Technical, Harlow, Essex, England.

Douce, R., and D. A. Day. 1985. Higher Plant Respiration (Encyclopedia of Plant Physiology, New Series, Vol. 18). Springer-Verlag, New York.

Schulze, E.-D., and M. M. Caldwell (eds.). 1994. Ecophysiology of Photosynthesis (Ecological Studies, Vol. 100). Springer-Verlag, Berlin.

7 Water Dynamics in Plants

Outline

Objectives

1. Describe the molecular structure of water, concentrating particularly on covalent bond force and hydrogen bonding.

2. Describe the impact of water's molecular structure on physical attributes of water, including cohesive and adhesive forces, heat behaviors, and electrical constants.
3. Explain important aspects of aqueous solution thermodynamics, focusing on potential energy.
4. Explain the definition of water potential in relation to plant systems.
5. Describe important physiochemical aspects of water in relation to cellular function.
6. Differentiate among the various cellular components of tissue water potential: osmotic, pressure, matric, gravimetric.
7. Explain how cell wall elasticity influences cell water potential.
8. Outline the basic hydrodynamics of the water transport system in plants.
9. Describe the soil–plant–air continuum and other possible mechanisms of water transport in the xylem.
10. Consider factors that influence each component of the water transport system in plants: root uptake, hydraulic conductance, capacitance, cavitation, transpiration.
11. Discuss the various technologies used to measure water potential in plants.
12. Discuss the pressure–volume curve and the types of data that can be derived from this relationship.
13. Summarize by presenting an integrated view of plant water relations.

I. INTRODUCTION

Water, a molecule that some call "the cradle of life," is of paramount importance to plant growth and function. All resource exchanges between a plant and its environment occur through the medium of water. Cell expansion and the physical/chemical integrity of the cell wall are water dependent. The molecular structure of proteins, polysaccharides, and other hydrophilic molecules depends on the polar properties of water and the resulting sheath of hydration. Cohesive and adhesive forces regulate the movement of water through transport systems. Chemically, water serves as an excellent solvent and acts frequently as a reactant. Considering the many significances of water to plants, it is not surprising that the amount of water available to plants constitutes an important potential stress in many natural habitats and managed systems.

All plants experience some sort of water limitation or overabundance. In natural systems, limited water availability can result from low precipitation, soil with low water holding capacity, excessive salinity, cold soil temperature, low atmospheric vapor pressure, high air and soil temperature, or a combination of these or other factors. In agricultural systems, the availability of water can be a primary limitation to yield. In fact, drought was predicted to cause 80% of the crop yield reduction (compared to potential) nationwide (Boyer 1982). Also, the forestry industry is strongly affected by water availability because tree ring diameter growth (wood production) is related directly to water availability and transpiration (Waring and Running 1978). This is the reason that ancient logs can be used to ascertain prehistoric climatic conditions.

This chapter serves as an introduction to most of the important aspects of water relations that have particular relevance to stress physiology. Biophysical, cellular, and hydrodynamic aspects of water in plants are covered. Furthermore, the methods of measuring cellular water relations and water flow dynamics are discussed. It is not possible to cover all aspects of plant water relations completely in one chapter. Therefore, we outline the most important aspects of water relations to plants, to develop the background understanding needed to interpret the mechanisms by which plants can compensate for limitation and overabundance (Chapters 8 and 9). Supplementary readings, reviews, and books on water relations are provided to reinforce and expand on topics covered in this chapter.

II. BIOPHYSICAL CHARACTERISTICS OF WATER

A. Molecular Structure

The molecular structure of water is dominated by two atoms that have an extreme difference in electronegativity. Oxygen, a highly electronegative atom, covalently bonds to hydrogen, an atom with low electronegativity, resulting in electrons predominantly orbiting oxygen. The consequence of this atomic interaction is strong dipolarity in the water molecule, the potential for hydrogen dissociation, and hydrogen bonding among water molecules (Figure 7.1). In water molecules hydrogen atoms are covalently bonded (sharing

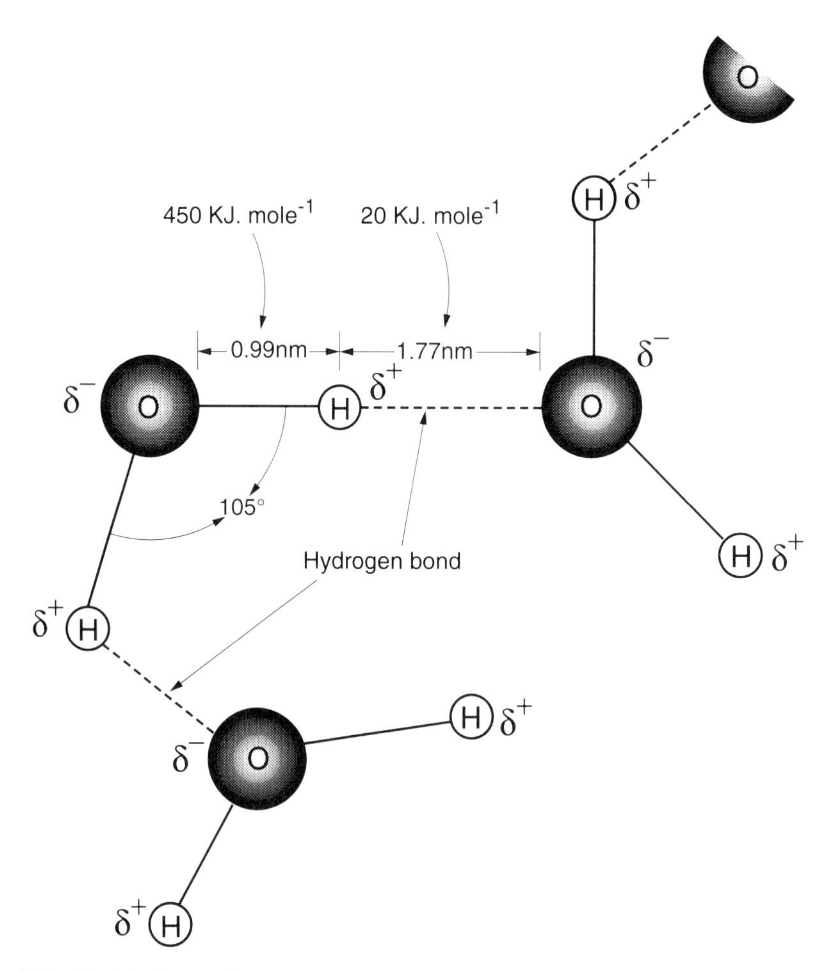

Figure 7.1 Model of the molecular structure of water molecules, including the hydrogen bond between two molecules. δ^+ and δ^- refer to charges developed on hydrogen and oxygen atoms due to their large difference in electronegativity. Contact angles are plotted as well as bond free energy. (Redrawn from Nobel 1991.)

one pair of electrons per hydrogen) to an oxygen atom with a force of 450 kJ mol^{-1}. Hydrogen bonding among several water molecules occurs between the hydrogen of one molecule and the oxygen of another with a force of merely 20 kJ mol^{-1}. Therefore, the hydrogen bonding between molecules is over 200 times weaker than the covalent hydrogen bond within the water molecule. Water molecules come together in large masses that are connected together by hydrogen bonds for short periods of time. These "clusters" of water molecules constantly break and rejoin into new clusters (termed *flickering clusters*). The strong dipolarity, hydrogen bonding capacity, and the flickering cluster properties of water confer a number of unique physical and chemical characteristics to water which are important to cellular metabolism.

B. Cohesive Properties

Hydrogen bonding among water molecules, and the resultant flickering clusters, hold masses of water molecules together. This cohesive force results in a relatively high surface tension (σ) for water in comparison to other liquids ($\sigma = 7.28 \times 10^{-2}$ N m^{-1} for water in air at 20°C). In addition, the strong dipolar nature of water influences its interaction with other surfaces. This results in strong adhesion of water to hydrophilic surfaces such as cell walls and macromolecules. The strength of cohesive and adhesive forces is important for the movement of water in small spaces such as the xylem elements of plants.

Capillarity is the ability of water to move vertically within a tube (such as a straw, xylem conduit, or soil pore). The height of capillary movement will depend on the relationship between the downward forces (gravity) and the upward forces (a result of cohesion and adhesion). Water in a tube will form a contact angle (α) with the surface of the tube (Figure 7.2), the value of which depends on the nature of the wall surface (hydrophobic, hydrophilic, smooth, pitted, etc.). The upward force in the capillary tube will be the product of surface tension, the cosine of the contact angle, and the bore circumference (Figure 7.2). The downward forces depend on the pressure on the system, a product of the water mass and gravitational force and the height of the water column. The height of water movement can be calculated by equating the upward and downward forces and solving for height (Nobel 1991). In a vessel element with an inner diameter of 20 μm, the column of water would rise about 0.75 m at atmospheric pressure. For the column of water to reach the height of many trees, approximately 30 m, the vessel lumen diameter must be approximately 5 nm. Vessel elements are normally much wider than 5 nm, commonly ranging from 20 to 50 μm. Therefore, capillarity alone could not account for the amount of water movement in the xylem, but it can account for some of the vertical movement of water in small statured plants. When considering the microscopic structure of vessel elements it is possible that capillarity may have a greater influence on water movement in the vascular tissues than is suggested by the analysis above. Openings in the xylem vessel and tracheid cell walls (bordered pits, scalariforn, etc.) lead to extensive interconnections between vessel elements. This may result in a network of tiny capillaries, which could act as a sponge to wick water up the vascular tissue and thus cause water to rise to the height of most trees. However, it is unlikely that this uptake mechanism could deliver a significant proportion of water transpired by tree canopies.

C. Heat Properties

The facility for hydrogen bonding among water molecules also results in unique heat transfer properties in comparison to other liquids. For example, water has one of the high-

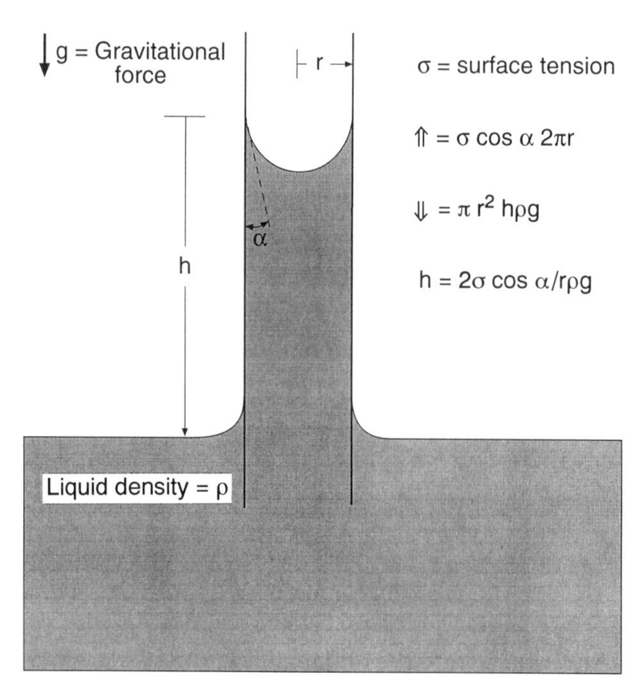

Figure 7.2 Diagram of the upward and downward forces that affect the height of water movement in a capillary tube. α represents the contact angle between water and the bore of the capillary. *h*, height of the water column, *r*, radius of the capillary inner bore; ρ, density of the water solution; *g*, gravitational force. (From Nobel 1991.)

est energy requirements for vaporization among liquids at atmospheric temperature and pressure (heat of vaporization, H_{vap} = 2.257 MJ kg^{-1} or 44 kJ mol^{-1} at 25°C). The high H_{vap} value of water results in a large transfer of heat from a leaf when transpiration is occurring, which results in a potentially effective surface cooling mechanism.

Latent heat transfer is the movement of heat between surfaces and water phases during evaporation or condensation. Since water requires a large amount of heat to break the hydrogen bonds holding the water molecules together (heat of vaporization), that heat is derived from the surface upon which the water is evaporating. Thus, the process of evaporation cools the wet surface and transfers heat energy to the water vapor. Water evaporates from cell surfaces inside the leaf and diffuses out the stomata. The evaporation causes heat to be transferred from the cell surfaces to the water vapor. Therefore, latent heat exchange by transpiration is an important cooling mechanism for leaves. Leaves can also gain heat by the same mechanism (but in reverse) when water condenses on the leaf surface. The effectiveness and significance of latent heat exchange to leaves is discussed further in Chapter 11.

In addition to the high heat of vaporization, water has relatively high phase transition temperatures. To illustrate this concept, phase transitions for molecules with 10 protons and 10 electrons can be compared. Water has the highest melting and freezing point (Figure 7.3). In fact, it is the only molecule represented that can be in a liquid, solid, and gaseous form at normal planetary temperatures. This physical characteristic of water makes this molecule particularly suitable as the matrix for life.

Water requires a large amount of heat to be warmed. Specific heat is defined as the

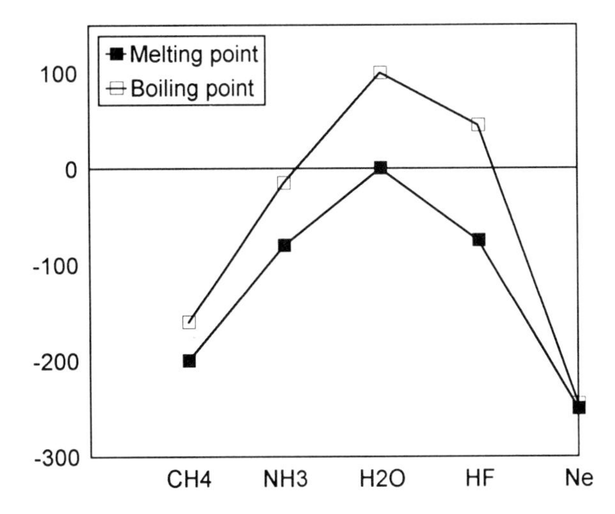

Figure 7.3 Melting and boiling points of molecules with 10 protons. (From Nobel 1991.)

amount of energy required to increase the temperature of 1 kilogram of material 1°C at standard temperature and pressure. The specific heat of water ($H_{sp} = 4182$ J kg^{-1} K^{-1}) is large compared to many other liquids. As a result of water's large specific heat, the mass of water in plants can absorb or lose a relatively large amount of heat without changing the temperature of the tissues. Thus water serves as an excellent thermal buffer for plants. This concept has been used frequently in the agricultural trade to limit the influence of short-term freezes on crops. A water spray is applied to crops as a buffer between air temperature and tissue temperature.

D. Dielectric Constant

The dielectric nature of a liquid refers to the attractiveness of cations and anions to each other when dissolved in liquid. If the dielectric constant is high, dissolved ions tend to stay separated. This relationship is defined by *Coulomb's law:*

$$\text{electrical force} = \frac{Q_1 Q_2}{4\pi E_0 D r^2} \tag{7.1}$$

where Q_1 and Q_2 are the charges of two ions, E_0 is a proportionality constant, D is the dielectric constant, and r is the distance between ions. Water has an extremely high dielectric constant ($D = 80.2$ at 20°C) compared to other liquids (e.g., hexane: $D = 1.87$ at 20°C). For example, table salt (NaCl) dissociates in water to Na^+ and Cl^-, and the attraction of Na^+ for Cl^- in water is 43 times less than in hexane. As a result, there will be 43 times more free ions in solution when equivalent normalities of table salt are added to water compared with hexane. This high dielectric property of water makes water an excellent solvent for polar substances and allows for abundant free ions in the cell solution.

E. Chemical Properties

The chemical potential of any compound is a thermodynamic description of its *Gibbs free energy,* the amount of energy available to do work. Studies of free energy normally relate the changes in free energy within a system, going through a transition from one state to another or one location to another. Therefore, *work* in a biological system refers to such activities as creating or breaking molecular bonds, cellular and organelle movements, moving ions through cellular components, or moving water up the xylem. The change in free energy that occurs when a system changes states or locations represents the amount of work done during the change. Thus, analyses of the free energy of water in plants will relate to such activities as the movement of water from soil to leaf, movement of water between the cell wall and the cytosol, changes in ion concentrations in the cytoplasm or vacuolar water, or any other change in the status or location of water in plants. The free energy of a system is the sum total of the free energy of all components in the system and is termed the *chemical potential* of the system. Chemical potential of an aqueous solution determines its (1) equilibrium state and (2) direction of transport. Water will always move from a region of high to low chemical potential as long as its movement is not prevented by impervious barriers. Movement will continue until chemical equilibrium is attained. In this chapter we restrict our discussion of chemical potential to an aqueous solution in the soil, plant, or atmosphere.

The chemical potential of an aqueous solution can be closely approximated by

$$\mu = \mu^* + \frac{RT\ \ln(a_w)}{V_w} + PV_w + Z_wFE + Mgh \tag{7.2}$$

where μ is the total chemical potential, μ^* a reference potential, R the universal gas constant, a_w the chemical activity of water, V_w the partial molar volume of water, P the pressure on the system, Z_w the charge number of water, F the Faraday constant, E the electrical potential, M the mass of water per mole, g the gravitational force, and h the vertical height of the solution. The reference potential (μ^*) is included in equation (7.2) because potential energy is a relative value. The reference potential contains an unknown constant and is not determinable. Therefore, it is a theoretical value. In most studies of plant–water relations the chemical potentials of two states are being compared, both of which have the same reference potential value. Thus the reference potential cancels out in these comparisons. In physiological studies of plant water relations, μ^* is considered to be the standard state, which is defined as the chemical potential of pure water at the same temperature and pressure as the solution.

The second component of the chemical potential equation [$RT\ \ln(a^w)/V_w$] represents the component of solution potential influenced by the activity of water. The activity of a substance *(a)* can be considered an adjusted concentration and is related to concentration *(c)* by an activity coefficient ($a = \gamma c$). The addition of any solutes (osmotically active molecules) to the water solution will decrease the activity of water. The solutes decrease the partial molar volume of water (V_w), displace the water molecules from each other, and decrease the kinetic energy of the water molecules that adhere to the solutes. The increase in solutes is represented as the mole fraction of solutes (Π). Since any increase in dissolved ions decreases the activity of water, an increase in Π decreases the osmotic potential of the solution. It is more convenient to express the osmotic potential of plants in terms of solute concentration rather than in terms of water's chemical activity. This convenient simplification can be used because of the direct relationship between osmotic potential and water's chemical activity [$RT\ \ln(a_w) = -V_w\ \Pi$]. The *van't Hoff approximation*

$$\Pi_s = RT\ c_s \qquad \text{MPa} \tag{7.3}$$

can be used to define the influence of solutes on the activity of water for dilute ideal solutions and ideal solutes (i.e., $\gamma_w = 1$), where c_s is the concentration of all solutes and Π_s is the osmotic potential of the aqueous solution.

The van't Hoff approximation will not accurately define the activity of water when colloidal or mucous materials are present, as these affect the activity coefficient of water. Therefore, when the water content of cells becomes low, or mucus levels are high (such as in cacti) the activity of water is lower than that predicted by the van't Hoff relationship. Extractions from plant tissues commonly have a solute concentration of 0.5 osmol kg^{-1} of extract (about 500 mol of osmotically active ions m^{-3}), which corresponds to an osmotic pressure of 1.22 MPa [0.002437 m^3 MPa mol^{-1})(500 mol m^{-3})], while that of seawater is 2.5 MPa.

The chemical potential of water is also influenced by the pressure (P) on the system. The free energy associated with pressure is the product of the partial molar volume (V_w) and pressure [see equation (7.2)]. The partial molar volume of a substance is the change in volume resulting from a given change in the amount (normally 1 mol) of a compound. This is normally equal to the volume of a mole of the substance, but when mixtures of different compounds are involved, the partial molar volume can differ from the volume for a mole of a pure substance. The partial molar volume of water is 18.05 cm^3 mol^{-1} at 20°C. Since pressure can be referred to as energy per unit volume (J m^{-3}), the partial molar volume has the units of m^3 mol^{-1}. Any increase in pressure will result in an increase in the free energy of the system (J mol^{-1}).

The electrochemical potential of a compound [Z_wFE in equation (7.2)] is a product of the Faraday (F), the electrical potential (E), and the charge number of the material (Z_w). This aspect of free energy is very important in discussions of ions and electrochemical potential across membranes because it influences the energetics regulating passive transport of ions across membranes. However, water has a charge number of zero, so this aspect of chemical activity of water is not significant to the movement of water or its thermodynamics.

The chemical potential of water is also influenced by gravitational forces. The impact of gravitational factors on the chemical activity of water will depend on the height (h) of the water above the ground, the acceleration of gravity (g), and the mass of water per mole (M_w). The higher the water column, the greater the influence of gravity. Since gravitational acceleration (about 9.8 m s^{-2}) and the mass per mole of water (0.018016 kg mol^{-1}) are constants, the effect of gravity on the chemical potential of water will be directly related to height.

The discussion of chemical potential of a material above assumes constant temperature. Since temperature influences the kinetic energy of water, and free energy is composed partly of kinetic energy, the Gibbs free energy of a solution will decrease with a decrease in solution temperature. Therefore, the tendency of water to move from one compartment to another may be affected by the temperature of the compartments. The impact of temperature on the free energy of water is particularly important to plants in environments where soil temperature lags behind leaf temperature. Significant limitation to water uptake by plants from soil can result because low soil temperature reduces the kinetic energy of water and makes it less likely to diffuse into the warmer tissues of the plants.

A convenient description of the free energy of water (μ_w) is commonly used to describe water relations in plants:

$$\mu_w = \mu_w^* - V_w\Pi_s + V_wP + m_wgh \qquad \text{MPa} \tag{7.4}$$

This relationship does not include the electrochemical potential [equation (7.2)] because the charge number of water is zero. Also, osmotic potential is used to define the chemical activity of water [see equation (7.3)]. Slatyer (1967) formally defined the difference between the chemical activity of water and the standard state, divided by the partial mole fraction of water, as the water potential:

$$\Psi = \frac{\mu_w - \mu_w^*}{V_w} = P - \Pi_s + \rho_w gh \qquad \text{MPa} \tag{7.5}$$

where Ψ is the water potential and ρ_w is the density of water (m_w/V_w); the other components were defined earlier.

The water potential equation is useful for describing the relative movement of water and the relative importances of osmotic, pressure, and gravitational factors on tissue water potential. This term can be compared to the amount of work that would be required to move water from a solution state to that of pure water. Since the reference is for pure water, pure water has a Ψ value of zero. The water potential of any solution containing solutes will have a negative water potential. The units for water potential are pressure units (energy/area), of which several are in common usage [atmospheres (atm), bars (bar), pascal (Pa), megapascal (MPa)]. Here we will use only MPa to describe the water potential of plants.

The water potential of any tissue is composed of four components and can be defined by

$$\Psi = \Psi_t - \Psi_s - \Psi_g + \Psi_m \qquad \text{MPa} \tag{7.6}$$

The pressure potential [$\Psi_t = P$ in equation (7.5)] is a result of the intracellular hydrostatic pressure (turgor pressure), the osmotic potential [$\Psi_s = -RTc_s$ in equation (7.5)] is a result of solute concentration in the cell, the gravimetric potential [$\Psi_g = \rho_w gh$ in equation (7.5)] is a result of the height of the tissues, and the matric potential (Ψ_m) is a result of the interaction between water molecules and surfaces in the cell (adhesive forces). Pressure and osmotic potential are considered the most important component of cellular water potential because gravimetric potential is small, in most plants, and matric potential is important only in a few situations.

III. CELLULAR ASPECTS OF WATER POTENTIAL IN PLANTS

A. Apoplast Versus Symplast

Traditionally, plant physiologists have considered water contained in plants to be divided between two main compartments. The apoplastic water, originally defined by Münch in 1930, is all that water outside the plasma membranes. This includes all water in the cell wall, intercellular spaces, and xylem conduits. Symplastic water is that within the plasma membrane and includes all the water in the cytoplasm and the central vacuole. Although there is considerable variation in the fraction of total water in the symplast or the apoplast, commonly a majority (70%+) of the tissue water is in the symplast. The relative symplastic volume is different among organs (leaves, stems, roots, etc.) because of the ratio of vascular tissue (xylem and phloem) to parenchymatic or meristematic tissue. In leaves 75 to 90% of total water is in the symplast, while in stems the symplast may have less than 50%. The symplastic fraction also varies among species (Roberts et al. 1980, Robichaux et al. 1986) and among developmental stages (Nilsen et al. 1984).

Most of the symplastic water is contained in the vacuole (80 to 90%). The vacuolar water contains a number of solutes, such as inorganic ions, organic acids, and anthocyanins. Since little to no pressure difference occurs across the tonoplast of the vacuole, the sum of osmotic and matric potentials must be equal between the cell protoplast (the cytosol) and the vacuolar water. The cytosol contains large numbers of macromolecules and membrane-bound organelles as well as solutes; thus the protoplasm must have a matric potential component. In contrast, vacuolar water has few to no large macromolecules or internal surfaces; consequently, vacuolar matric potential is negligible. Thus, the osmotic potential of the vacuole must be lower than that of the protoplasm. In some unusual cases (e.g., in cacti) the central vacuole may also have mucilage, which adds a matric potential component to the vacuolar water potential and can cause extracted vacuolar water to have twice the viscosity of pure water. It is important to understand that the osmotic potential of the central vacuole causes the development of positive hydrostatic pressure in the cell.

Water moves freely between the symplastic and apoplastic compartments, depending on the water potential of the two fractions. In fact, due to the presence of the casparian strip in young rootlets, water must make the transition through the symplast before traveling through the apoplast to the rest of the plant. Since apoplastic water is located outside the plasma membrane, cellular hydrostatic pressure does not influence apoplastic water potential in a positive manner. Apoplastic water potential is dominated by the negative influence of its solutes and the negative influence of tension developed in both the cell walls (due to a high Ψ_m) of the vascular tissue and the forces of transpiration (discussed later).

Apoplastic water has traditionally been thought of as a dilute solution freely moving in cells walls, between cells, and in the xylem. However, recent studies have elucidated some important physiochemical properties of apoplastic solutes that affect our understanding of water dynamics in plants (Canny 1995). The flow in the xylem conduits is leaky and its magnitude depends on conduit diameter. The water that exits the xylem penetrates the symplast faster than do some solutes. This results in some high solute concentrations at specific locations in the apoplast called *sumps*. The solutes diffuse away from the sumps at rates that are much slower than the rate of solute diffusion in water. Therefore, there are physical or chemical forces acting on the apoplast to retain solutes in sumps. Ion concentrations in the small vessels in leaves or roots can often be high (up to 200 m*M*). All this information clearly shows that the apoplastic water is not a uniform dilute solution, and the variation in apoplast solute concentration may have a significant impact on water movement throughout plants.

B. Osmotic Potential

As defined earlier, osmotic potential is based on the concentration of solutes in water. Extracts of water from xylem contain little solute content and do not account for a significant portion of tissue osmotic potential. However, intracellular water can contain large quantities of solutes, creating an osmotic potential at the turgor loss point as low as −5.0 MPa (twice that of seawater) in some cases (Meinzer et al. 1986).

There are four main classes of osmotically active solutes that can significantly affect tissue osmotic potential: inorganic ions, carbohydrates, nonprotein amino acids, and organic acids. Two of the four classes, nonprotein amino acids and carbohydrates, are compatible with the protoplasm. The other two can reach high concentrations only in the vacuole. High concentrations of inorganic ions in the protoplasm can cause protein dysfunction and otherwise interfere with metabolic processes. High concentrations of organic acids will alter the pH of the cytosol and negatively affect cellular metabolism.

The most abundant inorganic ion associated with changes in osmotic potential is potassium. Also, in some halophytes, high sodium concentration can result in the development of low osmotic potential. If inorganic ion concentrations constitute a high proportion of molecules affecting low osmotic potential, they must be isolated in the vacuole. The influences of cytosolic-compatible solutes and matric surfaces in the cytosol on cytosolic water potential must equal the influence of vacuolar inorganic solutes on the water potential of the vacuole. Thus, when vacuolar inorganic solutes increase so must cytosolic osmotic moieties (assuming no change in cytosolic matric potential). Several mono- and disaccharides constitute a large proportion of osmotically active moieties in the cytosol. These include sucrose, glucose, fructose, and raffinose. The most common nonprotein amino acids associated with osmotic activity in the cytosol are glycine-betaine (NNN-trimethyl glycine), and proline. Organic acids such as malic or citric acid often reach high concentrations in the vacuole, particularly in CAM plants.

C. Pressure Potential

The pressure component of cellular water potential is referred to as the *turgor potential* (Ψ_t). This is a result of hydrostatic pressure in cells which occurs when cell pressure balances the difference in water potential between the environment around cells and the cytoplasm. Turgor potential in cells is normally between 0 and 1.0 MPa. Negative turgor potential $\Psi_t < 0$) is unusual except in cell walls, where large matric forces can cause a strong negative tension on water in the wall. Normally, if the water potential of the environment around cells causes water to move out of the cells so that turgor potential temporarily dips below zero, the cell shrinks away from the wall, goes into a plasmolyzed state, and turgor potential remains at zero. Plasmolysis is generally detrimental to cells because cell membranes can tear at plasmodesmata during plasmolysis.

The magnitude of turgor potential in a cell is dependent on the elasticity of the cell wall (E). Cells that have elastic walls (a low modulus of elasticity) are able to maintain positive turgor potential over a wide range of cell water potential. Conversely, those cells with low E will have a greater loss of turgor potential as water potential decreases.

D. Gravimetric Potential

Gravimetric potential (Ψ_g) increases 0.1 MPa for each 10-m increase in plant height. Therefore, the maximum gravimetric potential in plants (30-m-tall tree) could be approximately 0.3 MPa. Commonly, leaf water potential ranges from -0.5 to -5.0 MPa, making gravimetric potential a small portion of total water potential. Furthermore, in many considerations of water movement, such as that across the plasma membrane, Ψ_g can be ignored because there is little change in height during the transition. However, gravimetric potential should be considered for analyses of the difference in water potential between roots and apecies of trees.

E. Matric Potential

One component of water potential that is not usually included in descriptions of plant water potential is matric potential (Ψ_m). Normally, this component is not significant for cells at high tissue hydration. However, if tissue hydration is very low (about 60% of maximum) or there are large quantities of mucilage (polysaccharides) or colloids in the tissues (Nobel 1991, Nobel et al. 1992), matric potential must be considered in order to balance the water potential equation. Matric potential is actually a special form of osmotic potential. Instead of relatively small ions regulating osmotic potential, large polar macromolecules or polar surfaces composed of such molecules affect matric potential. Large polar molecules or surfaces will interact with the polar nature of water, forming hydrogen bonds with water molecules, thereby reducing the kinetic energy of water (Passioura 1980). This interaction results in a layer of water molecules surrounding macromolecules (sheath of hydration) or covering surfaces that has a lower kinetic energy than the surrounding free water molecules. Interactions between polar macromolecules or polar surfaces with water cause a change in the activity coefficient of water (γ_w) which affects the chemical activity of water molecules that comprise the sheath of hydration. Normally, matric potential is not used in analyses of intracellular water, but this water potential component may become important in considerations of soil water, water in the cell wall, and in cells with high concentrations of mucilage because of the high proportion of surface area to volume in these systems.

IV. HYDRODYNAMICS OF THE PLANT SYSTEM

A. Soil–Plant–Air Continuum

Plant hydrodynamics is the consideration of all aspects of water movement from the soil to the atmosphere through the plant system. This topic includes the flow dynamics, transfers between segments of the continuum, resistances met along the way, and temporary storage and delivery characteristics of water in plants. Similar to a plumbing system, the movement of water from the soil to the atmosphere occurs in a continuous column of water from the soil solution to the atmosphere through the xylem (Figure 7.4). Any break in the water stream prevents water movement through the xylem. The column of water in plant vascular tissue is maintained by the cohesive and adhesive properties of water molecules. The main driving force for water movement through the plant is a gradient of free energy for water in the soil solution to water vapor in the atmosphere (Passioura, 1982).

B. Root Accumulation of Water

Most of the water accumulated by plants travels from the soil solution into the root system. There are some cases in which an alternative water absorption technique occurs. For example, ground-dwelling bromeliads (*Tillandsia*) accumulate most of their required water from fog through leaf cuticles (Rundel 1982), and some shrubs can use an osmotic mechanism to accumulate water from nearly saturated air (Mooney et al. 1980). However, these alternative routes are rare and are found only in species inhabiting extreme climates.

It was originally postulated that the majority of soil water enters root systems through root hairs or fine roots because they lack suberized coatings and have a large surface area (Dittmer 1937). However, several studies have shown that suberized roots can be the sites

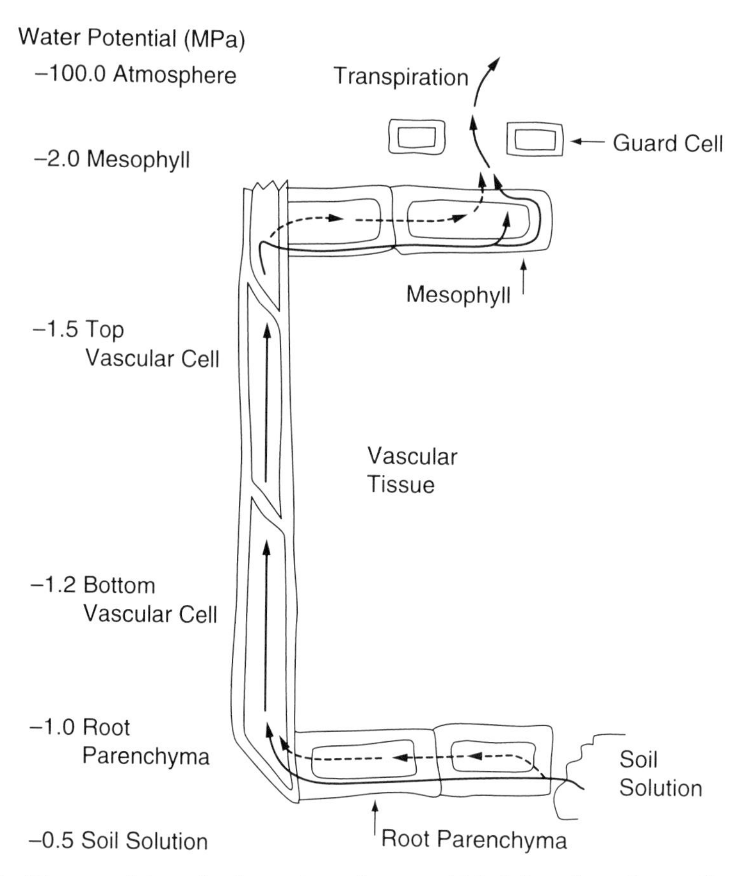

Figure 7.4 Diagram of the soil–plant–air continuum, which defines the pathway of water movement in plants. The pathway indicated by a solid line represents a purely apoplastic rout. The dashed lines represents water movement through the symplast of the root and leaf parenchyma cells.

of most water accumulation. Plants with only suberized roots can accumulate enough water for normal growth (Nightingale 1935), and if nonsuberized roots are removed, accumulation of water continues at a normal rate (Chung and Kramer 1975). Water diffuses into root cells because root cells have a more negative water potential than the soil solution. Therefore, any factors that reduce the water potential of the soil solution (high soil matric interactions, high soil salinity, frozen soil solution) will inhibit the movement of water into roots. When roots are not in contact with liquid water (no capillary water in the soil), they may be surrounded by an atmosphere with saturated vapor pressure (due to the presence of hydrostatically held water in the soil). Under this condition, no water will move into roots because the water potential of saturated air is far below that of the liquid water in root cells.

The pathway by which water moves from the epidermal root cells to the stele at the center of the root is complex (Figure 7.5). An apoplastic route traverses the walls of the cortical cells. A symplastic route involves water movement through the cytosol from one cell to the next. There are many more resistance barriers along the symplastic pathway (such as plasma membranes) than the apoplastic route. A third possibility is a vacuolar pathway in which water passes into the vacuole of the epidermal cells, then travels

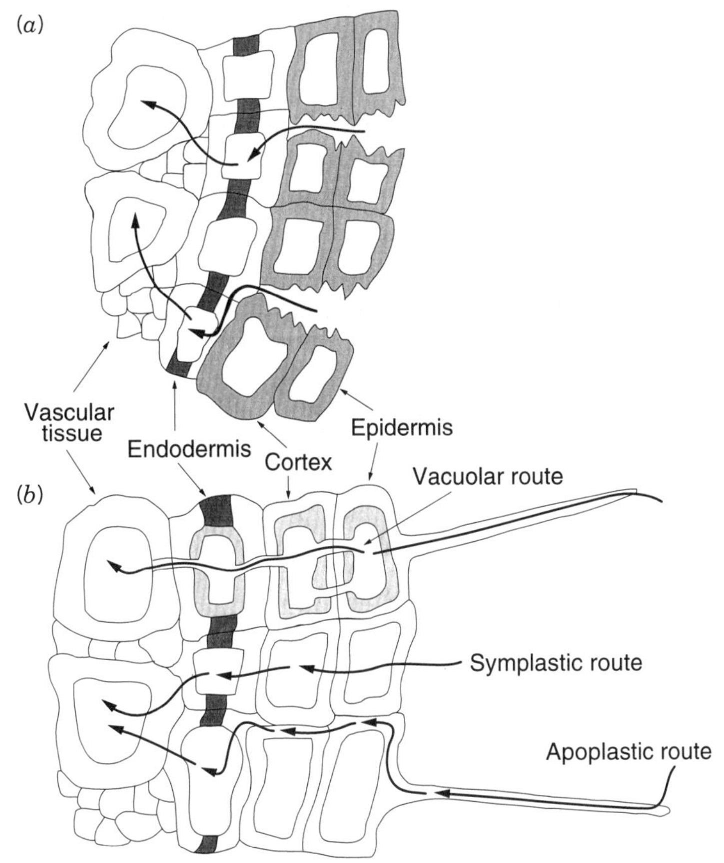

Figure 7.5 Several possible routes for water to traverse from the soil solution to the vascular system in plant roots. (A) Water moves into a suberized root through fissures in the root cortex. (B) Water moves into a nonsuberized root in the root hair zone by vacuolar, symplastic, and apoplastic routes.

through plasmodesmata from one cell vacuole to another. Water probably travels by all three routes on its way to the casparian strip. However, the apoplastic pathway is thought to be the main pathway for transporting water into plants because this route has the least resistance among the three possible routes.

In the endodermis tissue, just inside the root cortex, is a continuous section of cell wall impregnated with suberin. This suberin layer is hydrophobic and effectively diverts water moving through the cortical apoplast into the symplast. Thus, water must enter the cytosol at the endodermis before entering the water transport tissues inside the stele. However, in older roots that have cracks extending past the endodermis, water can enter the appoplast and travel to the stele without passing thorough the symplast.

C. Alternatives to the Soil–Plant–Air Continuum Model

The soil–plant–air continuum (SPAC) model is presently the prevailing explanation for water movement in plant vascular tissue. However, there are several bits of information that disagree with the SPAC model. For example, the SPAC model depends on a difference in tension (negative hydrostatic pressure) between the xylem at the top and base of trees. When xylem tension is measured with a sensitive pressure sensor (the pressure probe), no difference between xylem tension at the top and bottom of trees is found. If there is no difference in xylem pressure along the length of the vascular tissue, and water still travels from higher to lower water potential, differences in solute potential between the leaf and root may be important.

Three other possible processes may influence water movement in plant vascular systems (Zimmerman et al. 1993). Earlier in the chapter, capillary forces were discussed. Capillary forces in the lumen of vessel elements could not create enough lift force to move adequate amounts of water up tall trees. However, the hydrated microcapillaries of vessel element cell walls can develop forces capable of lifting water 100 m. The flow rates in microcapillaries would be very small, preventing this mechanism from moving enough water to account for transpiration. Although microcapillary water may not serve as the rout of water flow during maximum transpiration, it may serve as a water resource during times when flow is limited in vessel elements. Essentially, microcapillary water could refill vessel elements that have low water content or are air filled due to excessive drought.

Solute concentration may also have an important effect on water flow in plants. Traditionally, xylem fluid has been considered to be a dilute solution. It is possible that xylem sap has been thought to be dilute because solutes are rapidly redistributed into live cells of the wood following cutting. In addition, we have noted earlier that sumps with high solute concentration exist in certain locations of the vascular tissue. Water movement could be propagated up a long vascular system by short segments of xylem that contain axial osmotic gradients. Under such a scenario water would move through the vascular tissue by traveling progressively along steps that each have an osmotic gradient. For the osmotic gradients to be maintained, an axial barrier to solute movement must occur with each step, and vessel elements must be sheathed by a solute barrier. Such barriers have been identified in only a few species (Zimmerman et al. 1993).

Several lines of evidence favor the concept of osmotically induced water movement in the xylem. First, it is hard to understand root pressure extending into stems without the existence of solute barrier sheaths. Second, this mechanism would explain the inability to find a gradient in hydrostatic tension along plant shoots. Third, this model would explain why vascular flow is easily interrupted by any slight bending of branches.

It is also possible that interfacial forces between water and the vessel element structure influences water movement in the xylem. Small air bubbles have been found attached to the walls of vessel elements (called *Jamin's chains*). Although this was reported initially as a natural aspect of vascular tissue, many scientist believed it was an artifact of severing the transpiration stream. Recently, the pressure probe has proven that air bubbles are present on the inner walls of vessel elements. If we assume that the inner walls of the lignified vessels are covered with tiny bubbles, this raises the possibility of water movement by *Marangoni convection*. Marangoni convection is gravity independent and occurs because of a gradient in interfacial tension along a water–gas interface. The flow of water molecules along the interface between vascular fluid and air bubble will cause a reverse direction flow in the lumen of the vascular element. If there are a large enough number of air bubbles, lumen fluid flow can be substantial.

It is likely that some combination of SPAC, osmotic, capillary, and Marangoni convection processes regulate the flow of water in xylem vessels. It is no longer appropriate to predict water flow solely on the basis of the SPAC model. Consideration of alternative models for water flow in plants may enable a greater understanding of the many seemingly anomalous measurements of plant water dynamics.

D. Hydraulic Conductance

Once water has entered into the xylem tissue, the gradient of water potential between the upper leaf surfaces and root will create tension (negative pressure) in the water column. Consequently, water will be pulled up the xylem conduits by some combination of the processes discussed above. Under some conditions, positive pressure can develop in the root vascular system. If stomata are closed and the soil solution has a high water potential (close to zero), the lower osmotic potential (due to a higher concentration of solutes) of the root tissue can cause movement of water into the root system from the soil, causing positive root pressure. This potential to push water up the stem occurs only under particular conditions and cannot account for the bulk of water movement in plants.

On the way through the root, stem, and leaf vascular system, water movement is impeded by various resistances. The relationship between flow of water and resistance in a tube can be estimated by the *Hagen-Poiseuille* (Nobel 1991) *relationship,* as modified by Wiederman:

$$\text{flux} = \frac{r^2}{8n}\frac{dP}{dx} \tag{7.7}$$

where the volume flux rate of water in a tube is determined by the radius (r) of the inner lumen of the tube, the viscosity (n) of water (a constant in xylem tissues), and the pressure gradient ($-dP/dx$) from one end of the tube to the other. Therefore, the volume flux rate of water in the xylem will increase with increasing radius of the xylem and with increasing gradient in water potential along the xylem conduits. The flux rate across a specific cross-sectional area defines the total flux of water through the tube. The total cross-sectional area of transport tissues in the roots, stems, and leaves are different, but the total volume of water flux rate is very similar. Therefore, the total flux of water in the xylem is a constant among transporting organs when the plant is at steady state. However, the particular water flux through any one section of the transport system will not be the same among all parts of the transport organs.

The hydraulic conductance (K) of a xylem section is defined as the ratio of water flux rate (J) to water potential gradient ($\Delta\Psi$) across the long direction of the xylem section:

$$K = \frac{J}{\Delta\Psi} \tag{7.8}$$

Therefore, increasing hydraulic conductance can occur either by decreasing the water potential gradient or by increasing the flux rate of water in the xylem.

The volumetric water movement in soil can also be described in relation to a gradient in hydrostatic pressure, in a similar manner:

$$\text{volumetric flux} = \text{flux}^{\text{soil}} = -K^{\text{soil}}\Delta\Psi^{\text{soil}} \tag{7.9}$$

where K^{soil} is the hydraulic conductance of the soil.

The hydraulic conductance in soil spaces is less than that in xylem-conducting elements because interstitial soil spaces do not approximate cylinders. Thus, factors that regulate the resistance to flow in soil spaces are more complex than those determining the resistance to flow of water in xylem conduits.

Resistance to water movement in the xylem depends on the cross-sectional diameter of the vascular system, the individual diameter of trachieds and vessel elements, the presence of embolisms (air pockets in xylem cells), and the nature of interconnections between the xylem cells. Water meets resistance as it moves through roots, stems, and leaves on its way to the atmosphere. In general, root resistance is larger than that of either other organ and has been estimated to account for 50 to 70% of the total plant resistance to water flow (Neuman et al. 1974).

E. Capacitance

Each organ of the vascular system has a capacitance which regulates the relationship between water that is stored and water that moves through the xylem. In plant physiological systems, capacitance is defined as the change in tissue water content for any change in water potential (units = m^3 MPa^{-1}). Therefore, tissues with a high capacitance deliver large quantities of water for any small change in water potential. It must be noted that this is not a consideration of the water potential gradient across the tissue but a temporal change in tissue water potential. The water involved in capacitance of a tissue may be located in the symplastic and/or the apoplastic fraction. Commonly, plants with high capacitance store large amounts of water in the symplast of parenchyma-dominated tissues.

Capacitance and resistance concepts can be combined into a time function. That time function describes how fast the water potential will change within a tissue as water is transferred from capacitance tissue to the xylem stream:

$$\frac{d\Psi}{dt} = RC \qquad \text{s} \tag{7.10}$$

where $\Delta\Psi/dt$ is the rate of change in water potential, R is the resistance to water movement in the tissue, and C is the capacitance of the tissue. Plants that have large stem capacitance, such as succulent trees (Nilsen et al. 1990), will have a longer delay between initiating transpiration (water vapor flow out the leaf) and decreasing water potential than that for plants with small capacitance.

F. Cavitation

The column of water moving from the root cell to the substomatal chamber in the leaf must be continuous. The water in the xylem is metastable, because it is under negative tension, which means that the xylem water is likely to change phases when under high tension. Therefore, under particularly low xylem water potential, small gas bubbles can arise in the xylem stream. If these small gas bubbles coalesce into a larger bubble, the bubble can block the flow of water in the xylem conduit (Figure 7.6). In fact, up to 50% of the transpiring area of a branch segment can be blocked by embolisms during the summer (Clark and Gibbs 1957, Tyree and Sperry 1989). However, even under such high embolism rates, adequate transpiration may occur. During extensive cavitation the xylem stream moves into a higher proportion of trachieds compared to vessel elements because

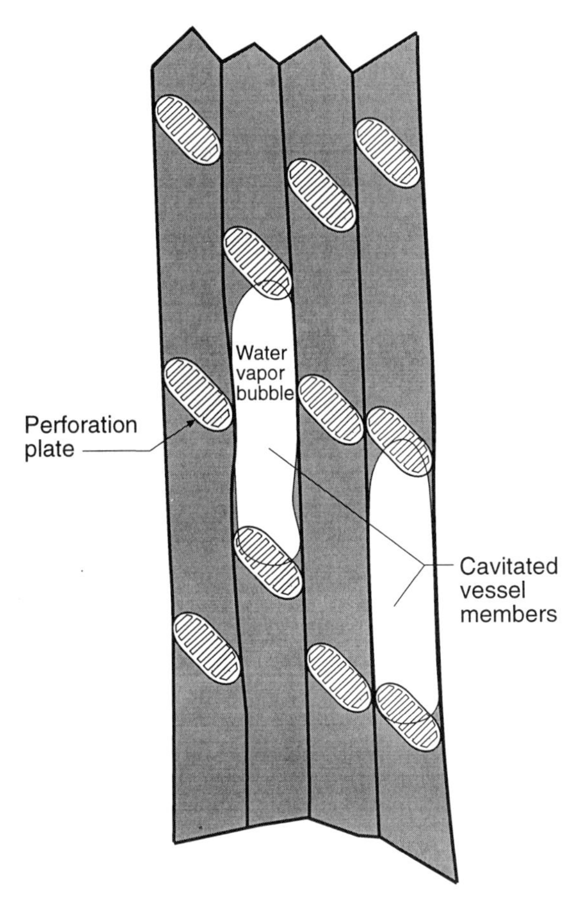

Figure 7.6 Diagram of the consequence of xylem embolism to water flow in the vascular tissue. Water flow is blocked in those cells that are embolized (cavitated vessel members), but water can still flow through surrounding noncavitated cells.

embolisms occur first in vessel elements and embolisms are restricted to the cell in which they occur. Consequently, cavitation in vessel elements causes increased resistance to water flow as water preferentially flows in trachieds (trachieds have a smaller vessel diameter than vessels) rather than vessels.

Water-stress-induced cavitation is thought to occur by an air seeding mechanism. Bubbles of air are sucked into a vessel element through intervessel or intertracheid pits from adjacent air-filled zones. The higher the tension in the vessel element, the more likely bubbles will pass through the vessel pits. The likelihood of tension-induced cavitation is also dependent on the permeability of the interconduit pits to the air–water interface. The introduction of small bubbles into the conduit changes the cohesion of water in the vessel, causing an embolism. This mechanism infers that the diameter of vessel elements does not affect the vulnerability to tension-induced cavitation (Sperry 1986, Tyree and Sperry 1989, Sperry and Tyree 1988).

Embolisms in the xylem stream can also occur during periods of freezing temperature (Sperry and Sullivan 1992). This is because the solubility of gases such as O_2, N_2, and CO_2 decreases in the xylem stream at low temperature. Therefore, gases come out of solution and form bubbles. Many of the gas bubbles may redissolve (if small enough) when the xylem temperature warms. However, if tension is placed on the xylem stream and bubbles reach a critical size, they will grow and form an embolism. The critical tension for bubble growth is related inversely to bubble radius:

$$\Psi_{crit} = \frac{2\sigma}{r} \qquad \text{MPa} \tag{7.11}$$

where Ψ_{crit} is the critical tension for bubble growth (MPa), σ the surface tension of water (7.28×10^{-8} MPa m^{-2} at 0°C), and r the bubble radius. Thus in freeze–thaw-induced cavitation, larger-diameter vessels will allow the growth of larger bubbles and suffer embolisms at lower tension than will smaller-diameter vessels (Sperry and Sullivan 1992).

G. Transpiration

The final step in the passage of water from the soil through the plant into the atmosphere is that from the leaf mesophyll cell wall to the turbulent air above the leaf. On their way from the leaf to the air, water molecules evaporate from the cell wall and diffuse through the substomatal chamber, the stomatal aperture, and the boundary layer above the leaf (Figure 7.7). To simplify the processes controlling diffusion of water from the leaf mesophyll to the atmosphere, we will assume that there are no temperature or pressure differ-

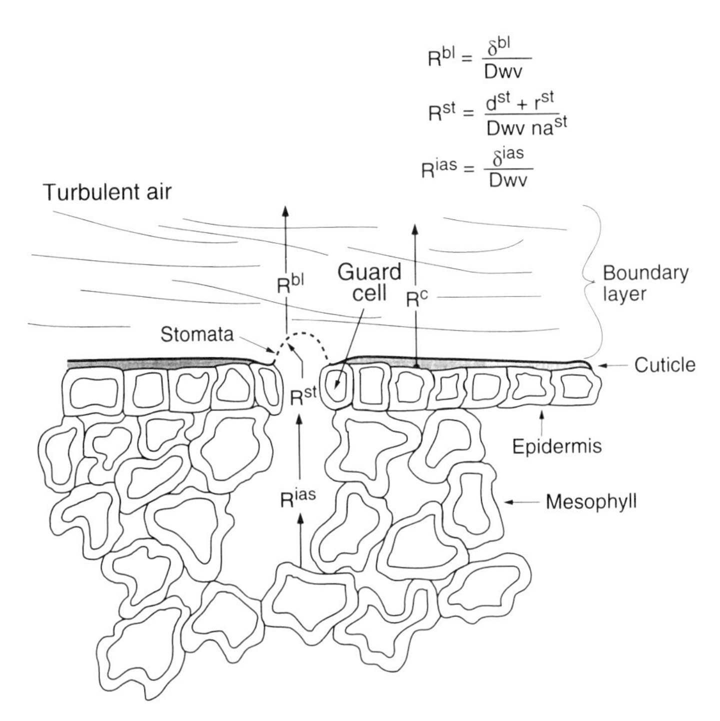

Figure 7.7 Diagrammatic representation of the resistance to water movement from the mesophyll cell wall to the turbulent air above the leaf. R^{bl}, R^{st}, R^{ias}, R^{c} represent resistance to the water vapor movement through the boundary layer, stomata, intercellular air spaces, and cuticle, respectively. δ^{bl} and δ^{ias} refer to the distance for water vapor to diffuse in the boundary layer and intercellular spaces, respectively. d^{st}, depth of the stomata; r^{st}, the radius of a dome of saturated air over the stomata; D_{wv}, the diffusion coefficient for water vapor in air; n, the number of stomata per leaf surface area; a^{st}, the area of the stomatal aperture.

ence along the various steps of the route that may alter the rate of diffusion (there will be a difference in temperature between the boundary layer and the turbulent air). The resistance to water diffusion across the substomatal space will be dependent on the linear distance from the cell wall to the base of the stomata:

$$R_{wv}^{ias} = \frac{\partial^{ias}}{D_{wv}} \qquad \mathrm{s\,m^{-2}} \tag{7.12}$$

where R_{wv}^{ias} is the resistance to water diffusion in the substomatal space, ∂^{ias} the linear distance from the cell wall to the base of the stomata, and D_{wv} the diffusion coefficient for water vapor in air.

The magnitude of substomatal resistance to vapor diffusion is singularly dependent on the size of the substomatal chamber. Plants with large, spacious, substomatal chambers will have the largest substomatal resistance to water vapor diffusion.

After the water molecules have traversed the substomatal chamber, they must travel through the stomata. The length of the route through the stomata depends on the depth of the guard cells (d^{st}) and the radius of the dome of saturated air above the stomata (r^{st}). The dome of saturated air above the stomata has a radius approximately equal to the radius of the stomatal aperture. Since the stomatal aperture is an ellipse, the r^{st} can be defined as $r^{st} = a^{st}/\pi$, where a^{st} is the area of a stomatal aperture. The resistance to water vapor is also dependent on the number of stomata per leaf area (n^{st}):

$$R_{wv}^{st} = \frac{d^{st} + \mathrm{r}^{st}}{D_{wv} n^{st} a^{st}} \qquad \mathrm{s\,m^{-2}} \tag{7.13}$$

Commonly, there are between 4000 and 100,000 stomata per cm^2 of leaf surface. The stomata are usually located on the lower leaf surface only (hypostomatous), rarely only on the top leaf surface (hyperstomatous), and occasionally on both leaf surfaces (ampistomatous). Resistance to water vapor movement can also vary considerably among positions on one leaf. Stomata can be open in local concentrations on the leaf (patches).

The depth of guard cells, radius of saturated air above the stomata, number of stomata, location of stomata, and patchiness are all morphological characteristics of a leaf established during development. The stomatal aperture and its patchiness is the only component of stomatal resistance that can be varied by physiological means after leaf maturation.

The final space for the water molecules to traverse on their route to the turbulent air is that between the stomatal dome of saturated air and the outer edge of the leaf boundary layer. As just described for substomatal resistance, the boundary layer resistance is singularly dependent on the linear distance that the water vapor must diffuse across. The resistance to water vapor diffusion across the boundary layer (R_{wv}^{bl}) will be dependent on the length (l) of the leaf (in the direction of prevailing wind) and wind velocity (v):

$$R_{wv}^{bl} = 4\left(\frac{l}{v}\right)^{0.5} \tag{7.14}$$

Larger leaves will have a larger boundary layer unless wind velocity is greater than 10.0 m/s. Above 10.0 m/s the boundary layer is near zero for normal-sized leaves. The amount of lobeing or splitting of leaf margins could affect boundary layer thickness, as could the cupping or rolling of leaf margins. Furthermore, extensive trichomes (leaf hairs) or leaf fluttering could influence the boundary layer thickness. Actually, the multiplier (4) in equation (7.14) is derived from many wind-tunnel experiments and is lower than that theorized for large planar surfaces (multiplier of 6). In fact, Gates (1980) suggested that 3.7 would be more appropriate for leaves. Also, the dependence upon $l^{0.5}$ is an approximation, because actual wind-tunnel determinations of this component range from $l^{0.4}$ to $l^{0.7}$. A similar approximation is applied to the velocity term. Thus equation (7.14) should be considered an approximation, and specific relationships may need to be developed for the specific leaf morphology of each species. Examples of the relationships between leaf shape, drag, and boundary layer thickness are given in Gates (1980), Nobel (1991), and Vogel (1981).

Total leaf resistance to vapor diffusion can be estimated as the sum of the substomatal, stomatal, and boundary layer resistances. Water can also evaporate from guard cell walls, as these have thin cuticle coverings. However, most of the epidermal cells are covered with a thick hydrophobic cuticle which creates a large resistance to water flow. Cuticular transpiration can account for a significant amount of total water loss when stomata are nearly closed or when the leaf has cracks or is damaged (Figure 7.7). Some pollutants, such as sulfur dioxide or ozone, may cause a reduction in cuticle thickness and decrease cuticular resistance. Since the resistance to water vapor diffusion out the stomata is much lower than cuticular resistance, the loss of vapor out the stomata dominates total water vapor diffusion from the leaf.

Modern considerations of vapor diffusion from leaves into the air (transpiration) utilize the concept of conductance rather than resistance. Conductance (g_{wv}) is the reciprocal of resistance and is proportional to the flux (J_{wv}). Conductance is preferred over resistance because conductance is more representative of how leaves are affecting water vapor diffusion, is easily expressed in molar units (mol m^{-2} s^{-1}), and takes into account mass flow of water vapor, which is not accounted for by traditional resistance calculations (Cowan 1977). See Pearcy et al. (1989) for a description of conductance calculations.

V. MEASUREMENT OF PLANT WATER RELATIONS

There are a diversity of techniques to measure various aspects of plant water relations. Some techniques are most appropriate for small tissue samples, others for field measurements of whole organs or plants, and others for greenhouse studies. Also, each technique has advantages for measuring particular aspects of plant hydrodynamics, or particular components of plant–water potential. In this section we review many, but not all, of the techniques employed to measure plant water relations. The discussion is separated into those techniques used to measure cellular water relations components and those used to measure hydrodynamic aspects of water flow. This division is used as a classification technique and is not meant to suggest that hydrodynamics and cellular water potential are unconnected. Quite the contrary, these two aspects of water relations in plants are highly linked into a dynamic whole. Although some general discussions of instrumentation will be covered, specific technical aspects of modern instrumentation can be obtained from other sources (see the Supplementary Reading).

A. Cellular Water Potential

1. Relative Water Content. Tissue water content can vary greatly among organs, developmental stages, seasons, habitats, and species. A simple calculation of water content, such as percentage of total weight, is not sensitive enough to evaluate cellular water relations characteristics. In some plants significant reduction in cellular water potential can occur with less than 1% decrease in the tissue water content, while in other plants a much larger amount of water loss is required to cause an equivalent effect on cellular water potential. However, a relative degree of tissue hydration can be calculated by comparing the current hydration of a tissue to its maximum potential hydration:

$$\text{relative water content} = \text{RWC} = \frac{\text{FW} - \text{DW}}{\text{TW} - \text{DW}} \tag{7.15}$$

where TW is the turgid weight (maximum potential hydration), DW is the dry weight, and FW is the current fresh weight of the tissue.

A RWC of 1 would represent tissues at their maximum hydration. The amount of water deficit (WD) or the amount of water less than full hydration can be calculated as 1 − RWC. The closer the RWC is to 1 (WD close to zero), the more favorable the hydration of the tissue. Fully hydrated weight is determined by placing the tissue in a saturation chamber until a constant weight is established. This technique can oversaturate the tissue more than would occur under maximum hydration in the field. Overhydration results in an underestimation of effective (in relation to full turgor pressure) RWC and can cause cell distention, which interferes with the analysis of cellular water potential components (Meinzer et al. 1986). Oversaturation during RWC determination is most likely to occur for tissues from plants of dry environments.

2. Incipient Plasmolysis. Cell water potential can be determined by a microscopic technique. Tissue sections are sliced and floated upon solutions in microscope slide wells with a coverslip. The wells are filled with salt solutions of known water potential. Following equilibration the cells are examined for plasmolysis (the pulling of the plasma membrane away from the cell wall). If plasmolysis has occurred, enough water has left the cell symplast to remove any hydrostatic pressure completely. Before the solution at which the first plasmolysis is observed, the penultimate salt solution is closest to the osmotic potential of the tissue. That is the water potential at which the bathing solution matches the cell and there is no turgor pressure.

The incipient plasmolysis technique is useful for determining osmotic potential only in small tissue sections. In some cases larger leaf sections can be used if the cytoplasm can be observed through the cuticle.

3. Solution Equilibration. A common technique for determining the water potential of tissues in physiological studies employs relative weight gain of samples from water uptake when placed in solutions of variable osmotic potential (Figure 7.8). Replicate samples (cores from tubers, leaf punches, leaflets, rootlets, etc.) from the tissues in question are weighed and placed into the salt solutions. The salt concentration is used to calculate the osmotic potential of the solution (van't Hoff relation). Since the salt solution has no hydrostatic pressure, the osmotic potential of the water solution equals the total water potential of the solution. After an adequate equilibration period, samples are removed from the solutions, patted dry, and reweighed. All those samples that gained weight had a water potential less than the solution, while those that lost weight had a water potential greater than the solution. The tissue water potential can be deter-

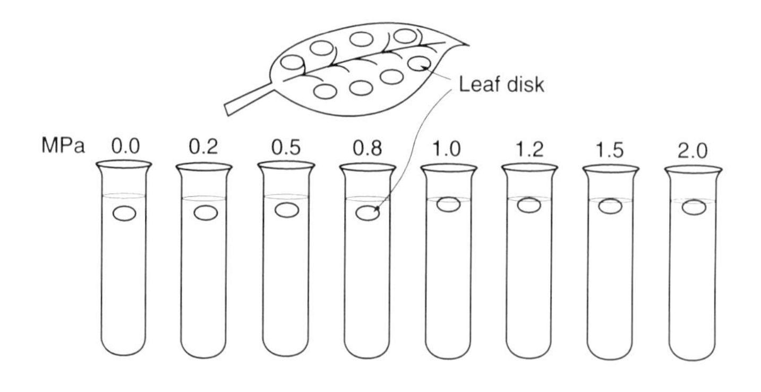

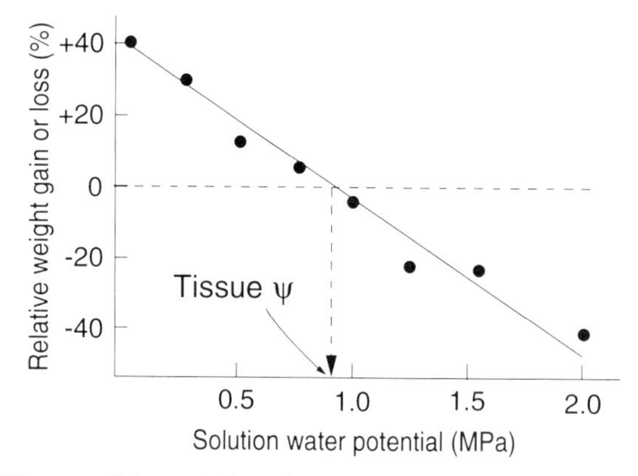

Figure 7.8 Diagram of the technique for determining water potential by the solution equilibration technique. If the tissue water potential is lower than the bathing solution, the tissue will gain weight. If the tissue water potential is similar to that of the bathing solution there will be little gain or loss of water.

mined by plotting the relative weight gain of samples at various solution water potentials. The point at which no weight is lost or gained is the tissue water potential (Figure 7.8).

A similar technique for determining tissue water potential is to suspend the tissue samples over solutions of known osmotic potential. The vapor pressure and water potential thereof above the solution can be calculated. The tissue samples will come to equilibrium with the water potential corresponding to the vapor pressure above the solution. This technique is also used to calibrate relative humidity sensors and thermocouple psychrometers.

Solution or vapor pressure equilibration requires the capability to collect multiple samples of each tissue. This requires a large tissue because the weight gain or loss may be determined inaccurately in small samples. In addition, the time required for equilibration may be up to several hours, depending on the nature of the samples. Furthermore, evaporation must be eliminated during sample handling because this will lower the tissue water potential.

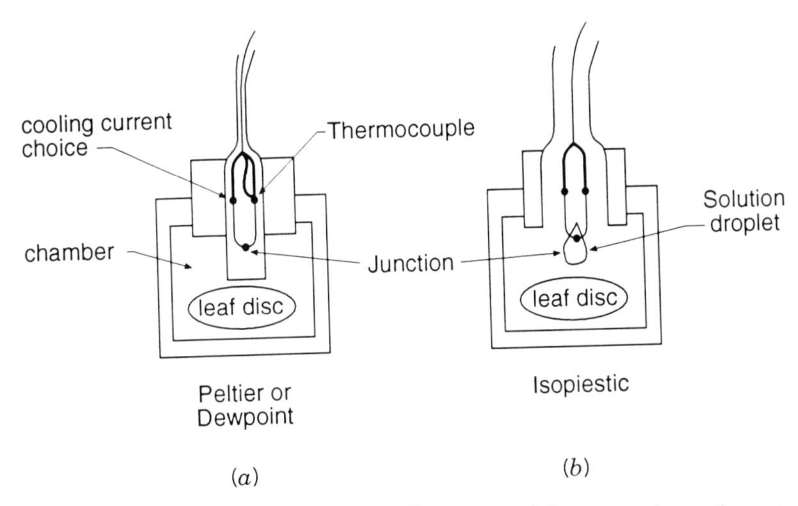

Figure 7.9 Diagrammatic representation of two types of thermocouple psychrometers.

4. Thermocouple Psychrometry. Thermocouple psychrometry is related to the vapor pressure equilibration technique. Small punches of leaf material (or similar small samples) are enclosed in small chambers (Figure 7.9). Water will evaporate from the sample until the vapor pressure of the chamber comes to a water potential equal to that of the tissue. Of course, the ratio of tissue volume to chamber space is critical to this determination. Since water is leaving the tissue into the chamber, the tissue water potential will decrease until equilibration occurs. An appropriate ratio of tissue to chamber size needs to be determined to minimize its effect on tissue water potential. Since the equilibrated atmosphere in the chamber represents the water potential of the sample, the sample water potential can be determined by measuring the vapor pressure of the chamber atmosphere and using that value to determine water potential

$$\Psi_{\text{total}} = \frac{RT(\ln e_{\text{chamber}})}{V_{\text{wv}}} \tag{7.16}$$

where e_{chamber} is the vapor pressure in the chamber following equilibration.

The vapor pressure in the chamber following equilibration can be determined by three methods. In one case a psychometric technique is used, which calculates relative humidity (RH) from the temperature depression that evaporation of pure water causes on a thermocouple. The *wet bulb depression* (List 1968) can be measured by bringing the thermocouple temperature below the dew point of the chamber atmosphere, which causes condensation on the thermocouple (the *Peltier effect*). The wet bulb depression can also be measured by placing a drop of pure water on the thermocouple just prior to measurement (Richards and Ogata 1958). The Peltier technique is favored because of the ease of operation. Relative humidity can then be used to calculate vapor pressure of the chamber by multiplying the saturated vapor pressure of the chamber (based on chamber temperature) by relative humidity (used in the decimal form).

In a second technique the dew point of the chamber atmosphere is measured instead of the wet bulb depression. In this case, termed *dew point hygrometry,* rapid sequential cooling currents are applied to the thermocouple junction to keep the junction at the dew point of the atmosphere. Dew point hygrometry has several advantages over psychometric techniques because the signal is larger and more constant than that from a thermocouple psychrometer. In some cases thermocouple psychrometer systems are capable of operating in the dew point mode.

In a third case, the *isopiestic point* is utilized to assay chamber vapor pressure. Two solutions of known Ψ are used to saturate the thermocouple junction after the sample has equilibrated with the chamber. The thermocouple signal is determined for each solution. Since the relation between Ψ and the thermocouple output is linear, a straight line can be drawn between these two points. Normally, the Ψ value of one solution is near that of the sample and the other is distilled water. The isopiestic point is that extrapolated point on the line where the thermocouple output is zero, which is the point of no net vapor movement between the solution bathing the thermocouple and the atmosphere in the chamber. Thus the isopiestic point represents the tissue water potential.

All techniques of thermocouple psychrometry must be carried out at constant temperature; therefore, the most dependable measurements of tissue water potential by thermocouple psychrometry are done in the laboratory. Some researchers have obtained constant conditions in the field by establishing a water bath in a cooler buried in the ground the day before sample collections, into which they place their thermocouple psychrometers. However, the thermocouples must have been calibrated at the same temperature as they are used in the field. This is because the calibration curves (relating Ψ to the thermocouple signal) developed for each thermocouple psychrometer individually are temperature dependent.

5. Pressure Chamber Technique. Probably the most elegant technique, because of its simplicity, is the pressure chamber technique for measuring water potential (Scholander et al. 1965). This technique is based on the assumption that tension in the xylem stream is equilibrated with the water potential of the leaf cells. The SPAC model implies that tension on the xylem stream increases as it gets closer to the leaf mesophyll cells because this gradient in water potential is the driving force for water movement. However, the xylem tension may be the same throughout the vascular tissues if alternative processes (besides SPAC) are contributing to water flow. The water potential of the xylem stream is dependent only on the negative pressure (tension) in vessel elements. There is an influence of matric and osmotic potential on apoplastic water at the surfaces of the vessel walls of the xylem, but this does not significantly affect the tension in the xylem stream (Turner 1981). Apoplastic osmotic potentials are commonly above -0.1 MPa, and apoplastic matric potential does not normally affect apoplastic water potential until the Ψ is below -14 MPa (Tyree and Jarvis 1982).

To determine the magnitude of xylem tension, a small sample (leaf or shoot) is excised from the plant and placed into a sealed bag to prevent evaporation. The excision should be sharp (with a razor blade or similarly sharp instrument), and the petiole or stem should not be recut, as this will result in an inflated value of Ψ. The xylem water will rebound into the cut stem a distance directly proportional to the tension the xylem water was under just before the excision. The sample is then inverted into a sealed chamber with the cut vascular tissue exposed (Figure 7.10). The chamber pressure is increased slowly until the xylem water just returns to the cut surface. At this point the chamber pressure is equal and opposite to the negative tension that was present in the

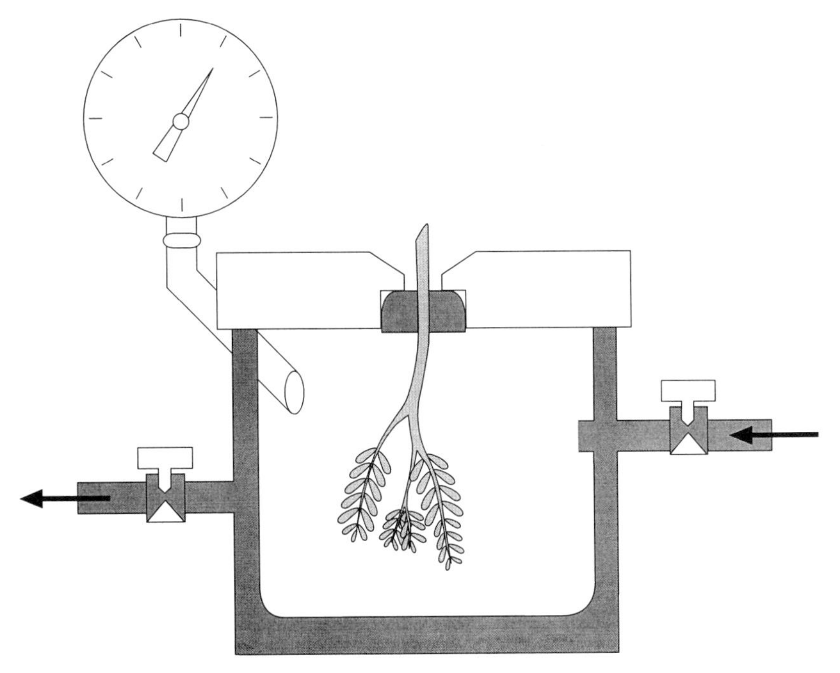

Figure 7.10 Diagram of a pressure chamber used to determine water potential of shoots. The chamber is pressurized until water in the cut end of the shoot is pushed back to the cut. This pressure is equal to and opposite the leaf bulk water potential.

shoot at the time of cutting. The simplicity of this technique permits the determination of tissue water potential in a short time, allows for a large sample size, and can easily be determined in field situations.

Some tissues are inappropriate for this method. In some herbaceous plants it is not possible to distinguish between the appearance of xylem or phloem sap at the cut surface because the vascular bundles have both tissues. In dicots, the phloem tissues can often be removed near the excision to enable clear determination of xylem water appearance. In addition, false readings are frequent in plants that have ducts with resin or milky juice. Also, plants with extensive aerenchyma make pressure chamber measurements difficult. The rate of pressurization, extent of petiole or stem extrusion from the chamber, and constriction intensity of the gasket holding the tissue are all critical to the accuracy of the measurement. Commonly, it takes considerable experience with the tissue and the apparatus before one is confident of the results.

6. Nuclear Magnetic Resonance. Most measurements of plant water relations require a destructive sampling technique. As a result, it has been very difficult to examine dynamic processes of tissue water relations over time in the same tissue. Nuclear magnetic resonance (NMR) has been used for many years in the medical field as a technique of imaging dynamic physiological processes. Recently, applications of this technology to studies of plant water relations have been instituted. In a loose sense, NMR is dependent on energizing protons and examining the dynamics of their return or relaxation to stable state. The spin-spin relaxation time of protons in biological molecules is known to be affected by many possible changes in the conformation of water molecules (Araulo et al. 1993), by changes in the relationship between water and membranes, and by the hydration influences of water with proteins (Mathur-DeVre 1984, MacFall et al. 1987). Therefore, the relaxation time for protons in the cell should be an indicator of interactions between cellular water and membranes and macromolecules. A relationship between NMR relaxation time and chilling sensitivity has been demonstrated (Kaku and Iwaya-Inoue 1987), as well as a differentiation between water in tumors versus water in noncancerous cells (Ling and Tucker 1980, Kaku and Iwaya-Inoue 1990). NMR imaging of water in sections through xylem have been used successfully to indicate the magnitude and location of occlusions related to fungal infection (MacFall et al. 1994). Although the specific plant biochemical processes which are indicated by the changes in proton relaxation times are not well defined, this technique may prove to be an excellent nondestructive window into the physiological status of water in cells.

7. Cellular Water Potential Components

Turgor Potential. The turgor potential of the symplast can be measured directly by the pressure probe technique (Figure 7.11). A section of the tissue is excised and placed on a microscope stage (Cosgrove and Durachko 1986). The microscope is fitted with an oil-filled microcapillary that is connected to a pressure transducer. The microcapillary is inserted into one cell on the microscope stage. The pressure transducer will then be able to determine the internal hydrostatic pressure in the cells. Although this technique can be quite accurate and can be used to determine the relationship between cell volume and hy-

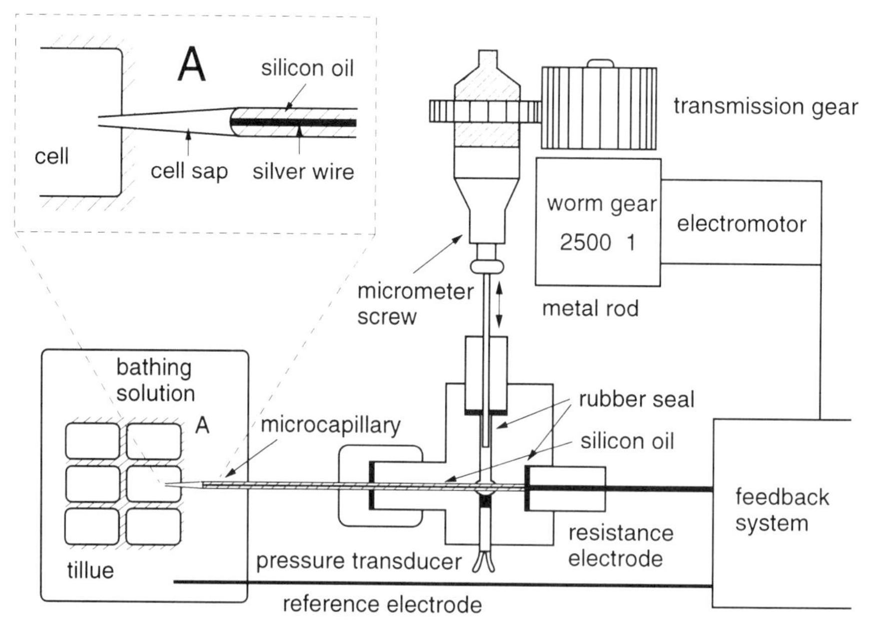

Figure 7.11 Diagram of a pressure probe system for determining cellular hydrostatic pressure. The microcapillary tube is gradually inserted into individual cells. The cell pressure is then measured by the pressure transducer.

drostatic pressure (needed to calculate elastic modulus of the cell wall), microtechnique skills are required. This technique is difficult to use in the field, but in greenhouse situations the pressure probe device has been adapted for use on whole organs in situ (Hüsken et al. 1978).

Turgor potential can be estimated from the difference between Ψ_{total} and Ψ_s. Thermocouple psychrometry is often used for this purpose. Replicate samples are taken from the tissue. One of the replicates is placed immediately into a thermocouple psychrometer while the other is frozen in liquid nitrogen. Freezing in liquid nitrogen breaks the plasma membrane and dissipates any hydrostatic pressure. The frozen and thawed replicate is then placed in another thermocouple psychrometer. The difference between the frozen and nonfrozen samples estimates the turgor potential. However, when the tissue is frozen, the symplastic and apoplastic water fractions mix, potentially diluting the symplast by as much as 20% (magnitude of dilution is dependent on the ratio of symplastic to apoplastic water). This will result in an underestimate of turgor pressure. A similar problem exists for the use of a French press to determine sap osmotic potential (see the section "Osmotic Potential"). The French press combines the apoplastic water (high osmotic potential) with the symplastic water (low osmotic potential) to make an estimate of symplastic osmotic potential (discussed in the next section). Turgor potential can also be determined by the moisture release technique (discussed later).

Osmotic Potential. As already mentioned, the osmotic potential can be estimated by freezing tissues in liquid nitrogen and using thermocouple psychrometry techniques to measure the thawed water potential. Tissue osmotic potential can also be assayed by freezing-point depression. The freezing point of solutions is dependent on the quantity of dissolved solutes. The freezing point will decrease as the osmotic concentration of a solution increases. A relationship can be established between freezing-point depression and the solution osmotic potential (1.86°C $osmolal^{-1}$). A hydraulic press (Figure 7.12) that serves to apply pressures up to 1000 MPa onto tissue samples is used to express sap from the tissues. Originally, this technique was developed by microbiologists to remove cellular fluid from unicellular organisms. Several different commercial osmometers can be used to measure the freezing-point depression of expressed cell sap. However, just as in the psychometric technique, the result is an overestimate of Ψ_s because symplastic and apoplastic water have mixed during the cell sap extraction.

Matric Potential. Matric potential is infrequently measured in plant tissues because it is normally considered to be an insignificant component of symplastic water potential. However, matric potential can be important in cell wall water potential. In fact, determinations of matric potential have revealed a direct association between the thickness of the cell wall and matric potential (Boyer 1967). Normally, matric potentials are determined by removing solutes from cell extractions (Boyer 1967). This technique has yielded estimates of approximately 0.1 to 0.07 MPa. Higher matric potential may be found for cells that have large quantities of polysaccharides (Nobel 1991).

Pressure–Volume Curves. The best procedure for determining components of cellular water potential is through the use of a graphical technique called the *moisture release*

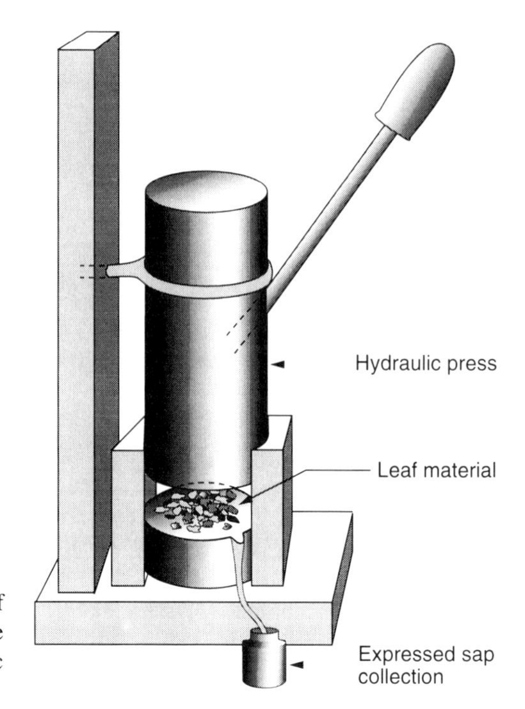

Figure 7.12 Diagrammatic representation of a French press used to extract cell sap for the determination of estimated cellular osmotic and matric potential.

curve (pressure–volume curve). A pressure–volume curve describes the relationship between tissue water content and the components of tissue water potential. The osmotic potential of a tissue can be related to tissue water deficit as

$$\frac{1}{\Psi_s} = \frac{WD}{\phi\rho_s RTN_s} \tag{7.17}$$

where WD is the water deficit, Ψ_s the osmotic potential, ϕ an osmotic coefficient, ρ_s the density of water in the symplast, N_s the total moles of solute in the sample, R a gas constant, and T the tissue temperature in°C.

The derivation of this relationship has been demonstrated elegantly by Tyree and Hammel (1972), Tyree and Jarvis (1982), and many others. Utilizing equation (7.17), a pressure–volume relationship can be adequately defined by the following equation: $1/P = (V_o - V_e)/(RTN_s - \Psi_t)$, where P is the equilibrium pressure in the pressure chamber, V_o the original volume of water, V_e the total volume of water expressed, and Ψ_t the tissue turgor pressure. This relationship can be used to determine turgor and osmotic potential as well as other tissue water potential components. After a certain amount of water has been removed from the shoot, the Ψ_t term becomes zero and the equation reduces to $1/P = (V_o - V_e)\ RTN_s = 1/\Psi_s$. This equation defines the line on Figure 7.13 which describes the linear change in $1/\Psi_s$ versus WD due to osmotic components alone.

The pressure–volume curve technique begins with an excised shoot or leaf. The shoot is weighed and placed into a saturation chamber so that the water content will rise to that of full hydration. Recall that saturation of the sample can cause cell distention, and erroneous results can result, particularly in species of dry regions (Meinzer et al. 1986). Fol-

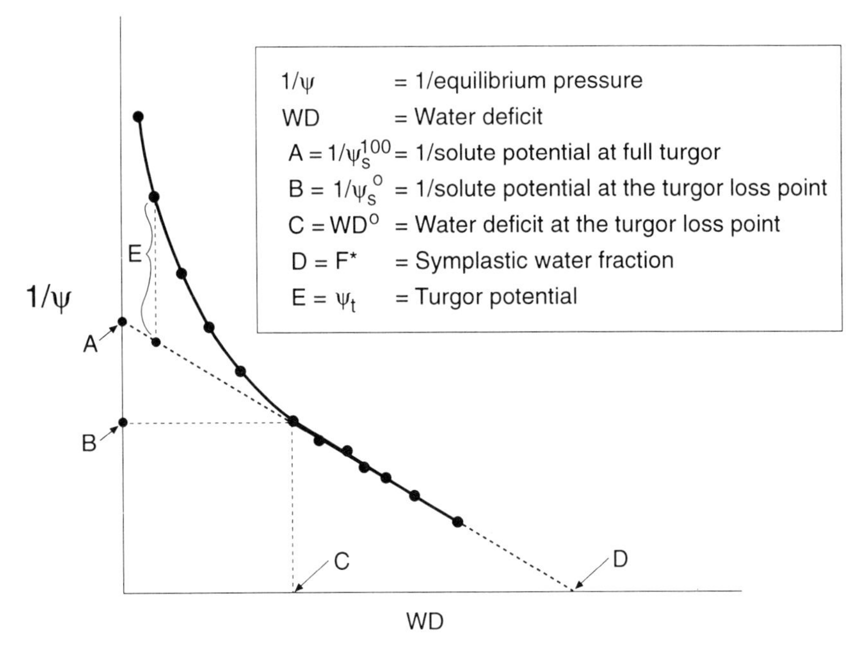

Figure 7.13 Pressure–volume curve used to determine the components of plant water potential. The box represents the types of tissue-water relations information that can be obtained from a pressure–volume curve.

lowing full hydration the shoot is weighed again and inverted into a pressure chamber with a moist lining. The chamber must have high humidity to prevent evaporation from the tissues during measurement (or the tissues can be enclosed in a balloon or plastic bag before enclosing in the chamber). The pressure is incremented in a stepwise fashion (usually 100 psi per increment), and the volume of exuded water is collected continuously at each pressure step. After each pressure increment, and water collection, a new equilibrium water potential is recorded. This requires lowering the pressure below the current step, waiting for a new equilibrium, and determining the new water potential of the stem or leaf. Following 10 to 15 increments and subsequent water volume collections, the stem or leaf is removed from the chamber, weighed, dried in a forced-air oven, and weighed again. The pressure–volume curve can be presented graphically in several ways, but the type II plot (inverse of the water potential vs the tissue water deficit) is easiest to interpret.

When the data are plotted with $1/\Psi_{total}$ versus WD (Figure 7.13), a first-order linear regression formulated through the last four or five points can be used to determine the osmotic potential at any tissue water deficit. The y intercept defines the osmotic potential at full turgidity (Ψ_s^{100}), and the x intercept defines the relative apoplastic water content (F^*). The curvilinear divergence of the points, at low WD, from the aforementioned linear regression represents the influence of Ψ_t on Ψ_{total}. The Ψ_{total} at the point where the impact of Ψ_t becomes zero is the osmotic potential at the turgor loss point (Ψ_s^0). The rate by which the curvilinear portion of the pressure–volume curve approaches the linear regression line represents the bulk elastic modulus (E) of the tissue. The steeper this line, the higher the bulk elastic modulus and the stiffer the cell walls. This relationship changes at lower water deficit; thus bulk elastic modulus usually decreases with increasing water deficit and decreasing turgor potential (Roberts and Knoerr 1978).

The moisture release relationship can also be used to determine tissue capacitance. In this case, tissue water potential (not its reciprocal) is plotted against tissue water deficit. The change in water potential for any given change in water volume is the tissue capacitance (Nobel and Jordon 1983).

Once the relationships between water potential, WD, osmotic potential, and turgor potential (obtained from the pressure–volume curve) are known, simple measurements of tissue water potential can be used to determine Ψ_t or other cellular water relations components (Nilsen et al. 1984). In this way the turgor potential or other tissue water potential components can be determined over a diurnal cycle or during developmental stages simply by measuring tissue water potential. The application of pressure–volume relationships to multiple water potential samples has resulted in a greatly improved understanding of plant water relations. However, one must be aware that developmental, diurnal, or seasonal changes in tissue water relations can alter the pressure–volume relationships.

The pressure–volume technique was first developed through studies of mangroves (Scholander et al. 1965) and conifers (Tyree and Hammel 1972). However, this technique is now applied to a diversity of species and organs with a wide range of volumetric water content and morphology. Several alternative techniques have been developed for assaying the water-deficit component of this relationship.

1. *The Hammel Technique.* This is the classical technique first described by Scholander et al. (1965) and improved by Cheung et al. (1975). The shoot is permanently inserted into the chamber. Exudate is collected from the cut end as the pressure in the chamber is incremented in steps (the technique described above). The method used to collect exudate is critical. Most commonly, small vials stuffed with tissue are preweighed and inserted over the cut end of the shoot. Following the equilibration period the vial is removed and weighed to determine water exuded. There must be no gas leak from the pressurized chamber, or the exudate volume will be underestimated. One must always weigh the shoot at the end of the experiment and determine if all the exudate was collected. If there is a large discrepancy (>10%) between the weight of collected exudate and the weight lost by the shoot, the pressure–volume curve (PV curve) analysis will be inaccurate.

The problems with this technique are that compressed air must be used instead of compressed nitrogen or the tissue will die from lack of oxygen. Each PV curve will take 5 to 6 h, so a multiple-chamber system is imperative for a large-enough sample size. Much care must be taken to prevent leaks so that the exudate can be collected accurately. The advantages are that there is less chance to damage the tissue and the results will be comparable to the classic technique.

2. *The Richards Method.* This method was first described by Talbot et al. (1975) and described further by Hinckley et al. (1980). In small samples there will not be enough water pushed out the stem or petiole to measure volume exuded accurately. Furthermore, any small gas leak from the pressure chamber will evaporate the exuded water. Therefore, after an equilibration pressure has been attained, the pressure is released and the sample is removed from the chamber and allowed to sit on the lab bench to desiccate for some time. The pressure must be released slowly or the tissue will freeze. After a prescribed time the sample is weighed and placed back into the chamber for a new water potential measurement. Water is being removed by evaporation out of the cells rather than by pushing out the xylem, and WD is measured by weighing the tissue rather than by collecting and weighing the exudate.

The advantages to this technique are that many samples can be done simultaneously, there is no error in collecting exuded water, water is removed by a more natural technique, and smaller samples such as individual leaves can be used. The disadvantages of this technique are potential loss of tissues during handling, a rapid release in pressure will kill the tissue by freezing, and damage to the petiole can occur because of repeated insertions into the pressure chamber. The results from this technique do not exactly match those from the Hammel technique. An excellent comparisons of the techniques is available in the literature (Ritchie and Rodin 1985).

3. *Thermocouple Psychrometry.* Pressure–volume curves can also be determined with thermocouple psychrometry. In this case, samples are punched out of leaves as they desiccate. The weights of the leaf punches are determined and related to the weight for a punch at full hydration. Then thermocouple psychrometry is used to determine water potential. Following the water potential measurement the sample is dried and weighed to determine the tissue WD. Many leaves in different hydration states are used to formulate the PV curve. Therefore, this technique depends on a uniform leaf area/weight ratio among all leaves. This technique is advantageous for samples that cannot be placed in a pressure chamber (fruits, very small leaves, seeds, etc).

The determination of a PV curve is an excellent technique for studying tissue water relations. The technique enables the researcher to determine the turgor and osmotic potentials at any water deficit. Furthermore, one can determine the bulk elastic modulus and capacitance for the tissue. This technique has become the mainstay of water relations analysis in plants.

B. Hydrodynamic Aspects of Plant Water Potential

1. Hydraulic Conductance. The water potential of leaves can decrease even though there is ample water in the soil. This is because the demand on the system, transpiration, is causing more water to leave the leaf than the vascular system can replace. Thus, the hydraulic conductance of the water pathway from the root surface to the leaf has a strong impact on leaf water potential. Hydraulic conductance (K) is defined as the total water flow rate through a defined segment of xylem divided by the change in water potential across the pathway [equation (7.8)]. To measure hydraulic conductance, the gradient in water potential and volumetric water flow must be known.

If hydraulic conductance of a stem section is to be determined in situ, the difference between the Ψ_{total} value of sunlit canopy leaves and Ψ_{total} value of a nontranspiring leaf (near the base of the stem section) can be used as the $\Delta\Psi$ value across the stem section. A nontranspiring leaf is established by covering the leaf with a plastic or foil enclosure, usually the night before. If the total plant hydraulic conductance is to be determined, the $\Delta\Psi$ value is that between the Ψ_{total} of the roots and the Ψ_{total} of sunlit canopy leaves. Root water potential can be estimated by determining the predawn plant water potential. This is only an estimate because the water potential at the root surface can decrease between predawn and the time of hydraulic conductance determination. Root water potential is estimated in some studies by measuring soil suction with tensiometers (Saliendra and Meinzer 1989).

The total flux (J) of water through the plant can be measured several ways. Direct measurement of stem water flow is possible with the use of heat pulse stem flow gauges (Figure 7.14). Essentially, two thermocouples are placed at two different points along a stem or branch. Between the two thermocouples a heat source rapidly heats the xylem stream. The heat will be carried to the upper sensor by flow of water in the

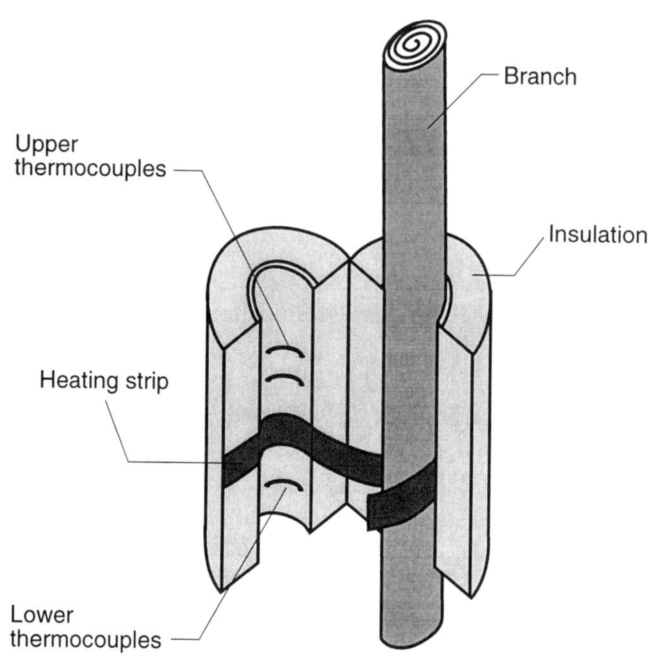

Figure 7.14 Stem flow gauge system employed to determine the water flow rate in the xylem tissues of a stem.

xylem, and to the lower thermocouple by thermal transfer within the tissues. The difference in the arrival time of the heat pulse to the two thermocouples can be used to calculate water flux (Cohen et al. 1983). Total plant water flow can also be measured by determining the transpiration rate with a porometer (described below). Transpiration per leaf area is multiplied by total leaf area to calculate total flow. If the leaf population is uniformly aged and in a uniform microclimate, there is excellent agreement between stem flow gauges and porometry measurements (Schulze et al. 1985). However, such a stipulation rarely occurs in natural systems. A third technique for measuring total water flow is by using energy balance equations (Abdul-Jabar et al. 1984). These indirect calculations require measurement of total incoming radiation, leaf absorptive and emissivity properties, atmospheric vapor pressure, and the difference between leaf and air temperature.

Any measurement of hydraulic conductance on intact plants will involve an estimation of both the water potential gradient and the total water flux. Therefore, the accuracy of these techniques is limited by the errors in measuring water potentials, transpiration, or stemwater flow. However, hydraulic conductance of portions of the xylem stream can be determined on excised stem sections with a higher degree of accuracy. Samples of the stem or root can be excised from the plant. One end of the stem or root is placed in a water solution while a known pressure gradient is established across the tissue (Calkin et al. 1985). The pressure gradient can be established by pressurizing the water immersed end in a pressure chamber or by placing a negative suction on the dry end (Figure 7.15). The quantity of fluid that moves through the stem or root section will depend on the hydraulic conductance of the tissue.

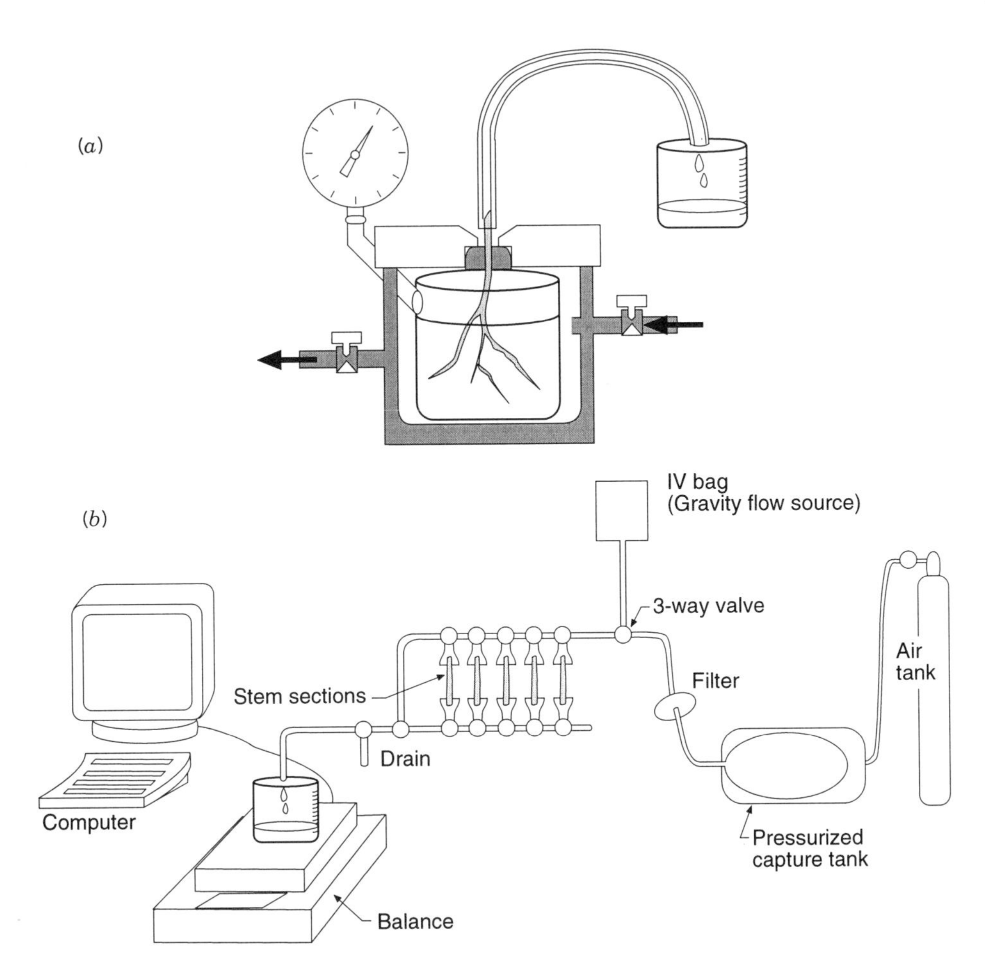

Figure 7.15 (A) Diagrammatic representation of the apparatus used to measure hydraulic conductance in roots using a pressure chamber. (B) Diagrammatic representation of the apparatus used to measure hydraulic conductance of stem sections. The gravity flow source is used to measure hydraulic conductance of each individual stem section. The pressurized tank is used to flush the stems of embolisms. Measurements of hydraulic conductance following flushing indicate the maximum attainable hydraulic conductance for each stem section.

2. Transpiration and Leaf Conductance. Transpiration of leaves is an important parameter of plant water relations because it defines the rate at which water is moving out of the plant. Water-use patterns are critical to making management plans for agricultural systems and in understanding the water cycle in ecosystems or individual plants. Transpiration rate depends on the gradient of water vapor pressure between the substomatal spaces and the turbulent air above the leaf (VPG) and the resistances to water flow through the leaf. It has been approximated by using the equation

$$\mathrm{TR} = J_{\mathrm{wv}} = \Delta w g_l \tag{7.18}$$

where TR is the transpiration $= J_{\mathrm{wv}}$, Δw the water vapor gradient between the leaf and atmosphere in moles, and g_l the leaf conductance.

It is assumed that the intercellular relative humidity is close to 100%. Thus leaf temperature can be used to determine the intercellular vapor pressure (from established tables of temperature versus saturated vapor pressure), and Δw can be determined by measuring air vapor pressure away from the plant and subtracting that of the leaf. Since the flux rate (transpiration) is equal to the product of VPG and leaf conductance, measurement of one (transpiration or conductance) can be used to calculate the other.

Three techniques have been employed to measure either transpiration or conductance. Transient porometers measure leaf conductance (Figure 7.16). A closed chamber is created that can be clamped over a leaf section. The amount of time it takes to reach a certain vapor pressure in the chamber is directly proportional to leaf conductance. This technique is dependent on accurate calibration of the chamber. Normally, several plates with different pore fields are used as a calibration technique. The conductance of the plates can be calculated from the pore number and diameter. Moist filter paper is clamped below the plates and the rate of humidity increase in the chamber is determined. A calibration curve is made between plate conductance and the rate of humidity increase in the chamber.

Transient porometers suffer from several problems. First, the conductance calibration is temperature sensitive, so a family of calibrations must be done to cover the possible range of leaf temperatures in the field. Second, stomata are sensitive to humidity. Therefore, the change in humidity, which the transient porometer depends on to measure conductance, can cause the stomata to change conductance. Third, many of the earlier transient porometers were built out of materials that absorbed and release water from their surfaces. Furthermore, if a large enough temperature difference exists between the chamber air and the leaf, water could condense in the chamber or on the leaf surface. These problems make results from transient porometers highly suspect. Certainly, the earlier models should not be used. Newer models that use updated materials and highly sensitive humidity sensors may be more accurate.

Null-balance porometers have an advantage over transient porometers because they establish a constant relative humidity over the leaf (Figure 7.17). A leaf is enclosed in a

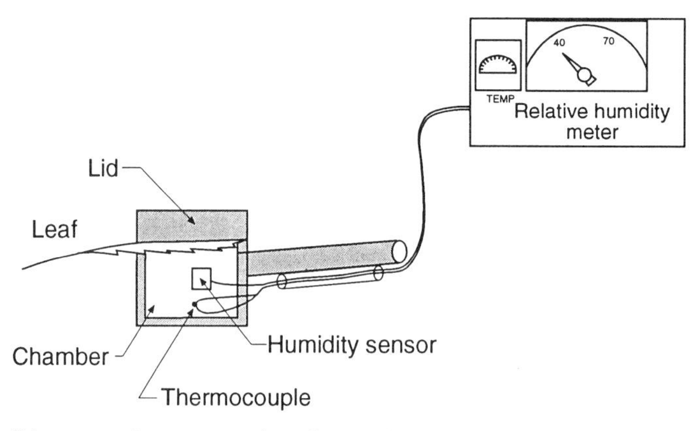

Figure 7.16 Diagrammatic representation of a transient porometer used to determine leaf conductance and transpiration. The lid is closed over a leaf and the data logger records the rate of change in air relative humidity and temperature. The faster the rate of change in relative humidity, the greater the amount of water added to the chamber by the leaf.

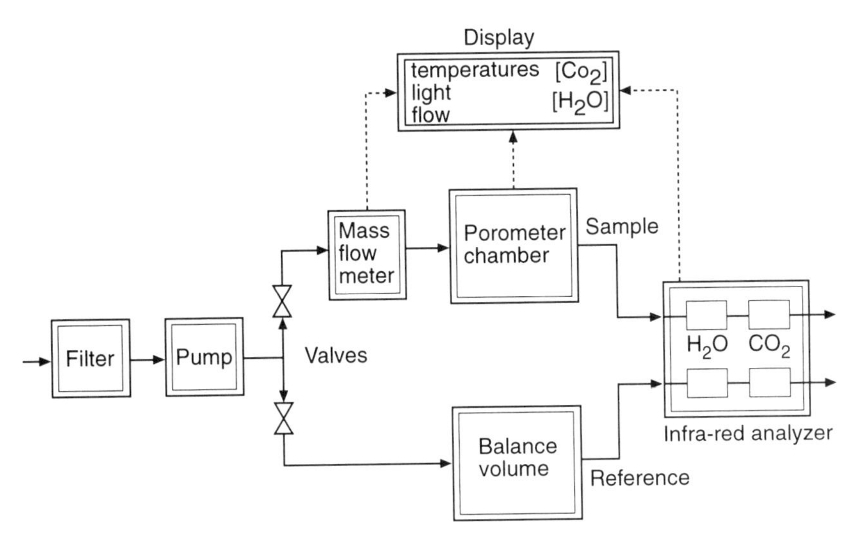

Figure 7.17 Pneumatic diagram of a null-balance porometer used to measure stomatal conductance and transpiration in many leaf morphologies. (Redrawn from Pearcy et al. 1989.)

chamber, and a stream of dry air (0% humidity) is passed over the leaf. The volumetric rate of the dry airstream is controlled by a mass flow controller and adjusted to maintain a specific humidity selected by the user. Normally, the humidity selected is that of the ambient humidity, which is determined with an open, empty porometer. The addition of water to the chamber by the leaf surface can be calculated from the chamber humidity, leaf area, chamber volume, and the volumetric flow rate of dry air. Leaf conductance can then be calculated from the transpiration rate in the chamber and VPG. Leaf and chamber air temperature must be determined because the determination of VPG depends on this difference. Therefore, many manufacturers of null porometers provide a jacket to insulate the chamber from external factors, causing temperature variation, to keep a constant temperature while measuring. Stomatal conductance can be determined if the boundary layer is calculated (usually by measuring leaf cutouts of moist filter paper) and subtracted from leaf conductance.

Null-balance porometers have several advantages over transient porometers. They maintain constant humidity during each measurement. Temperature change during measurement is minimized because less time is taken for each measurement, there is an internal fan to cool the chamber, and there is a insulation jacket. Modern null-balance porometers have highly accurate humidity sensors (dew point mirror or vaisala humicap capacitor), and the chambers are composed of materials that minimize water sorption. The dry airflow is controlled by mass flow controllers, and all calculations are achieved by an on-board microprocessor, which makes the measurement less subjective. A wide array of chamber adapters are available for use with many types of leaf morphology. These many advantages make null-balance porometers the method of choice for measuring stomatal conductance. Of course, transpiration and conductance can also be measured through gas exchange techniques.

Null-balance porometers have one limitation, due to the potential change in temperature of the leaf surface during measurement. The internal environment of the leaf chamber does not match that of the microenvironment around the other leaves on the plant.

The differences are due to the removal of boundary layer in the porometer chamber and a change in the VPG between the leaf and air. Thus the transpiration rates calculated by the null-balance porometer are not identical to those of the leaves in their natural microenvironment. However, there is no reason why the leaf conductance differs from that of the rest of the leaves. If the chamber humidity is the same as that of the ambient air, one can calculate the actual transpiration rate by using the leaf temperature of other leaves on the plant, calculating ambient VPG, and using leaf conductance from the null-balance porometer to recalculate transpiration rate.

The constant-flow porometer is a third instrument used to measure leaf conductance and transpiration (Figure 7.18). In this case a known flow rate of dry air is passed over a leaf in a porometer chamber. This differs from the null-balance porometer because no adjustment of dry air flow rate is used to establish a selected humidity. Therefore, one has to change the area of leaf to approximate a certain humidity in the chamber or the chamber humidity will be different among measurements. This technique suffers from having different humidities among measurements but is better than the transient system because the humidity is stable in each measurement. Other constant-flow porometers can now measure H_2O and CO_2 before and after the chamber. In this way transpiration and conductance are calculated from the change in humidity across the chamber. This results in humidities closer to ambient, but the detectors must be accurate as the change in humidity may be small across the chamber.

VI. SUMMARY: INTEGRATED VIEW OF WATER RELATIONS

We have separated our discussion of plant water relations into biophysical, cellular, and hydrodynamic aspects. This categorization is only for convenience, because all these as-

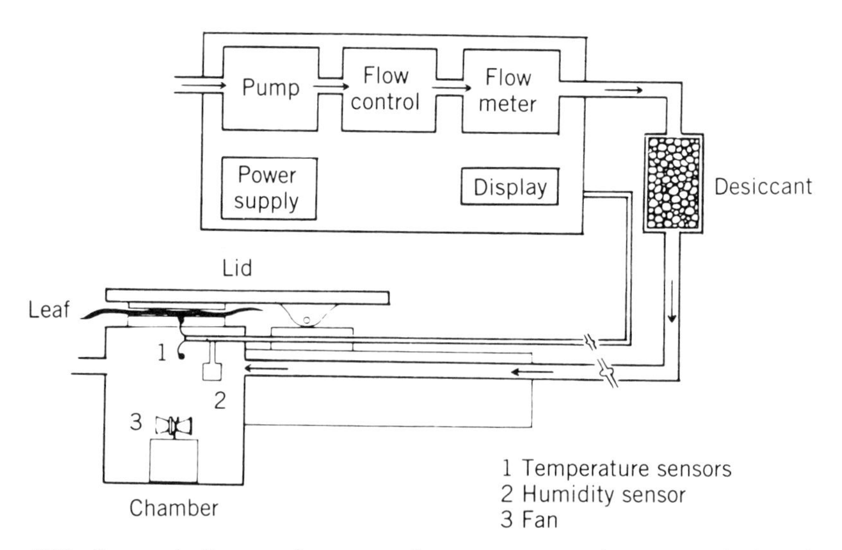

Figure 7.18 Pneumatic diagram of a constant-flow porometer used to measure leaf conductance, transpiration, and carbon dioxide accumulation. (Redrawn from Pearcy et al. 1989.)

pects of water relations interact. An increase in stem hydraulic resistance can cause a decrease in shoot water potential, a decrease in leaf conductance, and a faster diurnal change in shoot water potential. The interactions between water relations components can be observed by treating the soil plant air continuum as an electrical circuit (Figure 7.19). Resistances to flow are located in roots, stems, and leaves. As the water flows from the root to the atmosphere, it is impeded by resistances and interacts with capacitors. If the capacitance is large (large amount of stored water in tissues), the water potential will decrease slowly in each segment of the pathway following stomatal opening, and there will be a large delay between the initiation of transpiration and root uptake of water from the soil.

The rate of water potential decrease in the leaf tissues is initially due to an increase in water deficit of the leaf. The increasing water deficit develops because the loss of water from leaf mesophyll by transpiration is not completely compensated by the hydraulic flow of water through the plant to the leaves. The leaf water deficit causes a decrease in leaf mesophyll turgor potential and an increase in osmotic concentration. The rate at which turgor is lost depends on the bulk modulus of elasticity. The higher the bulk elastic modulus, the faster the decrease in turgor pressure. Stomata sense the changes in tissue water potential and reduce stomatal aperture at a specified tissue water potential. This reduced stomatal conductance reduces transpiration; hydraulic transport of water increases relative to transpiration, causing a decrease in cellular water deficit and an increase in cell turgor pressure.

These are only a few of the possible interactive scenarios that one can imagine be-

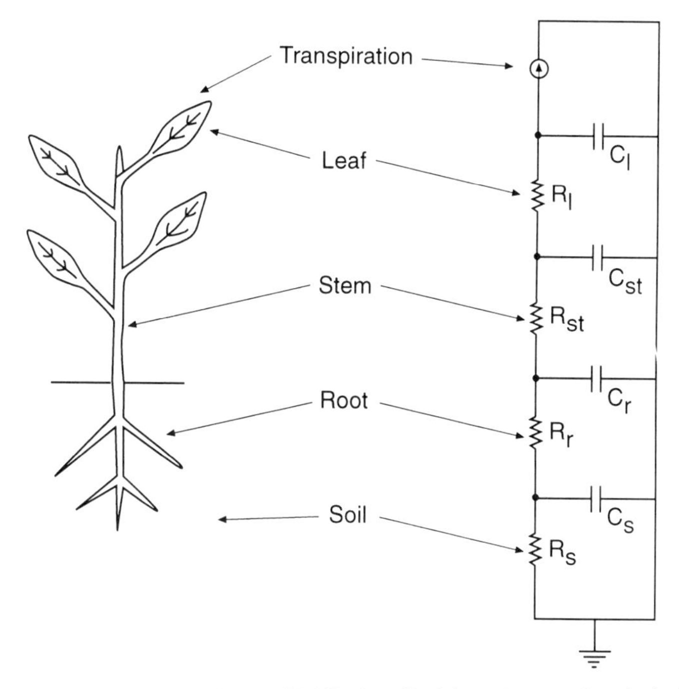

Figure 7.19 Plant water flow pathway (SPAC) described by using an electrical circuit analogy. *R* refers to resistance and *C* refers to capacitance.

tween the various aspects of plant water relations. A simple measure of tissue water potential is entirely inadequate to understand the water relations behavior of plants. One must take an integrated view of water flow dynamics, morphological relationships between root surface area and leaf area, stomatal regulation, and cellular water relations characteristics in order to understand the water relations characteristics of a species.

STUDY-REVIEW OUTLINE

Biophysical Characteristics of Water

Cohesive Properties

1. The cohesive properties of water, due to hydrogen bonding among water molecules, creates the characteristically high surface tension of water solutions.
2. The high surface tension and strong adhesion properties of water cause capillary rise of water, but this cannot account for the magnitude of water movement in most plants.

Heat Properties

1. Water solutions have a high heat of vaporization and specific heat compared to other nonaqueous solutions.
2. The large specific heat causes water to be an excellent thermal buffer, and the large heat of vaporization causes transpiration to serve as an effective cooling mechanism for leaves.

Dielectric Constant

1. The dielectric constant defines the capacity of a solvent to keep solutes separated.
2. Water has a high dielectric constant; therefore, aqueous solutions are excellent solvents for polar molecules and ions.

Chemical Properties

1. The chemical potential of a solution in relation to its surroundings determines the rate and direction of its diffusion. The chemical potential of water is defined as its water potential.
2. In plant physiological applications water potential is composed of pressure potential, osmotic potential, gravimetric potential, and matric potential.
3. Pressure potential is due to hydrostatic pressure of the solution. For solutions in cells, this is termed the turgor pressure or turgor potential.
4. Osmotic potential occurs as a result of dissolved polar substances or ions in solution.
5. Matric potential is due to the adhesive characteristic of water when in contact with surfaces or large macromolecules.
6. Gravimetric potential is mainly dependent on the height of the solution above a reference point (usually, sea level).

Cellular Aspects of Water Potential in Plants

1. Water potential in plant cells is defined as the difference between the solution chemical potential and a reference potential (that of pure water) divided by the partial molar volume of water.
2. Cellular water potential is usually influenced only by hydrostatic pressure and osmotic concentration (gravimetric and matric potential are less important).
3. Water in plants is divided between the symplastic fraction (that inside the plasma membrane) and the apoplastic fraction (that in the vascular system, cell walls, and intercellular spaces).
4. Water moves freely between the symplast and apoplast, and their water potentials are usually equal. However, the main factors contributing to the water potential of the apoplast (negative pressure) and the symplast (osmotic concentration, positive pressure) are different.

Osmotic Potential

1. Four compounds can compose most of the osmotic component in the symplast: inorganic ions, organic acids, nonprotein amino acids, simple carbohydrates.
2. Inorganic ions and organic acids can be at high concentration in the vacuolar water but not in the cytoplasm.
3. Nonprotein amino acids (glycine-betaine, proline) and simple carbohydrates (sucrose, raffinose) can come to a high concentration in the cytoplasm.
4. The osmotic potential of the vacuole is lower than that of the cytoplasm because the cytoplasmic water potential is also influenced by matric potential, and total water potential must be equal between cytoplasm and vacuole.

Pressure Potential

1. The hydrostatic pressure of the vacuole and cytoplasm are equivalent and positive.
2. Negative hydrostatic pressure can develop in the cell walls because of high matric potential. Negative hydrostatic pressure also occurs in the vascular system but not in the symplast.
3. The quantity of turgor pressure in the cell is also dependent on the cell wall elasticity.

Gravimetric Potential

1. The gravimetric potential is only important to some tall trees because gravimetric potential increases by only 0.1 MPa per 10 m of height.

Matric Potential

1. Matric potential is important to water in cell walls because of the high concentration of charged surfaces.
2. The only time when matric potential is important to cellular water potential is when the tissue is dehydrated or there is extensive mucus in the vacuole.

Hydrodynamics of the Plant System

Soil–Plant–Air Continuum

1. Water moves through the plant in a continuous column from root surface to mesophyll cells.
2. The driving force for water movement is the difference between atmospheric water potential and that of the soil. This is the basis of the soil–plant–air continuum model.
3. Alternative forces such as capillary, osmotic, or Marangoni convection may also influence water movement in the xylem.

Root Accumulation of Water

1. Most water is accumulated through suberized roots.
2. There are three possible routes for water to enter the vascular system: (a) apoplastic route: water moves through the cell walls of many cells until reaching the endodermis; (b) symplastic route: water enters the symplast and travels through the symplast to the vascular tissues; and (c) vacuolar route: water enters the vacuole and travels from vacuole to vacuole through plasmodesmata until entering the vascular apoplastic water.
3. Water probably follows all three routes (possibly most travels through the apoplast).

Hydraulic Conductance

1. Hydraulic conductance can be defined as the ratio of the volumetric water flux through and the water potential gradient across the hydraulic tissue.
2. The resistance to water flow in vascular tissue depends on the diameter of vessel elements, the nature of interconnections between the cells, and the cross-sectional area of the vascular tissue.

3. In general, root resistance is greatest among plant organs. Thus, root hydraulic conductance is lowest among the vascular organs.

Capacitance

1. Capacitance is defined as the amount of water released from a segment for any given change in water potential of that segment.
2. Capacitance and resistance can be combined into a time function, which indicates the rate of tissue water potential decrease following the initiation of transpiration.

Cavitation

1. Cavitation occurs when an air embolism lodges in a xylem cell. Normally, the embolisms first appear in the vessel elements and later in the trachieds.
2. A cavitation will cause a rerouting of the water stream and can increase xylem hydraulic resistance because water is shunted into vessels of smaller diameter. However, cavitation does not normally cause severe desiccation of canopy tissues.

Transpiration

1. Transpiration is the evaporation of water from the mesophyll cell walls and its movement out the leaf, through the stomata, and into the turbulent air.
2. Four main resistances can be defined for transpiration.
   a. Resistance due to the size of the substomatal chamber.
   b. Resistance due to the size of the stomatal pore and the morphology of the stomata.
   c. Resistance due to the thickness of the boundary layer.
   d. Resistance due to the hydrophobic cuticle on the surface of leaves.
3. Transpiration can occur through the stomata (a majority of transpiration) or through the cuticle (important only when stomata are closed or the leaf is damaged).
4. Boundary layer thickness is dependent on the length of the leaf in the direction of the wind, and the wind velocity.
5. Conductance, 1/resistance, is used primarily to describe the impact of leaves on transpiration because it can be expressed in molar units, and is directly (not inversely) proportional to transpiration.

Measurement of Plant Water Relations

Tissue Water Relations

1. The relative water content (RWC) of a tissue is the fraction of turgid water weight that is remaining in the tissue.
2. The water deficit (WD), the fraction of water missing from a tissue compared to full hydration, is commonly used in water relations research.
3. Incipient plasmolysis is equal to the osmotic potential of cells.
4. This technique is good for working with tissue slices, epidermal peels, or cell suspensions.
5. Solution equilibration is a technique for determining the water potential of tissues by measuring the relative weight gain or loss in an array of different osmotic solutions.
6. Solution equilibration is inappropriate outside the lab because multiple samples are necessary from each tissue, and constant temperature is necessary to stabilize the water potentials of the solutions.
7. Vapor pressure equilibration is similar to solution equilibration except that the tissue is suspended over a known salt solution.
8. In thermocouple psychrometry a small piece of tissue is placed into a small chamber, and the vapor pressure in the atmosphere of the chamber equilibrates with the sample. A thermocouple system is then used to determine the atmospheric vapor pressure.

9. Three techniques can be utilized to determine atmospheric vapor pressure.
   a. Psychrometry in which the thermocouple junction receives a cooling current to make water condense on the junction. The wet–dry temperature difference is used to calculate vapor pressure.
   b. Dew point hygrometry uses frequent small cooling pulses to ascertain the wet–dry temperature difference.
   c. The isopiestic point technique places the thermocouple junction in known salt solutions after atmospheric equilibration. The line plotting thermocouple response versus solution water potential can be used to determine tissue water potential.
10. Pressure chamber techniques are the simplest method for measuring water potential and most applicable to the field.
11. Shoots or leaves are inverted in a pressure chamber and the pressure is increased until the cut surface becomes wet. At this point, the pressure in the chamber equals the inverse of the xylem pressure potential.

Tissue Water Potential Components

1. Turgor potential can be measured with a pressure probe system, but this is normally limited to microscopic and laboratory applications.
2. Turgor potential can be estimated by measuring tissue water potential and tissue osmotic potential and subtracting. However, many osmotic potential determinations overestimate osmotic potential and thus turgor potential is underestimated.
3. The best way to determine turgor potential relationships of plants in field situations is by the pressure–volume technique.
4. Frozen tissues in thermocouple psychrometers is a method for measuring osmotic potential because freezing relieves any turgor potential.
5. Similarly, a French press can be used to express sap from tissues. The osmotic potential of the expressed sap can be measured in a osmometer by the freezing-point depression technique.
6. Measurements of osmotic potential by expressing sap or freezing tissues is inaccurate because the cytosol is diluted by the apoplastic water.
7. The best technique for measuring osmotic potential is the pressure–volume curve because this technique only measures symplastic water potential components.
8. Matric potential can be measured by taking ions out of an expressed sap sample and then measuring water potential of the solution.
9. The pressure–volume curve is a plot of the relationship between equilibrium pressure and relative volume of water exuded from a tissue. A type II plot in which 1/equilibrium pressure is plotted against water deficit is preferred.
10. This pressure–volume curve plot can be used to determine the osmotic and turgor potential at any water deficit or water potential.
11. The pressure–volume curve can also be used to estimate bulk elastic modulus and tissue capacitance.
12. There are several techniques for performing a pressure–volume curve. The particular technique is selected based on criteria such as sample size, need for large sample numbers, and the nature of the tissue.

Hydrodynamic Aspects of Water Relations

1. Hydraulic conductance, the ratio of water flow to water potential gradient, can be measured on intact plants by measuring the total water flux and the water potential gradient.
2. Water potential gradient in stems can be determined as the difference between a transpiring and a nontranspiring leaf.
3. Water potential gradients for roots can be determined as the difference between soil water potential and the water potential of a nontranspiring leaf near the bottom of the plant.

4. Water flux through the xylem can be measured by either (a) stemflow gauges, (b) leaf transpiration measurements, or (c) estimations from energy budget equations.
5. Hydraulic conductance of stem or root sections can be measured on excised sections by immersing one end in water and either pressurizing the immersed end or placing a suction on the dry end.
6. Transpiration is the product of vapor pressure gradient (between inner leaf spaces and turbulent air) and leaf conductance. Thus a measurement of VPG and one of the other two characteristics allows calculation of the third.
7. Transient porometers measure leaf conductance by determining the rate of change of vapor pressure in a closed chamber. This system is inappropriate because changing humidity affects stomatal conductance.
8. Null-balance porometers maintain a constant humidity in the measurement chamber by passing a controlled rate of dry air across the leaf surface.
9. Constant-flow porometers have intermediate effectiveness because they can maintain a constant humidity, but the humidity is not regulated to match ambient conditions.
10. A recalculation of transpiration must be done from ambient leaf temperature and vapor pressure, because the leaf temperature is slightly different in the measurement chamber (in comparison to ambient) and the boundary layer is minimized in the null-balance or constant-flow porometer.

SELF-STUDY QUESTIONS

1. Two spherical chambers (A and B) are linked together by a small tube that has a semipermeable membrane blocking the movement of water from one chamber to the next. At time 1 the two chambers have an equal volume of solution with an equal concentration of dissolved substances. What would be the result of the following perturbations to the system?

(a) Increasing the solute concentration in chamber A.

(b) Increasing the temperature of the solution in chamber A relative to that in chamber B.

(c) Raising the altitude of chamber A versus chamber B.

(d) Reducing the pressure of chamber A relate to chamber B.

(e) Filling the solution in chamber A with a high concentration of mucilage.

2. What are the possible routes that water can take to enter the xylem of roots, and which of these is most likely to dominate?

3. What is the critical difference between the definitions of hydraulic conductance and capacitance?

4. What factors regulate the vulnerability of xylem to cavitation during water limitation and freezing?

5. Which components of the vapor diffusion pathway out of leaves is affected by a change in the difference in temperature between the leaf and the turbulent air? In what way are these components effected?

6. Design a leaf that maximizes or minimizes transpiration based on the leaf features alone.

7. Compare the advantages and disadvantages of measuring tissue water potential by the incipient plasmolysis, solution equilibration, thermocouple psychrometry, and pressure chamber techniques.

8. Under what conditions would one utilize expressed sap or pressure–volume curves for determining osmotic potential?
9. Diagram a fictitious pressure–volume curve and label the important components on the curve.
10. Explain how the change in water potential with change in water deficit obtained by a pressure–volume curve can be used to determine both capacitance and bulk elastic modulus.
11. What are the critical differences among the following techniques for measuring leaf conductance: transient porometry, null-balance porometry, constant-flow porometer?

SUPPLEMENTARY READING

Boyer, J. S. 1985. Water transport. Annual Review of Plant Physiology 36:473–516.

Boyer, J. S. 1969. Measurement of the water status of plants. Annual Review of Plant Physiology 20:351–365.

Jarvis, P. G., and T. A. Mansfield (eds.). 1981. Stomatal Physiology. Cambridge University Press, Cambridge.

Jones, H. G. 1983. Plants and Microclimate: A Quantitative Approach to Environmental Plant Physiology. Cambridge University Press, Cambridge.

Kramer, P. J. 1983. The Water Relations of Plants. Academic Press, New York.

Larcher, W. 1980. Physiolgical Plant Ecology, 2nd ed. Springer-Verlag, Berlin.

Nobel, P. S. 1991. Physiochemical and Environmental Plant Physiology. W. H. Freeman, San Francisco.

Ritchie, G. A., and T. M. Hinckley. 1975. The pressure chamber as an instrument for ecological research. Advances in Ecological Research 9:165–254.

Slatyer, R. O. 1967. Plant Water Relations. Academic Press, New York.

Slavic, B. 1974. Methods of Studying Plant Water Relations. Academia Publishing House, Prague, and Springer-Verlag, Berlin.

Turner, N. C. 1987. The use of the pressure chamber in studies of plant water status. Proceedings of the International Conference of the Measurement of Soil and Plant Water Status. 2:13–24.

8 Water Limitation

Outline

Objectives

1. Explain important aspects of the soil environment that regulate water availability to plants.
2. Identify other ways in which plants could experience a water deficit.
3. Study the relative sensitivity of various physiological functions to water limitation.
4. Explain the various ways in which water deficit affects plant and cell physiological function.
5. Characterize the mechanisms by which plants compensate for water deficit.
6. Explain drought escape mechanisms, including developmental regulation and dormancy mechanisms.
7. Describe drought tolerance mechanisms, in which plants experience low turgor poten-

tial, including osmotic adjustment, bulk cell elasticity, cell volume water deficit relationships, and changes in proportional symplastic water fraction.

8. Describe desiccation-tolerant species and explain the mechanisms used to avoid the detrimental affects of desiccation.
9. Classify the mechanisms of water deficit tolerance with a high water potential. Include a discussion of mechanisms of reducing water loss and enhancing water accumulation.
10. Describe the interaction of growth regulators with water stress responses concentrating on ABA and ethylene.
11. Pay special attention to water use efficiency covering physiological and environmental regulation of stomata and the concept of stomatal optimization.

I. INTRODUCTION

In Chapter 7 the general characteristics of plant water relations were covered. It is very important to have an excellent understanding of the interactive components of plant water relations before considering plant physiological, behavioral, or structural responses to water limitation. In this chapter we utilize the concepts and methods outlined in Chapter 7 to understand the mechanisms by which plants compensate for water limitation.

Water relations of plants involve the interactions among all organs and their interaction with the environment. One cannot study the impact of plant water stress in one cell type or one molecule without knowing the basic water relations of the entire system. Interestingly, the study of water relations in plants is similar to the study of ecological properties of ecosystems (called *systems ecology*). Both disciplines are concerned with fluxes, residence times, pool sizes, and bioenergetics affecting movements of resources and materials. Just as in systems ecology, the regulation of the overall water dynamics in a plant is not dependent on one component of the system. There are many feedback and feedforward interlinking relationships. Therefore, the answers to questions about water stress phenomena are not going to come from the study of one component of the system but rather from the study of the dynamic nature of water throughout the system.

Our approach to this subject is first to present the underlying situations that can cause limitations in water resources to plants. A sensitivity analysis and specific effects section follows, to address which specific water relations components are critical to which physiological functions. The bulk of this chapter covers the many mechanisms by which plants can compensate for water limitation. Water use efficiency and the mechanism of stomatal aperture regulation are not discussed in this chapter because they are covered extensively in Chapter 6.

II. AVAILABILITY OF WATER IN THE ENVIRONMENT

A. Soil-Induced Water Limitation

A great majority of all water in plants is accumulated through the soil or the root medium. Only in rare cases of vascular plants is water accumulated from other sources (Rundel 1982). Therefore, we will consider the soil (or growth medium) as the primary source of water for plants. Water in the growth medium is extremely variable in both space and time, and its availability depends on many physical and biological factors.

Several physical characteristics of soils affect the quantity and quality of water available to plants. Soil texture defines the proportion of various soil particle sizes in the bulk soil. Clay soils are dominated by small particles with a large surface area/volume ratio, while sands are dominated by large particles with a small surface area/volume ratio. The distribution of particle size classes in loam soils is intermediate between clay and sandy soils. Following precipitation, water will drain through a soil profile rapidly in sandy soils, leaving little capillary water behind compared with that in clay soils. However, the soil solution that remains in sand-dominated soils after precipitation will have a higher water potential than that in clay-dominated soils because of the lower matric potential in soils with larger particle (and pore) sizes. Therefore, sandy soil will retain a smaller amount of precipitation than that of clay soils, but the soil solution will have a higher water potential in sandy compared with clay soils. As the quantity of water decreases in soils, there is an impact of soil partical size on the functional relationship between soil water potential and soil water content (Figure 8.1A). Soil water potential decreases at a faster rate as soil water content decreases in sand-dominated soils compared with clay-dominated soils. In summary, soil texture influences the quantity of water retained after precipitation, the initial water potential of the soil solution, and the functional response between soil water potential and soil water content.

This discussion of the effect that sand- versus clay-dominated soil texture has on soil water brings out several important aspects of soil water availability. The total amount of available water in a soil profile is dependent on the quantity of water in the profile and the quality (water potential) of that water. Thus shallow sandy soils have low water availability because the quantity of water is small, while deep hydrated clay soils may have little

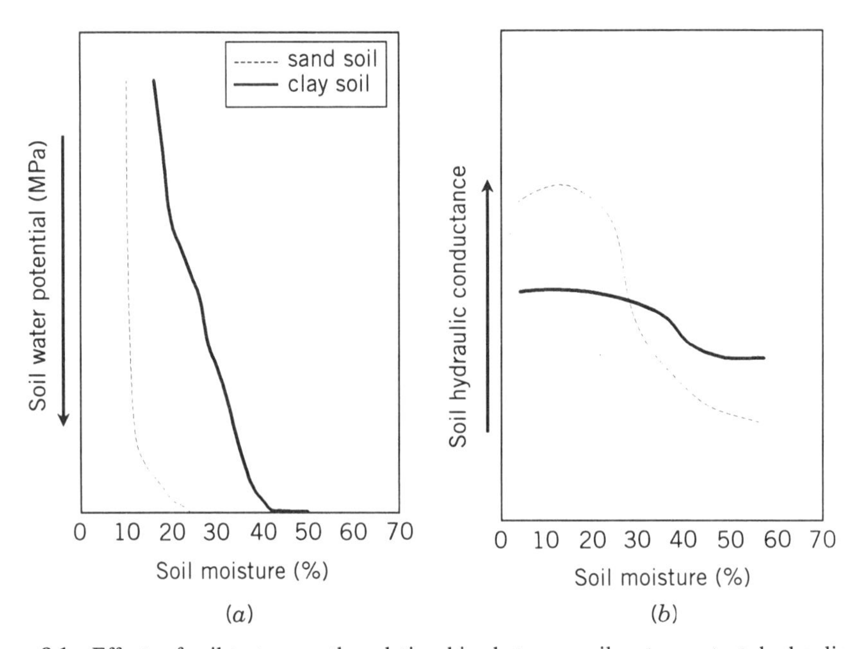

Figure 8.1 Effects of soil texture on the relationships between soil water content, hydraulic conductance, and soil water potential. (A) Relationship between soil moisture content as a percentage of total soil weight and soil water potential for sand- and clay-dominated soils. (B) Relationship between soil hydraulic conductance and soil moisture content for sand- and clay-dominated soils.

water available because a large matric potential is decreasing the soil solution water potential. Although the relative sizes of soil particles are important factors regulating water availability, they are by no means the only factors.

Other aspects of soil biophysics that affect water availability include organic matter, stratification, aggregation, compaction, salinity, microbial activity, and the action of invertebrates. High organic matter content increases soil water storage capacity. This is because organic matter can cause aggregation in clay soils, which increases pore size and improves water penetration. In addition, organic matter adheres to sand particles, reducing pore space and increasing water-holding capacity. Salinity affects soil osmotic potential and can affect the size of pore spaces by affecting soil structure (aggregation patterns of soil particles). The higher the salinity, the lower the soil water potential. If a soil is stratified, this can influence the downward percolation of soil water, potentially retaining water in lenses of saturated soil and leaving other soil layers dry. Soil compaction can reduce percolation and increase runoff, thereby drastically reducing soil water holding capacity. Soil invertebrates can have a massive impact on soil porosity and aggregation. In particular, invertebrates (such as earthworms) that ingest soil can increase aggregation because their eliminated material is more highly aggregated than that they ingested.

The rate of liquid moisture delivery to the root surface (soil hydraulic conductance) is also critical to the concept of soil water availability (Passioura 1988). The hydraulic conductance of soil is dependent on the gradient of water potential between the soil and root, the size of the soil pores, the nature of soil aggregates, whether the root is growing in a pore of its own making, and the path length to the root surface. Hydraulic conductance of soils is also sensitive to dissolved ions. In particular, cations influence the swelling of clay particles and can result in decreased soil conductivity. In general, at high moisture content (high soil water potential) hydraulic conductivity is greater in sandy soils because of the impact of large pore spaces. Under this scenario, most water is not bound to soil pore surfaces and thus is free to flow to zones of lower water potential. However, clay-dominated soils have a higher hydraulic conductivity than sandy soils at lower soil water potential because under these conditions clay soils have a higher ratio of free to bound (adhering to soil pore surfaces) water than do sandy soils (Figure 8.1B).

Water availability to plants is also dependent on the architecture of plant root systems. Diffuse root systems make use of a larger volume of surface soil and thus a larger water volume in surface soil layers than do tap-rooted species. Tap-rooted species utilize soil water in deeper soil layers, which may have a different water holding capacity than surface soils. Phreatophytes (an extreme form of tap-rooted species) have deep root systems that proliferate in the capillary fringe (phreatic zone) just above the groundwater (Figure 8.2). Thus, phreatophytes have a seasonally, relatively stable water resource compared to surface-rooted species. Roots from phreatophytes commonly travel 5 to 10 m deep into the soil profile to reach the groundwater phreatic zone, and mesquite roots have been recorded at a depth of 53 m (Phillips 1963). Every species has evolved a root system that is tailored to its particular habitat and its water use needs. In any environment there are likely to be several different root architectures among species which utilize a spectrum of soil water resource. In chaparral systems for example, grasses utilize surface water, some shrubs utilize water from moderate depths in the soil profile, and other shrubs and trees utilize water in the deep regions of the soil profile. Many of these tap-rooted shrubs in rocky environments have roots that penetrate deep into the soil profile along fissures in bedrock.

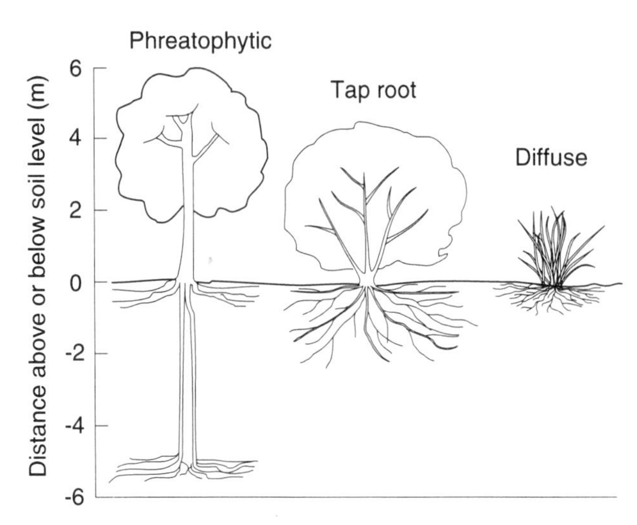

Figure 8.2 Diagram of the root architectures for species with diffuse, tap, and phreatophytic root systems.

Soil profiles can be composed of various layers that have different water-holding capacities. Root systems that proliferate in the upper part of a soil profile will be affected by different water availability relationships than roots which proliferate in deeper soil layers. Surface soils often have a greater water-holding capacity (capillary water) than deep soil layers, due to higher organic matter. However, deeper soil layers may have pockets in which water is prevented from flowing out (in fissures along boulders). Furthermore, since fewer species may be utilizing the deeper soil layers, more water may be available due to lower competition from other species.

Immediately following watering or precipitation, some fraction of the water enters the soil by percolation (the other fractions are runoff and evaporation). The fraction of precipitation that percolates into surface soil layers depends on soil wettability, soil porosity, whether the soil is cracked, and the extent of compaction. The percolating water fraction moves through the soil profile due to gravity. The rate of movement depends on the nature of pores in the system. However, after gravity has caused the water to pass through a soil zone, that water retained against gravity is located in capillary porosity. The quantity of water retained in capillary pores is called the *field capacity.* If there is not enough precipitation to bring the entire soil to field capacity, part of the soil remains dry. The dry soil may constitute patches in the profile that have a localized slower percolation rate or wettability (Figure 8.3), or (if the soil is extremely uniform) the lower soil layers will be dry. Therefore, those plants with roots in the upper soil layers or in zones of higher percolation (e.g., deep fissures along rock faces) will have a greater water availability.

The discussion above indicates that there are many physical and biological properties of the soil that affect water availability to roots. However, some generalities can be made. The potential quantity of water accumulated will depend on the magnitude of root proliferation in the soil volume and the abundance of competitive roots. The rate at which water moves toward the roots depends on the soil water potential (mostly osmotic potential near the root), the root water potential, and the hydraulic conductance of the soil (dependent on matric potential and pore geometry). We have discussed only a short outline of

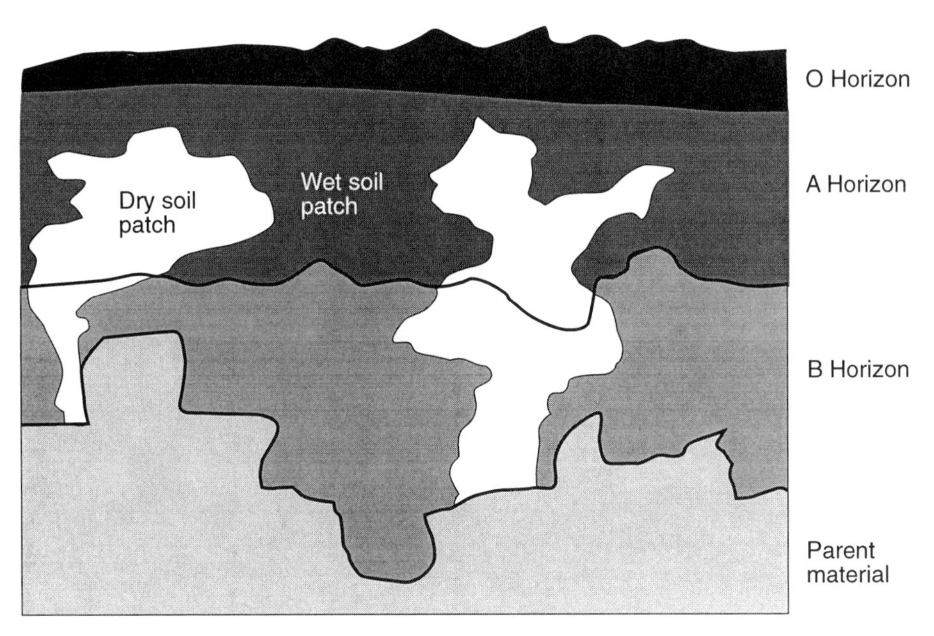

Figure 8.3 Diagram of a soil with patches of variable soil texture and wetability. The variability in soil wetability results in a heterogeneous soil moisture profile following precipitation.

soil and root system characteristics that influence water availability to plants. Extensive reviews of these processes can be found in soil science texts or general plant water relations books (see the Supplementary Reading).

B. Causes of Water Limitation

Limited water availability to the cells of plants can be due to physical and climatic characteristics of the environment, the soil–precipitation relationship, the soil–plant relationship, the atmosphere–plant relationship, excessive demand by the plant, or any combination of these. Obviously, low precipitation is a cause of water limitation. Precipitation ranges from over 2000 cm y^{-1} to a mere few mm y^{-1}. The driest desert of the planet is in western South America (Atacama desert), where little rainfall has been recorded over the past 100 years (Rundel et al. 1991). In a very general sense, plant success (as measured by community productivity) increases with increasing precipitation (Lieth 1973).

The amount of precipitation that enters a site is dependent on regional topography and climatic conditions. Topographic characteristics that affect precipitation are often related to elevation and the movement of air over that elevation gradient. In general, precipitation increases, then decreases, as elevation increases. This is because as the mass of air rises, its temperature decreases and approaches the dew point, inducing precipitation. As the air continues to rise, moisture is taken out of the air by precipitation at middle elevations, resulting in drier air at higher elevation. The direction of air movement in relation to the elevation relief also affects precipitation. Precipitation is highest on the windward side of the elevation gradient and lowest on the leeward side. This is because as the air decreases in elevation, its temperature increases above the dew point, which inhibits precipitation. This *rain shadow* can affect large areas of flat land on the leeward side of mountains, such as that of the short grass prairie on the leeward side of the Rocky Mountains in the United States.

The relative aspect (compass direction of the site), degree of slope, and position of a site on a slope can affect the water balance of that site. These factors are not related directly to precipitation, but rather, they affect runoff and evaporation. Southerly aspects (in the northern hemisphere) have higher evaporation rates than northerly aspects because of the solar angle. Thus, slopes that face south have less water available because of higher water loss by evaporation. In addition, the actual position of the site on the slope affects site water retention. For example, the valley bottom and toe slope (the slope closest to the valley) have higher water retention than that of the ridge top and shoulder slope (the slope nearest the ridge). Forest scientists have developed an index that determines the site water-holding quality based on aspect (compass direction), slope steepness, and slope position. The *forest site quality index* (FSQI) is small when the site has low water retention and high potential evaporation (steep southerly position near the ridge top) and large when potential evaporation is low and water retention is high (toe or valley position with shallow slope and northerly exposure). The FSQI has been shown to be highly correlated with forest productivity and species composition and is used in forest management decisions.

Many other climatic factors regulate the quantity of precipitation. For example, general atmospheric airmass circulation patterns (Hadley cells) can determine geographic location of deserts. Those latitudes with descending air (30° N or S) are likely to have desert conditions. Oceanic currents also affect regional precipitation patterns. Cold currents along coastal regions can cause a Mediterranean climate, while warm coastal currents can create regions of higher precipitation. These are just a few of the factors regulating site precipitation patterns; a more complete discussion of the global processes that regulate regional climate can be found in several general texts on ecology (see the Supplementary Reading).

The seasonality of precipitation is also a critical factor regulating plant response to water stress. Many of the world's ecosystems have climates with distinctly seasonal precipitation that has shaped the evolution of water relations in plants. For example, the Mediterranean climates of the world are characterized by a relatively wet winter and an extensive summer drought. Thus, when temperature is below optimum for growth, water is available, and when temperatures are optimum for growth, water is limiting. Seasonal drought is also characteristic of tropical dry woodland and grassland biomes. In addition, random (stochastic) occurrences of low precipitation also occur in regions of relatively high precipitation. Ecosystems such as the mountain coniferous forest and the eastern deciduous forest experience stochastic drought periods during particular climatic patterns. The regularity of drought induces evolutionary adaptations to water limitation which may be very different from those induced in plants by stochastic drought events.

Assuming that there is adequate precipitation for plant growth, water limitation can still occur because of the demand for water resulting from the interaction between plants and the atmosphere. In many crops a midday wilt is a common occurrence even shortly after irrigation. In natural systems, a midday stomatal closure occurs, although there is adequate water available in the soil. An atmosphere around the transpiring surface with a low vapor pressure will cause a large amount of transpiration if stomata are open. This large water loss can exceed the capacity of the plant's hydraulic system to replace it, resulting in a water deficit. The mechanisms by which plants cope with water limitation induced by an atmospheric–plant interaction or a soil–plant interaction are different (Schulze 1986).

There is also a synergistic relationship between dry air conditions and low soil moisture in many habitats. Deserts with the lowest global precipitation also have low atmospheric vapor pressure. In Mediterranean climates, the summer drought is a consequence of little precipitation, high temperatures, and low atmospheric vapor pressure. Therefore, studies of plant responses to drought must consider both the direct effect of low soil water and the effects of low atmospheric vapor pressure on plant water potential.

In natural and agricultural situations, total annual or seasonal precipitation (irrigation) is often used as an index of water availability. This index suffers from not taking into consideration the influences of soil characteristics on water availability. Therefore, many authors have chosen to measure soil water content directly. Soil water content can be measured by percent of dry weight, tensiometers, neutron probes, psychrometry, or time-domain reflectometry. Such measures of soil water content are unable to measure water availability to the plant directly because they do not take into account soil hydraulic conductance or root distributions. In fact, one could measure high soil water content even though water is not available to plants because the plant roots are undeveloped or hydraulic conductance is low.

Several indices of site water availability have been used to determine the extent of drought. We mentioned FSQI, a qualitative index of site water availability. Such an index is good for determining site water availability before the site is covered with vegetation. However, after the community is developed, total community transpiration can have an overriding affect on site water balance. Studies have shown that the predawn water potential (plant water potential before sunrise) for plants on a north slope is often lower than that for plants on the south slope, which is opposite to what the FSQI would predict. This is because the northerly slopes of Mediterranean shrublands and deciduous forests are covered with broadleaf shrubs with relatively high transpiration rates, while the southerly slopes have a community of lower leaf area composed of evergreen species with lower transpiration rates. Furthermore, the FSQI can only apply to forested systems with elevation relief and cannot be used in a flat topography or an agricultural system.

Indices of site water availability that include the impact of precipitation and atmospheric vapor pressure have been used. Site water balance indices are based on the difference between annual precipitation and annual potential evapotranspiration. Total precipitation is measured directly and potential evaporation can be measured directly or calculated from energy budget equations. The problem with these annual indices is that they do not consider the seasonal aspects of water balance. A site may have a positive water balance at a time when species are dormant, or visa versa. Classically, monthly comparisons of precipitation and evaporation plotted over an annual cycle (climate diagrams) have been used to represent seasonal changes in site water balance. An integrated index of site water balance that is weighted on the basis of plant phenological activity has been developed for desert ecosystems (Comstock and Ehleringer 1992). This site water balance index (ω) ranks sites based on the mean evaporative demand during the most likely growing seasons. The index is appropriate only when leaf and air temperature are similar and atmospheric relative humidity is low.

$$\omega = \frac{\frac{1}{P_{\text{total}}} \sum_{\text{Jan}}^{\text{Dec}} \left(e_{a,\text{sat}} \frac{P}{E_p} \right)}{\sum_{\text{Jan}}^{\text{Dec}} \frac{P}{E_p}} \tag{8.1}$$

where P_{total} is the total atmospheric pressure, $e_{a,\text{sat}}$ the mean monthly air saturated vapor pressure, P the total monthly precipitation, and E_p the total monthly evapotranspiration. Values are summed over the months during active plant growth.

As a more direct measure of site water availability, plant physiologists frequently use the predawn leaf water potential as an index. This value is normally at equilibrium with the soil profile because stomata have been closed all night and thus predawn water potential should equal site soil water potential in the rhizosphere. The predawn water potential does not include the effects of low atmospheric vapor pressure on site water availability or the effects of soil hydraulic conductance on the rate of water delivery to roots. In addition, there are several species that have open stomata at night or otherwise do not equilibrate with soil water potential during the evening.

Whatever index is selected to represent site water relations needs to be related to plant water relations in order to be a robust index for water stress studies. This is because certain plants have optimal physiology at lower site water balances than those of other species. This is due to evolutionary factors that have tailored the components of water relations to operate in specific ranges of plant water potential. Shifting drought-adapted plants to another environment in which the plants can maintain higher water potentials may not necessarily improve growth and productivity.

III. ANALYSIS OF SENSITIVITY TO WATER LIMITATION

The linkage between water availability and deleterious influences on physiological processes is easy to document. However, the specific aspects of tissue water relations that cause the deleterious impacts are enigmatic. Many physiological characteristics are correlated with the water potential of the mesophyll tissue. However, the correlations are species specific. Thus one species may have severely reduced photosynthesis at tissue water potentials of −2.0 MPa (e.g., soybean), while that of another species (such as *Larrea tridentata,* creosote bush) is not affected. This phenomenon was first presented and dealt with in the early 1970s (Hsiao 1973, Hsiao et al. 1976). There is a general hierarchy of sensitivities among general physiological activities (Figure 8.4). Most sensitive are cell expansion, cell wall synthesis, protein synthesis, protochlorophyll formation, and nitrate reduction.

The water relations characteristic that is most associated with cell growth and other sensitive physiological processes is the change in water potential rather than the absolute value of tissue water potential. Frequently, a decrease in cell water potential of only 0.1 MPa can cause a decrease in the cell enlargement rate and result in reduced cell size in shoots and roots. Among the various components of water potential, turgor potential (Ψ_t) decreases most rapidly with any change in tissue water potential. Thus, Ψ_t was identified as the best indicator of water stress (Hsiao 1973, Hanson and Hitz 1982). Many studies

```
Process                                     ΔΨ (MPa)
___________________________0_____0.5_____1.0_____1.5_____2.0_____2.5

Cell Growth.......................... --------.

Cell Wall Synthesis...................  --------

Protein Synthesis.....................  -------

Protochlorophyll formation.........       --------

Nitrate Reductase..................         -------

Stomatal Closure...................             ------------
        some xerophytes............                              ------------

CO2 Assimilation...................               -------------
        some xerophytes............                                ---------

Stem hydraulic conductance.......                       ------------

Proline accumulation...............                      ---------------

Sugar accumulation.................                      -----------------
```

Figure 8.4 Table listing relative sensitivities for major physiological functions to water limitation. Water limitation is defined as the change in tissue water potential. $\Delta\Psi$, the change in tissue water potential from some initial to some final state. This index does not concern the absolute value of tissue Ψ. (Redrawn and modified from Hsiao 1973.)

have shown a correlation between turgor potential and physiological function, but few studies have tried to evaluate the mechanism by which turgor potential is regulating physiological function.

Turgor pressure is not always associated with changes in physiological function induced by water limitation. Turgor pressure of corn tissues at different developmental stages responded differently to water limitation (Boyer 1970). When the elongation rate of differentiating cells was inhibited by withholding water, cell expansion decreased along with a decrease in $\Delta\Psi_t$ of differentiated cells. Yet there was no change in turgor pressure of nondifferentiated (juvenile) tissues (Figure 8.5). Thus, as the water limitation occurred, cell size decreased and turgor pressure remained constant in differentiating cells (Boyer 1970).

How much turgor pressure is required for optimal physiology? This is a hard question to answer because turgor pressure varies between 1.0 and 0.5 MPa among species during normal water availability conditions. Turgor potential may decrease from these values at dawn to approximately zero at midday in many species. Furthermore, some species normally have turgor pressure close to zero over most of the daylight hours during the entire growing season (Nilsen et al. 1984) without any indication of physiological dysfunction. Under these conditions cell expansion and growth occur at night when turgor pressure is maximum.

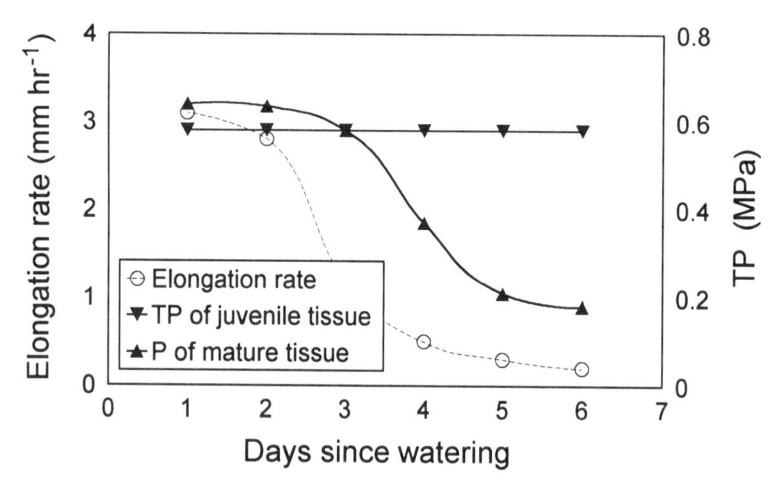

Figure 8.5 Relative elongation rates of differentiated and undifferentiated soybean leaf tissue during a water-withholding experiment. Notice that the shoot elongation rate decreases in accordance with the turgor potential of the mature differentiated tissues, but no such relationship is measured for undifferentiated tissues. (Redrawn from data in Boyer 1970.)

Generally (although there is some disagreement among scientists) turgor pressure is still accepted as the best indicator of water stress in plants. The specific mechanisms by which turgor regulates physiological function probably relate to cell walls and membranes. Since cell expansion is dependent on cell pressure and the cell wall yield threshold (see Chapter 2), there can be no cell expansion without turgor pressure greater than the yield threshold for cell expansion. Cell wall synthesis is highly related to cell expansion; therefore, it is reasonable to expect these two functions to be most sensitive to reductions in turgor pressure. After the cell has matured, no more cell expansion will occur, yet cell physiology remains sensitive to turgor pressure. Studies with algal systems have indicated that slight changes in turgor pressure decrease membrane permeability to water and ions (Zimmerman and Steudle 1975). Plant cell membrane structure, and the spatial arrangement of enzymes, transport channels, cellulose synthesis rosettes, and receptor proteins may be dependent on turgor pressure. Thus, when turgor pressure decreases, the spatial relationships of these proteins change, and membrane function is disrupted (Hsiao 1973; see Chapter 3).

IV. EFFECTS OF REDUCED WATER POTENTIAL ON PHYSIOLOGY

A reduction in mesophyll water potential (with or without a reduction in turgor pressure) can affect the physiology of cells in several ways.

1. Reduced water potential reduces the chemical activity of water and thereby modifies the structure of water in the cell.
2. A lower chemical activity of water can cause a change in the structure of the sheath of hydration around proteins and thereby reduce their efficacy.
3. The relationship among intracellular membranes of chloroplast, nucleus, mito-

chondria, endoplasmic reticulum, tonoplast, plasmalemma, and others will change because the cellular positions of these membranes will change.

4. A loss of turgor may cause a change in the spatial position of transport channels and membrane enzymes, and decrease membrane thickness.
5. A change in cell pressure and the resultant cell wall shrinkage may constrict the entrances to plasmodesmata.
6. The concentration of molecules in specific regions may change due to the loss of water in some subcellular locations.

Many of the phenomena listed are general factors affecting physiology of cells. Therefore, we would expect to see the effects of changing water potential in all aspects of plant physiology. Thousands of articles are available in the literature concerning the many correlations between specific physiological function and plant water limitation. We will review a few of the most important effects while keeping the general effects in mind.

A. Effects on Growth

When water limitation is large enough to increase tissue water deficit, there will be a reduction in turgor pressure. Since cell expansion is dependent on cell Ψ_t, developing cells will expand less and cell size will be smaller under these conditions. The turgor potential of differentiating cells remains constant during decreasing water potential even though severe reductions in cell expansion may occur (Boyer 1970, Barlow 1986). Therefore, the impact of reduced turgor is transmitted from mature cells to developing cells.

The consequence of reduced cell size to growth pattern of the whole plant is dependent on the timing of the water limitation in relation to the phenology of the plant. If water limitation occurs in the beginning of the growth cycle, leaf area will be reduced and carbon gain throughout the growing season will be reduced because of smaller leaves. Other secondary impacts will result from the reduction in leaf area, including changes in water and nutrient use patterns. If reduced turgor occurs during inflorescence development, the number of flowers is reduced and possibly all reproductive effort may be aborted, yet there will be little to no impact on plant mass. If water limitation occurs during fruit maturation, inflorescences will develop normally and entire plant mass will be unaffected, but seed fill may be inhibited and fruit abscission may be enhanced (O'Toole and Chang 1979). In addition, some plants are sensitive to water limitation during specific seasons (periods of active growth) but not during other seasons (Nilsen and Muller 1981b).

The critical water potential for inhibition of cell expansion is different among species and among organs within plants (e.g., corn roots and leaves have different critical water potentials). For example, changes in water potentials of -0.2 to -0.4 MPa (Boyer 1970) cause cessation of leaf expansion in sunflower, while the threshold for corn leaf expansion is a change of -0.7 MPa and that of soybean is a change of -1.2 MPa (Acevedo et al. 1979).

B. Effects on Cell Ultrastructure

The general impacts of water limitation on the structure and function of membranes is also observable in the ultrastructure of cells. Mild water limitation may disrupt the structure of microbodies, releasing hydrolyzing enzymes into the cytoplasm. The presence of these lipases and proteases further disrupts the normal structure of all cytosolic membranes. If the tonoplast is degraded, the vacuolar fluid can empty into the cytosol (Fellows and Boyer 1978). Since the vacuolar fluid may contain relatively high concentrations of solutes, damage to cytosolic proteins will probably result from a breached tonoplast.

Chloroplast and mitochondrial structure can be affected by severe water limitation. Increases in alkaline lipases inside chloroplasts during water limitation have been correlated with a degradation of thylacoid structure (Viera de Silva et al. 1974, Giles et al. 1976). Increases in plastoglobule (derived from thylacoid membrane) frequency and size (Figure 8.6) is associated with decreased water potential in some species (Poljenkoff-Mayber 1981, Steinmüller and Tevini 1985). Separation of thylacoid membranes from the appressed membranes observed in chloroplast from water-stressed plants may be due to dis-

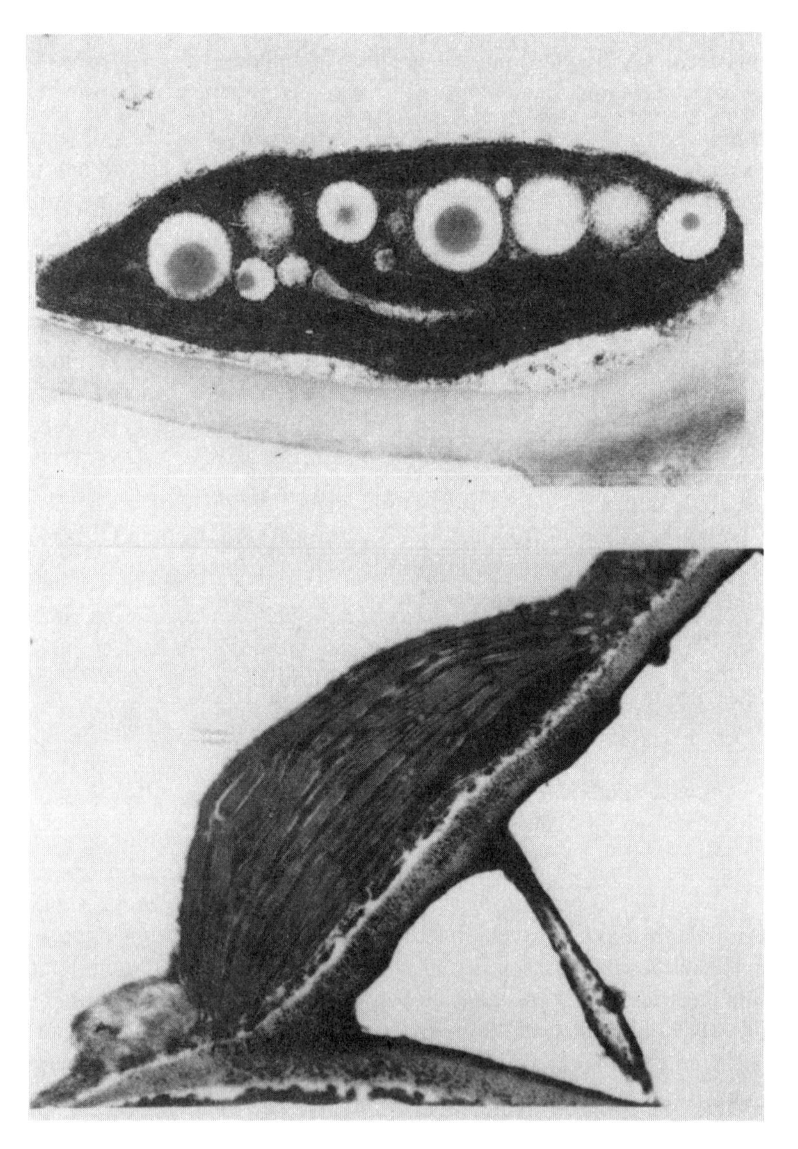

Figure 8.6 TEM photographs of a chloroplast from *Rhododendron maximum* with (A) and without plastoglobules (B).

lodging of photosystem II complexes (Powles 1984). In addition, severe water limitation can cause the appearance of fat droplets in the cytosol and a large rearrangement of chromatin around the nucleoli (Poljenkoff-Mayber 1981).

C. Effects on Photosynthesis

Water limitation has numerous effects on photosynthesis that have been reported extensively during the past 50 years. Although there are such a large diversity of effects and examples thereof, some generalities can be stated. The initial impact of water limitation on photosynthesis is usually stomatal closure. Stomata may close because of a root signal (Davies and Zhang 1991), probably abscisic acid, or because of low turgor pressure in the guard cells (Rashke 1975, Collatz et al. 1991). Stomata also close in response to increasing vapor pressure gradient between the leaf and air (VPG), although this may not be associated with a change in water potential (Turner et al. 1984).

Stomatal closure induced by water limitation causes a depletion of carbon dioxide in the intercellular spaces (C_i). This is termed stomatal inhibition of photosynthesis. Once C_i has decreased relative to oxygen, photorespiration is stimulated. If the light intensity is high enough, photorespiration cannot utilize all the energy-containing products of the electron transport system. Under such a scenario, photoinhibition can occur (Björkman et al. 1981), resulting potentially in a buildup of free radicals in the chloroplast. The impact of water stress on the light-harvesting systems and electron transport chains during periods of stomatal closure and high light intensity may be most important on photosynthesis. This subject is covered extensively in Chapter 10 (the impacts of irradiance) and Chapter 13 (interactive effects of multiple stresses).

Nonstomatal inhibition of photosynthesis (due to photoinhibition or other mechanisms) normally accounts for a larger proportion of photosynthetic inhibition as water potential becomes lower (Ogren and Öquist 1985). However, during the initial phases of water limitation, stomatal closure and nonstomatal inhibition occur concurrently. In fact, there is evidence in several species that nonstomatal inhibition may occur first, causing a temporary increase in C_i, which causes stomata to close (Briggs et al. 1986).

The mechanisms by which water limitation directly inhibits the photosynthetic apparatus is controversial. In some taxa the main nonstomatal impact is that of photoinhibition (Björkman and Powles 1984), while in other taxa the activity of rubisco dominates nonstomatal inhibition (Ogren and Öquist 1985). Reduced turgor may increase the permeability of the outer chloroplast envelope, which results in a change in chloroplast pH and ion concentrations. The change in ion concentrations and pH can affect secondarily the activity of rubisco. Some studies have found that photosynthetic enzymes are relatively immune to the deleterious effects of water stress (Björkman et al. 1980b, Mayoral et al. 1981), while others indicate direct impact on photosynthetic enzymes (O'Toole et al. 1976). The degradation of chlorophyll increases, and the concentration of the chlorophyll (in particular, the *a*/*b* binding protein complexes) decrease during water stress. Therefore, light harvesting and electron transport associated with photosystem II is preferentially decreased (compared to that of photosystem I) by water stress (Björkman et al. 1981).

D. Effects on Dark Respiration and Carbohydrate Metabolism

As water limitation progresses, photosynthesis decreases before that of respiration; consequently, the ratio between photosynthesis and respiration decreases. The decrease in the ratio of photosynthesis to respiration, and the potential increase in both photorespiration and dark respiration during water stress, have caused many authors to believe that water limitation could cause plant starvation. However, it is more likely that the plant will suffer greater damage to the shoot system from metabolic effects of water limitation other than carbohydrate deprivation.

Simple sugar concentration in some plant tissues may increase with water stress because starch stored in chloroplasts is mobilized. Loss of starch is a common correlation with water limitation, but simple carbohydrates do not always increase in accordance with the reduction in starch. Simple sugars that are derived from starch mobilization are used by many physiological processes. Therefore, it is not surprising that the buildup in the simple sugar pool does not account quantitatively for the reduction in starch during water stress.

Carbohydrate translocation also decreases during water limitation during the day but may increase relative to well-watered plants at night (Bunce 1982). The decrease in sucrose translocation is not due to specific effects on phloem loading processes. In fact, phloem loading is relatively resistant to water limitation (Sung and Krieg 1979). The cause of reduced photosynthate translocation is the change in source–sink relationships during water stress. Low CO_2 assimilation by leaves and increased respiration in mesophyll cells of leaves decreases the gradient of sucrose between the source leaves and the photosynthate sinks. The reduced gradient from source to sink causes a reduction in carbohydrate flow in the phloem.

Patterns of resource allocation change during water limitation. In many species, a majority of growth occurs in root tissues instead of in leaf tissues. Thus, there is a decrease in the root/shoot ratio. The timing of water limitation relative to the developmental status of the plant affects allocation patterns. Water stress that occurs during early growth phases causes large shifts in root/shoot ratio. In contrast, water stress during reproductive phases (late in the growing season) has little to no effect on root/shoot ratio, but flowering and seed set are reduced, or fruit abortion is increased.

E. Effects on Nitrogen Metabolism

Nitrate and ammonia accumulation decrease during water limitation. The flow of nitrogen from roots to leaves slows and higher concentrations of nitrate and ammonia build up in water-stressed roots than in the roots of well-watered plants (Nilsen and Muller 1981a). The higher concentration of nitrogen ions in the roots of water-stressed plants inhibits the accumulation of nitrogen from the soil. Therefore, the reduction of nitrogen accumulation is not due to specific effects of water stress on the transport proteins or accumulation mechanisms; rather, the changes in nitrogen use and flow result in conditions in the plant that inhibit nitrogen accumulation kinetics.

Water limitation is associated with an increase in protein hydrolysis and a decrease in protein synthesis. In adition, a decrease in polyribosome abundance is correlated with the decreased protein synthesis. Coincident with the decrease in total protein is an increase in free amino acids. Much of the amino acid accumulation is due to the reduction in protein synthesis, but in some cases biosynthesis of particular nonprotein amino acids is stimulated (e.g., betaine, proline).

Some studies have indicated that an increase in ribonuclease occurs during water limitation. There is little evidence that mRNA translation is affected (Shah and Loomis 1965); thus the cytosolic mRNA pool most probably decreases due to an increase in cytosolic ribonuclease activity (Todd 1972).

F. Generalized Response to Water Stress

A generalized progression of events can be suggested for the impact of water limitation on most plants. During the initial stages of water limitation, turgor pressure decreases, causing a reduction in cell expansion and a change in the distribution of growth regulators (discussed in Chapter 4). The decreased tissue turgor potential along with an increase in leaf free abscisic acid causes stomatal constriction. The stomatal constriction reduces the flow of water through the system and decreases intercellular carbon dioxide. The lower C_i can stimulate a reopening of stomata if water availability does not decrease rapidly. However, if turgor continues to decline, stomata will continue to close and photorespiration will increase. Decreased carbon flow into the leaf (reduced photosynthesis) will cause a mobilization of starch and potentially an increase in respiration. Nonstomatal reductions to photosynthesis occur, which limits the C_i depletion. Following the adjustment of cell expansion, stomatal aperture, and photosynthesis, further reductions in water potential cause impacts on cytoplasmic physiology such as photoinhibition. Protein synthesis decreases and nonprotein amino acids increase. Only under extreme water limitation will ultrastructural abnormalities occur.

Many authors have commented on the general similarities between the physiological impacts of water, light, heat, and salinity stress. Water limitation is often associated with increasing leaf temperature because transpiration has been reduced, eradicating a major leaf-cooling process. In addition, heat-shock proteins can be induced by water limitation. Water limitation often causes intense leaf movement in a similar manner as plants undergoing high-light conditions. This is because leaves that are suffering from water limitation have a low C_i and are sensitive to irradiance. Furthermore, salinity causes a low soil or root medium water potential, and separating the specific ion effects from the water limitation effects can be difficult. Although a centralized general response to stress has been suggested, and there are many similarities among cellular, molecular, and anatomical responses to resource limitation, each resource (light, water, heat) also has unique impacts on growth and development that must be considered.

V. MECHANISMS BY WHICH PLANTS COMPENSATE FOR WATER LIMITATION

The mechanisms by which plants can compensate for water limitation are as varied as the number of water-stress-tolerant plants. Commonly, these mechanisms are categorized into those that result in avoiding the dry condition and those that result in tolerating the reduced water availability (Turner 1986). Tolerance mechanisms can be divided into those that maintain a high water potential and those that result in a significant drop in tissue water potential (Figure 8.7). None of these categories are mutually exclusive. In fact, most of the individual mechanisms are dependent on other mechanisms to solve the water limitation problem.

In any community suffering from a temporal limitation of water there will be a series of species surviving through the use of any number or combination of mechanisms to compensate for the low water availability (Nilsen et al. 1984). Commonly, the environments with limited water availability have co-occurring high heat, light, or salinity. Therefore, it is often difficult to separate the physiological and structural responses of the plants to the individual environmental conditions without manipulative experiments. Al-

ADAPTATIONS TO WATER STRESS

I: Drought Escape

A. Rapid phenological development
B. Developmental plasticity (deciduousness)
C. Extended dormancy

II: Drought Tolerance With Low Water Potential

A. Maintenance of Turgor Potential

1. Osmotic adjustment
2. Increase or decrease in elastic modulus
3. Decrease in cell volume
4. Decrease in symplast volume vs apoplast volume

B. Desiccation Tolerance

1. Protoplasmic tolerance
2. Few Plasmodesmata

III: Drought Tolerance with High Water Potential

A. Reduction in water loss

1. Decreasing leaf conductance
2. Decreasing canopy leaf area
3. Leaf temperature moderation

B. Enhancement of Water Accumulation

1. Increased root density and or depth
2. Increased hydraulic conductance
3. Increased capacitance
4. Hydraulic lift

Figure 8.7 Various mechanisms by which plants can tolerate or avoid water limitation. This is a revision and extension of that presented in Turner 1986.

though the various mechanisms attributed to water limitation will be discussed individually, one must recognize that the mechanisms we discuss are also involved in compensating for heat, light, and salinity overabundance.

A. Escape from Water-Limiting Conditions

Mechanisms that concern escape from water limitation normally relate to a developmental aspect of the life cycle in annuals, or seasonal developmental phenology in perennials. There are many cases of rapid maturation and seed set in desert annuals (Mulroy and Rundel 1977). Several of these desert annuals produce only one pair of mature leaves before flowering and setting seed (Figure 8.8). The time from seed germination to seed set for these species may be as short as a few weeks (Kemp 1983). The requirements for germination of seeds of desert annuals is frequently several successive imbibitions (Bentley 1974). In this way the germinating seed has a higher likelihood of encountering moist conditions that remain for an adequate time period. Rapid annual development is com-

Figure 8.8 One of many desert taxa that develop rapidly after the advent of several precipitation events in the Sonoran desert of California.

mon in species from other low-precipitation areas, such as Mediterranean and other seasonally dry regions of the world.

Species can also avoid periods of water limitation by surviving as a subterranean perennating organ until moist conditions occur. These tuber, corm, or bulb species commonly have a high water storage capacity in the perennating organ. Normally, only two leaves are produced (such as in *Welwitchia* sp. of the namibian desert), and flowering is initiated rapidly and lasts for a short duration.

Perennial species can also utilize developmental changes to avoid seasonal periods of water limitation. The drought deciduous phenomenon in Mediterranean and desert climates is an excellent example. Leaf populations are developed in the winter months when water availability is relatively high. As plant predawn water potential decreases in late spring and summer (due to a combination of increased evaporative demand and low precipitation) most leaves abscise. Drought deciduous species survive the period of water stress in a dormant state (Nilsen and Muller 1982).

B. Drought Tolerance with Low Water Potential

The ability of a plant species to tolerate low water availability means that the plant can continue metabolic processes during periods of water limitation. In some species, metabolic processes continue even though tissue water potential decreases. In other species, tissue water potential does not decrease significantly when water limitation occurs. Both of these possibilities represent a tolerance of water limitation. The mechanisms used by species to continue metabolism under low tissue water potential are different from those used by plants which maintain a high water potential during water limitation. Frequently, species utilize a combination of techniques to moderate decreases in tissue water potential and maintain metabolic activity at lower water potential.

1. Osmotic Potential Adjustment. Tissue water potential is the added result of osmotic potential and turgor pressure (see Chapter 7 for a review). Thus when tissue water potential decreases, a change in osmotic potential may be able to maintain turgor pressure at a water potential that would otherwise result in turgor loss. Since turgor pressure is considered of primary importance to plant cell metabolism, a great deal of research has been done on osmotic adjustment.

Changes in the concentration of solutes is associated with several environmental perturbations. Increasing environmental salinity (soil medium, salt spray, etc.) can result in increased osmotic concentration, particularly in halophytes. Osmotic concentration can also increase during soil water depletion. In fact, increasing osmotic concentration can be induced similarly in halophytes by withholding water or increasing salt concentration in the root zone. Furthermore, osmotic concentration can increase in tissues subjected to cold temperature, which results in protection from freezing because tissue freezing point is lowered as solute concentration increases.

As explained in Chapter 7, four constituents of cells account for a major portion of solutes that concentrate during osmotic adjustment. The energetic requirement (respiratory cost) of adjusting osmotic concentration with the various constituents is different. If a suite of ions such as potassium (K^+), calcium (Ca^{2+}), and sodium (Na^+) are used to adjust osmotic concentration, these ions must be isolated in the vacuole because a high concentration of ions would be toxic to the cytoplasm. Energetically, this is an inexpensive method for plants to adjust osmotic potential because no carbon skeletons are required to formulate the osmotic constituent. On the other hand, energy is required to transport ions into the vacuole. Osmotic adjustment with cytosol-soluble carbohydrates requires extensive amounts of reduced carbon skeleton but is compatible with the cytoplasm and thus does not need to use energy resources for transport into the vacuole. However, high levels of simple sugars in the cytoplasm can stimulate respiration and thereby be an added cost to cellular carbon balance.

Osmotic adjustment with organic acids is moderately expensive for a plant, because organic acids require carbon skeletons and must be transported into the vacuole. In addition, the vacuolar pool of organic acids can be partly derived from Calvin cycle intermediates which could reduce carbon fixation by depleting the pool of Calvin cycle intermediates.

The use of nonprotein amino acids such as betaine (*N,N,N*-trimethyl glycine) and proline for adjusting osmotic potential is relatively expensive compared to the other osmotic moieties, because an induced biosynthetic pathway is required to produce the osmotica. In the case of betaine a two-step pathway (Figure 8.9) is required to convert choline to betaine (McCue and Hanson 1990). Therefore, osmotic adjustment with nonprotein amino acids requires both carbon skeletons and ATP for the biosynthetic pathway.

Normally, osmotic adjustment occurs through the use of several different osmotic moieties in the cytoplasm and vacuole. The vacuolar ion concentration may increase simultaneously with the cytoplasmic nonprotein amino acid pool. Frequently, the increase in nonprotein amino acids is only a small part of total symplasmic osmotic adjustment; the rest is accounted for by vacuolar ions. Many more ions are needed (than amino acids in the cytoplasm) to adjust the vacuolar osmotic potential because a large portion of the symplastic water is in the vacuole.

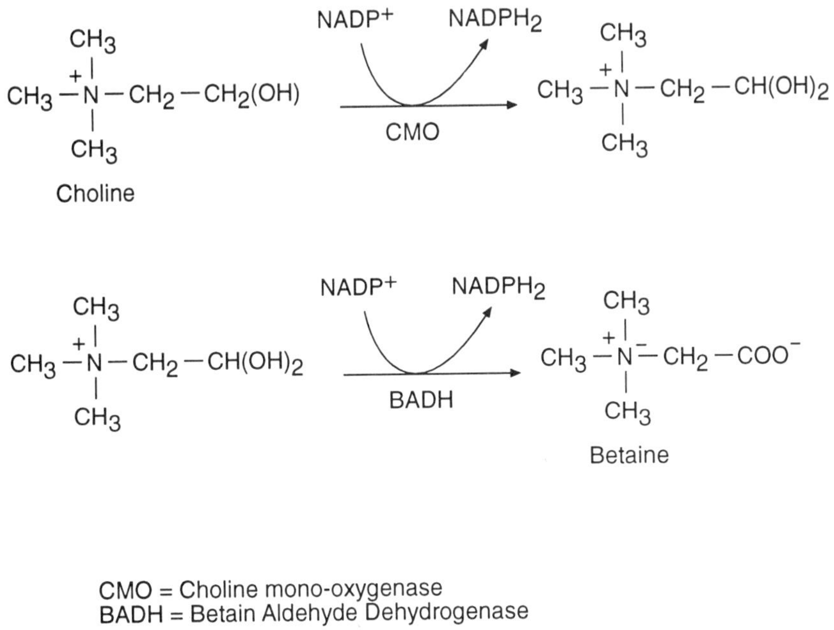

CH_3
$CH_3-\overset{+}{N}-CH_2-CH_2(OH)$ → ($NADP^+$ → $NADPH_2$; CMO) → $CH_3-\overset{+}{N}-CH_2-CH(OH)_2$
CH_3

Choline

CH_3
$CH_3-\overset{+}{N}-CH_2-CH(OH)_2$ → ($NADP^+$ → $NADPH_2$; BADH) → $CH_3-\overset{+}{N}^{-}-CH_2-COO^-$
CH_3

Betaine

CMO = Choline mono-oxygenase
BADH = Betain Aldehyde Dehydrogenase

Figure 8.9 Two-step pathway required to convert choline to betain during osmotic adjustment. The enzyme that regulates the first step is choline monooxygenase (CMO), and the enzyme that regulates the second step is betain aldehyde dehydrogenase (BADH).

The osmotic relationships of a tissue can be adjusted in several ways. Water loss from the tissues will cause an increase in osmotic potential, but this is not osmotic adjustment. Osmotic adjustment can also occur by a change in the total number of osmotically active compounds (type 1 adjustment). This can be observed in the results of a pressure–volume curve when the osmotic potential at full hydration (Ψ_s^{100}) decreases (becomes more negative) but the relative symplastic volume (F^*) remains the same (Figure 8.10A). This is considered "true" osmotic adjustment because the total number of solutes per cell has increased. This type of osmotic adjustment results in a changing relationship between Ψ_s^{100} and Ψ_s^0 (osmotic potential at full turgor and at the turgor loss point). As the number of solutes increase in the cell, the difference between these two values becomes smaller.

Osmotic adjustment can also occur by changing the ratio of symplasmic and apoplastic water (type 2 adjustment). Since the symplasm contains a majority of the solutes, a decrease in symplastic volume relative to the apoplast will cause an increase in effective osmotic concentration. A pressure–volume curve would reflect this type of osmotic adjustment when there is little to no change in Ψ_s^{100} and a decrease in F^* (Figure 8.13).

A third type of osmotic adjustment occurs when there is a decrease in both Ψ_s^{100} and F^* (type 3 adjustment). The osmotic lines of pressure-volume curves are parallel as this type of osmotic adjustment occurs (Figure 8.10B). Type 3 osmotic adjustment results in a consistent relationship between Ψ_s^{100} and Ψ_s^0 as osmotic adjustment occurs.

True osmotic adjustment (type 1) will predominate in nongrowing systems where differentiated tissues require a physiological osmotic adjustment. Type 2 osmotic adjustment is likely to predominate when growing systems are building new tissues under a limited

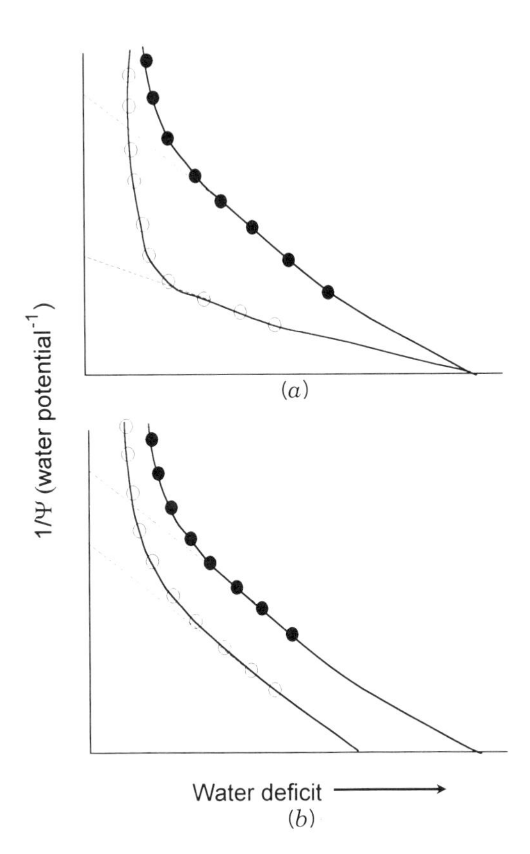

Figure 8.10 Two possible ways that osmotic adjustment can occur in plants as demonstrated by pressure–volume curve relationships. (A) This is true osmotic adjustment when the number of osmotica increase and there is no change in symplastic water fraction. (B) The case where solutes are increasing and symplastic fraction is decreasing in a coordinated manner.

water availability (discussed later in the Chapter). Type 3 osmotic adjustment can occur in growing systems in which the new tissues have used a balanced change in physiology and anatomy to maintain turgor potential at lower tissue water potential.

Osmotic potential varies in plant tissues on a number of temporal scales. During cell expansion, osmotic potential characteristically decreases, which enables continued influx of water for turgor maintenance and cell expansion. Thus, an increase in total solutes is required to maintain turgor during cell enlargement in all tissues (Morgan 1984). Seasonal changes in osmotic relationships commonly occur during slow changes in water availability (Dileanis and Groenveld 1989). Leaves that are produced late in the growth season often retain the capacity to osmoregulate longer than those produced early in the growing season. It is often difficult to separate osmotic changes, due to a changing leaf population from environmental impacts on osmotic adjustment in one leaf population, because leaves continue to be produced throughout the changing environmental conditions. Studies on short flush species (species that produce new clipping production during relatively short periods of the growing season) have shown that turgor is maintained in differentiating tissues by changes in osmotic potential and cell wall elasticity (see Chapter 7), but in mature tissues by osmotic adjustment alone (Nilsen et al. 1983).

Osmotic adjustment can also occur over a diurnal period (Nilsen et al. 1984). Diurnal adjustment of osmotic potential may be large enough to maintain turgor pressure during relatively large diurnal changes in plant water potential (up to 3 MPa). In some cases, without the diurnal osmotic adjustment, turgor pressure would reach zero in the mesophyll cells by late morning (Nilsen et al. 1984).

Osmotic adjustment can be an effective mechanism for maintaining turgor pressure during mild water stress. The maintenance of turgor pressure will allow stomata to stay open and carbon gain to proceed at water potentials that would have prohibited these functions if osmotic adjustment had not occurred. However, if a water shortage remains for extended periods, osmotic adjustment will not be able to overcome the deleterious effects of water limitation. Measurements of changing osmotic potential at full hydration are in the range 1 to 3 MPa. This limitation to osmotic adjustment may be due to cytoplasmic tolerance of vacuolar salinity and the effectiveness of nonprotein amino acids or carbohydrates to protect enzymes, and other proteins, from ion toxicity.

Osmotic adjustment through increasing ion concentration also is a very effective mechanism of maintaining cell turgor for plants growing in saline habitats, because ions are abundant. In fact, many members of the Chenopodeaceae, commonly saline-tolerant species, use both ions (including Na^+) and nonprotein amino acids for osmotic adjustment.

2. Tissue Elasticity. Water potential drops rapidly with increasing water deficit at the beginning of a pressure–volume curve. This is due primarily to a loss of turgor pressure. The rate at which the turgor pressure decreases with increasing water deficit was defined as the elastic modulus (Chapter 7). A shallow slope of the curvilinear portion in the P–V relationship indicates flexible walls that are able to maintain a larger quantity of turgor to lower water potential.

Both a smaller E and a larger E can be considered as mechanisms for tolerating water limitation. A decrease in E would allow the maintenance of turgor at relatively low water potentials and high tissue water deficit compared to tissues with a higher E. The smaller E results in higher turgor at any water deficit compared to tissue with a higher E value (Figure 8.11A). Therefore, the advantage of adjusting E down is to maintain higher turgor potential at lower tissue water contents.

In contrast, an increase in E could be responsible for minimizing the change in tissue water content during water deficit. A larger E results in a relatively rapid decrease in turgor pressure as water deficit increases, but this is associated with a smaller change in tissue water content as water potential decreases (Figure 8.11B). Therefore, an increase in E stabilizes tissue water content rather than stabilizing turgor pressure. Furthermore, a large E also causes a rapid decrease in tissue water potential with an increase in water deficit. This has been postulated as a mechanism to maintain a large difference between soil and tissue water potential to enhance water uptake from the soil without a significant change in tissue water content. Although many studies support the importance of decreasing E to water stress tolerance, as of now there is little evidence to support the importance of increasing E to water stress tolerance.

Changes in elastic modulus during water limitation are variable among species. In one study of four species of the Owens desert, three of four species decreased elastic modulus during water stress (Dileanis and Groeneveld 1989). Among 10 species growing in a single canyon in the Sonoran desert, there was no indication of changing elastic modulus between wet and dry seasons (Nilsen et al. 1984). One of two species native to Hawaiian lava flow habitats decreased E during water stress, while the other had no change in E (Robichaux 1984). In general, there is more evidence that a decrease in E accompanies a drought period rather than an increase in E.

The E value is not constant across a wide range of turgor pressure for most tissues. Thus cell wall elasticity is sensitive to turgor pressure. The slope of E versus turgor pressure (E_{vat}) is commonly negative. This can be seen in a pressure–volume curve as an increase in the slope of the curvilinear portion with increasing water deficit (Figure 8.12A).

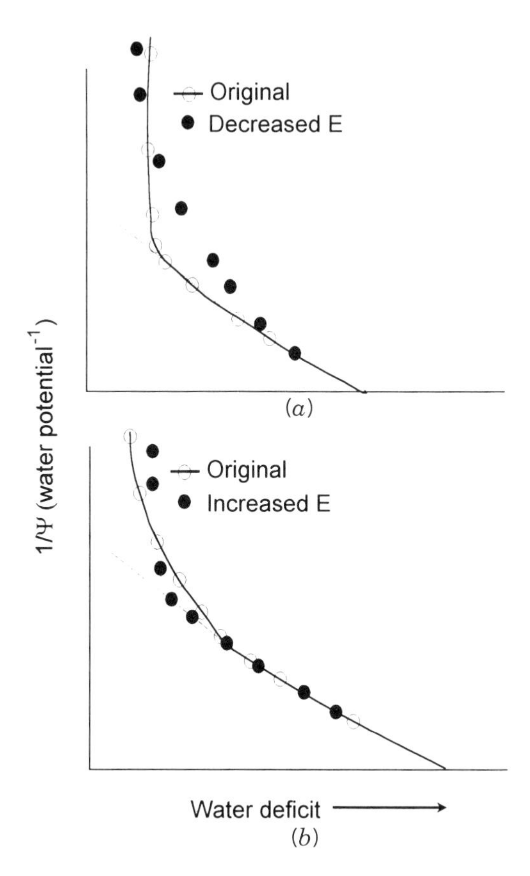

Figure 8.11 Consequences of lower and higher volumetric elastic modulus *(E)* to changes in turgor potential, water deficit, and water potential. (A) Decrease in *E* with no change in osmotic potential or symplastic water fraction. (B) Increase in *E* with no change in osmotic relations or symplastic water fraction.

If E_{vat} is small, turgor pressure is likely to be maintained by elastic modulus. However, if E_{vat} is large, other factors, such as osmotic potential, are most important to maintaining turgor pressure during water limitation (Figure 8.12B). Studies of E_{vat} have found that the relationship between E and turgor is not necessarily linear (Roberts et al. 1981). However, the significance of the different functional relationships between E and turgor potential have yet to be explored.

3. Symplastic Water Fraction. The apportionment of water among compartments in tissues can be considered a mechanism to tolerate water limitation with a low water potential. The most basic separation of water is between the apoplast and the symplast. When the symplastic water fraction decreases relative to the apoplastic fraction, solutes are held in a smaller proportion of tissue water. This increases the osmotic concentration without requiring more solutes and may assist in maintaining turgor potential at lower water potentials (Figure 8.13). Commonly, the ratio of symplastic to apoplastic water is established during the development of the tissue; thus reapportionment of water between symplasm and apoplast requires the construction of new tissue. Several studies have shown an association between decreasing symplastic water fraction and water limitation (Pavlic 1984, Calkin and Pearcy 1984). However, in several studies symplastic water fraction increased with a decrease environmental water availability (Ritchie and Shula 1984, Meinzer et al. 1986). This relationship between water limitation and symplastic water

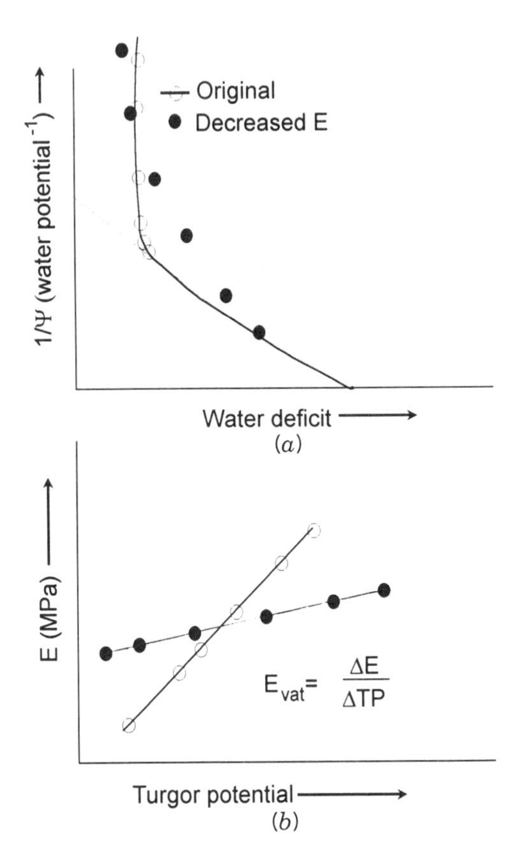

Figure 8.12 Consequences of a change in the slope of elastic modulus against turgor pressure (E_{vat}) to turgor maintenance. (A) Pressure–volume curves representing two samples with a different E_{vat}. (B) Relationship between E and turgor pressure for the two pressure–volume curves shown in (A).

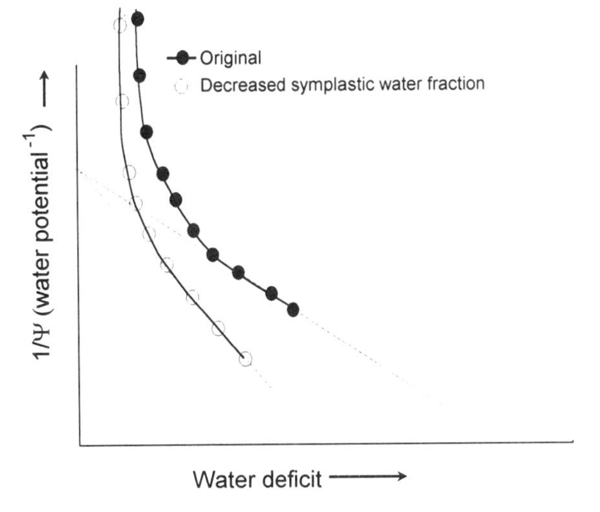

Figure 8.13 Consequence of a decrease in symplastic water fraction on the osmotic potential relationships of cells as demonstrated by pressure–volume curves. The result is a change in osmotic potential without any change in the number of osmotically active solutes in the cytosol.

fraction can only be explained by the transfer of solutes from the apoplast to the symplast causing water to flow into the symplast.

During the process of seed germination the apportionment of water between symplast and apoplast may have significant consequences to embryo water relations. In cucumber seeds, the water potential gradient during early seed development (20 to 40 days after anthesis) is from seed to pericarp, which causes a reduction in seed water content. Later in development the water potential gradient is in the other direction, resulting in the seed gaining water content and turgor pressure for cell elongation during germination. Changes in the allocation of water to symplast and apoplast is the basis for the changing gradient of water potential from seed to pericarp (Welbaum and Bradford 1988).

Movement of water from symplast to apoplast or from one tissue to another can be induced by changes in mucopolysaccharide concentrations. In *Opuntia ficus-indica,* water is retained in the chlorenchyma (part of the cortical tissue) at low water potential because of a buildup of mucopolysaccharides in the water storage parenchyma (Goldstein et al. 1991). Desiccation can also be induced during cold conditions when ice forms in intercellular spaces and draws water out of the symplasm. In a manner similar to that stated above, *Opuntia humifusa* increases mucopolysaccharides in the apoplast, which causes the retention of water in the symplast after intracellular freezing occurs (Loik and Nobel 1991).

4. Cavitation Susceptibility. At low xylem water potential, embolisms may develop in xylem cells. When a xylem cell has an embolism, cavitation occurs, which precludes water movement through that cell. As the percentage of cells that are cavitated increases in a stem, the hydraulic conductance decreases. Therefore, species that are able to resist the formation of embolisms will be able to continue transpiration at lower water potentials than will species that are susceptible to embolism.

Xylem susceptibility to embolism is different among species and among developmental stages within one species (Crombie et al. 1985, Tyree and Sperry 1989). Since xylem anatomy and morphology are critical to constraints placed on hydraulic conductance by cavitation, models of cavitation susceptibility relate to xylem morphology. Water-stress-induced embolisms are most likely developed as a result of air seeding from adjacent cells through pits in xylem element cell walls. The meniscus inside a pit between an air-filled cell and a water-filled xylem cell can withstand a certain amount of pressure differential (ΔP) before succumbing. When the pressure differential between the air-filled cell and the water-filled xylem cell exceeds ΔP (referred to as the *bubble pressure*), small air bubbles are injected into the xylem stream (Sperry and Tyree 1988). The bubbles disrupt the cohesion of water in the xylem and the water retracts in both directions, causing an embolism (Figure 8.14). The magnitude of bubble pressure is dependent on the surface tension of xylem water (σ_{xylem}) and the diameter of the pits in the cell walls (D):

$$\Delta P = 4\,\frac{\sigma_{xylem}}{D} \tag{8.2}$$

Therefore, those species with small-diameter pits will have the lowest susceptibility for tension-induced cavitation (Tyree and Dixon 1986), and those species with sclariform or other larger pits have a higher susceptibility to cavitation. For this reason cavitation begins in the early wood (early wood has more sclariform pitted cells) and is less likely to develop in late wood. Furthermore, species from moist habitats have a higher vulnerabil-

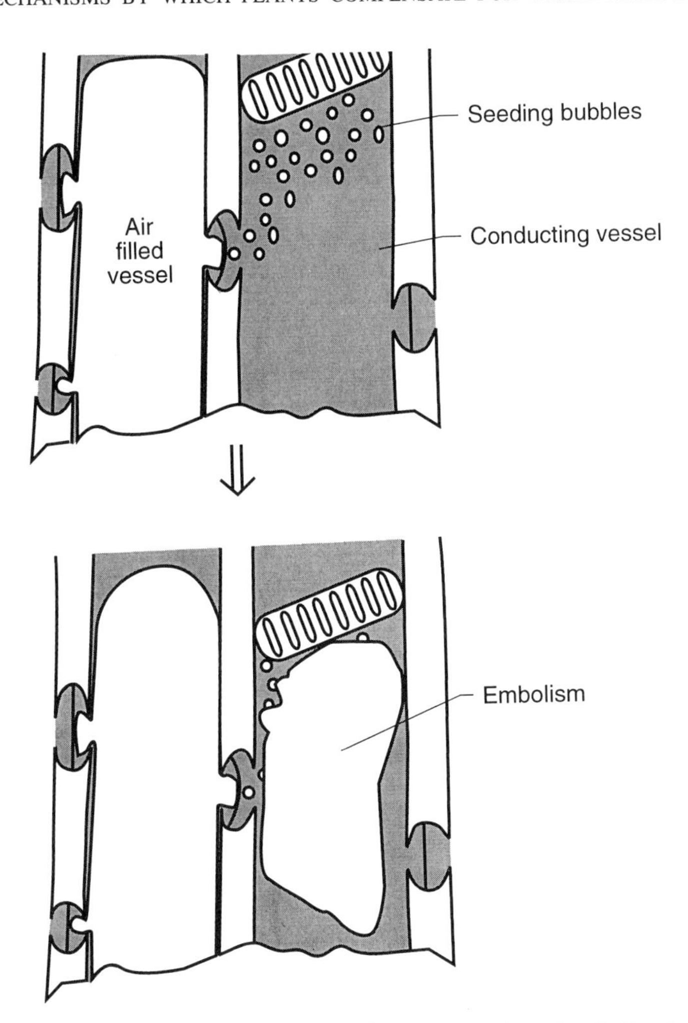

Figure 8.14 Diagram of the "air seeding" model for the induction of embolism in xylem vessel elements due to low water potential. Small bubbles of air are drawn into the fluid-filled cells through vessel pores that contact an air-filled vessel. The small bubbles can coalesce and eventually form an embolism.

ity to cavitation compared than do species from drought-prone environments (Sperry et al. 1988, Kolb and Davis 1993).

Xylem vulnerability to cavitation can also be thought of as a mechanism to reduce water use during moderate dry periods. That is because the loss of some xylem conduits will reduce hydraulic conductance and cause reduced leaf water potential, which will induce stomatal closure and lower transpiration. The result is a conservation of soil water resources. In addition, water that is redirected from the cavitated elements enters into a smaller cross sectional area of functional xylem conduits, which will decreases tension in the xylem. This can result in improved tissue water potential in the leaves. Therefore, small amounts of cavitation can be beneficial to plants under soil water limitation (Tyree and Yang 1990, Lo Gullo and Salleo 1992, Sperry 1995).

In perennial plants, individual branches may have different vulnerabilities to embolism. This is due to the fact that the branches are of different developmental stages and were developed under varying climatic conditions. During extreme water stress, some branches will suffer extensive cavitation before others. When those vulnerable branches become heavily cavitated, water is redirected to the rest of the plant. Axis splitting in desert plants may be related to the same mechanism of water conservation by "sacrificing" vulnerable sections of the plant during periods of particularly low water availability (Jones 1984).

*5. **Desiccation Tolerance.*** Several groups of species have the ability to tolerate almost complete desiccation and regain metabolic activity rapidly upon rehydration. Commonly, these species are lower vascular plants. For example, sphagnum mosses, lichens, several algae, and some ferns are capable of desiccation tolerance (Gaff 1980). In these plants, water evaporates from the tissues, cytoplasm shrinks in volume, and the plasmamembrane pulls away from the cell wall. This results in a highly concentrated cytoplasm and large air spaces inside the cell wall (Figure 8.15).

Three criteria must be met by the tissues of desiccation-tolerant species. First, cellular metabolism must be so designed that high concentration does not affect basic cellular machinery. Second, there must be relatively infrequent connections between cells by plasmodesmata because cell shrinkage would break the connections. Third, the cell wall must be able to withstand extensive dehydration without losing structure.

There is considerable variation in desiccation tolerance among species. Many species require several days after rehydration to recover maximal physiological activity. A few species, such as *Tortula ruralis* (resurrection fern), are able to resume maximum metabolic activity within 30 minutes of rehydration. It is clear that there are very few plasmodesmata between cells of desiccation-tolerant plants. The mechanisms by which the dehydrated cytoplasm can remain viable and rapidly rejuvenate upon rewetting are not well understood, but the concept of "glassy cytoplasm" may be important. Species that

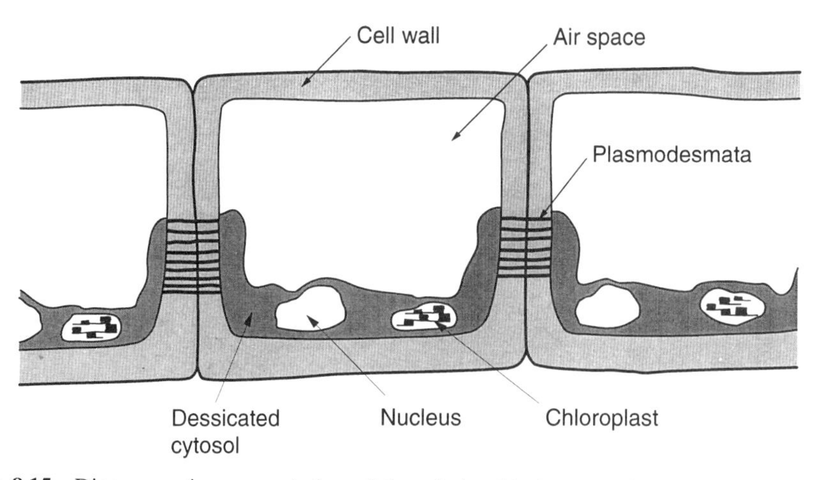

Figure 8.15 Diagrammatic representation of the relationship between the plasmamembrane and cell wall for a resurrection fern in the desiccated state. Note that the cytoplasm has pulled away from the cells wall except in the cell wall regions containing fields of plasmodesmata.

experience desiccation may concentrate specific low-molecular-weight carbohydrates such as raphanose. At low water content, high concentrations of raphanose convert the cytoplasm into a "glass" state, which may protect organelles from damage (Koster 1991). This concept is covered in detail in Chapter 14.

C. Drought Tolerance with High Water Potential

It is possible for species to resist damage during water limitation by maintaining a high tissue water potential. The ability to maintain a high water potential when soil water potential decreases or atmospheric demand increases must include a mechanism of decoupling the plant from normal hydraulic flow dynamics. Such mechanisms can occur at the leaf, stem, or root level. Species that utilize these methods of decoupling their water flow from the environment do not normally utilize osmotic adjustment or changes in cell wall elasticity to tolerate water limitation.

1. Reduction of Water Loss. If a plant can reduce total water use by the canopy, soil water will be conserved. Changes in leaf conductance properties can result in decreased water use. In particular, changes in the relationship between atmospheric vapor pressure deficit and leaf conductance can lower transpiration during periods of high atmospheric demand for water (Figure 8.16). Decreased conductance at high atmospheric vapor pressure will reduce the negative impact of atmospheric demand for water on cellular water potential (Fanjul and Jones 1982).

Leaf conductance decreases linearly with increasing vapor pressure gradient between the leaf and air (VPG) in both herbaceous and woody species. Even though transpiration may increase over this range of VPG (Turner et al. 1984), water loss increases less than it would have if stomatal conductance did not decrease. Thus, the decrease in leaf conduc-

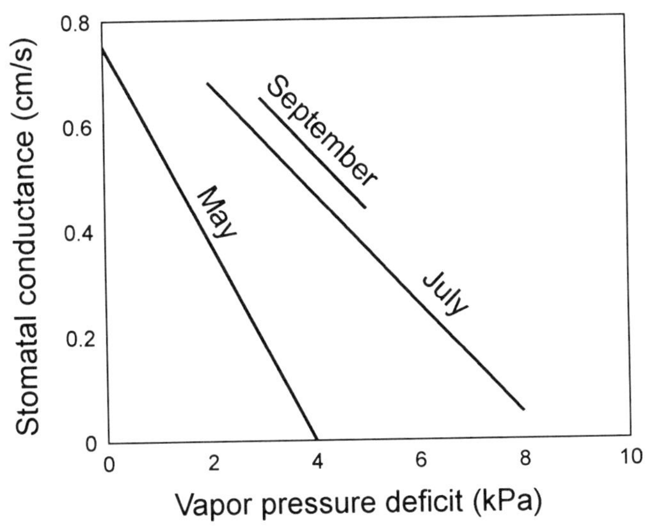

Figure 8.16 Relationship between stomatal conductance to water vapor and atmospheric vapor pressure deficit in *Prosopis glandulosa*. Notice the change in slope of the relationship as the season progresses from May to July. (Redrawn from Nilsen et al. 1983.)

tance moderates the impact of VPG on leaf water potential. It is often difficult to determine if VPG has a direct impact on stomatal conductance or if its impact is directly upon guard cell water potential. Studies of epidermal peels have demonstrated a rapid and reversible impact of dry air on the aperture of a single stoma (Lange et al. 1971). The ability of individual stomata to respond directly to vapor pressure when the rest of the epidermal peel is fully hydrated suggests a direct impact of VPG on stomatal aperture (reviewed in Chapter 6).

In many species the slope of the relationship between VPG and leaf conductance becomes more shallow during seasons characterized by greater water deficit (Figure 8.16). For example, in desert taxa the relationship between VPG and leaf conductance becomes shallow (lower conductance at any VPG) during the dry seasons of the year (Forseth and Ehleringer 1983). The shallow response of conductance to VPG during the desert summer is most likely induced by lower tissue turgor potentials (Forseth and Ehleringer 1983, Nilsen et al. 1984).

Species from different habitats have different relationships between VPG and conductance. Those from coastal habitats have a greater sensitivity to VPG than those from dry habitats (Mooney and Chu 1983, Nilsen 1995). When a large diversity of species are evaluated, those species with a high ratio of leaf area to root area have a high sensitivity to VPG, while those with a large root/low leaf area ratio have less sensitivity to VPG (Bunce 1986). Thus, the relative sensitivity of conductance to VPG may depend on the dynamics of flow limitations for water in conducting tissue.

Midday closure of stomata occurs in many species growing in climates with high atmospheric vapor pressure deficit. During midday, stomatal conductance decreases, which reduces the atmospheric demand on a plant's water conducting system. In the afternoon, when vapor pressure decreases (due to decreasing temperature), stomatal conductance increases again. This diurnal pattern of conductance is also accentuated during periods of low water availability (Nilsen and Muller 1981b) and high temperature (Roessler and Monson 1985). Midday stomatal closure may be a direct result of optimal stomatal regulation by plants (discussed later in the chapter).

Stomatal closure may also reduce transpiration in response to low soil moisture. In this case, when the roots dry out they send a signal to the guard cells, which induces stomatal closure (Bates and Hall 1981). The signal molecule is probably abscisic acid (ABA), because stomata are extremely sensitive to abscisic acid. Physiological regulation of stomatal aperture, including the effects of ABA and cytokinins, was covered in Chapter 6.

Water loss can also be reduced if the plant decreases the total transpiring surface. Many species of dry habitats such as Mediterranean or desert regions are drought deciduous. In Mediterranean habitats many species that lose leaves during the onset of drought have a dimorphic leaf population. For example, *Salvia melifera,* a chaparral-drought deciduous shrub in the Lamiaceae, loses over 90% of its leaves at the onset of the dry season. The 10% that remain are smaller leaves in shoot axes, and they are rotated to expose the white underside (Gill and Mahall 1986). *Lotus scoparius,* a suffrutescent summer deciduous chaparral shrub, preferentially drops larger leaves during the beginning of the dry season and retains smaller leaves until later in the dry season, when all leaves abscise (Nilsen and Muller 1981b). In fact, dimorphic leaf populations are common on Mediterranean species (Margaris 1977, Orshan 1972, Westman 1981). The dry season leaves are smaller and less frequent, which reduces the canopy transpiration rate and conserves water while maintaining metabolic activity.

In desert habitats, many species become aphyllous during dry periods. Many of these species have green photosynthetic stems (Gibson 1983). The stems have a vertical orientation which reduces the potential for photoinhibition during high light and low water availability (Ehleringer and Cooper 1992). Furthermore, the stems have lower conductance and higher water use efficiency than those of leaves (Comstock and Ehleringer 1988, Nilsen 1992a). In fact, the $\delta^{13}C$ of stems is characteristically lower than that of leaves on the same plant (Ehleringer et al. 1987b), which indicates that stems have a higher water use efficiency than leaves. The photosynthetic stems can continue to provide carbon for the plant and severely reduce the total water use by the canopy (compared to the same canopy with leaves). Therefore, a high water potential can be maintained in stem photosynthetic species relative to other desert species.

Water use efficiency (WUE), discussed in Chapter 6, is an important aspect of plant adaptation to drought. An increase in WUE indicates that more carbon can be accumulated for growth with the use of less water. Thus, an increase in WUE is a water conservation measure. In Chapter 6 the impact of photosynthetic biochemistry (C_3, C_4, and CAM) on water use efficiency was discussed. The differences between water use efficiency of these three physiological types is reflected in their distributions across the landscape (Teeri and Stow 1976, Teeri et al. 1978). However, the question "What is the importance of increasing WUE to C_3 plants during water stress?" is also significant. Commonly, when a plant experiences a negative change in water potential, stomata close, reducing transpiration and carbon accumulation. Initially, transpiration is reduced more than photosynthesis, C_i decreases, and WUE increases. An increase in WUE also occurs when stomata close in response to increased VPG. Is this change in WUE important to plant survival during drought, or is it simply a consequence of the physical diffusion properties of gases and limitations to photosynthesis?

In any functional analysis of water use efficiency the components of the carbon gain–water loss relationship need to be defined. In agricultural systems yield (Y) of a crop is often related to total evaporation from the field (plant transpiration plus evaporation) to produce a water use efficiency index. If the water loss term is restricted to transpiration, the water use index is referred to as a *transpiration ratio.* In various crops there is a relatively constant transpiration ratio within physiological types. For example, CAM crops have a WUE of 20 to 35×10^{-3} (g dry matter produced/g water transpired). That value for C_4 crops is 2.75 to 4.0×10^{-3}, and for C_3 crops it is 1.0 to 2.5×10^{-3}. On the other hand, *instantaneous water use efficiency* refers to a comparison of CO_2 assimilated versus H_2O transpired by individual leaves or canopies (usually on a molar basis).

Studies of instantaneous WUE by gas exchange techniques are difficult in field situations because environmental conditions around the leaves are always changing, so it is hard to get an average plant water use efficiency. Even in pot studies, the use of gas exchange measurements to determine whole plant WUE are labor intensive. However, ranking genotypes or species based on relative WUE is accomplished consistently and precisely by measuring carbon isotopic composition of the tissue (Chapter 5). Recent studies have shown that WUE is positively associated with increases in bread wheat plant weight. Genotypes with a greater WUE produced more biomass under well-watered or drought conditions (Bahman et al. 1993). However, grain yield decreased in barley with increasing water use efficiency (Acevedo 1993). The relationship between WUE and yield is complex and will vary for specific species in specific habitats. Therefore, simply breeding for high WUE in crops may not necessarily translate into higher yield in low-water-availability sites.

The relative significance of water use efficiency to plants has been studied primarily at the theoretical level. A general model describing water use efficiency (Cowan 1977) is

$$\frac{P_n}{E} = \frac{C_a[1 - (C_i/C_a)]}{1.6(e_l - e_a)} \quad (8.3)$$

where P_n is the net photosynthesis rate, E the transpiration rate, p_a the partial pressure of CO_2 in the atmosphere, C_i the partial pressure of CO_2 in the intercellular spaces, e_l the vapor pressure in the leaf, and e_a the atmospheric vapor pressure. Since water use efficiency is relatively stable for any species, this relationship can be simplified as

$$\frac{P_n}{E} = \frac{K}{e_l - e_a} \quad (8.4)$$

where K is a species-specific constant.

These relationships serve well to describe the instantaneous water use efficiency, but they are inadequate for describing a longer-term view of water use efficiency. There are other processes in plants that result in carbon losses (volatilization. etc.), and these will lower water use efficiency. In addition, loss of water through the cuticle is an additional water loss pathway that is not taken into account in the models of water use efficiency described above. A model that accounts for these two processes is

$$\frac{P_n}{E} = \frac{C_a[1 - (C_i/C_a)](1 - \Phi_c)}{1.6(e_l - e_a)(1 - \Phi_w)} \quad (8.5)$$

where Φ_c is the carbon dioxide lost from the leaf by sources other than net photosynthesis and Φ_w is the water lost through the cuticle.

The influence of cuticular conductance on the relationship between water use efficiency and net photosynthesis can result in an optimum stomatal aperture (Figure 8.17). At low cuticular conductance (high resistance) water use efficiency increases as stomata close. In contrast, at higher cuticular conductance (lower resistance) a distinct optimum water use efficiency is reached at moderately low stomatal aperture. A very similar scenario can be developed for the influence of boundary layer resistance of the functional relationship between water use efficiency and stomatal conductance. That is, the lower the boundary layer resistance, the more likely there will be optimal water use efficiency.

Any given set of environmental conditions in combination with a specific leaf morphology will potentially result in a specific stomatal conductance that will result in optimum water use efficiency. But environmental conditions are not stable over the season, day, or even seconds in a subcanopy environment. This led researchers to consider if there is a theoretical optimal stomatal behavior that maximizes water use efficiency in the face of changing environmental conditions (Cowan 1982). The development of this theory is dependent on the relationships between transpiration, photosynthesis, and their second derivative (referred to as λ). As stomata open transpiration progressively increases faster than net photosynthesis (Figure 8.18A). The second derivative (λ) of this relationship (the instantaneous change in transpiration/the instantaneous change in net photosynthesis) increases gradually over a wide range of stomatal conductance, then increases rapidly at a critical stomatal conductance (Figure 8.18B). Optimal stomatal behavior predicts that λ should be kept below that critical stomatal conductance, and λ will decrease to smaller values when water is less available. Thus, as the climatic conditions change over a day cycle, a constant λ will result in maximum water use efficiency. Also, under times of low water availability, λ will be relatively small (but constant), to maximize water use efficiency.

One result of optimal stomatal behavior theory is an explanation of midday stomatal

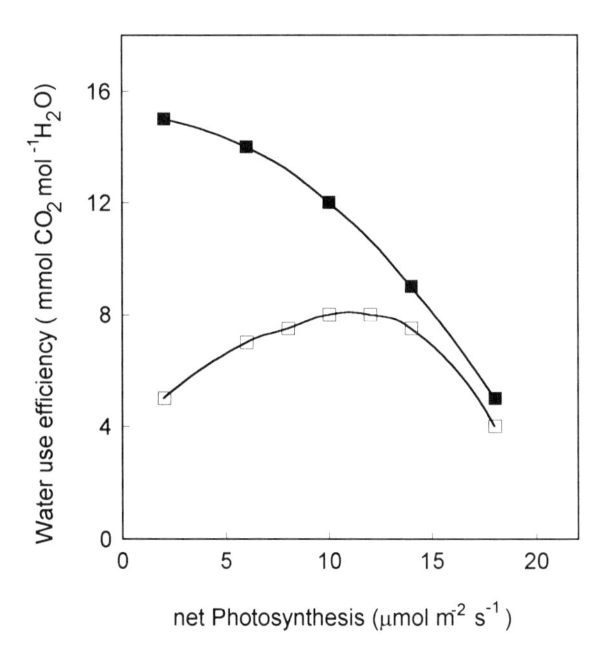

Figure 8.17 Plot of the relationship between water use efficiency and net photosynthesis. The two different lines represent different cuticular resistance against water vapor movement. Filled squares represent a situation with high cuticular resistance, and open squares represent a situation with low cuticular conductance. In this case the reduction in photosynthesis is due to a reduction in stomatal conductance. (Redrawn from Jones 1992.)

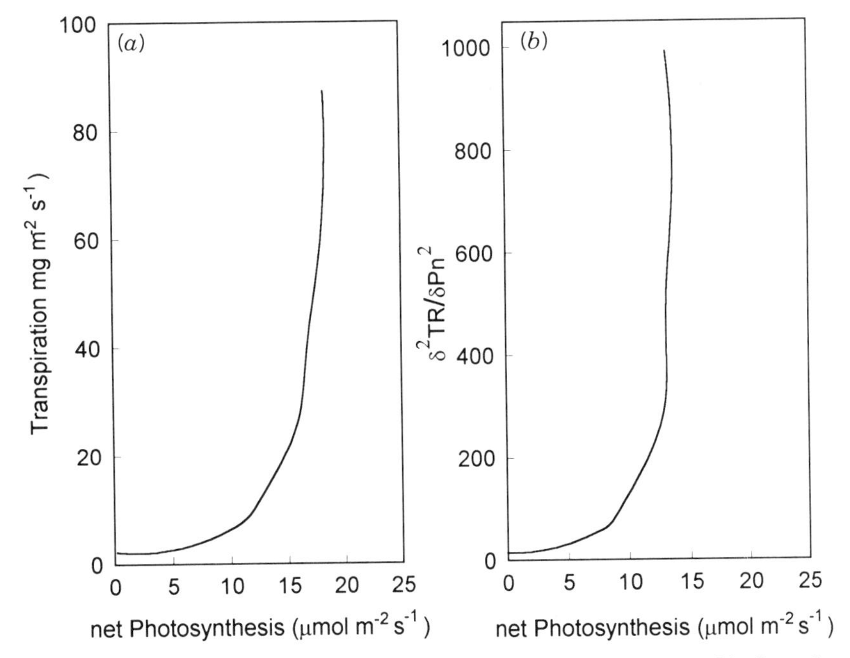

Figure 8.18 (A) Functional relationship between leaf transpiration rate *(E)* and leaf net photosynthesis (P_n). (B) Functional relationship between the second derivative of the ratio between leaf transpiration and leaf net photosynthesis (λ), and stomatal conductance. Notice the dramatic rise in λ above a net photosynthetic rate of 10 (μmol m^{-2} s^{-1}). (Redrawn from Jones 1992.)

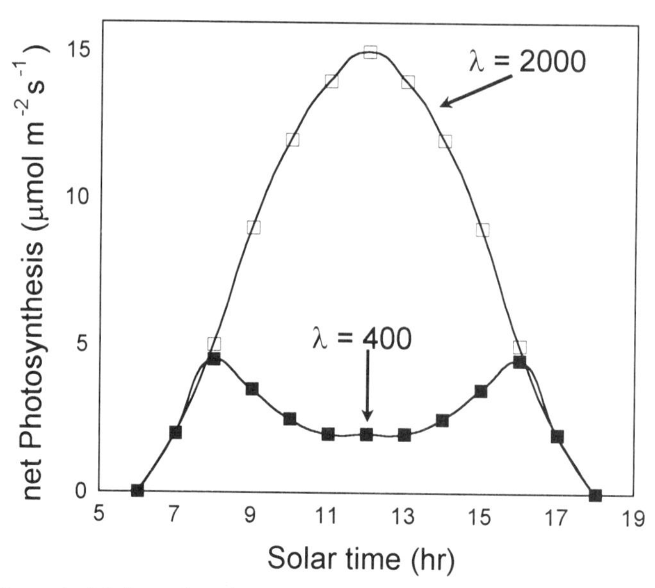

Figure 8.19 Theoretical daily cycles of net photosynthesis (P_n) with a relatively high λ and a relatively low λ. (Redrawn from Jones 1992.)

closure. At a high value of λ, a diurnal cycle of photosynthesis will rise in the morning, stabilize over the afternoon, and decrease in the evening (Figure 8.19). In contrast, a low value of λ will result in a significant midday depression of stomatal conductance and net photosynthesis. Midday depression of stomatal conductance is known to occur in many species only when water is relatively limiting. Consequently, field results agree with the theory that λ is decreased in plants to maximize water use efficiency during periods of low water availability. Actually, if one calculates the advantage received by optimal stomatal behavior in comparison to constant stomatal conductance, the difference is a gain of approximately 10% in water use efficiency. The question that ecological physiologists have to answer now is how important is that 10% gain in WUE.

2. Enhancement of Water Accumulation. In addition to optimally regulating transpiration, species can maintain high water potential by utilizing larger proportions of soil water. Root systems that travel deep into the soil (including *Phreatophytes*) to exploit water resources unused by most species are commonly found in drought-prone areas. Deep root systems also have the capacity of mining deep soil water and releasing it in surface soils. This hydraulic lift occurs at night when the phreatic roots are saturated with water and the surface roots are in dry soil. The water potential gradient between the soil and root pulls water out of the root into the soil (Caldwell and Richards 1989). Thus, water used by transpiration from surface soils during the day can be replaced by hydraulic lift from the phreatic zone at night.

Other morphological characteristics of root systems also are important for soil water absorption. For example, cacti have extensive root systems along the upper surface of the soil. This geographically wide distribution of roots ensures a strong possibility that rainwater will be absorbed by the cactus. During periods of extended drought, some cacti sever connections with the root system because in dry soil this extensive surface of root would tend to draw water out of the plant.

As discussed earlier in this Chapter, some species can absorb water by organs other than roots (Rundel et al. 1991). For example, *Tillandsia* species in the Peruvian and Ecuadorian Andes absorb most of their water from fog because there is little to no rainfall in the Atacama desert. These terrestrial bromeliads are distributed in a narrow band corresponding to the elevation of maximum fog duration. Also in the Atacama desert resides a species that utilizes salt crystals to extract water from the air (Rundel et al. 1991). Epiphytic bromeliads of tropical seasonally dry forests can utilize the water contained in a water catchment basin formed by their leaf bases. Although there are a few rare species that utilize organs other than roots to accumulate water, most species must depend on soil water resources.

Hemiparasitic species utilize water resources from their host. To accomplish this, the water potential of the hemiparasite must always be lower than that of the host xylem stream, so that water will divert from the host vessel elements into the parasite conduction system. Mistletoes growing on desert species have higher conductance and lower water potential values than those of the host (Ehleringer et al. 1985).

When water is unavailable in the soil, plants can utilize their own water resources to maintain transpiration and minimize the reduction in water potential. High capacitance in roots or stems is found in several desert taxa and is identified by the succulent nature of the root system or stem (Nilsen et al. 1990, Holbrook 1995). Water is stored in various tissues of stems or roots during periods of high soil moisture and humid atmosphere. During the dry season, transpiration can continue because water is being diverted from the capacitance of the plant (Holbrook and Sinclair 1992) to the leaf. In Baja California, a series of stem succulent trees employ this technique to maintain high water potential during the dry season (Nilsen et al. 1990). Those species with the highest ratio of capacitance to leaf area have the smallest difference between predawn and midday water potential (Figure 8.20) among species in any habitat, Those species with high capacitance are buffered

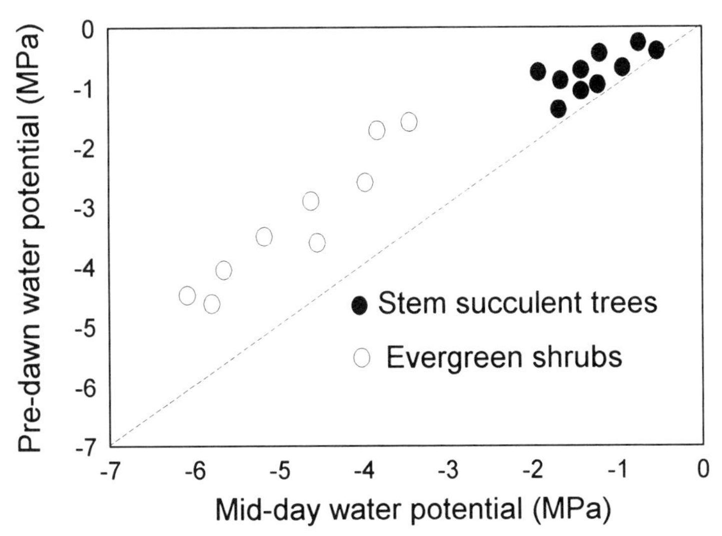

Figure 8.20 Plot of the predawn against midday leaf water potential for species in a desert wash of Baja California over an annual cycle. The dashed line represents no difference between predawn and midday water potential. Notice that values for stem succulent species are close to the no-difference line, while evergreen and deciduous shrubs have a greater difference between predawn and midday water potential. (Redrawn from data in Nilsen et al. 1990.)

from short-term changes in environmental water availability. Furthermore, those species with the greatest capacitance have leaves with low solute concentration and high elastic modulus (i.e., they loose turgor pressure at relatively high water potential).

It is important to recognize that the value of the capacitor is dependent on the quantity of water retained and the ability of the tissue to release water (quality of the capacitor). Succulent stems that retain water in cells with extensive mucopolysaccharides do not have a large amount of water available. This is because mucopolysaccharides contain only a small volume of water per volume of cell (Loik and Nobel 1991).

VI. SUMMARY

All terrestrial plants, whether in agricultural or natural systems, experience a temporally and spatially heterogeneous water environment. Depending on the evolutionary history of the particular species, some location or time will have an inadequate water availability for optimal plant performance. The inadequacy could stem from a low ratio between the supply of water and the absorptive surfaces, an excessive demand by the transpiring surfaces, or both. Without compensation for the inadequacy of water supply, a multitude of physiological consequences may occur.

The first processes affected by water limitation are cell expansion, nitrate reduction, and protein synthesis. Reductions in cell expansion are followed by a cascade of impacts on fundamental physiological processes such as membrane function and transcription. These impacts on fundamental cell physiology need not occur in all plants when they experience water supply limitation. In fact, there are a wide array of mechanisms utilized by plants to compensate for water limitation and avoid the deleterious effects.

In general, mechanisms utilized to avoid the detrimental impacts of water limitation on plants occur in suites of characters in each species. Although adaptations to water stress can be classified as avoidance or tolerance, species will often utilize a combination of mechanisms. For example, osmotic adjustment may occur in leaf tissues to maintain turgor at low water potential, at the same time that lower conductance and leaf fall reduce water use. Furthermore, not all species in a given environment will employ the same suite of characters. Some may adjust by changes in timing of developmental stages, while others will adjust hydrodynamic aspects of water flow, or water potential properties of cells.

Exploration for genetic regulation over water stress adaptation mechanisms will be very difficult to accomplish. Water stress adaptations involve suites of characters regulated by a number of interacting gene systems in any particular species. Thus, it is unlikely that one gene or gene family will be identified as that which confers water stress tolerance in any species. Nevertheless, water stress is important to the growth and evolution of plants and has stimulated a strong interest in searching for molecular regulation of water stress adaptation (Chapter 14).

STUDY-REVIEW OUTLINE

Availability of Water in the Environment

Soil-Induced Water Limitation

1. Soil water availability is dependent on the quantity and quality of water in the soil solution.
2. Sandy soils have a lower retention of total water than do clay-dominated soils, but the water in sandy soils has a higher water potential at any soil water content than in clay soils.
3. Organic matter increases soil water holding capacity by its effects on soil porosity.

4. Increasing soil salinity decreases water availability to plants because it lowers soil osmotic potential.
5. Other factors that affect soil water quality and quantity are soil stratification, compaction, and the action of invertebrates.
6. Soil hydraulic conductance regulates the rate at which water flows to the root surface through soil pores.
7. The factors that regulate hydraulic conductance are soil texture, soil water potential gradient between the root and soil pores, the soil texture, and the nature of soil aggregates.
8. Root systems of plants have a diversity of possible architectures that regulate what fractions of the heterogeneous soil water resource each plant species utilizes.
9. The species composing any community will have a diversity of root architectures, resulting in water extraction from various soil locations.

Non-Soil-Induced Water Limitation

1. Water limitation in plants can be induced by atmospheric conditions as well as low precipitation.
2. Topographical location and regional climate regulate precipitation at any one site. Precipitation can vary from 2000 to a few millimeters per year.
3. Commonly, low precipitation co-occurs with high light, heat, and evaporative demand.
4. Many indices have been developed to rank sites based on relative water balance. However, the best measure of water availability to plants is predawn water potential.
5. Plants can wilt in high light or at midday because the vascular architecture is inadequate to supply the water needs of the leaves even though there is adequate water in the soil.

Sensitivity Analysis

1. The water potential characteristic best associated with water stress is the change in tissue water potential, not the absolute water potential.
2. The physiological processes most sensitive to changes in tissue water potential are cell expansion, cell wall growth, protein synthesis, and nitrate reduction.
3. The water potential component that changes most rapidly with changing water content of a tissue is turgor potential.
4. Turgor potential is considered the best indicator of plant water stress.

Effects of Reduced Water Potential on Physiology

1. Reduced water potential reduces the chemical activity of water and thereby modifies the structure of water in the cell.
2. A lower chemical activity of water can cause a change in the structure of the sheath of hydration around proteins and thereby reduce their efficacy.
3. The relationship among intracellular membranes of chloroplast, nucleus, mitochondria, endoplasmic reticulum, tonoplast, plasmalemma, and others will change because the cellular positions of these membranes will change.
4. A loss of turgor may cause a change in the spatial position of transport channels, membrane enzymes, and decrease membrane thickness.
5. A change in cell pressure and the resultant cell wall shrinkage may constrict the entrances to plasmodesmata.
6. The concentration of molecules in specific regions may change due to the loss of water in some subcellular locations.

Effects on Growth

1. Cell growth (expansion) is one of the most sensitive physiological processes to decreasing turgor pressure.

2. Cell expansion can occur only when turgor pressure is greater than the cell wall yield threshold.
3. Cell wall yield thresholds are different among species, organs, and developmental stages.

Effect on Cell Ultrastructure

1. Mild water limitation may disrupt the structure of microbodies.
2. Other observable changes in ultrastructure such as chloroplastic plastoglobules or nuclear membrane disorganization occur only under extensive water-limiting conditions.

Effects on Photosynthesis

1. Stomatal limitation of photosynthesis occurs when water limitation induces stomatal closure, reducing gas exchange and decreasing intercellular carbon dioxide concentration.
2. Nonstomatal inhibition of photosynthesis by water stress includes direct impacts on light reaction components as well as specific effects on rubisco activity.

Effects on Dark Respiration

1. Mild water stress stimulates respiration in many cases.
2. The increase in respiration with water stress may be a result of starch mobilization, causing an increase in respiration substrate.
3. During initial stages of water stress respiration increases and photosynthesis decreases, reducing the net carbon gain of leaves.
4. Photosynthate translocation is reduced during water stress because the leaves are not as strong a source for photosynthates as they were under optimal water conditions.

Effects on Nitrogen Metabolism

1. The rates of nitrate and ammonia accumulation from the soil solution decrease during water stress.
2. The decrease in nitrogen accumulation is not associated with specific effects of water stress on root nitrate or ammonia uptake mechanisms or enzyme kinetics.
3. Since nitrate reduction and protein synthesis is reduced during water stress, inorganic nitrogen pools build up in tissues and inhibit nitrogen accumulation.

Generalized Response to Water Stress

1. Many of the impacts of water stress on metabolism are the same as those due to other stresses such as heat, cold, light, salinity.
2. Limiting one resource can often affect the availability of other resources.
3. Mechanisms that plants use to respond to water limitation often also provide protection against other stresses, such as salinity or cold temperature.

Mechanisms by Which Plants Compensate for Water Limitation

1. There are many possible combinations of factors that plants use to compensate for water limitation.
2. In any community different species experiencing the same environmental water balance will utilize different suits of factors to adjust to the same site water balance.

Escape from Water-Limiting Conditions

1. Escaping water stress usually refers to developmental mechanisms to complete the life cycle when water availability is high and remain in a dormant state when water availability is low.
2. Short-lived plants that flower after producing only two leaves are examples of water stress avoiders.

3. Other examples of drought avoidance are subterranean perennial plants and drought deciduous shrubs.

Drought Tolerance with Low Water Potential

1. These are species that are able to tolerate periods of low water availability without becoming dormant but still experiencing low water potential.

Osmotic Potential Adjustment

1. When the osmotic potential of tissues decreases, turgor potential can be maintained at lower water potentials.
2. True osmotic adjustment (type 1) occurs when the actual number of osmotically active solutes increases in the symplasm without any change in the proportion of water in the symplasm.
3. Type 3 osmotic adjustment occurs when both the absolute number of dissolved solutes changes and the symplastic water fraction changes.
4. Type 2 osmotic adjustment occurs when the proportion of water in the symplasm decreases without any change in the absolute number of dissolved solutes.

Tissue Elasticity

1. Both an increase and a decrease in tissue elasticity have been suggested as mechanisms of tolerating low water availability.
2. If tissue elasticity decreases, the cells can maintain turgor potential at larger cell water deficits.
3. If tissue elasticity increases, cells loose turgor at smaller water deficits but can maintain turgor potential to lower water potential. This serves to maintain tissue water content at a relatively stable value.
4. Most often (based on field studies), if tissue elasticity changes during water stress, it decreases.

Symplastic Water Fraction

1. A decrease in the proportion of water in the symplasm can improve tissue turgor potential.
2. The adjustment in symplastic water fraction normally occurs by the construction of new tissue.
3. The addition of mucilage to the apoplast can reduce the symplastic water fraction.

Cavitation Susceptibility

1. Cavitation of vessel elements in the vascular system interrupts the flow of water from roots to leaves.
2. Those species with a xylem structure that inhibits the formation of embolisms will be protected against cavitation and can maintain water flow in the vascular tissue at low water potential.
3. Conversely, some scientists have argued that plants that have a vascular system that is susceptible to cavitation will conserve water use during periods of low water availability.

Desiccation Tolerance

1. Some species, particularly lower vascular plants, are tolerant of very low water content in their tissues.
2. These species must have few plasmodesmata, be tolerant of concentrated cytoplasm, and have a desiccation-tolerant cell wall.

Drought Tolerance with High Water Potential

1. Drought tolerance with high water potential occurs when species are able to maintain metabolism during water limitation without experiencing low water potential in their tissues.

Reduction of Water Loss

1. Species that are able to reduce transpiration but retain carbon gain (increase water use efficiency) can maintain high water potential during water limitation.
2. Since low atmospheric vapor pressure usually accompanies water stress, those species that can change the relationship between leaf conductance and VPD during water stress can maintain a higher water potential than those species without such a regulation mechanism.
3. Optimal stomatal behavior theory suggests that stomatal conductance is adjusted during climatic changes in order to keep λ relatively constant. λ expresses the second differential of the ratio between transpiration and photosynthesis.
4. When water availability is low, λ is low, and this promotes a midday depression in stomatal conductance. The midday depression maximizes water use efficiency at low water availability.
5. If the total transpiring area can be decreased by preferential abscission of large leaves, total canopy transpiration is reduced but carbon gain is maintained.
6. Complete leaf loss can occur in species that have photosynthetic stems to maintain a high water potential and continue carbon gain.

Enhancement of Water Accumulation

1. Absorption of water from the atmosphere by organs other than roots can assist in maintaining a high water potential when soil water potential decreases.
2. Some species decouple themselves from their roots when their roots are in very dry soil.
3. Internal tissue capacitance can serve as an alternative source of water for plants.
4. Hydraulic lift of deep water to surface soils increases water availability to some plants.

SELF-STUDY QUESTIONS

1. Define the characteristics of soil that regulate the availability of water to plants.
2. What combination of site characteristics would be likely to minimize water availability to plants in a mountain deciduous forest region and a flat grassland region?
3. How would you develop an index that would accurately define the availability of water to plants without measuring the plant?
4. Rank the sensitivity of the following physiological functions to reductions in water potential: stomatal conductance, cell expansion, nitrate reduction, photosynthesis, protein synthesis, cell ultrastructure.
5. In what specific ways might a reduction in water potential affect cytosolic physiological processes?
6. Differentiate between nonstomatal and stomatal limitations to photosynthesis by water stress. By what mechanism does each affect photosynthesis?
7. What are the four osmotically active compounds plants use to adjust tissue osmotic potential? Rank these on the basis of carbon and energy cost to the plant.
8. Differentiate between the type 1, 2, and 3 mechanisms of osmotic adjustment that occur in plants.
9. Which do you think is a more likely mechanism for withstanding water stress: an increase or a decrease in E?

10. What is the difference in the likelihood that large or small xylem vessels will succumb to embolism during water stress? Why?
11. What measurement of a plant's water potential could you take to indicate whether the plant has a high or low capacitance?
12. What are some phenological or life-cycle characteristics of plants that are associated with water stress avoidance or tolerance?

SUPPLEMENTARY READING

Brady, N. C. 1974. The Nature and Properties of Soils, 8th ed. Macmillan Publishing Co., New York.

Koxlowski, T. T. 1973. Water Deficits and Plant Growth, Volume III, Plant Response and Control of Water Balance. Academic Press, New York.

Lange, O., P. S. Nobel, C. B. Osmond, and H. Ziegler (eds.). 1982. Encyclopedia of Plant Physiology, Vol. 12B, Physiological Plant Ecology II. Water Relations and Carbon Assimilation. Springer-Verlag, Berlin.

Smith, J. A. C. , and H. Griffiths. 1993. Water Deficits: Plant Responses from Cell to Community (Environmental Plant Biology Series). Bios Scientific Publishers, Oxford.

9 Flooding

Outline

Objectives

1. Define flooding conditions and the terminology that refers to plants adapted to flooding conditions.
2. Describe the prevalence of flooding conditions in natural and agricultural systems.
3. Explain the environmental characteristics of sites that are flooded or prone to flooding.
4. Describe the detrimental effects of flooding on plants, including anaerobic, nutrient, and mechanical impacts.
5. Cover the various anatomical mechanisms that plants have evolved to compensate for the detrimental aspects of flooding conditions.
6. Explain the physiological adaptations against anaerobic conditions that develop during flooding conditions.

I. INTRODUCTION

The flood of 1994 along the Mississippi and Missouri river drainages of the United States is a vivid image in the minds of many people. Flooding is a dramatic event for the Yellow River in China, where more than 1500 floods have been recorded, with the 1887 flood killing more than 900,000 people (Briggs 1973). In fact, no continent of the globe is exempt from the effects of flooding. The most common definition of flooding is when a region that is usually dry is inundated with water. Most people recognize that flooding events can cause major damage to natural ecosystems, agricultural fields, and human habitation around the world. However, flooded conditions are also a natural part of many ecosystems of the world. Even desert systems have washes that are occasionally flooded. Therefore, the definition of flooding in relation to plant physiology and ecology requires a more subtle view than that used above.

Ecosystems that are normally affected by flooding, called *wetlands,* are diverse in species composition and ecosystem function (Maltby and Turner 1983). This inherent diversity of wetland systems is a consequence of variation in spatial and temporal flooding conditions. There is considerable discussion about the definition of wetlands because of the importance of that term to legislation that regulates land use in wetland systems. Definitions of wetland systems are based on criteria such as flooding, anaerobic conditions, species composition, or a combination of these. The particular set of criteria change among definitions of wetlands, depending on the interests of those who are stating the definition. For example, the U.S. Army Corps of Engineers use an entirely plant-based definition; ". . . those areas that are inundated or saturated by surface or ground water at a frequency and duration sufficient to support, and that under normal circumstances do support, a prevalence of vegetation typically adapted for life in saturated soil conditions . . ." (Maltby 1991). Another definition of international significance is from the Ramsar Convention on Wetlands of International Importance Especially as Waterfowl Habitat, in which wetlands were defined as ". . . areas of marsh, fen, peat land or water, whether natural or artificial, permanent or temporary, with water that is static or flowing, fresh, brackish, or salt, including areas of marine water the depth of which at low tide does not exceed 6 meters" (Maltby 1991). Anyone is entitled to be confused by the breadth and diversity of wetland definitions. Here we will define a *wetland* as a region that experiences saturated soil conditions, at or above the root zone of most species, during at least a month of the growing season. This definition is based on the time required for flooding conditions to affect significantly the metabolism and growth of plants. Under these conditions plants in the region will have mechanisms to compensate for flooded conditions.

Plants found in wetland systems are as or more diverse than the wetland systems themselves. If a plant has any mechanisms to compensate for flooded conditions, whether anatomical, behavioral, or physiological, the plant is termed a *hydrophyte.* Hydrophytic plants that must have flooded conditions to grow and complete their life cycles are termed *obligate hydrophytes.* In contrast, *facultative hydrophytes* are those species that can compensate for flooded conditions but do not require flooded conditions for their optimal growth and reproduction. Wetland systems that are temporarily flooded are likely to have a high proportion of facultative hydrophytes, while constantly flooded systems are likely to have a high proportion of obligate hydrophytes. Regions that have extremely rare flooding events are likely to be dominated by nonhydrophytes, which have no mechanisms to compensate for flooding conditions. These species will suffer during extended flooding conditions. Rheophytes are plant species that reside in aquatic systems with flowing water, such as rivers and streams. In agricultural systems, a majority of crops are sensitive to prolonged flooded conditions, although some exceptional cases of crops such as rice are hydrophytes. Since many crops are sensitive to flooding and a large amount of agricultural land is located on river floodplains, flooding is an important stress in agricultural systems worldwide.

II. NATURAL WETLAND SYSTEMS

Wetland ecosystems as defined above account for at least 6% of the world's terrestrial habitats (Maltby 1991). This does not include the transitional zones between wetlands and terrestrial systems, which can be extensive. Although wetlands occupy a relatively small proportion of the world's land surface, they have a disproportionate importance to ecosystem processes. Wetland systems have been touted as having many important "ecosystem services," a few of which are discussed below. Water that passes through wetland systems is modified, or purified, before it reaches the groundwater. Wetland systems store water for recharging the groundwater. Wetland systems can have very high productivity, particularly in the case of salt marshes. Floods that pass through wetlands are attenuated by plants. The wetland system also captures sediments from floods. Many vertebrate species utilize wetland systems as nurseries for their young or as migration stopover points. However, wetland systems have been severely affected and reduced in acreage by land development around the world. For these and other more anthropogenic reasons, most nations of the world have developed legislation to minimize development in wetland systems.

Wetlands and the species that inhabit them are located in all biomes of the world. The classification of wetlands can be based upon hydrologic regime, geographical position, water chemistry, soil characteristics, historical genesis, and size. This diversity of defining criteria encompasses areas of less than a hectare to thousands of hectares, areas with the world's highest and lowest productivities, areas that are acid to alkaline, areas that are forested to barren, areas that are fresh to extremely saline, and areas from mountaintops to below sea level. Therefore, it is not surprising that general patterns of wetland ecosystem function are not easy to describe. However, as discussed later in this chapter, there are some general attributes of wetland systems that shape the morphology, anatomy, and physiology of hydrophytes.

Wetlands can be generally grouped on the basis of whether they are flooded permanently or temporarily (Figure 9.1). Here we discuss bogs, swamps, and rice paddies as examples of relatively permanently flooded regions, and we discuss marshes, floodplains, and irrigated fields as temporarily flooded regions. These are by no means the entire diversity of wetland systems, but they include the majority of wetland ecosystems and hydrophytic species.

Swamps are what most people consider to be wetlands because they are constantly inundated with water. Swamps are located at the margins of many lakes, in still backwater areas of rivers, and in parts of floodplains. Some specific examples are cypress swamps of the southern United States, maleleuca swamp forests of New Guinea, and mangrove forests of southern Florida and other tropical regions. Swamps (or glades) dominated by herbaceous species, such as those of the Florida everglades or the papyrus swamps of Africa, will be considered as marshes in this treatment. Although swamps are restricted to

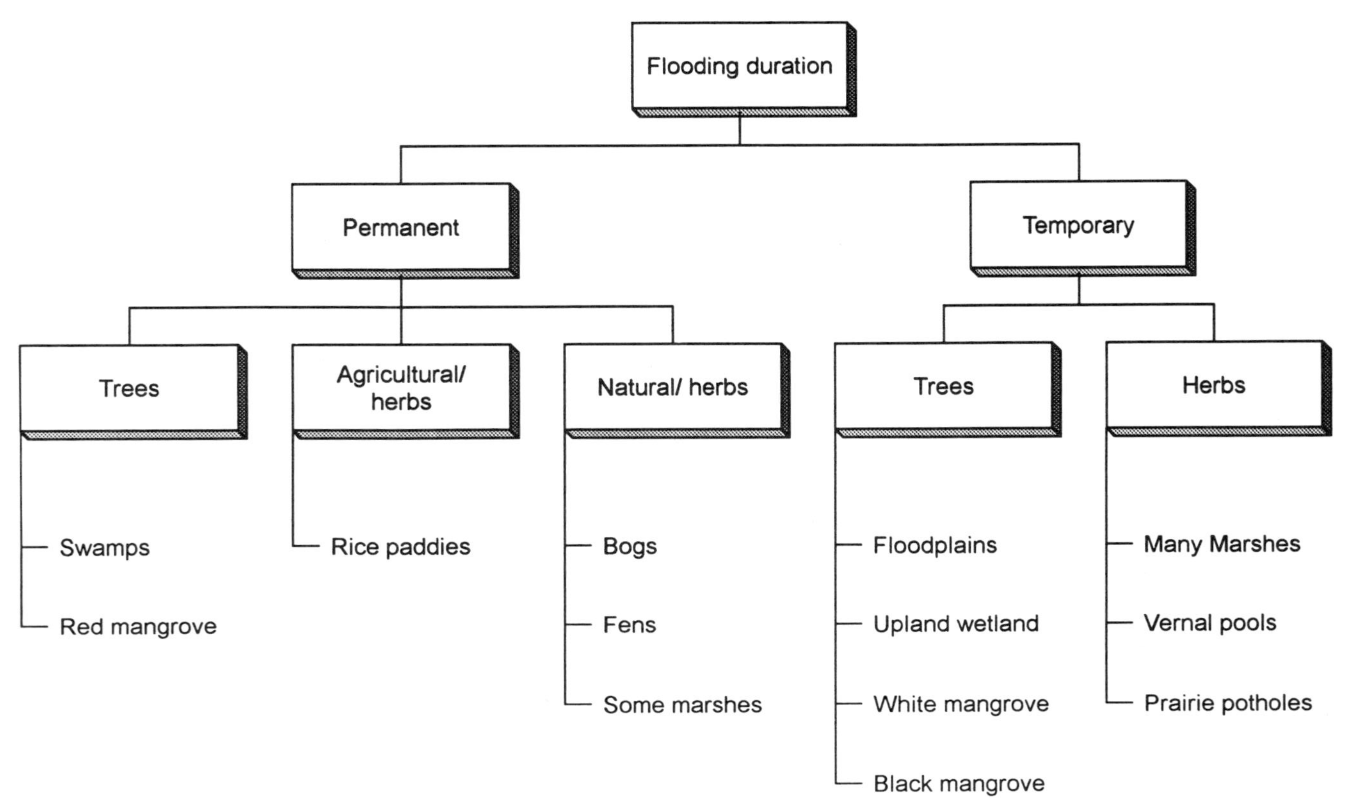

Figure 9.1 Flowchart representing a systematic ordering of wetlands based on the longevity of flooded conditions and the types of plant growth forms in the wetland.

relatively small proportions of North America and Europe, some regions of Asia, such as Indonesia, have more that a quarter of their land area as swamp. Mangrove forests are the world's most abundant swamplands, covering more that 14 million hectares (Maltby 1991). The greatest amount of mangrove swamp is located in the Indian Ocean to Western Pacific region, where most coastal regions are dominated by species of mangrove (Chapman 1984). Three species of mangrove dominate North American mangrove swamps, but there are approximately 70 species of mangrove trees and shrubs worldwide (Duke 1995). There is strong spatial zonation of mangrove species within mangrove swamps (Figure 9.2). Outermost are red mangrove (*Rhizophora*), which are most tolerant of the tidal oceanic conditions. Further inland, black mangroves (*Avicennia*) are taller and more resistant to nonflooded conditions. There is some question whether mangrove species actually build coastline or simply reduce erosion, but red mangroves can advance seaward at a rate of 100 m per year in some situations (Macintosh 1983). Different species of mangrove are located in specific zones of the swamp, depending on their tolerance of salinity, anaerobic conditions, and tidal action.

In North America the most abundant swamp is that dominated by cypress trees. The Okefenokee swamp in southern Georgia, the Great Dismal Swamp of Virginia and North Carolina, the Bijou of Louisiana, and cypress strands of Florida are example regions of cypress swamp. These regions are dominated by bald cypress (*Taxodium distichum*) and associated shrubs or trees. Few herbaceous species are present in these systems.

Marshes are characterized by temporary to permanent flooding and are dominated by herbaceous species. Marshes can be located at the edges of lakes, the edges of tem-

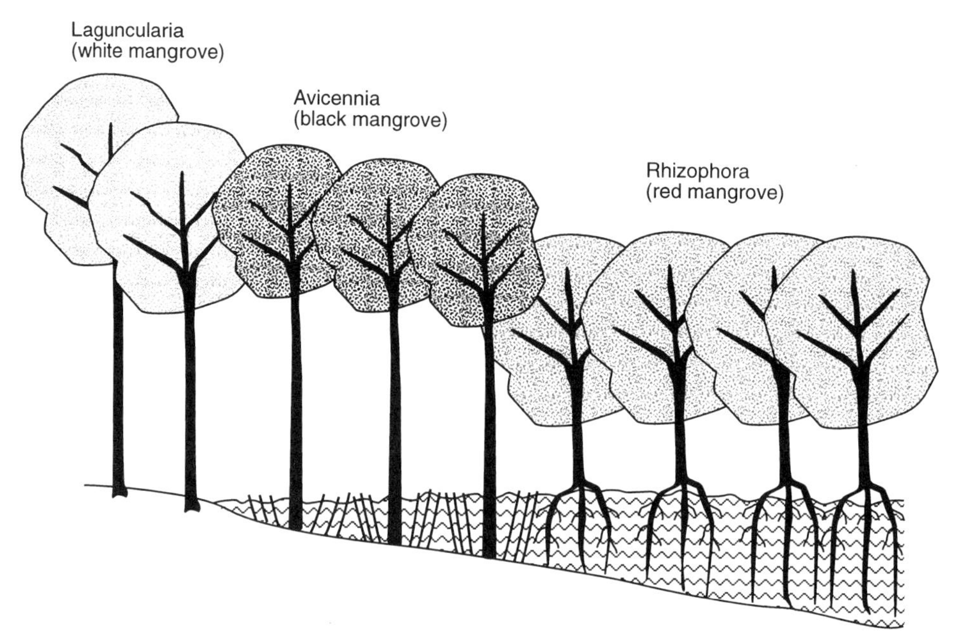

Figure 9.2 Diagrammatic representation of the zonation of various mangroves in a coastal mangrove swamp. Notice the prop roots on red mangrove and the pneumataphores on black mangroves.

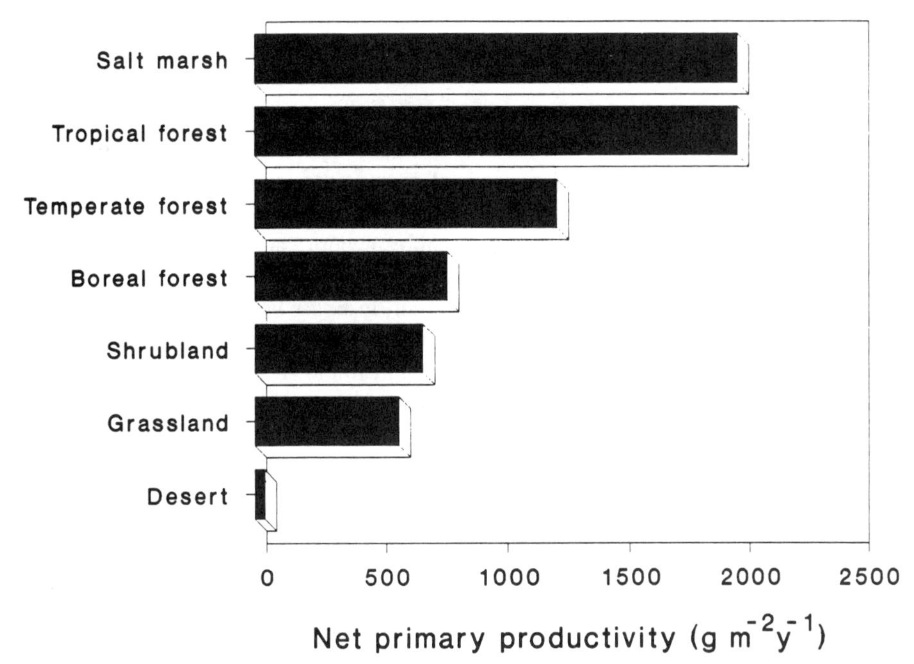

Figure 9.3 Compares the representative productivity of various ecosystem types in the world. The data for this figure are from Table 5.2 in Whittaker, 1975.

porary pools such as vernal pools in California or prairie potholes in Minnesota, stream backwaters, or ocean margins. Saltwater marshes are among the most highly productive ecosystems of the world (Figure 9.3) and dominate temperate and Arctic region shoreline (particularly in eastern North America and Europe). Within saltwater marshes there is species zonation in which *Spartina* dominates in areas closest to the ocean, and *Sueda* and *Salicornia* dominate in the highest salinity areas near mean high tide level. Freshwater marshes constitute approximately 90% of freshwater wetlands in the United States. They are distributed throughout North America, occurring in almost all ecosystems and latitudes. Dominant species of freshwater marshes are grasses, sedges, and rushes, such as those in the genera *Phragmites, Scirpus, Juncus, Carex, Eleocharis,* and *Typha* (cattail). The latter is a particularly important freshwater marsh genus, of which three species (*T. latifolia, T. angustifolia,* and *T. glauca*) dominate marshes throughout North America. Several species in freshwater habitats are tolerant of extreme variation in water availability. For example, species of *Isoetes* in Californian vernal pools are nearly submerged in one season and experience drought in another season. For this reason, several species of *Isoetes* have particularly flexible photosynthetic physiology (Keeley 1991).

Peat lands are characterized by sites in which the soil is very acidic and high in organic matter that is usually derived from *Sphagnum* moss. Since *Sphagnum* mosses are characteristic of temperate, alpine, and arctic areas, peat lands have traditionally been considered to be high-latitude or high-altitude ecosystems. However, there are tropical peat lands, most on the borders of the South China Sea, where they are associated with swamps and marshes of tropical lakes. Even though tropical peat land does exist, it has been estimated that about 90% of the world's peat lands are in the temperate and boreal regions.

Bogs are a type of peat land characterized by a high water table maintained by precipitation and poor drainage. The soil is commonly covered with *Sphagnum* peat moss and is very acidic, due to the slow decomposition of organic material. Nutrients are very low, particularly phosphorus, and soils are anaerobic. Dominant herbaceous genera are *Sphagnum, Carex,* and *Juncus,* but many small shrubby genera can be present. In particular, species of the family Ericaceae, such as *Vaccinium macrocarpon* (cranberry), which are clonal and have small leathery leaves are common in bogs. Fens are differentiated from bogs because they have water resources other than precipitation, such as groundwater or streams. Fens are characterized by higher nutrient availability and therefore have a larger diversity of species. Plants that inhabit bogs and fens have adaptive mechanisms to cope with low nutrient availability, acidic soil, and anaerobic conditions. These areas are also the regions in which insectivorous plants such as species of *Drosera* and *Saraccenia* are found.

The areas near rivers and streams that are temporarily flooded are classified as floodplains (Kozlowski 1984). These areas are exposed to flooded conditions in a stochastic manner. There may be several years between floods, or floods may happen on a regular basis. Following a flood, semipermanent areas of water can be left in depressions such as oxbows. In lower reaches of rivers floodplains commonly support bottomland hardwood forest. The trees of these forests have mechanisms for coping with short-term inundation but cannot tolerate permanent flooding. The largest remaining area of bottomland forest in the United States is in the lower reaches of the Mississippi River. Although this forest still accounts for 23.5 million hectares, it is much reduced because of the channelization of rivers in North America. Other areas of the world retain a large fraction of forested floodplain, such as in Africa where nearly one-half of all wetland is forested floodplain (Drijver and Marchland 1985).

Bottomland forest species are diverse and include herbs, shrubs, and trees. Species composition of bottomland forest is dependent on the frequency, duration, and intensity of flooding. Woody taxa have developed mechanisms to compensate for temporary anaerobic conditions and the mechanical impacts of violent floods. Nutrients are relatively high in relation to other wetlands because of the infrequent anaerobic conditions and the silt deposited during occasional flooding.

Agricultural fields are not exempt from flooding. In fact, some fields are constructed to promote flooding, as this is important for the crop. Irrigation techniques can often cause temporary flooding, as can seasons of high precipitation. Since bottomland forest has soil with good nutrition and texture, these are preferred regions for agriculture. Floodplains are also selected for agriculture in regions of low precipitation because of the access these areas have to irrigation water. Therefore, many agricultural fields developed in natural floodplains are predestined to flooding. Levees are built to minimize this problem in some areas, but flooding still occurs with some regularity. Since most crops are sensitive to the effects of flooding, flooded or saturated soils can cause damage even when present for only short periods. In fact, corn seedlings experience the detrimental effects of flooding after less than 1 h. The sensitivity of crops to flooding has generated considerable interest in engineering mechanisms of flood tolerance in crops (see Chapter 14 for more details).

III. ENVIRONMENTAL ATTRIBUTES OF FLOODED CONDITIONS

Prolonged saturation of wetland sediments or upland soil in which plants grow has a significant impact on both abiotic and biotic attributes of the rhizosphere. Populations of soil microorganisms shift in composition and abundance, which subsequently changes soil solution chemistry because of differential metabolism by the new soil flora. Both of these changes (soil microflora and soil solution chemistry) affect root and therefore plant performance. Moreover, the salinity (or other chemical attributes) of floodwater will affect biological activity in the rhizosphere.

The most important detrimental characteristic of a flooded ecosystem for plants is the resulting oxygen partial pressure in the root zone. This is important because (1) roots are particularly sensitive to anaerobic conditions, and (2) anaerobic conditions support a unique microbial community compared with aerobic conditions, and this severely affects nutrient relations of the soil. Low oxygen is such a general characteristic of prolonged flooding that many of the definitions of wetlands involve a measurement of anaerobic conditions in the soil for some defined time of the growing season.

Oxygen concentration of sediments has traditionally been determined by using oxygen electrodes (Armstrong 1994). In general, oxygen electrodes are based on polargraphic measurement of oxygen. In essence, a negative electrode (often platinum, silver, or gold) is attached to the negative pole of a battery. This electrode will become negatively charged (a cathode) because electrons will tend to move from the battery to the end of the electrode. When a second electrode (usually silver) is attached to the positive pole of a battery, electrons will flow toward the battery and cause the electrode to be positively charged (an anode). A film of electrolyte (commonly KCl solution, which ionizes to K^+ and Cl^-) completes the circuit between the two electrodes. When oxygen is present it is reduced at the cathode and a flow of electrons is initiated from the cathode to the anode. The electrical current produced by oxygen reduction is related to the quantity of oxygen in the test solution. The ease of dipping an electrode in a flooded sediment has made measurement of oxygen partial pressure a mainstay of environmental analysis in wetlands. To measure the oxygen concentration at different depths in the sediment, wells can be constructed that access only certain depths of the soil profile. At equilibrium with the atmosphere, water can contain 227 mmol oxygen m^{-3} at 20°C. A sediment can be considered hypoxic when its oxygen content is below 50 mmol m^{-3}, and a flooded soil is anoxic when its oxygen content is not detectable. Endogenous oxygen in tissues of plants can also be determined by inter- and intracellular oxygen microprobes (Armstrong et al. 1994a).

Why is oxygen partial pressure in the root zone of a flooded ecosystem so low? There are two primary processes that drive oxygen partial pressure to low values in flooded soils. First, oxygen partial pressure in the soil is dependent on diffusion of atmospheric oxygen into the soil profile. If the soil is highly porous, there can be a greater gas volume in the soil and a larger reserve of oxygen. Oxygen diffuses from the atmosphere into the soil pore spaces because roots and microorganisms are removing oxygen by respiration and creating a lower oxygen partial pressure than that in the atmosphere. When root and microbial respiration deplete oxygen in a flooded soil, the rate of oxygen diffusion from the atmosphere is reduced (compared with its diffusion in soil gas phases) because of the high resistance to its diffusion in water compared with air. Most gases diffuse 10,000 times faster in air than in water. Second, the solubility of oxygen in water is particularly low. In fact, there will be at least a 30-fold decrease in oxygen concentration between the gaseous and aquatic phases at the interface between the atmosphere and a flooded soil. Thus, the delivery rate of oxygen to roots in flooded soils is low because (1) oxygen has a low solubility in water, and (2) the diffusion rate of oxygen in water is relatively slow.

In a flooded soil at steady state with the atmosphere, an aerobic zone (hypoxic) will exist but it will only extend between 1 and 20 mm into the soil (Armstrong et al. 1991a). Furthermore, the depth of hypoxia in sediments and the extent of anoxia will vary on a spatial and temporal scale. Soon after a flooding event, anaerobic conditions in the soil can develop in a few hours or faster at higher temperature. As the length of time a soil is flooded increases, the depth of the hypoxic zone decreases. In tidal marshes soil anaerobic conditions can vary seasonally and daily due to the frequency and intensity of tides. Since a lack of oxygen has dramatic effects on respiration, the temporal and spatial distribution of anaerobic soils in wetland systems has a major impact on the evolution of physiological and anatomical adaptations in wetland plants.

As mentioned earlier in the chapter, low oxygen concentration in the soil solution has ramifications for the composition and abundance of microorganism populations. Flooding of soils causes a major change in the activity of specific classes of bacteria. Shortly after the onset of flooding (1 to 2 days), aerobic bacteria in soils will have depleted the soil solution of oxygen, creating anaerobic conditions. After anaerobic conditions develop, obligate aerobic bacteria become inactive and facultative, and obligate anaerobic bacteria become active. The anaerobic bacteria utilize molecules other than oxygen to accept electrons in their energy-acquisition, oxidation–reduction reactions. Oxidation reactions are not defined by the presence of oxygen. Rather, they are defined as those reactions that lose electrons, and reduction reactions gain electrons. This transfer of electrons (or electron flow) is a ubiquitous part of all biological activity. In fact, the electrochemical potential (or electron availability) of a biologically mediated chemical system (such as a soil solution) will be determined by the composition of aerobic/anaerobic microorganisms. The redox potential of a solution measures the solutions electrical potential. A large positive redox potential indicates an oxidizing environment (aerobic bacteria are dominant), while a strongly negative redox potential indicates a reducing environment (anaerobic bacteria dominate). If the redox potential is known, one can predict the nature of many chemical and biological processes that regulate the activity and availability of most major nutrients and trace metals in soils.

Redox potential is most often measured with an electrode composed of a platinum wire and a calomel electrode to complete the electrochemical circuit. In certain situations, particularly in oxidized soils, there can be instability, or drift, in the electrode reading. In oxidized soils chemical and biological reactions are not so dependent on the intensity of reduction, and direct measurements of oxygen concentration are more likely to reflect soil chemical processes. Therefore, platinum–calomel electrodes are not very useful in aerobic soils. However, in anaerobic systems (such as in flooded soils) redox potentials are important and platinum–calomel electrodes provide stable and useful information. In fact, under anoxic conditions, a number of important nutrient transformations are dependent on the intensity of reduction in the soil solution.

After anaerobic conditions are established in a flooded soil, obligate and facultative anaerobic bacteria dominate the microflora of the soil. These organisms are able to flourish in anoxic sediments because they utilize other molecules besides oxygen as electron acceptors for their energy-producing metabolism. Over time the activity of anaerobic bacteria will cause the redox potential to decrease. At a redox potential of approximately 225 millivolts (corrected to a pH of 7) nitrate (NO_3^-) availability will decrease to zero. At a redox potential of approximately 200 mV the manganate form of manganese (MnO_2) is reduced to the manganous form (Mn^{2+}). The ferric iron [$Fe(OH)_3$] is reduced to the more soluble ferrous form (Fe^{2+}) at approximately 100 mV. At around −150 mV sulfates (SO_4^{2-}) are reduced to sulfide sulfur (H_2S), and at approximately −200 mV, methane is formed from the reduction of carbon dioxide and certain organic acids (Gambrell et al. 1991). Therefore, in anaerobic soils, redox potential can indicate the status of a number of important nutrients in the soil solution.

In addition to affecting the availability of major nutrients in the soil, prolonged flooding has other impacts on soil chemistry. Anaerobic bacteria are much less efficient than aerobic bacteria at extracting energy from soil organic matter. This results in two general effects on soils. First, flooded soils have larger amounts of organic matter because anaerobic bacteria do not decompose organic matter as fast as do aerobic bacteria. Therefore, organic matter has a longer residence time in wetland soils than in upland soils. Second, the end products of anaerobic consumption of organic matter are not the same as those produced by aerobic respiration. For example, respiration by aerobic microbes results in mostly carbon dioxide, nitrate, sulfur, and some residual humics. In contrast, respiration by anaerobic microorganisms can result in the production of carbon dioxide, methane, low-molecular-weight organic acids, hydrogen, ammonia, amines, hydrogen sulfide, or other chemicals, depending on the redox potential. Thus, the soluble organic molecules in anaerobic soils are of a different composition than those in aerobic soils.

In most wetland systems there is a thin layer of aerobic conditions at the surface of the soil. This results in an aerobic–anaerobic interface. This interface may be particularly important for transformation of major nutrients such as nitrogen. In aerobic soils, nitrification (ammonia conversion to nitrate) is the dominant nitrogen transformation process and is accomplished by aerobic bacteria (Figure 9.4). In contrast, denitrification (conversion of nitrate to nitrous oxide or dinitrogen) occurs through the action of anaerobic bacteria. Ammonia that is mineralized in anaerobic sediments diffuses up into the aerobic zone, where it is nitrified to nitrate. The nitrate then diffuses down into the anaerobic sediments, where it is

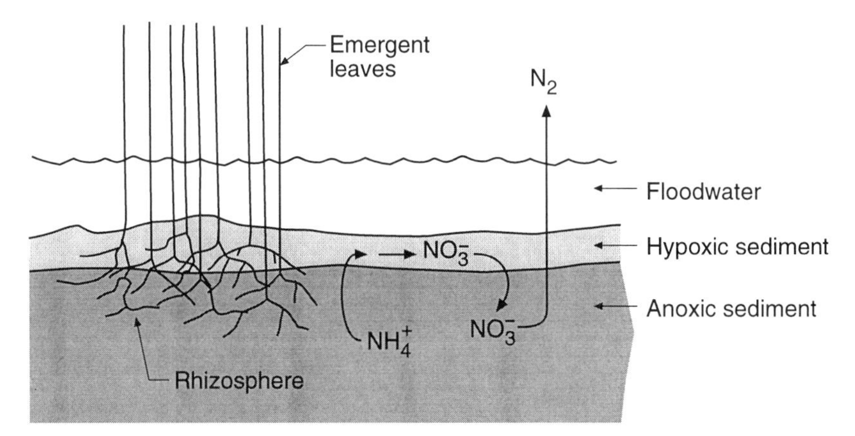

Figure 9.4 Diagrammatic representation of nitrogen transformations in a marsh ecosystem. Notice the importance of the aerobic and anaerobic layers in the overall transformation process from ammonium ion (NH_4^+) to dinitrogen (N_2).

denitrified into N_2 or N_2O, which diffuse up into the atmosphere. Therefore, wetland systems do not store large amounts of nitrogen in their sediments, and nitrogen becomes an important limiting nutrient because mineralized nitrogen is lost by the nitrification/denitrification processes set up by the anaerobic/aerobic interface (Patrick 1982). The transfer process of nutrients between the anaerobic and aerobic interface will become important when we consider the interface between roots of hydrophytes and the soil solution.

Some anaerobic microorganisms generate gases as a by-product of their energy acquisition mechanism. For example, *Desulovibrio* species reduce sulfates to hydrogen sulfide, and other microorganisms produce methane. Hydrogen sulfide, methane, and other reduced gases can dominate the gas phase of the soil solution in a flooded ecosystem and will volatilize, giving the commonly encountered "swamp gas" smell of wetlands.

Salinity is another important attribute of the environmental conditions of wetlands. In particular, coastal marshes and mangrove swamps are affected by variable salinities. Salinity in salt marshes and mangrove swamps may reach as high as several times that of seawater. High soil-solution ion concentration has two types of detrimental impacts on plants. In the first case, excessive ions in solution decrease the soil-solution osmotic potential and affect water availability to plants. The second impact is due to the concentration of specific ions in the solution. The negative specific ion impact on plants will depend on the ion concentration and the chemistry of the specific ions that are causing high salinity. Salinity in wetland systems is normally due to oceanic salt sources; thus the most influential ions are sodium and chlorine. Ecosystems at the interface of terrestrial and marine environments have been present for millions of years. It is logical to expect that plants inhabiting salt marshes and mangrove swamps will have adaptations to cope with high salinities, because of natural selection or other microevolutionary pressures. In fact, many plants from salt marshes or mangrove swamps are able to complete their life cycles in an environment with between 0.5 and 6% salt content on a dry weight basis. This is the definition of a halophyte. In contrast, glycophytes, such as most forest, shrubland, and grassland plants, cannot complete their life cycles under such salinities (Flowers 1975).

Salinity can also develop in agricultural systems where some type of flooding irrigation is used. During flooding with fresh water, sediments filter out on the basis of size. The small particles that are deposited last can clog the soil pores, inhibiting water infiltration. As a consequence, most of the water is evaporated, leaving the dissolved salts behind. After prolonged flood irrigation, the soil, and therefore the irrigation water, becomes saline. Salinity in an agricultural setting is particularly important because most crops are glycophytes. Although there are important interactions between flooding and salinity, this chapter focuses on the effects of flooding. Salinity is not covered in this chapter, but it is important to remember that salinity can be a major factor influencing the suite of traits developed by wetland plants to compensate for flooding conditions in salt marshes, mangrove swamps, and agricultural fields.

There are also some mechanical attributes of flooded systems that can be detrimental to plants. The actions of tides and the movement of flowing water create strains in plants that are rooted in wetland ecosystems. This is particularly true of species that reside in floodplains (the rheophytes). Floodplains are characterized by occasional floods that may have sudden rushes of water over the short term, creating acute strain on woody species. In contrast, flooded conditions can also provide support for a number of herbaceous species. Emergent or predominantly submerged species are common in moderately deep wetlands such as marshes and pond edges. Many of these species require the water for mechanical support and would lie pendant on the ground in the absence of flooding.

IV. DETRIMENTAL EFFECTS OF FLOODING ON PLANT FUNCTION

A. Effects on Respiration

The most immediate effect of anaerobic soil conditions on plants is a reduction in aerobic respiration in roots. The switch to anaerobic conditions in sediments around roots, upon extended flooding, causes root cells to switch to anaerobic respiration, which is much less efficient than aerobic respiration (Figure 9.5). For each glucose molecule that enters aerobic respiration, 36 ATP molecules can be produced, while only 2 ATP are produced per glucose by anaerobic respiration. Thus, the ATP/ADP ratio, the concentration of ATP, and the energy charge of root cells are reduced under flooded conditions (Pradet and Raymond 1983). Most ATP produced by aerobic respiration occurs in mitochondria by oxidative phosphorylation; however, all ATP produced in anaerobic respiration is dependent on glycolysis and fermentation that both occur in the cytosol. Since the significance of mitochondria to the energy state of cells is reduced under anaerobic conditions, it is not surprising that the ultrastructure and function of mitochondria are affected rapidly after anoxia begins. For example, in wheat, noticeable swelling of mitochondria, a reduction of internal cristae, and an increased transparency of the mitochondrial matrix occur after 1½ h of anoxia (Vartepetian 1991). Upon transferring seedlings back into aerobic condi-

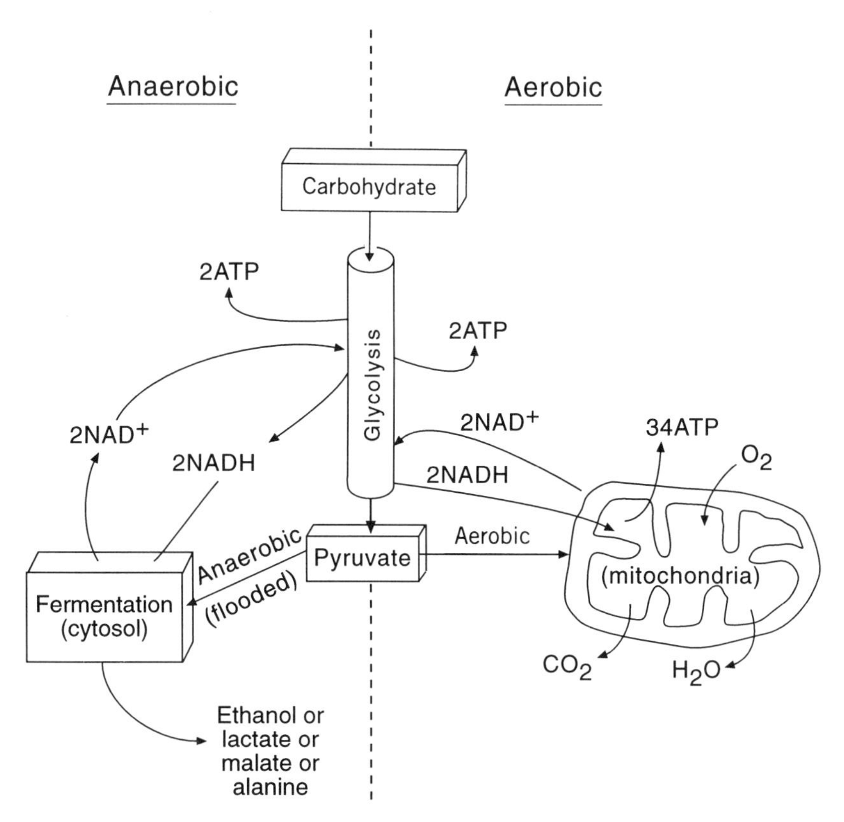

Figure 9.5 Flow diagram for aerobic and anaerobic respiration. Notice the pivotal position that pyruvate occupies in relation to fermentation and oxidative respiration.

tions, the ratio of ATP/ADP increases rapidly and the structure of mitochondria returns to normal shortly thereafter. It takes more than 9 h of anoxia to permanently alter the morphology and destroy the oxidative phosphorylation of mitochondria in sensitive species.

Anaerobic respiration does not completely oxidize glucose to carbon dioxide and water as does aerobic respiration. Therefore, some form of reduced carbon (usually ethanol in plants) is an end product of anaerobic respiration (Figure 9.5). In order for anaerobic respiration to produce enough ATP for cell survival, a potentially excessive amount of ethanol may be produced in the cytosol and stored in the central vacuole. Ethanol concentration can build up to toxic levels during anaerobic conditions unless mechanisms exist to detoxify or remove ethanol from root cells.

As discussed in Chapter 6, plants have a unique respiration pathway called *cyanide-resistant respiration.* This respiration pathway is particularly germane to anaerobic conditions because oxygen is not required as the electron acceptor. There have been many suggestions about the significance of cyanide-resistant respiration (also called *alternative respiration pathway*). Lambers (1980) suggests the following list of possibilities for the significance of the alternative respiration pathway:

1. To generate heat in fruits of many species, and in inflorescences of species in the Araceae family, without generating ATP.
2. To allow respiration to continue in fruits that have cyanogenic glycosides (a mechanism for defense against herbivores).
3. To eliminate excess reducing power.
4. To affect ion accumulation rates.
5. To be a mechanism of osmoregulation.
6. To oxidize excess carbohydrates that are not required as a source of carbon skeletons.

The evidence in support of any one of these significances is weak, but it is important to notice that this alternative electron transport system uses reducing power and oxidizes carbohydrates without producing any ATP. Under anaerobic conditions, fermentation occurs in the cytosol in order to regenerate oxidized NADH so that glycolysis can continue. Fermentation will oxidize NADH, and in the process will produce ethanol as an end product. Since the alternative pathway oxidizes NADH in the mitochondria without producing ethanol, a high activity of cyanide-resistant respiration may reduce the rate of fermentation, reduce the rate of ethanol buildup, and generate enough oxidized NAD to keep glycolysis active.

B. Effects on Photosynthesis and Carbon Balance

In many flooded situations plant shoots are not submerged and are not experiencing anaerobic conditions directly, but shoots respond to the metabolic conditions of roots. This relationship has been shown many times by split-root studies in which a small fraction of roots experiencing stress affects the function of shoots. One of the first responses of photosynthesis to root zone flooding is stomatal closure in sensitive species. This phenomenon has been observed in a number of different species, including important crops such as tomato, wheat, pepper, and bean (Pezeshki 1994). In accordance with the reduced stomatal conductance, photosynthesis of sensitive species decreases rapidly following

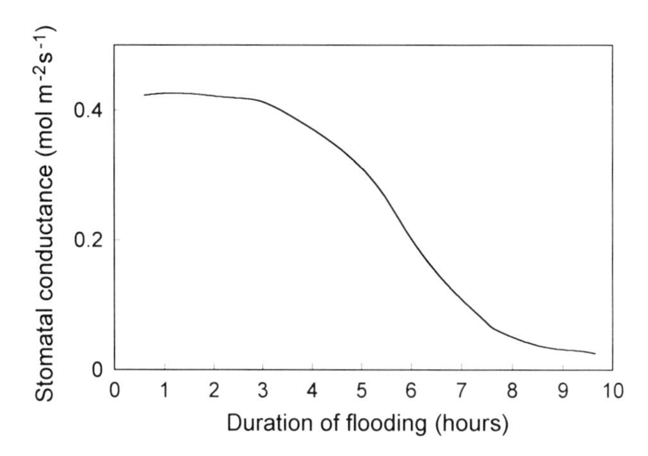

Figure 9.6 Representation of the influence of flooding on stomatal conductance of many species that are sensitive to flooding.

flooding. For example, photosynthesis in *Quercus falcata* seedlings decreased to zero after 3 days of flooding (Pezeshki and Chambers 1985). Although there have been several studies that demonstrate a close correlation between stomatal closure and reduced photosynthesis after flooding (Figure 9.6), nonstomatal inhibition can be involved in photosynthetic inhibition by flooding. The precise mechanism by which nonstomatal inhibition of leaf photosynthesis occurs during root zone flooding is not well understood. However, secondary effects of flooding on photosynthesis have been demonstrated. For example, inhibition of photosynthetic enzymes by hydrogen sulfide has been found to occur (Takemoto and Nobel 1986). Furthermore, it is likely that changes in carbohydrate translocation, buildup of inhibitory growth regulators (ethylene and ABA), or the small adenylate pool in the root cell cytoplasm may also be responsible for photosynthetic inhibition during flooding.

Roots of plants under flooded conditions require a large amount of carbohydrate because of the inefficiency of anaerobic respiration compared to aerobic respiration. In addition, flooding induces a complete shutdown of most protein synthesis, and about 20 stress-induced proteins are synthesized (see Chapter 14 for more details). These anaerobically induced proteins are mostly glycolytic proteins. Therefore, anaerobic respiration is strongly increased by flooding, as is the requirement for carbohydrates. As a result, root tissues rapidly become depleted in carbohydrate unless the plant has stored an excessive amount of carbohydrates before flooding occurred. Several authors have described this situation as "carbohydrate starvation" during flooding (Setter et al. 1987). Carbohydrate starvation is enhanced further in roots because translocation of carbohydrates from leaves to roots is inhibited during flooding conditions (Brandle 1991).

Nonstomatal inhibition of photosynthesis may also be a secondary reaction to the buildup of inhibitory growth regulators in leaves (Yamamoto and Kozlowski 1987a, 1987b). During flooded conditions translocation of cytokinins from roots to leaves is reduced, while translocation of ABA and ethylene from roots to leaves increases (Zhang and Davis 1987). In particular, hypoxia in roots, induced by flooding, causes an increase in ethylene synthesis and translocation (Raskin and Kende 1984) due to a radial gradient in oxygen concentration within the root (Figure 9.7A). Ethylene synthesis stops in anoxic conditions because the last enzyme in the ethylene biosynthesis pathway (ACC-oxidase) requires oxygen (Figure 9.7B). If ACC-oxidase is inhibited, why does ethylene synthesis

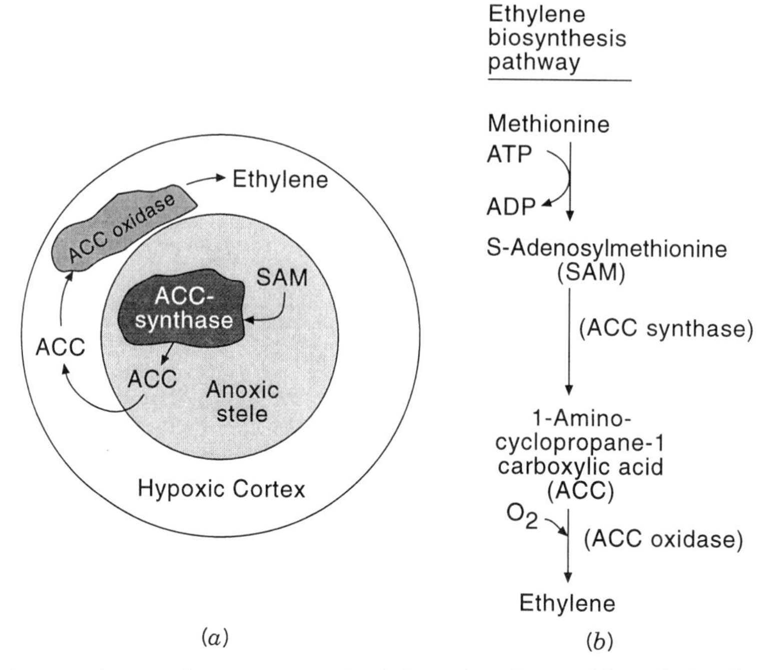

Figure 9.7 (A) Diagram of a root cross section in hypoxic sediments. The stele (core) of the root becomes anoxic while the cortex remains hypoxic. (B) Flow diagram for ethylene biosynthesis. Note that ACC-synthase can perform its reaction in an anaerobic environment while ACC-oxidase requires oxygen.

increase in roots during flooding? The answer to this question lies in the variation of oxygen concentration among root tissues. Another enzyme in the ethylene biosynthesis pathway, ACC-synthase, regulates ethylene synthesis but does not require oxygen. Although there are several ACC-synthase genes, the transcription of at least one has been shown to increase in the stele of roots during anoxic conditions (Zarambinski and Theologis 1993). The stele of roots becomes anoxic before the peripheral cortical tissues because radial diffusion of oxygen into the roots is inhibited by the presence of respiring cortical cells and the impedance of cell walls. Since ACC-synthase activity increases preferentially in the center of roots (anoxic) compared with outer tissues when flooding occurs (more hypoxic), ACC diffuses to outer cells, where ACC-oxidase can complete the ethylene synthesis pathway (Armstrong et al. 1994b). Ethylene also induces the formation of aerenchyma (gas passages through the plant from roots to leaves). The increase in ethylene synthesis near the stele and the induction of aerenchyma (discussed later) combine to stimulate ethylene transport to leaves. Further discussion of the relationships between flooding and phytohormones can be found in Chapter 4.

C. Effects on Water Relations

Intuitively, one would not think that flooding would affect the water relations of plants detrimentally. After all, water is very abundant in flooded conditions. Of course, the salinity of the water will affect plant–water relations. An osmotic drought induced by floodwater that has a higher salinity than root tissues is significant in saltwater marshes, mangrove swamps, and other saline wetlands. However, a few studies have also shown a detrimental effect of freshwater-flooded systems on plant–water relations. As stated in the photosynthesis section of this chapter, one of the first reactions to flooding is stomatal closure. Flooding increases atmospheric humidity which, along with stomatal closure, reduces transpiration. Stomatal closure in response to flooding is not induced by leaf water deficit. A root signal, probably a combination of reduced cytokinin synthesis and transport along with an increased transport of ABA and ethylene, induces stomatal closure. Since transpiration is reduced during flooded conditions, the amount of water uptake from roots need not be as high to maintain plant water balance as it would be in a nonflooded condition.

Several studies have demonstrated a decrease in the permeability of roots to water in flooded conditions. In fact, wilting can be a transient response to flooded conditions (Jackson 1956). Wilting in flooded plants has been attributed to death of the root system that was built during aerobic conditions, which severely limits water accumulation. Another effect of ethylene is the induction of adventitious roots in flooded conditions. Adventitious roots develop in hypoxic conditions near the surface of the soil. Water transport is enhanced by these roots in comparison to the roots in the anoxic sediments. Stomata may reopen following prolonged flooding, because root function has been transferred from the old aerobic roots to the newly developed adventitious roots (Regehr et al. 1975).

The generalized train of events by which flooding affects water relations is the following. First, flooding induces stomatal closure because of a change in growth regulator translocation to leaves. Lower stomatal conductance and higher humidity place a reduced demand on the root system for water acquisition. Under continued anaerobic conditions, root water uptake is inhibited, but this will not affect plant water balance because of the low transpiratory demand. If the specific plant is capable of producing adventitious roots, ethylene will induce their production and adventitious roots will elongate before the old root system dies. There may be a short time when wilting occurs, between the time when a large enough adventitious root system has been developed and the time when aerobic roots loose their permeability to water and die. Following successful deployment of adventitious roots, stomatal conductance may increase again. Root hydraulic conductance is likely to be different when water uptake is dependent on adventitious roots (in comparison to hydraulic conductance of the aerobic roots), because the cellular structure of adventitious roots may contain an extensive volume of aerenchyma, for gas transport, at the expense of xylem water-conducting tissue.

D. Effects on Nutrient Relations

Flooding will significantly affect the nutrient relationships of plants. In fact, there are three main areas of plant nutrition that will be affected by flooded conditions: (1) reduced transport of water will reduce transport of nutrients to leaf tissues; (2) anaerobic conditions decrease the adenylate pool in root cells, which will reduce the abundance of ATP for active nutrient uptake mechanisms; and (3) anaerobic conditions in the soil change the availability of a number of important macronutrients. The first mechanism of inhibition relates to the previous discussion of water relations under flooded conditions. Although the delivery rate of water to the shoots has some impact on nutrient delivery rate, the selective permeability of root cells to ions is more critical to the nutrition of plants.

Under flooded conditions oxygen partial pressure in roots will decrease. Different tissues respond to specific oxygen partial pressures differently. This fact has stimulated development of the concept *critical oxygen pressure* (COP) (Saglio et al. 1984), the oxygen partial pressure at 25°C that induces a decrease in cellular respiration and a steady increase in the respiratory quotient (mol CO_2 evolved/mol O_2 absorbed). The COP could be higher than atmospheric oxygen partial pressure, as in the case of maize roots. This is because the resistance to radial oxygen diffusion in roots results in a core of cells that are anoxic even at atmospheric oxygen concentrations around the roots. Young roots and root tips have the highest COP among roots because of their high respiration rate per unit volume of tissue and the lack of air spaces between cells (particularly in nonwetland species). Therefore, after flooding, the COP is reached first in the root fraction most involved with nutrient acquisition (young roots and root hairs).

The concept of COP applies directly to oxygen consumption, but it can also be related to other metabolic processes. For example, in maize roots the COP for energy metabolism (ATP/ADP ratio and adenylate pool) corresponds closely to the COP for respiration. In contrast, the COP for root extension is higher than either of the other two (Saglio et al. 1984). Therefore, root extension requires metabolic processes in addition to respiration. In terms of energy-dependent nutrient uptake, it is likely that this process decreases at an oxygen partial pressure lower than the COP for respiration, because the pool of available ATP decreases below this COP. It is well known that sustained influx of ions into root cells is dependent on a proton efflux pump located in the root cell plasma membrane. This H^+ ATPase may be dependent on either the absolute concentration of ATP or the ratio of adenylates (ATP/ADP). Whatever the metabolic factor is that regulates the H^+-ATPase, it is well known that anaerobic respiration is inadequate for supporting cation efflux (Cheeseman and Hanson 1979). For example, in just 5 min of anoxia the electrogenic component of plasma membrane potential was reduced from -160 mV to -100 mV, which completely interrupted active K^+ transport but did not affect passive transport of K^+ (Cheeseman and Hanson 1979). Other metabolic processes probably compete with the plasma membrane ATPases for the available cytoplasmic ATP, thereby inhibiting the ATP-dependent proton pump.

Some of the first reviews of flooding impacts on plants noted a rapid onset of chlorosis in plants (particularly crops) when exposed to flooding (Kramer 1951). Similarly, there was a decrease in tissue concentrations of N, P, and K that coincided with chlorosis. This nutrient limitation could be due to a buildup of anaerobically derived toxins that inhibit nutrient uptake, a decrease in nutrient availability in anaerobic sediments, an inhibition of nutrient uptake mechanisms, or all three. An elegant set of experiments done in the 1970s resolved these issues. When studying barley plants it was observed that the onset of chlorosis occurred shortly after anaerobic conditions developed (2 h after inundation), long before toxic compounds built up or nutrients were unavailable in sediments (Drew and Sisworo 1979). In fact, if all nutrient levels are kept constant in a sterile hydroponic system (no bacterial conversions of nitrogen forms), and nitrogen is bubbled through the solution, barley plants will become chlorotic even though adequate nutrients in the proper form are present (Drew 1991). Therefore, the primary influence of flooding on root accumulation of nutrients is that on the active transport proteins in the root membranes.

Flooding-induced anaerobic conditions in the soil around roots also change the availability of nutrients to plants. As mentioned previously, anaerobic conditions create a reducing environment in soils, causing the reduction of many oxidized plant nutrients. For example, nitrate abundance will be low and soil nitrogen will be dominated by ammonium ions. Furthermore, soil pH will become particularly acidic, which will reduce the availability of phosphate. Extended anaerobic conditions will lead to a reduction of iron forms into a particularly soluble form of iron (ferrous iron), which can have detrimental effects on plant roots. In addition, sulfates are reduced to sulfides, making them unavailable to plants. Thus, in protracted flooding conditions many macronutrients are converted into reduced forms that are unavailable or potentially toxic to plants. Changing nutrient availability in sediments augments the detrimental impact of anaerobic conditions on ATP-dependent transport mechanisms and its consequent impact on plant nutrition.

E. Mechanical Effects

Plants that reside in environments that have stochastic flooded conditions are exposed to significant mechanical damage. In floodplains along rivers, occasional floods will carry extensive quantities of silt that will be deposited on the floodplain, burying the bases of trees and shrubs. Furthermore, the action of rapidly flowing floodwater can be particularly damaging, ripping away large branches or tearing poorly rooted plants out of the ground. The trees that dominate desert washes are exposed to violent flash floods many times during their lifetimes.

In contrast to the potential mechanical damage of flash floods in desert washes and floodwater in floodplain ecosystems, water provides a supportive matrix to plants growing in regions with long-term flooding. Water lilies and other herbaceous floating-leaved or emergent species rely on the presence of water to support their leaves on or above the water for photosynthesis. Other species float on top of the water during their entire life cycles. However, for most nonhydrophytes, flooding causes some form of mechanical pressure.

V. ADAPTATION TO FLOODING OF PLANTS IN NATURAL ECOSYSTEMS

A. Avoidance of Anaerobic Cells by Adjusting Gas Diffusion

1. Root Location and Morphology. In previous sections we have illuminated the most detrimental aspect of flooded environments, anaerobic conditions. The reason for anaerobic conditions is the low solubility and high resistance to diffusion of oxygen in an aqueous solution. Therefore, if plants can change the pathway of diffusion and the matrix in which oxygen diffuses to roots, this will circumvent one of the main detrimental aspects of flooded habitats. Recall that in flooded habitats the upper few centimeters of soil are often not completely anaerobic. Shallow-rooted plants will have their roots in a hypoxic condition rather than an anoxic state, and therefore can retain partial aerobic respiration. The effectiveness of shallow rooting as an adaptation to flooding is limited because the potential volume of sediments that roots can exploit is restricted; thus shoot growth is also restricted. In addition, surface rooting predisposes plants to mechanical damage (uprooting) by fast-flowing floodwaters. The advantages of surface rooting in flooded conditions can be enhanced by having thin roots. Roots of fine diameter have less resistance to radial oxygen diffusion, and thus do not develop an anaerobic core as would occur in thicker roots. Furthermore, a greater surface area of roots per root volume will enhance exploitation of the limited rhizosphere volume. Surface rooting has been identified as an adaptive strategy for coping with flooding in a large number of wetland species (Justin and Armstrong 1987).

Plants adapted to wetland environments may have root systems that penetrate deeply in the soil even though the lower layers of the soil are anaerobic. Deep penetration of roots in wetland ecosystems can be attributed to higher porosity of roots to gas diffusion. The higher porosity is in part due to the presence of large air spaces between cells, but even without these air spaces the porosity of some wetland roots is high. Differential cell configuration may be an important mechanism of enhancing root porosity. Cubic cell configuration, in contrast to hexagonal cell packing, may be responsible for the inherently high porosity of some wetland species' roots compared to those of nonwetland species. Cubic cell packing can result in a maximum of 21.4% porosity, whereas that of hexagonal packing is only 9.3% (Justin and Armstrong 1987).

Alternative structures on stems and roots can assist in gas transport to roots. In black mangroves (*Avecinnia* species), extensions of roots, termed *pneumataphores,* grow vertically out of the soil to a height above floodwaters. When black mangroves experience flooded conditions, oxygen from the atmosphere enters the pneumataphores and diffuses to the roots. Diffusion occurs through the pneumataphore because there are large spaces between cells for gas movement in the pneumataphore, and the root cells drive their oxygen concentration down to values well below that in the atmosphere. In many floodplain genera, such as *Salix, Alnus,* and *Nyssa,* branches and stems develop small hydrophobic cracks in their bark called *lenticels.* Lenticels are also sites through which oxygen can enter the bases of stems and diffuse toward roots. Cypress trees are well known for producing large extensions of root tissue that extend above the water line called *cypress knees.* Although cypress knees resemble pneumataphores (in concept, not in structure), they have not been shown to be a site of oxygen diffusion to roots.

Flooding also induces the development of roots from the bases of stems, adventitious roots. These roots are usually derived from localized excessive phellogen activity that produces strongly hydrophobic lenticels on the surface of the stem. If the lenticels are below the water level, roots will emerge. However, if the lenticels are above the waterline, they will enhance gas diffusion into the stem but will not produce roots. When these flood-adapted adventitious roots enter the soil they do not suffer complete anoxia because (1) they enter surface soil that is hypoxic, (2) they have large gas spaces containing a reservoir of oxygen in their tissues, and (3) they are connected to the stem close to the atmosphere, which allows oxygen to diffuse into their tissues. In some cases adventitious roots, which can penetrate deeper than the fine surface roots, will stabilize plants (that have surface root systems) in wetland systems in addition to enhancing water and nutrient uptake (Hook 1984).

2. Anatomical Changes: Aerenchyma. There are many hydrophytes that do not develop extraneous gas transport structures, yet are able to maintain oxygenated roots and rhizomes in anaerobic soils. These species utilize internal anatomical alterations to enhance the diffusion of oxygen to roots and the diffusion of CO_2 from the roots to the atmosphere. *Aerenchyma* is a specialized form of cortical tissue found in wetland species and to a limited extent in nonhydrophytes when they are flooded. Aerenchyma can develop in roots, stems, and leaves by two developmental sequences. In one case, cortical cells grow to relatively large size and then die, leaving behind large spaces that form continuous gas-filled columns through the stem to roots and rhizomes (Figure 9.8B). This "*lysogenic*" mechanism of aerenchyma formation is common in grasses (*Poaceae*) and sedges (*Cyperaceae*). In some cases, aerenchyma occurs when cortical cells simply pull apart from each other without the formation of new cells. This "*schizogenic*" aerenchyma is common in

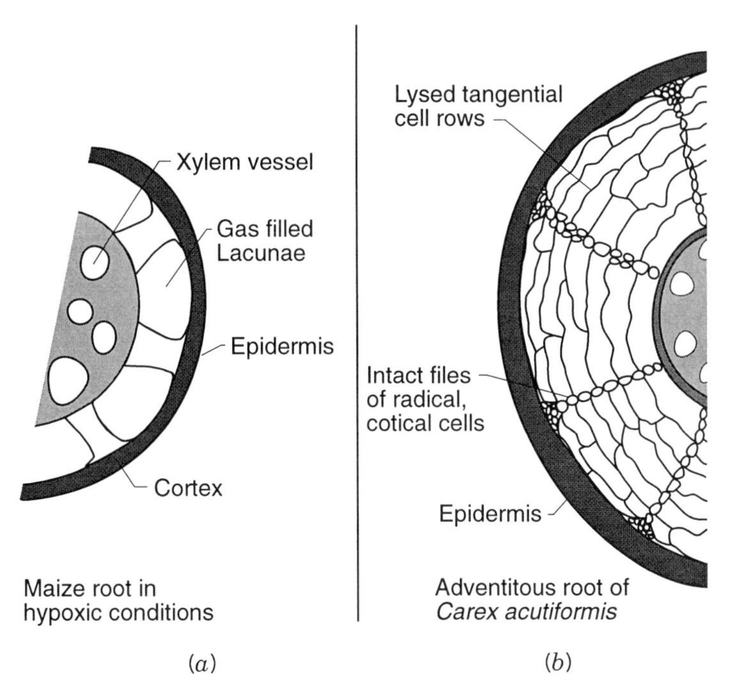

Figure 9.8 (A) Schizogenic formation of aerenchyma in the cortex of *Maize* roots when they are flooded. (B) Lysogenic aerenchyma formation in the roots of *Carex acutiformis*.

members of the *Polygonaceae* (Figure 9.8A). In a few cases aerenchyma develops in the pith rather than in the cortex. Wherever or however the aerenchyma forms, the result is a continuous gas-filled pathway and higher porosity from leaves or stems located in the atmosphere to roots and rhizomes located in anaerobic sediments.

There are two significances of aerenchyma to wetland plants. First, the resulting higher porosity when aerenchyma is present greatly reduces the resistance to radial transfer of gases within the root. Second, when aerenchyma is present, the oxygen demand per unit volume of root decreases, because the density of respiring cells in the cortex is low. Thus plants with aerenchyma in their roots are able to have a greater surface area of root with a smaller respiratory demand. The former significance is probably the most important. Aerenchyma also increases the diffusion of oxygen from the atmosphere into roots and its subsequent diffusion out of roots into the sediments producing an aerated sheath around the roots. Other gases, such as carbon dioxide, methane (Dacey and Klug 1979), and dinitrogen, can also be transported through aerenchyma (Ueckert et al. 1990). The carbon dioxide transported in aerenchyma has been shown to enhance gross photosynthesis in *Typha latifolia* because chloroplasts in the cortical cells lining leaf aerenchyma can use aerenchyma gases for photosynthesis (Constable and Longstreth 1994).

Although aerenchyma will be very effective for gas transport in herbaceous species or young roots of woody species, secondary growth reduces the effectiveness of aerenchyma. Secondary growth in roots begins in the pericycle of the stele. After secondary growth occurs, the primary, apical-corticular, aerenchyma (produced in the cortex) will be cut off from the gases coming through the shoot. Thus, for many plants secondary wood growth in roots may be the most detrimental aspect of growth for flood tolerance.

In rare cases, secondary cortex could also become aerenchymatous and reestablish the connection between roots and gases from the shoots (Laan et al. 1990). For example, most of the stem diameter of the legume *Aeschynomene aspera* is aerenchymatous tissue, which is secondary xylem. The absence of pit membranes in this type of aerenchymatous xylem indicates that gases may diffuse freely among vessel elements (see Sculthorpe 1967, pp. 65–67). Normally, flood tolerance of large woody tree species cannot be dependent on aerenchyma formation in primary tissues because of the extensive secondary growth in roots. Bottomland trees also have invested a great amount of resources into the large and relatively deep root system. These roots must have a long life in order to attain a benefit that exceeds their construction cost. Bottomland trees cannot simply replace their extensive root system with adventitious roots during prolonged flooding. To tolerate flooded conditions, large wetland tree species may have to depend on reduced metabolic rates, mechanisms of detoxifying end products of anaerobic respiration, and large supplies of carbohydrate reserve to fuel glycolysis and fermentation.

3. Convective Gas Flow Mechanisms. There are limitations to the effectiveness of aerenchyma and diffusion for delivering gases to the rhizosphere. For oxygen to penetrate more than 300 cm deep into the rhizosphere by diffusion, the roots must have maximum porosity, very low oxygen demand, and release little oxygen out into the rhizosphere. Yet there are many wetland species that have roots which penetrate 500 cm or more into the soil or survive in water depths of several meters. For adequate oxygen to get to these deep root systems of wetland plants there must be a pressurized delivery system. Several different types of such systems, called *convective gas flow systems,* have been discovered in wetland plants. The first convective gas flow system was reported in *Nuphar,* a type of water lily in which petioles extend down into an anoxic root zone several meters below the water surface (Dacey 1980 1981), and similar processes have been found in many other wetland species (Armstrong et al. 1991b).

Convective gas flow systems can be separated into two basic classes. Non-throughflow convective processes occur when most of the carbon dioxide released by respiration in the roots remains dissolved. The dissolved carbon dioxide may diffuse out of the roots and into sediments, diffuse into the transpiration stream, or be fixed into an organic molecule such as malate, instead of becoming gaseous and diffusing up the aerenchyma channel. Since the carbon dioxide does not become gaseous, this creates a pressure deficit (oxygen removal by respiration without replacement by carbon dioxide) which will draw or "suck" atmospheric gases into the plant toward the rhizosphere (Figure 9.9A). A certain degree of non-throughflow convection probably always occurs in flooded plants with aerenchyma, but it is normally not very effective for increasing gas diffusion to the rhizosphere. This non-throughflow system is limited because the pressure deficit is greatest at the leaves and decreases to zero in the roots. In addition, along the aerenchyma pathway, activities such as photosynthesis or anaerobic respiration in lower stems or leaves can create enough gas to negate the negative pressure developed in hypoxic roots. Analyses of potential non-throughflow convection in deepwater rice indicate that it should be possible to ignore non-throughflow convective mechanisms completely and still be able to account for all the oxygen diffusion required by roots (Beckett et al. 1988).

Non-throughflow convection has been known in mangroves for some time (Scholander et al. 1955). In fact, a significant pressure deficit (5 to 6 kPa) can be developed in leaves of mangrove species by carbon dioxide dissolution during inundation by tides. It is likely that convective processes in white and black mangroves occur by two mechanisms. When

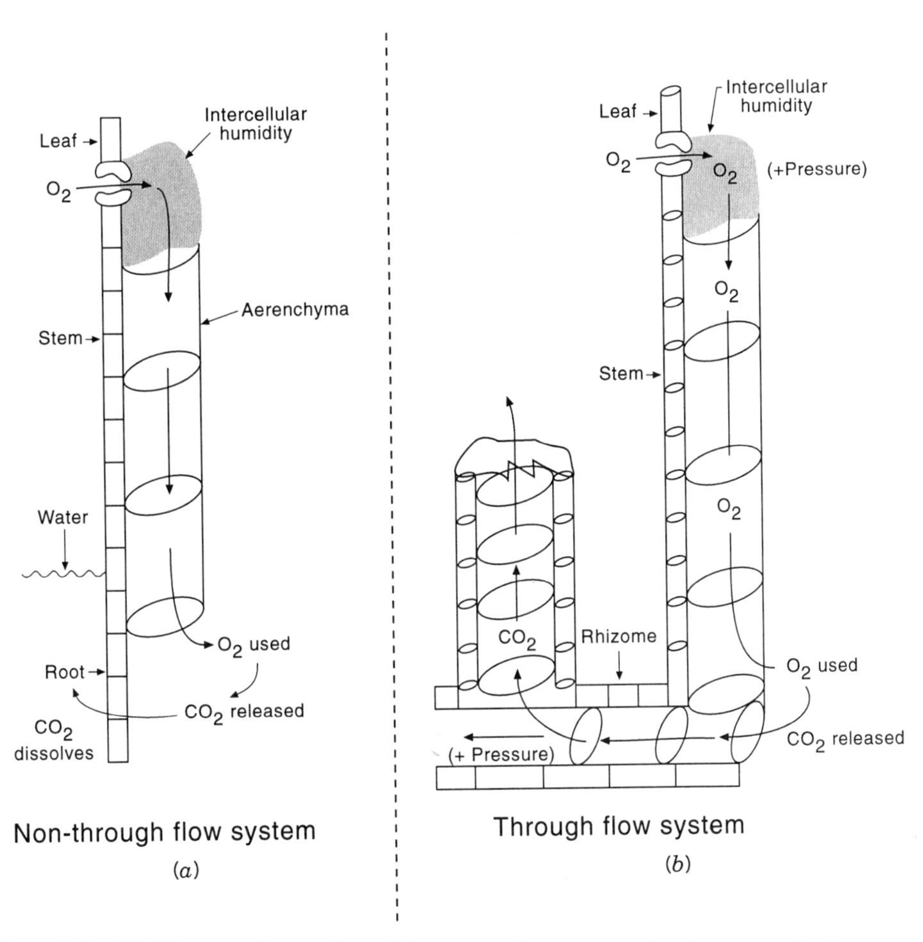

Figure 9.9 (A) Diagram of a non-throughflow convection system whereby oxygen in the atmosphere is transported to roots of some wetland plants. (B) Diagram of a throughflow convection system used by some wetland species to transport gases to and from the roots and rhizomes.

tides move out, atmospheric air rushes in to refill aerenchyma spaces. This is not particularly effective at supplying oxygen to roots because the atmosphere contains a fairly low partial pressure of oxygen. Following reinundation with tides, the roots develop a small negative pressure and oxygen diffuses to roots by non-throughflow convective processes. Neither the non-throughflow convection nor the air influx between tides is thought to be particularly effective at supplying oxygen to respiring roots during tidal inundation (Curran et al. 1986). In fact, it may simply be the ability of aerenchyma to store oxygen that is most significant to root respiration during tidal inundation of mangroves.

In contrast to non-throughflow convection, throughflow convection can be generated in at least three ways: (1) by humidity-induced convection (HIC), (2) by thermal transpiration convection (TTC), and (3) by venturi-induced suction (VIS). In all these cases, there is no change in flow velocity along the diffusion path, and flow rates can be relatively large (15 cm^3 min^{-1} in *Nuphar* petioles). The first two cases of throughflow convection (HIC and TTC) have been found in many wetland species, while the last has been found in only one species. A partition, with micropores, must be located between the pressurized transport system and the atmosphere for either HIC or TTC. This partition is the epidermal layer of leaves and the micropores are stomata.

In the case of HIC, the area close to the micropores inside the gas transport system (substomatal spaces) must be humidified constantly. The humidity inside the stomatal chamber, coming from the mesophyll cells, creates a lower oxygen and nitrogen concentration inside the leaf than that outside (these gases dissolve in the humidified air). Since the depth of stomata is small (less than or equal to 10 μm), oxygen and nitrogen will rapidly diffuse inward, creating a pressurized substomatal chamber. The pressurized substomatal gases will convect back to the atmosphere by the route of least resistance, through the aerenchyma. The gases travel through rhizomes (if present) and are vented out dead culms, another leaf, or another route in the same pressurized leaf (Figure 9.9B). The difference between HIC and non-throughflow convection is that (1) the pressure along any part of the aerenchyma in HIC is constant, and (2) the internal gases of aerenchyma in HIC can have a concentration of oxygen equal to that in the atmosphere. The pressure in the aerenchyma of species that utilize HIC is often between 200 and 1300 Pa (Brix et al. 1992). The effectiveness of HIC depends on the pore size (stomatal aperture). If the stomatal aperture is less than 1 μm HIC will be maximal. Between a pore size of 1 and 3 μm HIC will operate, but the pressures developed in substomatal spaces will be lower. Significant flows of gases can be attainable up to a 3-μm-diameter pore size even with relatively low pressure differentials (Armstrong and Armstrong 1994).

Similarly to HIC, TTC requires a humidified internal space. This mechanism (sometimes called *thermoosmosis*) is induced when leaf temperatures increase over air temperature. Gases diffuse from the atmosphere into the leaf because their solubility in the humidified substomatal chambers is greater (due to the higher temperature) than that in air with a cooler temperature. As a result, internal pressures are developed and convection down the aerenchyma occurs. As water lily leaves float on top of a water surface, a large temperature gradient can develop between the leaf and the air. Thermal transpiration convection (TTC) is commonly thought to be the main mechanism convecting gases to the roots or rhizomes of floating-leaved species (Grosse and Bauch 1991). Therefore, TTC is most effective in plants with large leaves that develop high internal temperature, and HIC is more effective in emergent macrophytes that may not have temperature differentials between leaf and air. In addition, both HIC and TTC are most pronounced during daylight hours, when temperature differentials develop and stomata are open.

The third internal gas convection mechanism (VIS) has distinctly different properties to those of the previous two mechanisms. No microporous surface is required, pressurization does not depend on a humidified internal chamber, and there is no diurnal change in gas pressure or flow as there is with HIC and TTC. The force that initiates VIS is wind blowing across dead and broken flowering stalks of *Phragmites australis*. Broken flowering stalks have exposed aerenchyma channels. The wind blowing across those channels creates a suction close to the cut surface which draws atmospheric gases into the aerenchyma (Figure 9.10). The magnitude of VIS flow will depend upon the height of the broken stalks (the closer to the ground the higher the flow) and on the wind speed squared. Thus, gusts of wind close to the water surface are important to VIS.

Whichever mechanism is used by wetland plants to convect gases (particularly oxygen) from the atmosphere into rhizomes or roots, this will enhance the plant's success in flooding ecosystems. Convective systems work best in species that have rhizomes, because the rhizome aerenchyma spaces will have oxygen concentrations close to those in

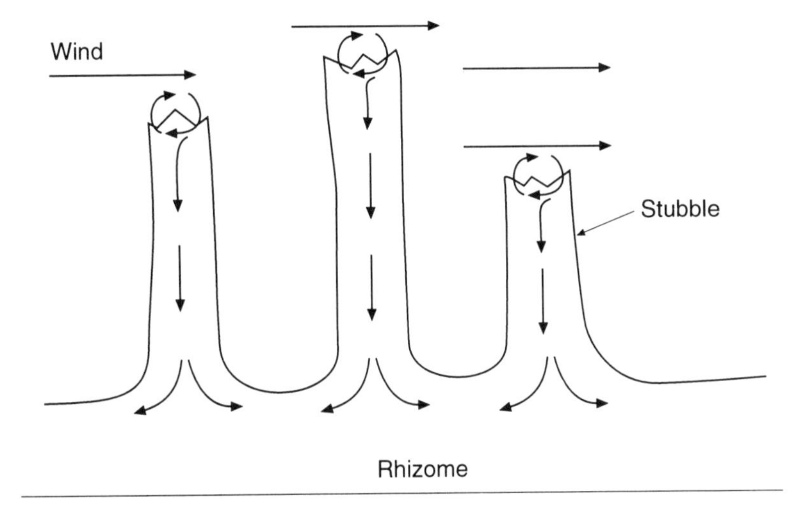

Venturi induced suction

Figure 9.10 Diagrammatic representation of venturi-induced suction. Wind blowing across the broken flowering stalks (stubble) of *Phragmites* creates a small suction, pulling atmospheric gases into the stubble and then into the rhizome.

the atmosphere and there will be a large concentration gradient of oxygen from rhizomes to roots. In fact, in many cases the rhizospheres surrounding the roots and rhizomes of wetland plants are hypoxic even though they reside in anaerobic sediments. Convective gas flow mechanisms allow roots to penetrate deeper into wetland soils and allow for the creation of a zone of partially aerobic sediments around roots and rhizomes. Oxygenation of the rhizosphere has a profound impact on nutrient relationships near the root surface.

4. Rhizospheric Oxygenation. When nonwetland plants are flooded and sediments become anaerobic, roots are exposed to the reducing environment and all the nutrient changes that go along with anoxia. However, some leakage of oxygen occurs out of roots into the rhizosphere in all plants no matter what the root porosity, which results in a hypoxic sheath around roots. In wetland plants that have aerenchyma and a convective gas flow mechanism, roots can penetrate into deep anaerobic sediments and still maintain a sizable aerobic sheath. The result is an aerobic microflora in immediate contact with roots, and the oxidized forms of iron, manganese, sulfur, and nitrogen will be present in the aerobic root sheath. In fact, an aerobic rhizosphere is often recognized by the presence of precipitated hydrated iron oxides around roots. The aerobic rhizosphere is most likely as important an adaptation to wetland ecosystems as that of aerenchyma.

In terms of mineral nutrition, an aerobic rhizosphere serves to protect against uptake of toxic amounts of ferrous iron or other heavy metals that are abundant in anaerobic sediments. For example, heavy metals such as zinc, arsenic, and cadmium can be complexed to the iron oxides and rendered insoluble near the aerobic root surface (Otte et al. 1991). In an anaerobic sediment these heavy metals would be soluble. An oxygenated rhizosphere will also support aerobic bacteria such as nitrifying bacteria. The interface between the aerobic rhizosphere (harboring nitrifying bacteria) and the anaerobic sediments (harboring denitrifying bacteria) has a significant impact on wetland ecosystem nitrogen cycling (see previous sections). In terms of root activity the nitrifying bacteria, living in an aerobic root sheath, oxidize ammonium ion, mineralized in anaerobic sediments, releasing nitrate close to the root surface for accumulation and use by the root. This nitrogen flow between aerobic and anaerobic layers can be substantial and is used commercially by constructing wetlands (usually freshwater marshes) as tertiary treatment facilities for sewage and agricultural effluent. The urea from the pretreated water enters into anaerobic sediments, where it is converted into ammonium ion. Nitrifying bacteria in the rhizosphere convert the ammonia into nitrate, and the denitrifying bacteria of anaerobic sediments convert the nitrate to dinitrogen, which goes to the atmosphere.

B. Tolerance by Metabolic Adaptation

1. Regulation of Respiration. Just as with adaptations to other stresses, there is an amalgam of mechanisms used by any plant to cope with flooding stress. Anatomical adaptations must be reinforced by metabolic adaptations. The consequences of anaerobic respiration for cell metabolism is one of the factors most likely to inhibit physiological processes. The switch from aerobic to anaerobic respiration severely curtails energy availability, requires a large amount of carbohydrate reserve, produces potentially detrimental endproducts, and causes cytoplasmic acidosis. The most dramatic metabolic impact of anoxia on cell metabolism is an almost complete disappearance of aerobic proteins and replacment by anaerobically induced proteins. These proteins are mostly glycolytic proteins, and they are transcribed and translated in high quantity (discussed further in Chapter 14). Thus, when root cells become anaerobic, glycolysis and fermentation are strongly increased, potentially resulting in a high concentration of ethanol. For roots to tolerate flooded conditions without dying, root cells must be able to keep the concentration of ethanol at a tolerable level, yet produce enough ATP to keep cells functional. In wetland plant rhizomes, ethanol concentrations may never reach a toxic concentration in anaerobic sediments because the porous tissues easily release ethanol into the aqueous surroundings. However, if the specific conditions of the habitat and plant can result in toxic buildup of ethanol, there are two types of adaptations that can compensate for this end product toxicity.

One metabolic adaptation that plants could use to avoid the production of ethanol by the glycolytic pathway is to use an alternative glycolytic pathway with an end product other than ethanol. Plants can utilize several different fermentation pathways that reduce NADH to NAD^+. The end products could be lactate, malate, succinate, or ethanol. Alanine can also be an end product of fermentation without producing ethanol, but NADH will not be oxidized to NAD^+ in this case (Figure 9.11). The particular end product of anaerobic respiration is partly dependent on pH. At a pH above neutrality, lactate fermentation is dominant, and as pH decreases (due partially to lactate fermentation), ethanolic fermentation is induced. Two enzymes are important for the conversion of pyruvate into ethanol. Pyruvate decarboxylase (PDC) converts pyruvate into acetaldehyde, and alcohol dehydrogenase (ADH) converts acetaldehyde into ethanol. Under acidic conditions ADH is stimulated. If the quantity of ADH is reduced in roots under anaerobic conditions, this will force fermentation to produce end products other than ethanol. The disadvantage of using alternative pathways for fermentation is that they are less efficient at oxidizing NADH than ethanol fermentation. Thus, the alternative fermentation pathways restrict the effectiveness of glycolysis compared with that of ethanolic fermentation. There is little

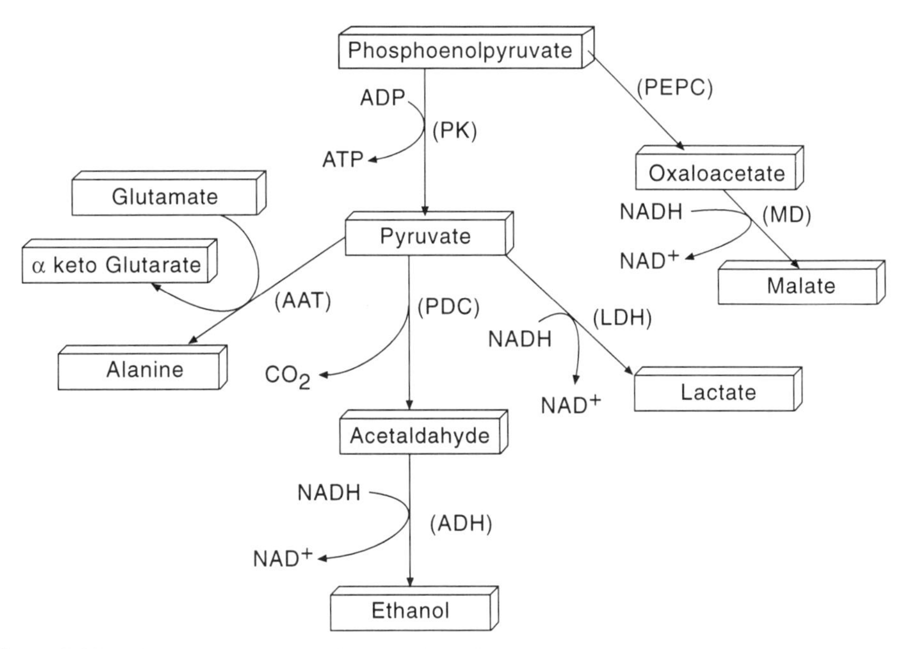

Figure 9.11 Diagrammatic respiration of a few of the fermentation pathways in plant cells. PK, pyruvate kinase; PDC, pyruvate decarboxylase; ADH, alcohol dehydrogenase; LDH, lactate dehydrogenase; PEPC, phosphoenol pyruvate carboxylase; MD, malate dehydrogenase; AAT, alanine aminotransferase.

current support for the down-regulation of ADH as an adaptation to flooded conditions. In fact, under anaerobic conditions several ADH genes are up-regulated in flood-tolerant and flood-sensitive species. In general, wetland species preferentially produce ethanol over other end products during anaerobic respiration; however, several mechanisms, such as transport of ethanol to aerated tissues, leakage of ethanol to the surroundings, or transport of ethanol in xylem stream to leaves, must be used to avoid ethanol buildup in root tissues.

The second adaptation against excessive ethanol buildup concerns cytosolic pH. During anaerobic conditions cytosol pH decreases, which inhibits lactate dehydrogenase (LDH) and stimulates ADH, leading to exclusive ethanol production. If pH can be maintained near neutrality, LDH can be active and less ethanol will build up. In flood-sensitive species, cytosolic pH drops rapidly when roots are hypoxic or anoxic. The rapid drop in cytosolic pH, called *acidosis,* is thought to be one of the main reasons why cells die in response to flooding. In flood-tolerant plants the pH drop may be counteracted by an alkalinization process. The formation of α-aminobutyric acid, the accumulation of amides, and possibly the accumulation of argenine may be part of the alkalinization process of wetland roots during anaerobic conditions (Crawford et al. 1994)

2. Preparatory Energy Storage. Anaerobic respiration produces a small amount of ATP per glucose molecule in comparison to aerobic respiration. Therefore, a large amount of carbohydrate is required to produce enough ATP for maintaining cell function in anoxic or hypoxic conditions. Carbohydrates could be provided for anaerobic respiration from stored sources or by translocation from photosynthetic leaves. As stated above, carbohydrate translocation is very sensitive to anaerobic conditions and plants cannot sustain a high carbohydrate flow to roots during flooding, except in some wetland species with extensive aerenchyma. Consequently, the ability to store large amounts of carbohydrate in roots or rhizomes may enable a longer tolerance of anaerobic conditions.

Only certain belowground tissues will be effective for carbohydrate storage. Thickened roots, stems, and seeds are effective at storing carbohydrates, but they are particularly sensitive to anoxia. Thick stems and roots preclude oxygen transport into core cells and lack extensive aerenchyma, so they are particularly sensitive to flooding and cannot effectively prolong their survival of flooding by storing carbohydrates. Large seeds may be able to store enough carbohydrate to elongate hypocotyl and radical, but they cannot develop roots or leaves in anaerobic conditions (except for a few genera such as *Oryza* and *Echinocloa*). Starchy seeds are fairly tolerant of hypoxia and can germinate at 0.01% oxygen, and seeds that store fat can germinate at oxygen concentrations as low as 2% (Armstrong et al. 1994b). However, following germination, growth is severely inhibited under hypoxic conditions. In many wetland species, seeds germinate when floating, in aerobic conditions between tidal floods, or on the tree before release (red mangrove has viviparous seeds).

Rhizomes, belowground stems adapted for vegetative reproduction, can be effective organs for storing carbohydrates for flood tolerance in wetland plants. Rhizomes of wetland plants can store a large amount of oxygen in aerenchyma when plants are dormant aboveground as a hedge against anaerobic conditions, and they can have as much as 50% of their dry weight in starch. In some cases (e.g., *Phragmites australis*), carbohydrates are stored in the rhizome in large amounts as sucrose instead of starch. The general seasonal pattern of carbohydrate concentration in rhizomes of wetland plants is (1) an increase to high levels at the end of the growing season, (2) a small decrease during the dormant season, and (3) a sharp drop at the beginning of the growing season. These results agree with the theory that carbohydrates are stored at the end of the growing season, used for anaerobic respiration to maintain viable buds during dormancy, and rapidly metabolized for the growth of shoots at the onset of the growing season. When new shoots are produced, convective gas transport will be renewed.

Lipids could also serve as a reserve for anaerobic respiration because they can be broken down into subunits that can enter glycolysis or enhance fermentation. However, lipids are not stored by wetland rhizomes during the dormant season. The high oxygen demand for processing lipids into respiration probably precludes the effectiveness of storing lipids for anaerobic respiration during the winter.

3. Lipid Metabolism and Mitochondrial Activity. In all the discussions above, we have assumed that mitochondria are inactive during flooded conditions because of the lack of oxygen as a terminal electron acceptor. In sensitive plants and in many tolerant plants, internal mitochondrial membranes disappear and mitochondria become engorged with lipids. However, in some species (particularly *Echinocloa phylopogon*), TCA intermediates are produced, acetate is readily metabolized, and germination is inhibited by cyanide, all of which suggest that mitochondria still function under anaerobic conditions (Kennedy et al. 1991). In fact, several (but not all) TCA cycle enzymes are active during anaerobic conditions (Figure 9.12). When mitochondria are isolated from plants grown in either anaerobic or aerobic conditions, they exhibit good respiratory activity and respiratory control. Mitochondria isolated from anaerobically grown plants had 70% of the respira-

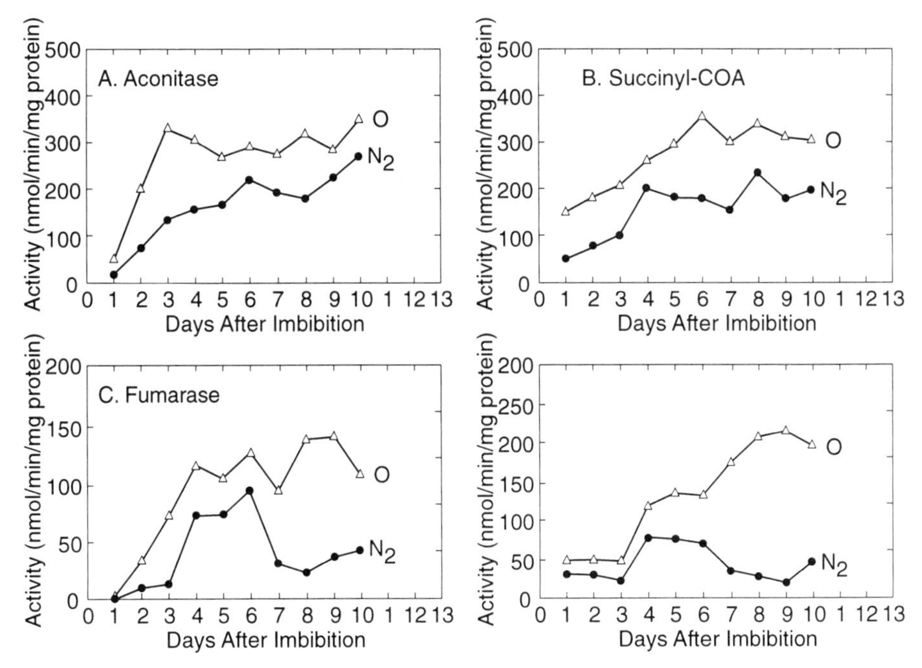

Figure 9.12 Relative activity rates of several enzymes from the Krebs cycle of respiration in the presence of oxygen (O_2) or the absence of oxygen (N_2). (Modified from data in Kennedy et al. 1991.)

tory activity of those isolated from aerobically grown plants. One of the most significant differences between mitochondria from aerobically and anaerobically grown plants was their mitochondrial absorption spectra. Mitochondria from plants grown in anaerobic conditions had a spectrum with a unique peak at 577, which was not present in mitochondria from aerobically grown seedlings. The different mitochondrial spectrum from anaerobically grown seedlings indicated that the composition of cytochromes was different between mitochondria from the two seedling types. Subsequently, quantitative differences in cytochome abundance were found. Aerobically grown mitochondria had relatively low cytochome *c*, while anaerobically grown mitochondria had relatively low cytochrome *a* and *a*3 (the cytochromes associated with oxygen reduction). Interestingly, a cytochrome usually restricted to bacteria, cytochrome *d*, was found in both types of mitochondria. It is possible that cytochrome *d* may be serving as an alternative terminal electron acceptor, or enhancing the use of nitrate, nitrite, or sulfite as an alternative terminal electron acceptor. *Echinocloa phyllopogon* contains a relatively high amount of nitrate in root tissues (compared to other crops), most of which is reduced during the first 24 h of anoxia. Much is yet to be learned about alternative uses of mitochondria during anaerobic conditions in this and other species.

Lipids accumulate in many cellular locations during anoxia, including mitochondria, chloroplasts, and liposomes. The buildup of lipids is commonly thought to be a detrimental aspect of anaerobic conditions derived from the large amount of oxygen needed to maintain membrane function. Since the supply of oxygen is limited, the amount of cellular membrane decreases and lipids accumulate. Lipids accumulate dramatically in barnyard grass, such that up to 14% of leaf dry mass is lipid under anaerobic conditions. However, there may be some advantages to lipid buildup during anaerobic conditions. Lipid synthesis in the absence of oxygen would reoxidize NADH to NAD^{+}. Thus lipid synthesis would produce oxidized NAD without the production of ethanol by fermentation. Lipid accumulation in leaves of barnyard grass results from the carboxylation of acetyl-CoA. The acetyl-CoA is derived from carbohydrates that have undergone glycolysis. Thus, during anaerobic conditions there could be a significant flow of carbon from carbohydrate to lipids which produces ATP by glycolysis and oxidizes NADH but does not generate extensive amounts of ethanol and does not affect cell pH. In addition, the stored lipids may serve as an important energy source when aerobic conditions return.

4. Effects on Growth Regulators, Particularly Ethylene. The impact of flooding on several important phytohormones was discussed extensively in Chapter 4. Here it suffices to review these impacts to reinforce the association between flooding, anaerobic conditions, and growth regulators. Among the four classes of growth regulators, ethylene increases in concentration most dramatically during flooding. The increase in ethylene is due partly to an increased synthesis of ethylene, partly to inhibition of ethylene movement out of roots, and partly to plant uptake of ethylene generated in anaerobic sediments. Ethylene synthesis increases under hypoxic conditions when ACC-synthase concentrations increase, which stimulates ACC synthesis. ACC then diffuses to aerated parts of the root or leaf and is converted into ethylene by ACC-oxidase. Ethylene is far less soluble in water than in air. Therefore, more ethylene is retained inside plant tissues when flooding occurs and ethylene concentrations increase. Irrespective of the causes for elevated ethylene synthesis, it is important to know the important effects of ethylene on plants under flooded conditions.

Increased ethylene is associated with adventitious root formation. Adventitious roots are probably formed in response to increased IAA concentration, but ethylene may make tissues more sensitive to the impact of IAA. Several lines of evidence support this concept. In horticultural applications, stem cuttings are coated with an auxin to stimulate adventitious root formation. No adventitious root formation will occur if the coleoptile and axillary bud are removed from seedlings exposed to ethylene. Since the axillary bud is the main source of auxin for seedlings, ethylene alone may not be able to cause the initiation of adventitious root buds. The following theory for adventitious root formation in flooded plants has been suggested (Voesenek and Van Der Veen 1994). During flooding the basipedal transport of auxin is inhibited, resulting in an increase in auxin at the base of the stem. Endogenous ethylene concentration increases rapidly after flooding for the above-mentioned reasons. High ethylene concentration makes the tissues sensitive to auxin, which stimulates the formation of lenticels and adventitious roots at the bases of stems just above the waterline.

Elevated ethylene has also been associated with several other important adaptations of plants to flooded conditions. There is strong evidence that ethylene is related to the development of aerenchyma in many plants. In most wetland plants, the development of aerenchyma and highly porous stems and roots is genetically controlled. In such plants aerenchyma develops in the presence or absence of anaerobic conditions. However, in nonwetland species such as corn, aerenchyma occurs only during flooded anaerobic conditions. Corn produces lysogenic aerenchyma during flooding. Application of exogenous ethylene will also induce lysogenic aerenchyma in corn under aerobic conditions (Drew et al. 1979). More recent studies demonstrate that hypoxia induces ethylene synthesis in corn roots, and this is related directly to aerenchyma formation (Brailsford et al. 1993). In addition, hypoxia and ethylene concentration are correlated with an increase in cellulase activity. This may be significant for lysogenic aerenchyma formation, because cell wall breakdown may be needed to form intracellular space. In contrast, there are data which show that cells collapse during lysogenic aerenchyma formation without cellulase activity or cell wall breakdown (Konings and Lambers 1991). If the latter evidence is correct, the activity of cellulase is immaterial to the development of lysogenic aerenchyma in corn.

In these discussions of flooding tolerance we have not considered submergence extensively. There are many situations in which floodplain species may become submerged by floodwaters for variable lengths of time. During complete submergence, tissues become hypoxic and photosynthesis is severely reduced. If petioles and shoots cannot extend above the water surface, the plant will suffer extreme carbohydrate starvation. Shoot elongation will not occur under anoxic conditions except in a few cases such as rice coleoptiles and leaves of *Typha* species. However, hypoxic conditions induce rapid shoot and petiole elongation in submerged plants. For example, deepwater rice shoots can elongate at a rate of up to 250 mm per 24 h when submerged, eventually leading to a total plant height of over 7 m (Voesenek and Van Der Veen 1994). The advantages of rapid shoot elongation during submergence are (1) to enhance gas exchange (both oxygen convection and carbon dioxide fixation), (2) to promote wind- or insect-mediated pollination, and (3) to release end products from anaerobic metabolism or gases from anaerobic sediments that have infiltrated into roots, such as methane and hydrogen sulfide. Ethylene has been shown to enhance shoot elongation rates in many emergent and submerged aquatic macrophytes. Shoot elongation in submerged wetland species may also involve giberellins and auxins. In both cases the sensitivity of the tissue to these phytohormones is enhanced by high ethylene concentration. For further information about stress and phytohormones, see Chapter 4.

5. Photosynthetic Adaptations to Flooding. Photosynthetic adaptations to flooding in wetland plants are most dramatic in submerged aquatic macrophytes (SAMs). These are species in which all or some of their leaves are submerged during part or all of their life cycles. If the species has both emergent and submerged leaves on the same plant, the emergent leaves are entire while the submerged leaves are compound. Many submerged leaves also have a particularly thin cuticle and mesophyll tissues, although the morphology of submerged leaves varies among species and habitats. The compound morphology, thin cuticle, and thin mesophyll are some of the structural mechanisms that minimize resistance to diffusion of inorganic carbon from an aqueous solution into photosynthetic cells.

Most wetland plants are C_3 species. This is not surprising because C_4 and CAM photosynthesis are considered adaptations against high heat and low water availability, and these alternative photosynthetic physiologies are found most abundantly in plants from dry or hot environments. However, in submerged environments dissolved carbon dioxide can become scarce, and bicarbonate ion can become abundant. Since rubisco reduces CO_2 and pep-carboxylase reduces bicarbonate ion, there may be an advantage for submerged aquatic macrophytes to employ pep-carboxylase preferentially as the primary carboxylating enzyme (Keeley and Morton 1982).

Submerged aquatic macrophytic plants (SAMs) in freshwater systems experience a diversity of stresses: (1) a low concentration of inorganic carbon, particularly CO_2; (2) a limited availability of nutrients such as nitrogen and phosphorus; (3) oxygen buildup in tissues that can inhibit photosynthesis; and (4) limited light availability because of photon absorption by water or turbidity in the water. All these stresses taken together severely limit photosynthesis in submerged leaves. Adaptations in anatomy and physiology of SAMs have in some cases maximized photosynthesis under these stringent conditions.

Morphological adaptations (related to photosynthesis) of submerged leaves can be arranged in two basic groups, based on whether they inhabit eutrophic or oligotrophic lakes. Oligotrophic lake environments are characterized by moderate to acid pH (5 to 7), low inorganic carbon content (0.01 to 0.25 mol m^{-3}), low electrical conductivity (<50 $\mu S\ cm^{-1}$), low inorganic nitrogen (1-3 g N cm^{-3} sediment), and relatively high light intensity. In contrast, eutrophic lake waters have higher pH (7 to 9), higher dissolved inorganic carbon (2 to 5 mol m^{-3}), higher electrical conductivity (>200 $\mu S\ cm^{-1}$), higher inorganic nitrogen (10 to 20 g N cm^{-3} sediment), and relatively low light intensity (Keeley 1991). Leaves of SAM species in oligotrophic lakes are mostly cylindrical with low surface/volume ratio, and they are hollow due to extensive aerenchyma. These leaves have a thick cuticle, are displayed in a rosette phyllotaxy, and may have extended longevity (evergreen). In contrast, submerged leaves of SAM species, in the relatively higher nutrition and lower light intensities of eutrophic lakes, are finely dissected and reside near the surface of the water. These finely dissected leaves have a high surface/volume ratio and very little internal air space. The cuticle is relatively thin, and leaves are of short longevity (deciduous).

The differences in leaf anatomy between SAM species from oligotrophic and eutrophic lakes correlates with their unique photosynthetic processes. Species of eutrophic lakes have a much higher photosynthetic rate than species in oligotrophic lakes. The higher photosynthesis is due to higher nutrition and higher leaf surface area on plants in eutrophic lakes. Species in eutrophic lakes often are capable of accumulating bicarbonate ion for photosynthesis (at the high pH of eutrophic lakes bicarbonate ion is abundant) and utilizing C_4 photosynthesis, which is a rare occurrence in plants from oligotrophic lakes. CAM photosynthesis is well developed in plants of oligotrophic lakes, and these plants can utilize CO_2 produced in sediments for photosynthesis. In fact, 40 to 100% of total carbon fixed may be derived from the sediments (Keeley 1991). Having large ovate leaves, a rosette growth form, and large internal air spaces all favor accumulation of carbon dioxide from sediments. Furthermore, utilizing CAM photosynthesis elongates the diel time of photosynthesis and prevents leakage of CO_2 out of leaves during decarboxylations. Therefore, the anatomy and physiology of submerged leaves on SAM species in oligotrophic lakes are coordinated to maximize CO_2 accumulation, while those for submerged leaves in eutrophic lakes are designed to maximize photosynthesis and maximize the accumulation of bicarbonate ion.

VI. SUMMARY

Flooding is considered a detrimental occurrence because of its impact on agriculture and human habitation. This opinion is reinforced by development of agricultural fields, towns and cities on floodplains, or other naturally flooding areas. However, flooding is a common occurrence in over 6% of the world's terrestrial ecosystems. These naturally flooded areas, wetlands, support a diversity of species and serve many important ecosystem services. Wetlands mitigate floods from rivers and streams, they recharge groundwater, they serve as a filtration device for groundwater purity, they are a sediment catchment basin, they serve as waterfowl habitat, and they have many different possibilities for the recreational industry. Even though wetlands have many ecosystem and industrial services, the amount of wetland has been severely eroded by development throughout the world.

This chapter has been focused on the detrimental impacts of flooded conditions and how plants can cope with those flooded conditions. Flooding can cause anaerobic sediments, reducing sediments, acidic or alkaline sediments, low nutrient availability, and mechanical damage due to floodwaters and siltation. Hydrophytes are species that have evolved some adaptations to compensate for temporary or permanent flooding. Anatomical and metabolic changes combine to produce various suites of adaptations that allow hydrophytes to grow and reproduce in a diversity of flooded conditions.

The most detrimental aspect of flooded environments, whether natural or agricultural, is low-oxygen partial pressure in the sediments. Roots in anoxic or hypoxic conditions switch from aerobic respiration to anaerobic respiration. Along with this switch is a down-regulation of aerobic protein synthesis and an up-regulation of approximately 20 anaerobic proteins. Many of the anaerobically induced proteins are glycolytic and fermentation enzymes such as alcohol dehydrogenase (ADH). High rates of cytosolic glycolysis and fermentation can cause acidification of the cytoplasm and a potential buildup of toxic products such as ethanol.

Anatomical mechanisms of coping with anaerobic conditions in emergent wetland plants concern an increase in organ porosity and a greater ability to access aerobic or partially aerobic (hypoxic sediments). Organ porosity can be increased by the development of aerenchyma, which allows the movement of oxygen from the atmosphere to roots via an internal air passage. Diffusion of oxygen through aerenchyma can be enhanced by various convection methods. Further enhancement of oxygen diffusion to roots can be attained by lenticels on the bases of stems and pneumataphores extending from roots up into the atmosphere.

Under many flooding conditions the top few centimeters of soil are partially aerobic (hypoxic), while deeper sediments are anaerobic (anoxic). Roots that reside in these upper sediments are able to maintain a portion of their potential aerobic respiration capacity. Thus, surface rooting is a common scenario for wetland species. Adventitious roots, induced by changes in phytohormone concentration, proliferate at or above the water line and grow into surface sediments. Following flooding, some hydrophytes stop using roots that are now in deeper anoxic soils and switch to using only surface-growing adventitious roots for water and nutrient uptake. The relatively high porosity of adventitious roots allows them to penetrate deeper in sediments and yet retain some aerobic respiration. A disadvantage of using surface roots to combat anaerobic sediments in floodplains is that plants are likely to be uprooted when they experience the common sudden intense floods that are characteristic of these habitats.

Metabolic adaptation to low oxygen concentrations involves an up-regulation of anaerobic respiration. Glycolysis and fermentation must occur at a fairly rapid rates to provide enough energy for maintenance and growth, because these pathways are inefficient at producing ATP compared with aerobic respiration. To support glycolysis, a large reserve of carbohydrates is imperative for tolerating extended anaerobic respiration of roots. The requirement for a carbohydrate pool in roots and rhizomes under anaerobic conditions is amplified because translocation of photosynthate from leaves is inhibited under flooded conditions. Hydrophytes must also be able to compensate for high ethanol concentrations and cytoplasmic acidosis in root cells, which result from extensive anaerobic respiration. Wetland plants lose a lot of the ethanol produced by fermentation by diffusion into the aqueous matrix through radially porous roots. However, some species also minimize ethanol production by continuing to use lactate fermentation or other alternative pathways that oxidize NADH but produce by-products other than ethanol. In addition, synthesis of fats can oxidize NADH and keep glycolysis active without producing ethanol. To overcome cytosol acidosis that comes from extensive fermentation (particularly lactate fermentation), many wetland plants utilize various cytosolic alkalinization mechanisms. These and other metabolic adjustments are utilized by various wetland plants in combination with anatomical adaptations to cope with anaerobic sediments and other detrimental attributes of flooded conditions.

Although there are many anatomical, morphological, and metabolic adaptations to flooding, no plants can survive to reproduce in a completely anaerobic environment. When roots are exposed to anaerobic conditions, ramifications extend to aboveground parts through communication molecules and by changing source/sink relationships. Therefore, any soil flooding for extended periods will detrimentally affect a large majority of plants, particularly crops. There are only a few crops that are able to tolerate extended flooding and retain yield. Since flooding occurs in agricultural settings, and flooding reduces yield in most crops, studies of molecular adaptations to anaerobic conditions in wetland plants will provide potentially beneficial information for crop breeding.

STUDY-REVIEW OUTLINE

Introduction and Definitions

1. Wetlands occupy 6% of terrestrial land area and are defined on the basis of flood duration, sediment redox potential, sediment oxygen concentration, and species composition. Many definitions of wetlands are possible using any number or combinations of these factors.
2. Hydrophytes are species that have evolved morphological, anatomical, or biochemical adaptations for coping with flooded ecosystems. Hydrophytes are able to grow and reproduce in wetland ecosystems.

Natural Wetland Systems

1. A classification of the diverse ecosystems grouped as wetlands can be based on the frequency of flooding. Permanently flooded ecosystems are swamps, bogs, rice paddies, and some marshes, and temporarily flooded ecosystems are floodplains, some marshes, and irrigated fields.
2. Swamps are permanently flooded ecosystems dominated by trees and shrubs. Good examples are mangrove swamps and cypress swamps.
3. Marshes can be permanently or temporarily flooded ecosystems that are dominated by herbaceous species. Some important examples are prairie potholes of Minnesota, vernal pools of California, stream backwaters, and coastal marshes.
4. Bogs and fens are types of peat lands which are dominated by *Sphagnum* moss, herbaceous plants, and clonal shrubs (often in the *Ericaceae*). Bog and fen soils are extremely acidic and have low nutrient availability.
5. Floodplains are areas near rivers and streams that are flooded only when the waters of the river rise. These areas are often dominated by trees and are termed bottomland forest. The largest remaining bottomland forest in the United States is that in the floodplain of the southern Mississippi River. Large bottomland forest regions still remain on other continents, such as in Africa.
6. Agricultural fields are not exempt from flooding. Irrigation can flood agricultural fields for the short term, but this does not last long enough to affect crops. In contrast, natural flooding by excessive precipitation or rising river waters can cause extensive damage to crops, particularly when agricultural fields are built on floodplains.

Environmental Attributes of Flooded Conditions

1. The most important detrimental characteristic of flooded conditions is low oxygen partial pressure in sediments.
2. When a soil is flooded, oxygen cannot diffuse fast enough through water to compensate for root and soil microorganism respiration. Consequently, anaerobic soils develop.
3. A few centimeters of surface soil can retain some oxygen (hypoxic), in comparison to deeper layers, which have no oxygen (anoxic).
4. Anaerobic sediments affect microorganism activity, which will cause a reduced nutritional condition in the sediment.
5. The low redox potential of flooded soils causes a decreased availability of oxidized nutrients such as nitrate.
6. Salinity can also be an important aspect of coastal wetlands such as marshes and mangrove swamps. Halophytes are species that are able to complete their life cycles in saline conditions.
7. There are also mechanical stresses in some flooded ecosystems, such as rapid water movement and siltation, which can affect plant performance.

Detrimental Effects of Flooding on Plant Function

Effects on Respiration

1. Anaerobic respiration occurs in the cytosol, not in mitochondria, and is less efficient than aerobic respiration.
2. During anaerobic respiration, the adenylate pool in the cytoplasm decreases and the ratio of ATP/ADP decreases.
3. Since anaerobic respiration does not completely oxidize glucose to carbon dioxide and water, organic end products (ethanol, lactate, etc.) will increase during anaerobic respiration.
4. An alternative electron transport process in plants (cyanide-resistant respiration) may be able to oxidize NADH without producing toxic end products.

Effects on Photosynthesis

1. The first impact of root-zone flooding on leaf photosynthesis is a decrease in stomatal conductance.
2. Decreasing cytokinin synthesis and transport, as well as increasing ABA, are probably the metabolic cues for decreasing stomatal conductance during flooding.
3. Nonstomatal inhibition of photosynthesis also occurs during flooding in sensitive plants.
4. Nonstomatal inhibition during flooding may be due to a buildup of photosynthetic end products (inhibited carbohydrate translocation), inhibition of photosynthetic enzymes by toxic gases coming from anaerobic sediments, the low adenylate pool, or a buildup of inhibitory phytohormones.

Effects on Water Relations

1. Flooding with salt water creates an osmotic drought for some sensitive plants. Saline wetlands may also cause specific ion toxicity in sensitive plants.
2. Flooding with fresh water may decrease root permeability to water and root hydraulic conductance. As a result, short-term wilting may occur in some species in flooded conditions.
3. Adventitious roots, formed in response to flooding, are more permeable to soil water and reside in hypoxic rather than anoxic sediments. Upon the growth of adventitious roots, water transport to shoots is once again enhanced.
4. Adventitious roots have lower hydraulic conductance than aerobic roots so total water flow in plants will be less when dependent on adventitious roots.

Effects on Nutrient Relations

1. Flooding reduces transpiration and therefore reduces the delivery rate of nutrients, dissolved in the xylem, to the shoots. This mechanism is unlikely to contribute to significant nutrient limitation during flooding.
2. Anaerobic respiration is unable to provide adequate ATP for active nutrient-uptake mechanisms. This inhibition severely limits nutrient accumulation in flooded conditions.
3. When the sediment oxygen concentration decreases below the critical oxygen pressure (COP), respiration is inhibited and nutrient uptake slows.
4. A rapid onset of chlorosis occurs in sensitive plants after flooding because of reduced nutrient uptake and iron toxicity.
5. Inhibition of active nutrient accumulation processes is the single most important factor influencing plant nutrient relations during flooding.
6. Anaerobic conditions in sediments reduce the availability of oxidized nutrients to roots. In particular, oxidized forms of nitrogen, sulfur, and phosphorus are severely reduced in anaerobic soils. This mechanism can affect nutrient accumulation during prolonged flooding conditions.

Mechanical Effects

1. In floodplain ecosystems, large or acute floods can cause significant damage to plants by the forces of rapidly moving water.
2. A second mechanical impact of flooding can be extensive siltation, which can bury stems and branches and erode leaves.

Adaptations to Flooding of Plants in Natural Ecosystems

Avoidance of Anaerobic Cells by Adjusting Gas Diffusion

Root Location and Morphology

1. Wetland plants often have shallow roots, to avoid deeper anaerobic sediments and utilize surface hypoxic sediments. However, shallow roots make these species particularly susceptible to uprooting during floods.
2. Flooding induces adventitious roots, at or just below the water surface, which grow into surface sediments and take advantage of hypoxic conditions. Adventitious roots also have high oxygen porosity and are connected to the stem near the atmosphere.
3. Cubic cell configuration in roots increases root porosity to oxygen and allows roots to penetrate deeper into flooded sediments.
4. Alternative structures of roots and stems can enhance oxygen transport to roots in anaerobic soils. Stem lenticels decrease the resistance to oxygen diffusion into stems. Pneumataphores, vertical root extensions, provide a porous pathway for oxygen to the roots of black mangroves.

Anatomical Changes: Aerenchyma

1. Intercellular gas transport can be enhanced by the development of aerenchyma, a tissue with large gas-filled passages.
2. Aerenchyma can be developed lysogenically by the breakdown of cortical cells, providing a gas connection between leaves and roots. Aerenchyma can also be produced schizogenically by cortical cells pulling away from each other to create high tissue porosity to oxygen.
3. Aerenchyma can be significant in two ways: (1) in increasing porosity for oxygen transport from leaves to roots, and (2) in decreasing the volume-specific demand for oxygen in roots. The first is more important for wetland species.
4. Leakage of oxygen out of roots with aerenchyma creates a partially aerated sheath around wetland roots.
5. Aerenchyma can also transport other gases, such as methane, from sediments, nitrogen from the atmosphere, and carbon dioxide from roots to leaves.

6. Secondary growth reduces the effectiveness of aerenchyma. This is a major limitation to woody plant development in constantly flooded ecosystems. However, there are a few species that have aerenchymatous secondary xylem.

Convective Gas Flow Mechanisms

1. Convective gas flow mechanisms utilize pressure-enhanced diffusion of oxygen from leaves to roots.
2. Non-throughflow convection occurs when the carbon dioxide respired by roots dissolves and is carried away in the aqueous phase, creating a negative pressure in the roots. This causes gases to move from the atmosphere through leaves and stems to roots.
3. Non-throughflow convection requires a humid substomatal chamber and involves a decreasing pressure gradient from leaves to roots. This system is inefficient at transporting large amounts of atmospheric gases to roots.
4. The best case for non-throughflow convection is that of red mangroves.
5. Throughflow convection also requires a humidified substomatal chamber and small stomatal aperture, but pressure is maintained at a constant level from leaf to root.
6. The gas pathway for throughflow convection starts at the leaves, travels through rhizomes, and exits by a different leaf or out broken inflorescences.
7. Throughflow convection can be based on (1) humidity-induced convection, (2) thermal transpiration convection, or (3) venturi-induced suction.
8. Thermal transpiration convection occurs when leaf temperature is greater than air temperature and pressure is developed in humid substomatal chambers. This mechanism is particularly effective for transporting gases in water lilies.
9. Venturi-induced suction has been found in only one species; it occurs as wind moves across broken stem bases.
10. Humidity-induced convection occurs in many wetland species and works best with rhizomes, in which aerenchyma can contain oxygen concentrations equal to that of the atmosphere.
11. Thermal- and humidity-induced convection are both most pronounced during daylight hours. However, there is no diurnal pattern to venturi-induced suction.

Rhizospheric Oxygenation

1. Wetland plants with aerenchyma are able to oxygenate a sheath of sediments around their roots.
2. The hypoxic sediments around roots increase the availability of oxidized nutrients by the action of aerobic microorganisms in the oxygenated rhizospheric sheath.
3. An oxygenated rhizosphere can often be recognized by a sheath of red iron oxides around roots in an otherwise gray anaerobic sediment.

Tolerance by Metabolic Adaptation

Regulation of Respiration

1. Flooding induces the translation of many enzymes associated with anaerobic respiration. This allows ATP production to maintain cell function in the absence of oxygen.
2. An alternative fermentation pathway to produce lactate, malate, or alanine instead of ethanol is employed in some wetland plants. But most wetland plants utilize ethanolic fermentation under anaerobic conditions.
3. Alcohol dehydrogenase is stimulated during flooding conditions and results in a buildup of ethanol.
4. Most ethanol is removed from anaerobic cells in wetland plants by the aquatic medium or the transpiration stream.
5. Wetland plants are able to compensate for cytosolic acidosis during anaerobic respiration by various alkalinization mechanisms.

Preparatory Energy Storage

1. To maintain adequate anaerobic respiration for cell function, a large amount of carbohydrate is needed. Sensitive plants can suffer carbohydrate starvation during flooding.
2. Wetland plants can store large amounts of carbohydrates in rhizomes to aid anaerobic respiration when leaves are incapable of providing oxygen to their roots.

Lipid Metabolism and Mitochondrial Activity

1. In most species mitochondria lose internal structure and lipids accumulate in the mitochondrial matrix during flooding.
2. Lipids may accumulate because their synthesis serves to oxidize NADH to NAD^+ without the synthesis of ethanol. Thus, there is a flow of carbohydrate through glycolysis to fats, which produces ATP, oxidizes NADH, but does not produce ethanol.
3. The buildup of lipids may also serve as an important energy reserve when flooding conditions conclude.
4. In some wetland species mitochondria retain their electron transport activity under flooded conditions. Electron transport in these species when under oxygen deprivation may be dependent on cytochrome *d*, which is normally found in bacteria but not in plants.

Effects on Growth Regulators (reviewed in Chapter 4)

1. Ethylene synthesis increases during flooding, which induces aerenchyma formation and also possibly induces the formation of adventitious roots.
2. Ethylene causes tissues to become more sensitive to the actions of other growth regulators, such as indoleacetic acid.
3. Ethylene also induces shoot and leaf elongation. This is important for plants that become totally inundated during flooding.

Photosynthetic Adaptations of Submerged Aquatic Macrophytes

1. SAMs can have both submerged and emergent leaves. The morphologies of these two leaf types can be very different.
2. Leaf anatomy and morphology of submerged leaves are designed to maximize the absorption of inorganic carbon from the water solution.
3. In oligotrophic lakes nutrients are limiting, and the most abundant form of inorganic carbon is carbon dioxide. Plants growing in this habitat have cylindrical, hollow leaves with a thick cuticle.
4. In eutrophic lakes nutrients are relatively abundant, but light intensity may be low. Plants growing in this habitat have finely divided compound leaves which are carried close to the surface.
5. The photosynthetic physiology of SAM plants adapted to oligotrophic lakes is often dominated by crassulacean acid metabolism (CAM). In addition, these species gain most of their carbon by fixing respired carbon dioxide evolved from the sediments.
6. Plants that grow in eutrophic lake systems perform mostly C_3 photosynthesis, although some C_4 photosynthesis occurs. Photosynthesis in these species is relatively high and may often utilize bicarbonate ion.

SELF-STUDY QUESTIONS

1. Construct a key that enables one to separate the major wetland ecosystems on the basis of three or four critical ecosystem factors.

2. What are the most significant environmental attributes of flooded ecosystems that stimulate morphological and metabolic adaptation in wetland plants?
3. Differentiate between the significance of hypoxia and anoxia to plants in a long-term-flooded ecosystem.
4. Describe the flow of nitrogen through various molecules that creates low nitrogen availability in wetland systems. What factors would regulate the intensity of nitrogen deprivation?
5. In what ways is the switch from aerobic to anaerobic respiration in roots detrimental to plants in flooded systems?
6. Differentiate between the processes that cause stomatal versus nonstomatal inhibition of photosynthesis in flooded conditions.
7. Why is it that a plant can wilt in a flooded soil when there is no lack of water?
8. What is the definition of critical oxygen pressure, and what is its significance to nutrient accumulation by plants in flooded habitats?
9. Why are many wetland plants rooted in surface soils only?
10. Differentiate between lysogenic and schizogenic formation of aerenchyma.
11. Why is throughflow convection more effective than non-throughflow convection in providing oxygen to root tissues?
12. Why is thermal transpiration convection most appropriate for water lilies, and why is humidity-induced convection most appropriate for sedges and reeds?
13. How does non-throughflow convection operate in relation to tides in black and white mangroves?
14. What are the advantages and disadvantages of utilizing alternative fermentation pathways in roots under anaerobic conditions?
15. How do mitochondria retain activity in anaerobic tissues of some wetland species?
16. What is the significance of lipid synthesis to root metabolism under anaerobic conditions?
17. Why is CAM photosynthesis abundant in submerged aquatic macrophytes?

SUPPLEMENTARY READING

Blom, C. W. P. M. 1990. Adaptations of plants to flooding. Aquatic Botany Special Issue 38(no. 1):entire issue.

Cooper, P. F. and B. C. Findlater (eds.). 1990. The Use of Constructed Wetlands in Water Pollution Control. Pergamon Press, Oxford.

Crawford, R. M. M. (ed.). 1987. Plant Life in Aquatic and Amphibious Habitats (British Ecological Society, Special Publication No. 5). Blackwell Scientific Publications, Oxford.

Jackson, M. B., D. D. Davies, and H. Lambers (eds.). 1991. Plant Life Under Oxygen Deprivation: Ecology Physiology and Biochemistry. SPB Publishing, The Hague.

Kozlowski, T. T. 1984. Flooding and Plant Growth. Academic Press, San Diego, CA.

Lugo, A. E., and S. C. Snedekar. 1974. The ecology of mangroves. Annual Review of Ecology and Systematics 5:39–64.

Mitsch, W. J., and J. G. Gosselink. 1986. Wetlands. Van Nostrand Reinhold, New York.

Schlesinger, W. H. 1991. Biogeochemistry: An Analysis of Global Change. Academic Press, San Diego, CA.

Sculthorp, C. D. 1967. The Biology of Aquatic Vascular Plants. Edward Arnold, London.

10 Irradiance

Outline

Objectives

1. Describe the factors that regulate the quantity, spectral quality, duration, and heterogeneity of the light environment experienced by plants.
2. Explain the various technologies used to measure attributes of the light environment. Include a discussion of radiometers, spectroradiometers, specific spectral sensors, and canopy photographs.
3. Describe the potential impacts of the radiation environment on plants. Do not include a extensive discussion of energy budgets, as this is covered in Chapter 11. Include photosynthetic effects, physiological sensors, phytochrome, and UV impacts.
4. Describe the various morphological, anatomical, and physiological ways that plants of diverse habitats adapt to high-light environments.
5. Discuss the mechanisms by which plants have adapted to environments with low light intensity.
6. Describe acclimation in response to light environment changes. Include a comparison of the relative ability of high- and low-light-adapted plants to acclimate.

7. Define the environmental conditions that result in temporal and spatial heterogeneity of light availability.
8. Describe the impact of heterogeneity of light intensity at the level of a plant canopy.
9. Discuss the impact of light heterogeneity on leaf-level processes. Focus on non-steady-state photosynthesis and transpiration responses to subcanopy light flecks.
10. Summarize the varied responses of light quantity, spectral quality, and heterogeneity on the evolution and performance of wild and crop species.

I. INTRODUCTION

Plants are affected in a complex manner by irradiance at all times. The architecture of plants is dependent on the quantity, direction, duration, and quality of light. Several physiological sensing systems respond to light quality and thereby regulate major developmental stages and physiological processes in plant cells. Photosynthesis is dependent on light for energy and for induction of enzymatic processes. In addition, major aspects of a plant's life cycle, such as seed germination and flowering, are dependent on light quality and quantity. The importance of light to the basic factors that regulate growth rate, development, structure, function, and behavior, makes adaptation and acclimation to light environment critical to a plant's survival in any ecosystem.

Attributes of light vary among the ecosystems of the world. In some situations plants grow at high latitude, where photoperiods vary greatly among seasons and ultraviolet (UV) irradiance can be relatively high. Other plants live in alpine conditions where light intensity is high but the photoperiod may be relatively constant (equatorial mountains), or in a subcanopy environment in which light intensity can be low but very patchy. Aquatic plants may experience low light intensities with unusual spectral quality because water absorbs some wavelength bands more effectively than others. Due to these extreme variations in light "resource" availability and quality, plants have evolved many mechanisms to ensure the capture of adequate energy and to minimize the risk of damage by excess light intensity.

At the onset of this discussion of plant adaptations to light environments as we would like to make it clear that responses to light conditions are not independent from responses to other environmental factors. This is particularly true for the interaction of light and temperature (Chapter 11). High irradiance affects plant processes directly through its impact on factors such as enzyme activity and photosynthesis, but high light also affects plant physiological processes indirectly by its impact on thermal attributes of the tissues. The capacity of plants to respond to light may also depend on the current water status or nutrient status of plants. Therefore, the interactions among different resource states of plants may regulate the manner by which plants respond to light.

This chapter is designed to cover the environmental variation in irradiance (see Chapter 4), irradiance measurement, and plant adaptation and acclimation to environments with various irradiance attributes. A discussion of the impact of irradiance on plant energy budgets is discussed in Chapter 11 (high temperature and energy budget); however, some of these concepts overlap between chapters.

II. ENVIRONMENTAL VARIATION

The total amount of irradiance impinging on the atmosphere is relatively stable ($\pm 3.5\%$) and is referred to as the solar constant (1368 W m^{-2}). However, the amount of irradiance entering the biosphere is much less than that of the solar constant ($\Phi_{s(\mathrm{dir})}$) and of different spectral quality than the irradiance impinging on the outer atmosphere. Solar irradiance is altered during its pathway through the atmosphere, due to the scattering and absorptive attributes of atmospheric constituents.

Solar irradiance is scattered in the atmosphere by suspended particles, volatile molecules, and clouds. A portion of scattered light is reflected out into space, while the rest becomes *diffuse irradiance*. Diffuse irradiance is sometimes referred to as *sky irradiance* because this fraction of irradiance lacks directionality, in contrast to direct beam irradiance from the sun (direct irradiance passing unimpeded through the atmosphere). Solar irradiance scattered by relatively small particles may cause a spectral shift in the irradiance, but most scattering effects have little influence on the spectral quality of diffuse light.

The total quantity of direct solar irradiance that passes through the atmosphere when the atmosphere is free of dust or clouds, depends on the optical airmass (m). The optical airmass is a relative term that defines the thickness of the atmospheric pathway for light at the angle of the sun in comparison to light coming directly down from straight overhead (the zenith). Thus, m is dependent on topographical elevation and the solar elevation angle (β). At high elevation or when the solar angle is high (sun is overhead) m is small and a higher proportion of direct solar irradiance penetrates the atmosphere compared with low-elevation sites or low solar angles (Figure 10.1).

The intensity of direct solar irradiance that reaches the biosphere at any location is also dependent on the transmissivity of the atmosphere. Clouds, dust, water, and other gaseous materials reduce transmission of solar irradiance. If all solar irradiance penetrated the atmosphere, transmissivity would be 1.0, but actual values range from 0.55 to 0.70. Direct solar irradiance on a horizontal surface is dependent on the solar angle (β), the solar constant ($\Phi_{s(\mathrm{dir})}$), transmissivity (τ), and the atmospheric optical mass (m).

$$I_{s(\mathrm{dir})} = \Phi_{s(\mathrm{dir})}\tau^{m} \cos \beta \tag{10.1}$$

Direct beam solar irradiance flux density at horizontal surfaces in the biosphere varies from zero at the poles in winter to approximately 40 MJ m^{-2} day^{-1} at midsummer just above 40° latitude. However, total solar radiant flux density is the sum of direct beam and diffuse solar radiant flux density. It is important to note here that *irradiance* is a term commonly used to describe radiant flux (W m^{-2}) between 200 and 3000 nm. The proportion of total irradiance that is diffuse, on a clear day at noon, is between 10 and 30%. However, close to dawn and dusk (low solar angle), and when clouds are present, diffuse irradiance accounts for a higher proportion (up to 100%) of total irradiance (Figure 10.1).

The spectral quality of solar irradiance is also affected during its path through the atmosphere. At the surface of the atmosphere the spectral distribution of energy in solar irradiance is close to that of blackbody irradiance (irradiance of platinum at 2700°C). However, absorptive characteristics of molecules in the atmosphere, particularly water, carbon dioxide, and ozone, alter the spectral quality of solar irradiance (Figure 10.2). Total solar irradiance can be subdivided into the wavelength bands of ultraviolet (200 to 400 nm), visible (400 to 700 nm), and shortwave infrared (700 to 3000 nm). Ozone in the stratosphere absorbs UV irradiance in the 200 to 380 nm wavebands, reducing the quantity of UV light hitting the earth's surface. Water, carbon dioxide, methane (CH_4), and several pollutants in the atmosphere (sulfur and nitrous oxides) absorb shortwave infrared (IR) irradiance. Thus, the chemical composition of the atmosphere screens much of the UV and IR irradiance while letting much of visible irradiance penetrate. It is not surprising that the optic systems of animals and photosynthesis have both evolved to use the visible waveband of total solar irradiance.

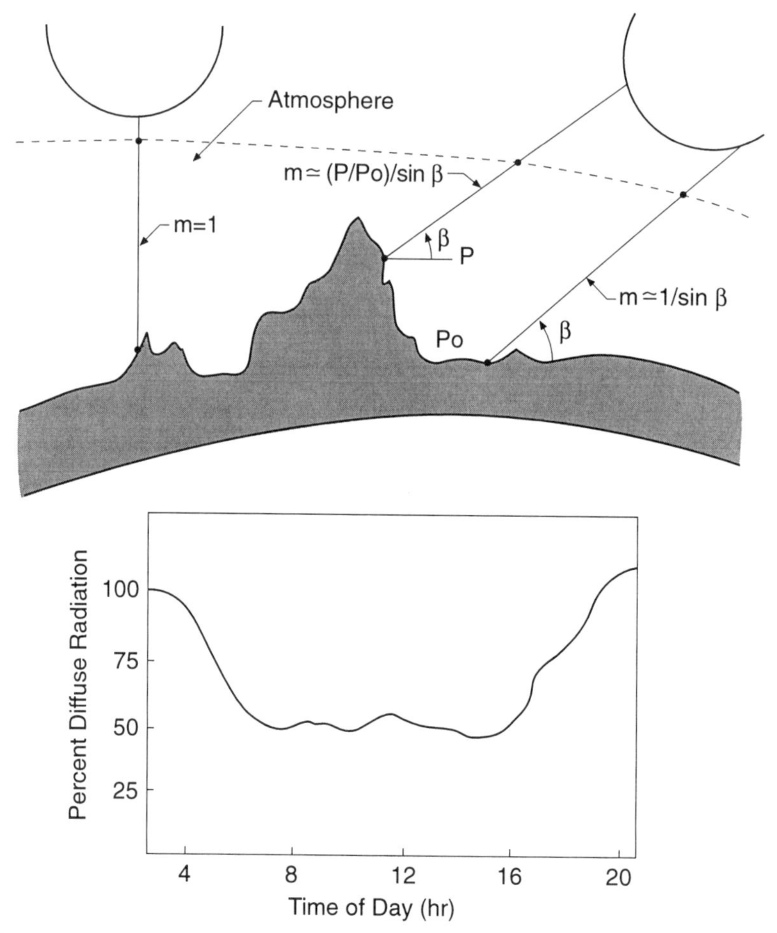

Figure 10.1 (A) Schematic representation of light scattering by atmospheric components and penetrating through the atmosphere. (B) Proportion of total irradiance accounted for by diffuse irradiance at 51.5° N latitude. (Modified from Jones 1992.)

The duration of available irradiance also varies on a diurnal cycle in the biosphere because of planetary motions. Day length, or *photoperiod,* is dependent on the season and the latitude (Figure 10.3). Photoperiod has profound effects on total daily irradiance and the relative proportions of direct and diffuse irradiance because solar angle varies with photoperiod. Several important subcellular physiological sensors, which regulate transitions between important growth and development phases of organisms, are dependent on photoperiod.

Characteristics of the terrain also affect the irradiance environment. Surface reflectance, or *albedo,* can affect total irradiance. In sites with high albedo (highly reflective), net irradiance on the plant will be relatively high because most of the reflected irradiance can affect the vegetation before returning to the atmosphere. The aspect and slope of the terrain can also regulate the amount of direct solar irradiance hitting the surface of the soil or vegetation. In the northern hemisphere, south-facing aspects receive the greatest irradiance. Depending on the latitude and season, specific slopes that are perpendicular to the direct solar beam will receive greater irradiance than other slopes. Therefore, total irradiance on vegetation is affected by attributes of the atmosphere and topographical conditions (slope, albedo, aspect, texture, etc.).

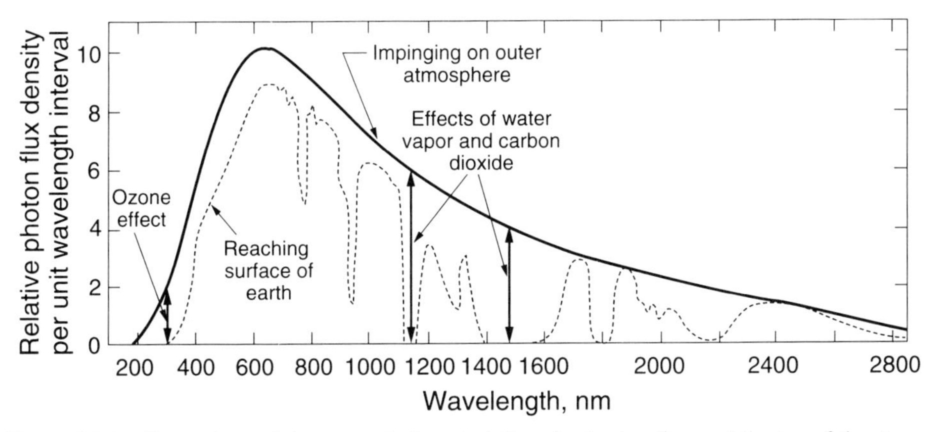

Figure 10.2 Comparison of the spectral characteristics of solar irradiance at the top of the atmosphere (solid lines) and at the Earth's surface (dashed line). Regions of absorption by atmospheric gases are indicated. (Combined drawing from information in Nobel 1991 and Jones 1992.)

Throughout this discussion of irradiance availability we have been assuming that there is an even distribution of irradiance impinging on a plant. However, large amounts of temporal and spatial heterogeneity in irradiance hitting the plant surface may occur. The distribution of clouds is not uniform, so that the daily pattern of atmospheric transmissivity will not be uniform. On some days the pattern of irradiance will be close to that predicted by equation (10.1), while on other days irradiance will change drastically, due to moving clouds in the atmosphere (Figure 10.4). Mountain environments in which background irradiance is high and fast-moving storm fronts are common have frequent stochastic reductions in irradiance due to clouds (*cloud flecks*).

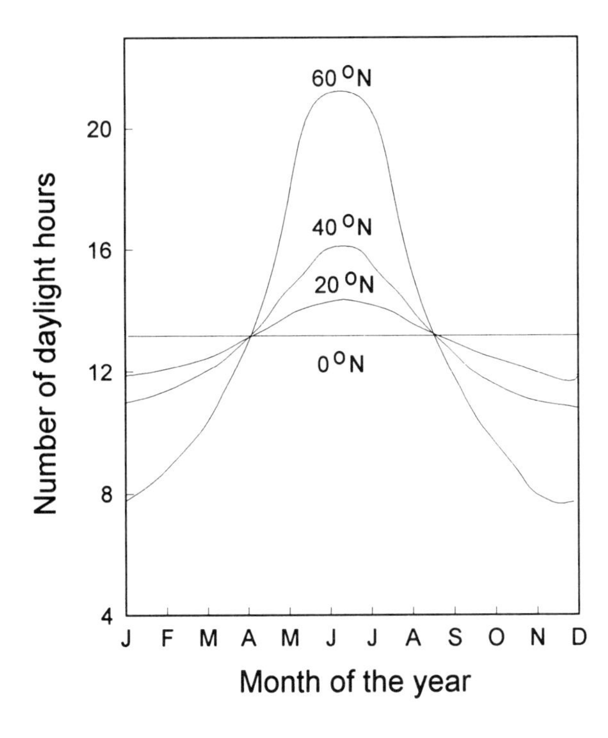

Figure 10.3 Number of hours a day during the year at different latitudes. (Modified from Downs and Hellmers 1975.)

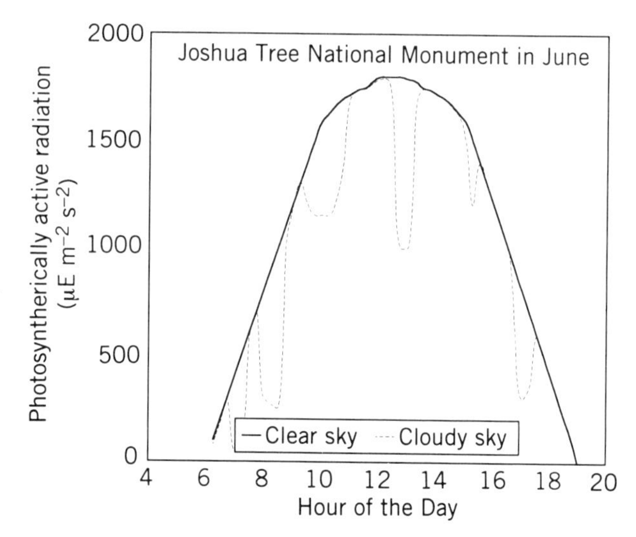

Figure 10.4 Diurnal cycles of irradiance from a high-elevation desert site in Joshua Tree National Monument, California. The solid line refers to a day in July with a clear sky. The dashed line represents a day in July when clouds were present.

Irradiance impinging on the surface of a tree or shrub canopy will be attenuated as it enters the canopy. Some light will scatter by reflecting off leaves and stems, some light will be absorbed by leaves and stems, some light will pass through gaps in the canopy as direct beam irradiance, and some light will pass through leaves. Light attenuation in the canopy is dependent on the density of the canopy and the angle of leaf blades. The interior of the canopy will have a higher proportion of diffuse light, a higher patchiness of direct light, and the diffuse light will have a different spectral quality from that of light hitting outer canopy leaves. Thus, leaves that are located within the canopy will have a very different light environment than that of leaves on the outer surface of the canopy. This heterogeneous light of the canopy interior makes it difficult to model the irradiance environment for all leaves on a plant. In addition, there are many epiphytic plants that live in the interior of other species' canopies, particularly in tropical forests. Epiphytes and climbing vines that reside at the top of a tropical forest canopy will receive more irradiance with a different spectral pattern than will plants or leaves of vines residing lower in the canopy (Figure 10.5).

There are also many species located on a forest floor whose light environment is modified by canopy trees. The light environment of subcanopy plants is dominated by diffuse irradiance that is depleted in visible (particularly blues and reds) compared with solar irradiance, due to light passing through canopy leaves. The intensity of the diffuse irradiance will vary depending on the leaf area index of the canopy. For example, common measurements of photosynthetically active diffuse irradiance under a temperate deciduous forest may be 100 to 250 μE m^{-2} s^{-1}, compared with 1 to 2 μE m^{-2} s^{-1} under a tropical forest. In addition, the type of species that form the canopy can affect the spatial distribution, frequency, and duration of direct irradiance that makes it to the subcanopy environment (Figure 10.6). Anyone who has walked in a forest has noticed the patches of light that highlight small locations of the subcanopy. These "sun flecks" occur because direct irradiance has penetrated through a gap in the canopy. Under temperate broadleaf forests, sunflecks are of longer duration and higher frequency than in the subcanopy of tropical forests (Figure 10.6).

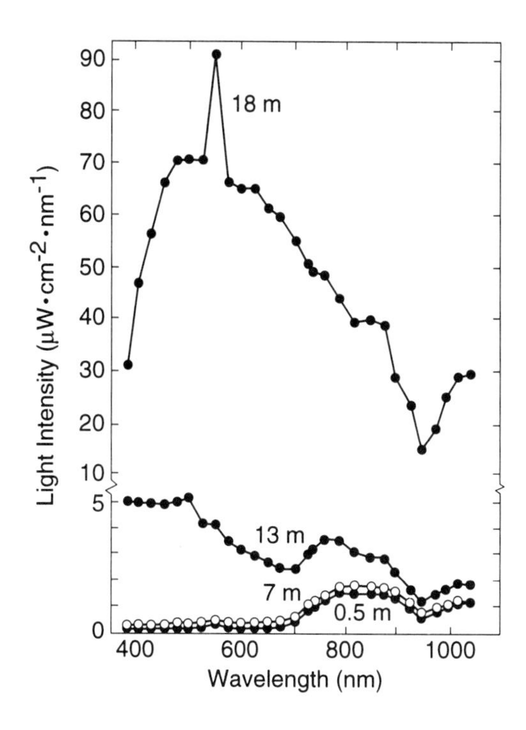

Figure 10.5 Changes in the intensity and spectral distribution of light at different heights in a tropical canopy. (From Johnson and Attwood 1970.)

Non-Steady State Light Patterns

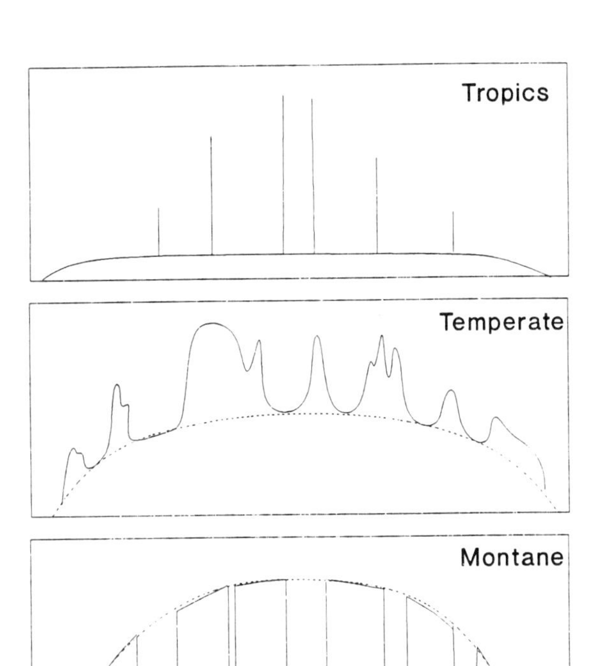

Figure 10.6 Characteristic daily pattern of irradiance during July for a shrub growing in a montane grassland, a shrub growing under a temperate deciduous forest, and a shrub growing under a tropical evergreen forest. Each trace represents those data for light intensity found in many studies.

III. MEASUREMENT

A. Total Irradiance

The total quantity of irradiance impinging on a surface is termed the incident radiant flux density or irradiance and is measured in W m^{-2}. Irradiance includes the energy of all wavelengths of light between 200 and 3000 nm. If one determines the irradiance of particular wavelengths, the term *spectral irradiance* is used and the SI (System International) unit is W m^{-2} nm^{-1}. Light has both wavelength and particle properties. If light is considered to be in discrete packages of energy, the photon is used to refer to that package of light energy, and quantum is the amount of energy in the package. When particle characteristics of light are used, the standard unit is the *einstein,* which equals a mole of photons (6.02×10^{23} photons). Although this unit has at times been used to refer to the amount of energy in a mole of photons, the correct and accepted usage is the number of particles in a mole of photons. The energy contained in an einstein is not uniform because the spectral characteristics of the photons (amount of energy in each photon) is not uniform. *Photon flux* refers to the number of photons per unit time absorbed or emitted by a surface (E s^{-1}), and the incident photon flux density is the number of photons per unit area per time (E m^{-2} s^{-1} or mol m^{-2} s^{-1}). In studies where the energy of incident irradiance is evaluated, the units W m^{-2} are used, and when studies concern photosynthetic properties, photon flux density is used. *Photosynthetically active irradiance* (PAR) refers to the irradiance impinging on a surface in the restricted wavebands that are directly involved with photosynthesis (400 to 700 nm). Normally, studies of photosynthetic effects of irradiance utilize the incident *photosynthetic photon flux density* (PPFD) of PAR and report units as mol m^{-2} s^{-1} or E m^{-2} s^{-1}. Those interested in the energy contained in PPFD must measure the photosynthetic irradiance (PI) and report units as W m^{-2}.

There are three general types of irradiance sensors. Thermoelectric sensors utilize the influence of irradiance on the temperature of different colored surfaces. Thermocouples, thermopiles, or thermisters are attached to a surface painted with black absorbent paint and compared with a reference temperature measurement of a white reflective surface. The sensor utilizes the difference in temperature of the two surfaces to determine irradiance. Thermoelectric sensors are designed to respond equally to irradiance of many wavelengths and can be used to measure the sum of solar and long-wave irradiance.

Photoelectric sensors are based on the differential responses of metals to irradiance. The photovoltaic effect is an electric field developed between two adjacent thin layers of materials. If this electrical field is short circuited, the current in the short circuit is linearly related to the photon flux on the surface. There are a number of different materials used to establish the electric field. Each material used has a different spectral response. For example, the long-wavelength cutoff point (point of low sensitivity) for silicon photocells and gallium arsenide phosphide (GaAsP) photocells is 1100 and 700 nm, respectively. Since the voltage output from a photocell is inherently nonlinear, photoelectric cells utilize a current–voltage converter to establish a linear response between voltage and irradiance.

The third method of measuring irradiance is by chemical reaction. Photochemically induced color changes have been calibrated by exposure to known light intensities or doses. Blueprint paper or dye solutions have been used successfully to measure total irradiance impinging on surfaces. One should be warned that these photochemical reactions are wavelength sensitive, so they are measuring only a portion of total irradiance.

Total solar irradiance is commonly measured with pyranometers. These sensors measure both direct solar irradiance hitting the sensor and indirect irradiance (W m^{-2}). Pyraheliometers are designed to have a narrow opening so that they measure only direct irradiance. Pyranometers and pyraheliometers do not measure long-wavelength irradiance (that greater than 3000 nm), but radiometers measure both solar and long-wavelength irradiance. Most pyranometers employ a thermoelectric mechanism. The sensor is commonly covered with a glass dome, which restricts the impact of turbulent air on thermal responses and screens out long and very short wavelengths. Spectral responses of pyranometers are usually between 300 and 3000 nm. There are pyranometers based on silicon photocell technology. However, these silicon cell pyranometers have a spectral response considerably shorter than that of solar irradiance and cannot be utilized under canopies where the irradiance has a spectral quality shifted to longer wavelengths.

Net pyraradiometers measure the difference between incoming and outgoing irradiance. They are based on two thermoelectric cells facing in opposite directions and well insulated from each other. This contrasts with total radiometers, which also have two thermoelectric cells, but the output is summed, resulting in a combined measurement of incoming and outgoing irradiance.

B. Photosynthetically Active Irradiance

As mentioned above, the appropriate spectral quality of light for photosynthesis is PAR. Thus, sensors that determine available irradiance for photosynthesis must measure a narrow band of wavelengths (400 to 700 nm). The original filtered silicon photocell for measuring PAR was developed in the early 1970s (Biggs et al. 1971), and modified versions are still in use today (the quantum sensor), making this the standard for PAR measurement. The cosine-corrected spectral absorptance of this sensor is very close to the ideal photon response curve. PAR sensors made today are very stable over time (less than 2% change per year), rugged, and easily portable to remote field sites. Specialized versions of the small disk PAR sensor have been constructed for underwater use (spherical sensor) and subcanopy measurements (1-m-long sensor).

The heterogeneity of light impinging on plants cannot be measured with one or a few quantum sensors. Rather, many sensors are required to represent the spatial and temporal variation of light in a subcanopy situation. Initially, techniques based on many quantum sensors were established (Hutchison et al. 1980) in which multiple sensors were placed on a cable and pulled through a section of canopy. Light intensity was constantly recorded on all sensors. Gutschick et al. (1985) initiated the use of many small (2 mm in diameter) and light (10-mg) GaAsP photocells mounted directly on leaves and interfaced with a computer through amplifiers and analog-digital converters. The GaAsP photocells have some inherent spectral inadequacies compared to the standard quantum sensors, resulting in up to 10 to 15% error (Chazdon and Fetcher, 1984) and should not be used when only a few sensors are required. However, if many sensors are required, the GaAsP photocells should suffice (keeping in mind the potential deviation from quantum sensor measurements) because the cost of many quantum sensors is prohibitive.

C. Spectral Properties of Irradiance

Irradiance impinging on the top of a canopy has a different quality from that reaching the bottom of a canopy. This and other situations can result in selective screening of wavelengths that standard radiation sensors cannot discern. To evaluate the spectral quality of light, a spectroradimeter is required. This device is similar to a spectrophotometer in the fact that the detection of light can be tuned to specific wavelengths. Spectroradiometers are based on a monochrometer that separates specific wavelengths of light, a photodetector, and a mechanism to adjust the photodetector output to the different energies of specific wavelengths. Spectroradimeters can provide a complete spectrograph of the irradiance impinging on a surface. Although it is possible to take spectroradimeters out into the field, they are expensive and not commonly used in studies of plant response to light. More commonly the researcher would like to determine the relative intensity of several specific wavebands, such as 660 and 730 (those causing phytochrome responses). Detection of narrow bands of wavelength can be obtained by using glass filters that have specific wavelength cutoff points. These filters are highly absorbent at shorter wavelengths and transmit wavelengths greater than the specific cutoff point. Therefore, by combining the measurement from a photocell covered with each of two specific filters, one can obtain irradiance measurement for specific waveband widths (usually about 25 to 50 nm).

Ultraviolet and long infrared irradiance are at opposite ends of the solar spectrum. As discussed earlier, ultraviolet light is defined as that irradiance between 200 and 400 nm. This light can be divided into three general ranges (UV-A = 320 to 400 nm, UV-B = 280 to 320 nm, and UV-C = 200 to 280 nm.) Stratospheric ozone concentrations effectively block all UV-C radiation. Thus, only UV-B and UV-A penetrate to the Earth's surface. One can measure the amount of UV-A and or UV-B irradiance by placing over a pyraradiometer specific filters that have high wavelength cutoff points that bracket either one of these two wavebands. Most commonly, scientists are interested in the UV-B waveband because this component of solar irradiance is more detrimental to plant systems than UV-A and is increasing due to changes in atmospheric ozone composition. UV-B monitors are available with permanently installed filters over their pyraradiometer.

Any object over 0^0 K will emit infrared irradiance. The amount of infrared irradiance emitted by a surface is proportional to the fourth power of the surface temperature (blackbody or Stefan–Boltzmann law):

$$E_{\mathrm{IR}} = e\sigma T^4 \tag{10.2}$$

where E_{IR} is the emmitted infrared radiation, σ the Stefan–Boltzmann constant (5.670×10^{-8} W m^{-2} K^{-4}), e the object emisivity, and T the temperature in kelvin.

Hand-held infrared thermometers that utilize the Stefan–Boltzmann law have been developed to assay the temperature of leaves nondestructively. Simply pointing the IR thermometer at the leaf or stem surface will record the surface temperature and infrared light emittance. Infrared light also can be directly measured inexpensively by placing a filter over a pyraradiometer that has a high wavelength cutoff of 700 nm.

D. Image Analysis of the Light Environment

The light intensity in a canopy and below a canopy is not uniform. One can measure light intensity at many points and average those values or represent them in a histogram to describe spatial variation in light intensity. However, those patterns of irradiance intensity change over the diurnal cycle and the season because of planetary movements. Diurnal cycles of irradiance can indicate temporal changes at one point but cannot represent the dynamic interaction between temporal and spatial properties of light on plants. The most robust measure of heterogeneous light environments has been through the use of canopy hemispherical photography (Anderson 1964). Fish-eye photographs (taken with a 4- or 8-mm lens) are taken of the canopy on a relatively cloudy day (to prevent sun glare) as an indirect estimate of the irradiance received at the point where the photograph was taken (Figure 10.7). The image on the negative is then captured by an image grabbing interface

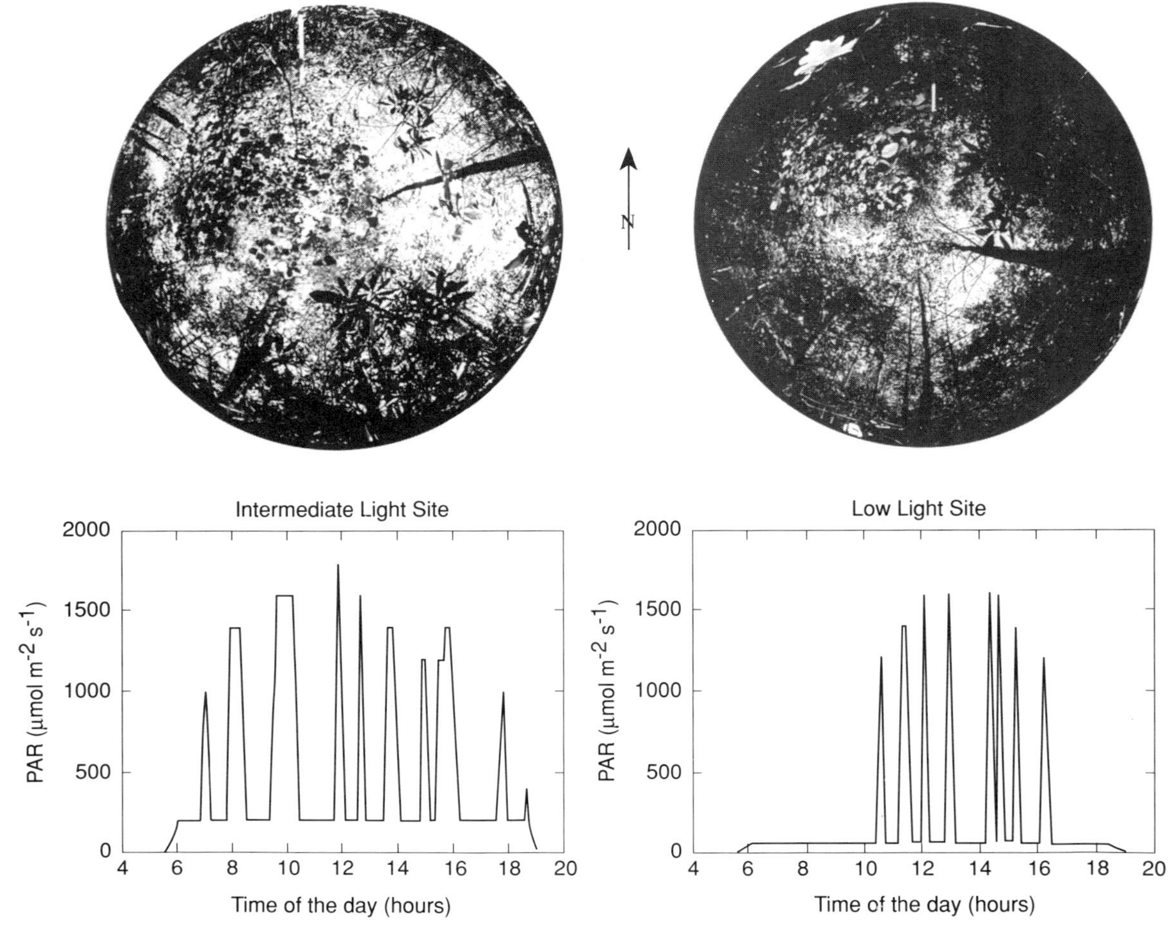

Figure 10.7 (A) Canopy photograph taken under two different temperate forest canopies in July. (B) Trace of irradiance that was derived by interpreting the canopy photographs, shown in part A, in relation to the solar track across the sky in July.

and saved in digital form on a computer. There are several software programs that will essentially overlay a solar track on the image and calculate the patterns of irradiance on the surface over a diurnal cycle (Rich 1989). The determination of total diffuse and direct irradiance on the surface by this method can be done for any day of the growing season or summed for the entire growing season. Instruments that measure integrated light intensity without a photographic technique are also available, but these are unable to represent the spatial and temporal heterogeneity of the subcanopy light environment.

IV. EFFECTS OF IRRADIANCE ON PLANTS

Irradiance that impinges on plants has multiple impacts on plant growth, development, and physiology. The intensity, duration, direction, and spectral quality of light can influence plant growth. The influences of irradiance intensity predominantly affect the energy balance of a tissue and its resulting temperature (discussed in Chapter 11). During periods of intense irradiance and/or high tissue irradiance absorption, tissue temperature can increase drastically compared with air temperature. It is not unusual for leaf temperature to exceed air temperature by as much as 10°C (Figure 10.8). Other physiological processes require less irradiance. For example, photosynthesis uses only a small part of total irradiance (less than 1%), and photoreceptors that regulate plant development require very small quantities of irradiance compared with that required for net photosynthesis.

A. Effects on Photosynthesis

A small portion (about 1%) of visible irradiance hitting a leaf is utilized by photosynthesis. Energy of photons that hit the various pigments in the chloroplast is collected by chlorophyll molecules in light-harvesting protein complexes and is transferred to electrons by the Hill reaction (see Chapter 6). Although leaves absorb most of the visible irradiance that hits them (commonly 80 to 95% absorptance), only a fraction of absorbed visible light provides energy for photosynthesis. The rest is absorbed by a diversity of other

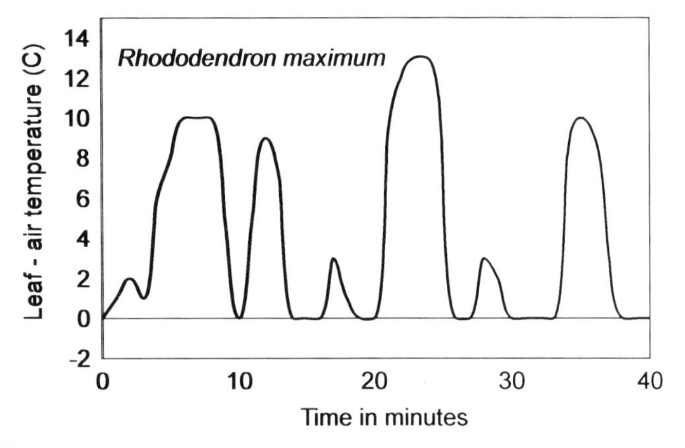

Figure 10.8 Representation of the leaf temperature for *Rhododendron maximum* compared with air temperature on a clear day in July. The measured leaf is on a plant growing under a temperate forest canopy.

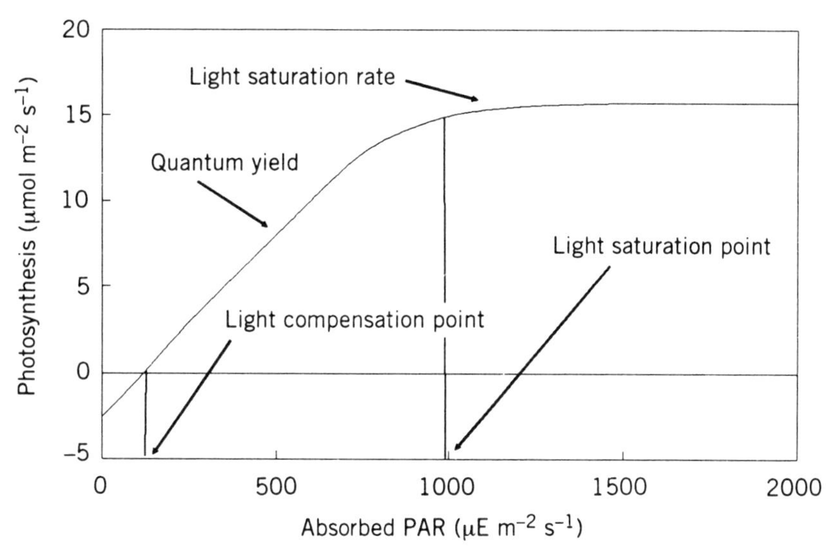

Figure 10.9 Diagrammatic representation of the response of photosynthesis to light intensity, termed a "light response curve." Critical points on the light response curve are highlighted by arrows and labels.

compounds. The small amount of visible irradiance that is transmitted through leaves is enriched in wavelengths near 550 nm compared with other visible wavelengths.

Photosynthesis increases with increasing absorbed irradiance in a predictable manner. The photosynthetic light response curve (Figure 10.9) has general properties that are uniform among species. The initial part of a light response curve is linear because light is the dominant limiting factor. All light response curves should include a measurement in the dark to indicate the dark respiration rate. The next important point is the *light compensation point,* that light intensity which is required for net photosynthesis to equal zero (gross photosynthesis counterbalances the sum of dark respiration and photorespiration). The slope of the linear increase in P_n with increasing light at relatively low light intensity (100 to 200 $\mu E\ m^{-2}\ s^{-1}$) is termed *quantum yield.* This term represents the increase in carbon gain (yield) for any increase in energy absorbed (quantum gain). When resources other than light become limiting, the light response curve loses its linearity and becomes horizontal. The light intensity corresponding to the point where P_n does not increase with increasing light is the *light saturation point.* Quantum yield is dependent on the capacity of the light-harvesting apparatus in relation to the electron transport chain. In most C_3 plants quantum yield is relatively constant (about 0.55 μmol CO_2 mol photons). Temperature and its impact on photorespiration cause changes in quantum yield in C_3 plants. Quantum yield varies significantly among species with different photosynthetic pathways (discussed further in Chapter 6). Light-saturated P_n is dependent on the capacity of the electron transport chain to produce ATP and NADPH, the availability of P_i, the capacity of the Calvin cycle, rubisco activity level, nitrogen concentration, and the rate at which triose phosphates are used by the cell. Since it is easier to interpret changes in quantum yield than changes in the light-saturated P_n rate, comparisons among species and/or sites are often made on the basis of quantum yield. If other diagnostic gas exchange response curves, such as a CO_2 response curve, are utilized along with the light response curve, the physiological processes that are regulating changes in light-saturated photosynthesis can also be determined (Sharkey 1985).

B. Physiological Sensors

Plants are able to utilize specific wavelengths of light to regulate major developmental changes. For example, seed germination, bending of stems and branches, initiation of flowering, internode elongation, and plastid replication are some of the developmental processes regulated by irradiance. Irradiance elicits a physiological response through the action of chromophores, which are molecules that absorb irradiance of a specific spectra and cause physiological change. The best known chromophores are those that absorb in visible wavelengths. We discuss only the phytochrome response, although there are several other chromophore systems (notably the blue light responses) that regulate important processes such as stomatal opening.

1. Phytochrome. As discussed in Chapter 4, phytochrome is a system of chromophores that can elicit many developmental responses in plants. The molecular structure of this chromophore is probably a dimer of two similar proteins (Figure 10.10). Each of these two proteins (I and II) is composed of two disk structures (A and B). The A domain of each protein contains the chromophore, while the B domain of these proteins probably interacts with the cell membrane (Tokutomi et al. 1989). The chromophore in the A domain can be in two forms, each of which absorbs light maximally at different wavelengths (Figure 4.2).

The spectral property of this dimer can be converted between two forms (Figure 4.3), which was discussed in Chapter 4. When light impinges on a population of phytochrome dimers, a photo equilibrium is established between the two forms. When red light is used to irradiate a plant containing phytochrome, a spectral equilibrium is established in which 85% of the phytochrome is in the P_{fr} form. Similarly, if only far-red light is used, not all phytochrome will be converted to P_r (97%). The inability to convert all phytochrome to one or the other form is due to an overlap in the absorption spectra of the two forms. Re-

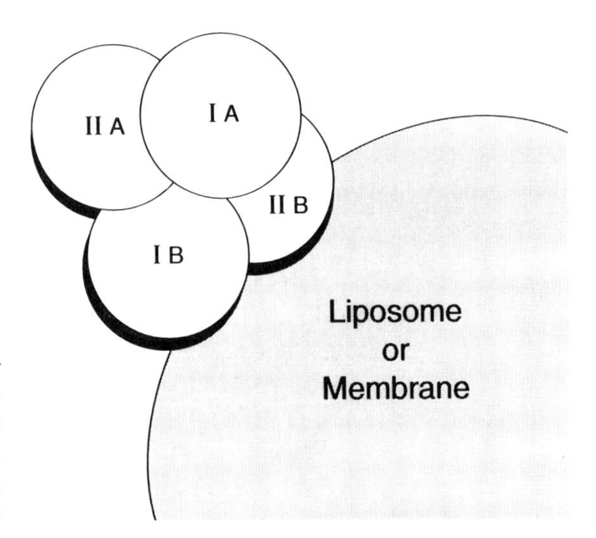

Figure 10.10 Most likely molecular model for the structure of phytochrome when adhered to a liposome or a membrane. Each disk (IA, IB, IIA, IIB) corresponds to a protein. (Redrawn from Tokutomi et al. 1989.)

call from Chapter 4 that the P_r/P_{fr} equilibrium is further affected by darkness because P_{fr} is slowly converted to P_r during the dark. In Chapter 4, general aspects of phytochrome structure, interconversions, and relationships with other phytohormones were discussed. In this Chapter we concentrate on the molecular mode of action and specific plant responses to light quality mediated through phytochrome.

The subcellular location of the two forms of phytochrome (P_r and P_{fr}) is different in green plants (nonetiolated form of phytochrome). P_r is located diffusely in the cytoplasm. When the cell is irradiated with red light, the P_{fr} aggregate around small ovoid structures, probably liposomes. There is also definitive evidence that P_{fr} can be associated with membranes, while P_r does not. Since P_{fr} seems to be the active form, the association of phytochrome with membranes may be a critical component of its physiological regulation capabilities.

One mode of action for phytochrome depends on the interaction between phytochrome and membrane permeability to calcium. When cells are irradiated with visible light, the spectral quality of this light causes the photoequilibrium of phytochrome to be strongly dominated by the P_{fr} form. P_{fr} binds to membranes, causing them to be more permeable to calcium. The increased calcium content in the cytoplasm binds with calmodulin (a sensitive calcium-binding protein). The calmodulin-Ca complex then binds to specific enzymes, activating them and stimulating a morphogenic response.

Phytochrome can also affect the transcription of genes for important photosynthetic enzymes (Figure 10.11). For example, high P_{fr} content in the cytoplasm activates a regulatory protein which enters the nucleus and binds to a light-activated element in the promoter region of the rubcs (low-molecular-weight component of rubisco) and cab (light-harvesting complex protein of photosystem II). Binding of the regulatory protein to the promoter stimulates transcription of these important genes (Schäfer et al. 1986). Therefore, phytochrome proteins affect plant physiology and development through impacts on membrane function, enzyme activity, and DNA transcription.

What are the environmental factors that affect phytochrome? This topic was also cov-

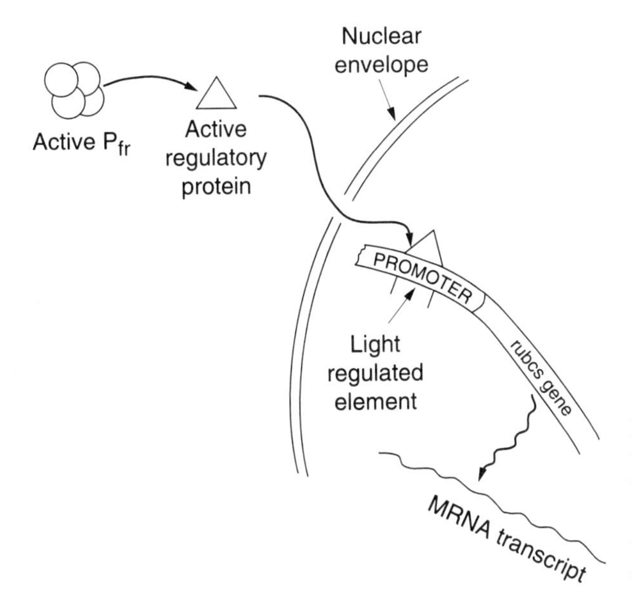

Figure 10.11 Molecular model for the interaction between phytochrome in the cytosol and transcription of the *rubcs* gene (gene which transcribes the small subunit of rubisco) in the nucleus.

ered in Chapter 4, but it is important to refresh the readers mind about causes for environmental variation in red and infrared light. The spectral equilibrium of the two phytochrome forms is dependent on the presence and spectral quality of light in the environment. When plants are in the dark, the slow reversion of P_{fr} to P_r and decomposition of P_{fr} causes the spectral equilibrium of phytochrome to become dominated by P_r. This slow conversion from P_{fr} to P_r dominance is the mechanism by which plants monitor the length of the night and regulate photoperiodic events. Immediately after a dark-grown plant is irradiated with sunlight, P_{fr} dominates the phytochrome pool. This is because the quantity of red light in sunlight is significantly more than far-red light. As long as sunlight is transmitted through the atmosphere, its spectral quality will cause a dominance by P_{fr}. However, if sunlight is transmitted through the leaves of a canopy, or water column, irradiance becomes enriched in far-red light in comparison to red light because chlorophyll absorbs red and transmits far-red light (Table 4.3). Those species who survive on the forest floor may need to respond to irradiance in a different manner than those at the top of the canopy because the subcanopy diffuse light is deficient in red light. In aquatic systems, submerged plants receive irradiance with a severely altered spectral quality compared to terrestrial plants. Aquatic species utilize a greater diversity of pigments to absorb irradiance for photosynthesis than terrestrial plants because of the spectral alteration of irradiance induced by the water column. Similarly, photomorphogenic responses of aquatic organisms regulated by P_{fr} may not function because water absorbs a large proportion of red light.

*2. **Physiological Effects of Phytochrome.*** A great majority of research on phytochrome-regulated phenomena have used the etiolated pea seedling or various algal systems. The results from these experiments cannot be applied directly to phytochrome-mediated responses of terrestrial plants in natural situations because the form of phytochrome in etiolated plants is not the same as that found in plants that have experienced light. However, by combining all the studies on a number of plant species, including some in natural situations, several important classes of phytochrome-mediated responses to irradiance can be identified.

Phototropic effects of phytochrome are growth responses of plants toward or away from the source of light. Bending of coleoptiles, reorientation of algal cells, and growth of etiolated seedlings toward the source of light are examples of phototropic responses. These and other types of phototropic responses are widely distributed among higher and lower vascular plants and thus are a general characteristic of plant metabolism. Phototropic responses are relatively slow morphological events (occurring over hours to days) and may have a significant lag time between initiation and response. This phototropic stimulation of shoot growth toward the light source is of great importance to plants growing in low-light environments. In temperate forests, after a canopy gap has been created, the branches of trees surrounding the gap preferentially grow toward the light source filling the gap. Young seedlings just emerging from the soil are at a critical stage for plant survival. If young seedlings grow toward the highest source of light (phytochrome mediated), the adult is more likely to have a positive carbon balance in the future.

Circadian rhythms in plants are cyclic phenomena that oscillate on a 24-h cycle. In much the same pattern as an electromagnetic wave, these rhythmic responses reach their peak activity every 24 h. When a plant that expresses a circadian rhythm is placed in a permanently dark environment, the cyclic period may elongate above or shorten below 24 h. However, the oscillations continue in the absence of light stimulation (they may dampen over time). This continual peak and trough physiological activity in the absence of light is due to an endogenous oscillator (a physiological clock). In a natural situation this endogenous oscillator is synchronized with the 24-h cycle by solar irradiance.

Phytochrome has a significant effect on the physiological clock mechanism. This can be shown clearly because the periodicity of circadian rhythms in dark-grown plants can be phase shifted (changing the timing of the peaks without changing the frequency of peaks) by exposing the plants to a short burst of red light. Exposure to short bursts of far-red light does not cause phase shifting. Some important circadian rhythms in plants are tendril movements in climbing vines, flower opening, stomatal opening, photosynthetic capacity, and respiration rate.

Photoperiodic responses of plants are also mediated by phytochrome. Induction of flowering, leaf abscission, and initiation of dormancy are a few of the photoperiodic effects on plant growth mediated by phytochrome (Table 4.2). The timing of these major developmental changes in relation to plant life cycle and climate is critical to plant survival. Phytochrome is also involved in plant response to seasonal changes in day length. These photoperiodic effects are regulated by the length of the dark period rather than the length of the light period. Long-day responses are regulated by short nights, and short-day responses are regulated by long nights. In fact, short-day physiology can be converted to long-day physiology by providing 1 minute (short flash) of red light in the middle of the night. The best studied response to seasonal photoperiod is induction of flowering because the regulation of flowering time is extremely important to the horticultural industry and to plant population biology. Various species respond differently to day length, so a classification scheme that separates long-day, short-day, day-neutral, or short-long-day plants was developed. All of these different patterns of flower induction regulation (except day-neutral) are dependent on the slow dark conversion of P_{fr} to P_r and the destruction rate of P_{fr}. Other seasonal developmental changes, such as dormancy and leaf abscission, are also regulated by the dark conversion of P_r to P_{fr}. It is very important to realize that this phytochrome-mediated regulation of developmental activity is influenced by other environmental characters, such as ambient temperature. That interaction can be critical to the successful interface between plant developmental stage and climatic conditions.

One example of an interaction between phytochrome-mediated leaf abscission and climate is summer deciduousness in Mediterranean chaparral species such as *Lotus scoparius*. The Mediterranean climates of the world (California, Europe, Australia, Chile, South Africa) have a predictably wet winter and spring and a strong regular drought during the summer and autumn. However, there is a stochastic element to the precipitation pattern, so that winter–spring droughts and summer–autumn rain storms occur. Summer deciduous chaparral plants, such as *Lotus scoparius* and *Salvia melifera,* lose most or all of their leaves during the summer to avoid low water potential during summer droughts (Nilsen and Muller 1981b, Gill and Mahall 1986). This growth response has traditionally been referred to as drought deciduousness because leaf fall occurs at the onset of the summer, coinciding with the Mediterranean climate seasonal drought. Leaf abscission occurs during periods of long photoperiod and renewed growth, while leaf production coincides with short photoperiods. Several experiments in which the effects of low water potential on leaf abscission during long and short photoperiods showed that leaf abscission in response to drought was inhibited under short photoperiods. Phytochrome was implicated because the short-day resistance to leaf fall during drought could be reversed with a short flash of light during the middle of the night (Nilsen and Muller 1981b). It was later shown that long photoperiods and water-stress-induced dormancy could not be broken by rehydration until short photoperiods return (Nilsen and Muller 1982, Comstock and Ehleringer 1986). In the absence of phytochrome regulation of the onset of dormancy, several leaf populations would be produced and lost during one year because of the stochastic nature of precipitation in Mediterranean habitats. Such behavior would severely reduce whole-plant net carbon gain.

Photomorphogenic responses are those in which light regulates the morphology of plants. Such responses can be classified into high-intensity responses [or high-fluence (HF) responses] and responses that require low light intensity [low-fluence (LF)]. Low-fluence responses require at least 0.1 μmol m^{-2} irradiance and need to convert only 0.02% of total phytochrome to the P_{fr} form. Some LF responses are lettuce seed germination stimulation, shoot elongation promotion, and chlorophyll accumulation. Most LF responses have been tied to phytochrome based on differential impacts of red and far-red light on the response selected. HF responses require light intensity greater than 10 mmol m^{-2}, and this light intensity must be sustained for hours. Some HF responses are a synthesis of anthocyanins and the induction of flowering in henbane. HF responses are probably regulated by a combination of chromophores, including phytochrome, a blue light receptor (cryptochrome), and chlorophyll.

The quality of light can also strongly affect plant morphology. For example, plants respond to the ratio of red to far-red light (FR/R) by changing stem elongation rate. Those plants adapted to high-light environments respond to the ratio of phytochrome far red to the total phytochrome population (P_{fr}/P_{total}) differently from those plants adapted to low-light sites. As P_{fr}/P_{total} decreases (due to light passing through a canopy or water column), the elongation rate of stems on high-light-adapted plants increases (Figure 10.12). Thus when high-light-adapted plants grow under a canopy, their shoot elongation rate increases, making it more likely that they will reach the canopy where there is high light. In contrast, shoot elongation of low-light-adapted plants is not affected by R/FR, so these species have a more compact growth structure under forest canopies than that of high-light-adapted plants.

Photonastic responses are movements of plant parts in response to light, but not directionally. Some of these responses do not involve permanent elongation of cells and are re-

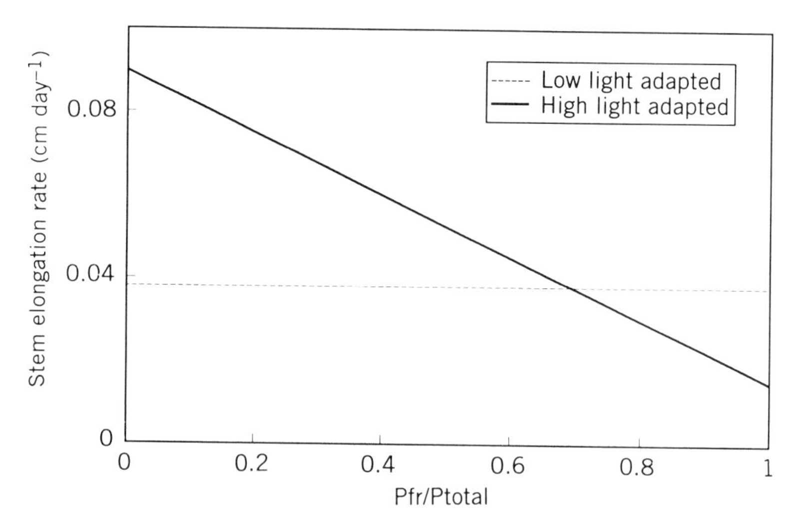

Figure 10.12 Differential growth response to the ratio of far-red light to total radiation (P_{fr}/P_{total}) of high-light- and low- light adapted plants. (Redrawn from data in Smith 1981, and Morgan and Smith 1979.)

versible. An example of a reversible photonastic response is leaf cupping in many grass taxa, where changes in turgor of bulliform cells causes cupping and uncupping of leaves. Some photonastic organ movements do involve differential cell elongation rates. For example, diaheliotropic leaf movements where leaves track the sun across the sky involve differential cell elongation (discussed in Chapter 11). The physiological mechanism that regulates many photonastic responses, such as "sleep" movements of legume leaves, have long been known to follow a circadian rhythm, and therefore are under the influence of phytochrome and are controlled by cryptochromic photoreceptors.

C. Effects of Shortwave (UV) Irradiance

Shortwave irradiance, also called *ultraviolet irradiance,* is that light energy with wavelength between 200 and 400 nm. Only about 7% of total solar irradiance hitting Earth's atmosphere is ultraviolet (UV), and less penetrates to the biosphere. Atmospheric molecules, particularly ozone, in the stratosphere absorb or scatter UV irradiance. However, each photon of UV light has high energy compared to photons of other solar wavebands (the shorter the wavelength, the higher the energy), which makes UV light highly photochemically active. The impact of UV light on plants outweighs the relatively low abundance of this irradiance compared to visible irradiance.

As discussed earlier, the UV spectrum can be divided into three main bands (Figure 10.13). UV-C is ultrashort wavelength irradiance (200 to 280 nm) with very high energy. The atmosphere effectively removes all UV-C from solar irradiance before it penetrates to the biosphere such that the solar spectrum is truncated at about 290 nm. Nucleic acids and proteins have their peak absorptance in the UV-C waveband, making them very sensitive to alteration or damage by this irradiance. In fact, germicidal lamps, used for sterilizing surfaces, emit irradiance in the UV-C waveband. Also, molecular biologists regularly utilize lamps containing high amounts of UV-C to destroy any foreign DNA on their microfuge tubes and transfer pipettes before doing an experiment.

A portion of UV-B irradiance (280 to 320 nm) does penetrate the atmosphere and af-

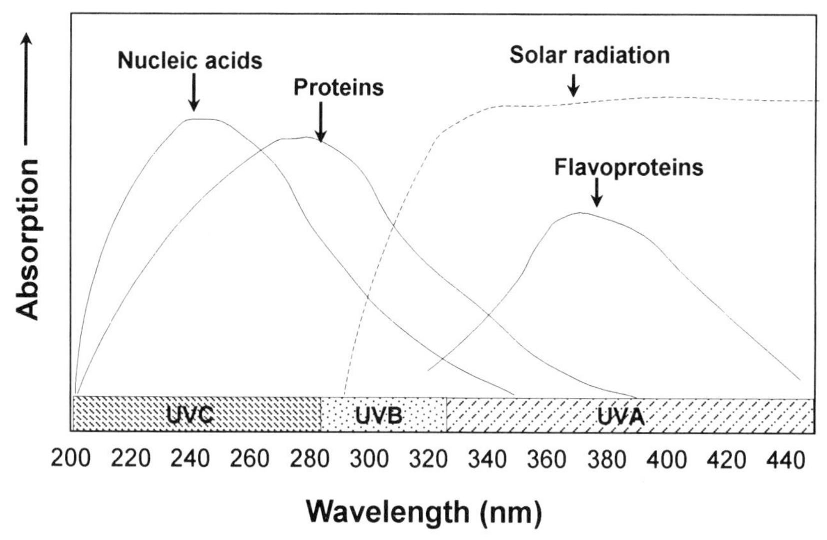

Figure 10.13 Ultraviolet spectrum in relation to the spectrum of solar radiation and the absorption spectra for nucleic acids, proteins, and flavoproteins. (Modified from Caldwell 1979.)

fect plants (Basiouny et al. 1978). Nucleic acids and proteins absorb less UV-B than UV-C, and UV-B has less photochemical activity than UV-C. Thus, a greater dose of UV-B than of UV-C is required to cause the same photochemical damage to nucleic acids and proteins. Eukaryotic cells are equipped with DNA repair machinery that can extract denatured sections, or polymers of pyrimidines, and resynthesize functional DNA. Prokaryotes do not have these complex repair systems and they are more sensitive to UV-B than eukaryotes.

Irradiance between 320 and 400 nm is UV-A, which is only weakly absorbed by proteins and nucleic acids The energy of UV-A photons is less than that of any other UV photons. Thus, UV-A photons have only limited photochemical activity. The atmosphere does not influence the transit of UV-A photons. Therefore, any changes in stratospheric ozone will not change the transmittance of the UV-A to the biosphere. Plants absorb a large majority of the UV-A that hits their leaves by compounds such as flavoproteins and chromophores (such as phytochrome). In fact, some important enzymes such as rubisco are activated by UV-A light. UV-A light is not damaging to plants, and pretreatment by UV-A can actually increase tolerance to UV-B.

The UV climate of the biosphere is not uniform. On a long-term basis (decades) the quantity of UV irradiance impinging on Earth's atmosphere varies about 10 to 12% on a 12-year cycle due to sunspot activity. Stochastic events can also change global UV-B irradiance. Volcanic emissions of chlorine gas causes a short-term increase in UV-B. This is because chlorinated compounds in the stratosphere cause photochemical destruction of ozone, thereby reducing the UV-B screening effect of stratospheric ozone. In recent history, anthropogenic emissions of chlorinated compounds such as CFCs (chlorinated flurocarbons), used in refrigeration systems and as an aerosol propellant, have caused a reduction in stratospheric ozone with a corresponding increase in UV-B irradiance on plants.

The intensity of UV-B irradiance also increases with latitude because the stratospheric ozone layer is thinner toward the poles. Therefore, arctic plants receive a higher dose of UV-B than do tropical plants. In a similar manner, the UV-B dose increases with elevation. The UV-B dose does not change significantly with aspect of slope (of the site or the leaves) because more than half the UV-B irradiance is diffuse (a result of scattering in the atmosphere) and does not have a large direct beam component as does total solar irradiance.

Studies concerning the impact of UV on plants have focused on UV-B because UV-A does not have damaging photochemical activity, and no UV-C irradiance penetrates the atmosphere. In contrast, the proportion of solar UV-B penetrating the atmosphere is variable in time and space, and UV-B has potentially damaging photochemical activity. A reduction of only 15% of atmospheric ozone can increase UV-B irradiance by about 50%, which will have a significant effect on plants (Caldwell, 1979).

Plants can be classified as "resistant" or "sensitive" to UV-B irradiance. Those plants that are resistant are not completely immune to the impact of UV-B, but they are able to tolerate a reasonable increase in UV-B (25%) without suffering physiological damage. The resistant variety or species can withstand some increase in UV-B dose because they have evolved mechanisms to screen UV-B before this irradiance enters the mesophyll cells. Some UV-B protection results from a thick cuticle because waxes reflect and absorb a portion of UV-B. However, a proliferation of other UV-B absorptive compounds, such as flavoproteins and flavenoids, are required for adequate UV-B resistance.

UV-B irradiance generally has a negative impact on plants. Small changes in UV-B dose can have dramatic effects on the physiology, morphology, and growth in sensitive plants. These same effects occur in resistant plants, but it takes a larger dose of UV-B to initiate the effect. Resistant species often respond to small doses of UV-B and UV-A by increasing leaf thickness, cuticle thickness, and flavenoid concentration, affording some protection against a further increase in UV-B dose. Some of the specific detrimental effects of UV-B are epidermal cell lesions, surface bronzing and chlorosis, decreased photosynthesis, decreased growth, and decreased yield (Teramura 1983, Björn and Åkerlund 1984, Murali et al. 1988, Tevini and Teramura 1989).

UV-B irradiance has a generally negative impact on photosynthetic processes. Only occasional studies demonstrate a positive effect, such as increased biosynthesis of chlorophyll in low light and moderate UV-B dose. The most sensitive component of photosynthesis to UV-B is photosystem II, and C_4 plants are more susceptible than C_3 plants. In resistant species such as peanut, there is no impact of UV-B on photosystem II (Basiouny et al. 1978). In relation to PSII inhibition, chloroplast ultrastructure is also affected by high UV-B. Several studies have shown a loss of membrane stacking, which is probably the result of PSII disruption. Stomatal conductance is not affected significantly by UV-B except in some especially sensitive plants. In fact, UV-A stimulates stomatal opening in many species. If stomata close in response to changing UV-B, this is probably a result of increasing C_i due to the inhibition of CO_2 reduction by photosynthesis. There is some evidence that a UV-B dose can affect the quantity of rubisco because the soluble protein pool of sensitive plants increases significantly during moderate UV-B dose and high light intensity. However, under low light intensity, a moderate UV-B dose reduces the soluble protein in leaf tissue severely (Vu et al. 1982a). Since a large proportion of total soluble protein is rubisco (more than 50%) a reduction in total soluble protein is likely to indicate a reduction in the pool of rubisco. Some studies have demonstrated the impact of UV-B on the activity state of rubisco when plants are grown under relatively low visible light intensity (Vu et al. 1982b). It has been well documented that photosynthesis is more sensitive to UV irradiance when plants are grown under low light compared to high light. Since leaves enhance electron transport and photosystem II under low light and this is the most sensitive photosynthetic component to UV-B, it stands to reason that such plants would be more sensitive to UV-B irradiance. Plants grown under high light have thicker leaves, more cuticle, and emphasize CO_2 reduction capability over PSII capacity. Photosynthetic pigments are also affected by UV-B. Chlorophyll and carotenoid pigment destruction increases with increasing UV-B dose in sensitive plants. The impact of UV-B on the chlorophyll *a*/*b* ratio is not generally consistent among plants and may represent different levels of sensitivity.

Growth and yield of crops and native plants is generally reduced for sensitive plants with moderate doses of UV-B. The first observation made on UV-B effects on crop growth was that seedlings exposed to moderate UV-B dose at low background irradiance are stunted compared to those receiving no UV-B (Mylar treated). The stunting was due to a reduction in internodal length, not a reduction in nodes produced per time. Thus UV-B was affecting an endogenous regulator of cell expansion. Plant biomass is effected by UV-B differently among species. In some cases growth is stimulated, and in other cases growth is inhibited or there is no impact (Caldwell et al. 1975). However, it is clear that inhibition of growth and yield in crops is dependent on background visible irradiance. As light intensity decreases, the deleterious effects of UV-B on growth increases.

V. ADAPTATION TO HIGH AND LOW LIGHT

Anyone who has spent any time in natural ecosystems knows that there is a tremendous variety of irradiance environments. Some habitats have high and uniform irradiance (deserts, arctic, and crops, for example), some have low and variable irradiance (subcanopy habitats), and others have high irradiance with intermittent low-light (mountain) environments. In a classic sense, plants have been divided into "sun" and "shade" species. Sun plants are able to grow well in high-light environments but suffer in low-light sites. In contrast, shade plants grow and reproduce in low-light conditions but are quickly damaged in high-light conditions. In a general sense, early succession species and species from high-light environments are sun plants (Table 10.1). Late succession species, which must begin their life cycle in a shaded environment, and subcanopy species are shade species. Therefore, the distribution of sun and shade species is segregated in space and time. Of course, phenotypic acclimation to a changed irradiance environment, or selection for ecotypes that are more or less tolerant of high light, are possible ways that one species can survive in sites with a different irradiance environment than that in which the species

TABLE 10.1 General Differences between Sun and Shade Leaves (Leaves of Plants Adapted to High and Low Light Intensity)

Characteristic	Sun Leaves	Shade Leaves
Morphological features		
Leaf area	−	+
Leaf thickness	+	−
Cell abundance	+	−
Cuticle thickness	+	−
Density of stomata	+	−
Specific leaf weight	+	−
Cell ultrastructural features		
Cell size	−	+
Cell wall thickness	+	−
Chloroplast frequency	+	−
Chloroplast orientation	vertical	horizontal
Proportion of stacked membrane	−	+
Starch grains in the chloroplasts	+	−
Chemical features		
Caloric content	+	−
Water content of fresh tissue	−	+
Cell sap concentration	+	−
Lipids	+	−
Anthocyanins, flavenoids	+	−
Chlorophyll content	+	−
Chlorophyll *a/b*	−	+
Light-harvesting complex (PSII)	−	+
Physiological functions		
Quantum yield	−	+ (or equal)
Light-saturated photosynthesis	+	−
Light compensation point	+	−
Light saturation point	+	−
Photoinhibition likelihood	−	+
Dark respiration	+	−
Photorespiration	+	−
Transpiration	+	−

Source: Data are from many sources but Boardman (1977) summarized these differences in a formal manner.

evolved. In this section we discuss the morphological and physiological ways that plants have adapted to habitats with different light environments.

A. Morphological, Anatomical, and Behavioral Adaptations

The intensity of irradiance affects all aspects of plant structure. However, the effect of irradiance on leaf structure and plant architecture are the best studied phenomena. Species adapted to high irradiance generally have thicker and smaller leaves with a thicker cuticle than those of species from low-light environments. Frequently, high-light-adapted species have stomata on both leaf surfaces rather than just on the bottom of the leaf. The smaller, thicker leaves serve to reduce energy load on the leaf and moderate leaf temperature. In fact, high-light-adapted leaves are often cupped or slightly curled, which further reduces their leaf area and energy load. A thicker cuticle absorbs the higher quantity of UV in high-light sites and reduces water loss. The thicker leaf also results in a higher area of the mesophyll in relation to the area of the epidermis (A_{mes}/A). This increases the potential mesophyll cell surface for diffusion of CO_2 into chloroplasts and reduces intercellular air space resistance for diffusion. In contrast, leaves on plants adapted to low-irradiance environments are larger, thinner, have a thinner cuticle, and are fewer in number per branch tip than that for high-light-adapted plants. This leaf morphology on shade-adapted species maximizes photon penetration into the leaf and potential absorption of photons in an environment of low photon flux density (Figure 10.14).

The anatomical characteristics of leaves in high-light-adapted plants are designed to reduce irradiance absorption. For example, many high-light-adapted plants have a surface layer of trichomes, which, when dead, serve as an excellent reflective surface. As light penetrates leaves, some is directed through the cell wall between palisade cells and focused on the mesophyll cells. Plants adapted to high-light conditions increase light focusing, while the anatomy of low-light-adapted plants maximizes light scattering in the interior of leaves. If light is scattered in the leaf, there is a higher chance that the photons will be absorbed instead of transmitting through the leaf.

The architecture of mature woody plants adapted to high-light sites is also different from those adapted to low-light regimes. The central growth characteristic of plants that makes them unique from other kingdoms is the modular nature of plant expansion. Growth modules such as leaves or stems can be produced and abscissed as the plant responds to environmental variation. Natural selection can stimulate the modification of certain modules of plants, thereby affecting survival in a specific environment. Leaves have evolved into spines, insect-trapping devices, phylodes, or floral accentuating structures by this evolutionary process. In addition, modules on different parts of a plant can adjust individually to their immediate environment, allowing plants to maximize their use of a heterogeneous environment. The branching patterns of plants in high-light sites are denser, but leaf display is often just at the outer portion of the canopy. In low-light-adapted plants, branch density is minimal (plants look leggy) and the leaf canopy is thick. The minimal branch density and the thickness of the leaf canopy in low-light-adapted plants are thought to maximize photon absorption in an environment with low photon flux density. However, plants that grow at the interface between a high-light and a low-light environment may have high-light attributes (thin canopy, heavy branching, high-light leaves) on one side of the plant and low-light attributes on the other side.

Leaf population demography is often dramatically different between plants adapted to high and low photon flux density. Leaves survive longer on plants in low-light sites (Fig-

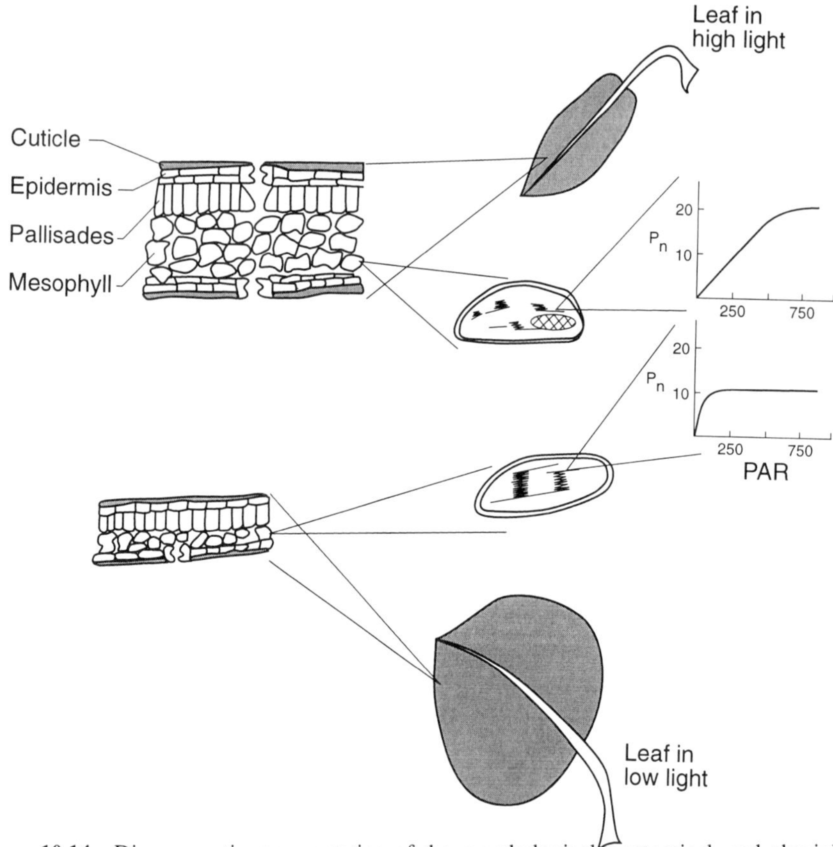

Figure 10.14 Diagrammatic representation of the morphological, anatomical, and physiological changes in shade- and sun-adapted leaves.

ure 10.15). Extended longevity is required because photon flux is limiting photosynthesis and it takes longer for the leaf to replace the cost of producing and maintaining the leaf (Williams et al. 1989). However, increased leaf longevity also carries susceptibility to herbivory. Long-lived leaves must be protected from herbivores, and thus low-light leaves must invest some of their carbon resources in a protective mechanism against herbivory.

Leaf angle can be thought of as a type of leaf behavior. Leaf angle is often more vertical in plants adapted to high-light sites. A horizontal leaf angle for plants in low light maximizes absorption of diffuse irradiance. Daily movement of leaf angle (diaheliotropism and paraheliotropism) is a characteristic of high-light plants. Temporal and reversible leaf movements are discussed in Chapter 11 because this topic relates specifically to leaf energy budget.

Light intensity also affects the hydraulic system of plants. In high-light sites stem hydraulic conductance is commonly higher than in plants adapted to low light. Furthermore, the relationship between atmospheric vapor pressure and leaf conductance is stronger in high-light sites. Leaves on plants in low-light sites are often not responsive to atmospheric vapor pressure, particularly when grown under low-light conditions. Stomatal frequency is lower and leaves are larger in low-light sites, both of which affect leaf temperature regulation. If leaves on plants adapted to low light are placed in a high-light environment, leaf temperature increases rapidly over air temperature; this increases the vapor pressure gradient between leaf and air and stimulates transpiration. Low-light-adapted plants have a high chance of wilting in high light if stomata do not close even if water is abundant in the soil, because transpiration is limited by hydraulic resistance of the vascular tissues.

B. Photosynthetic Physiology

Why is high- or low-light intensity a stress to plants? Excessive photon flux density on a leaf surface will affect leaf temperature and potentially bring leaf temperature above that which inhibits physiological function. However, if anatomical and morphological adaptations of leaves limit total photon absorption, leaf temperature can be maintained within a nondestructive range. A second problem of high photon flux density is that the light-harvesting apparatus of photosynthesis may absorb more light energy than can be processed by the electron transport chain. This creates an imbalance in photosynthetic physiology that can be damaging. In contrast, low-light situations do not induce problems of leaf overheating, so leaves can be large and have high absorptivities. On the other hand, photon flux density can be low enough to reduce light harvesting severely, causing a low potential for carbon reduction in photosynthesis. This results in a delicate balance between canopy carbon gain and respiratory requirements of the whole plant. Therefore, low-light adaptation involve maximizing light absorptive properties of leaves and canopies and minimizing respiratory costs of the whole plant.

The best way to determine if a plant's photosynthesis is adapted to high-or low-light environments is to measure a light response curve (Figure 10.16). Plants adapted to high-

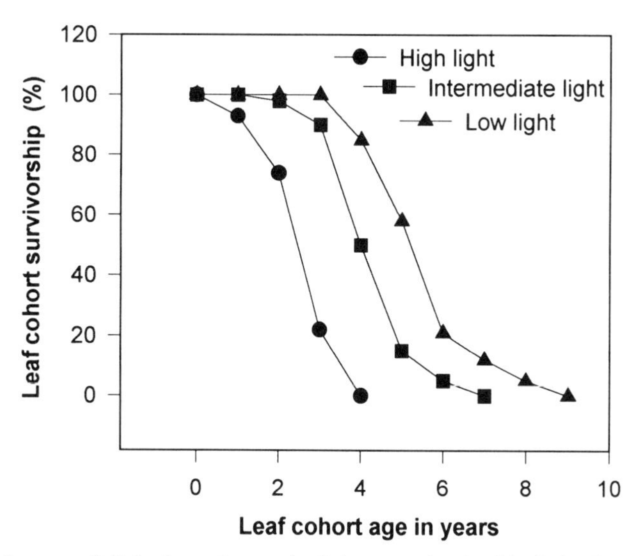

Figure 10.15 Impact of light intensity on leaf demography in *Rhododendron maximum*. The high-, low-, and intermediate-light sites correspond to the nature of the canopy over the *R. maximum* shrubs. (Redrawn from Nilsen et al. 1987a.)

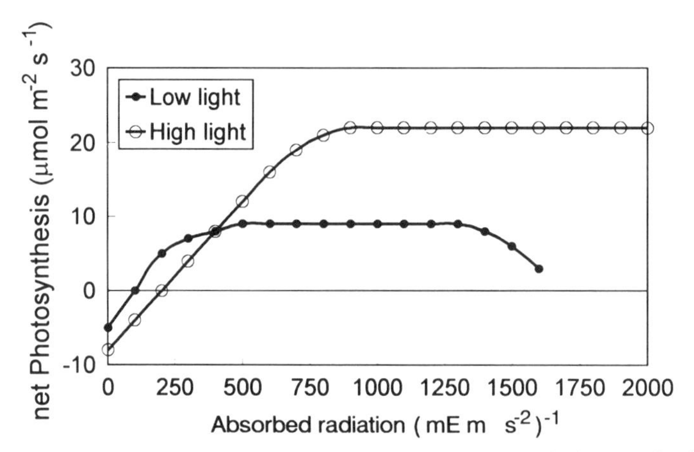

Figure 10.16 Comparison of a representative light response curve of photosynthesis for leaves adapted to high light or leaves adapted to low light. Similar relationships have been found in a number of species.

light sites will have relatively high dark respiration, low quantum yield, high-light saturation point, and a high-light-saturated photosynthetic rate compared with plants adapted to low-light intensity (Bjorkman et al. 1988). In accordance with these differences in the light response curve, there are changes in the pools of chlorophyll molecules. Low-light-adapted leaves have less total chlorophyll per leaf area (because they are thinner) and a smaller chlorophyll *a*/*b* ratio. Since most of chlorophyll *b* is in the light-harvesting component of photosystem II, changes in the chlorophyll *a*/*b* ratio reflect changes in the relative abundance of light-harvesting complexes PSII and PSI. If a representative molecule from each protein complex is measured relative to total chlorophyll content, the specific impact on LHCII can be shown (Figure 10.17). The concentration of the quinone pool as-

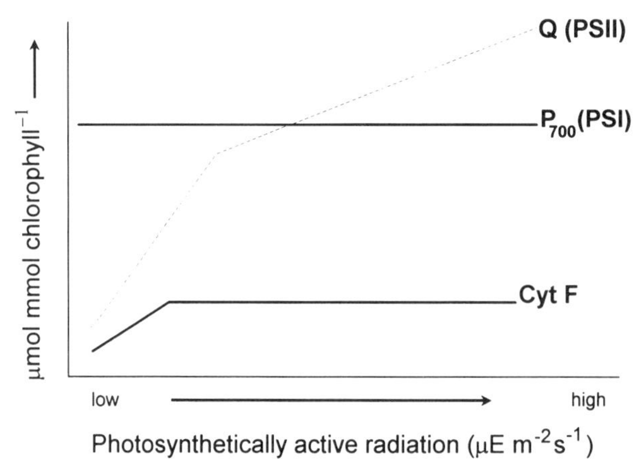

Figure 10.17 Representation of the manner by which specific components of the photosystem core complexes change in relationship to chlorophyll concentration in increasing light intensities. Similar responses have been found in a number of species.

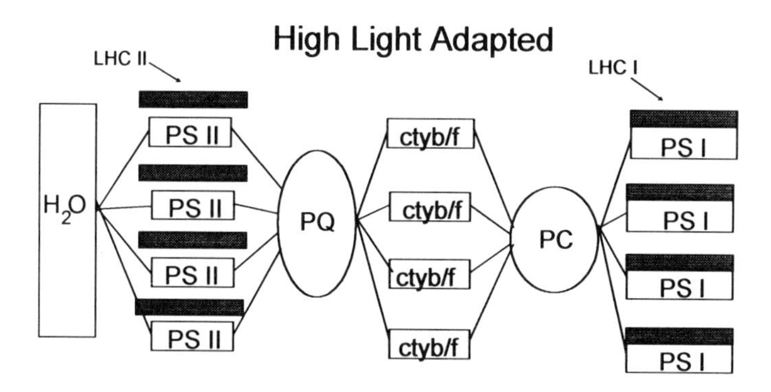

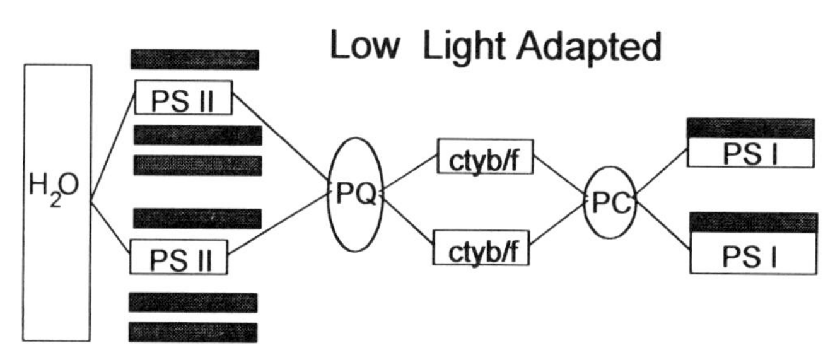

Figure 10.18 Diagrammatic representation of relative pool sizes for components of the photosynthetic apparatus in plants adapted to high-light sites and those adapted to low-light sites.

sociated with the PSII core complex decreases dramatically at low light intensity, while that of P_{700} (an indicator of PSI core complex abundance) is unchanged. This indicates an increase in the abundance of LHCII in relation to the PSII core protein complex in conditions of low-light intensity (Anderson et al. 1988).

A diagram of the relative pool sizes of the various components in the light-harvesting apparatus and electron transport chain can help clarify adaptation to low-light environments (Figure 10.18). Plants adapted to low-light sites have maximized light-harvesting capacity by increasing the number of LHCII per PSII core complex and minimizing the number of protein complexes related to electron flow. Therefore, if these plants are exposed to high light, light energy will be absorbed at a rate which the electron transport system cannot accommodate. In contrast, plants adapted to high light have minimized light-harvesting capacity by reducing the ratio of LHCII to PSII but increased the investment in electron transport capacity. When high-light-adapted plants are exposed to low-light conditions, the plants may not have the capacity to generate enough ATP or NADPH to continue photosynthesis at a rate that is adequate to compensate for respiration.

Differences between the light-harvesting capacity or electron transport activity in plants adapted to high- or low-light environments must be balanced with changes in the biochemical components of CO_2 reduction. The primary carboxylation enzyme in C_3 plants, rubisco (ribulase biphosphate carboxylase-oxygenase), is light activated. Therefore, in the absence of any light (when no new ATP and NADPH are being generated by the photochemical reactions of photosynthesis), CO_2 reduction is shut down. The amount of light required to activate rubisco is a very small quantity of light compared to that in low-light environments. It is unlikely that the difference in light intensity between low- and high-light sites has a direct impact on rubisco activity. The total quantity of rubisco and other photosynthetic enzymes must be higher in high-light plants because these plants have a relatively large capacity to phosphorylate ADP and reduce NADP. To keep energy flowing through the photochemical reactions in high-light situations, ADP and NADP pools must be regenerated at an appropriate rate by the enzymatic processes of the stroma.

Differences between the biochemical processes for CO_2 reduction in C_3 and C_4 species affect light response curves. Due to the extra amount of ATP required for C_4 physiology compared with C_3 photosynthesis, the quantum yield of C_3 plants is higher than C_4 plants at moderate leaf temperature. However, light-saturated photosynthesis is higher in C_4 plants compared with C_3 plants because photorespiration is inhibited in bundle sheath cells. The C_4 physiology is therefore predisposed to high-light habitats, and the frequency of C_4 species in low-light habitats is small compared with that of C_3 plants.

Photoinhibition occurs in all plants when the photochemical processes of photosynthesis do not have enough ADP or NADP to enable continued flow of energy through the electron transport system. Under these conditions, electron flow is blocked and molecules that are designed for oxidation/reduction reactions become reduced. Electrons of the chlorophyll in the reaction center become highly excited and can cause damage to the supporting proteins unless the energy is dissipated from the reaction center. Much of this potential damage occurs on the photosystem II core complex, which is designed both to absorb light energy and be inactivated by excess light. This incongruity seems counterproductive at first glance, but it is critical to high-light adaptation. Light inactivation of PSII is due to amino acid damage in the D1 protein induced by excess light energy absorption and reduced electron transport capacity. The amino acid damage cannot be repaired in place, so the damaged D1 proteins are cleaved out of the PSII core complex, decomposed by a protease in the stroma, resynthesized by the chloroplast translation machinery, and reattached to a PSII core complex. This D1 protein turnover is a general characteristic of chloroplasts in any light environment. As the light intensity increases, the number of D1 proteins damaged increases more rapidly than the repair mechanism. Thus, at high light intensity and low electron flow, many D1 proteins are damaged and disconnected from PSII core complexes, which is one type of photoinhibition. Many environmental factors can affect the turnover rate (either degredation or reassembly and repair) of D1 protein (Demming-Adams and Adams 1992). However, it is very clear that if a plant is adapted to low-light environments (high photon absorption efficiency, low electron transport), these plants are predisposed to extensive photoinhibition.

Photoinhibition is normally a temporary symptom of excessive irradiance because of the repair mechanism for the D1 protein of PSII. In contrast, permanent damage to the photosynthetic machinery can occur if oxidations of other parts of the system are extensive. Photo-oxidation is the process by which photosynthesis is permanently damaged by oxidations, which cannot be repaired by chloroplast biochemistry and is observed in leaves as a bleaching of chlorophyll. This permanent damage occurs when photorespiration, photoinhibition, and other energy-dissipating mechanisms are not adequate for dissipating excess, absorbed light energy. Photooxidation can occur during situations where the environment has caused stomata to close (water stress, cold temperature) during periods of high irradiance. When the products of the electron transport chain are not consumed as fast as they are being produced, excess energy is absorbed and must be dissipated before damage such as photo-oxidation reduces photosynthesis permanently.

Leaves have several mechanisms of dissipating excess energy before and after it has

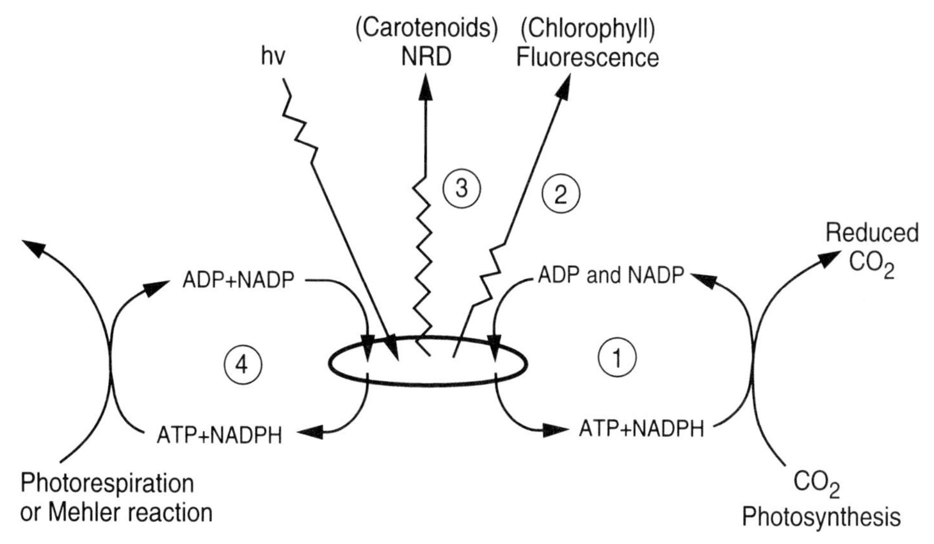

Figure 10.19 Diagrammatic representation of the four ways that light energy can be dissipated after absorption by chlorophyll: (1) photosynthetic use of ATP and NADPH for CO_2 reduction, (2) fluorescence of chlorophyll, (3) nonradiative dissipation of heat energy by the carotenoid system (xanthaphylls), and (4) use of ATP and or NADPH by nonphotosynthetic reactions such as photorespiration and the Mehler reaction.

been absorbed by the leaf in addition to that by photoinhibition and photorespiration (Björkman and Demming-Adams 1994). Changes in leaf reflective properties and leaf angle are effective mechanisms of avoiding light absorption (discussed in Chapter 11). There are four mechanisms by which leaves dissipate photon energy absorbed by chlorophyll (Figure 10.19). Conversion of photon energy by photosynthesis into reduced carbon molecules is the first of these mechanisms. The second is by re-emission of photon energy (fluorescence) from chlorophyll. The third is by energy (ATP) consumption by metabolic processes that do not result in reduced carbon (for example: photorespiration, ion transport, Mehler reaction). The fourth is by dissipating absorbed light energy as heat in a bed of pigments (Thayer and Björkman 1992), which is termed *thermal* or *nonradiative dissipation* (NRD).

Nonradiative dissipation (NRD) of energy is related to both chlorophyll and carotenoid pigments in both PSI and PSII. The electron transport chain utilizes absorbed energy to create a high proton gradient in the lumen of the thylakoid. If ATP is being consumed by metabolic processes at an adequate rate (in proportion to the proton gradient being established across the thylakoid membrane), the inner thylakoid proton concentration comes to a steady state. However, if metabolic use of ATP decreases relative to proton flow into the thylakoid lumen, the proton concentration increases and the pH of the inner thylakoid lumen decreases. The pH change is thought to induce a conformational change in the thylakoid membrane, altering the association of pigments with their proteins and inducing a thermal deactivation of chlorophyll. This both converts light energy to heat in chlorophyll and reduces light absorption. Carotenoids (particularly violaxanthin, zeaxanthin, and antheraxanthin) have long been thought to afford protection to the thylakoid membrane from damage due to excessive irradiance. Recent evidence suggests that when there is an overabundance of excitation energy in PSI or PSII, violaxanthin is deepoxidized to zeax-

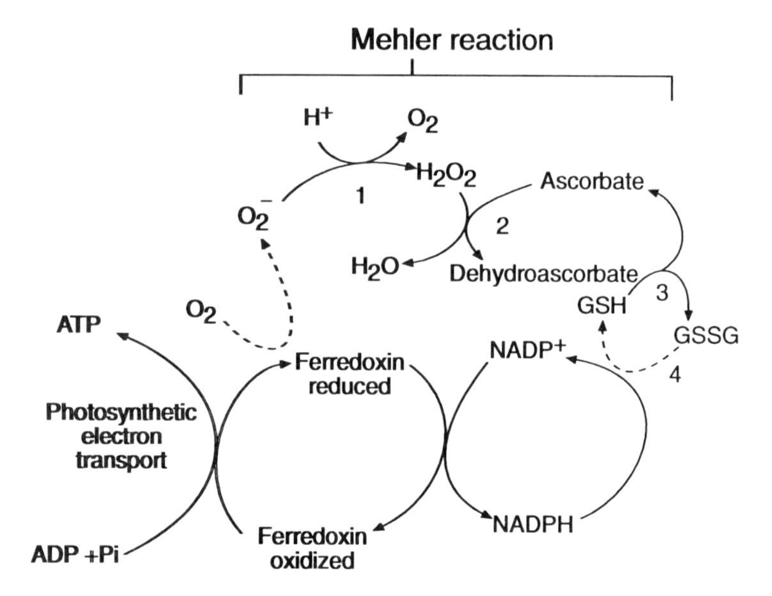

Figure 10.20 Biochemical pathways of the Mehler peroxidase ascorbate reactions in the chloroplast. (See Appendix A for a list of abbreviations.)

anthin via antheraxanthin, with zeaxanthin epoxidase, which dissipates energy as heat (Thayer and Björkman 1992).

The Mehler peroxidase ascorbate reaction (Figure 10.20) also is an important metabolic mechanism for dissipating excess energy through the consumption of NADPH that has been reduced by energy flowing through PSI. The critical aspects of this reaction for protection against excessive absorbed light energy is the use of energy and electrons from PSI to reduce O_2, leading to H_2O_2, which in turn reacts with ascorbate to produce H_2O and O_2. Since ATP is not used but NADPH is utilized, this reaction secondarily may enhance the proton concentration in the thylakoid engaging NRD by xanthaphylls.

The Mehler peroxidase reaction is also significant in relation to free radicals of oxygen. If other mechanisms of utilizing electron flow are inadequate, and electron flow is continuing, a probable destination for electrons is molecular oxygen, which results in the superoxide radical (O_2^-). Superoxide can be spontaneously converted into more dangerous forms of oxygen free radicals. Therefore, a buildup of superoxide will be detrimental to chloroplast function. Superoxide can be converted to hydrogen peroxide by superoxide dismutase (SOD) in the chloroplast. This is the first step leading to the Mehler peroxidase reaction sequence, which eventually releases water and oxidizes NADPH and utilizes electrons from PSI. The effectiveness of this reaction system for utilizing excess electron flow depends on SOD. There are several forms of SOD in higher plant chloroplasts that can be separated based on the metal cofactors. One form (Cu/Zn SOD) is the most abundant stromal form and is likely to dominate the conversion of superoxide to hydrogen peroxide. Hydrogen peroxide is converted to water in a reaction mediated by ascorbate peroxidase. In addition to hydrogen peroxide, ascorbate is a substrate and the products are monodehydroascorbate radical and dehydroascorbate. Both products can be reduced by different pathways in which the ultimate electron donor is NADPH. Recent evidence suggests that Mn SOD and one form of ascorbate peroxidase may both be bound to thylakoid membranes close to PSI core complexes. This would maximize the efficiency of this alternative electron flow pathway in chloroplasts. In this process electrons have flowed from PSI to water, superoxide has been converted to molecular oxygen, and NADPH has been oxidized.

C. Leaf Longevity and Light Intensity

Leaf longevity is regulated by genetic as well as environmental factors. Clearly, there is a dramatic difference between leaf longevity of deciduous and evergreen species. The term *evergreen* species refers to the presence of leaves at all times of the year, but it does not refer to specific leaf longevity. Among evergreen species leaf retention is variable. In some evergreen species, particularly those in the tropics, leaves are being produced and lost year round, and leaf longevity may be only a few months. Some evergreen species retain one cohort of leaves just until the next leaf population is produced, which may be less than one year. In contrast, some evergreen species retain leaves for extended periods (several years), and they have several different annual cohorts of leaves in the canopy at any one time. The dynamic regulation of leaf age structure in a canopy is often determined by light intensity. For example, leaves of *Rhododendron maximum* are evergreen, lasting for 3 to 7 years depending on the light environment in which the plant is growing. In high-light sites many smaller leaves are produced on each shoot each year, but they survive for only 2 to 3 years. In contrast, plants growing in low light produce fewer, larger leaves (compared with leaves in high-light sites) on each branch that remain for 5 to 7 years (Nilsen 1986). Similarly, leaves on plants in low-nutrient or low-temperature sites tend to remain active for longer than leaves on plants from high-nutrient and high-temperature sites. Leaf survivorship in each of these cases may be related to the cost of leaf production and maintenance compared with the carbon gain possibilities for the specific environment.

In low-light sites photosynthesis is inhibited by low light, and leaves have developed physiological mechanisms to maximize the efficiency of light absorption. If one calculates the cost (in ATP units) to produce leaves in high or low light, the values come out to be very similar on a dry weight basis. It is true that leaves of high-light sites have a higher maintenance cost than those from low-light sites because of higher respiration. However, the ratio of construction cost plus maintenance expenses to possible carbon gain is higher in leaves adapted to low-light sites. It takes longer for a leaf to replace its construction cost in low- compared to high-light sites. Thus, the leaves of low-light-adapted plants have an extended longevity, which compensates for the high construction cost/daily carbon gain ratio.

Increased leaf longevity carries with it an increased potential for damage by other potential stresses of the environment. For example, the possibility of encountering an extended drought increases with increased leaf longevity. The detrimental effects of UV-B dose and pollutant dose increases with the length of exposure. Furthermore, the potential for damage by herbivores and or other pathogens increases with leaf longevity. Leaves that have extended longevity frequently have well-developed mechanisms for defense against these potential stresses (Mooney and Gulmon 1982). The production of defensive compounds or anatomical structures to avoid damage, such as that by UV irradiance, adds additional energy cost to producing leaves and further exacerbates the construction cost/net photosynthesis ratio.

VI. ACCLIMATION TO AN ALTERED LIGHT ENVIRONMENT

Photosynthetic adaptation to light regime is an excellent case for studying adaptation and acclimation (see Chapter 1). Plants that reside in low-light environments have maximized photon absorption relative to electron transport and CO_2 reduction and have a characteristic light response curve. These characters contrast with those of species in high-light environments. In many cases populations of one species may grow in high-light environments, while other populations of the same species grow in low-light environments. One excellent example of this is *Solanum dulcamara,* in which populations can be found in a forest (forest clone = low light) or in a field habitat (field clone = high light). Specific genotypes (clones) from the forest population and field population were grown in two common gardens, high and low light intensity (Gauhl 1976). The light response curves of each clone was determined in each common garden setting (Figure 10.21). It was clear

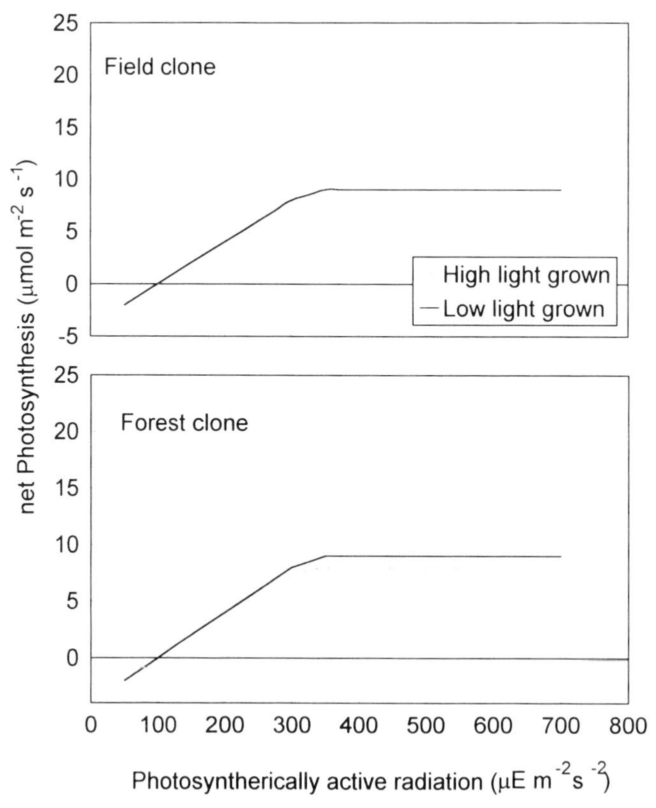

Figure 10.21 Comparison of light response curves for *Solanum dulcamara* from a high-light site (field clone) or a low-light site (forest clone) grown under high or low light availability. (Redrawn from Gaul 1976.)

from this study that ecotypes (genetically selected populations) adapted to high-light conditions can acclimate to low light. In contrast, low-light-adapted ecotypes cannot acclimate to high-light conditions.

VII. ADAPTATION TO IRRADIANCE HETEROGENEITY

Earlier we stated that the light environment under a forest canopy or in mountain meadows may be heterogeneous in space and time because of clouds or the presence of a canopy (Figure 10.6). Plants growing in forest subcanopies spend a majority of their time under a low-light regime. In fact, much of the information on shade adaptation and acclimation is derived from studies on forest subcanopy plants. However, we just stated that the physiology of leaves adapted to low light maximizes light absorption efficiency and that such leaves may not have the machinery to dissipate excess energy under high-light conditions. Low-light-adapted leaves are less able to acclimate to high-light conditions and suffer photoinhibition and eventually, photooxidation. Are the short-duration "flecks" of sunlight, to which subcanopy plants are often exposed, damaging to photosynthesis? To answer this question we have to examine the patterns by which photosynthesis and leaf conductance respond to natural *sun flecks* or experimentally induced *light flecks*.

Under forest canopies and inside the canopy itself sun patches are extremely diverse. In an effort to categorize those patches on the forest floor, an index has been developed based on the height and width of the canopy gap that creates the high-light condition. If the ratio of gap diameter to gap height (GDR) is less than 0.01, the resulting light area is classified as a sun fleck. Sun patches have a GDR of between 0.01 and 0.05, and sun gaps are characterized by a GDR greater than 0.05. Therefore, sun flecks are created by small canopy gaps relatively high in the canopy.

The nature of particular sun flecks can be compared based on four main criteria. The first is the maximum intensity of irradiance in the sun fleck. Those sun flecks derived from the top of tall canopies have a strong penumbra effect. This effect makes the edges of the sun fleck fuzzy and reduces the intensity of irradiance in the sun fleck. Sun flecks below a redwood forest have strong penumbra effects and have relatively low irradiance compared to sun flecks below or within a crop canopy. Sun flecks in the canopy of a crop such as soybean are of short duration, frequent, and have high intensity with sharp boundaries. The second distinguishing character of a sun fleck is the total integrated amount of photon flux density over the lifetime of the sun fleck at a particular microsite. The total PFD is dependent on the light intensity and the duration of the sun fleck, which is the third distinguishing character. Fourth, the impact of sun flecks is dependent on the amount of time that has passed since the last sun fleck.

The natural distribution of sun flecks has been mostly studied primarily below forest canopies (Chazdon 1988). Sun flecks tend to be bunched into small portions of the day cycle. The total PFD of all sun flecks in a day tends to be determined by one or a few long sun flecks with high PFD (Gildner and Larson 1992). There is a large variation in the total amount of daily irradiance accounted for by light in sun flecks (10 to 80%). Some of that variation is due to differences in canopy structure or microclimate. However, the variation among microsites in one forest may be as large as that between forests. It is always true that the diffuse light below a forest or in a canopy is fairly consistent, and all variation in the light regime is dominated by the effects of sun flecks.

If a heterogeneous light environment is a common characteristic of subcanopy sites and inner canopy leaves, how important are the high-light periods to the total daily carbon gain of the leaf. There have been only a few studies of the significance that sun flecks play for total carbon gain. In several studies of subcanopy tropical forests, sun flecks accounted for 30 to 60% of total daily carbon gain (Pearcy and Calkin 1983, Pearcy 1987). In contrast, sun flecks accounted for only 10 to 20% of total carbon gain in subcanopy species of temperate forests (Weber et al. 1985). The difference may be due to the intensity of the diffuse irradiance. In a study of *Rhododendron maximum* under temperate forest canopies, sun flecks accounted for only a small proportion of total daily carbon gain for plants under a sparse deciduous canopy. However, when grown under a dense evergreen canopy, with lower diffuse irradiance than under the deciduous canopy, sun flecks accounted for a majority of daily carbon gain (Nilsen, unpublished data).

Since sun flecks can have a significant impact on total daily carbon gain, improving the efficiency of sun fleck use by photosynthesis would be an adaptation to this heterogeneous light environment. Photosynthesis responds rapidly (within 1 s) to a sun fleck (Figure 10.22). This initial rise is followed by a slower increase. The magnitude of photosynthetic response is dependent on the immediate previous history of activity in the leaf. This is called the induction state and is measured by the photosynthetic rate 60 s after the sun fleck is started (IS_{60}). There are actually two induction states, a short-term induction state (within 1 s) and a long-term induction state (over 5 to 60 s). The short-term induction state is important only in high-light-adapted leaves (Tinoco-Ojanjuran and Pearcy 1993). Full induction of photosynthesis to its maximum value requires over 30 min, which is longer than most sun flecks. However, full induction can be reached if frequent sun flecks occur with short rest periods between the sun flecks. The induction of photosynthesis by

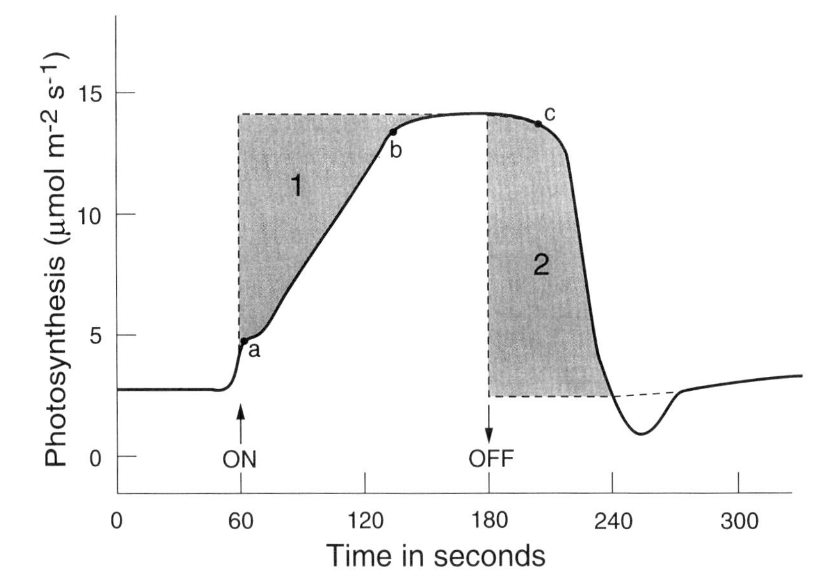

Figure 10.22 Rapid response of photosynthesis to a short-term flash of light (light fleck). Point a refers to the fast induction response, point b refers to the induction state at 60 s (IS_{60}), and point c represents postillumination photosynthesis. Shaded area 1 represents less than potential photosynthesis due to the requirement for induction. Area 2 represents greater than expected carbon gain due to postillumination photosynthesis. The dashed line represents the expected photosynthesis if there is no time lag between light changes and photosynthetic response (expected photosynthetic response to a light fleck).

frequent short sun flecks is a characteristic of all plants measured (Pearcy and Seeman 1990). The rest periods between sun flecks in natural systems are variable, so leaves are likely to be at various induction states at the beginning of any sun fleck.

The induction state also influences an unusual phenomenon in sun fleck physiology, postillumination photosynthesis. When a sun fleck ends, photosynthesis does not stop immediately. In fact, photosynthesis at the induced rate will continue for a minute or more after a sun fleck of 20 s has ended. The postillumination photosynthesis is due to changes in the pools of photosynthetic intermediates (Figure 10.23). During the rest period of low-light the PGA pool increases because there is a limited amount of energy resources coming from the photosystems. When the sun fleck starts, the PGA pool decreases rapidly and the pools of RUBP and triose phosphate increase to a high level. When the sun fleck stops, photosynthesis continues because of the abundance of RUBP, and the PGA pool is reestablished. Therefore, the RUBP pool is acting as a capacitor for photosynthesis to continue CO_2 reduction, even after the light intensity has decreased. The postillumination photosynthesis has been shown to contribute significantly to total carbon gain induced by sun flecks.

The time for induction of photosynthesis reduces the use of the initial phase of the high-light period, and postillumination extends photosynthesis past the end of a sun fleck. Do these two processes counterbalance each other, or does one have an overriding impact on the effectiveness by which leaves can utilize a sun fleck? The efficiency by which a leaf utilizes sun flecks [light use efficiency (LUE)] can be calculated from the ratio of actual carbon gain compared to expected carbon gain during a sun fleck. One can calculate the expected carbon gain from a sun fleck by assuming that photosynthesis increases and decreases immediately at the beginning and end of the sun fleck. When carbon gain from a sun fleck is actually measured, it can be considerably greater than the carbon gain that would be expected based on the previous assumption (Figure 10.24). Thus, greater than 100% efficient use of sun fleck light by photosynthesis (LUE $>$ 100) can be attained. The difference between the observed and expected carbon gain during a sun fleck is due to

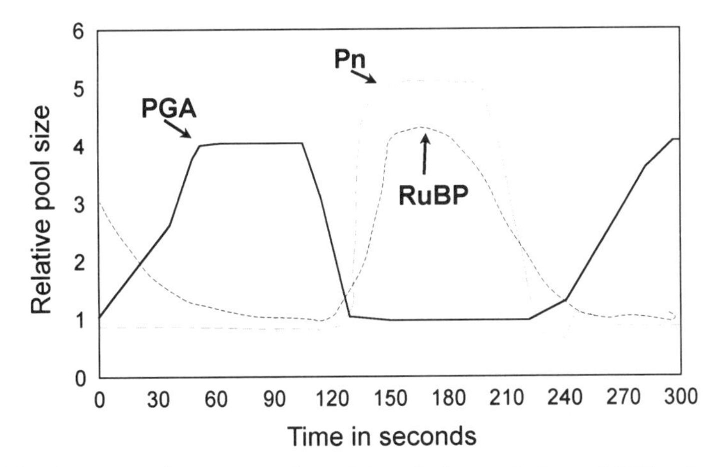

Figure 10.23 Diagrammatic representation of the relative pool sizes of PGA (phosphoglyceric acid) and RUBP (ribulose bis-phosphate) before, during, and after a sun fleck. Also plotted is the photosynthetic response (P_n) to the light fleck. On and off arrows refer to the time when the light is turned on or off.

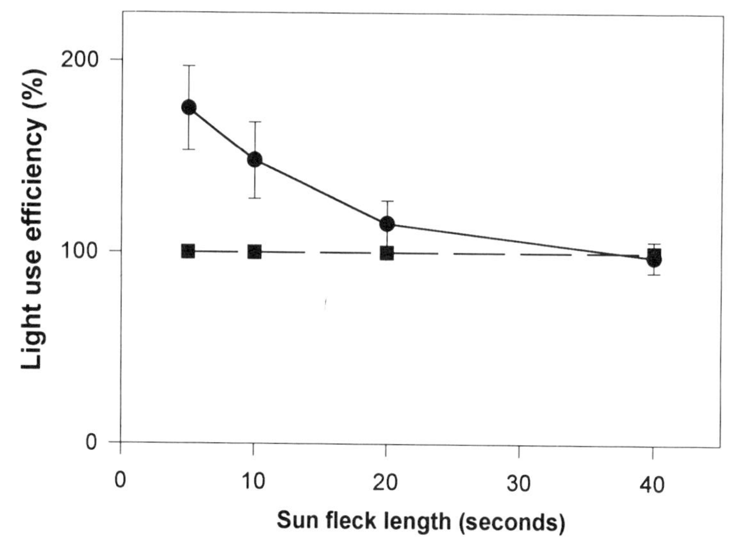

Figure 10.24 Changes in LUE (sun fleck light use efficiency; solid line), and the actual percentage of maximum postillumination carbon gain (dashed line), as the duration of a sun fleck increases. Note that postillumination carbon gain remains constant but LUE decreases. (Redrawn from data and text in Pearcy et al. 1994.)

the impact of postillumination photosynthesis. As the duration of a sun fleck gets longer, the contribution of postillumination carbon gain to total carbon gain decreases. This is a relative decrease since postillumination carbon gain is relatively constant while total carbon gain increases with sun fleck duration. Therefore, if a leaf experiences many short sun flecks, postillumination photosynthesis can greatly increase total carbon gain. If sun flecks are few and of long duration, postillumination photosynthesis will not add considerably to carbon gain during sun flecks. This conclusion has been supported in one study where it was found that as total daily PFD increased in redwood forest subcanopy sites (due to an increase in total light in sun flecks), actual daily carbon gain became an increasingly smaller proportion of expected daily carbon gain (Pfitsch and Pearcy 1989).

There are several other important responses to sun flecks other than photosynthesis. Leaf temperature increases during sun flecks because of the higher solar irradiance impinging on the leaf during a sun fleck and the generally large leaves of plants growing in low-light sites. Although the impact of these heat flecks has not been studied completely, they will affect physiological activity. For example, respiration will increase due to increased leaf temperature. In one study of *Rhododendron maximum*, the thermal optimum for photosynthesis was not the same as the temperature of leaves in diffuse light. Only during sun flecks does leaf temperature reach the optimum for photosynthesis (Bao and Nilsen 1988), which is 6 to 10°C over air temperature. Sun flecks also induce changes in leaf conductance that will affect transpiration and whole plant–water relations.

Stomatal conductance (g_s) responds more slowly to sun flecks than photosynthesis (Figure 10.25). Only in sun flecks that last more than 2 to 3 minutes will g_s reach its maximum potential. Also, g_s decreases slower after the end of a sun fleck than it rose at the onset of photosynthesis. In general, g_s does not limit carbon gain in a sun fleck unless the sun fleck is greater than 10 minutes in duration. The rate at which stomata open and close

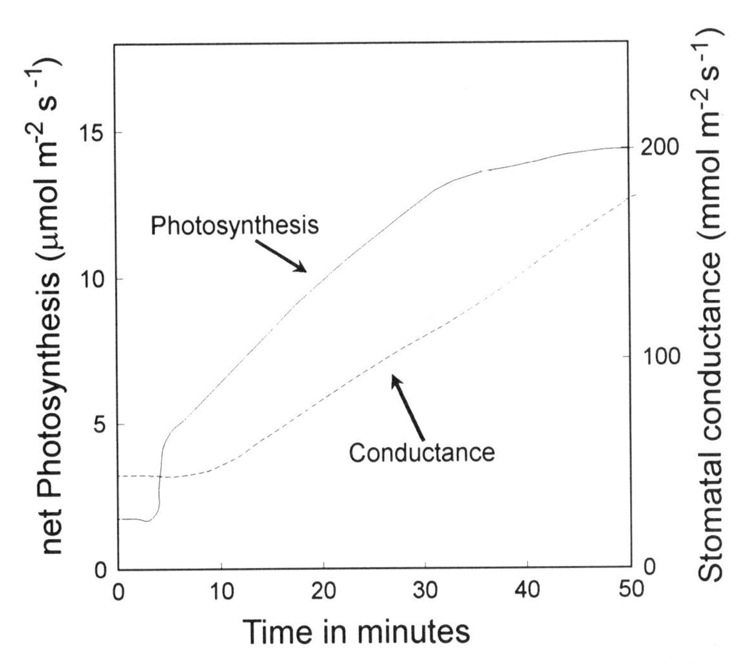

Figure 10.25 Timing for photosynthesis (P_n) and stomatal conductance (g_{st}) during a light fleck. (From Pearcy et al. 1994.)

in response to short-duration high light is species specific. Stomata of species adapted to low-light sites often open faster than those adapted to high-light sites. In the case of a tropical gap species, *Piper auretum,* stomata open and close faster in shade-grown plants, causing greater stomatal limitation to photosynthesis during sun flecks than that of high-light-grown plants (Tinoco-Ojanjuran and Pearcy 1992).

The relationship between photosynthesis, leaf conductance, and leaf water potential is different among species experiencing relatively long periods (10 to 12 min) of high or low light. Several studies (Knapp and Smith 1987, 1989, 1990; Knapp 1992) have suggested that subcanopy species can be classified into three main groups (Figure 10.26). Thermal avoiders are those species that have relatively low conductance during the period between sun flecks (Figure 10.26A). When a sun fleck hits the leaf, photosynthesis increases rapidly but is soon limited by stomatal conductance. Stomatal conductance and transpiration increase to high values during sun flecks in this class of species, which moderates the potential increase in leaf temperature. These species, classified as *thermal avoiders,* must have an ample supply of water in the soil, a high capacitance, or a high hydraulic resistance; otherwise, leaf water potential will decrease drastically during and shortly after the sun fleck. Those species that maintain relatively high leaf conductance (near their maximum g_{st} value) during the low-light period are classified as *water spenders* (Figure 10.26B). This classification was selected because under low light, photosynthesis in these species is limited by light intensity rather than CO_2 diffusion. Thus, having wide-open stomata under low-light conditions is a luxury consumption of water by transpiration, which results in a severely reduced water use efficiency. Photosynthesis increases rapidly at the onset of a sun fleck in water spenders and is not limited by leaf conductance until relatively late in the sun fleck duration. However, the relatively high g_{st}

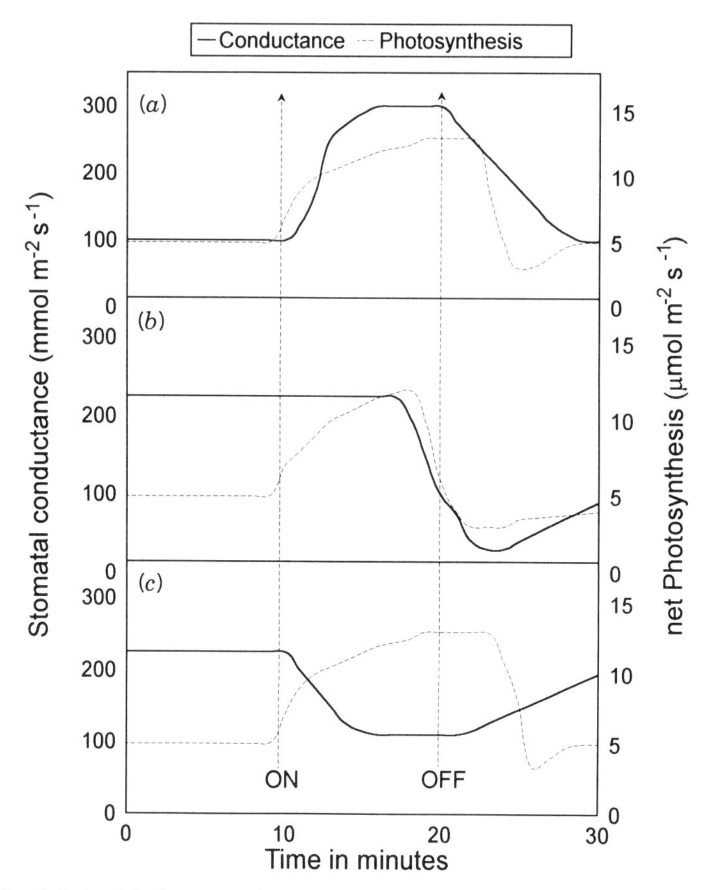

Figure 10.26 Relationship between photosynthesis and leaf conductance during sun flecks for three classes of species: (A) the thermal avoider pattern, in which stomatal conductance increases in a delayed fashion after photosynthesis; (B) The water spender pattern, in which high conductance can cause wilting which results in stomatal closure and photosynthesis inhibition; and (C) the thermal tolerance pattern, in which stomata close at the onset of a sun fleck, inducing high temperature in leaves. (Data gathered from Knapp and Smith 1987, 1989.)

value and the large vapor pressure gradient between the leaf and air (that develops in a sun fleck) together cause high transpiration. The high transpiration can lead to wilting, a common phenomenon for leaves of water spenders in sun flecks. The third class of species has relatively high g_{st} during the low-light period, like the water spenders, but g_{st} decreases during sun flecks (Figure 10.26C). This group of species is designated as *thermal tolerators* because leaf temperature increases dramatically in a sun fleck due to the low g_{st}. Photosynthesis increases in sun flecks for thermal tolerators, but transpiration could decrease in the sun fleck because of the low g_{st}. This results in a maximization of water use efficiency during the sun fleck. Although this classification scheme is currently theoretical, there is some evidence for species falling into one or the other class. In general, trees and shrubs tend to be thermal avoiders, while herbaceous species tend to be water spenders or thermal tolerators. However, growth form is not necessarily tied to one specific type of response (Knapp 1992).

There is little doubt that sun flecks provide a major amount of light variation in canopies and subcanopy environments, sun-fleck light can account for a large proportion of total daily carbon gain, and plants have adapted to sun flecks in a number of different ways. However, are sun flecks a stress for subcanopy plants or inner canopy leaves? If leaves that are adapted to low diffuse light experience a sun fleck of short duration (with low total PFD), this should not be considered a stress to the leaf because photosynthesis is induced and little to no photoinhibition occurs. When low-light-adapted plants are exposed to sun patches or sun gaps in which the high-light periods are long and have high PFD, an imbalance develops between metabolite use and energy absorbed. The result is photoinhibition or photooxidation. The prevalence of photoinhibition will be enhanced in water spenders, which develop low water potentials during long sun flecks. Stomatal closure in response to low turgor pressure will reduce C_i and promote photoinhibition. The potential for photoinhibition is also regulated by the ability of leaves to dissipate energy as heat through the xanthophyll epoxidation cycle (NRD). Shade-acclimated leaves have a smaller capacity for NRD of light energy and are more likely to suffer photoinhibition than sun-acclimated leaves. There is also an important interaction between leaf temperature and photoinhibition in sun flecks. When temperature raises quickly, NRD is less effective and photoinhibition is more likely. Thus if a shade leaf has high leaf conductance (water spenders), this may moderate leaf temperature, enhance NRD, and postpone photoinhibition. Such an adaptation in shade-adapted plants operates well only if leaf water potential is maintained high because of high hydraulic conductance or high capacitance.

VIII. SUMMARY

Light intensity and quality are essential characteristics of the environment that affect the physiology of cells and the growth and development of whole plants. Excesses, deficiencies, and inappropriate spectral distributions of light can be considered stresses to the integrated plant system. Plant physiology, growth, and development have evolved to utilize the particular spectral quality of light that results after sunlight penetrates the atmosphere. Most of this penetrated light is in the visible range and that is the light utilized for most physiological sensors (red and blue light chromophores) and photosynthesis. Ultraviolet light can have many detrimental impacts on plants (particularly UV-B), and infrared irradiance affects the thermal balance of leaves. Leaves have developed several morphological and behavioral mechanisms to alleviate the stress of excessive UV-B and excessive infrared irradiance.

Assuming that the spectral distribution of light is appropriate for proper physiological function, the intensity of light can still constitute a stress to plants. Plants have adapted through genetic means to environments with either high-or low-light attributes. Adaptation to high-or low-light habitats involves morphological changes (such as leaf size and thickness), behavioral changes (leaf angle, deciduousness, leaf movements), and physiological changes (ratio of photon absorption to electron transport capacity). Once a plant species has adapted to a particular light regime there is still the possibility of phenotypic response to an altered light regime. Acclimation of leaf structure and function to an altered light environment is more likely when a sun plant is adapting to shaded environment. However, acclimation of shade-adapted plants to high light does occur to some extent by induction of enzyme activity.

Natural light intensity is variable in space and time. In particular, the light environment under canopies or within a canopy is characterized by a background of constant diffuse irradiance punctuated by short flashes of high-intensity irradiance (sun flecks). The short flashes of high light (lasting a few seconds to several hours) may constitute a significant stress to leaf physiology of shade-adapted plants. Photoinhibition and photo-oxidation are likely to occur if there is an imbalance between light absorption and the use of metabolic intermediates of the Calvin–Benson cycle. However, subcanopy plants have evolved several mechanisms, including photosynthetic induction and postillumination photosynthesis, to maximally utilize the light provided during sun flecks for carbon gain. In fact, much of the total daily carbon gain by a subcanopy plant may be due to photosynthesis during and shortly after sun flecks. When sun flecks are of long duration, approaching that of sun patches or sun gaps, photoinhibition and photo-oxidation are likely to occur if the absorbed energy cannot be dissipated from the leaf.

The dynamics of physiological response to light intensity and quality is an underlying character of plant physiology. The survival of species in the wide diversity of habitats in the biosphere is in large part due to adaptations and acclimation to the variation in quality and intensity of light on a spatial and temporal scale. Understanding the way that plants cope with the stress of excessive irradiance, light deficiency, sun flecks, and/or inappropriate spectral distribution of light is a critical component of any stress physiologist's background.

STUDY-REVIEW OUTLINE

Environmental Variation

1. The solar radiation received by Earth's atmosphere is relatively constant, but that received by the biosphere is variable.
2. Atmospheric constituents scatter and absorb some of the solar radiation. Scattering does not affect spectral quality, but absorption of irradiance by the atmosphere does alter the spectral quality of light hitting Earth.
3. Diffuse irradiance is not directional and is due to scattering of light by the atmosphere. Diffuse radiation can account from 10 to 100% of radiation, depending on the site, time of day, or season of the year.
4. Atmospheric components absorb all UV-C, much of UV-B, and most of IR irradiance, but almost all visible light penetrates to the biosphere.
5. The spectral quality, intensity, and heterogeneity of light is altered by the presence of a canopy.

Measurement

1. Irradiance can be measured with a thermoelectric, thermovoltaic, or chemical method. Thermoelectric sensors work by determining the difference in temperature between a black and a white surface with the use of thermocouples, thermisters, or thermopiles.
2. Total irradiance is termed incident radiant flux and is measured with pyranometers. A pyraradiometer measures total irradiance between 200 and 3000 nm wavelengths.
3. Net radiometers measure the difference in incoming and outgoing radiation and total radiometers measure the sum of incoming and outgoing irradiance.
4. Pyraheliometers measure total direct radiation and do not measure diffuse radiation.
5. Photosynthetically active radiation (PAR) is the quantity of photons hitting a surface within the 400 to 700 nm waveband. Such light is measured with quantum sensors.
6. The complete (200 to 3000 nm) spectral characteristics of light must be determined with a spectroradimeter. However, if one needs to know the abundance of light in a specific waveband, filters can be used to screen out light in other wavebands.
7. The impact of canopies on light environment can be determined with an image analysis

method. Photographs of the canopy are taken and then the image is analyzed in a computer system.

Effects of Irradiance on Plants

1. Most irradiance absorbed by plant leaves affects the thermal properties of the leaves, and only about 1% of that is used by photosynthesis.
2. Large leaves are uncoupled from air temperature and may have large differences between air and leaf temperature when exposed to high light.

Effects on Photosynthesis

1. The influence of light on photosynthesis can best be described by a light response curve.
2. Quantum yield is the linear increase in photosynthesis with increasing light intensity at relatively low light intensity.
3. The light compensation point is the amount of irradiance required to compensate for the sum of dark respiration and photorespiration.
4. The light saturation point is that light intensity at which light is not the only limiting factor for photosynthesis and photosynthesis does not increase linearly with increasing light intensity.
5. The light saturation rate is the maximum photosynthetic rate attained at high light intensity.

Physiological Sensors

1. Some important chromophores are phytochrome, blue light chromophore, and UV-absorbing chromophores. Sensors that respond to infrared light are common in animals but unknown in plants.
2. Phytochrome is a dimer of two types of proteins. One lobe of each protein associates with membranes, while the other lobe contains the active sites.
3. The phytochrome dimer can be converted between a form that absorbs red light (P_r) and a form that absorbs far red light (P_{fr}).
4. The far-red absorbing form of phytochrome is the active form, which induces physiological response.
5. Changes in the spectral distribution of light or the length of the night influences the relative abundance of P_r and P_{fr}.

Physiological Effects of Phytochrome

1. One molecular model of phytochrome action suggests that phytochrome changes membrane permeability to calcium, which activates calmodulin. Active calmodulin then activates specific enzymes initiating the physiological response.
2. A second molecular model of phytochrome action indicates that phytochrome activates a regulatory protein that enters the nucleus, binds to a promoter region, and stimulates gene transcription.
3. Some phytochrome effects are phototropic in which organs grow toward the light.
4. Photoperiodic effects mediated by phytochrome regulate morphological changes such as flower induction and leaf abscission.
5. Photomorphogenic responses are changes in morphology as a consequence of phytochrome-mediated physiology.
6. Photonastic movements such as leaf "sleep" movements are not well understood and may not involve a phytochrome mechanism.

Effects of Shortwave (UV) Irradiance

1. Ultraviolet irradiation can be divided into three main wavebands with different photochemical capacity. UV-C has high photochemical capacity and is absorbed by nucleic acids and some

proteins. UV-B has less photochemical activity but is also absorbed by nucleic acids, proteins, and flavoproteins. UV-A has little photochemical activity, and nucleic acids and proteins absorb little of this radiation.
2. Studies of UV radiation on plants focus on UV-B because UV-C is completely absorbed by the atmosphere, and UV-A has no detrimental impact on plant function.
3. The amount of UV-B radiation increases at high latitude and at high elevation.
4. UV-B dose is dependent on the intensity of UV-B and the length of exposure to UV-B.
5. UV-B has a generally negative impact on plants, lowering growth, yield, and photosynthesis.
6. A major site of damage by UV-B is photosystem II. Therefore, plants in low-light that have accentuated photosystem II will be very susceptible to UV-B damage.
7. To avoid UV-B damage, leaves can develop thick cuticles and a proliferation of flavenoid compounds.

Adaptation to High and Low Light

Morphological, Anatomical, and Behavioral Adaptations

1. In low light, leaves are thinner, have thinner cuticle, and are larger than leaves on plants adapted to high light.
2. Leaves have a greater longevity in species adapted to low-light environments. The increased longevity increases lifetime carbon gain but requires an investment in defense.
3. Leaves on high-light-adapted plants are vertically oriented rather than perfectly horizontal and can have a reflective surface.
4. Plants growing in high-light sites can have higher hydraulic conductance than plants growing in low-light sites.

Photosynthetic Physiology

1. Leaves of plants adapted to high-light have a lower quantum yield, a higher light compensation point, a higher light saturation point, and a higher light-saturated photosynthetic rate than those of leaves of low-light-adapted plants.
2. In low-light-adapted leaves, light harvesting is enhanced by increasing the amount of LHCII compared with PSII core complex, having chloroplast horizontal, and having extensive appressed membrane.
3. In high-light plants electron transport is enhanced by having large pools of plastoquinone, plastocyanin, and photosystems.
4. The biochemical properties of leaf photosynthesis must be balanced with the electron transport capacity.
5. Photoinhibition, damage of the D1 protein of PSII, causes a reduction in electron transport capacity but also reduces light absorption. Therefore, photoinhibition is one line of defense against the damaging nature of excess absorbed irradiance.
6. The xanthophylls are important for nonradiative dissipation of energy when photosynthesis and photorespiration cannot rid the photosystems of excess energy.
7. The Mehler peroxidase ascorbate reaction oxidizes NADPH, utilizing some of the electron flow in photosystems when excess energy is present. This Mehler reaction also causes a buildup of H^+ ions in the thylakoid, which stimulates NRD by carotenoids.

Acclimation to an Altered Light Environment

1. Acclimation is a phenological response of morphology or physiology to changes in climatic resources such as light.
2. Because plants have modular growth, they are able to abscise old organs and replace them with new organs which may be more tailored to the new resource conditions. Modules can also acclimate on one part of the plant differently than on other parts of the plant.

3. Plants adapted to low light have less ability to acclimate photosynthetic processes that plants adapted to high-light sites.

Adaptation to Irradiance Heterogeneity

1. The light intensity under or within a canopy is distinctly variable, due to sun flecks.
2. Important attributes of sun flecks are (a) their duration, (b) their maximum intensity, (c) their total photon flux density (PFD), (d) and the amount of time since the last sun fleck.
3. Sun flecks occur under all canopies but they are of longer duration under temperate canopies than under tropical canopies.
4. The higher the gap in the canopy, the greater the penumbra effect in the sun fleck and the fuzzier the edges of the sun fleck become.
5. Carbon gain during sun flecks can account for a large majority of carbon gained by leaves.
6. Photosynthesis does not immediately reach its maximum in a sun fleck; the maximum rate must be induced by extended periods in the sun fleck or by many frequent sun flecks.
7. After the sun fleck is over, photosynthesis continues because of the buildup of intermediates for the Calvin–Benson cycle.
8. Stomatal conductance increases more slowly than photosynthesis at the beginning of a sun fleck and decreases more slowly than photosynthesis after the sun fleck.
9. Thermal avoiders increase both conductance and photosynthesis in a sun fleck.
10. Thermal tolerators decrease conductance and increase photosynthesis in a sun fleck, which maximizes water use efficiency but causes significant increases in temperature.
11. Water spenders have high conductance and may wilt at the end of sun flecks.

SELF-STUDY QUESTIONS

1. What regulates the spectral quality of light that affects a plant in an open field versus a plant in the subcanopy of a redwood forest?
2. Define the difference between a net radiometer, pyraheliometer, and a pyraradiometer.
3. How is the influence of irradiance on black-and-white surfaces used to measure irradiance?
4. What are the advantages and disadvantages of using quantum sensors or light GaAsP sensors?
5. What are the three subdivisions of UV irradiance, and which are damaging or helpful for cell physiology?
6. Diagram a light response curve and label the important characteristics of the curve.
7. What is the molecular model for phytochrome, and how does phytochrome affect transcription or calmodulin-mediated enzyme reactivity?
8. Differentiate between phototropic, photonastic, and photomorphogenic effects of phytochrome.
9. In what way is UV light a stress to plants?
10. How can plants increase their resistance to UV irradiance?
11. What are four ways in which the morphology and anatomy of leaves adapt to habitats with different light intensity?

12. How do the light-harvesting components of photosynthesis compare between high-light- and low-light-adapted plants. A diagram of light response curves for leaves of high- and low-light-adapted plants may be helpful.

13. What is the molecular mechanism for photoinhibition? Under what conditions is photoinhibition likely to occur?

14. How are plants able to rid themselves of excess irradiance once absorbed by chlorophyll molecules?

15. Why can the Mehler peroxidase ascorbate reaction reduce the severity of damage to photosynthetic reactions when excess energy is being absorbed by chlorophyll?

16. What are the costs and benefits for leaves with extended longevity in sites of low light intensity?

17. Compare the capacity of high- or low-light-adapted plants to acclimate to a change (either higher or lower light) in their light environment.

18. What types of experiments are used to determine if a physiological characteristics is due to genetic or phenotypic causes?

19. Diagram a characteristic response of photosynthesis and stomatal conductance to a 90-s sun fleck.

20. Under what conditions are sun flecks damaging to plants adapted to low-light environments?

21. Describe the three possible types of photosynthesis and conductance responses to sun or shade flecks. Include the consequences of this behavior to leaf temperature and leaf water potential.

SUPPLEMENTARY READING

Furuya, M. (ed.) 1987. Phytochrome and Photoregulation in Plants. Academic Press, Tokyo.

Krause, G. H., and E. Weiss. 1991. Chlorophyll fluorescence and photosynthesis: The Basics. Annual Review of Plant Physiology and Plant Molecular Biology 42:313–349.

Moses. P. B., and N. -H. Chua. 1988. Light switches for plant genes. Scientific American 258 (April):88–93.

Powles, S. B. 1984. Photoinhibition of photosynthesis induced by visible light. Annual Review of Plant Physiology 35:15–44.

Smith, H. 1975. Light quality, photoreception, and plant strategy. Annual Review of Plant Physiology 33:481–518.

11 High Temperature and Energy Balance

Outline

Objectives

1. Introduce the variation in thermal environments among different habitats and different parts of plants.
2. Discuss energy balance relationships in plants.
3. Cover each physical factor of thermodynamics of plants that influences tissue temperature.

4. Discuss the various time functions that occur when plant tissues respond to thermal changes.
5. Present the various techniques used to measure plant tissue temperature.
6. Discuss measurements of physiological response to tissue temperature, including the LD_{50} concept.
7. Discuss the various detrimental effects of high temperature on plant metabolism. Include impacts on photosynthesis, respiration, membrane function, and enzyme kinetics.
8. Present and evaluate the various mechanisms by which plants moderate leaf temperature. Include morphological, behavioral, and physiological mechanisms.
9. Cover variation in tissue high-temperature tolerance among species, and evaluate the various physiological mechanisms that occur in plants which enhance tolerance of high tissue temperature.

I. INTRODUCTION

Terrestrial and aquatic environments around the world experience a wide range of temperature conditions. Air temperature in some environments, such as Death Valley, California, reaches close to 60°C during the summer, and remains above 50°C day and night for months. In contrast, some environments, such as those in Antarctica, have air temperatures above freezing for only small windows of time, and frequently have air temperatures of −50°C. Furthermore, any one climatic area may have a unique spatial or temporal variation in air temperature. Temperature climates have wide variation in seasonal (often a 40°C difference between summer and winter temperatures), and daily (often 20°C) temperature. Tropical alpine habitats may have little to no seasonal variation in air temperature but extensive daily variation (−5 to 30°C each day). Thermal attributes of a particular environment can also have large (regional) and small (microsite) spatial patterns. North and south slopes of a hillside in the temperate zone can have a difference of 25°C in air temperature, while the difference between the air temperature of north and south slopes in arctic ecosystems may be only a few degrees C. There is a strong increase in air temperature near the soil surface compared with a few centimeters above the soil. Canopy sunflecks cause localized hot spots that move across the forest floor. Even though there is a wide variety of thermal attributes among communities, there are vascular plants that can tolerate almost the entire range of thermal conditions in terrestrial environments, including those of Antarctica and Death Valley.

Plants are poikilotherms (except for a few rare exceptions), thus temperatures in their tissues are reflective of their thermal environment. Due to the wide variation in ambient temperatures among environments where plants reside and the poikilothermic nature of plants, it is logical to expect a wide range of metabolic, morphological, and anatomical adaptations to thermal conditions. The wide variance of spatial and temporal thermal attributes among habitats can also cause stress in plants, particularly when plants are moved into environments that are different from those in which they evolved. Such is often the case for crops.

Plant adaptation and acclimation to thermal conditions have important applied significance, in addition to their significance to the basic function of plant metabolism. As mentioned above, crops are often cultured in habitats that are foreign from those in which they evolved. Agricultural scientists are constantly working to develop varieties of crops that are able to maintain yield in habitats that are at the extremes of thermal tolerance of specific crops. The development of some of these varieties has expanded the environmental tolerances of crops, and the range of appropriate climatic conditions for crop production has increased. It is important to understand the mechanisms by which plants naturally tolerate environments with extreme temperatures, because this information can lead to efficient development of new crop varieties by breeding for specific adaptive traits, or by genetic engineering of those traits.

Understanding thermal tolerances of native species has also applied value in managing natural systems. There is a plethora of evidence that thermal characteristics of the globe are changing. Global warming will occur with different intensities in different locations; however, the most sensitive situations may be high-altitude communities and the southern margin of northerly distributed species. Recent evidence clearly shows that a change in air temperature will have an impact on the survival of particular species in alpine habitats and will result in changing species distributions. Through an understanding of the mechanisms by which plants tolerate and acclimate to a changing thermal environment, managers of natural systems may be able to develop plans to prepare for this impending new thermal environment.

In any environment, the temperature of all organs of plants will not be the same (Figure 11.1). One important gradient in temperature within plants is the vertical gradient from shoot apices to root apices. In general, during midday, temperature is highest at shoot apices, decreases toward lower parts of shoots, and increases close to the ground surface. This thermal gradient is due to the impact of light on temperature of outer branches and solar heating of the soil surface. At midnight, roots may have the highest temperature among organs, and temperature may decrease from the soil surface to the outer branches and shoot tips. This nocturnal temperature gradient within plants is due to residual heat in the soil from solar heating and radiation of heat to the environment by outer branches. The most stable temperatures over the season and day are those in the root zone, and the most variable temperatures are experienced at the soil surface and at shoot apices (including outer canopy exposed leaves).

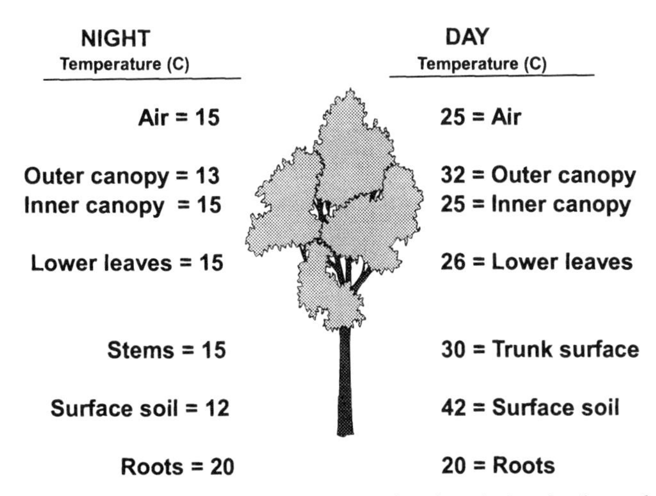

Figure 11.1 Variation in temperature among organs of a plant during the day and night. Note that at night roots have the highest temperature, while during the day the highest temperatures are in the base of the stem near the soil surface.

Different organs of plants also have different sensitivities to temperature extremes. The most sensitive locations on plants are apical meristems of roots and shoots. These tissues are particularly sensitive to high temperature because of the intricate processes involved in active cell division. Since root apical meristems are located in the most thermally buffered environment, most attention is directed toward thermal tolerance of shoot tips, leaves, inflorescences, and axial buds. In addition to apical meristems, leaves are also very sensitive to thermal variation, due to their lack of any significant surface buffer tissue. For example, phloem and xylem have surface bark or cortical tissues that can serve as a thermal buffer. In some alpine tropical species such as tall monocarpic plants sheathing petioles buffer the daily temperature change in cambial tissues (Rada et al. 1985). Therefore, the stems of these species are less affected than leaves by changes in air temperature because stem temperature can remain fairly constant even though air temperature varies considerably each day. It is not true that all stems tend to be homeothermic compared to leaves. In fact, stems of most species have temperatures fairly close to air temperature.

The variation in temperature among different organs, and the variation in temperature among the same organ on different species growing at the same site, is evidence that tissue temperature is not determined by ambient air temperature alone. In rare cases, tissue temperature of plants can be influenced by endogenous respiratory activity. In particular, the inflorescence of species in the *Araceae* can have temperatures significantly over air temperature because of a highly active alternative respiration pathway. However, in contrast to many animals, internal metabolism does not have a significant influence on tissue temperature for most plant species. Tissue temperature is regulated primarily by organ morphology, evaporation, and the electromagnetic radiation balance.

The *energy balance* of an organ (leaves, for example) refers to a steady-state situation in which energy absorption by a tissue is equal to energy loss by that tissue (Campbell 1977). All plant organs must come to this steady-state condition after an alteration in exogenous or endogenous energy content. For example, subcanopy leaves exposed to diffuse radiation will have an equilibrium leaf temperature that is a result of the steady-state balance of energy input and output. When those leaves experience a sudden increase in light intensity (sun fleck), more energy is absorbed than is currently counterbalanced by energy dissipation processes. As a result, energy builds up in the tissue, temperature rises, and energy loss mechanisms may increase. If energy loss processes can increase to compensate for the increased absorbed energy, leaf temperature may return to the pre-sunfleck value. However, if energy loss processes cannot increase adequately, leaf temperature will increase to a new steady-state level in the sun fleck. Therefore, morphological, anatomical, behavioral, or physiological factors that affect energy gain or loss properties of tissues are centrally important to tissue temperature regulation. The rest of this chapter is devoted to explaining how tissue temperature (particularly leaves) is regulated by tissue energy balance, how plants adjust their thermotolerance, and what mechanisms plants utilize to keep tissue temperature within a tolerable range.

II. PHYSICAL FACTORS THAT REGULATE LEAF TEMPERATURE

A. Atmospheric Temperature

Ambient temperature around a particular plant organ has a direct effect on organ temperature. This is because plants are poikilotherms, and their internal metabolism has a minimal impact on tissue temperature. Some tissues will have a temperature that is very close to air temperature at all times ($\Delta T = 0$). These organs are classified as "coupled" to air temperature. Such tissues, or organs, are relatively immune from the impacts of changing energy absorption on leaf temperature. They have the ability to change energy dissipation rates rapidly in response to changes in energy absorption. Other tissues are "decoupled" from air temperature and will have a variable ΔT, depending on energy relationships of the environment. These tissues have a limited ability to change the rates of some energy loss mechanisms when changes in energy absorption occurs. Some leaves can have as much as a 10 to 15°C difference between air and leaf temperature under certain conditions (Bao and Nilsen 1988). However, most decoupled leaves will have 5°C or less difference between air and leaf temperature.

In the case of leaves, ΔT also has a secondary impact on transpiration. Transpiration is partly dependent on the vapor pressure difference between the inner leaf spaces and the atmosphere (see Chapter 7). As ΔT increases, the VPG increases, which causes an increase in transpiration (as long as stomatal conductance remains unchanged). Transpiration is itself one of the mechanisms by which leaves can dissipate energy (see latent heat exchange, described below). Therefore, there is a feedback relationship between increasing ΔT of leaves and increasing energy dissipation by latent heat exchange.

B. Net Radiation Energy Balance

1. Solar Radiation Balance. One of the most important aspects of tissue energy balance is energy absorbed from radiation impinging on the tissue surface. Radiation comes to the leaf surface from a series of different sources (Figure 11.2). Direct and diffuse solar radiation sum to produce total radiation impinging on a leaf from the sky. Diffuse radiation from the sky is that which is scattered by particles or clouds in the atmosphere. Reflected radiation (a portion of diffuse radiation) is solar radiation that hits surfaces (soil, trunks, branches, leaves, etc.) near leaves and is reflected toward leaves. Reflected radiation will increase radiation absorbed by leaves and can account for 10 to 30% of the total radiation impinging on a leaf surface. The fraction of shortwave radiation (not infrared radiation)

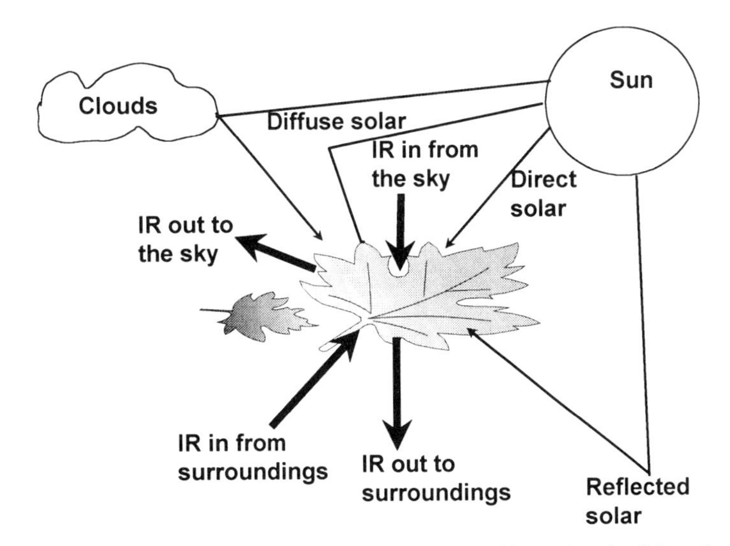

Figure 11.2 Various energy sources that are absorbed by and emitted from leaves.

reflected from surfaces is termed the *albedo*. The albedo of ground surfaces can vary from about 0.60 for snow to 0.10 for peat soils (Rosenberg et al. 1983). The higher the albedo, the higher the significance of reflected radiation to energy absorbed by leaves. Reflectance of the surroundings is further modified by the angle of incidence between the radiation source and the reflective surface. The smaller the angle of incidence, the greater the reflectance.

The total amount of absorbed radiation can be calculated as

$$\text{absorbed radiation} = a(\Phi^{\text{direct}} + \Phi^{\text{diffuse}}) + ar(\Phi^{\text{direct}} + \Phi^{\text{diffuse}}) \quad \text{or} \quad = a(1 + r)\Phi \qquad (11.1)$$

where a is the absorptance, which is the fraction of total radiant energy flux (Φ) absorbed by the leaf, Φ^{direct} the direct solar radiant energy flux, Φ^{diffuse} the total diffuse radiant energy flux, r the fraction of Φ reflected from surroundings onto the leaf surface, and Φ the total of diffuse and direct radiance flux. This equation is an approximation because the spectral quality of reflected or sky diffuse light can be different; therefore, the two forms of diffuse light can contain different amounts of energy. Furthermore, leaf absorptance can be dependent on the spectral distribution of radiance. Additionally, the amount of reflected radiation can depend on the wavelength distribution of direct radiation.

The characteristic of plant leaves that has the greatest impact on absorption of solar and reflected radiance is leaf absorptance (a). Leaf absorptance varies with the wavelength of light impinging on the surface. For example, in the visible wavebands (400 to 700 nm) the value of a is often between 0.8 and 0.9. Absorptance of near-infrared radiation (800 to 1500 nm) is relatively low (0.1), but absorption of longer-wave radiation (1500 to 3000 nm) is also approximately 0.9. Inasmuch as leaf absorptance has a dramatic effect on radiation absorption, energy balance, and photosynthesis, the ability of plants to adjust their absorptance by morphological or behavioral mechanisms will have a great affect on tissue temperature and metabolism. Leaf absorptance can be as low as 0.4 in some species under some conditions. The mechanisms that regulate absorptance of leaves are discussed later in the chapter.

2. Infrared Radiation Balance. Leaves also absorb radiation emitted by surrounding surfaces (leaves, trunks, etc.) and the atmosphere. *Infrared* or *thermal radiation* is emitted by any object with a temperature above 0^0 K (absolute zero). Over 99% of radiation emitted from the surroundings of a plant occurs at wavelengths above 4000 nm; thus the surroundings are emitting radiation with wavelengths far into the infrared. The radiation emitted from the sky is derived mostly from water and carbon dioxide, which also emit long-wave radiation between 5000 and 13,000 nm. The concentration of these gases varies in the atmosphere, and that variation in gas concentration regulates the amount of emitted radiation. Clear cloudless skys at night have a low long-wave emittance. The presence of clouds or dust particles in the atmosphere will increase long-wave emissions from the atmosphere.

The amount of long-wave radiation (infrared radiation or IR) emitted by any object can be described by the Stephen–Boltzmann relationship. This relationship shows that the amount of long-wave radiation emitted by a blackbody radiator (perfect radiator) is proportional to the fourth power of its surface temperature:

$$\text{infrared radiation emitted} = \sigma T^4 \qquad (11.2)$$

where σ is the Stephen–Boltzmann coefficient and T is the absolute surface temperature of the radiator. Therefore, a small increase in the surface temperature of the radiator will greatly increase the amount of IR emitted.

Leaves and other plant surfaces absorb infrared radiation that is being emitted from all surrounding surfaces and from the sky. The amount of infrared radiation emitted by the surroundings will depend on their collective temperature, and a sky temperature can be determined from the amount of infrared radiation emitted from the atmosphere (see Chapter 10). The amount of infrared absorption can be calculated based on leaf absorption properties and infrared radiation from the plant surroundings:

$$\text{infrared radiation absorbed} = a_{IR}\sigma[(T_{surr})^4 + (T_{sky})^4] \quad (11.3)$$

where: a_{IR} is the fraction of the infrared radiation impinging on a leaf surface that is absorbed by the leaf, T_{surr} the mean temperature of surrounding surfaces, and T_{sky} the effective sky temperature. Leaf absorption of IR is dependent upon its IR absorptance coefficient. The absorptance coefficient for IR is normally between 0.94 and 0.98, indicating that a great majority of IR around leaves is absorbed.

Leaves and other plant surfaces also emit IR. The amount of IR emitted will depend on the fourth power of the leaf temperature (based on the Stephen–Boltzmann law). Emittance of IR from any surface depends on the surface temperature as modified by an emissivity coefficient (e). Emissivity of a perfect radiator is defined as 1. Therefore, there is no emissivity coefficient in equation (11.2). Leaves and plant surfaces are not perfect radiators; therefore, they have a value of e that is less than 1. Because the emission rate of IR is the reverse of its absorption, it is logical to assume that the same factors regulating a_{IR} also influence e_{IR}. Consequently, e_{IR} is normally very similar to a_{IR}, (about 0.96). The Stephen–Boltzmann relationship can also be used to describe emitted radiation from leaves, but a factor of 2 must be added for most leaves because of their two-sided structure:

$$\text{infrared radiation emitted by a leaf} = 2e_{IR}\sigma(T^{leaf})^4 \quad (11.4)$$

Recall that the total radiation absorbed by a leaf (or other plant surface) is the sum of solar and IR radiation absorption. The following equation can be used to describe the net radiation balance of leaves:

$$\text{net radiation} = a(1 + r)\Phi + a_{IR}\sigma[T_{surr})^4 + (T_{sky})^4] - 2e_{IR}\sigma(T^{leaf})^4 \quad (11.5)$$

Most of the radiation absorbed by leaves is lost by emission of IR. However, during the daylight hours when solar radiation is large there will be a positive net radiation balance. During the night, net radiation may be negative, particularly in situations where sky temperature is low (clear nights). Although there will be a positive or negative net radiation balance, the energy balance of a leaf must equal zero. During the day, when net radiation is positive, other mechanisms of energy loss become important for leaf temperature maintenance. In the absence of these other energy loss processes, leaf temperature will rise, increasing energy loss by IR emission and reestablishing an energy balance. Several other energy loss mechanisms, important to leaves, are discussed below.

C. Latent Heat Exchange

When water evaporates it requires energy, which is taken from the evaporation surface. The removal of energy by water evaporation is termed *latent heat exchange*. Most of the water that evaporates from plants does so in transpiration. Therefore, when stomata open and transpiration is occurring, energy is being lost to the surroundings. The amount of energy lost through latent heat exchange can be calculated from the transpiration rate by

$$\text{latent heat exchange} = J_{wv} H_{wv} \tag{11.6}$$

where J_{wv} is the transpiration (usually between 0 and 10 mmol m^{-2} s^{-1}), and H_{wv} is the heat of vaporization for water at the temperature of the leaf (at 25°C, $H_{wv} = 44.0$ kJ mol^{-1}).

During the day, when gas exchange is occurring, loss of energy by latent heat exchange is inevitable. However, leaves can adjust the amount of energy lost by latent heat exchange by adjusting the stomatal aperture. Therefore, in high-radiation environments, latent heat exchange may be an important mechanism of cooling leaves. The effectiveness of latent heat exchange as a leaf-cooling mechanism is reduced in environments where water is also limiting. For example, in desert habitats or in coastal marine habitats, radiation is high during the day but water has limited availability. Most desert plants have low conductance to minimize water loss (a water conservation mechanism), which also reduces the magnitude of latent heat exchange. Under these situations another heat loss mechanism must effectively remove energy from the leaf in order to moderate leaf temperature.

D. Conduction/Boundary Layer Resistance

Energy can be dissipated from leaves by conduction and convection. Conduction occurs when energy in the form of heat is transferred from one relatively warm surface to the contacting adjacent cooler surface. Therefore, leaves that have a temperature greater than the air contacting the leaf surface will conduct heat energy into the air. Convection occurs when heat is removed from the surface of an organ by moving water or air. There are two types of convection: free and forced, of which the latter is most relevant to leaves. Forced convection is caused by wind removing the heated air outside the leaf boundary layer (the layer of unstirred air on the leaf surface that was heated by conduction). As wind speed increases, the amount of forced-air convection increases. The amount of heat dissipated by conduction and convection in combination can be estimated based on the following equation:

$$\text{energy lost by convection and conduction} = 2K^{air}\frac{T^{leaf} - T^{air}}{\delta^{bl}} \tag{11.7}$$

where K^{air} is the thermal conductivity coefficient in air and δ^{bl} is the thickness of the boundary layer of unstirred air around the leaf. Note that the equation has a multiplier of 2 to take into account the two surfaces of leaves.

Thus, conduction and convection of energy from leaves depends on the difference in leaf temperature between the leaf and air (ΔT) and the thickness of the boundary layer. The larger the ΔT and the smaller the δ^{bl}, the larger will be conduction and convection. If leaves are coupled to air temperature (ΔT is close to zero), the most important factor regulating conduction and convection is the thickness of the boundary layer.

Boundary layer thickness has been shown to depend on the length of the leaf in the direction of the wind and the wind speed:

$$\delta^{bl} = 4\sqrt{\frac{l}{v}} \tag{11.8}$$

This equation was developed by iterative experimentation in wind tunnels. The multiplier of 4 is an appropriate factor for leaves in field-turbulent air, even though the theoretical boundary layer displacement on a flat plate is 6 (Nobel 1991 and citations therein). Furthermore, the square-root component for leaf length $(l)^{0.5}$ is for a flat plate and does not always give the best results for leaves. Exponents between 0.3 and 0.5 are appropriate for the length component (Gates and Papain 1971), and exponents of 0.5 to 0.7 are appropriate for the wind velocity component (Nobel 1991) of equation (11.8). In general, boundary layers of leaves can range between 0.5 and 10 mm for leaves with lengths of 0.002 and 0.5 m (length in the direction of the wind) at low wind velocity (0.1 m s^{-1}). If wind velocity rises to 10 m s^{-1} or greater, the leaf boundary layer will be less than 1 mm at a leaf length greater that 0.5 m.

Boundary layer thickness and ΔT clearly interact through their respective influences on conduction and convection. If the boundary layer is relatively large (large leaf and low wind velocity), the leaf is relatively decoupled from air temperature and ΔT will increase. Thus, at times of variable wind speed, leaf temperature may rise at low wind speed and drop at high wind speed. However, during this time, conduction and convection of heat energy may remain constant because of the counterbalancing effects of ΔT and δ^{bl} on the magnitude of conduction and convection.

E. Other Influences on Leaf Temperature

Leaf temperature is regulated primarily by the influences of net radiation, latent heat exchange, conduction and convection, and air temperature. However, there are other attributes of the environment and the plant that can modify these general regulators of leaf temperature. One aspect of the environment that can have a significant impact on plant tissue temperature is the soil surface. Surface soil commonly has a high absorptance; thus surface soil temperature can become relatively high when exposed to solar radiation. High soil surface temperature will lead to heat being emitted by the soil surface. The emitted heat can affect the long-wave radiation load on leaves located near the base of plants. Plants with basal rosettes can be particularly susceptible to soil-radiated heat.

Although internally generated heat from catabolic reactions such as respiration is usually insignificant compared to other regulators of plant leaf temperature, there are a few cases in which internal metabolism does affect tissue temperature. There is no doubt that inflorescence of species in the Araceae utilizes cellular respiration, which heats the stigma surface in order to attract pollinators. Except for this case, temperature modulation by endogenous respiration is insignificant for plants.

Heat can also be generated in an environment by anthropogenic factors. Habitation changes the albedo of a habitat and thus alters the amount of reflected radiation. Effluent from reactors or other municipal industry will alter the thermal environment of downstream habitats. Cities can cause thermal inversions that would otherwise not occur.

Changes in the thermal environment by anthropogenic mechanisms is a major topic in environmental sciences.

F. Time-Course Responses of Leaf Temperature

Responses between air temperature, radiation balance, transpiration, and thermal conductance are not instantaneous. There is a non-steady-state time dynamic which results from the interaction of these factors. A thermal time constant (τ), which refers to the speed by which tissue temperature follows changes in air temperature, can be calculated for most plant tissues. Except for the largest leaves (greater than 0.5 m in length), τ will be less than 1 min (Jones 1992). Other tissues, such as fruits or large stems, may have values of τ greater than 500 min (Monteith 1981). Furthermore, the thermal conductivity in plant tissues is relatively slow; thus the internal temperatures of large organs will lag behind that of surface tissues. For example, in cactus stems one side of the stem can be 10°C different from the other side of the stem, and the central tissues of the stem can be cooler than either external surface (Nobel 1978). Similarly, the stems of *Espeletzia* (a high-elevation, tropical, monocarpic plant with thick stems) can have temperatures that lag behind changes in air temperature by 12 to 14 h (Rada et al. 1985). The τ for any particular tissue will depend on the thickness of the tissue as well as its outer morphological attributes (presence of spines or trichomes, height off the ground, compass direction, etc.).

There are several different dynamics for the interaction between air temperature change and tissue temperature. In the case of step changes, the air temperature changes instantaneously and the tissue temperature lags behind (Figure 11.3A). Instantaneous environmental temperature change can occur in submerged leaves when the water temperature changes drastically due to a change in water flow dynamics of a stream. A step change in temperature is uncommon in terrestrial leaves, although very rapid changes in air temperature can occur from the effect of sun flecks on leaf temperature. A ramp change dynamic occurs when air temperature gradually changes and tissue temperature lags behind at slower rate (Figure 11.3B). This is a very common dynamic for leaves, stems, or fruits as air temperature changes gradually over a diurnal cycle. If τ is particularly large and there is a sizable delay between air and tissue temperature changes over a diurnal cycle, a harmonic relationship can develop (Figure 11.3C). In a harmonic relationship both the tissue and air temperature are following a sine-wave-like dynamic, but they are out of phase. In particularly thick stems, the amplitude of the wave pattern of stem temperature will decrease with depth in the trunk tissues.

G. Techniques for Measuring Thermal Attributes of Tissues

Tissue temperature is commonly measured by various types of thermocouples or thermistors. To understand the operation of thermocouples and thermistors, one must understand the operation of a resistance thermometer. The resistance thermometer is based on the fact that all metals change electrical resistance with a change in temperature. Therefore, temperature can be measured simply by determining the electrical resistance within a uniform metal. Although each metal has a coefficient of resistance, and many metals could be used for resistance thermometers, platinum is usually the metal of choice. The platinum wire must be encased to protect the metal and can be at a remote location from the readout device. This device measures resistance to electrical flow in the platinum wire; thus the magnitude of the temperature-dependent signal will depend on the length of wire

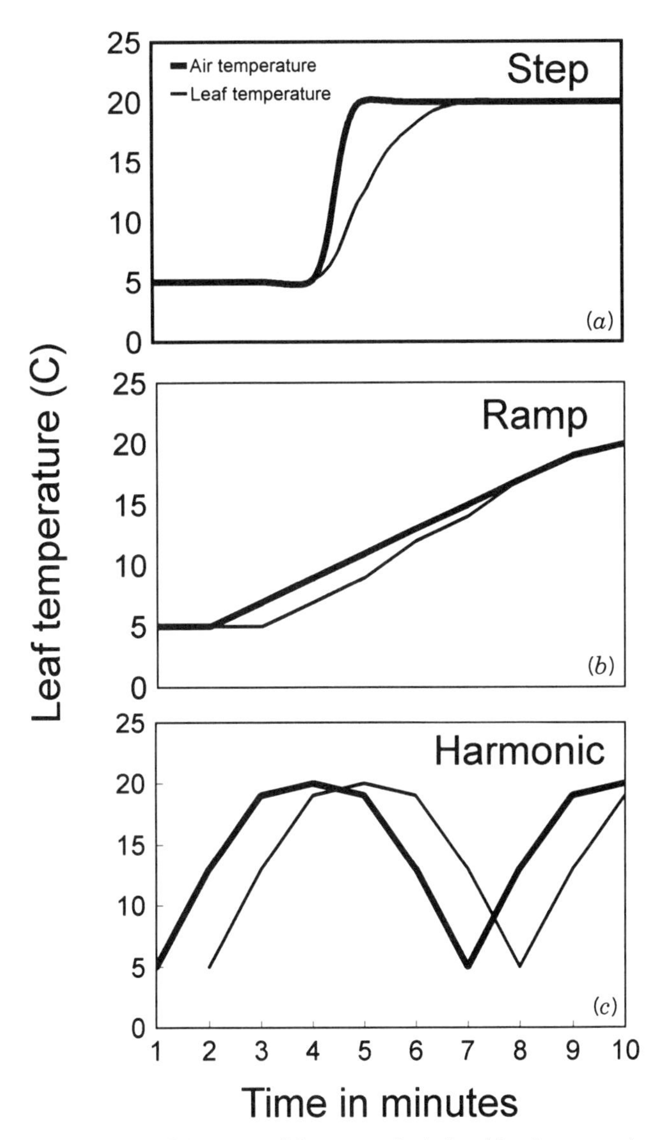

Figure 11.3 Time dynamics of three possible types of relationships between air and tissue temperatures. (Revised and redrawn from Jones 1992.)

(which limits the distance between the sensor and the remote detection device). As a result, resistance thermometers are usually located fairly close to their detection device. The high expense, small distance between sensor and detector, large size, and relatively slow response make resistance thermometers less attractive to scientists as a tool for measuring plant tissue temperature dynamics.

Thermistors are very similar to resistance thermometers (in that resistance is measured), but they utilize a semiconductor (metallic oxide) rather than a pure metal. The size and shape of the semiconductor can be variable. There are small bead thermistors or large flat disks. The rate of response is dependent on the size of the thermistor; thus small bead thermistors are favored for determining tissue temperature because of their fast response time. Bead thermistors are incorporated into many instruments and can be used as remote sensors in a data acquisition system. Thermistors are not the most abundantly used device for measuring tissue temperature because they are relatively expensive and fragile compared with thermocouples.

Thermocouples are different from thermistors and resistance thermometers because a thermocouple signal depends on the electrical properties of two dissimilar metals. Two wires, each made of a different metal, are joined at their ends (both ends). When these two junctions are at different temperatures a current will flow (Figure 11.4A) through the circuit. The magnitude of the electrical force gradient (voltage) in the circuit and the direction of electron flow are directly proportional to the difference between the temperatures of the two junctions. The term *thermocouple* comes from the fact that two metals are coupled together and the electrical result is dependent on temperature. Different metal junctions will produce a different relationship between junction temperature difference and voltage. The coefficient that defines the voltage gradient produced per unit change in temperature for any pair of metals is called the *Seebeck coefficient* (after Johann Seebeck, the discoverer of the thermocouple circuit). The most common pair of metals used in determining plant tissue temperatures is copper–constantan, which has a Seebeck constant of 38.7 $\mu V\ C^{-1}$ (when one junction is maintained at 0°C). Many other possible pairs of metals can be used, each of which has a different Seebeck coefficient and a different range of temperature sensitivities.

Thermocouples can easily be used to determine the absolute temperature of a specific tissue when one of the two junctions is maintained at 0°C (the reference junction) and the other junction is embedded or attached in the tissue (Figure 11.4B). The reference junction can be established by keeping it in a slurry of distilled water ice (usually in a Thermos) or by utilizing an electronic reference (the most popular technique). A microvoltmeter can then be attached to the positive (copper) and negative (constantan) leads to measure the voltage signal from the thermocouple circuit. Thermocouples are very popular for measuring tissue temperature because they are inexpensive, can be constructed easily in the lab or field, and the signal (temperature-dependent voltage) is not affected by extension of the wires to remote locations.

A thermocouple junction is easily constructed by twisting the two different wires together and soldering (soldering has no effect of the Seebeck coefficient) or by butt-welding the two wires. If very thin gauge wire is used, a small junction can be inserted into leaf tissues or appressed to leaf surfaces (Figure 11.5A). In addition, the constantan wire can be extended past the junction (Figure 11.5B) and used to wrap a thermocouple around a needlelike leaf without influencing the sensing junction (the first twist of the junction wires). Furthermore, several different junctions can be connected in parallel to one pair of wires to get an average temperature of those junctions (Figure 11.5C). If the thermocouples are closely linked together in series, this will produce a larger signal output and is called a *thermopile*. Thermopiles are often employed in radiation sensors (see Chapter 10).

Thermocouple wire is available in various sizes. Thin wire, 36 gauge, is often used for leaf or tissue temperatures, while heavier-gauge wire (24) is often used for air and soil temperature or the extension wire from leaf thermocouples. Wire gauge selection is based on two important criteria. The first is the mass of tissue being measured. The wire thickness must be less than the thickness of the tissue. A thermocouple wire gauge of 24 is too thick for leaf temperature because this gauge is thicker than most leaves. The second cri-

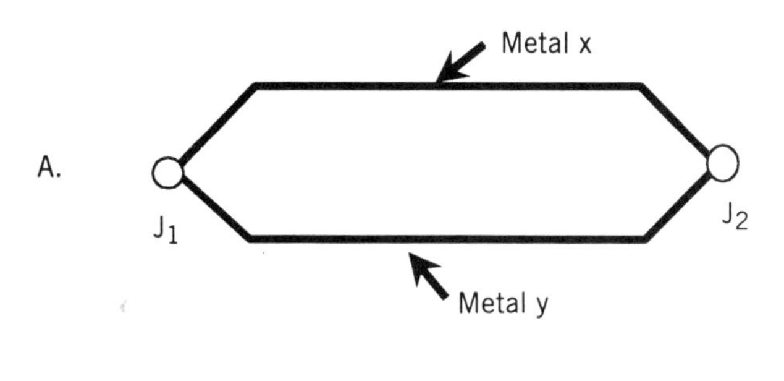

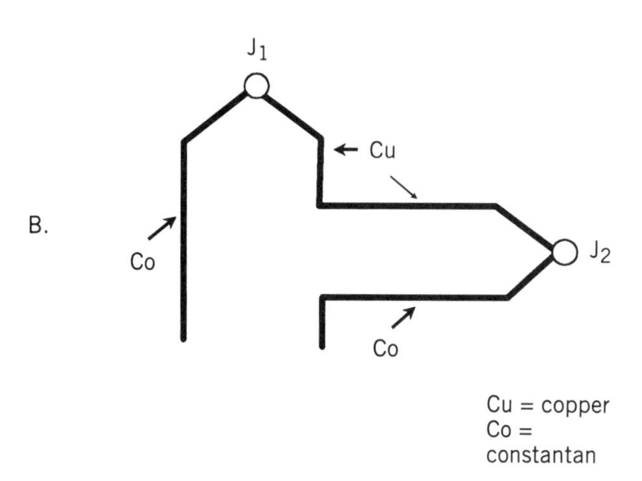

Figure 11.4 (A) Basic circuit for a thermocouple based on electron flow through differentially heated junctions (j1, j2) of two different metal wires. (B) Thermocouple circuit based on copper (Cu) and constantan (Co) thermocouple wire with a tissue sensing junction (j1) and a "cold" reference junction (j2).

terion is the speed of response for thermocouples (the time constant), which is dependent on wire gauge. Higher-gauge wires (thinner) have shorter time constants than that for thicker wires (24-gauge copper/constantan wire has a time constant of 0.9 s, while that for 16-gauge wire is 3.1 s). Short time constants are valuable when the timing of temperature variation is significant, while longer time constants are useful when average temperature is required. The gauge of a thermocouple can change from the sensing junction to the microvoltmeter. For this reason, leaf thermocouples are usually 10-cm-long 36-gauge wire connected to 24-gauge wire which extends to the data logger. Thermocouple connectors must be used to keep the integrity of the copper and constantan wire from the sensor to the voltmeter.

There are some aspects of using thermocouples that should be considered before implementation. One of the most critical considerations when using thermocouples to determine leaf temperature is how to attach the thermocouple to the tissue. In some studies the sensing junction is inserted into the lower surface of the leaf. This technique is used for

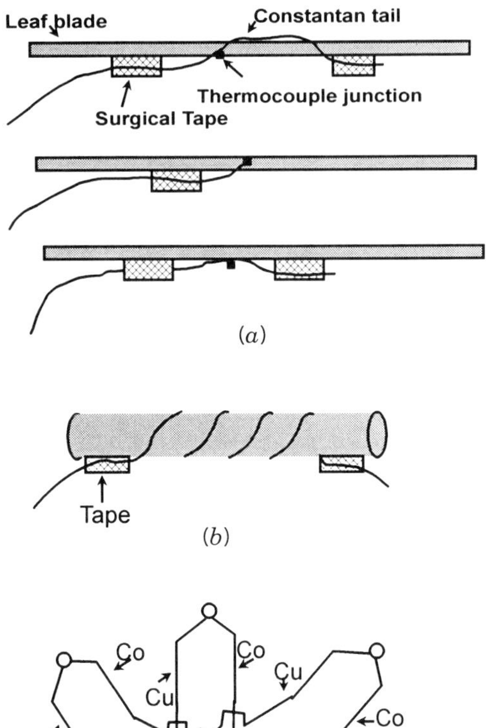

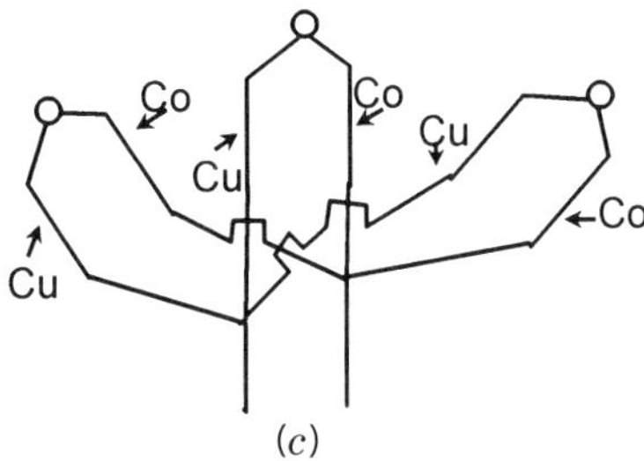

Figure 11.5 (A) Various ways that thermocouples can be attached to planar leaf surfaces. Note that affixed or inserted thermocouples have the sensing junction on the lower leaf surface. (B) Thermocouple design for use determining the temperature of round leaves or stems. (C) Thermocouples linked in parallel give an average signal of all the sensing junctions.

strong leaves but cannot be used for thin leaves such as those of soybean. Inserted leaf thermocouples may penetrate through the leaf and thus may be partly exposed to radiative heating. Therefore, it is important to let only the constantan wire penetrate the leaf surface and keep the copper constantan connection below the leaf surface. In other applications the sensing junction is affixed to the lower leaf surface with surgical tape, or the sensing junction can be attached to a plastic paper clip and clipped to the underside of leaves. Adhering thermocouples to leaves instead of inserting them may be important when determining leaf freezing point, because thermocouple insertion may create ice nucleation sites and disrupt an accurate measurement of leaf supercooling. A second important factor that can cause errors when using thermocouples is radiative heating. If direct light is hitting a sensing junction, the thermocouple will heat and not accurately measure the tissue surface or air temperature. The impact of radiation on thermocouple overheating is dependent on wire gauge. Lower-gauge wires are more easily heated by radiation. All precaution must be taken to keep thermocouple junctions out of direct light. In particular, air-temperature sensors (24 gauge) are highly susceptible to radiative heating.

Plant tissue temperature can also be measured remotely by the use of infrared thermometers. Earlier we stated that all objects above 0^0 K emit long-wave radiation. The amount of long-wave radiation emitted depends on the surface temperature of the object; thus, infrared emissions can be used to calculate the surface temperature of plant organs. The intensity of infrared radiation emission is converted back to surface temperature by using the Stephen–Boltzmann law [see equation (11.2)]. Infrared thermometers are gun-like devices that are pointed at a surface to detect surface temperature. The advantage of this system is that the leaf or other plant surface is not touched while temperature is measured. The disadvantages of this system is that the accuracy is limited to 0.5°C, small surfaces (individual leaves of a few centimeters in area) are difficult to measure, and small changes in emissivity of the surface cause large errors in predicted temperature. Therefore, infrared radiation thermometers are used in some experiments where the tissue is relatively large, emissivity is constant over time, and a small number of replicate measurements are needed.

III. DETRIMENTAL EFFECTS OF HIGH TEMPERATURE ON CELL METABOLISM

Before we discuss the potential detrimental impact of temperature on plant metabolism, it is instructive to review the mechanisms by which temperature affects metabolic rates. An overview of metabolism would show that enzymatic reactions and biochemical pathways are relatively sensitive to temperature, diffusion processes are moderately sensitive, while photochemical reactions are relatively insensitive to temperature changes. The impact of temperature on enzymatic reactions includes the effects on energy levels of substrates and thermal stability of enzymes. In general, reactions that have high minimum energy requirements have a strong thermal sensitivity.

Many metabolic reactions are regulated by the presence of an activation energy. Metabolic reactions can be regulated by physical processes such as temperature, because the average kinetic energy level of substrates is less than that of the activation energy (Figure 11.6). The number of substrate molecules above the activation energy can be described by the Boltzmann energy distribution:

$$n(E_a) = n^{-E_a/\Re T} \tag{11.9}$$

where n is the total number of molecules, E_a the activation energy, $\Re$ the gas constant, and T the absolute temperature of the system.

The rate of an enzymatic reaction is proportional to the number of molecules above the activation energy. Therefore, a rate constant for a reaction with a specific activation energy can be developed:

$$k = A^{-E_a/\Re T} \tag{11.10}$$

where k is the rate constant and A is an approximately constant function that depends on the type of metabolic process.

When the logarithms of this rate constant equation [equation (11.10)] are taken, the resulting equation is the *Arrhenius equation:*

$$\ln k = \ln A - \frac{E_a}{\Re T} \tag{11.11}$$

It can be deduced from the Arrhenius equation that the natural logarithm of the rate constant is proportional to $1/T$ with a slope of $-E_a/\Re$. Based on this relationship, a very use-

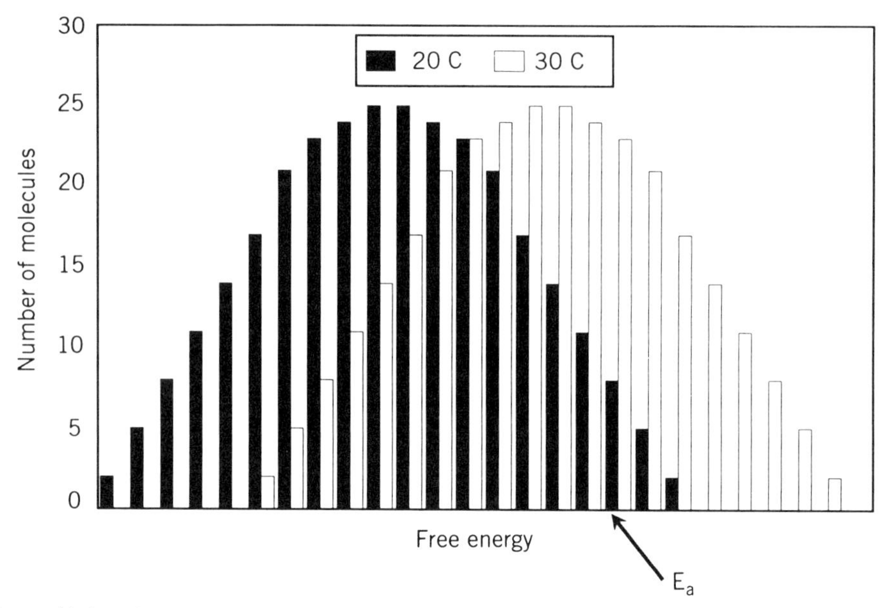

Figure 11.6 Diagram that represents a histogram of energy levels among a population of substrate molecules. E_a refers to the activation energy, that substrate energy level required for the reaction to occur. Notice that an increase in substrate temperature changes the distribution of energy levels of substrate molecules and increases the number of molecules with energy above the activation energy.

ful concept, the Q_{10}, has been developed for studying the impact of temperature on metabolic processes. The Q_{10} refers to a comparison of reaction rates between one temperature and a temperature 10°C higher. This is an arbitrary coefficient, but it is useful when comparing the temperature sensitivity of a metabolic process. The Q_{10} should be used with care because it is applicable only when comparing temperature effects in the logarithmically increasing phase. If other factors are influencing the reaction besides temperature, the Q_{10} will not be useful as a comparative index.

$$Q_{10} = \frac{A^{-E_a/\Re(T+10)}}{A^{-E_a/\Re T}} \quad \text{and} \quad Q_{10} = \exp\left[\frac{10E_a}{\Re T(T+10)}\right] \tag{11.12}$$

Based on equation (11.12), it is possible to calculate that if a reaction has an activation energy of approximately 50 kJ mol^{-1}, the Q_{10} will be approximately 2 (e.g., $2 = \exp[(10 \times 50{,}000)/(8.3 \times 293 \times 303)]$. Metabolic reactions with higher activation energies will have larger Q_{10}. The Q_{10} concept is often applied to metabolic pathways in addition to individual reactions. In this case the concept is used slightly inaccurately because many metabolic reactions are involved in a pathway with a diversity of different activation energies. The Q_{10} of a metabolic pathway will be dependent on one or a few reactions with large activation energies, or on particular reaction pathways in which other factors influencing substrate energy level are important. Respiration, for example, has a Q_{10} value close to 2. However, respiration is an extensive pathway and the Q_{10} value varies among developmental stages, organs, and times (ranges of Q_{10} for respiration from 1.5 to 2.5 are not uncommon).

Recall that a physiological process can be evaluated by the Q_{10} only in the logarithmic phase of a reaction response to temperature. However, most physiological process have a definite thermal optimum. The optimum temperature refers to that in which the process reaches its maximum potential. Above or below this optimum the reaction rate will decrease. At temperatures below the optimum the reaction rate decreases because fewer substrates have energy greater than the activation energy. The reaction stops increasing linearly somewhat below the optimum because other factors of the reaction become limiting, or a balance has been reached between the thermal behaviors of the two opposing reactions involved with this process. At temperatures above the optimum, inhibition occurs often because of thermal inhibition of enzyme function, or changes in membrane fluidity (if the process has enzymes that are associated with membranes). Thus any consideration of the detrimental impact of high temperature on physiological processes must concentrate on enzyme and membrane thermal stability.

A. Effects on Photosynthesis

As is the case with many elaborate metabolic pathways, tissue temperature has complex impacts on photosynthesis. Photochemical processes such as light harvesting and electron transport are relatively immune from direct effects of temperature. However, photochemical processes are not immune from thermal-induced feedback from the biochemical processes of photosynthesis or thermally induced changes in thylakoid membrane fluid. In fact, the overall pattern of net photosynthetic response to temperature is reminiscent of a temperature response curve for an individual enzyme (Figure 11.7). Net photosynthesis

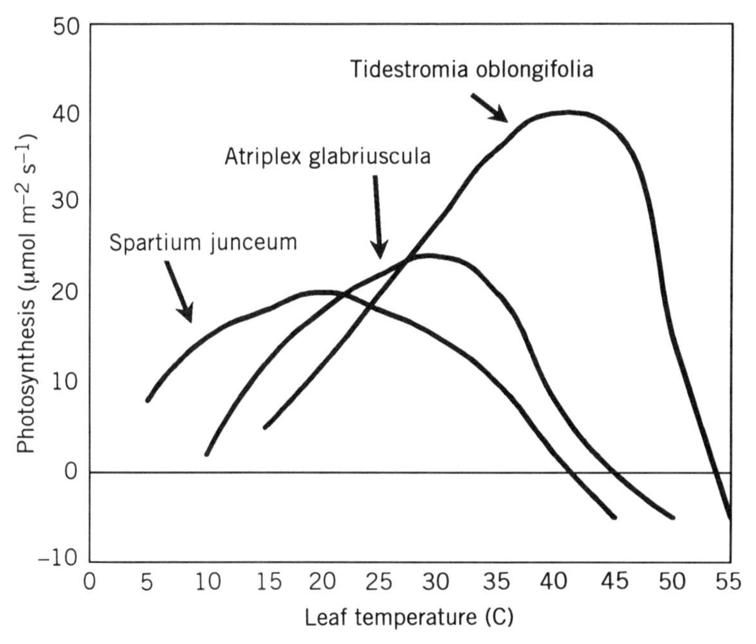

Figure 11.7 Representative temperature response curves of photosynthesis for several species. (Data gathered from Nilsen et al. 1993 and Osmond et al. 1980.)

increases with temperature increase, approximately linearly, at low temperatures. An optimum temperature is reached which can have a variable breadth, following which a strong thermal inhibition occurs. The thermal response of photosynthesis at low temperature is due to the kinetics of specific rate-limiting enzymes. In particular, rubisco and fructose bis-phosphate and phosphatase are very important for determining net photosynthetic rate at low temperature. Although photosynthetic regulation at low temperature is an important concept, the focus of this chapter is the consequences of high tissue temperature to photosynthetic processes.

The optimum temperature for photosynthesis is variable among species and among habitats or seasons for one species. Temperate plant species commonly have optimum temperatures for photosynthesis between 20 and 30°C, while that for desert species may be as high as 45°C (Björkman et al. 1975, Mooney et al. 1978). However, not all desert plants have high thermal optima for photosynthesis at all times of the year (Mooney et al. 1978, Nilsen and Sharifi 1994). Acclimation of the temperature optimum in response to seasonal changes in air temperature occurs in some species growing in environments where there are large ranges in temperature during the growing season. The largest acclimation of temperature optima occurs in desert evergreen species and is in the range of 10°C (Mooney et al. 1978). Also, at relatively high ambient temperature, photosynthesis is generally inhibited relative to a more moderate temperature regime. In other words, the photosynthetic rate at the optimum temperature for heat-acclimated tissue is lower than that of non-heat-adapted leaves. This point brings out the important interaction between affects of temperature on photosynthesis and respiration.

Although the direct effects of heat on respiration are discussed later in this chapter, it is prudent to consider the combined result of increasing temperature on photosynthesis and respiration at this point. As we discussed earlier in this chapter, respiration has a Q_{10} of approximately 2. Therefore, as temperature increases, respiration increases logarithmically. In most plant species, respiration is not inhibited until reaching temperatures of 40 to 45°C. In fact, respiration continues to increase after gross photosynthesis has reached its limitations by CO_2 diffusion and enzyme kinetics (Figure 11.8). High respiration and photorespiration rates at high temperature reduces net photosynthesis of C_3 relative to that at lower ambient temperature. Therefore, part of the reason that net photosynthesis decreases at high temperature is because of the disproportionate increase in respiration compared with gross photosynthesis. But this is not the primary reason why photosynthesis decreases precipitously at high temperatures.

One of the classic studies in environmental physiology of plants concerned the mechanisms that regulate high-temperature tolerance of photosynthesis in two desert species (Björkman et al. 1980a; see Berry and Raison 1981 for a review of this study): *Atriplex sabulosa* (a C_4 species from a cool coastal environment) and *Tidestromia oblongifolia* (a C_4 species from a desert environment). The temperature optimum for photosynthesis was 25°C for *Atriplex* and 45°C for *Tidestromia* (Figure 11.9A). The difference between the thermal tolerance of photosynthesis in these two species could have been due to a number of factors, including (1) respiration increasing quicker (in relation to gross photosynthesis) in one species than the other, (2) different thermal properties of key photosynthetic enzymes, (3) a different temperature for membrane damage induction; and/or (4) differential thermal stability of photochemical reactions.

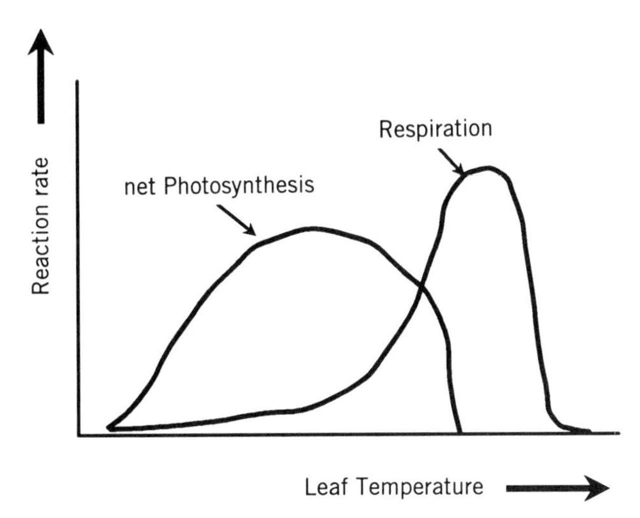

Figure 11.8 Relationship between influences of temperature on gross photosynthesis and respiration.

Thermal inhibition of photosynthesis for both *Atriplex sabulosa* and *Tidestromia oblongifolia* occurred far before thermal inhibition of respiration (Figure 11.9B). Therefore, some of the reduction in net photosynthesis of both species at high temperature was due to respiration, but this relationship did not explain the rapid inhibition of gross photosynthesis at high temperature. The thermal tolerance of critical enzymes in photosynthesis such as rubisco, ribulose 5-phosphate kinase, and NADP reductase were examined in the study. In some cases there was no difference in thermal tolerance of the enzyme between the two species (NADP reductase), and in other cases the difference in thermal tolerance of the enzyme did not correspond with the thermal tolerance of photosynthesis (rubisco). However, the thermal tolerance of some enzymes differed between species and matched the thermal tolerance of photosynthesis (ribulose 5-phosphate kinase). Therefore, part of the explanation for differential thermal tolerance of photosynthesis between these two species was due to the thermal tolerance patterns of particular enzymes. There was no evidence that thermal tolerance of the plasma membrane was directly associated with temperature inhibition of photosynthesis. In fact, cell membrane leakage occurred at temperatures at least 10°C higher than photosynthetic inhibition.

The factor most associated with thermal tolerance of photosynthesis in both species was photosystem II activity. Quantum yield of photosystem II–driven electron transport decreased at a temperature that corresponded exactly with photosynthetic inhibition of both species. In contrast, PSI electron transport was unaffected by temperature. What factors of PSII electron transport would be sensitive to temperature? There are two main possibilities. First, the fluidity of thylakoid membranes could have been affected at high temperature. It is likely that PSII core complexes or PSII light-harvesting complexes can be dislodged from thylakoid membranes when membrane fluidity changes. Second, PSII integrity is dependent on electron flow dynamics. Therefore, if high temperature disrupts

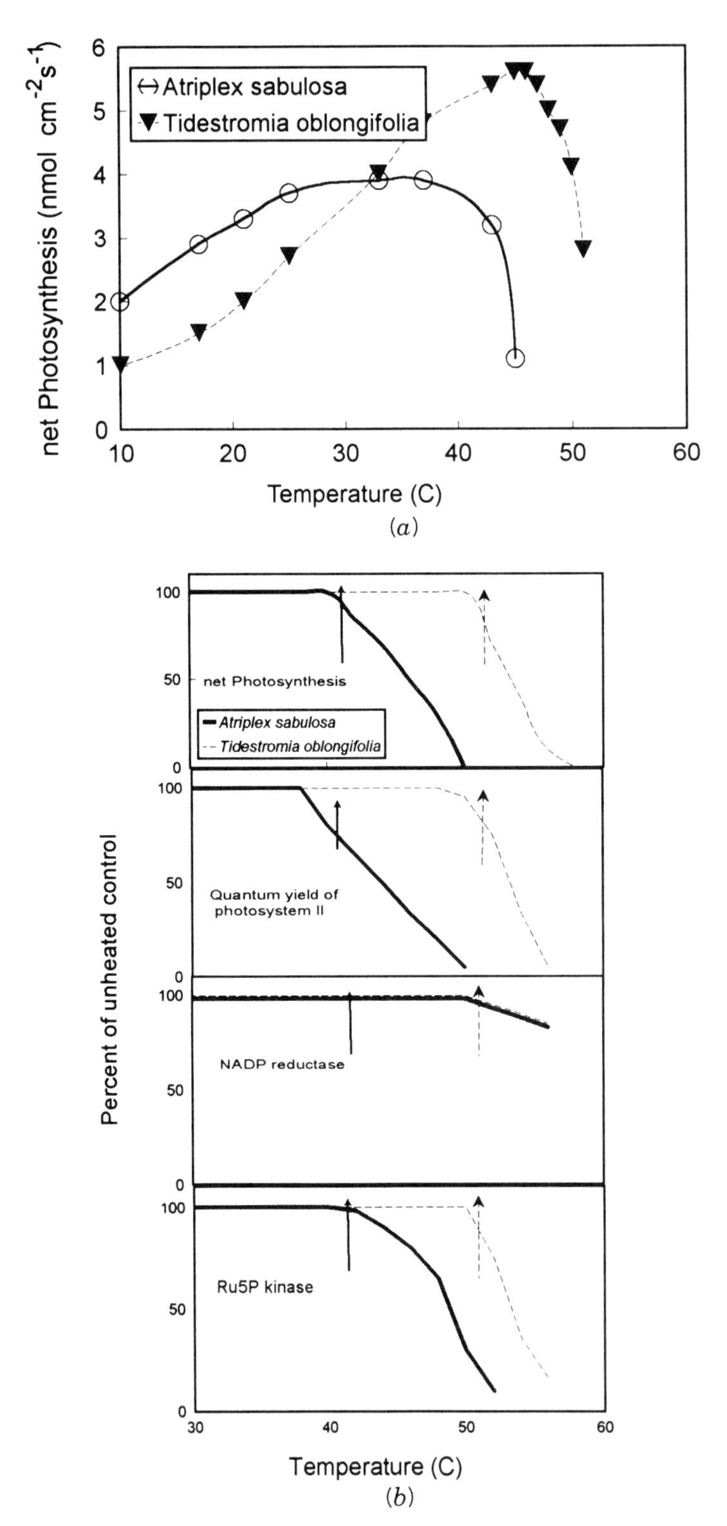

Figure 11.9 (A) Temperature response curve for photosynthesis of *Atriplex sabulosa* (a coastal temperate species) and *Tidestromia oblongifolia* (a desert species). (B) Temperature dependence of various processes in the leaves of *Atriplex sabulosa* (a coastal temperate species) and *Tidestromia oblongifolia* (a desert species). (From Björkman et al. 1978.)

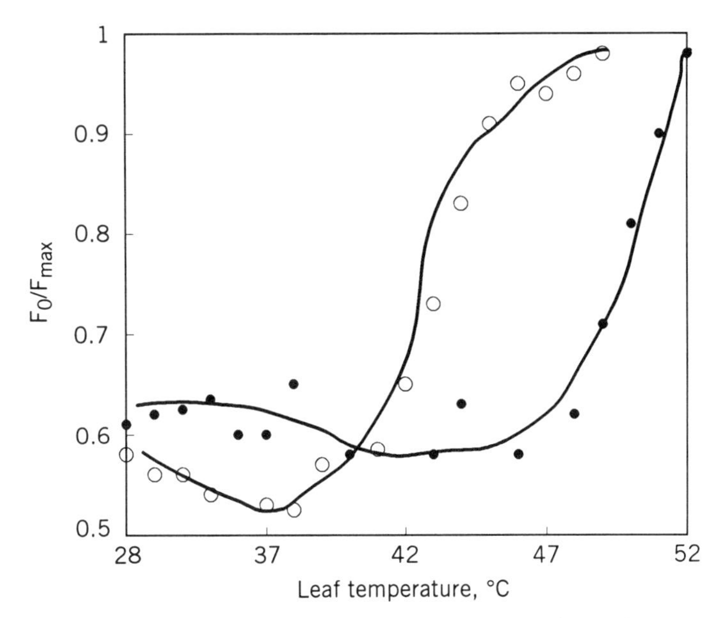

Figure 11.10 Changes in the ratio of basal fluorescence (F_0 to maximum fluorescence (F_{max}) in *Atriplex lentiformis* grown at high temperatures (45/30°C) or low temperatures (20/15°C). (From Pearcy et al. 1977.)

the metabolic process that delivers electrons to PSII (the Hill reaction), or the process that accepts electrons from PSII core complex (plastoquinone, or O_2), PSII is likely to dislodge from the thylakoid membrane in a dysfunctional state. The inhibition of PSII electron transport by high temperature is further supported by a sharp increase in the basal level of chlorophyll fluorescence (Figure 11.10) that corresponds to photosynthetic inhibition by high temperature (Pearcy et al. 1977, Bilger et al. 1984). Recent studies have shown that Hill reaction protein complexes may be stabilized by low-molecular-weight heat-shock proteins under high-heat conditions (Coleman and McConnaughay 1994). Differential expression of these heat-shock proteins may result in differential thermal tolerance of PSII. The interaction between HSPs and photosynthesis is discussed further in Chapter 14.

B. Effects on Respiration

Responses of respiration to heat among different species is relatively uniform, as exemplified by the fact that the Q_{10} for respiration of most species is relatively constant. However, the dynamics of high-temperature inhibition of respiration is different among species. In the case of *Atriplex sabulosa* and *Tidestromia oblongifolia,* there was a 5°C difference in the thermal inhibition point of respiration between the two species (Figure 11.9B).

There is some evidence that the cytochrome electron flow path of respiration is inhibited by lower temperatures than that which inhibits the alternative electron flow path. In *Glycine max* (soybean), cytochrome-mediated electron flow decreased by 50% after 4 days at 40°C, while the alternative pathway increased to 100% of its capacity (Leopold and Musgrave 1980). It is possible that cytochrome-mediated electron transport is sensitive to heat stress because this electron pathway is dependent on oxidative phosphorylation. If a heat-stressed tissue has a lower demand for ATP, this will inhibit the cytochrome electron transport pathway because a change in adenylate ratio (ATP/ADP) inhibits oxidative electron transport and stimulates the alternative electron flow pathway.

The impact of high temperature on respiration is also associated with membrane function. This relationship can be elucidated by comparing the temperature required to cause membrane leakiness (loss of membrane integrity) and the temperature required to initiate inhibition of respiration. Once again, in the case of *Atriplex sabulosa* and *Tidestromia oblongifolia,* there is an exact match between the thermal inhibition temperature for respiration and the temperature that induces membrane leakiness in both species (Figure 11.9B). Therefore, high-temperature inhibition of respiration is highly likely to be related to membrane dysfunction.

C. Effects on Membrane Function

It is clear that thermal inhibition of photosynthesis and respiration is in part regulated by the functional integrity of membranes. This is not surprising, because the main functions of chloroplasts and mitochondria are dependent on membranes. If high heat affects these important physiological process by its effect on membranes, what are the mechanisms by which heat destroys the appropriate function of membranes? High temperature influences membrane fluidity, induces membrane leakiness, and affects the relationship between integral proteins and the membrane. As discussed in Chapter 3, all membranes have a specific temperature at which they switch from the gel state to the liquid-crystalline state (T_c). The T_c value is related to the function of membranes but does not relate to the extreme conditions that are associated with complete membrane dysfunction.

Most membrane scientists agree that there are two extremes in the organization states of biological membranes. The *hyper-rigid state* occurs at very low temperatures and occurs at the interface of the gel-solid membrane state. The *hyperfluid limit* is pertinent to high-temperature stress because it occurs at high temperature when the fluidity of the liquid-crystalline state becomes too great. The mechanics of this particular state transition at high temperature are not well understood, but some experiments indicate that the lipid bilayer becomes thinner, becomes leaky, and the thermal stability of integral proteins is lost. For a detailed discussion of these relationships and other interactions between membranes and temperature, see Chapter 3.

D. Effects on Enzyme Function and Protein Synthesis

One of the most dramatic effects of high temperature on cell metabolism is the sudden precipitous drop in enzyme function at high temperature. This decrease in function is due to several possible mechanisms. First the tertiary shape of proteins, which regulates their function, is affected drastically by high temperature. *Denaturation,* the loss of tertiary structure at high temperature, is a consequence of breaking the weak molecular attractions (proton bonds, hydrophobic attractions, ionic interactions, etc.) that keep a polypeptide chain folded appropriately. Upon losing their tertiary shape, proteins are dysfunctional. Each enzyme system or membrane protein has a unique sensitivity to denaturation based on the particular array of molecular bonds that constitutes its tertiary structure. Those proteins that have simple tertiary structure maintained primarily by disulfide bonds are relatively immune to denaturation. In contrast, proteins that have complex structure, based on a diversity of relatively weak molecular bonds (such as most enzymes), will be especially sensitive to heat-induced denaturation. Thermal stability of enzymes can vary to some degree among species that grow in different environments. Although differential thermal stability of isozymes has been often found, this is usually not the most important factor regulating thermal tolerance of a species or metabolic process.

The metabolic rate of a process can be affected by a change in proteins synthesis or degradation, as well as a change in enzyme-specific activity. One of the most dramatic effects of rapid-onset high temperature (heat shock) is a complete loss of low-temperature proteins, which is replaced by a lower rate of protein synthesis focused on a small number of heat-induced proteins. In addition, ubiquitone-associated protein degradation increases at elevated temperatures. Soluble protein pools decrease rapidly in response to heat shock (except for the heat-shock proteins) due to decreased protein synthesis and increased protein degradation. Therefore, metabolic inhibition by high temperature is due to a multifaceted impact on enzyme concentration, gene transcription, mRNA translation, enzyme degredation, and enzyme specific activity.

IV. MECHANISMS OF MODERATING LEAF TEMPERATURE

A. Phenology

One of the main mechanisms plants can utilize to cope with high tissue temperatures is by seasonal changes in the abundance of sensitive organs. Leaves and inflorescences are the organs most sensitive to high temperature. In regions with seasonally high air temperature, many species utilize leaf abscission to avoid periods of high heat. This summer deciduousness is most prominent in ecosystems where high air temperature coincides with high radiation and low temperature. The combination of high air temperature, high radiation, and low water availability is particularly suitable for greatly elevated tissue temperature. Leaves will absorb large amounts of radiation and be unable to utilize latent heat exchange as a mechanism for energy dissipation because stomata are relatively closed. As a consequence, leaf temperature is likely to rise to a high value (well above air temperature) because of the synergistic impact of high air temperature, excessive radiation absorption, and low leaf transpiration. Many species in chaparral, desert, and seasonal dry tropics lose their leaves at the onset of the dry and hot summer (Nilsen and Muller 1981).

It is also common for summer deciduous species of the Mediterranean and desert regions to have photosynthetic stems (Nilsen 1995). Photosynthetic stems have a number of advantages over leaves during high light, high air temperature, and low water conditions. First, stem photosynthesis has a higher water use efficiency than leaf photosynthesis. Stems also act as a water conservation mechanism because carbon gain can continue during the summer when leaf photosynthesis is precluded by environmental conditions. Second, stems have a larger mass, so their temperature is buffered compared to thin leaves. This character, and the vertical orientation of photosynthetic stems, keeps stem temperature equal to or below air temperature during the summer (Nilsen et al. 1996). The impact of vertical orientation on tissue temperature is discussed further later in the chapter.

B. Morphological Changes

Many plants also avoid seasons of high heat without forsaking their leaf population. This is accomplished by changing the morphological attributes of their leaf population just before the initiation of the hot and dry season. Plants with this capacity are said to have dimorphic leaf populations. Species with dimorphic leaf populations are found in Mediterranean, desert, and tropical habitats (Margaris 1975, Mooney et al. 1977, Nilsen et al. 1986). To moderate temperature in the secondary leaf population, leaves are often smaller and narrower (reduce radiation load), more vertically oriented, and more reflective than the cool season leaves (Figure 11.11). In some instances, both large and small leaves are present on the branch during the cool season. At the onset of the hot season, the large leaves of *Lotus scoparius* and *Cytisus scoparius* abscise, leaving only small vertically oriented leaves during the summer (Nilsen and Muller 1982, Nilsen et al. 1993).

In addition to reducing radiation load under hot conditions by retaining only small leaves, many plant species also decrease leaf absorptance in response to seasonal or environmental radiation load. The best documented example of changing leaf absorptance in response to season or habitat energy balance is that for *Encelia* species of the American deserts (Ehleringer and Björkman 1978). Various members of the genus *Encelia* inhabit desert and Mediterranean habitats in southwestern North America and northwestern Chile. Those species that reside in desert habitats have leaves with a lower absorptance than species from the moister Mediterranean habitat. Differences in absorptance among species is due to the presence of dead reflective trichomes on the upper leaf surface. Desert species with leaf trichomes are able to maintain lower leaf temperatures under conditions of high radiation and low water availability compared to Mediterranean species under the same climatic conditions. This relationship is the same along an aridity gradient in California and Chile (Ehleringer et al. 1981). In fact, a mutant form of *Encelia farinosa* was located in Death Valley, California that did not have extensive leaf trichomes as does the wild type of this desert species (Ehleringer 1983). The mutant *E. farinosa* had much higher leaf absorptance than that of leaves from the wild-type *E. farinosa* (Figure 11.12A). Although the mutant form had high leaf absorptance, it could maintain leaf temperatures similar to those of the wild type only by having alternative methods of reducing leaf energy balance, such as high transpiration rates and steep leaf angles.

Seasonal changes in radiation intensity and ambient temperature were also correlated with changes in leaf absorptance of the desert species of *Encelia*. During the hot summer, leaves that had high absorptance (which developed in the relatively moist and cool spring) changed to low absorptance in the summer as leaf trichomes matured and died on leaf surfaces (Figure 11.12B). In general, changes in trichome reflectance are an important mechanism for reducing leaf absorptance in a diversity of species. In cacti, the presence of spines reduces absorptance properties of the succulent stem. In the California chaparral, one deciduous species (*Salvia mellifera*) is able to maintain a small population of relatively large leaves during the hot and dry summer. This species can maintain these leaves during the summer conditions (high radiation, high temperature, and low water availability) because the leaves rotate to expose their under surface, which is covered with dead reflective trichomes. If a summer storm occurs, the leaves of *Salvia mellifera* turn over to expose their upper surface, which increases absorptance for photosynthesis (Gill and Mahall 1986).

C. Leaf Orientation

The orientation of leaves with respect to the solar angle and the solar azimuth is also a critical factor for regulating leaf temperature in hot environments. The more oblique the solar angle is to leaf lamina, the smaller the incident direct radiation on the leaf. In envi-

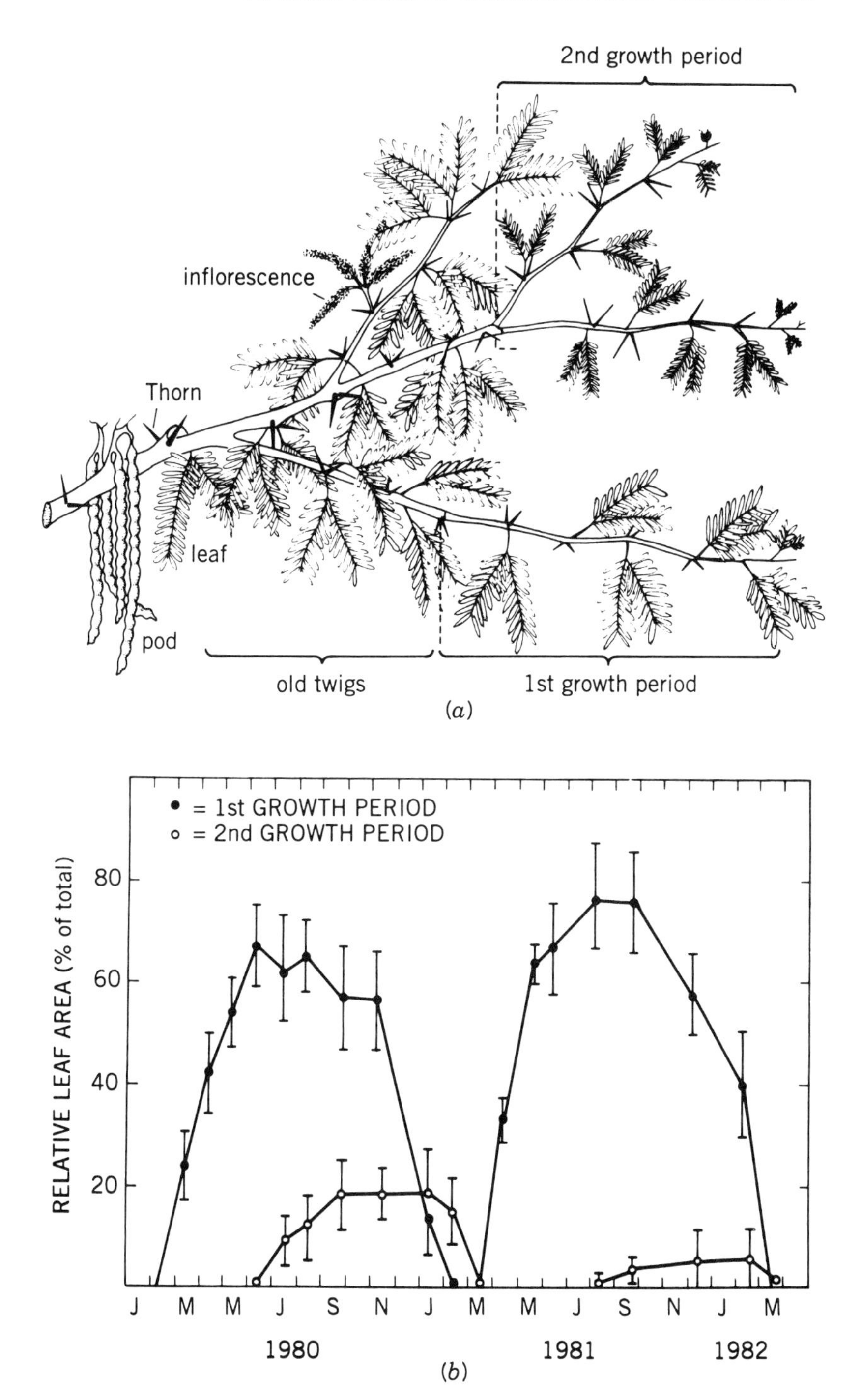

Figure 11.11 (A) Diagram of the branch of *Prosopis glandulosa* with cool season and hot season leaves. Notice the difference in leaf size between first- and second-growth period leaves. (B) Plot of the leaf area of cool and wet season leaves (first growth period) and leaves produced during the hot and dry summer (second growth period) on *Prosopis glandulosa* in the Sonoran desert of California. (Both plates contain material from Nilsen et al. 1986.)

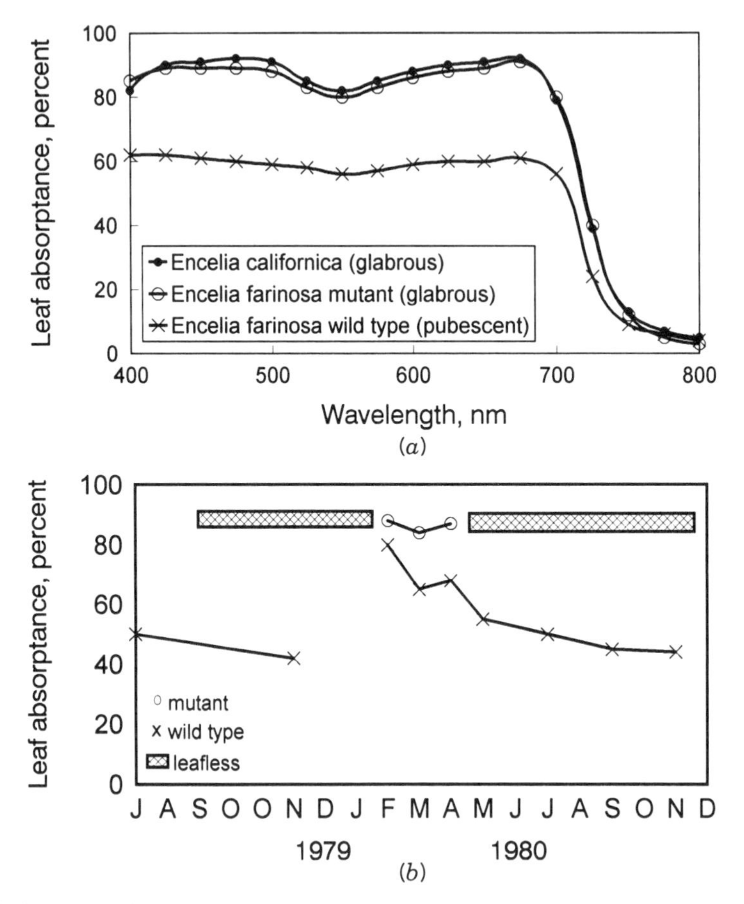

Figure 11.12 (A) Leaf absorptance spectra between 400 and 800 nm for leaves on *Encelia californica*, a glabrous mutant of *E. farinosa,* and a normal pubescent *E. farinosa.* (B) Seasonal time course of leaf absorptances to solar radiation in the 400 to 700 nm waveband for the mutant *E. farinosa* and the wild-type *E. farinosa.* (Used with permission from Ehleringer 1983.)

ronments that have most solar radiation in the form of diffuse radiation, leaf angle is not significant for maintaining leaf temperature. However, in environments where incident solar radiation is dominated by direct beam radiation, leaf angle and azimuth are important factors regulating leaf temperature. The angle of incidence between a leaf lamina and the solar direct beam radiation is calculated by the cosine of incidence:

$$\text{cosine of incidence} = \cos \alpha_l \sin \alpha_s + \sin \alpha_l \cos \alpha_s \cos(\beta_s - \beta_l) \qquad (11.13)$$

where α_l and α_s are the angles above the horizon of the solar direct beam (s) and the leaf lamina (l). The azimuths (compass direction corrected for magnetic deviation) of the sun and the leaf lamina are represented by β_s and β_l.

When the cosine of incidence is 1, the solar direct beam is perfectly perpendicular with the leaf lamina, and if the cosine of incidence is zero, the solar direct beam is perfectly parallel with the leaf lamina. Many evergreen species, such as *Simmondsia chinen-*

sis (jojoba), in desert habitats or other high-radiation sites have vertically oriented leaves. In fact, vertical orientation is an important mechanisms for moderating leaf temperature for different populations of one species among sites with different energy loads (Field et al. 1982).

The cosine of incidence for vertically oriented leaves (that are in a fixed position over time) is close to 1 in the morning and evening and close to zero at midday. Also, the diurnal cycle of cosine of incidence changes during the year because of the seasonal change in solar angle. Irrespective of the seasonal impact of solar angle on cosine of incidence, the maximum cosine of incidence on any daily cycle of the year, for a vertical leaf, will be in the morning and evening. This contrasts with a horizontal leaf, which will have its greatest cosine of incidence at the highest solar angle; near midday. Vertical orientation reduces the relative incident radiation during the midday hours (compared with horizontal leaves) when air temperature is highest and solar radiation is strongest (Figure 11.13). Thus, vertical leaf orientation minimizes leaf absorption of solar radiation during midday, when moderating leaf temperature may become critical.

Leaves on some early successional broadleaf species in grassland ecosystems are vertically oriented and held at specific azimuth directions. For example, the compass plant (*Silphium laciniatum*) has vertical leaves that are characteristically oriented in a north–south plane (Jurik et al. 1990). When leaves of the compass plant are manipulated in angle or azimuth, the carbon gain and water use efficiency decreases relative to the vertical north–south leaf orientation. Uniform leaf azimuth is not as important a mechanism for moderating leaf temperature as variation in leaf angle, because leaf populations that have a uniform azimuth are infrequent.

Leaf angle can change on a seasonal basis as the solar angle changes. It is not unusual in desert species for leaves to be horizontal in the relatively cool and moist winter and vertical during the hotter and drier summer. For example, the leaves of *Atriplex hymenolytra* change from horizontal to near vertical (70° inclination) between the winter and summer in Death Valley, California (Mooney et al. 1978). Similar changes in leaf angle can be found in Mediterranean shrubs, grassland forbs, and many other species of

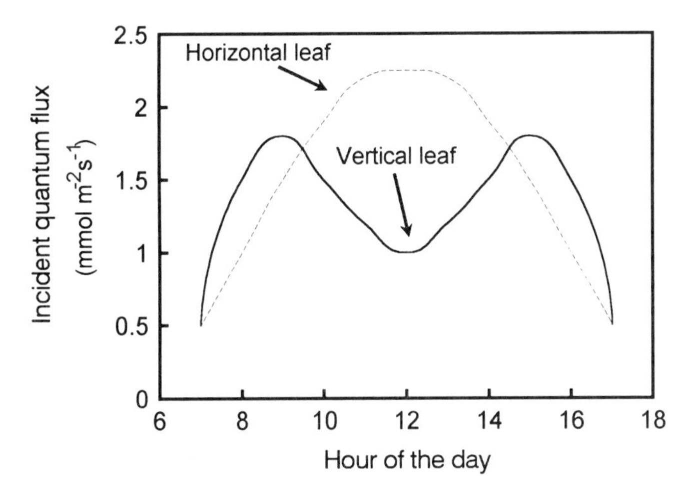

Figure 11.13 Representative plot of solar radiation incident on a horizontal and a vertical leaf for a summer day in the temperate zone of the northern hemisphere.

high-light environments. Transpiration is inhibited during the summer because of low water availability; thus leaf temperature is moderated by an increase in leaf angle and associated decrease in leaf absorptance.

Vertical orientation of the photosynthetic organ during hot and dry seasons of the year can also be accomplished by switching from leaf to stem photosynthesis. In desert and Mediterranean environments (as well as in some crops) species may have photosynthetic stems (reviewed by Nilsen 1995). Stems of these green twigged species are vertically oriented; thus the photosynthetic organ has a low cosine of incidence at midday during the summer months. Stem temperatures are rarely more than 1°C above or below air temperature (Nilsen and Sharifi 1994). The vertical orientation of stems is important for their appropriate function in high light and hot environments, because if the stems are held at a horizontal position, they suffer irradiance-induced photosynthetic damage (Ehleringer and Cooper 1992).

Diurnal changes in leaf angle may also be related to processes that regulate leaf temperature. Some types of diurnal changes in leaf orientation, such as "sleep movements" in beans and rapid leaf closure in *Mimosa*, have been recognized for a long time (Darwin 1881) but are not related to modifying radiation absorption by leaves. However, there are heliotropic leaf movements that relate directly to adjusting radiation load, moderating leaf temperature, and affecting photosynthesis and transpiration. Solar tracking leaves or diaheliotropic leaves move during the diurnal cycle to keep a perpendicular angle to the direct solar beam. Thus, the leaves on diaheliotropic plants maintain a constant cosine of incidence of 1, even though the solar angle is changing during the day (Figure 11.14). In contrast, paraheliotropic leaves maintain a cosine of incidence close to zero, minimizing diurnal radiation absorption. The diurnal pattern of radiation on a fixed horizontal surface is quite different from a dia- or paraheliotropic surface (Figure 11.15). Therefore, leaf movements during the day strongly affect radiation balance and leaf temperature.

Desert environments, with high temperature and radiation, should contain many species with heliotropic leaves if these movements are important for modifying leaf temperature. In fact, 75% of desert annual species in one survey demonstrated heliotropic leaf movements (Ehleringer and Forseth 1980). Heliotropic movements in desert species are affected by plant–water relations status. For example, the cosine of incidence for *Lupinus arizonicus* is approximately 0.5 when the leaf water potential is 1.0 MPa (Figure

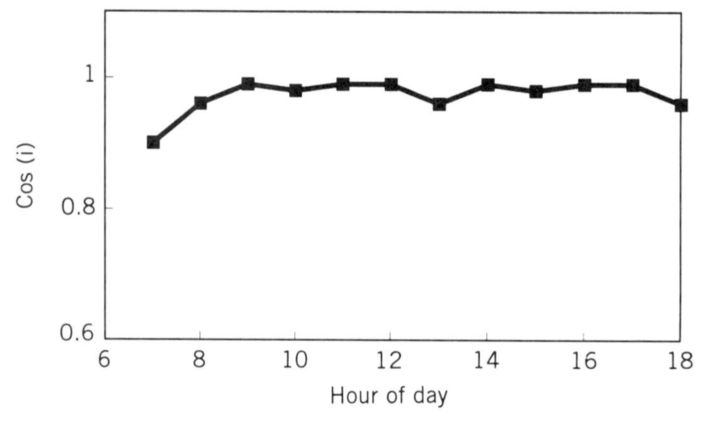

Figure 11.14 Diurnal of cosine of incidence for a leaf population of *Malvastrum rotundifolium*. (Data adapted from Ehleringer and Forseth 1980.)

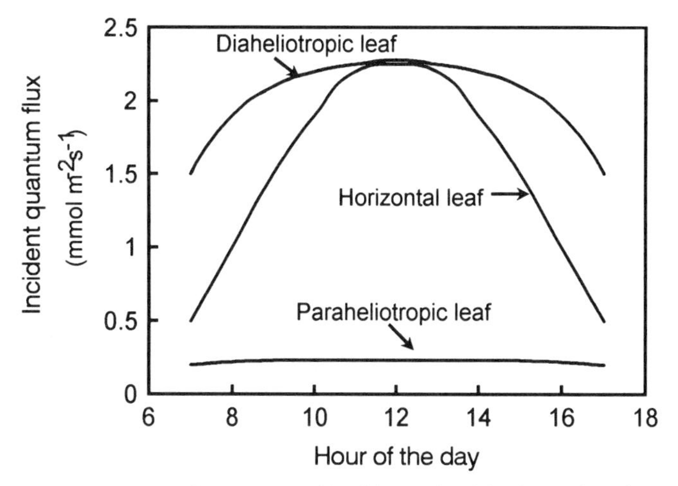

Figure 11.15 Radiation that strikes leaves with either a fixed horizontal surface, diaheliotropic movements, or paraheliotropic movement. (Redrawn and adapted from Ehleringer and Forseth 1980.)

11.16A). However, as leaf water potential decreases, the cosine of incidence also decreases (Forseth and Ehleringer 1982, 1983). Stomatal conductance decreases with decreasing water potential, reducing the amount of latent heat exchange from leaves of *L. arizonicus*. Therefore, paraheliotropism (low cosine of incidence) becomes more important for moderating leaf temperature at low water potential for *L. arizonicus*. In other species, such as the common bean, heliotropic leaves respond directly to air temperature as well as water potential (Figure 11.16B). At high air temperature, paraheliotropic movements in bean leaves increase in intensity.

Leaf movements may be combinations of nonheliotropic ("sleep movements") and heliotropic processes. For example, leaves of *Erythrina herbacea* have their tip pointed downward with no axial rotation at night. In the morning, leaflets rise continuing to a midrib orientation of 60°C just before midday. As the leaves are moving upward, the leaf rotates around the midrib by an average angle of 40°. These movements result in an initial increase in leaf solar radiance reception, followed by minimal reception at midday, and a second period of high solar reception in late afternoon (Herbert 1984). Another important point made by this article is that there is variation in leaflet orientation among leaves. This is particularly true on larger species in nondesert habitats.

Variation in leaf orientation within a canopy may be an important mechanism for maximizing solar radiation within a canopy. Outer leaves may be held at a vertical position relative to inner leaves in order to increase the radiation in the inner canopy. Phyllotaxi (location of petiole placement on the stem axis) may also be designed to maximize the amount of direct solar radiation received by a shoot complex. For example, phyllotaxi causes a minimization of overlap among leaves of different ages in *Rhododendron maximum, Acer rubrum,* and other species of deciduous forest understory. In addition, leaf angle becomes more pendent at lower positions of the branch in *Rhododendron maximum,* which also increases light interception by the whole shoot complex (E. T. Nilsen, personal observation). Therefore, leaf orientation may be serving an energy balance purpose by keeping leaf temperature moderate, and/or leaf orientation may be serving to maximize canopy carbon gain processes (Herbert 1991).

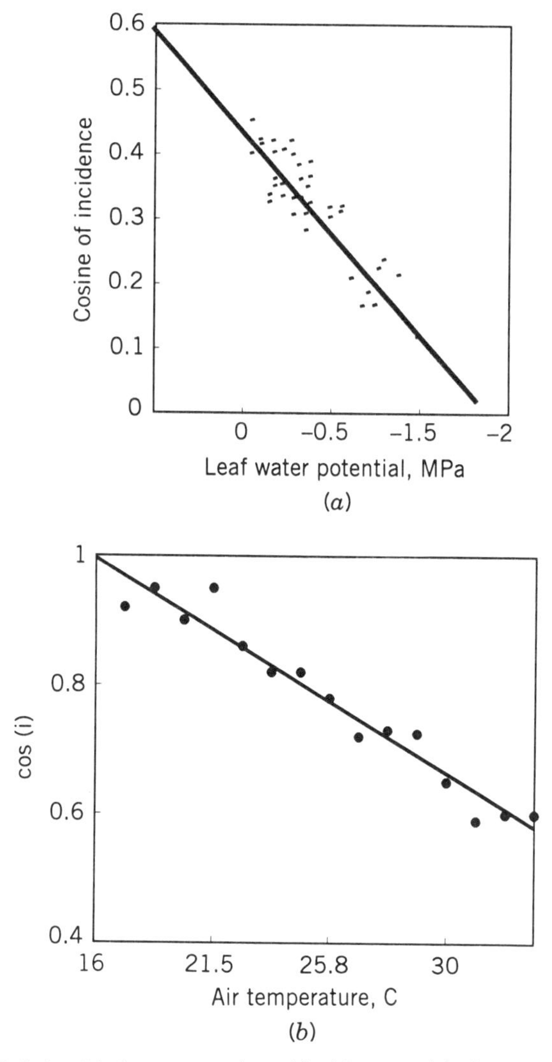

Figure 11.16 (A) Relationship between cosine of incidence and leaf water potential in *Lupinus arizonicus*. (Data adapted from Forseth and Ehleringer 1980.) (B) Relationship between temperature and the cosine of incidence for *Phaseolus vulgaris* cv. Blue Lake Bush. (From Fu and Ehleringer 1990.)

D. Transpiration

Latent heat exchange was discussed earlier as a mechanism of heat dissipation from leaves. This mechanism can be effective only in habitats with high water availability. Plants in desert or Mediterranean habitats, which are characterized by high heat and irradiance, usually cannot utilize transpiration as a mechanism for dissipating heat.

In many habitats, the season with the highest temperature of the year coincides with high irradiance and relatively low water availability. Water conservation becomes particularly important for evergreen species during these times. Thus, stomata of evergreen leaves must close in order to reduce water loss during the time of year when latent heat exchange has its greatest significance. Stomatal closure does not necessarily mean that la-

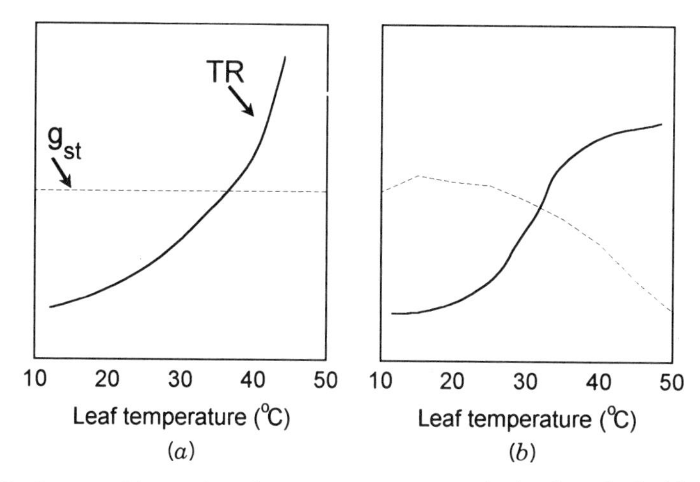

Figure 11.17 Impact of increasing air temperature on transpiration by a leaf with constant leaf conductance (A), or variable leaf conductance (B).

tent heat exchange is reduced because of the relationship between air temperature and vapor pressure. As air temperature increases, relative humidity of the air decreases, and the vapor pressure gradient (VPG) between leaf and air increases. This increase in VPG causes higher transpiration at any leaf conductance (Figure 11.17). However, as discussed in Chapter 8, there is often a direct relationship between VPG and leaf conductance as a water conservation mechanism. Therefore, a dilemma for plants in high-heat and low-water environments is to close stomata and conserve water and at the same time, compensate for high heat. This is probably a main reason why desert and Mediterranean species have elaborate morphological and behavioral techniques for modifying leaf temperature.

Some species in desert habitats have extensive water resources even in the hottest and driest periods of the year. Phreatophytes may be able to use latent heat exchange as a major leaf-cooling mechanisms. Evidence from *Prosopis glandulosa* indicates that latent heat exchange is not a major leaf cooling mechanism. Leaf temperature does not differ from air temperature, leaflets are relatively small, stomatal conductance is reduced during the summer, and stomata respond linearly to VPG (Nilsen et al. 1983). These observations suggest a pattern of association between air temperature, transpiration, and water relations. If stomata remain open during periods of high temperature and large VPG, transpiration will increase and a greater demand will be placed on the hydraulic system of the plant. It is highly likely that excessive transpiration during the hot and dry summer months would induce low water potentials in the stem, which may induce embolism and hydraulic collapse. Therefore, the hydraulic architecture of plants places a constraint on the potential use of transpiration as a cooling mechanism.

V. MECHANISMS OF TISSUE TEMPERATURE TOLERANCE

A. Variation in Tissue High-Temperature Tolerance

There are a diversity of heat tolerances among plants. The species that have been evaluated can be grouped as sensitive, facultatively tolerant, and heat tolerant. Sensitive species are those that are inhibited by relatively low temperatures and do not have the capacity to adjust their heat sensitivity in response to increasing ambient temperature. Sensitive species demonstrate heat injury at tissue temperatures between 30 and 40°C. This group includes most temperate, arctic, and alpine plants. Interestingly many pathogenic viruses and bacteria are also intolerant of temperatures above 45°C. Facultative, heat-tolerant species are those that can adjust cytoplasm heat tolerance in response to changing ambient thermal conditions. Commonly, these are species from high-light environments, such as early successional temperate species, desert species, and Mediterranean species. These species can acquire hardiness to heat extending up into the 50°C range. Some cacti can express heat tolerance to 60 or 65°C for limited times (30 min). A tissue temperature of 60 to 70°C for extended periods is an upper limit for plant species. Extreme heat tolerance is only an attribute of some prokaryotes and some plant seeds. For example, thermal vent bacteria and those bacteria residing in sulfur hot springs can withstand temperatures up to 90 to 110°C. Terrestrial plants cannot tolerate these temperature ranges unless they are in a seed state. For example, many fire-induced seeds can tolerate relatively high temperatures (120°C) for short periods and retain germinability. In fact, some postfire annual species require heat to melt outer coatings of wax in order to stimulate germination.

There are several ways of determining heat tolerance of tissues. The most general heat tolerance measure is thermostability of protoplasmic streaming. Protoplasmic streaming, or the circulation of protoplasm in plant cells, is based on the cellular cytoskeleton. When heat causes a breakdown in the cellular cytoskeleton, protoplasmic streaming is inhibited. Thus, observing protoplasm in cells on a temperature-controlled microscope stage is a basic measure of thermotolerance. Another basic measure of thermotolerance concerns the integrity of the plasma membrane. At high temperature the plasma membrane becomes superfluid and loses its differential permeability. In such a study, tissue sections are placed in distilled water and maintained at a range of temperatures. Those tubes held at a temperature above the tissue thermotolerance will increase in conductivity (due to cell contents leaking into the distilled water) relative to those tubes maintained at a temperature below the tissue thermotolerance (Figure 11.18). An LD_{50} (temperate at which 50%

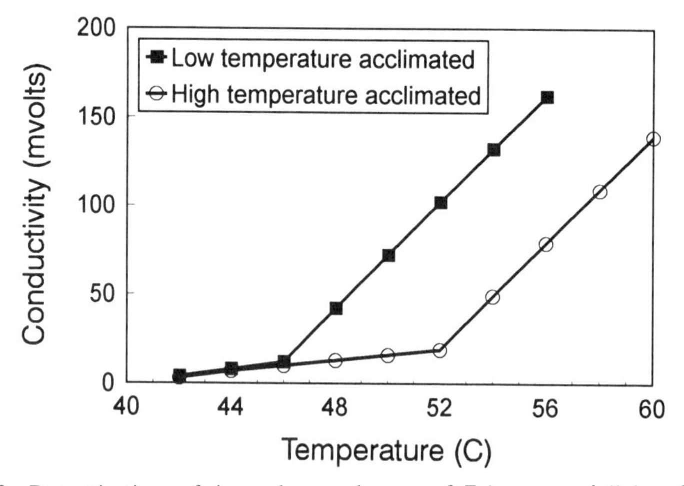

Figure 11.18 Determinations of tissue thermotolerance of *Eriogonum alnii* (an obligate heliophyte from temperate North America) by measurement of plasma membrane leakage. The greater the bathing solution conductivity, the greater the thermal damage. (Data from Hill-Braunchweig 1993).

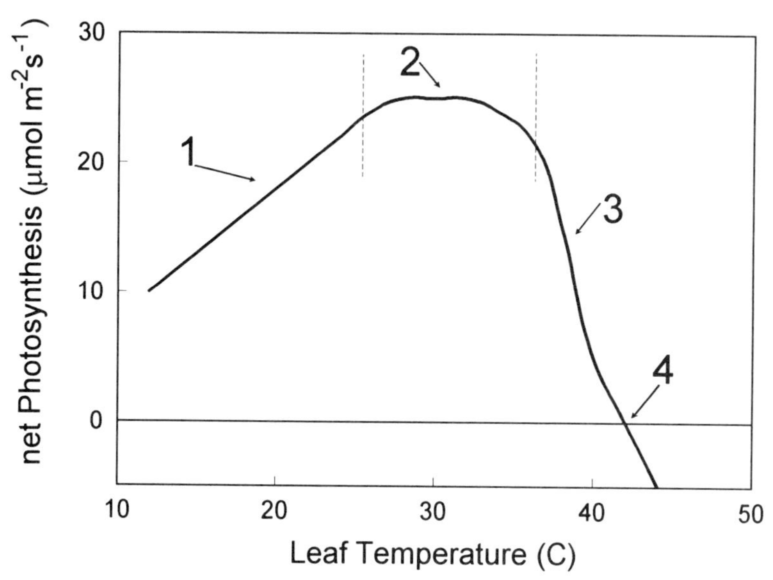

Figure 11.19 Generalized temperature response curve for an enzymatic reaction or a metabolic process. 1, temperature-dependent reaction rate; 2, thermal optimum range; 3, temperature inhibition; 4, thermal cutoff point.

of maximum conductivity occurs in the bathing solution) is often used to compare tissue thermotolerance among species.

Temperature response curves are often employed in studies on thermotolerance of particular enzymes or metabolic pathways. A temperature response curve plots metabolic or enzymatic rate against tissue temperature. Three basic regions of temperature response curves can be designated (Figure 11.19). In the lower-temperature range of this relationship, the metabolic process is limited primarily by temperature. Reaction rate increases linearly with an increase in temperature in this area of the temperature response curve. This is the range in which a Q_{10} can be calculated for the metabolic process. The thermal optimum range is the span of temperature in which the reaction rate is at or above 90% of its maximum. Temperature inhibition occurs at high temperature when the reaction velocity dips below 90% of its maximum. This region of the temperature response curve is usually characterized by a rapid decrease in reaction rate as tissue temperature increases. The last significant aspect of the temperature response curve is the thermal cutoff point, which is defined as that high temperature at which reaction rate reaches zero. In some studies an LD_{50} (high temperature at which the reaction decreases to 50% of its maximum) is utilized as a relativized index of thermal stability.

B. Mechanisms That Regulate Enzyme Heat Tolerance

Earlier in this Chapter we discussed thermotolerance of particular metabolic pathways, such as photosynthesis and respiration. In both cases, thermotolerance was regulated by a combination of enzyme thermotolerance, membrane thermostability, and protection and repair mechanisms. Rather than cover the thermotolerance of a diversity of metabolic pathways, in the rest of this Chapter we focus on the attributes of enzymes and membranes that regulate their thermostability.

Remember that earlier in this chapter it was explained that enzyme thermosensitivity can be observed at high temperature when a precipitous decrease in enzyme activity occurs. This decrease, called *denaturation,* occurs because the tertiary structure of the enzyme, which regulates the physical and chemical attributes of the active sites, is destroyed. Active site integrity is lost at high temperature because the weak molecular bonds, which maintain that active site conformation, are broken. Hydrophobic attractions, ionic interactions, and hydrogen bonds among amino acid side groups come apart, and the three-dimensional morphology of the enzyme is lost. Disulfide bonds are the strongest and most thermally stable of the many types of interactions that maintain the structure of enzymes. Therefore, if enzyme structure is maintained primarily by disulfide bonds, the enzyme may have high thermal stability. Although enzyme thermotolerance could theoretically be altered by changes in the types of molecular forces that maintain the enzymes tertiary structure, this type of thermotolerance has not been widely observed. This is probably due to the fact that DNA transcription and other protein synthesis processes have a sensitivity similar to that of tissue temperature, as many enzymes are involved in primary metabolism. Therefore, there is no selective advantage to having enzymes with high thermotolerance if the transcription and translation machinary required to synthesize the enzyme is not similarly thermotolerant. Regulation of thermotolerance in plants is mostly a function of membrane thermal stability and thermal protection devices such as heat-shock proteins.

C. Mechanisms That Regulate Membrane Heat Tolerance

Membrane thermostability is one of the primary factors regulating tissue thermotolerance in plants. For example, photosynthetic thermotolerance is due primarily to thermostability of thylakoid membranes. As discussed in Chapter 3, several aspects of membrane composition affect membrane thermostability in response to cold temperatures. Actually, most studies of membrane thermotolerance have dealt with maintenance of fluidity at low temperature (see Chapters 3 and 12 for further discussion). Comparatively few studies have focused on high-temperature tolerance by membranes. Seven possible mechanisms that affect membrane integrity have been suggested from studies on low-temperature tolerance (Harwood et al. 1994):

1. Changes in positional distribution of acyls in a lipid class.
2. Changes in *cis*/*trans* isomerization
3. Changes in fatty acyl chain length
4. Changes in fatty acid unsaturation
5. Other modifications of the acyl chains
6. Changes in the proportion of lipid classes
7. Changes in the lipid/protein ratio

The first two mechanisms are theoretically possible, but they have not been shown to occur in higher plants. Alterations in fatty acyl chain length have been found in a few species but the impact of changes in acyl chain length on membrane thermostability has not been separated from associated influences of unsaturation. Unsaturation has been shown to be a major mechanism by which membranes can tolerate low temperatures. Desaturases, membrane-bound enzymes that induce increased unsaturation, may greatly enhance membrane fluidity under cold temperatures. Many regulatory processes, such as the influence of oxygen or membrane fluidity on desaturase activity, have been studied and all involve changes at low temperature (see Chapter 12 for more information). The only evidence that increasing saturation of fatty acids will increase thermal tolerance comes from artificial hydrogenation of chloroplast membranes (cited and discussed in Chapter 3). Increasing hydrogenation greatly enhanced thermostability of thalakoid membrane (Vigh et al. 1989).

D. Heat-Shock Proteins in Plants

One of the most dramatic effects of sudden high heat on plant metabolism is a complete down-regulation of most genes and an up-regulation of heat-shock genes. This response to laboratory-applied sudden short-term heat shock is ubiquitous among organisms and has a powerful impact on basic metabolism. Two main modes of heat-shock protein (HSP) protection of metabolic processes are chaperone and protease actions. Chaperone-type heat-shock proteins serve to refold proteins that have been denatured by excessive heat, and protease-type HSPs degrade proteins that have been permanently damaged (and tagged by ubiquitone). There is also some recent evidence that constitutive HSPs may be able to protect some aspects of cell metabolism from high-heat damage. Verification of these laboratory-induced responses in field situations is a major thrust of modern research on HSPs. Thermal adaptation by plants to heat shock is covered extensively in Chapter 14 and is not repeated here.

VI. SUMMARY

Plants are able to inhabit successfully an enormous range of different thermal environments, extending from hot desert to arctic. Plants are also poikilotherms, thus are unable to maintain a constant internal temperature. Therefore, there is strong selection for adaptation to variable thermal environments among plants. Furthermore, thermal attributes of any particular habitat can vary both temporally and spatially on different scales. Plants must be able to acclimate to changing thermal conditions in order to survive in many habitats.

Tissue temperature moderation is frequently accomplished by changes in morphology and position of sensitive structures. Meristems sensitive to thermal extremes are often protected by sheathing tissues, which buffer temperature change. Leaf temperature is regulated by a complex interaction between absorbed radiation and energy emission mechanisms. Solar radiation absorption can be modified by changing leaf angle, changing leaf size, increasing trichome abundance, or changing the phyllotaxi of shoots. Energy emittance can be modified by changing transpiration, changing leaf morphology as it influences conductance, and altering infrared radiation emitance coefficients. Leaves on species that inhabit regions of high radiation utilize a suite of these mechanisms to moderate leaf temperature.

Tolerance to high temperature also is affected by a complex array of metabolic processes. The most important aspect of cell biology that affects thermal tolerance is membrane thermostability. At high temperature membranes enter into a superfluid state under which membrane thickness and association with proteins changes, holes can develop in the membrane, and differential permeability is lost. Membrane thermostability at high temperature can be modified by changes in fatty acid unsaturation, the position of fatty acids on the glycerol backbone (position 1 or 2), the composition of fatty acids, and the abundance and composition of sterols.

Metabolic thermotolerance is also related to enzyme thermal attributes. In most cases, enzymatic reactions are relatively sensitive to high temperature, at which denaturation occurs. The sensitivity of enzyme morphological conformation, and physiological activity at high heat is due to the weak molecular forces that maintain tertiary shape of proteins. If protein shape is dependent on disulfide bonds, enzymes will be more thermotolerant. However, an abundance of disulfide bonds in an enzyme decreases the ability of enzymes to change shape during their enhancement of metabolic reactions. Thus there is variation in thermotolerance among enzymes and among isozymes. The thermotolerance of some metabolic processes is regulated by the thermotolerance of particularly temperature-sensitive enzymes, while in other cases overall thermotolerance is regulated by membrane attributes rather than specific enzyme temperature characteristics.

Rapid onset of high tissue temperature severely reduces protein synthesis and selectively up-regulates heat-shock genes. Originally, this response was elicited in laboratory situations, but HSPs have been found to be induced in natural situations. Two main metabolic functions have been suggested for heat-shock proteins: chaperones and proteases.

In summary, thermal adaptation and acclimation in plants occurs at a number of different levels. At the level of the organ or the organisms, alterations in morphology or phyllotaxi can moderate leaf or canopy temperature. At the level of membranes, changes in membrane constituents affect thermal stability. At the level of enzymes, changes in isozyme abundance and changes in the nature of bonds that maintain tertiary structure of enzymes affect thermal attributes of particular reactions. At the level of genes, heat shock induces transcription of specific genes that enhance recovery from heat damage. Plants must be able to coordinate thermal adaptation and acclimates at all these levels. The future of research on high-temperature adaptation must focus on the interactions among levels of organization and constraints placed by specific levels of organization on thermal adaptation of other levels.

STUDY-REVIEW OUTLINE

Introduction

1. The diversity of terrestrial ecosystems contain a broad range of thermal patterns to which plants have adapted.
2. Plants are poikilothermic; thus they must compensate for this broad array of possible habitat thermal patterns by morphological, behavioral, or physiological means.
3. Understanding thermotolerance in plants has applied significance to crop production and natural systems management.
4. Meristems (apical and axial) and leaves are the most sensitive organs to high-temperature inhibition. The likelihood for thermal inhibition of a particular organ is determined by the organ sensitivity and environmental thermal patterns around the specific organ.
5. The temperature of an organ is determined by the energy balance established between the environment and the plant tissues. Energy must be balanced between the organ and its environment; therefore, changes in energy absorption affect energy dissipation and tissue temperature.

Physical Factors That Regulate Leaf Temperature

Atmospheric Temperature

1. Coupled leaves have a small difference between leaf and air temperature (ΔT), while ΔT is large for uncoupled leaves.

2. Commonly large leaves in relatively high light sites are uncoupled, while smaller leaves, or compound leaves, are commonly coupled to air temperature.
3. ΔT has a secondary impact on transpiration because ΔT influences the difference in vapor pressure between the atmosphere and leaf intercellular spaces.

Solar Radiation Balance

1. Leaves and other plant organs absorb most of their radiation from solar sources. Solar radiation impinges on leaves as diffuse radiation or direct beam radiation. Diffuse radiation can arrive at the leaf surface from the sky or reflected off surrounding surfaces.
2. Only a portion of solar radiation is absorbed, and this amount depends on leaf morphological attributes. Leaf absorptance is normally 0.8 to 0.9, except for leaves with extensive dead trichomes, whose absorptance is 0.4 to 0.6.

Infrared Radiation Balance

1. Leaves absorb from the surroundings and emit to the surroundings infrared (or thermal) radiation.
2. The amount of infrared radiation emitted by the surroundings toward leaves is proportional to the fourth power of the average surrounding temperature. Similarly, the amount of infrared radiation emitted by the sky is a measure of the sky temperature.
3. The amount of infrared radiation absorbed by a leaf is dependent on the leaf infrared absorptance (usually close to 0.98) and the sum of the surrounding and sky temperatures.
4. The amount of infrared radiation emitted by leaves is dependent on the leaf infrared emittance coefficient (usually close to 0.96) and the temperature of the leaf.
5. An equation for net radiation balance of leaves can be composed by adding the solar radiation absorption and infrared radiation absorption, then subtracting infrared radiation emission.

Latent Heat Exchange

1. When water evaporates, it removes energy from the evaporation surface. Thus, transpiration of water from leaves removes energy from the mesophyll cells.
2. The amount of energy released by evaporation from leaves (latent heat exchange) is determined by transpiration and the heat of vaporization for water.
3. Since stomatal conductance and transpiration are influenced by temperature, light, and humidity, these factors also affect latent heat exchange.
4. In high-light environments, latent heat exchange can become a major mechanism of energy dissipation from leaves.

Conduction/Boundary Layer Resistance

1. Energy can be dissipated from leaves by conduction directly into the air contacting the leaf surface. Thus leaves with a higher temperature than air will conduct some heat into the air.
2. There are two types of convection (free and forced). Forced-air convection occurs when wind moves air across the leaf and removes the air near the leaf surface that has been heated by conduction (the boundary layer).
3. The amount of heat that can be dissipated by convection and conduction is proportional to the difference between leaf and air temperature, and inversely proportional to the thickness of the boundary layer.
4. The thickness of the boundary layer is dependent on the length of the leaf in the direction of the wind, and inversely proportional to wind velocity.

Time-Course Responses of Leaf Temperature

1. Leaf response to changing energy environment is not immediate. A thermal time constant defines the rapidity with which leaves respond to temperature change.

2. In most cases thermal time coefficients are 1 min or less. However, in some cases (such as cactus stems) thermal time coefficients can be very long (500 min).
3. Thermal time coefficients for tissues depend on tissue thickness, tissue hydration, and tissue morphology.
4. Various time functions, such as ramp, step, and harmonic, can describe the way that tissue temperature responds to a changing energy environment.

Techniques for Measuring Thermal Attributes of Tissues

1. Resistance thermometers are those that relate the change in temperature with the change in electrical resistance within one metal, such as platinum.
2. Resistance thermometers are not often used in physiological studies because they are relatively large, must be close to the recording device, and are expensive.
3. Thermisters also measure resistance to electrical flow as a proxy for temperature. However, instead of a pure metal, thermisters utilize semiconductors such as metallic oxides.
4. Thermisters are cheaper than resistance thermometers, but thermisters cannot be located at a great distance from a recording device, and they are relatively fragile.
5. Thermocouples are the most popular temperature recording device in physiological, ecological, and agricultural applications.
6. The thermocouple signal is voltage (not resistance) and is generated by differential electron flow from two different metals joined at junctions held at different temperatures.
7. Infrared thermometers measure tissue surface temperature from a distance and therefore have no effect on the tissue.

Detrimental Effects of High Temperature on Cell Metabolism

1. Thermal sensitivity for metabolic processes from the most to least sensitive would be ranked: enzymatic, diffusion, and photochemical.
2. The Q_{10} value, based on the Arrhenius equation, defines metabolic responses to temperature within a temperature range that does not inhibit the metabolic process.
3. At temperatures above the optimum range of a metabolic reaction, inhibition occurs because of enzyme denaturation and changes in membrane fluidity.

Effects on Photosynthesis

1. The optimal temperature for photosynthesis is variable among species that inhabit regions with different thermal attributes.
2. As temperature increases in leaf tissues, respiration increases more dramatically than gross photosynthesis. Therefore, part of high-temperature inhibition of photosynthesis is due to high respiration and photorespiration rates.
3. The difference in thermal stability of photosynthesis among species is due partly to differential sensitivity of key enzymes to high temperature, and mostly to differential thermal stability of photosystem II.

Effects on Respiration

1. Respiration has a Q_{10} value that ranges between 1.5 and 2.5, which means that the rate of respiration approximately double during each 10°C increase in tissue temperature.
2. There is some evidence that cytochrome-mediated electron flow is inhibited at lower temperatures than the alternative electron transport chain. Thus, at high temperature most respiratory electron flow occurs through the alternative pathway.
3. Thermal inhibition of respiration may be closely linked to thermal stability of membranes.

Effects on Membrane Function and Integrity

1. Most of what is known about membrane atributes that affect membrane thermostability comes from studies of plant adaptation to low temperature.

2. High temperature influences membrane fluidity by causing a hyperfluid state, induces holes (leakiness), and disrupts the association between lipids and proteins.

Effects on Enzyme Function and Protein Synthesis

1. The sudden drop in metabolism at high temperature is a result of enzyme denaturation (the loss of function due to a loss of tertiary structure).
2. Simple proteins that are held in shape by mostly disulfide bonds are relatively immune from thermal inhibition. In contrast, those complex enzymes whose shape depends on a variety of weak molecular bonds are susceptible to thermal damage.
3. Thermal inhibition of metabolic reactions can also be due to the effects of high heat on protein synthesis, transcription, or protein degradation.

Mechanisms of Moderating Leaf Temperature

1. Leaf abscission during periods of high temperature and radiation is a mechanism of avoiding heat stress.
2. Frequently, summer deciduous species have photosynthetic stems which maintain carbon gain when leaves are unable to tolerate the heat and radiation load.
3. Narrow, compound, and small leaves are characteristic of species that survive in regions with high heat load. If the high-heat condition is seasonal, the leaf population may be dimorphic. Leaves that develop in the hot season are smaller and thinner than those that developed in the cool season.
4. Dead trichomes on the surface of leaves reduce radiation absorptance of leaves and moderate leaf temperature in high-light and high-heat environments.
5. Vertical leaves reduce radiation absorption during midday: the time of day when leaves are most likely to suffer heat damage.
6. Paraheliotropic leaves move during the day to minimize radiation impinging on their leaf surface.
7. The cosine of incidence indicates the angle of incidence between solar direct radiation and the leaf surface. A cosine of incidence of 1 represents a leaf that is perpendicular to the solar angle, and a cosine of incidence of zero represents a leaf that is parallel to the solar angle.
8. Transpiration is not usually a major mechanisms of leaf cooling because high heat is usually associated with high radiation and low water availability.

Mechanisms of Tissue Temperature Tolerance

1. Most plants can tolerate temperatures up to 40°C, and some can tolerate temperatures as high as 60°C. Only some bacteria can tolerate temperatures above 70°C for extended periods.
2. The highest temperature tolerance in plants is found in some cacti that can tolerate 60 to 65°C for a short time.
3. Thermal tolerance of tissue is measured by the sensitivity of cytoplasmic streaming or membrane leakiness to increasing temperature.
4. Variation in thermotolerance among enzymes is not usually due to differences in the abundance of disulfide bonds. Rather, temperature stability of transcription and translation have a greater influence on metabolic processes.
5. Metabolic thermotolerance is due primarily to membrane thermostability.
6. Membrane thermostability in response to high heat may be due primarily to the amount of saturation in fatty acids.
7. The most dramatic influence of heat on metabolism is the induction of heat-shock proteins.
8. HSPs may serve to refold denatured proteins, degrade damaged proteins, or protect other proteins from thermal damage.

SELF-STUDY QUESTIONS

1. Describe the various morphological attributes that you would expect for a leaf that is uncoupled with atmospheric air temperature.
2. What are the major routes of radiation absorption and emittance from plant tissues.
3. Which plant tissues are likely to suffer from high-heat stress the most? Why?
4. How important is endogenously produced heat (from respiration) to heat balance in plants?
5. What is the average absorptance of leaves? What characteristics of stems and leaves influence their absorptance of solar radiation?
6. Why is it that leaf temperature may be well below air temperature on a clear night, while leaf temperature will equal air temperature on a cloudy night?
7. What factors of the environment regulate the amount of infrared radiation absorbed by plant surfaces?
8. Describe the process that leads to latent heat exchange from leaf tissues.
9. Differentiate between conduction and convection in relation to leaf energy balance.
10. What factors regulate the thickness of a leaf boundary layer?
11. Differentiate between the mechanical operation of resistance thermometers, thermisters, thermocouples, and infrared thermometers.
12. What characteristics of thermocouples need to be considered before measuring the temperature of plant tissues?
13. In what way does environmental temperature relate to the activation energy of reactions?
14. Explain the meaning of Q_{10}. Is this a constant for any reaction? If not, why not?
15. Rank the significance of the thermostability of enzymes, photochemical reactions, and membranes to the mechanisms for thermal inhibition of photosynthesis. Do the same for respiration.
16. In what way does high heat impair the activity of the photochemical reactions of photosynthesis?
17. Describe several morphological attributes of leaves that can reduce leaf temperature in an environment with high heat and high light.
18. Diagram the expected diurnal cycle of cosine of incidence for a vertical leaf, a horizontal leaf, and a paraheliotropic leaf.
19. What type of plants, in what type of habitats, commonly have photosynthetic stems? Of what advantage are photosynthetic stems to the species you identified?
20. What are two significant factors for leaf movements to individual leaves and to the canopy of larger plants?

21. What is the upper thermal limit in most terrestrial higher plants? What is the most likely reason for this upper thermal tolerance?

22. Describe two important mechanisms for determining tissue thermotolerance in plants.

23. What factors regulate thermotolerance of enzymes? Of membranes?

24. What are two basic functions of heat-shock proteins in plants?

25. If you were to design a plant with ultimate tolerance of hot environments, what would be its morphological, behavioral, and physiological attributes?

SUPPLEMENTARY READING

Baker, N. R., and S. P. Long. 1986. Photosynthesis in Contrasting Environments. Elsevier, Amsterdam.

Campbell, G. S. 1977. An Introduction to Environmental Biophysics. Springer-Verlag, New York.

Cossins, A. R. 1994. Temperature Adaptation of Biological Membranes. Portland Press Proceedings, London.

Niklas, K. J. 1994. Plant Allometry: The Scaling of Form and Process. University of Chicago Press, Chicago.

Nobel, P. S. 1991. Physiochemical and Environmental Plant Physiology. Academic Press, San Diego, CA.

Pearcy, R. W., J. Ehleringer, H. A. Mooney, and P. W. Rundel. 1989. Plant Physiological Ecology Field methods and instrumentation. Chapman & Hall, New York.

Precht, H., J. Christophersen, H. Hansel, and W. Larcher. 1973. Temperature and Life. Springer-Verlag, New York.

Wyman, R. 1991. Global Climate Change and Life on Earth. Chapman & Hall, New York.

12 Low Temperature: Chilling and Freezing

Outline

OBJECTIVES

1. Define chilling and freezing stress, establish global distribution of low temperature, and address changes in global temperatures.
2. Discuss temperature limits of chilling-sensitive and chilling-insensitive plants and factors influencing response types.
3. Develop the concept of primary and secondary lesions as they pertain to cold stress and describe the development and expression of cold-induced injury symptoms.
4. Discuss the impact of chilling stress on the physiology and biochemistry of chilling-sensitive plants.
5. Develop the concept of differential sensitivity of cellular organelles to chilling stress.
6. Differentiate between chilling and freezing stress in plants and address different mechanisms whereby plants deal with freezing temperatures.
7. Discuss ice formation in plant cells and the consequences of ice development.
8. Discuss the physiological and biochemical ramifications of freezing stress in plants.
9. Describe morphological and physiological strategies that have developed in plants adapted to cold environments.

I. INTRODUCTION

Low temperature is a very arbitrary term but will be used in the context of chilling stress or temperatures above freezing that inflict injury on plants (from just above freezing to 15 to 20°C) and freezing stress that causes injury to plants at 0°C and below. Plants are poikilotherms, that is plants assume the temperature of their environment. Low temperature is critical to plants in that it is an important determinant in the normal distribution of natural plant communities and defines the range of distribution and growth of important agricultural crops, many of which are grown at or near the temperature boundary limits of their genetically determined survival abilities.

Globally, temperatures decline at higher latitudes, which results from a combination of lower solar elevation and changing seasons (summer–winter), resulting in longer nights. Reduced temperatures are also associated with increasing altitude and with distance from coastal regions. Local reductions in temperature can occur in valleys or low-lying areas in mountainous regions, particularly during periods when seasonal changes in temperatures are occurring, resulting in pockets of cold air being trapped under warmer air in low-lying valleys. It has been estimated that only 6.6% of the total continental area of Earth experiences temperatures that do not drop below +15°C (Amazon basin, Congo basin, and parts of southeast Asia), and only 25% of the continental surface is considered safe from frost (Sakai and Larcher 1987). Mean annual temperatures below 0°C, −10°C (survival limit for freezing sensitive plants), −20°C (survival limit of broad-leaved evergreen trees of maritime temperate climates), and −40°C (distribution limit for deep supercooling woody plants) are represented by 64, 48, 35, and 25% of Earth's landmass, respectively (Sakai and Larcher 1987).

We have historically accepted, rather than addressed, the effect of small climatic changes on plants. For example, it has been conjectured that a 1°C decrease in the world mean temperature would result in a 40% reduction in rice production and would most probably change the range of cultivation of such crops as well as the natural distribution of native plants. Obviously, increases in temperature, similarly, could influence cropping practices and plant distribution. The ongoing debate relative to whether global temperatures are increasing or declining is certainly pertinent to this discussion. Recent evidence suggests that temperatures are actually declining, particularly in regions of intense industrialization such as China, Europe, and the United States. Studies suggest that aerosols (sulfates, dust, soot) generated from industrial processes and volcanic activity (sulfates and particulates), dust generated from eroded soils and deserts, sulfates generated from phytoplankton as a result of fertilizer pollution of the oceans, and smoke generated from forest fires and burning of crop residues serve as nuclei to condense moisture to increase localized cloud cover. The ultimate effect is to reduce radiation energy reaching the surface of the earth during the day and essentially trapping that energy that would be normally dissipated back to the atmosphere at night. Figure 12.1 shows the relationship of cloud cover and temperature in the United States over a 90 year period. These findings suggest that regional reductions in temperature can override any warming effects due to the greenhouse effect and that such changes could have significant impacts on not only regional but global climates (Pearce, 1994). The implications should be obvious to the reader with respect to the distribution of terrestrial plants and the impact that could occur relative to agricultural production in heavily populated industrialized regions of the world.

II. CHILLING STRESS

Chilling stress is usually limited to plants native to or growing in tropical or subtropical regions of the world, which as referred to previously, represents only about 6.6% of the continental area. The temperature range for chilling stress in such plants ranges from just above freezing to 15 to 20°C in some chilling-sensitive plants. Plants vary greatly relative to their sensitivity to chilling stress. Chilling-sensitive plants have been defined as plants

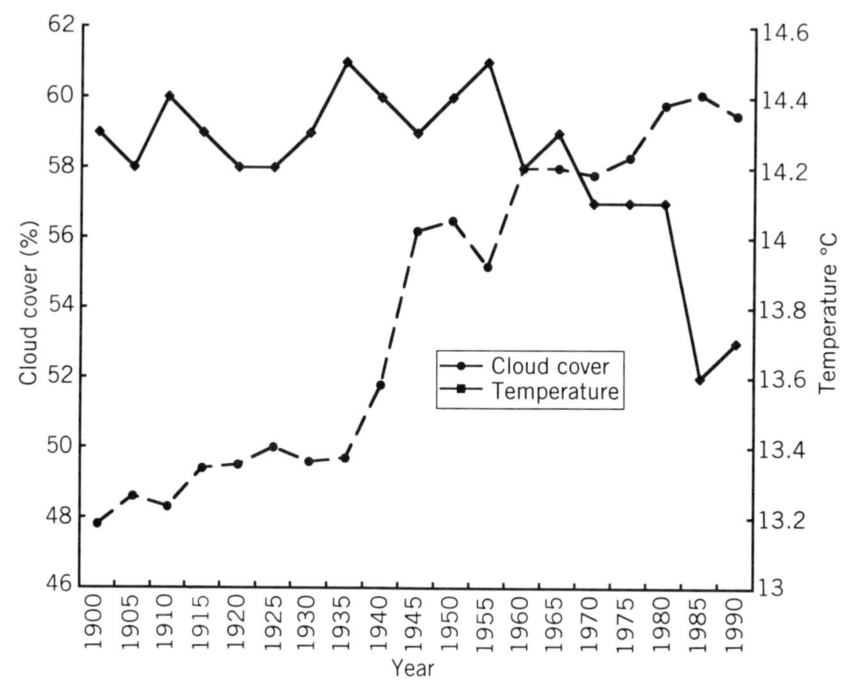

Figure 12.1 Annual temperature range and cloud cover in the United States. (Modified from Pearce 1994.)

that are killed or injured by temperatures above the freezing point of the tissue up to 15 to 20°C (Graham and Patterson 1982). Chilling-resistant plants are those able to grow at temperatures near 0°C. However, considerable variability exits among plants relative not only to genetic factors but to stage of development, metabolic status (dormancy or active growth), and conditions under which plants are growing during and after the chilling episode. Generally, plants are more sensitive to chilling under nondormant conditions (high metabolic activity), during younger stages of development, during the day or under high-light intensities, under drought stress (although drought hardening can increase chilling tolerance), and when nutrients are limiting (particularly K, which is involved with osmotic adjustment in plants).

The onset of symptom expression due to chilling injury in plants varies visually and temporally among plant species. Symptom expression may take hours to months, depending on the sensitivity of the plant and other environmental conditions alluded to above. In some plants, injury may be expressed during the chilling period or after the plant tissue has been rewarmed. Symptoms of chilling injury include cellular changes (changes in membrane structure and composition, decreased protoplasmic streaming, electrolyte leakage, and plasmolysis), altered metabolism (increased or reduced respiration, depending on severity of stress, production of abnormal metabolites due to anaerobic conditions), reduced plant growth and death, surface lesions on leaves and fruits, abnormal curling, lobbing and crinkling of leaves, water soaking of tissues, cracking, splitting and dieback of stems, internal discoloration (vascular browning), increased susceptibility to decay, failure to ripen normally (immature fruits lose the ability to ripen if chilled), and loss of vigor (potatoes lose the ability to sprout if chilled) (Saltveit and Morris 1990). Some of the more common symptoms of chilling stress are rapid wilting followed by water-soaked patches which develop into sunken pits that reflect cell and tissue collapse. Following warming the sunken pits usually dry up, leaving necrotic patches of tissue on the leaf surface. Chilling symptoms in fruit vary and include sunken pits in cucumber, browning of skins and degradation of the pulp tissue in bananas, and "blackheart" of pineapple (Wilson 1987).

A. Primary Versus Secondary Lesions

Scientists have long debated whether there is some primary event, and what the event is, that leads to chilling injury in plants. Several hypotheses have developed over the years that have tried to tie various physiological, biochemical, and biophysical events to a primary cause for the injury and symptomology observed in chilling-sensitive plants. Evidence has accumulated that chilling stress affects several functions in plants, including biochemical and biophysical structure of membranes, nucleic acid synthesis, changes in protein synthesis, enzyme conformation, affinities and activation energies, water and nutrient (particularly Ca) balances, cellular cytoskeletal structure, and photosynthetic and respiratory function. Which of these represent the *primary event* and which are *secondary events* leading to symptom expression is the basis for the debate. Raison and Orr (1990) propose a single primary event based on temperature transition in the molecular ordering of membrane lipids, which was expressed initially some 25 years ago (Lyons and Raison 1970). As pointed out in Chapter 3, chilling temperatures can cause membrane phase transition from a liquid- crystalline phase to the solid-gel phase. Such changes can decrease membrane permeability (fluidity) and could affect any of the above-mentioned responses of plants to chilling stress, and according to Raison and Orr (1990), relegate such responses to secondary events leading to symptom expression.

Figure 12.2 outlines the concept of primary and secondary events that lead to chilling injury in plants as expressed by Raison and Orr (1990). This model suggests that chilling injury can be divided into a single primary event and several secondary events. The primary event (probably, membrane phase transition) is initiated when temperature drops below a certain critical temperature. This temperature will vary with species and conditions under which the plant is grown. The primary event is then responsible for initiating numerous secondary events, but the order of initiation, if any, is not clear. If the level of chilling is not too great or too long and the plant is returned to warmer temperatures, the process can be reversed and the plant does not sustain injury. However, if the stress is maintained at too low a temperature and too long, and then returned to warmer temperatures, injury and cellular degradation are accelerated. The concept developed in this model is very useful in clarifying frequently loosely used terms, such as chilling-sensitive, chilling-insensitive, and chilling-tolerant or chilling-resistant plants. In this scheme, chilling-sensitive plants are injured at chilling temperatures and will die from such injuries, chilling-insensitive plants can grow and reproduce at temperatures near 0°C, while chilling tolerant or chilling-resistant terminology is reserved for those plants showing variation in time course in the expression of secondary events. As an example, chilling-sensitive plants vary temporarily relative to visual expression of injury (hours, weeks, months). The longer the time it takes to express the symptoms, the more tolerant/resistant the plant is to chilling stress.

Naylor (1983) argues against a single primary lesion as being responsible for initiating the cascade of events that lead to injury and symptom development. Instead, he envisions chilling stress affecting the physical concepts of coordination of metabolic pathways (rate

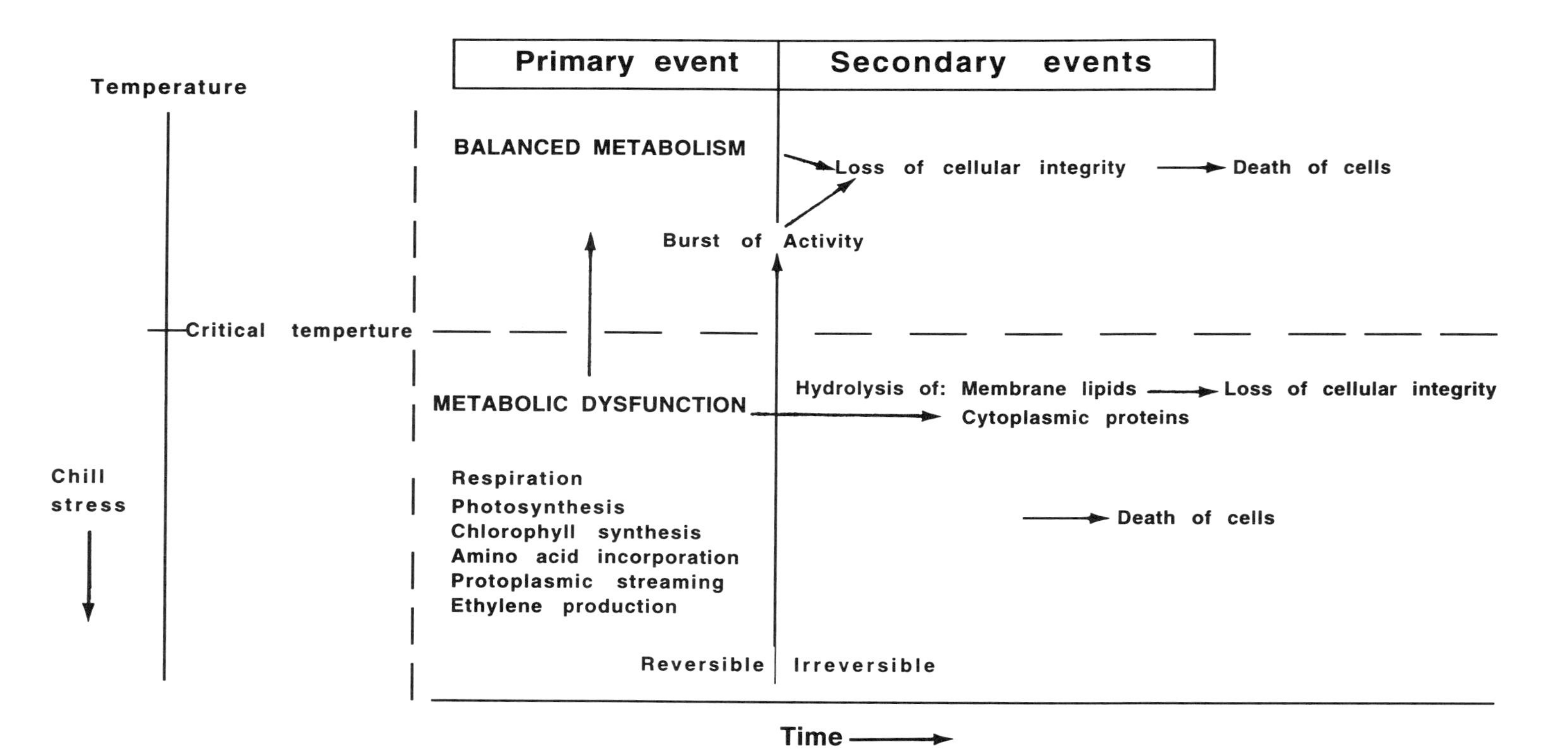

Figure 12.2 Relationship between "primary" and "secondary" events during chilling injury. (Modified from Raison and Orr 1990.)

TABLE 12.1 Classes of Weak Chemical Bonds

Class of Bond	Approximate Range of Bond Energies (Enthalpies) (kcal mol^{-1})
Van der Waals forces	1
Hydrogen bonds	3–7
Ionic bonds	5
Hydrophobic interactions	1–3
Covalent bonds	50–100

Source: Modified from Naylor 1983.

effects) and the stability of complex biological molecules (weak bond effects), either or both of which may be affected at different levels of organization and expression. In the case of *rate effects,* sensitivity of any one metabolic process, or enzyme within that process, to chilling temperatures could have a significant impact on the functioning of other enzymes in that process or substantially alter the balance among multiple metabolic processes. *Weak bond effects* may even seem more significant a factor in favor of multiple primary lesions when one compares the bonding energies of various types of chemical bonds (Table 12.1). Naylor (1983) points out that biochemical structure, including protein configuration, membrane structure, nucleic acid structure, and most biochemical interactions, require a high degree of specificity, which, if not totally, is highly dependent on weak chemical bonds. Thus when one considers the importance of just hydrogen bonding (bonding energies 10 to 12 times weaker than covalent bonds) in protein configuration, the structure and stability of membranes, and protein–membrane and protein–substrate interactions, it becomes clear that there may be numerous opportunities where biochemical functions could be affected simultaneously by chilling stress, making the identification of a single primary lesion difficult if not impossible. The ability of organisms to cope with chilling temperatures and other environmental stresses then lies with the development of physical and/or biochemical strategies that protect the organism against breaking of weak bonds.

B. Physiological and Biochemical Effects of Chilling Stress

Chilling stress can affect proteins both qualitatively and quantitatively. Generally, soluble proteins increase and proteins that have enzymatic functions can be either up- or down-regulated, depending on specific function. Table 12.2 summarizes some of the enzymes or enzymatic-controlled processes that are regulated (up or down) by chilling stress. As can be seen from this table, some of the proteins/enzymes regulated may have detrimental effects relative to the overall metabolism of the plant, while others may have a beneficial role relative to protecting the plant.

Generally, enzymes are more labile to high temperatures than they are to low temperature. However, complex enzymes possessing subunits such as pyruvate P_i dikinase and phosphofructokinase, which are involved in the carbon fixation reactions in C_4 plants and glycolysis, respectively, are inactivated by chilling temperatures as a result of converting them from tetramers to dimers. Another enzyme thought to be affected by chilling stress and one that is critical to osmotic relations in plants is K^+-mediated ATPase activity. As-

TABLE 12.2 Enzymes or Processes Up- or Down-regulated in Plants as a Result of Chilling Stress

Enzymes or Processes Up-regulated	Enzymes or Processes Down-Regulated
a. Phenolic synthesis	a. Carbon fixation reactions
(1) PAL (phenylalanine ammonium lyase)	(1) NADP-malate dehydrogenase (C_4 reactions, high light)
(2) CQT (hydroxycinnamoyl CoA quinate hydroxycinnamoyl transferase)	(2) pyruvate P_i dikinase (C_4 reactions, high light)
b. Respiratory enzymes	(3) PEP carboxylase (C_4 reactions)
(1) Glycolysis	(4) Rubisco (dark reactions)
(2) TCA cycle	(5) Fructose-1,6-bisphosphatase (dark reactions)
(3) PPP (pentose phospate pathway)	b. Light reactions
c. Cryoprotectants	(1) NADP reductase
(1) Invertase (starch to sucrose)	(2) Plastocyanin
(2) Proline	(3) CF1
(3) Putrescine	(4) Ca^{2+} ATPase
(4) Betaine	c. Respiration
d. Antioxidants	(1) NADP-malate dehydrogenase (TCA cycle)
(1) Glutathione	(2) 3-PGAL dehydrogenase (Glycolysis)
(2) Ascorbate	(3) Phosphofructokinase (Glycolysis)
e. Membrane lipids	(4) 6-Phosphogluconate dehydrogenase (PPP)
(1) Desaturases	(5) Glucose-6-P-dehydrogenase (PPP)
(2) Acetyl-CoA carboxylase	d. Nitrogen metabolism
f. Nonenzymatic proteins	(1) Gutamate dehydrogenase (NH4 to α-ketoglutarate)
(1) Ice nucleators	(2) Aspartic β-semialdehyde dehydrogenase (lysine and alanine)
(2) Cryoprotective proteins	e. Starch metabolism
g. Invertase (decreased inhibitor activity)	(1) Fructose-1,6-bisphosphatase
	(2) Starch synthase
	f. Antioxidants
	(1) SOD (superoxide dismutase loss of Zn and Cu)
	(2) Catalase
	(3) Ascorbate peroxidase

Source: Modified from Graham and Patterson 1982.

sociated with reduced activity of this protein is leakage of K^+ from cells exposed to chilling stress. It appears that this process is reversible if the stress is not prolonged or to severe (Palta and Weiss 1993). The importance of osmotic relations on chilling sensitivity and ATP, ADP, and AMP levels in bean plants is illustrated in Figure 12.3 (Wilson and McMurdo 1981). As is represented in this figure, the activity of all three adenine derivatives is drastically reduced when chilling and subsequent transfer to warmer temperatures are conducted at a relative humidity of 85%. When the same experiment is conducted at 100% RH, only AMP is substantially affected. Further, this experiment illustrates the important interaction of plant–water relations and chilling sensitivity.

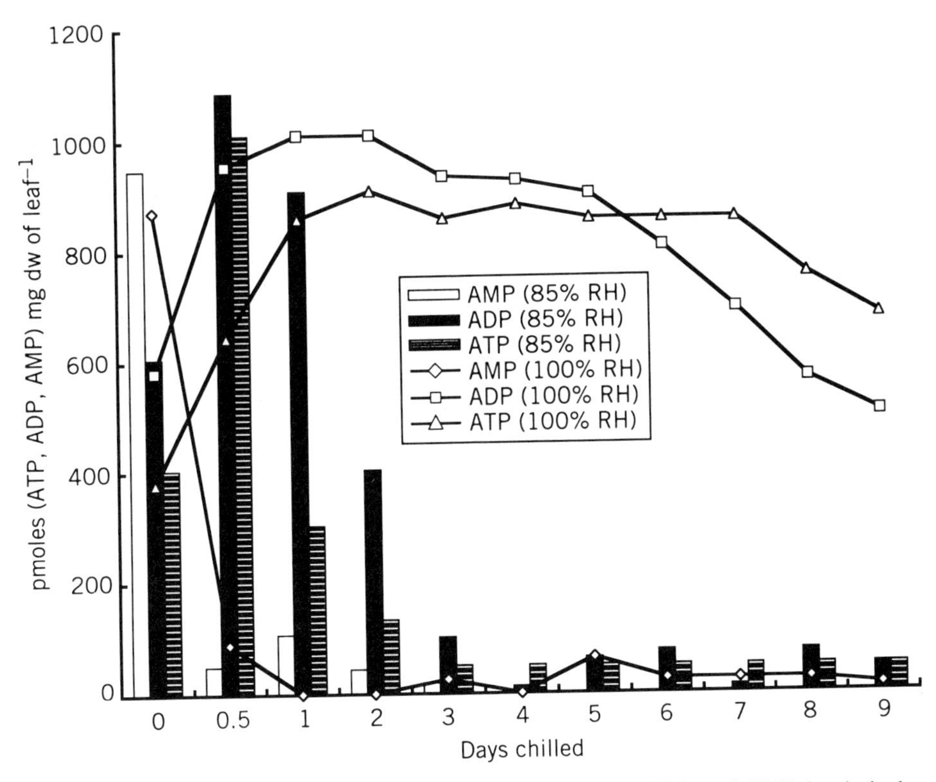

Figure 12.3 Effects of chilling injury and water loss on ATP, ADP, and AMP levels in bean leaves. (Modified from Wilson and McMurdo 1981.)

Chilling stress generally weakens hydrophobic interactions, exposes sulfhydryl groups, and can alter the lipid environment surrounding membrane-associated proteins which can ultimately lead to changes in the configuration of proteins and possibly enzyme kinetics. Chilling affects enzyme kinetics by altering the reaction velocity (V_{max}) and the affinity (K_m) of the enzyme for its substrate. In some cases K_m has been shown to decrease (> affinity for substrate) for enzymes such as PAL, CQT, PEP carboxylase, and rubisco, and the response may be either constitutive or induced. Figure 12.4 illustrates the effect of chilling (1.5°C) on PEP-carboxylase activity in alpine, temperate, and tropical plants compared to 20°C. Obviously, the relative activity for this enzyme was higher for plants adapted to cooler environments than was the activity for the tropical plants. This suggests differences in the enzyme or other mechanisms available to cooler-climate plants for protecting enzymes. Figure 12.5 compares the K_m values for chilling-sensitive tomato, chilling-insensitive wheat, and *Caltha*. Only the chilling-sensitive tomato responds to the chilling treatment (1.3°C), with an increase in PEP K_m which suggests reduced substrate affinity by the enzyme at chilling temperatures. Figure 12.6 illustrates the effects of chilling temperatures on carbon assimilation in a "typical" vascular plant. The effects of chilling influences total carbon gain but also influences the distribution of assimilates for growth, storage, and respiration (Levitt 1980).

Photosynthesis is inhibited in chilling-sensitive plants by affecting both the light and dark reactions of photosynthesis. Chilling is more injurious to plants if it occurs during

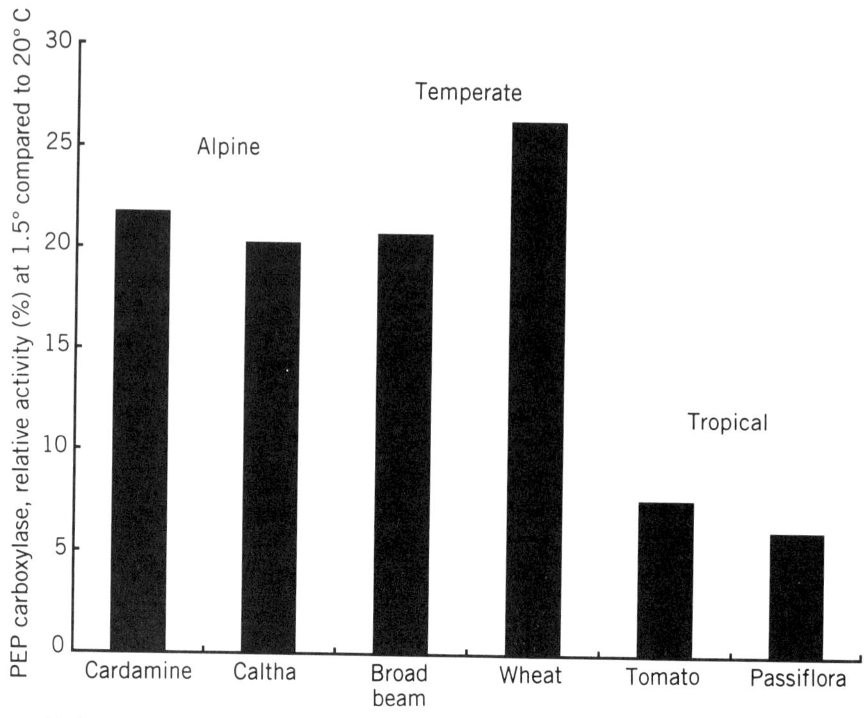

Figure 12.4 Relative activity of PEP-carboxylase at 15°C compared to 20°C for alpine, temperate and tropical plants. (Modified from Graham et al. 1979.)

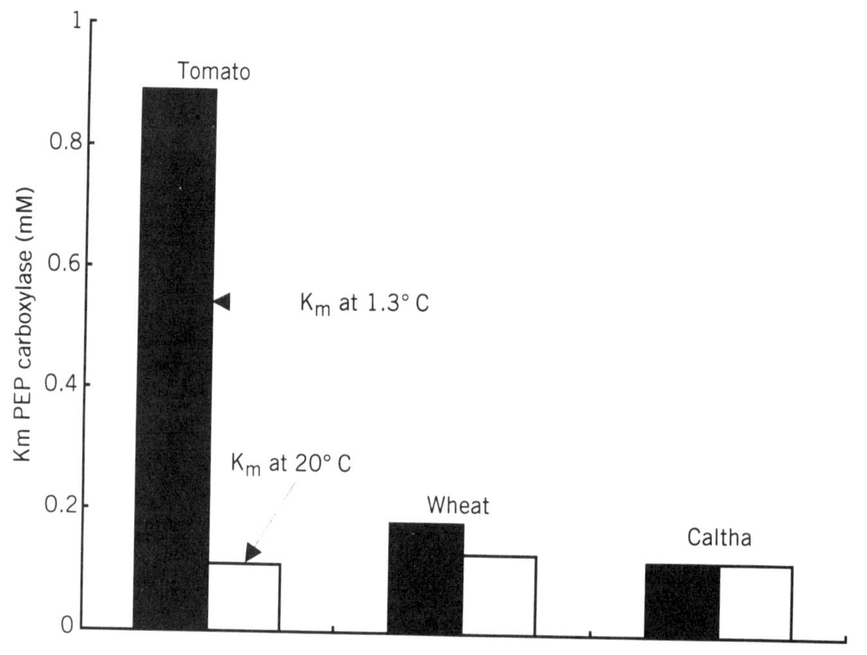

Figure 12.5 K_m values for PEP carboxylase prepared from cold-sensitive (tomato) and cold-tolerant (wheat and *Caltha*) plants. (Modified form Graham et al. 1979.)

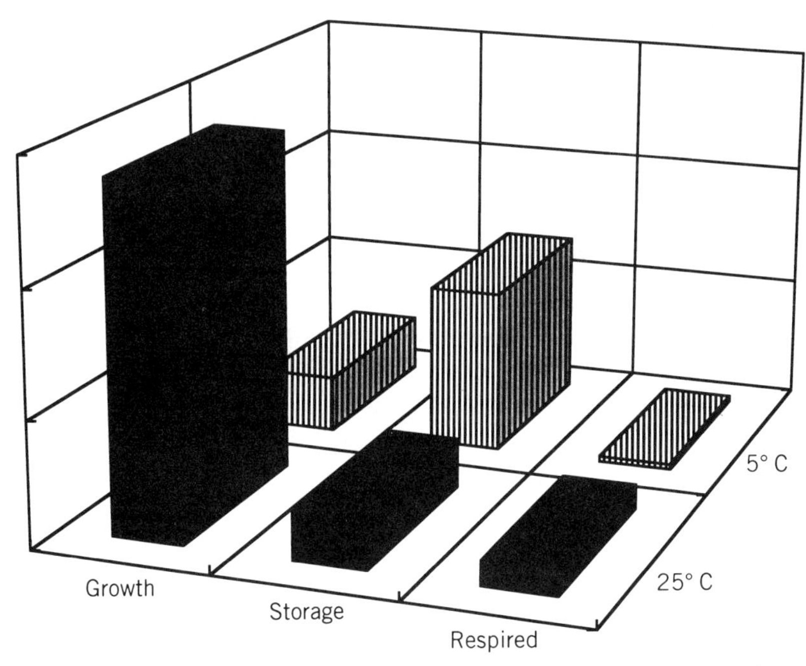

Figure 12.6 Relative proportions of carbon assimilated at high and low temperatures for a typical vascular plant. (Modified from Levitt 1980.)

the day or when the plant is experiencing drought stress. In the former case, injury is probably due to photoinhibition since chilling has disrupted the normal electron transport carrier system, which can lead to electrons being used to form free-radical species, which then facilitate degradation of membranes. This response may be reflected as a loss of chlorophyll in plants, as illustrated in Table 12.3. In this experiment, chilling-sensitive and chilling-insensitive plants were compared relative to chlorophyll content after being subjected to 17°C for 24 h and then compared to plants maintained at 24°C. Obviously, chilling-sensitive plants appear to be more susceptible to photo-oxidation than the insensitive plants based on chlorophyll retention. The amount of NADPH is generally reduced in the light reactions, which suggests a blockage of photosynthetic electron trans-

TABLE 12.3 Chlorophyll Content of the First Leaf or Cotyledons of Etiolated and Greened (at 17°C for 24 h) Chilling-Resistant and Chilling-Sensitive Plants

Chilling-Resistant Plant	Percent Chlorophyll	Chilling-Sensitive Plant	Percent Chlorophyll
Cauliflower	83	Pearl millet	28
Cabbage	62	Rice (Indica)	23
Radish	60	Cantaloupe	22
Barley	57	Cotton	21
Wheat	45	Sorghum	19
Oats	43	Maize	19

Source: Modified from McWilliam et al. 1979.

port which is probably at the oxidative side of photosystem II and appears to involve the D1 protein (Bowers 1994). As indicated previously, several photosynthetic enzymes are chilling sensitive, so the rate of enzyme activity is reduced, resulting in reduced metabolic pools such as 3-phosphoglyceric acid required for the dark reactions of photosynthesis. Thus reduction in carbon fixation occurs as well as transport of accumulated carbon, since in some plants starch accumulates in chloroplasts (Graham and Patterson 1982). This may reflect lower activity of starch-degrading enzymes or perhaps reduced capacity to transport carbohydrates to sieve cells.

Reduced photosynthetic rates may also reflect reduced transfer of CO_2 into the plant since under chilling-stress plants frequently become dehydrated and stomates close. Figure 12.7 illustrates how chill hardening can affect diffusive resistance in a chilling-sensitive plant such as bean and how the response compares to a chilling-insensitive plant such as pea (Wilson 1987). The experiment shows that stomatal closure is faster in non-hardened chilling-insensitive pea plants than either hardened or nonhardened chilling-sensitive bean. However, hardening of bean plants results in stomatal closure occurring sooner and for a longer period of time than for nonhardened plants. Such responses may be a disadvantage relative to reduced photosynthetic rates but are important relative to maintaining plant hydration, which is crucial to maintaining a low level of chilling injury.

Respiration rates can either increase or decrease, depending on the severity and length of time the plant is exposed to chilling stress. Short-term minimal stress tends to increase respiration rates, while long-term or extremely low temperatures leading to cell damage and death tend to reduce rates. Increased rates, as measured by O_2 consumption, have been demonstrated to be associated with the increased activity of the cyanide-insensitive branch from the normal electron transport chain. No ATP is generated from this pathway, and what its function may be relative to chilling stress is not known, but it could serve to recycle reduced pyridine nucleotides, which may be necessary for normal carbon metabolism (Markhart III 1986). In some instances, chilling-sensitive plants accumulate ethanol, lactate, acetaldehyde, and alanine, which may be indicative of reduced mitochondrial enzyme activity and increased glycolytic function. Accumulation of ethanol and lactate may in themselves be injurious to plants. Figure 12.8 illustrates how chilling stress affects the activation energy of succinate oxidase in mitochondria of chilling-sensitive and chilling-insensitive plants. Breaks in the Arrhenius plots indicate the temperature where membrane lipids are supposedly transformed, resulting in changes in enzyme conformation that changes the kinetics of enzyme activity. In this example, only chilling-sensitive plants exhibited the break in the Arrhenius plot. This could suggest that differences in mitochondrial lipid composition could explain the differences in chilling sensitivity between the plants studied. However, some chilling-insensitive plants have been shown to exhibit breaks in Arrhenius plots even though they exhibit a lack of sensitivity at the temperatures employed. This suggests that other mechanisms may be involved in the insensitive response.

Structural changes in organelles have been observed with chilling stress, and it appears that differences in organelles exist relative to chilling sensitivity. Table 12.4 shows the degree of injury exhibited by tomato cotyledon organelles after different chilling periods at 5°C. It would appear from this experiment that order of sensitivity is plastids > mitochondria > peroxisomes > nuclear envelope > tonoplast > plasmalemma. It is also of interest that microtubules disappear after 4 h of chilling treatment. There may also be some survival benefit to the observation that the tonoplast and plasmalemma exhibit lower sensitivities to chilling stress than the other organelles. In the case of the cell membrane,

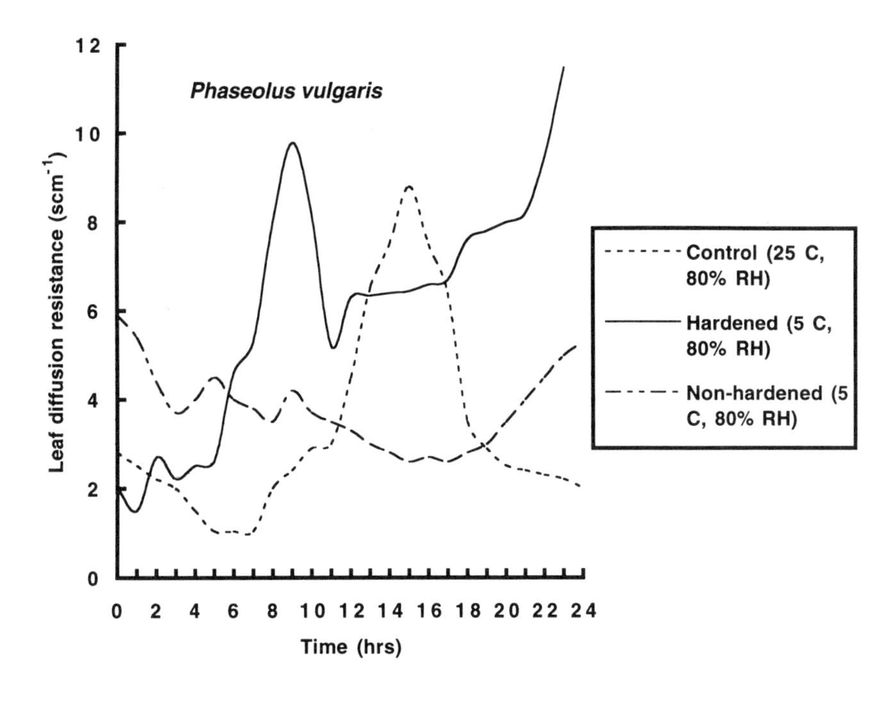

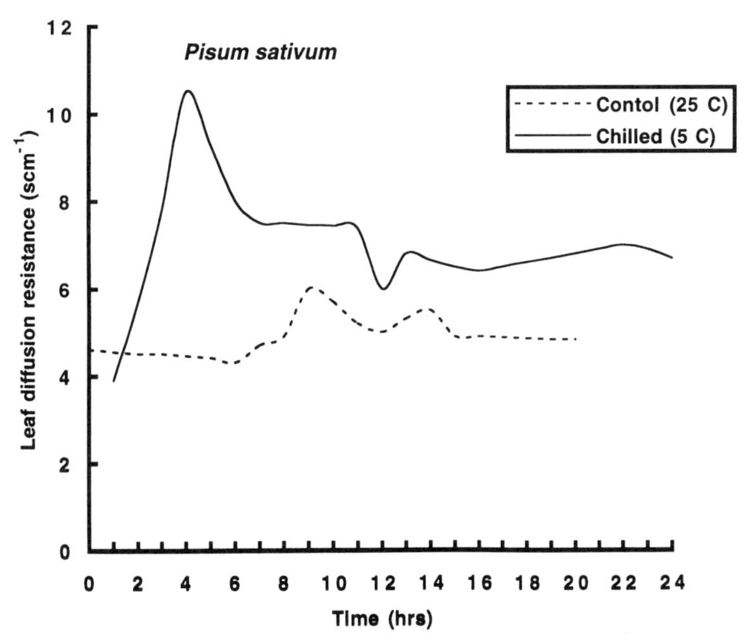

Figure 12.7 Comparison of leaf diffusion resistance of chill-hardened and nonhardened bean and pea plants during chilling at 5°C (80% RH) compared to controls maintained at 25°C (80% RH). (Modified from Wilson 1987.)

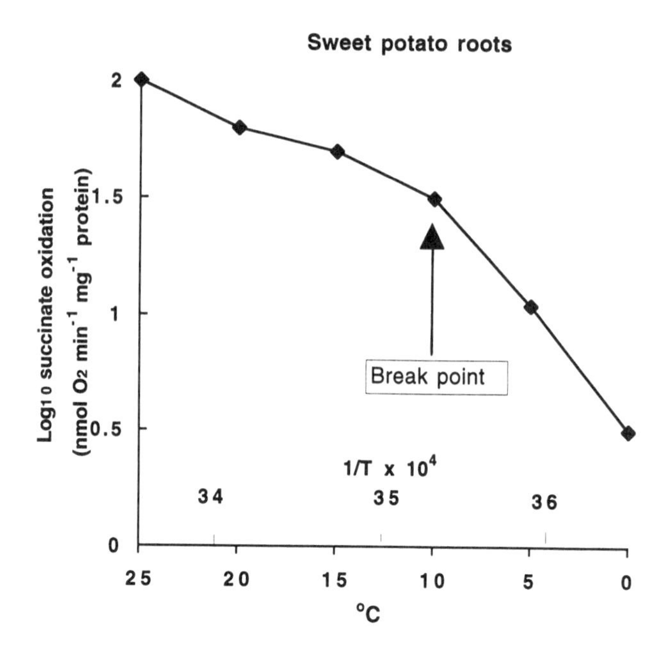

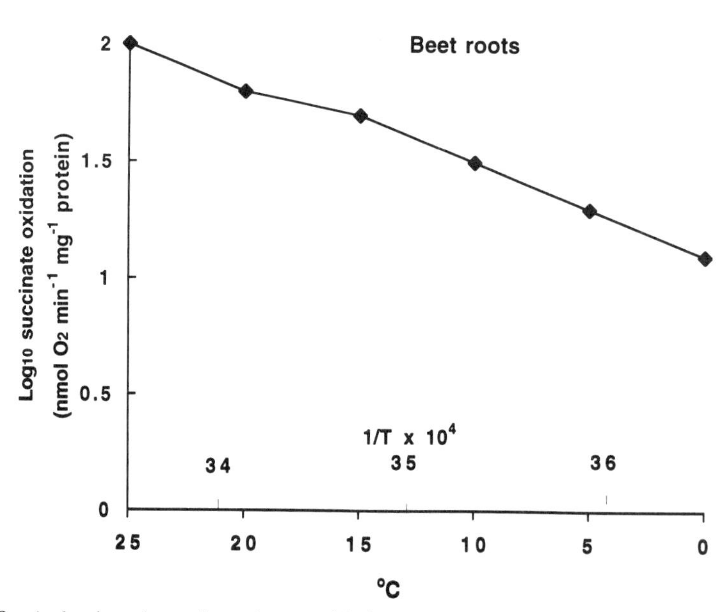

Figure 12.8 Arrhenius plots of succinate oxidation by plant mitochondria obtained from chill-sensitive (sweet potato) and chill-resistant (beet) plants. (Modified from Lyons and Raison 1970.)

TABLE 12.4 Dependency of Chilling Symptoms on Time in Various Compartments of Tomato Cotyledon Chilled at 5°C[a]

Chilling Time (h)	Plasma-lemma	Tono-plast	Mito-chondria	Plastids	Nuclear Envelope	Peroxi-somes	Micro-tubules
2	—	—	+	—	—	—	—
4	—	+	+	+	+	+	—
8	—	+	+	++	+	+	absent
12	—	+	++	+++	+	+	absent
16	+	+	+++	++++	+	++	absent
20	+	++	+++	++++	+++	++	absent
24	++	++	+++	++++	++++	++++	absent

Source: Modified from Ilker et al. 1979.

[a]—, no injury; +, slight; ++, moderate; +++, severe; ++++, extreme.

once it ruptures, there is little chance for the reconstitution of other cellular organelles once chilling is removed, and in the case of the tonoplast, its role in osmotic control is important with respect to chilling-induced dehydration. Both of these cellular structures could be viewed as last bastions for survival of the cell. Chilling-induced ion leakage of plasma membranes is illustrated for bean plants in Figure 12.9. In this experiment, bean

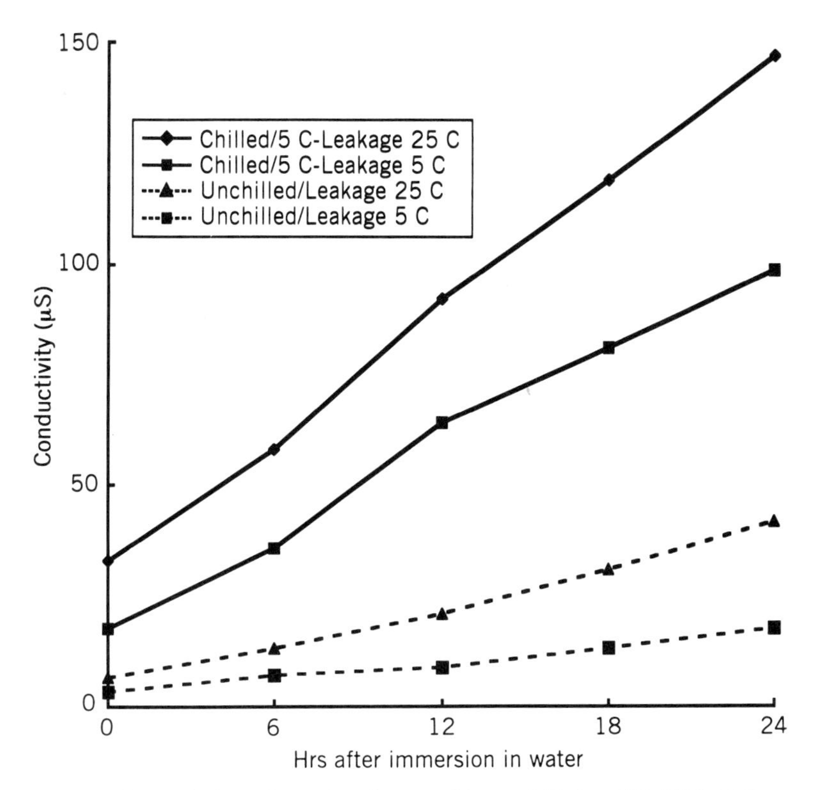

Figure 12.9 Leakage of electrolytes from leaves of bean chilled at 5°C (85% RH) compared to unchilled leaves placed in water at 25°C or 5°C. (Modified from Wilson and McMurdo 1981.)

leaves subjected to chilling at 5°C exhibited higher rates of ion leakage than nonstressed leaves. Leakage was also greater when measured at 25°C compared to 5°C. This response suggests that membrane aberrations may be the reason for the observed acceleration of injury symptoms when chilled plants are transferred from cold to warm temperatures.

Chilling-sensitive plants can be acclimated to low temperature by gradually decreasing the temperature over an interval of time or by subjecting plants to drought stress or in some cases ABA application (see Chapter 3). In the natural environment, day length may be key to initiating events that lead to acclimation. Herbaceous plants may only require reduced temperatures to initiate acclimation, while alpine and temperate species may require only photoperiod. Some species may rely on both temperature and photoperiod to activate the process. The process of acclimation (hardening) itself may involve a multitude of changes in the physiology and biochemistry of the plant. Figure 12.10 illustrates the hardening pattern in cauliflower leaves with seasonal changes in temperature. Note that as temperature declines, electrolyte leakage also declines, which is used as a measure of membrane permeability (Grout 1987). In some cases plants appear to accumulate carbohydrates (sucrose, raffinose, stachyose), sugar alcohols (sorbitol, ribitol, and inositol), proline, and glycine betaine that may serve to stabilize membranes (maintaining hydrophobic interactions and phospholipid headgroup stabilization) or proteins (by sustaining water around the proteins and maintaining the subunit conformation) associated with membranes as well as help to maintain favorable ionic and osmotic balances. In addition, enzyme kinetics (V_{max} and K_m) could be altered in favor of more efficient functioning under chilling stress. As indicated earlier, some enzymes are up-regulated and some are down-regulated, and in some instances new proteins have been observed to accumulate

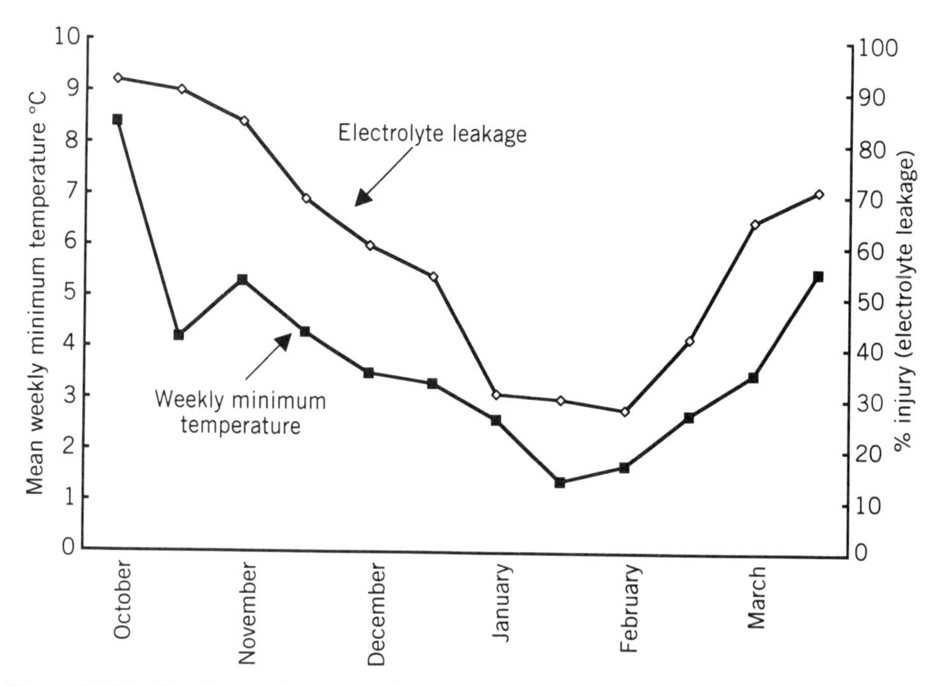

Figure 12.10 Correlation of mean weakly minimum temperature to hardening pattern in cauliflower as measured by electrolyte leakage. (Modified from Grout 1987.)

which may have a protective function. Some appear to have characteristics of heat-shock proteins, while others are different. Although the process of cold acclimation takes weeks or months to accomplish, deacclimation can be accomplished in a matter of days.

III. FREEZING STRESS

Two types of freezing processes are recognized in nature: radiation and advective freezing. The former usually occurs on cool clear nights when heat from the surfaces of plants is radiated back to the atmosphere, resulting in lower surface temperature than the surrounding atmosphere. Differential leaf surface and air temperature can lead to condensation of moisture on the leaf surface and can result in frost formation if the temperature is low enough. Advective freezing usually occurs when very cold air moving into an area from polar regions causes extremely low air temperatures that result in rapid freezing of plant parts (Bowers 1994).

As with chilling stress, plants differ in their ability to avoid or tolerate freezing stress. For example, some sensitive plants are killed at temperatures just below freezing (i.e., vegetable crops), while some cereal grains can acclimate to temperatures approaching −30°C, and many perennial woody species, that are fully acclimated, can tolerate temperatures as low as −196°C. However, the ability to tolerate or avoid freezing injury depends on other factors, such as rate of freezing and thawing, speed and depth of acclimation, plant tissue/organ differences, soil/plant–water relations, snow cover, nutrition, plant morphological characteristics, and growth habit. Freezing and thawing rates are very important. In nature, during spring and fall frost periods temperatures normally decline at a rate of 1 to 2°C/h. In the case of advective changes in temperature, much faster rates of decline can occur, and in the latter case can result in greater injury than that with slower rates of cooling. Figure 12.11 shows how early stages of seedling development differ in *Beta vulgaris* with respect to frost resistance (Cary 1975). Early stages of seedling development are more sensitive than later stages after true leaves have developed. Note the variance in frost resistance among tissues and the seasonal changes in tissue sensitivity.

Chilling-sensitive plants are killed or injured at temperatures between 0 and 15 to 20°C and are always killed at temperatures at 0°C or lower. Such plants normally occur in tropical or subtropical habitats. Plants that occupy temperate, high-latitude, or altitude environments generally can withstand freezing temperatures either by avoidance or by tolerance of freezing. One avoidance mechanism involves the process of supercooling, whereby cellular solutes accumulate in cells and lower the freezing temperature of the cytoplasm to well below the freezing point of pure water. This is accomplished by one of two ways: (1) metabolic synthesis of solutes such as sugars, polyols, and other osmotic molecules; or (2) by the movement of water from one tissue to another area that is less sensitive to freezing. This process leads to dehydration and accumulation of solutes in the cytoplasm, which ultimately lowers the freezing point. If carried too far, this process can lead to other problems related to dehydration stress. Another avoidance mechanism that has been shown to occur in some plants is the production of metabolic heat to elevate temperatures to prevent freezing. This mechanism has been observed in some species of the Areaceae and "snow plants" (Bowers 1994). How prevalent this process is among plants is not certain. Cacti and other succulents that are characterized by thick tissue mass and abundant water content may avoid freezing stress by accumulating

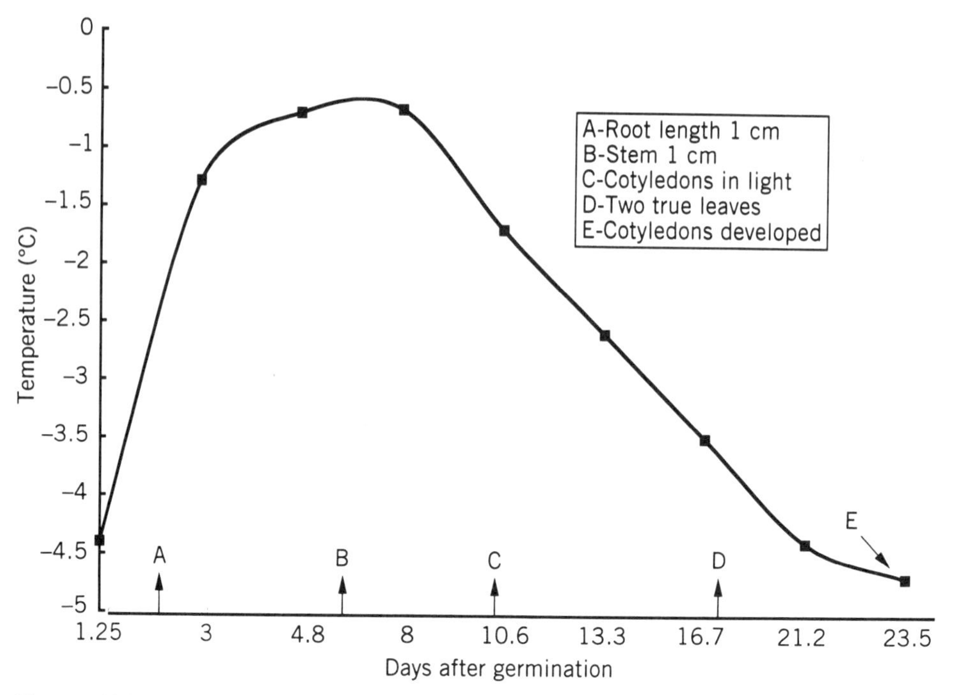

Figure 12.11 Frost resistance (LT50) during germination and early development of beet seedlings. (Modified from Cary 1975.)

residual heat during the day and slow dissipation of the stored heat during the cooling period, to the extent that the plant maintains a high enough temperature and long enough to avoid freezing. All of these mechanisms are attempts to avoid freezing–thawing stress.

Tolerance mechanisms include processes that allow ice to form in plant tissues without lethal consequences. This process usually involves extracellular ice formation (formation of ice in the apoplast of the cell or tissue, which includes cell wall space and intercellular space among cells and the xylem). Intracellular ice formation (ice formation within the cell) is generally lethal. However, if freezing–thawing occurs slowly, it may be possible for very fine structured ice crystals to form intracellularly and not be lethal. This appears to be an unusual occurrence, however.

Avoidance mechanisms seem to be more prevalent in herbaceous plants, while tolerance is more characteristic of woody perennials and evergreens. However, both mechanisms may coexist in the same plant but reside in different tissues. For example, vegetative and floral buds can supercool by moving water from the internal tissues to bud scales, where it freezes and causes no injury or reduced injury to those tissues. Similarly, the ray parenchyma of woody deciduous plants supercools, but the bark, including the phloem, tolerates ice formation under freezing conditions. Leaves of pine, spruce, cranberry, and citrus can supercool, but survival is primarily by tolerance to ice (Palta and Weiss 1993). Table 12.5 compares ice initiation temperatures of several woody perennials and herbaceous annuals. It appears that field-grown plants freeze just above the freezing point and that herbaceous annuals freeze at about 1°C below that of the woody perennials. This may suggest that an avoidance mechanism (supercooling) is operative in the herbaceous

TABLE 12.5 Temperature at Which Freezing Was Initiated in Plants under Field Conditions

Species	Mean Ice Initiation Temperature (°C)
Prunus persica	−1.6
Malus domestica	−1.3
Fagus sylvatica	−1.6
Cornus florida	−1.8
Ilex crenata	−1.6
Juniperus chinensis	−1.3
Pyrus communis	−2.1
Rubus spp.	−1.6
Pinus strobus	−1.2
Taxus cuspidata	−2.0
Lycoperiscon esculentum	−2.0
Zea mays	−2.5
Glycine max	−2.7
Phaseolus vulgaris	−2.7
Gossypium hirsutum	−2.5

Source: Modified from Ashworth and Kieft 1995.

plants as opposed to a tolerance mechanism (lower-temperature ice formation) in the woody plants. The woody perennials will survive the freezing temperatures, but the herbaceous annuals will be killed.

How, where, and when ice formation occurs in freezing-tolerant plants is critical to their survival and ability to limit injury. Acclimation to low temperature is an important component of the how and when part of the question of tolerance to freezing stress. Plants begin the process of acclimation by detecting environmental signals such as changes in photoperiod (short days) and changing temperature. Plants acclimate in ways not fully understood, but increases in the synthesis of organic solutes, specific proteins, and the retailoring plant membranes appear to be part of the process.

Ultimately, this process culminates in dormancy. The acclimation process has much to do with defining where ice forms in plants. Generally, the osmotic potential of the apoplast of cells and tissues is higher than the cytoplasm of cells. Thus, at freezing temperatures ice formation will begin in the apoplastic region first. This is particularly true if the freezing process is slow. Rapid freezing could result in symplastic cell freezing and death of the cell.

The initiation of ice formation requires ice nucleators (INs), which are frequently restricted to the apoplastic regions and possibly, tissue surface. Two classes of INs have been identified: heterogeneous and homogeneous nucleators. The former can include dust, specific species of bacteria, fungi, insects, wind, agitation, and even ice itself. Homogeneous ice nucleation refers to the temperature that pure water or a solution freezes without the help of heterogeneous nucleation and involves supercooling of the liquid or solution well below its freezing point. For example, pure water can supercool to −38°C before homogeneous IN causes it to freeze. Most IN occurs in plants via heterogeneous IN, probably as a result of normal bacterial flora on plant surfaces. However, it appears that plants may also possess intrinsic INs that appear to be chemically diverse (Brush et al. 1994). In rye plants, the composition and number of intrinsic INs varied with cold acclimation and day length. Plants grown at 5°C under short days had INs comprised predominately of proteins, while those grown at 20°C consisted of primarily carbohydrate and phospholipids. However, it appears that herbaceous plants, many of which are frost sensitive and important agriculturally, seldom contain significant number of intrinsic INs active at temperatures above −5°C (Lindow 1995 and references therein). Thus in such plants the primary determinant of frost damage at temperatures in the range 0 to −5°C are IN bacteria. Two species of plant epiphytic bacteria known to be INs are *Pseudomonas syringae* and *Erwinia herbicola*. The ability of IN bacteria to initiate ice formation in plants resides with their production of specific proteins that orient water molecules into an icelike lattice. The greater the density of IN bacteria on plant surfaces, the greater the degree of IN. Non-IN bacteria can compete with IN forms and reduce the incidence of IN. Control of IN bacteria through the use of antibiotics and non-IN bacteria that compete successfully with IN bacteria for plant surface area holds promise relative to protection of frost-sensitive plants to freezing temperatures.

Once freezing begins in the apoplast, water will move from the cytoplasm to the extracellular space down a vapor pressure gradient that leads to the growth of ice crystals until freezing equilibrium is reached between the liquid phase of the cytoplasm and the solid phase of the apoplast. The rate of ice formation is important relative to potential injury to the plant tissue. Generally, the faster the rate of ice formation, the more injury sustained. Two factors that can affect the rate of ice formation are membrane and cell wall porosity. Cell walls are comprised of microcapillaries of various sizes. Water within small microcapillaries freezes at a lower temperature than if it is associated with large capillaries (Ashworth 1993). Also, the smaller the pore size, the slower the rate of cellular water loss and thus ice formation. The greater the size and number of membrane pores also causes an increase in the rate of water loss. The membrane also serves as an effective barrier to intrusion of ice crystals into the cytoplasm. Once IN is initiated, ice formation usually begins in the xylem and can propagate throughout the vascular system at a rate as high as 60 cm/min (Chen et al. 1995 and references therein). Ice formation stops when it reaches a tissue with a higher temperature or encounters other barriers to its formation.

Ice formation is an exothermic reaction and at the time of transformation from liquid to solid heat is given off. In plants, two distinct exotherms occur, one at relatively high temperatures (high-temperature exotherm, HTE) which represents nonlethal extracellular ice formation and one at supercooling temperatures (low-temperature exotherm, LTE) that results from intracellular ice formation and results in cellular death. This has also been referred to as the *eutectic point*. The LTE may be characterized by multiple increases in temperature, which is thought to be a result of individual cells freezing at different temperatures. Figure 12.12 illustrates shifts in the HTE and LTE in lateral shoots of golden delicious apple trees. These results show clearly that the LTE shifts to lower temperatures as seasons change toward colder temperatures (Ketchie and Kammereck 1987). Exothermic reactions can be measured using temperature-sensing thermocouples and calorimetric measurements in studying ice formation in freezing-tolerant and freezing-sensitive plants.

The movement of water from the cell and the formation of extracellular ice leads to the dehydration of the cell and the accumulation of solutes within the cytoplasm. This has the advantage of lowering the freezing point of the cytoplasm, which in turn reduces the probability of intracellular freezing. The negative aspect of this process relates to the removal of water from the cell and the increase of ion concentrations, which could

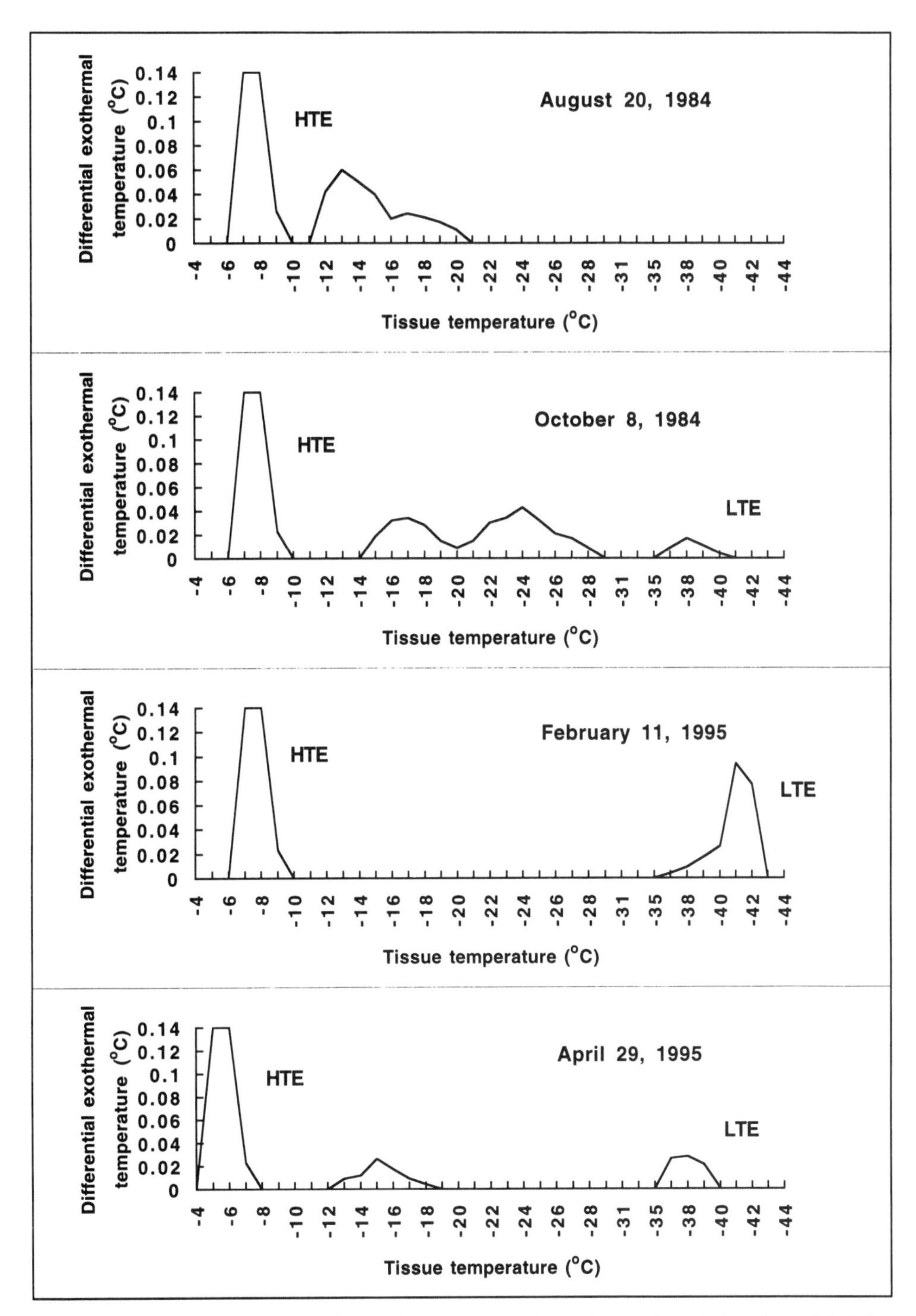

Figure 12.12 Seasonal changes in exotherm patterns in apple tree lateral branches. (Modified from Ketchie and Kammereck 1987.)

potentially affect membrane stabilization, protein structure, and ultimately, cell function. As water is removed from the cell and ice forms, the cell cytoplasm shrinks as a result of water loss and compression of the cell wall against the cytoplasm by the developing ice. Plants that form ice at relatively low temperatures, without appreciable supercooling, usually sustain less freezing injury than plants that supercool to low temperature. In the latter case, ice formation, when it occurs, is very rapid and usually intracellular, compared to ice formation at lower temperatures and extracellular. Resistance to freezing stress in plants may relate to the rigidity of the cell wall. A more rigid cell wall could ultimately result in less compression of the cytoplasm as ice crystals developed. This may have implications relative to the thawing process when injury is usually manifested. The cytoplasmic membrane is thought to be attached to the cell wall. The greater the shrinkage and compression of the cytoplasm, as a result of water loss and ice formation, respectively, the greater the potential for tearing and detachment of the membrane from the cell wall. In addition, when thawing occurs, in particularly rapid thawing, and the cell wall reassumes its original shape, the membrane could be damaged further by tearing it away from the cell wall as a result of rapid cell wall expansion. Slow thawing, on the other hand, could allow time for the membranes to reconstitute and "grow" with the expanding cell wall. To support this notion, synthesis of cell wall components that could increase cell wall rigidity and membrane attachment to cell walls have been shown to occur in plants undergoing low-temperature acclimation (Chen et al. 1995). Thus, in nature, episodes of rapid freezing and thawing are responsible for most of the injury that plants sustain in environments prone to freezing.

The physiological processes that are affected by freezing stress are in many ways similar to those affected by chilling stress but are accentuated. One of the primary differences relates to the development of intracellular ice and possible cavitation of cells that appears to lead to the death of cells and tissues (Ristic and Ashworth, 1993). Also, in freezing-tolerant plants the hardening process leads to a dormant condition that causes the down-regulation of all metabolic functions, including photosynthesis and respiration. During the early stages of hardening, photosynthesis and respiration increase, which probably represents the production of metabolites important in the hardening process (i.e., sugars, polyols, protective proteins, ice nucleators). However, during the latter stages of hardening, these processes decline. In freezing-sensitive plants, the inability to harden either through avoidance or tolerance mechanisms exposes membranes, which according to some represents the primary lesion leading to injury and metabolic processes to excessive stress to the extent that recovery is not possible.

Plants inhabiting extremely cold climates such as polar regions, the tundra, and alpine areas exhibit morphological and physiological characteristics that tend to provide an adaptive advantage for survival in such environments. Plants growing in such areas generally include herbaceous, mostly perennial rosette plants, tussock graminoids, cushion plants, and dwarf shrubs. Few woody trees grow in such areas because of permafrost and the low accessibility of roots to soil. Along with the cold temperatures, low moisture, high winds, and low nutrient levels are usually prevalent. The ability of plants to survive in these environments relates to their ability to adapt morphologically and physiologically to these environmental stresses. The short-prostrate stature of plants, small, thick ericoid pointed leaves, low biomass/leaf area, high root/shoot ratio, and the production of roots near the surface all represent important strategies developed by plants to deal with these stresses (Körner and Larcher 1988). Survival of plants in cold environments relates to their ability to utilize what thermal energy is available to support metabolic functions.

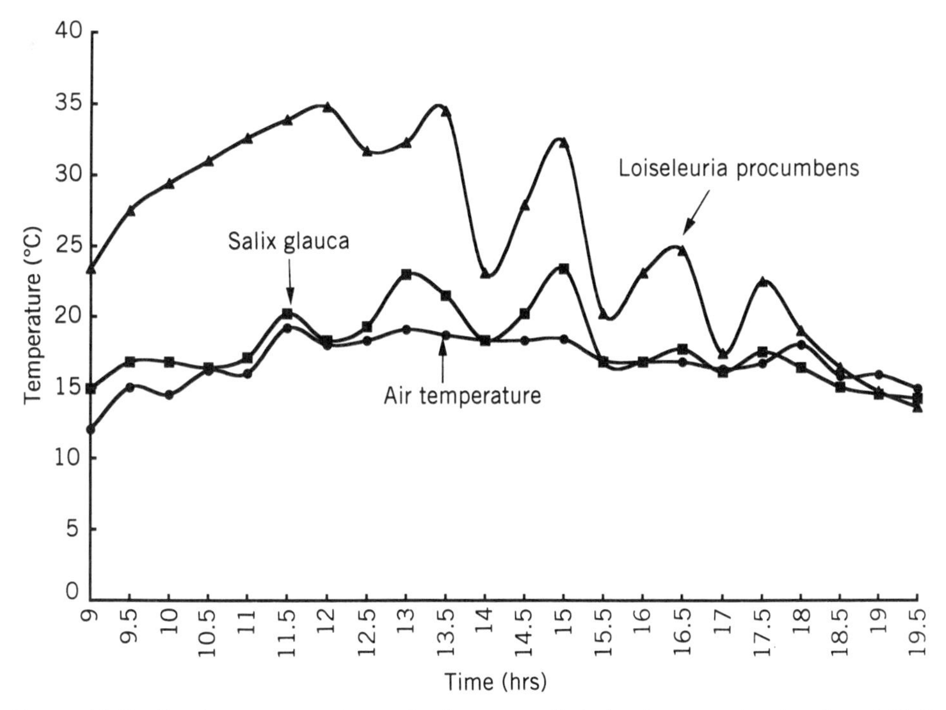

Figure 12.13 Leaf temperatures in stands of prostrate (*Loiseleuria procumbens*) and erect (*Salix glauca*) plants growing at 900 m in Norway. (Modified from Galuslaa 1984.)

The short, prostrate growth habit and reduced surface area afford the plant the opportunity to absorb heat energy radiated from the soil during the day and to avoid the cooling and desiccation associated with cold winds. Figure 12.13 illustrates the temperature differential between plant tissues and the atmosphere. Note that temperatures are higher in tissues closer to the soil, while root temperatures decline with depth. In some alpine regions, leaf canopy temperatures have been recorded to range from 25 to 35°C, with a maximum of 40°C, with an air temperature of 10 to 20°C during the same period. However, during the night, plant temperatures fall below atmospheric temperatures due to radiation cooling, but temperature differentials are less than during the day, ranging from 3 to 5 K.

It appears that cold-climate plants have developed efficient and opportunistic strategies that take advantage of the limited growth resources that are available to them. Much of this strategy relates to how they trap, store, respire, and allocate energy. Plants in these environments grow slowly but have photosynthetic rates that are comparable to plants growing in temperate and warmer environments. Some studies suggest that higher photosynthetic rates in cold climates is a result of elevated levels of rubisco (metabolic compensation). Most, if not all, are C_3 plants, but as a result of the cold climates in which they grow, photorespiration is kept to a minimum.

Cold-climate plants tend to store a large portion of their carbohydrate reserves in belowground organs and commit large percentages of photosynthate to maintenance and replenishment of roots and other belowground structures. This strategy is important

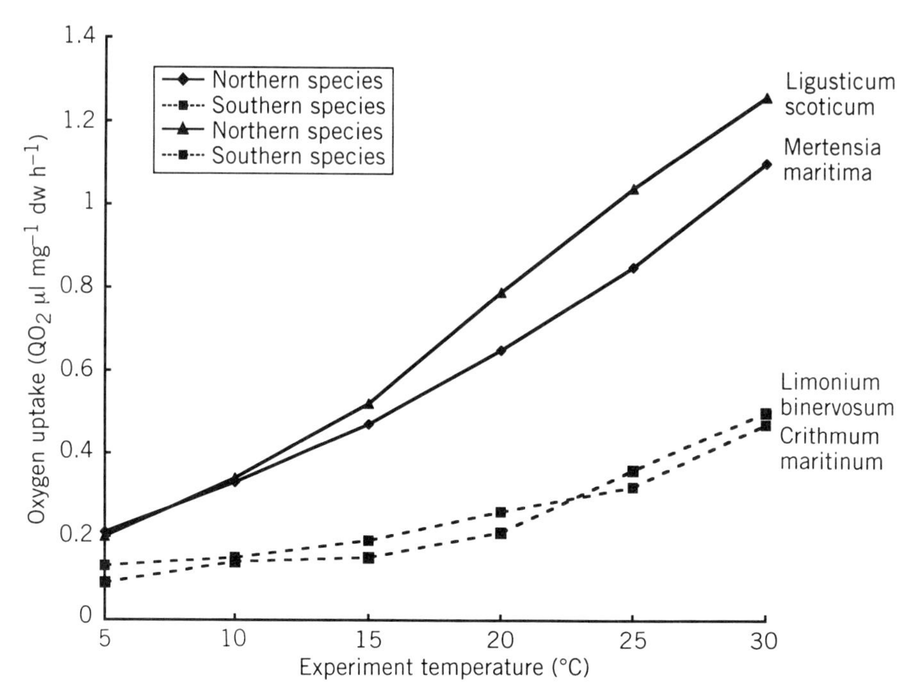

Figure 12.14 Influence of temperature on the respiration rate of two northern and two southern plant species. (Modified from Crawford and Palin 1981.)

because of the low soil nutrient levels and low water availability. In addition, when growing conditions are such that photosynthesis can occur, it is important that the plant can rapidly take advantage of the opportunities available. Thus cold-adapted plants typically have higher respiration rates than species growing in warmer climates as indicated in Figure 12.14. This affords the plant the opportunity to utilize stored reserves rapidly in support of growth under what are usually very short environmental windows conducive to growth. Although overall growth rates of cold-adapted plants are viewed as being slower than warm climate plants, comparison of graminoids growing in tundra and temperate grassland areas reveals that tundra plants during the time of limited growth have comparable growth rates of temperate grasses (Figure 12.15). Another advantage to storing large carbohydrate reserves in the roots relates to abnormally cold seasons when little to no growth occurs. The plant then relies on storage carbohydrate to sustain it during these times. In some cases, negative net carbon gains have been reported for plants in extreme cold climates, further illustrating the importance of large carbon stores.

Although cold climate areas are generally low in nutrients, particularly nitrogen and phosphorus, it appears that these elements are not limiting to growth of plants adapted to these environments. In fact, nitrogen levels are as high in these plants, on a weight basis, as other plants grown under what are usually considered sufficient conditions. This probably reflects the low growth potential of cold-adapted plants and the low biomass/element composition (Körner and Larcher 1988).

Reproductive strategies also differ among cold-adapted plants compared to temperate

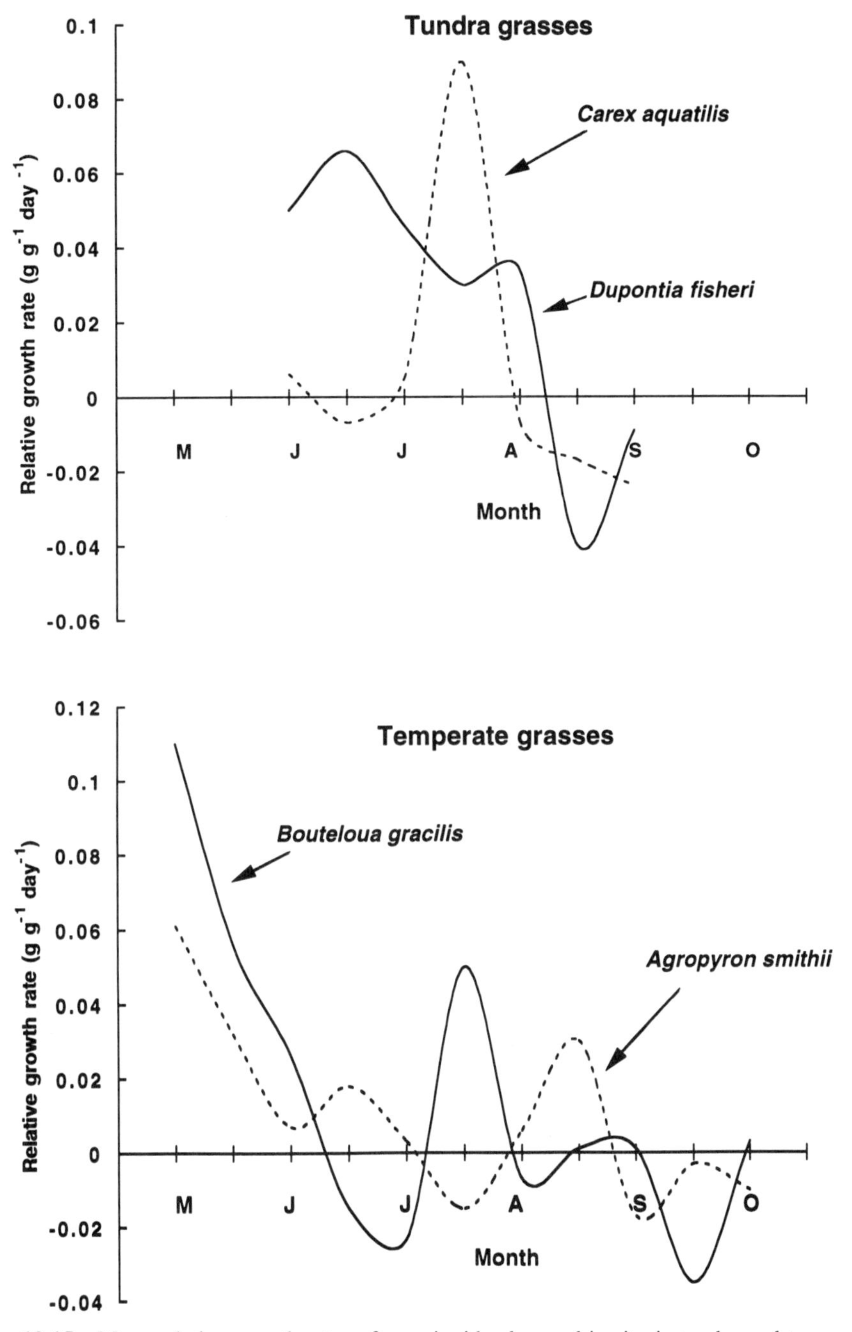

Figure 12.15 Mean relative growth rates of graminoids observed in situ in tundra and temperate grasslands. (Modified from Chapin et al. 1980.)

or warm climate plants. These strategies, like others these plants have evolved, reflect the need to conserve carbon in these harsh climates. It has been estimated that as much has 80% of the angiosperm species growing in some Scandinavian floras exhibit some type of asexual reproduction. The energy expense of sexual versus asexual reproduction has been estimated to be 10,000 times higher for sexual reproduction. Also a factor that may reflect on this trend is the lack of pollinators in these environments. When sexual reproduction occurs, dioecious plants appear to be more common. When flowers are produced, they are parabolic in shape, which tends to focus the sun's rays toward the center of the flower, which warms the reproductive structures and attracts pollinators. The flowers are usually dark in color and track the sun, which would lead to greater heat absorbance.

Thus it appears that cold-adapted plants have evolved a number of characteristics that allow them to survive extremely cold environments and actually restricts their distribution to colder climates in a manner similar to that which restricts the distribution of plants adapted to warm climates.

IV. SUMMARY

Plants have varying abilities to survive chilling (0°C to 15 to 20°C) and freezing temperatures (0°C and below). This is reflected in the global distribution of natural plant communities and the ability to sustain important food crops throughout the world. Chilling-sensitive plants are killed within the higher defined limits of chilling, while chilling-insensitive plants can tolerate temperatures near freezing. Chilling resistance and tolerance perhaps should be restricted to temporal differences in response to chilling stress. Freezing-sensitive plants are killed or injured at or just below freezing temperatures. Plants that are less sensitive resort to avoidance and/or tolerance mechanisms to survive. When plants are subjected to freezing stress, the aspects of where, how, and when ice is formed is paramount to the survival of specific plant species. Generally, intracellular ice formation is lethal to most plants.

The concept of primary and secondary events leading to chilling injury is important in defining temporal events that lead to plant injury. However, the identification of a single primary event leading to expression of secondary responses in plants may be complicated by the simultaneous expression of multiple biochemical/biophysical events that could be affected by low temperature.

Many aspects of the biochemistry and physiology of plants can be affected by cold stress. In some cases, depending on the plant, environmental conditions and the degree of cold stress, enzymes can be either up-regulated or down-regulated. Where up-regulation leads to the production of new proteins, metabolites that serve a protective function are frequently formed. When down-regulation occurs, important metabolic functions such as photosynthesis and respiration may be impaired. Such changes, however, could reflect the onset of dormancy, which ultimately benefits and allows plants to survive in freezing environments.

Plants typically adapted to cold environments have evolved growth habits, morphological characteristics, and efficient biochemical and physiological strategies that allow them to exist in such hostile environments. The intrinsic nature of these adaptations is expressed when these plants are taken from their native environments and grown under conditions seemingly more favorable for optimum growth.

STUDY-REVIEW OUTLINE

Chilling Stress

1. Chilling and freezing stress are distinguished by temperature ranges above or below 0°C that results in injury or death to sensitive plants.
2. Temperatures decline at higher latitudes, with increasing altitude and with distance from coastal regions.
3. Greenhouse gases and aerosols may be causing significant changes in global temperatures that could have a substantial impact on natural plant distribution and crop production regions.
4. Chilling-sensitive plants are generally restricted to tropical or subtropical regions, where temperatures range from 0 to 20°C. Chilling-resistant plant can grow near 0°C.
5. Plants can be acclimated (hardened) to low temperature by drought, declining temperatures, and/or changing day length.
6. Symptom expression in chilling-stressed plants differ with plant, degree of stress, and tissue. Symptoms may be expressed during chilling or after removal of the stress.
7. Cold-induced membrane phase transition has been identified by some as the primary lesion responsible for initiating events leading to symptom expression.
8. Weak bonding energy associated with protein configuration, membrane structure, nucleic acid structure, and biochemical interactions may represent multiple sites of cold injury in plants, making it difficult to identify a primary site of cold injury.
9. Under chilling stress, enzymes may either be up- or down-regulated, depending on plant species and degree of cold stress. Enzymes up-regulated may reflect acclimation processes, while down-regulated enzymes may reflect sensitive enzymes or those involved with dormancy induction.
10. Chilling stress can affect the velocity and affinity of enzyme activity. Cold and warm temperature–adapted plants respond differently to low temperatures relative to enzyme kinetics.
11. Plants subjected to high light intensity are more sensitive to chilling stress.
12. Cold stress can affect photosynthesis rates by inhibiting the light and dark reactions of photosynthesis.
13. Cold stress can cause reduced chlorophyll levels, blocks photosynthetic electron transport, reduces photosynthetic enzyme activity, and reduce stomatal conductance.
14. Respiration can either increase or decrease in plants subjected to chilling stress. If the stress is not severe, respiration increases and may reflect the acclimation process or increased activity of the cyanide-insensitive branch of normal electron transport.
15. Discontinuities in Arrhenius plots in which membrane-associated enzyme activity is measured against decreasing temperatures have been used as evidence for membrane phase transition as being the primary event involved in cold stress.
16. Membranes of cellular organelles have different sensitivities to cold stress.
17. Cold acclimation may be initiated in plants by photoperiod and/or declining temperature but differs with plant species.
18. Sugars, polyols, and proteins accumulate in cold-acclimated plants and appear to be important in low-temperature resistance.

Freezing Stress

1. Plants are subjected to radiation and advective freezing.
2. Plants can avoid freezing injury by supercooling, production of metabolic heat, or storing absorbed heat energy.
3. Plants tolerate freezing injury by forming extracellular ice.
4. Intracellular ice formation is generally lethal.
5. Freezing injury and lethality are related to freezing and thawing rates.
6. Extracellular ice formation can result in dehydration and ionic stress.
7. Freezing tolerance and avoidance mechanism may coexist in different parts of the same plant.

8. Ice formation is initiated by ice nucleators.
9. The nucleation of ice at temperatures closer to 0°C is less injurious to plants than nucleation at lower supercooling temperatures.
10. Factors affecting rate of ice formation include cooling rate, cell wall porosity, and membrane porosity.
11. High- and low-temperature exotherms occur in the freezing process. The former is thought to represent extracellular ice freezing, while the latter represents the freezing of intracellular water or possibly, bound water.
12. Plant differences or shifts in exothermic reactions with decrease in temperature may reflect resistance or acclimation to cold stress.
13. Freezing injury may be related to cell wall rigidity and the ability of membranes to re-form as the cell wall re-forms after the cell thaws.
14. Physiological, biochemical, and acclimation processes are similar in plants subjected to chilling and freezing stress, except in the latter the responses are accentuated and ice formation is involved.
15. Plants inhabiting cold environments have adapted morphological, physiological, and biochemical mechanisms that allow such plants to exist in cold, low-nutrient, and low-moisture environments.
16. Photosynthetic rates of cold-adapted plants approach those of temperate plant species.
17. Generally, respiration rates of cold-adapted plants are higher than those of plants that are adapted to warmer climates.
18. Plants adapted to cold environments have developed reproductive strategies that conserve and maximize energy utilization.

SELF-STUDY QUESTIONS

1. Define chilling and freezing stress.

2. How do plants differ from animals with respect to their abilities to deal with cold temperatures?

3. Describe the range of cold temperatures globally.

4. What evidence suggests that global cooling could be counteracting the greenhouse effect, at least in industrialized nations?

5. Define chilling-sensitive and chilling-resistant plants.

6. Outline the range of plant symptoms that can result from chilling stress.

7. What is the difference between a primary and a secondary lesion?

8. What is considered to be the primary lesion relative to chilling stress in plants?

9. What significance do weak bonding energies have with respect to chilling stress in plants?

10. Specify biochemical and structural components of cells that exhibit weak bonding energies.

11. How does chilling stress affect proteins (enzymes) both functionally and biochemically?

12. Name some enzymes that are up-regulated and down-regulated in response to chilling stress.

13. What physiological/biochemical significance can be attributed to question 12?
14. Describe how chilling stress can affect enzyme kinetics.
15. Summarize the effects of chilling stress on photosynthesis and respiration in plants.
16. What are Arrhenius plots, and how can they be used to determine how chilling stress can affect physiological and biochemical responses in plants?
17. Review the relative sensitivities of cell oganelle membranes to chilling stress. What possible significance might this have relative to cell survival?
18. Define cold acclimation (hardening) and indicate what is involved in the process from a physiological and biochemical viewpoint.
19. Define advective and radiation freezing as it applies to plants.
20. Define frost avoidance and tolerance and give examples of each.
21. Describe the process of ice formation in a frost-tolerant plant.
22. What are ice nucleators, and what is their importance in ice formation and frost tolerance in plants?
23. Aside from ice formation, what other stresses can occur as a result of extracellular ice formation in plants?
24. What factors affect the rate of ice formation, and why is rate important?
25. Explain where ice forms in cells and tissues and how rapidly it progresses throughout the plant.
26. What are high and low temperature exotherms, and what is the significance of determining them in plants?
27. Why is the thawing process considered to be critical relative to freezing injury in plants?
28. What roles may the cell wall and cytoplasmic membrane have in freezing injury in plants?
29. Outline growth habit and morphological adaptations characteristic of plants found in cold climates.
30. What other stresses besides cold are plants exposed to in cold climates?
31. What physiological and biochemical strategies do cold-adapted plants have that allow them to exist in cold climates?
32. Why can't some plants adapted to cold climates be equally successful or more so in warmer habitats?
33. How do rates of photosynthesis and respiration of cold climate plants compare with temperate climate plants?
34. Describe the allocation of carbon in cold-climate-adapted plants.
35. What reproductive strategies have cold-climate plants evolved, and what advantage to the plant are these strategies?

SUPPLEMENTARY READING

Crawford, R. M. M. 1989. Studies in Plant Survival: Ecological Case Histories of Plant Adaptation to Adversity (Studies in Ecology, Vol. II). Blackwell Scientific Publications, Boston.

Katterman, F. 1990. Environmental Injury to Plants, Academic Press, New York.

Li, P. H. 1989. Low Temperature Stress Physiology in Crops. CRC Press, Boca Raton, FL.

Sakai, A., and W. Larcher. 1987. Frost Survival of Plants. Springer-Verlag, New York.

13 Multiple Stress Interaction

Outline

Objectives

1. Discuss the importance of stressor interaction to the evolution of plants.
2. Describe several specific combinations of resources that are important regulators of plant growth form and metabolic function.
3. What are some of the synergistic stresses that enhance the detrimental impact of either stress on plants? For instance, what is the synergistic impact of low temperature and high light on photosynthesis?
4. Describe some specific cases of antagonistic stressors that relieve or ameliorate the detrimental affect of another stressor. Include as an example how cold acclimation enhances water stress tolerance.
5. Discuss some anthropogenic stressors that interact with natural stressors of the environment. For example, consider the effects of atmospheric CO_2 concentration, gaseous pollutants, and acid rain on plant response to natural environmental stress.

I. INTRODUCTION

Throughout this book we have considered individual stressors, their detrimental impacts on plants, and various characteristics of plant structure and function that ameliorate those detrimental impacts. However, natural systems are not characterized by variations in single factors. Quite the contrary, many environmental attributes vary simultaneously within and among sites. Some of these factors are mechanically linked (such as high light and heat), while other factors vary independently (e.g., heavy metals in soils and ozone in the troposphere). The response of plants to one factor may be influenced by variability in other attributes of the environment. Plants have been exposed to multiple-factor interaction throughout their evolution because of the nature of their environment (Figure 13.1). Therefore, the study of single-factor responses in plants is somewhat artificial.

Why is it that there are many studies on single-factor responses in plants and far fewer studies on the integration of multiple factors on plant stress physiology? One reason for this phenomenon is the basic nature of science. To determine the impacts of one factor on a system (plant in our case), all other factors must be kept stable. If other attributes of the environment are not stable, it is very difficult to discern the response of plants to the factor in question. All experimental design classes stress the importance of a control and background stability, even in the case of multiple-factor analysis. In addition, the statistical treatment of an experiment with one dependent and one independent variable is much simpler than that of an experiment with multiple independent factors. The number of replications and the complexity of statistical design increase exponentially with increases in independent variables. For example, if one factor is studied over four states, four treatment combinations, each with a set of replicates, are needed. However, if two factors are studied (with four states) along with their interaction, then 16 treatment combinations are

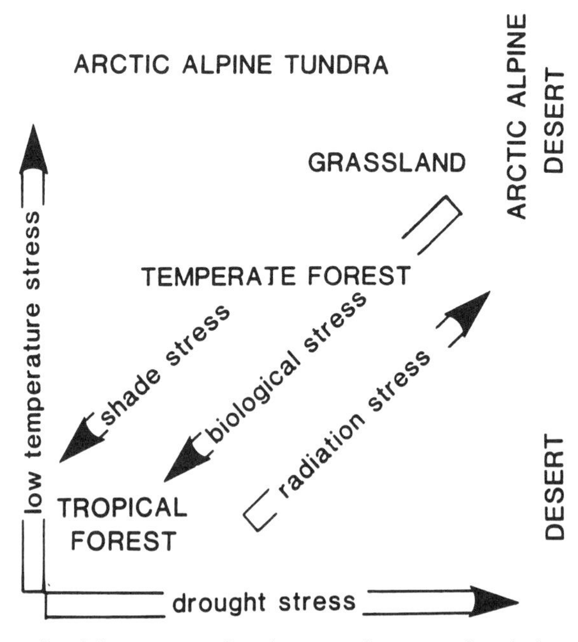

Figure 13.1 Schematic of the patterns of environmental stressors in relation to major vegetation types. The basic matrix is dependent on low-temperature stress and drought stress, while other stresses are embedded inside this matrix. (Adapted from Osmond et al. 1987 Figure 1.)

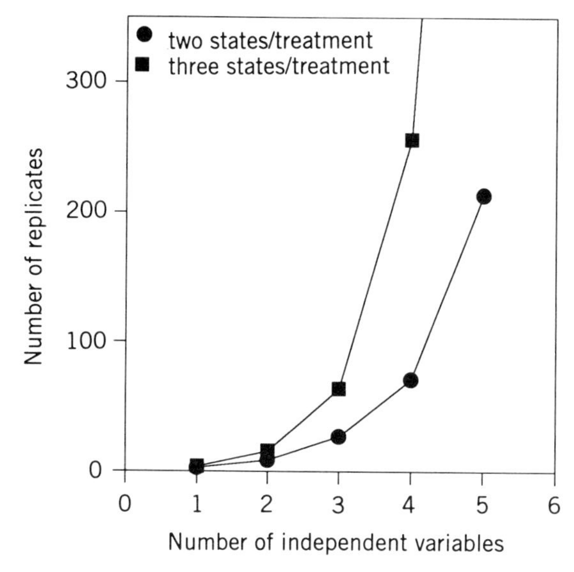

Figure 13.2 Relationship between the number of replicates (plots or experiments) in experimental designs (two or three states per treatment) with increasing numbers of independent variables. Independent variables in this case refer to the different stressors involved in the experiment. Note the exponential nature of the relationship.

required. For three factors, 64 combinations are needed (Figure 13.2). Because of the greater complexity in statistical treatment, experimental designs are more complicated, and greater replication (i.e., more degrees of freedom) may be required to discern treatment effects. The larger requirement for replications requires more space and funds, which may often be limiting factors in scientific research.

Although experiments on multifactor interaction may be difficult and expensive relative to single-factor experiments, information is available on stress factor interaction and plant performance. Some of that information has been gathered in experiments specifically designed to analyze complex interactions between stressors, while other bits of information come from descriptive measurements of natural processes. Before discussing specific cases of stressor interaction, we consider some theoretical implications of multiple stresses and plants.

A. Law of the Minimum

Resource limitation is one class of stressor that plants experience in all habitats. Any one of a number of different resources at low availability can limit maximum plant growth potential. However, it is likely that a number of resources may be limiting simultaneously in any environment. Is plant performance regulated by one most important (primary) limiting resource? If this were true, raising the availability of any other limiting resource would not improve plant growth because the primary limiting resource is constraining growth. This is *Liebig's law of the minimum,* which states that plant growth is limited by a single resource at any one time; only after the resource limitation is relieved can another resource influence plant growth (Figure 13.3). In contrast to Liebig's law of the minimum, one could argue that plants have evolved to compensate for resource imbalances (compensatory theory), and growth should be limited equally by all resources simultaneously (Bloom et al. 1985).

There are several experiments which demonstrate that more than one factor limits growth and reproduction at any one time. In Alaskan tundra, growth of most species responds to an increase in either light, nitrogen, or temperature (Chapin and Shaver 1985).

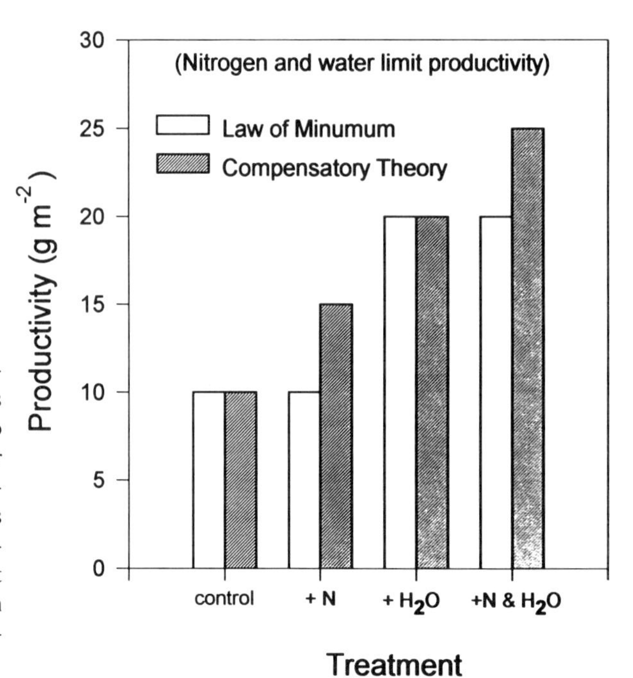

Figure 13.3 Diagrammatic representation of growth responses expected when plants respond to additions of nitrogen, water, or both. Open bars represent the predicted response when growth is regulated by the law of the minimum. Hatched bars represent the predicted response when growth is regulated by compensatory theory.

In desert systems, plants increase growth in response to both water and nitrogen amendments (Lauenroth et al. 1978, Sharifi et al. 1988). The photosynthetic rate of subcanopy plants is increased by increasing light and increasing water (Lipscomb 1991). In fact, results in most resource limitation studies in the field, greenhouse, or growth chambers suggest that plants are limited simultaneously by a number of resources.

Several important significances of compensatory theory in plants are suggested by Chapin et al. (1987). First, if there are multiple resources limiting growth, this suggests that the efficiency of resource acquisition has been maximized. If only one factor is limiting growth, this suggests that the plant has excess capacity for acquiring the other nonlimiting resources. The energy and materials used to develop the excess capacity for acquiring other resources could be directed toward the machinery needed to acquire the limiting resource and would result in increased growth. Second, if a plant is limited simultaneously by several resources, this means that increases in a number of different resources could increase growth and reproduction. This improves the chances that growth will be increased because any one of a number of different occurrences would enhance growth rather than only one type of occurrence. Third, there is evidence that when plants are limited by several resources, increases in two or more of those limiting resources has a synergistic impact on growth (Peace and Grubb 1982, Lipscomb 1991).

B. Functional Convergence

It has been argued by several authors that plants may function in an economical manner similar to the way a manager operates a firm (Bloom et al. 1985, Mooney and Gulmon 1982, Field 1991). Managers stop investment in one part of the company when investment in that part falls below the return of investment in another aspect of the company. In the case of plants, allocation of resources into biochemical capacity for carbon gain should be inhibited whenever a limitation in any resource prevents capitalization on the increase in carbon gain capacity. In other words, it is inefficient for plants to allocate a lot of nitrogen to synthesize rubisco if water availability is scarce and stomatal conductance is low (limitation by carbon dioxide diffusion).

Functional convergence is best defined as follows: Biochemical capacity for carbon reduction should be reduced whenever a limitation in the availability of any resource prevents the effective exploitation of that capacity. A good example of this relationship is that plants have evolved slow growth rates (low carbon gain) in environments that have low calcium availability (Chapin 1980), even though low calcium does not have a direct effect on carbon gain processes. Thus, functional convergence of metabolic processes suggests that biochemical capacity for carbon acquisition may reflect limitations by all resources (which directly or indirectly affect photosynthesis). As such, biochemical carbon gain capacity may be a master integrator of environmental stresses and would serve as the best indicator of the magnitude of environmental stress on plants.

Of course, plants are not company managers, but the analogy brings up important concepts of resource use efficiencies and their relation to plant evolution. It is not likely that evolution has been perfect in its modeling of plant genotypes such that all plants have the ability to balance their biochemical capacities precisely with the nature of resource limitations of their environment. There are certainly going to be genetic constraints on the ability of plants to modulate metabolism on an evolutionary scale. In addition, changes in allocation patterns to maximize resource use efficiency in a temporally dynamic environment will encounter metabolic constraints. However, the overwhelming evidence suggests that compensatory adjustment in metabolic processes in response to multiple resource limitation (within limits) is the rule rather than the exception.

C. Why Study the Effects of Multiple Stresses?

There are several important reasons (both applied and theoretical) for studying the interactive impact of stresses on plant function. First, it is clear from the discussion earlier that overall growth is regulated by multiple resource limitations rather than one primary limiting factor. Therefore, to understand the regulation of plant growth, the interaction among various stressors on plant function must be understood. Second, the economic analogy discussed previously clarified how natural selection might affect metabolic processes in plants. Multiple stress interaction (resource limitations) was a main factor leading to genetic adaptation. Therefore, to understand adaptation in plants, the influence of multiple stressors on plant metabolism must be understood. Third, some resource limitations and over-abundance lead to constraints on plant response to other resources. It would be difficult to interpret experiments on plant response to one resource if, unbeknown to the researchers, another resource is constraining plant response to the former. Fourth, evolutionary processes in natural systems has resulted in species that respond differently to multiple stress combinations. Therefore, shifts in the balance among resource limitations and/or environmental stresses may influence competitive hierarchies among coexisting species. In fact, plant responses to shifts in the balances among multiple resource limitations are central themes in recent theories on competition and succession in natural ecosystems (Tilman 1988).

There are also important factors for studying multiple stress interactions in applied sciences. There are many changes in agricultural and natural systems that occur because of the effects that human populations have on environments. To understand the consequences of these changes to agricultural productivity in a number of different climatic regions, it is important to understand the interaction between natural resources and artificial environmental perturbations. This was the rational for initiating the National Study of Atmospheric Pollution Impacts on Agricultural Production (NAPAP). Dramatic shifts in climate may occur in the future because of gaseous emissions into the atmosphere. Those changes in climate will have different effects on crops based on other stresses faced by specific crops in specific regions.

In the rest of this chapter we focus on two basic aspects of plant response to multiple stressors. The most important combinations of resource limitations that shape plant form and function will be discussed. Then examples of specific interactions (synergistic and antagonistic) will be discussed to develop some integrated views of plant stressors and how they react with anthropogenic perturbations of the environment. The examples we selected are not designed to be inclusive; rather, they are selected to elucidate particularly important patterns of relationships among environmental stressors.

II. EFFECTS OF MULTIPLE RESOURCE LIMITATIONS ON PLANT FUNCTION

The structure and function of plants have been sculpted over time by evolutionary processes, including adaptation to multiple resource limitations (Chapin 1991). The effects of multiple resource limitations on plants can be observed by examining how resources are allocated within plants in response to resource limitations. One of the clearest examples of resource allocation patterns and its association with maximizing growth in various environments is that of nitrogen distribution within and among tree and shrub canopies. In particular, this topic concerns the patterns of nitrogen distribution in canopies with varying light availability. Is nitrogen allocated among leaves of a canopy in a manner that optimizes potential productivity?

A. Canopy Resource Use Optimization (Pn_{max}–N relationship)

What is meant by *optimization of resource allocation?* A resource can be defined as optimally distributed if any change in its distribution will reduce canopy carbon gain. Two main characteristics of trees and shrubs exert a major impact on canopy carbon gain. First, the architecture of branching, morphology of leaves, and spatial placement and orientation of leaves will determine the pattern of light on and within the canopy. Species with sparse, widely spaced branches, and small leaves will have a different extinction coefficient of light through the canopy than that of densely branched canopies with large leaves. Subcanopy plants in high-light sites will have a different distribution of light in their canopy than that of subcanopy plants under a dense canopy. Deciduous plants will have a greater amount of within-canopy sun flecks than that of conifers. If functional convergence has optimized canopy carbon gain, each scenario suggested above should have a different distribution of carbon gain capacities among leaves in the canopy.

The second main factor that affects carbon gain optimization in canopies is the temporal and spatial distribution of nitrogen in the canopy. Several studies have demonstrated that potential carbon gain (measured as Pn_{max}, or the maximum potential photosynthesis)

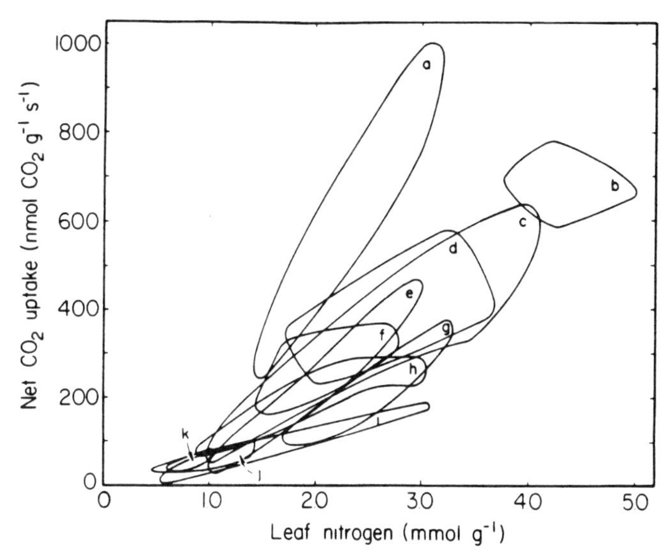

Figure 13.4 Relationship between photosynthesis and nitrogen for a wide variety of species. Each outline circumscribes all the data reported in a single study. Included are data from 33 C_3 species (b–k) and 11 C_4 species (a). The experimental variables used to induce variation in tissue nitrogen concentration were nitrogen availability, light availability, and leaf age. (From Chapin et al. 1987 and adapted from Field and Mooney 1986.)

is linearly related to tissue nitrogen concentration (Figure 13.4). There are two important points about this relationship. First, the relationship holds across species and ecosystem lines. Thus, the relationship between leaf nitrogen and maximum photosynthetic capacity is a universal aspect of plant metabolism. Second, the relationship persists when considering leaf aging (Field and Mooney 1983), light availability (Hirose et al. 1989), nutrient availability (Sage and Pearcy 1987), or intrinsic differences among species (Field and Mooney 1986). Third, the relationship between Pn_{max} and N is due to intrinsic aspects of basic plant metabolism. The importance of basic metabolic properties to the relationship between Pn_{max} and N is shown by the facts that the sensitivity of Pn_{max} to N is greater in C_4 than in C_3 species (Pearcy et al. 1982), greater in sun-adapted than in shade-adapted species (Seemann et al. 1987), and greater in mesophytic than schlerophytic leaves (Evans 1989).

The ultimate reason for this relationship is not due to the proportional amount of nitrogen in photosynthetic versus nonphotosynthetic components of cells. The determination of Pn_{max} is performed under conditions where the main limiting factor for photosynthesis is rubisco (see Chapter 6). Thus the relationship between Pn_{max} and nitrogen should be dependent on the amount of leaf nitrogen contained in rubisco. However, only about 5 to 30% of leaf nitrogen is contained in rubisco (Seemann et al. 1987), and this is not a high enough proportion to dominate the nitrogen sensitivity of Pn_{max}. Field (1991) outlines the most likely mechanism that regulates the Pn_{max}–N relationship: "Plants regulate investments in biochemical capacity for photosynthesis so that the limitation by factors other than biochemical capacity is held below a level that depends on the costs of increasing photosynthetic capacity." This explanation is dependent on compensatory theory and is based on the observations that (1) the Pn_{max}–N relationship is consistent across wide species and habitat differences, (2) the $Pn_{max}-N$ relationship is robust during various environmental stress conditions, and (3) the $Pn_{max}-N$ relationship is sensitive to the identity of limiting resource. Since the relationship between $Pn_{max}-N$ is universal and robust, the distribution of nitrogen in a canopy will regulate canopy potential carbon gain.

If the canopy has an optimized carbon gain, the distribution of light in the canopy (one resource limitation) must be coordinated with the distribution of nitrogen (a second limiting resource) in the canopy. Therefore, nitrogen within a canopy should be distributed in proportion to the distribution of absorbed radiation (Farquhar 1989). Assuming that there is a constant absorptance for leaves in the canopy, the Pn_{max}/leaf area and N/leaf area should both change proportionally with the mean light interception for leaves in a canopy. In most if not all studies of canopy resource distribution there is a parallel change in Pn_{max}/leaf area and N/leaf area among layers or groups of leaves (Field 1983, Hirose et al. 1989, Hollinger 1989, Ellsworth and Reich 1993, Kull and Niinemets 1993). It is clear that the distribution of nitrogen among leaves in a canopy is established in coordination with the distribution of light in the canopy. Therefore, the requirement to change nitrogen distribution patterns among leaves with changes in plant architecture or microclimate in order to maximize productivity has been a major process regulating the structure and function of plants.

Specifically in relation to environmental stress, the $Pn_{max}-N$ relationship is normally maintained to some degree in controlled experiments of plant response to individual stressors. Commonly, both N/leaf area and Pn_{max}/leaf area decrease in a parallel manner during drought stress (Walters and Reich 1989), nitrogen limitation (Sage and Pearcy 1987, Walters and Reich 1989), low light (Seemann et al. 1987), or pollutants (Reich and Schoettle 1988). In some cases the parallel change in N/leaf area and Pn_{max}/leaf area does not occur, particularly during phosphorus-limitation experiments (Reich and Schoettle 1988, Mulligan 1989). Therefore, the linkage between Pn_{max} and tissue nitrogen concentration is generally applicable to both the distribution of resources in a canopy and the response of whole plants to individual stresses.

B. Light and Heat Dynamics and Subcanopy Plants

In Chapter 10 the light dynamics of subcanopy environments was discussed, as was leaf temperature dynamics of subcanopy plants in Chapter 11. However, these two chapters dealt primarily with one stress factor, even though both light absorptance and temperature changes in leaves occur concurrently. Many researchers have shown that efficient use of sun flecks has a major impact on leaf carbon gain in subcanopy plants. Since light response is important to photosynthesis in subcanopy plants, and other factors such as leaf temperature and VPG change in a sun fleck, it is logical to assume that there will be interaction among the effects of these factors upon photosynthesis in sun flecks.

During the time between sun flecks, leaf photosynthetic machinery acclimates to low light intensity. Therefore, it requires a long sun fleck duration, or repeated sun flecks, to induce maximum photosynthesis. Does photosynthesis also acclimate to other environmental factors between sun flecks? There is very little information on photosynthetic response to leaf temperature dynamics during sun flecks. In fact, most studies of sun fleck dynamics maintain leaf temperature constant during the experiment. In one study on *Rhododendron maximum,* temperature response curves were done along with measures of temperature variation in response to sun flecks (Bao and Nilsen 1988). Leaf temperature was between 20 and 25°C between sun flecks and 28 to 34°C during sun flecks (Figure 13.5A). In comparison, the optimal temperature for photosynthesis was between 27 and 37°C (Figure 13.5B). Therefore, during the time between sun flecks, photosynthesis is

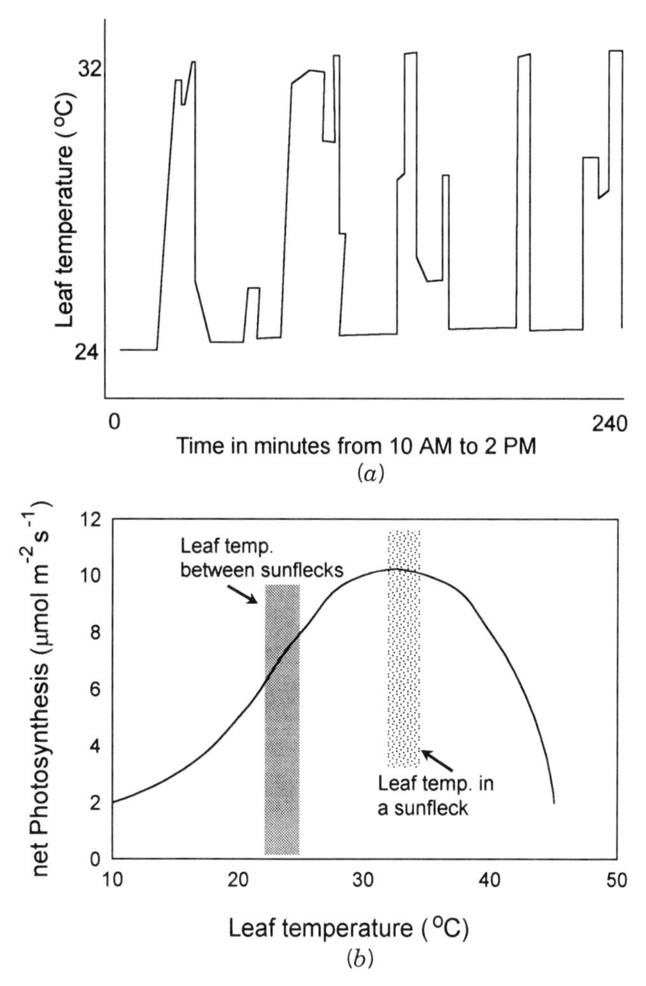

Figure 13.5 (*a*) Plot of leaf temperature for *Rhododendron maximum* growing in the subcanopy of a deciduous forest. High leaf temperatures occur during sunflecks. (*b*) Temperature response curve for *Rhododendron maximum* photosynthesis.

limited by temperature and light, while during a sun fleck, leaf temperature moves into the optimum for photosynthesis so that the leaf can take advantage of increased light.

There is also likely to be an interaction between light variation and vapor pressure gradient between the leaf and the air. During a sun fleck, leaf temperature may increase relative to the air above the leaf. When leaf temperature increases, the internal vapor pressure of the leaf increases in comparison to that in the air, which causes VPG to increase. The increase in VPG induces an increase in transpiration at any stomatal conductance. Thus, in a sun fleck, transpiration increases even if stomatal conductance remains constant. The greater water loss in a long sunfleck may induce lower leaf water potential, which may induce stomatal closure and wilting. The lower stomatal conductance may constrain the ability to capitalize on sun fleck light (Figure 10.26). The consequence of these dynamics between light, heat, and water were considered during the development of long-term sun fleck response theory in plants (see Chapter 10 for further information on water spenders, thermal avoiders, and thermal tolerators).

C. Herbivore Induction of Defensive Compounds Versus N and C Costs

Herbivores constitute a type of biotic stress in plants. Any plant in any habitat is likely to harbor any number of different herbivores. Trees will have stem borers, leaf miners, aphids, caterpillars, root nematodes, and any number of other classes of herbivores. Since herbivores cause a reduction in potential growth, the presence of herbivores is an important stress that induces adaptation and acclimation in plants. Does the induction of defensive mechanisms in plants constrain the ability of plants to respond to other resources?

There have been several theories developed about why specific plants allocate a greater portion of their resources to defense compared with other species. One theory predicts that the amount of investment in defense is determined by the apparency of plants to herbivores (Feeny 1976). Apparency was determined by the predictability of locating plants in space and time. Thus, late successional species that are long lived, or crops that are grown in monocultures, are more apparent to herbivors, while early successional plants that are ephemeral are less apparent. Apparent plants are more likely to have evolved effective defensive mechanisms (relatively resource expensive) against specialist herbivores. In contrast, unapparent plants are likely to depend on some weaker qualitative defenses against generalist herbivores. Thus, plant apparency was suggested to determine whether selection would favor expensive defenses against specialist herbivores or inexpensive defenses against generalist herbivores.

Upon closer examination it was observed that differences in plant apparency coincided with differences in environmental resource limitation (Coley et al. 1985). Those sites with high resource limitation (late successional, for example) have plants of low apparency, and sites with less resource limitation (early successional) have plants with greater apparency. One of the first studies to show that the influences of apparency and resource limitation on plant investment in defense could be separated was a study of defense investment in British plants (Grime 1977). Three groups of plants were defined, each of which contained a specific structural and functional set of attributes shaped by interactions between resource limitation and biotic stresses such as herbivory. Coley et al. (1985) explain that the apparency theory is not supported by several lines of evidence. First, apparency theory predicts different defensive mechanisms against specialist and generalist herbivores. This is not generally supported by empirical evidence. Second, there is no evidence that apparent species have a higher incidence of specialist herbivores compared to unapparent species.

Even though apparency theory does not solidly explain the distribution of herbivory defense mechanisms, it is likely to be an important corollary factor influencing resource limitation theory. Resource limitation selects for inherently slow growth rates. Under slow growth rate leaves must be retained for longer periods, and therefore the leaves experience longer exposure to potential herbivory and they must have greater investment in herbivory defense mechanisms. Thus, there is a strong interaction between environmental stress (resource limitation) and herbivory pressure that shapes the structure and function of plants (Nilsen et al. 1987a).

D. Adaptation to Multiple Resource Patches

Plants adapted to environments with multiple resource limitations may have evolved a different capacity to utilize short-term resource abundances compared to plants adapted to environments with adequate resources. Compensatory theory supports the view that evolutionary processes alter the genetics of plants in order to balance resource utilization, so that more than one resource is limiting growth simultaneously. Based on this concept there should be a difference in the way that plants from low-resource environments utilize resource pulses compared to plants from high-resource environments. These concepts have recently been studied with split-pot experiments. In a split-pot experiment, roots of plants are divided into two areas of low resource availability. After growth for a designated period, the resource availability of one section is increased dramatically. These studies have shown that plants from environments of low resource availability do not capitalize on resource pulses as well as do plants from high-resource environments (Crick and Grime 1987, Kuiper 1988).

The ability to utilize short-term resource patches is an aspect of phenotypic plasticity. During periods of low resource availability, physiological mechanisms of resource acquisition decrease and resource use efficiency increases. For example, under low water availability, stomata close and transpiration decreases, but water use efficiency increases. In addition, under low nutrient availability, the amount of nutrient transport proteins decrease in root cell membranes, and overall plant nutrient use efficiency increases. The rapidity in which plants can increase resource acquisition in response to a sudden resource pulse is an indication of phenotypic plasticity.

In general, plants that have evolved in environments of low resource availability have low phenotypic plasticity. This relationship has been proposed as an indication that there is an inverse relationship between adapting to low resource availability and phenotypic plasticity (Stewart and Nilsen 1995). This theory suggests that in low-resource habitats, resource patches are rare and of short duration. Selection processes would favor genotypes that ignored temporal resource patches because the time required to increase resource acquisition mechanisms and utilize those newly acquired resources in growth may be longer than the duration of resource availability. Thus, plants with high phenotypic plasticity would have low resource efficiency in low resource environments.

Many environments have patchy resource availability both above ground (light patches) and below ground (nutrient patches). It is highly likely that these two resource areas vary independent of each other. There are likely to be patches of high light in low-nutrient areas and patches of high light in high-nutrient areas. To maximize resource acquisition in any of these habitats, plants would need to vary their investment into aboveground or belowground tissues. In fact, recent theories on the function and evolution of plants have placed heavy emphasis on the relative abundances of resources above and below ground and its association with trade-offs between allocation of resources above or below ground (Tilman 1988, Shipley and Peters 1990). This is another case in which multiple resource variation is selected as a major factor regulating the morphological and physiological nature of plants.

III. SYNERGISTIC INTERACTIONS BETWEEN STRESSORS

There can be little doubt that the interaction between stressors (limiting resources included) has a major impact on the adaptation and acclimation of plants. Therefore, it is important to examine specific interactions. Synergistic interactions are those in which the combination of stressors induces greater detrimental impact on plants than that of either stressor alone.

A. Low Temperature and High Light

When plants experience low temperature, plant function can be affected detrimentally, particularly photosynthesis (see Chapter 12). The detrimental impact of cold temperature on photosynthesis is especially pronounced when a substantial amount of light accompanies the cold treatment. An equal amount of light under warmer conditions does not induce similar detrimental effects. In *Lemna minor*, chilling induces no detrimental effect if it occurs in the dark, but at light levels above 700 $\mu mol\ m^{-2}\ s^{-1}$ chilling induces inhibition of light-saturated and light-limited photosynthesis (Lindeman 1979). Thus, high light intensity and cold temperatures act synergistically to magnify their respective effects on photosynthesis.

Increased susceptibility of photosynthesis to light inhibition during low temperature has been related to the following processes: (1) When photosynthesis is inhibited by cold temperature, this predisposes leaves to develop excess excitation of photosystems; (2) cold temperature inhibits proteins synthesis and enzyme activity so that biochemical reactions of photosynthesis are slower [this also predisposes the photochemical reactions (not temperature inhibited) to having excess energy acquisition]; and (3) several of the alternative pathways for energy dissipation (Mehler peroxidase reaction, photorespiration) are inhibited at low temperature. In general, cold temperature inhibits biochemical utilization of photon energy and predisposes the photochemical reactions to damage by excess excitation.

Low temperature also induces photoinhibition and that inhibition is due to a reduction in photosystem II activity. The photoinhibition under low-temperature conditions may provide somewhat of a safeguard against damage due to the adverse effects of excess excitation energy (Oquist and Hunter 1989). In addition, cold acclimation induces the formation of several enzymes associated with oxygen free-radical scavenging, such as ascorbate peroxidase and oxidase, dehydroascorbate reductase, glutathione reductase, superoxide dismutase, as well as others (Guy and Carter 1984, Halliwell and Gutteridge 1985). Increased levels of these enzymes will hasten recovery from low-temperature and high-light damage when tissue temperature increases. Therefore, chilling-resistant species decouple electron flow by photoinhibition and synthesize enzymes that will scavange free radicals (produced during the cold treatment) upon rewarming.

In temperate vegetation there are few subcanopy evergreen shrubs. This is probably due to the interaction between cold temperature and high light. The light intensity in the subcanopy of a deciduous forest is highest during the winter. Therefore, evergreen leaves (particularly broad leaves, which absorb a lot of radiation) would have to experience high radiation and cold temperatures all winter. The most successful (widely distributed) broad-leaved evergreen shrub of the eastern temperate forest is *Rhododendron maximum*. Leaves of this species last up to five years (Figure 13.6), during which time they experience four winters when light intensity is high and air temperature is low (Nilsen et al. 1987b). If low temperature and high light is a particularly strong stressor combination for leaves, how do *R. maximum* leaves tolerate these conditions?

Sensitivity to low temperature and high light is expressed in *R. maximum* leaves when grown in very high light sites (no canopy present over the leaves). Under these conditions the leaves turn yellow and last for only two years. Chlorophyll concentration decreases significantly after the second winter (Nilsen and Bao 1987), and plastoglobules become abundant (Nilsen et al. 1988). Even though photochemical damage to leaves is easily observable in the highest-light sites after two years, no such damage is observed for five years in sites where there is a deciduous canopy over the *R. maximum* plants. The nega-

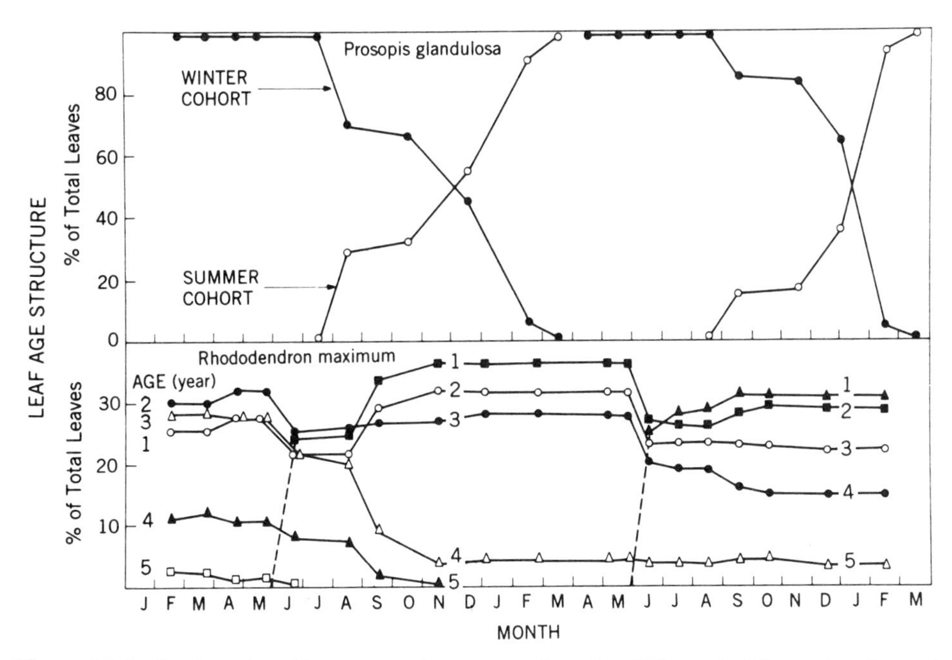

Figure 13.6 Leaf survivorship patterns for two species with different leaf longevity. *Prosopis glandulosa* is a deciduous tree of desert washes and *Rhododendron maximum* is an evergreen shrub of temperate forests. Notice that leaves remain on *R. maximum* branches for 5 years. (From Nilsen et al. 1987a.)

tive impact of light intensity on *R. maximum* leaves during the winter is lessened by the presence of thermonastic leaf movements (Figure 13.7).

During winter months leaves of *R. maximum* become pendant and below −2°C become curled. The steep angle and curled morphology reduces the total amount of radiation impinging on the leaves during cold conditions. When leaves are kept from performing these movements, photosynthetic potential is reduced during the following year. In fact, quantum yield decreases up to 50% in comparison to the control (Figure 13.8) if leaves are kept horizontal during the winter (Bao and Nilsen 1988). Those *Rhododendron* species that do not express thermonastic leaf movements are not cold hardy (Nilsen 1992b, Nilsen and Tolbert 1993). Thus, thermonastic leaf movements in *Rhododendron* species confer cold hardiness and allow the maintenance of photosynthetic potential by reducing absorbed radiation during winter months (Nilsen 1992b).

Another mechanism of compensating for the synergistic impact of low temperature and high light is by increasing the amount or effectiveness of nonradiative dissipation (NRD) of excitation energy (see Chapter 10 for a description of the biochemistry). Xanthophyll epoxidation utilizes energy absorbed by photosystem II and releases heat, thereby utilizing photon energy without passing electrons through the electron transport chain. This may prevent damage to the electron transport chain and reduce the buildup of free radicals in leaf tissues. Although this is probably a mechanism of preventing radiative damage during cold conditions, this topic is just now being investigated (Demming-Adams and Adams 1992). In general, the larger the amount of xanthophylls and the larger the proportion of xanthophylls in the A and Z form (compared to the V form), the greater the potential for NRD. In two species of conifers, both the xanthophyll concentration and

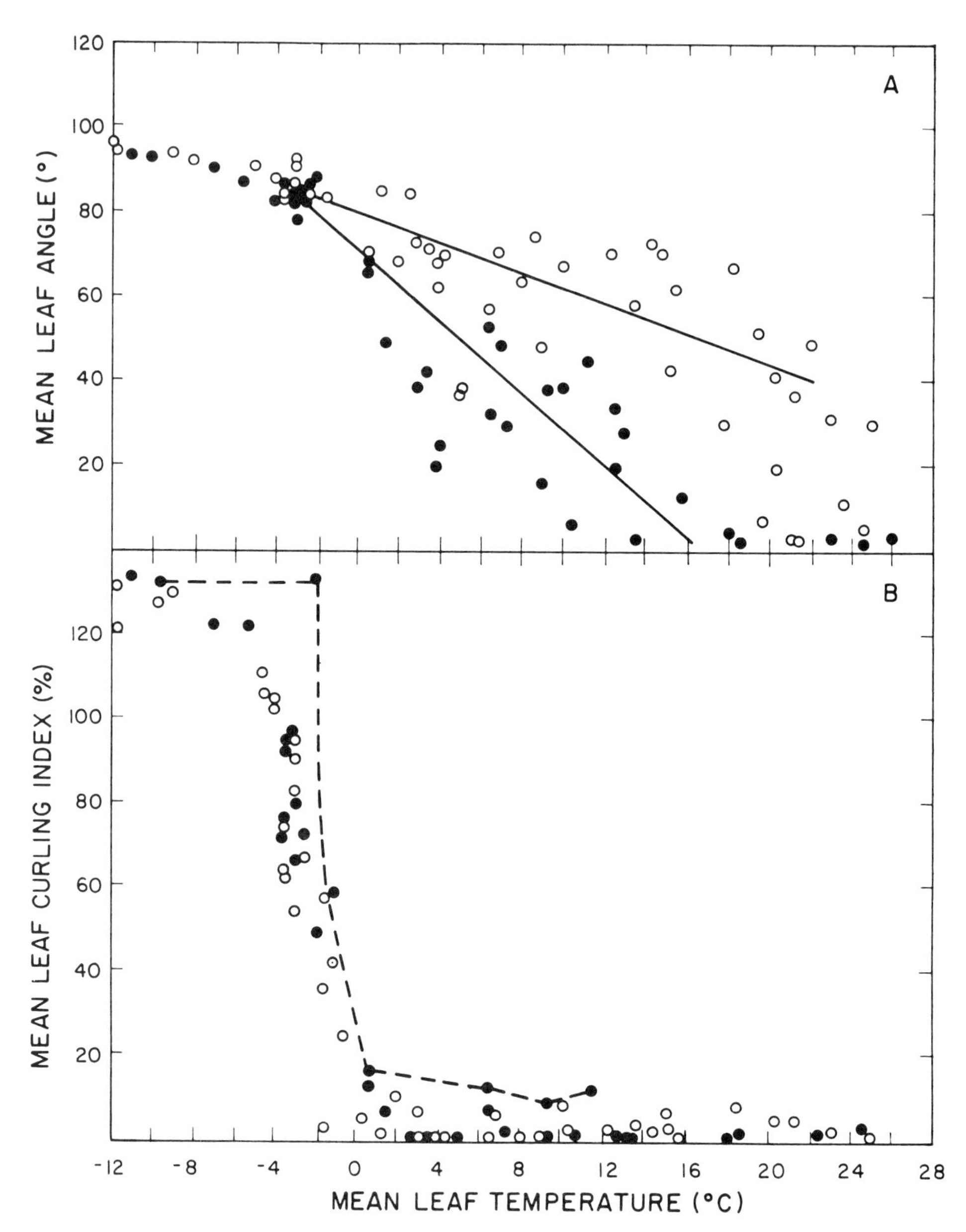

Figure 13.7 Relationship between leaf temperature and leaf orientation in *Rhododendron maximum*. (A) Dependency of leaf angle on leaf temperature. At lower temperature the leaves are more pendent. (B) Dependency of leaf curling intensity on leaf temperature. The dashed line represents the interaction for one particular leaf. Notice the sharp induction of leaf curling at approximately −2°C. (From Nilsen 1985.)

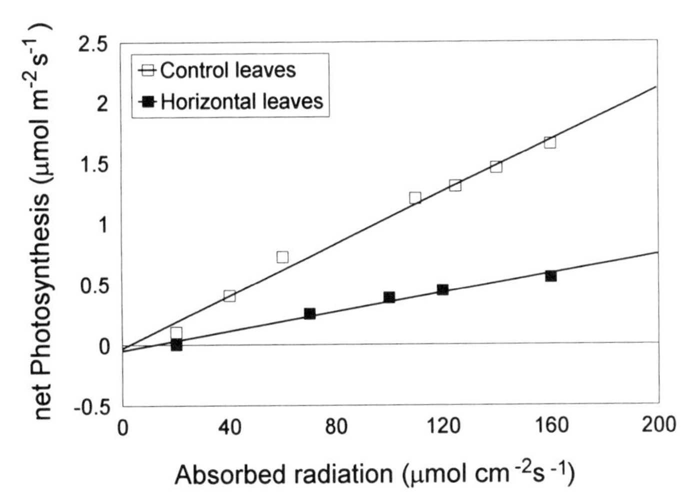

Figure 13.8 Relationship between net photosynthesis and absorbed radiation for *Rhododendron maximum* leaves at low light intensity. The slope of the line indicates quantum yield. Control leaves were able to move freely in response to temperature during the winter. Horizontal leaves were maintained in a horizontal position all winter and kept from performing any thermonastic leaf movements.

the ratio of A + Z/total xanthophyll were higher in leaves on a clear day in the winter compared with a clear day in the summer (Adams and Demming-Adams 1994).

B. Low Water or Low Light and Low Nutrients

The ability to accumulate nutrients from the environment is dependent on several characteristics of the plant and the soil solution. Nutrient accumulation depends on the concentration and availability of dissolved nutrients in the soil solution. Nutrient availability in the soil solution is contingent upon many characteristics of soil structure and chemistry, such as particle size, particle aggregation, cation exchange capacity, pH, and organic matter concentration. It is not the purpose of this chapter to go over all of the soil characteristics that regulate nutrient availability; however, we discuss some of the main factors that regulate nutrient concentration in the soil solution.

1. The structure and chemical composition of the humus, which is derived from the decomposition of plant material, has an important impact on nutrient availability. Therefore, the availability of nutrients in soil is partly regulated by the nature of organic matter coming into the soil from the vegetation of the region. In sites with low water availability, plant adaptation has resulted in leaves that have low quality for decomposition. Therefore, in drought-prone areas, soil nutrients can be limited because mineralization of nutrients from the humus is lower than in well-watered areas, and partly because of poor nutrient quality of the litter.

2. Nutrient accumulation is also dependent on the availability of water in the soil profile. A large majority of ions and low-molecular-weight organic compounds accumulated by plants are taken up from the soil solution. These compounds cannot be accumulated

from dry soil. Furthermore, the organisms that decompose and mineralize humus in the soil are restricted to moist areas. Desert regions can have low mineralization rates that are restricted to the time when soil is wet from a rainstorm.

3. The abundance of roots and the proportion of young versus old roots affect a plant's ability to absorb nutrients. The larger the rooting surface area and the higher the proportion of young roots, the greater the capacity for nutrient absorption.

4. Nutrient accumulation is dependent on the abundance and kinetics of transport proteins in the membranes of root cells.

5. Nutrient accumulation, particularly P accumulation, is dependent on the presence of mycorrhizal infection. Most species form symbiotic associations with mycorrhizae (endo- or ectomycorrhizae), and this association benefits plants by (a) increasing root surface area, (b) increasing the ability to accumulate particular nutrients because mycorrhizal hyphae can have a greater ability to accumulate specific nutrients compared with that of roots, and (c) mycorrhizae exude enzymes that decompose organic molecules containing specific elements, rendering the element available for accumulation by roots.

These are not all the factors that regulate nutrient uptake by plants, but the list serves the purpose of introducing the complexity of interacting factors that regulate nutrient accumulation. Low light and low water in the environment specifically interact with nutrient accumulation in several synergistic manners (Loomis and Worker 1963).

In drought-prone environments, plants are forced to operate at low stomatal conductance in order to reduce water use by transpiration. Increasing nitrogen or other nutrient investment in leaves may not result in a significant increase in photosynthesis when plants are under water limitation. Consequently, plants in drought-prone areas have low tissue nitrogen concentrations, and photosynthetic limitation is dominated by stomatal conductance. At any stomatal conductance an increase in tissue nitrogen concentration will increase photosynthesis and increase carbon gain per water used; however, that same increase in tissue nitrogen will reduce carbon gain per nitrogen investment because limitation by conductance increases in importance. Therefore, at constant water loss, increases in tissue nitrogen increases water use efficiency. Also, at constant nitrogen concentration decreases in conductance increase water use efficiency but decrease overall photosynthesis, which will decrease nitrogen use efficiency. This suggests that there is a trade-off between nitrogen use efficiency and water use efficiency (Figure 13.9). Therefore, if plants are maximizing nutrient use efficiency in response to nutrient limitation, water use efficiency cannot also be maximized.

During a drought the amount of water in the soil decreases and nutrient availability decreases. Therefore, in environments with low nutrient availability, prolonged drought exaggerates the limitation of nutrients. Also, the efficiency of transport proteins in root cells are affected negatively by water stress. Under water-limited conditions, the root-weight-specific nutrient accumulation decreases (Figure 13.10). Nutrient limitation often co-occurs with water stress because water stress inhibits soil solution nutrient concentration, mineralization rates, and plant nutrient accumulation processes.

Low-light conditions will also interact synergistically with nutrient limitation. It has been well demonstrated that nutrient accumulation by root cells is inhibited when plants are grown under low-light conditions (Figure 13.11). Low-light availability decreases the root/shoot ratio of plants, and therefore reduces nutrient accumulation capacity. In addition, reduced light reduces the flow of carbohydrates from leaves to roots; thus there is a limitation to the amount of energy available for nutrient accumulation. Under low-light

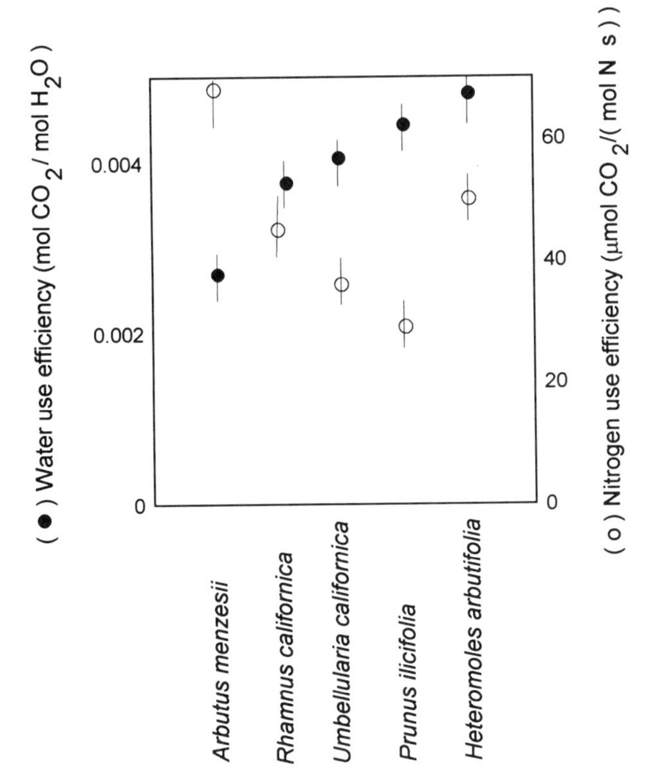

Figure 13.9 Comparison between water use efficiency and nitrogen use efficiency among several chaparral species. (From Field et al. 1983.)

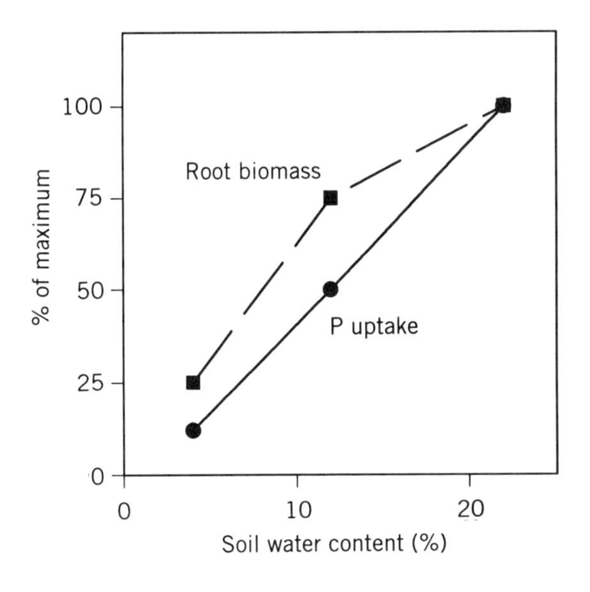

Figure 13.10 Relationship between root biomass and phosphorus uptake rate per root weight of tomato plants in response to soil water content. Both root biomass and phosphorus uptake are inhibited by drought. (Redrawn from Figure 6 in Chapin 1991.)

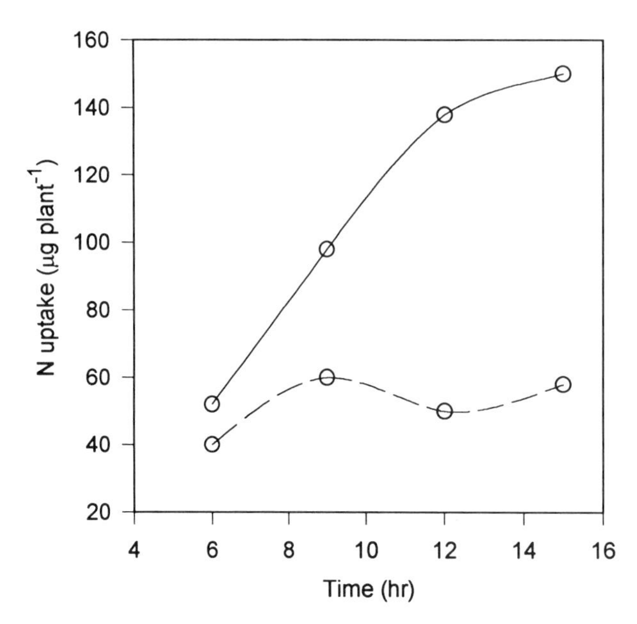

Figure 13.11 Impact of light intensity on time-dependent ammonium accumulation in young rice plants. The solid line represents plants grown under high light intensity, and the dashed line represents plants grown under low light intensity. (Adapted from Mengel and Viro 1978.)

conditions, photosynthesis increases little when there is an increase in tissue nitrogen content. Therefore, plants adapted to low-light conditions have less developed nutrient accumulation mechanisms than those of plants adapted to high-light sites.

In the case of nitrogen, plants can accumulate ammonium or nitrate from the soil solution. When most plants are given the choice between ammonium or nitrate, ammonium is absorbed preferentially. This is probably because the amount of energy needed to accumulate nitrate from the soil solution and reduce it to ammonium is greater than that needed for accumulating ammonia. In fact, those species that reduce nitrate to ammonium in the chloroplast utilize photosynthetically produced energy for the nitrate reduction. Under high-light conditions, the carbon cost for nitrate reduction is insignificant, but under low-light conditions the amount of carbon needed for nitrate reduction can be a significant drain on photosynthesis. Consequently, plants that are acclimated or adapted to low-light conditions utilize soil ammonium pools as their primary nitrogen source.

One of the important evolutionary aspects of plant response to resource limitation is a generalized reduction in growth rate. Therefore, plants adapted to drought-prone regions or low-light sites will have a low growth rate, a low requirement for nutrients, and a reduced potential for nutrient accumulation. Thus, there may be an antagonistic interaction between stresses in which adaptation to nutrient limitation also reduces the plant's susceptibility to water stress or low-light intensity. This follows from compensatory theory, which suggests that a balance of resource limitations should occur in any one site. In contrast, species that evolved in regions with adequate water and nutrients (and crops that have been engineered for such sites) will suffer a strong synergistic effect between nutrient limitation and water or light limitation.

IV. ANTAGONISTIC INTERACTIONS BETWEEN STRESSORS

There are some cases in which adaptation or acclimation to one stressor can provide protection against another simultaneous stressor. This is termed an *antagonistic interaction* between stressors. One example was mentioned in a previous paragraph, where adaptation to nutrient limitation results in a slow growth rate and consequently, decreased sensitivity to water stress. There are many other such cases, a few of which are discussed here.

A. Chilling and Water Stress

One of the main responses to chilling stress is the up-regulation of genes that code for cryoprotectant compounds. As a result, the abundance of these osmotically active compounds increases in the cytoplasm, which lowers the cell osmotic potential. This is a similar mechanism (osmotic adjustment) as that which confers some protection against water stress. Therefore, acclimation to chilling conditions may accord a degree of drought-stress tolerance. This interaction is utilized by foresters to prepare seedlings for transplanting. The major reason why seedlings do not survive following transplanting into logged sites is water stress. Pretreatment of seedlings with chilling stress greatly increases seedling transplant success even though the transplanted seedlings do not experience chilling conditions during their first summer of growth.

Not every case of acclimation to chilling will result in increased drought tolerance. For example, changes in membrane fluidity in response to cold temperature may have little impact on water stress tolerance. In addition, there are cases in which adaptation to cold-stress tolerance can actually increase sensitivity to drought stress. For example, in *Rhododendron maximum,* cold-stress tolerance is partly determined by leaf thermonastic movements (Nilsen 1991). For leaves to possess these thermonastic movements, they must have a high tissue water content. Therefore, the *Rhododendron* species that express cold-stress tolerance by thermonastic leaf movements are intolerant of low tissue water content and highly susceptible to water stress (Nilsen and Tolbert 1993).

B. Light Intensity and Water Stress

Plants acclimate and adapt to light limitation and light excess by changes in morphology, anatomy, and physiology. Those changes also affect plant response to water stress. An important question is: How is tolerance to water stress affected by changes in light intensity?

In one study, the interaction between water stress and light intensity was compared among several *Rhododendron* species and *Kalmia latifolia* (Lipscomb 1991). Photosynthetic capacity was reduced under low-light conditions and under low-water conditions in all species (Figure 13.12). The magnitude of water stress inhibition was greater for all species when under high-light conditions than when under low-light conditions. In all species tested, there was no difference in the water potential of low-water treatments under high or low light intensity, and stomatal conductance was determined primarily by water potential. Thus, under high-light conditions, decreasing stomatal conductance has a large impact on photosynthesis, while a similar change in stomatal conductance has a much reduced effect on photosynthesis under low-light conditions. Consequently (on the basis of photosynthetic inhibition), plants under high light intensity are more susceptible to water stress than are plants under low-light conditions.

A second reason why plants under low light are protected from damage due to water stress comes from the impact of closed stomata on excess energy absorption. If plants are in a high-light environment, they are predisposed to oxidative damage in the chloroplasts during periods of water stress due to the high light intensity. Plants under lower-light con-

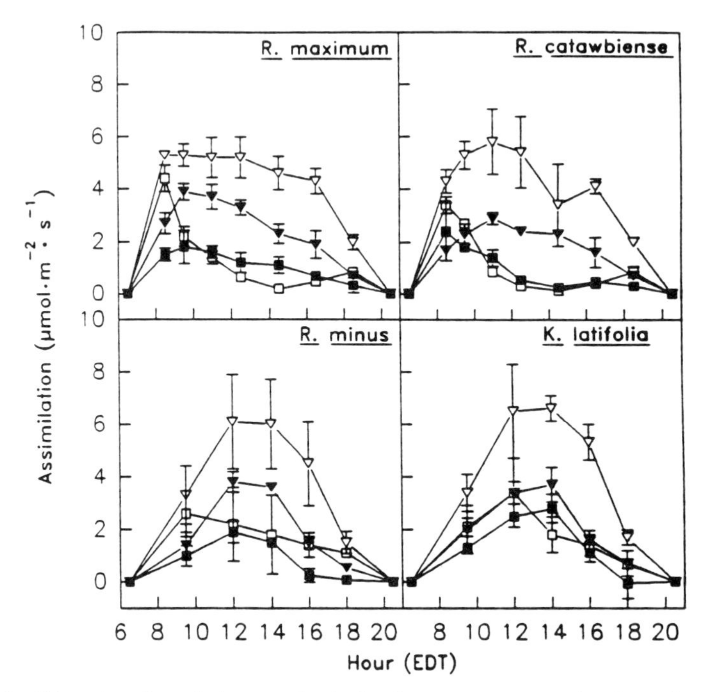

Figure 13.12 Diurnal cycles of photosynthesis for four evergreen species from the eastern temperate forest when grown in various combinations of light and water availability. Inverted triangles represent high water treatments, and squares represent low-water treatments. Filled symbols represent low light treatments, and open symbols represent high-light treatments. (From Lipscomb 1991.)

ditions can experience relatively prolonged drought without experiencing significant photo-oxidation.

V. ANTHROPOGENIC STRESSORS AND NATURAL RESOURCE LIMITATION

Due to the activity of human populations, environmental resource availability can be altered around plants. Fossil fuel combustion and other aspects of human habitation cause an increase in the abundance of sulfur dioxide (SO_2), ozone (O_3), carbon monoxide (CO), and nitrous oxides (NO_x), as well as many other gaseous species in the atmosphere. In particular, atmospheric carbon dioxide is increasing, due to the interactions of human populations and the biosphere. The increase in atmospheric carbon dioxide has been linked to a multitude of possible climatic impacts on the biosphere. These and many other environmental changes are occurring because of human population density, technological development, and their cumulative impacts on ecosystem processes. The abundance and concentration of pollutants changes spatially and temporally over the landscape due to population centers and the socioeconomic status of the population. Since natural resources and climatic conditions are not uniform (spatially and temporally) over the landscape, there is potential for interaction between natural stressors and anthropogenic effects, particularly those of airborne pollutants.

A. Increased Carbon Dioxide Concentration and Resource Limitations

The carbon dioxide concentration of the atmosphere has been increasing steadily, from about 270 ppm in 1870 to 350 ppm today, and is expected to reach between 600 and 800 ppm by the middle of the next century (Cipollini et al. 1993). The majority of this increase is attributed to increased fossil fuel combustion and decreased global areas of temperate and tropical forest. Plants generally respond to increasing carbon dioxide concentration by increasing growth, increasing water use efficiency, and decreasing photorespiration (Bazzaz 1990), although there are notable exceptions. For example, C_3 plants have a greater response to elevated CO_2 than do C_4 plants because C_4 plants are close to CO_2 saturation at current atmospheric CO_2 concentrations. Plants that reside in ecosystems that are already resource limited will respond differently than agricultural plants to increasing atmospheric CO_2 (Figure 13.13).

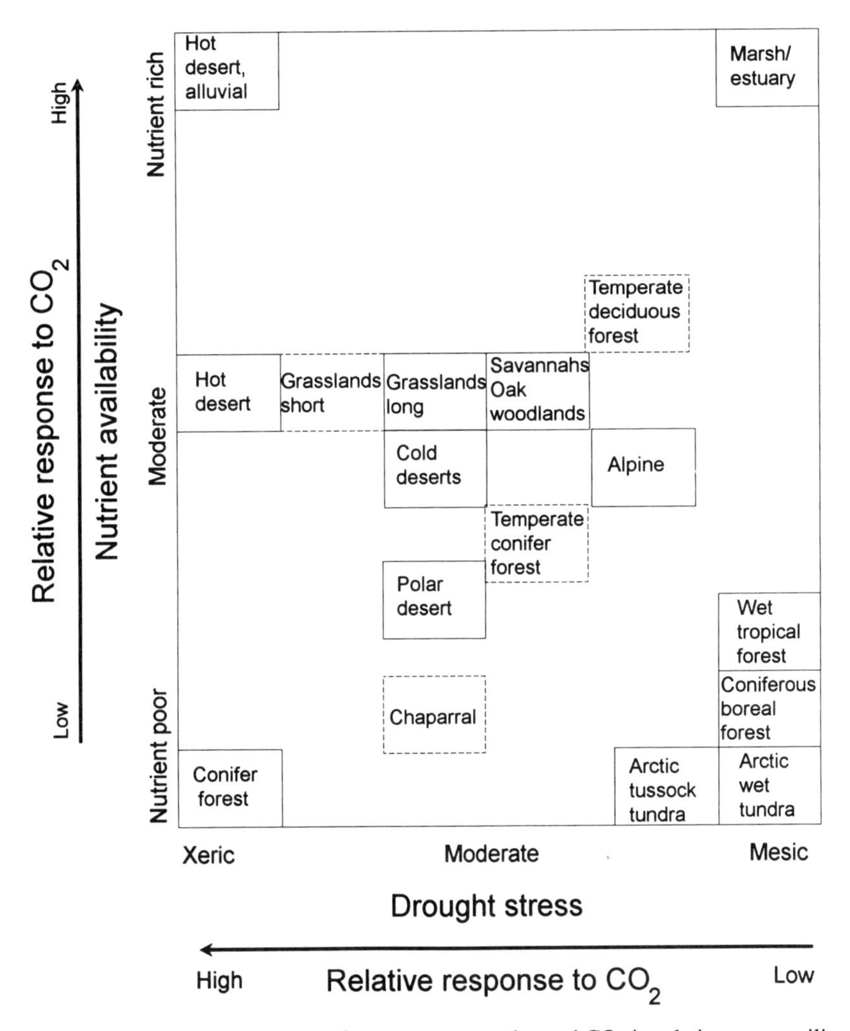

Figure 13.13 Hypothesized response of ecosystems to enhanced CO_2 in relation to prevailing nutrient and water availability. Notice that the largest response is expected in nutrient-rich and dry habitats, while the smallest response is expected in low-nutrient and mesic environments. Many of these ecosystems are yet to be investigated. (From Mooney et al. 1991.)

There is some controversy about the potential magnitude of global atmospheric CO_2 change because of disagreement about the rate of CO_2 dissolution in the oceans and the magnitude of plant productivity. It is likely that climatic changes as a consequence of atmospheric CO_2 rise may reduce global productivity and enhance atmospheric CO_2 rise. Emanuel et al. (1985) predict a 6.7 to 17% increase in desert area due to global warming (a definite response to CO_2 increases), which will reduce total global production. Increasing global temperature may increase decomposition rates and cause a shorter turn-over rate of organic carbon in detritus (Schlesinger 1984), which will also enhance CO_2 concentrations in the atmosphere. Higher temperatures in boreal systems may increase the release of isoprene and monoterpenes from conifers, and these compounds also result in increased CO_2 when oxidizing gases (O_3) are in the atmosphere. Therefore, it is extremely likely that atmospheric CO_2 will continue to rise, causing a change in climatic patterns, and affecting plant response to other limiting resources.

These generalized responses to elevated CO_2 are modified by resource availability (Goudriaan and deRuiter 1983). In particular (Figure 13.14), photosynthetic and growth responses to increased atmospheric carbon dioxide are dependent on nitrogen resources (Bazzaz and Miao 1993). Nitrogen availability is a limiting factor for photosynthesis and growth in many ecosystems, and this limitation will act as a constraint to the magnitude of growth response to elevated carbon dioxide (Thomas et al. 1994). An exacerbating factor is that anthropogenic emissions of nitrogen can account for 90% of atmospheric nitrogen–containing compounds (Lovett 1994) and a large proportion of nitrogen deposition.

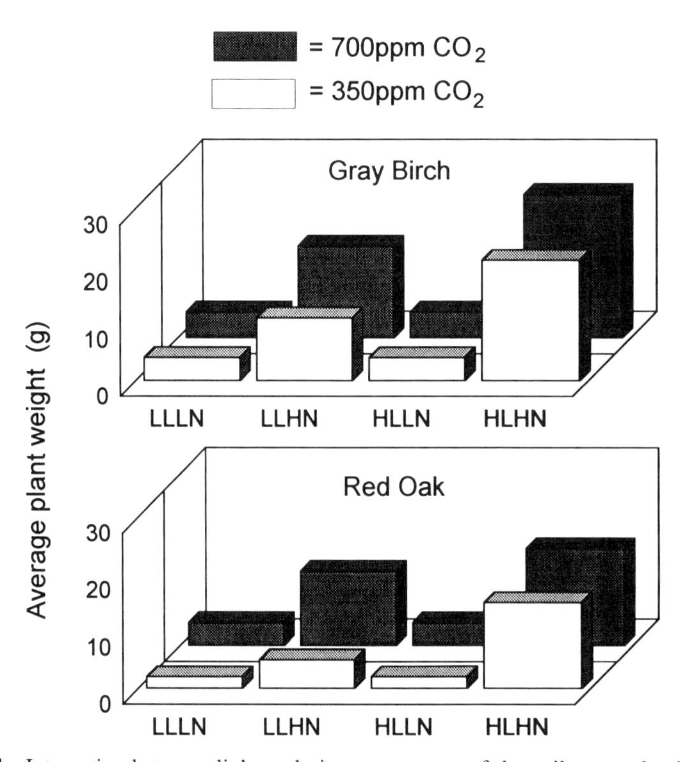

Figure 13.14 Interaction between light and nitrogen content of the soil on productivity response to elevated atmospheric carbon dioxide concentration. Adapted from Bazzaz and Miao 1993.

Therefore, there are areas in which increased atmospheric carbon dioxide coincide with increased nitrogen deposition. In areas receiving significant nitrate deposition in rainfall (dissolved nitrous oxides from pollution) there will be a larger growth response to elevated CO_2 than in areas free of nitrous oxide pollutants.

Increased atmospheric CO_2 can also improve nitrogen use efficiency. Under conditions of high-atmospheric carbon dioxide, there will be a higher concentration of CO_2 for rubisco to act upon. Increased CO_2 in the chloroplast leads to increased carboxylation efficiency because the CO_2/O2 ratio increases near the active site of rubisco; thus less rubisco is required to reduce the same amount of CO_2. Lower concentrations of rubisco and photorespiratory enzymes are needed because of the increased carboxylation efficiency. Thus, under elevated CO_2, nitrogen can be reallocated from rubisco to chlorophyll and electron transport proteins. The end result of the nitrogen reallocation is a higher nitrogen use efficiency. However, if rubisco concentrations decrease and nitrogen is not reallocated to alternative metabolic processes, tissue *N* concentration will decrease. In fact, elevated CO_2 treatment has been associated with lower tissue nitrogen concentration in some species (Larigauderie et al. 1988), but not in others (Sage et al. 1989), and plant C/N ratio generally increases under enriched CO_2 treatment. It is clear that the response of plants to anthropogenically induced, elevated atmospheric CO_2 will depend on the resource balance in the habitat and plant adaptation to that resource balance (Mooney et al. 1991).

B. Atmospheric Pollutants and Plant Resource Availability

A number of compounds such as sulfur dioxide, ozone, and nitrous oxide released into the atmosphere from anthropogenic sources have detrimental effects on plants. The intensity of these stressors (magnitude of detrimental effects) is dependent on their atmospheric concentration, climatic conditions, soil resource availability, and genetic makeup of the vegetation. Atmospheric concentration of pollutants vary both temporally and spatially in ecosystems. In particular, sulfur dioxide concentrations are focused in short time periods when wind conditions favor the transport of sulfur dioxide from smokestack effluents into the surrounding vegetation. The amount of sulfur dioxide that actually enters leaf cells is dependent on stomatal conductance at the time that atmospheric SO_2 is high. Therefore, internal pollutant dose is due to the interaction between factors that regulate stomatal conductance and factors that regulate atmospheric concentration of the pollutant. For example, the internal dose of sulfur dioxide is strongly dependent on atmospheric humidity (McLaughlin and Taylor 1981).

In addition to the interaction between climatic factors that regulate stomatal conductance and pollutant concentration, internal dose is affected by vegetation growth form and physiological status. Species that reside in low-nutrient environments have reduced photosynthesis and stomatal conductance compared to those species in high-nutrient environments. Therefore, the detrimental effects of pollutants on species of low-resource environments may be less than that for species in resource-abundant sites (Winner and Mooney 1980a). Furthermore, agricultural species will be affected more than many species in natural environments because agricultural species have been bred for maximum productivity, and this usually results in particularly high stomatal conductance.

Pollutants also affect stomatal conductance differently among species. In some cases fumigation with SO_2 induces stomatal closure (Winner and Mooney 1980b), and in other cases, stomata open in response to SO_2 fumigation (Black and Unsworth 1980). Some plants can induce defensive metabolism in response to ozone concentrations of intercellular spaces, whereas other plants cannot. This suggests that plant response to atmospheric pollution is variable, and that variability will affect the detrimental result of high atmospheric pollutant concentrations. If the inducible defensive mechanism requires specific resources, and those resources are limiting, the detrimental impact of pollutants on metabolisms may increase relative to situations when resources are abundant. The interaction between resource availability, genetic regulation of stomatal aperture, and atmospheric pollutant concentration is complex and specific for each species in each habitat. Understanding multiple-factor interaction is particularly important when determining the potential impact of atmospheric pollutants on vegetation.

C. Acid Rain and Cold Stress Tolerance

Atmospheric pollutants (particularly SO_2, and NO_x) interact with water in the atmosphere, which results in acidic precipitation (see Schlesinger 1991 for a review). Acid precipitation contains nitric and sulfuric acid, which ionize to release hydrogen ions and sulfates or nitrates in the wet deposition. Therefore, when pollutants dissolve in atmospheric water, nutrients are added to rainfall in the form of nitrate and sulfate. This results in a type of ecosystem fertilization. In addition to wet deposition of pollutants, dry deposition on leaves can also be a source of nutrients to ecosystems when rain passes through the canopy (Lovette 1994). The impacts of acid rain on ecosystem function are not clear because of the many interactions between factors that affect the influence of pH in rain on soil nutrient processes. However, there is little doubt that acid rain contains nutrients, and this could increase soil nutrient availability to plants.

In many environments plants acclimate to slowly decreasing temperatures by a reduction in metabolic activity and an up-regulation of mechanisms that protect tissues against freezing damage. Acclimation against potential freeze damage is important for evergreen species that inhabit cold regions, such as many conifers. In these regions, nutrient availability in late fall is usually low because cold soil temperature inhibits mineralization of organic nitrogen sources. If conifers experience a period of high nutrition in the fall, new foliage will be produced. New foliage is relatively susceptible to frost damage because acclimation has not occurred. Therefore, there may be an important interaction between acid rain and freezing tolerance in cold-climate conifers (Johnson 1988). Acid rain can cause an increase in soil nutrition late in the growing season. This induces fall growth in conifers that will be particularly susceptible to freeze damage. This particular interaction between the fertilization effect of acid rain and freeze tolerance has been suggested as one of the mechanisms causing the decline in spruce fir forests worldwide.

VI. SUMMARY

All plants in all ecosystems experience a diversity of abiotic and biotic conditions. Many of these factors affect the survival, reproduction, and growth of plants individually. However, there are also important effects on plants due to interactions among those factors. A significant part of the evolution of plant form and function is due to adaptive response to multiple-stressor interaction. Experimentation on interaction among stressors is difficult because the number of replications in an experimental design increases exponentially as the number of independent variables increases. This makes experiments on multiple factors time consuming, difficult to analyze statistically, and potentially expensive. Consequently, the number of studies on single stressors far exceeds the number of studies on multiple-stressor interaction.

An economical analogy can be used to explain the way that plants respond to combinations of limiting resources. Growth and reproduction are not regulated by one limiting factor of the environment as was proposed in Liebig's law of the minimum. In contrast, plants follow a compensatory pattern in which multiple resources are equally limiting of growth and reproduction. The economic analogy explains that optimal resource partitioning favors allocating resources among metabolic processes to balance the various limitations of the environment. Efficiency is optimized by not allocating excess resources to a metabolic pathway that will not result in increased growth or fitness.

Various stressors can interact in different ways. Synergistic interaction is when one factor enhances the detrimental effect of another factor. High light can cause photoinhibition in plants that do not have a large ability to dissipate energy absorbed by the photochemical reactions of photosynthesis. The magnitude of photoinhibition increases when leaves are under cool temperatures. Therefore, cold temperature and high light act in a synergistic way to regulate leaf photosynthesis. Another important synergistic interaction is that between low light and nutrient limitation. Under low-light conditions root nutrient uptake mechanisms are down-regulated, and the flow of carbohydrates needed to fuel the uptake enzymes is restricted. Therefore, nutrient accumulation is inhibited and this may become critical when the plants are nutrient limited by a dilute soil solution.

Antagonistic interactions are those in which the impact of one stressor reduces the detrimental impact of another stressor. Plants that are preacclimated by water stress can have increased tolerance to cold conditions, and plants acclimated to cold conditions have a greater tolerance of drought. If plants are grown under low light, the influence of low water potential on photosynthesis is less than that for plants under high-light conditions.

A relatively recent addition to stressors in our biosphere is that of pollutants. The concentration of gaseous pollutants is temporally and spatially dynamic. Some plants will receive a heavy dose and other plants will experience no atmospheric pollutants. Various plants that do experience heavy pollutant concentrations in the atmosphere around their leaves respond differently to the pollutant concentration. The variation in sensitivity to pollutants is a consequence of mechanisms that limit intercellular dose and detoxification mechanisms.

One of the most important gaseous pollutants released by human activities (CO_2) is also a natural component of the atmosphere. The documented rise in atmospheric CO_2 is a consequence of both fossil fuel combustion and a reduction in forest biomass. Although there is some controversy about particulars of the global carbon cycle and climate change, there can be no doubt that CO_2 concentrations are increasing and that global temperature is tightly linked to atmospheric CO_2. There will be an interaction between elevated CO_2 and air temperature because they are so strongly linked by biogeochemical processes. A general response of plants to elevated atmospheric CO_2 is an increase in growth and an increase in water use efficiency of C_3 plants. Plant response to elevated CO_2 is modified by nitrogen availability, light intensity, and plant growth form. Adaptation to low water or nutrient availability will limit a plant's ability to respond to elevated CO_2. In contrast, C_3 plants in agricultural fields are likely to have a dramatic response to elevated CO_2 because of abundant nutrients and water.

Plant response to other gaseous pollutants is also affected by climatic conditions and plant evolutionary history. If climatic conditions induce stomatal closure, the effective internal dose of a pollutant will decrease and the plant will be less affected. If plants have evolved to maximize water and nutrient use efficiency in environments with limited resource availability, they will have relatively low growth, photosynthesis, and conductance. When these plants experience high atmospheric pollutant concentration the impact will be small relative to that of plants that have high conductance. The impact of pollutants on plants can be enhanced by stomatal opening in response to the pollutant. Pollutants can also result in a fertilization effect in ecosystems that receive a large amount of dry and wet deposition. Increased nutrition in the fall (as a result of wet and dry deposition of soluble pollutants) causes the initiation of new growth in conifers. The new growth is susceptible to freezing. Therefore, autumnal fertilization through wet and dry fall of pollutants can decrease the tolerance of needles to freezing.

Multiple interaction between stresses is an important factor shaping the evolution of plants and defining the manner by which plants respond to environmental stresses. If plant response to a stressor is not consistent or nonexistent, this probably means that other stressors are restricting plant response. Multifactor interaction among environmental resources is important to adaptation and acclimation in plants, and experiments on these interactions should be a main thrust of future research in plant stress physiology.

STUDY-REVIEW OUTLINE

Introduction

1. The number of research studies on single-factor stresses is far greater than that for multiple stresses.
2. One of the reasons for the limited number of multiple-stress experiments is the statistical complexity of these experiments. The number of replications increases exponentially with an increase in the number of factors.
3. In any environment plants will experience a number of different resource limitations simultaneously. It is important to understand if one or all of these resource limitations will regulate plant performance.
4. Liebig's law of the minimum states that plant growth should be limited by one major limiting resource. In contrast, compensatory theory states that plants will balance their use of resources so that many different resources will be limiting. Most evidence supports the latter theory.
5. The efficiency of resource acquisition and allocation can be optimized by compensatory allocation patterns. Resources are not wasted on processes or structures that are not limiting growth.
6. Functional convergence states that allocation of resources to carbon gain processes should reduce whenever a limitation in any resource prevents capitalization on the investment.
7. Carbon gain may be the master integrator of environmental resource limitations because it reflects limitations by all resources.
8. Why study multiple stress interactions?
   a. Plants experience multiple stresses in all ecosystems.
   b. Plant growth and carbon gain are regulated by multiple-stress interactions, not only one limitation at a time.
   c. Some resource limitations restrict plant response to other resource limitations.
   d. Succession and competition among plant species may be regulated by the manner by which species respond to the balance of multiple limiting resources.
   e. The intensity of anthropogenic effects on plants (pollution, etc.) is modified partly by natural resource variation.

Effects of Multiple Resource Limitations on Plant Function

1. Optimization of resource allocation can be defined as a state in which any change in the resource distribution (within the plant) will cause a reduction in carbon gain.

2. Two factors that are important regulators of canopy carbon gain are light and tissue nitrogen concentration. Carbon gain decreases linearly at low light intensities in a canopy and as tissue nitrogen concentration decreases.
3. The main factor that regulates the distribution of nitrogen in a canopy is probably the distribution of light. A positive relationship exists in which high-light areas are populated with leaves that have high tissue nitrogen content, and tissue nitrogen decreases as light intensity decreases in the canopy.
4. In subcanopy environments in which carbon gain occurs mostly during sun flecks, there is likely to be a strong interaction between the influences of heat, vapor pressure, and light on leaf photosynthesis.
5. During a sun fleck, leaf temperature can rise considerably (10 to 15°C). If the temperature optima for leaves was close to the leaf temperature during inter-sun-fleck periods, photosynthesis would be inhibited by heat during a sun fleck and the efficiency of sun fleck light use would decrease.
6. The thermal optimum for photosynthesis in *R. maximum* leaves is equal to the leaf temperature in a sun fleck, and maximum carbon gain can occur during sun flecks. However, photosynthesis is inhibited by low temperature in between sun flecks.
7. There is an important interaction between resource allocation to defensive structures and/or compounds, and herbivory stress. The amount of investment in defense may depend on the apparency of plants.
8. Late successional species that are long lived, or species in monocultures, are more apparent to herbivores and thus should invest more resources into defense against herbivores (particularly specialist herbivores).
9. However, differences in environmental resource limitation coincide with differences in apparency.
10. Resource limitation selects for species with inherently slow growth rates. Slow growth rates and long retention of leaves predisposes plants to herbivore damage. Therefore, plants from regions of low resource availability commonly have proportionally large investments of resources into defense.
11. Plants from environments of low resource availability do not capitalize on high-resource patches as effectively as do plants from regions of high resource availability.
12. It is possible that there is an inverse relationship between adapting to low resource availability and possession of high phenoptypic plasticity. Plants with high phenotypic plasticity are able to utilize patches of high resource availability by increasing growth, while plants adapted to low resource availability have low phenotypic plasticity.
13. Above- and belowground resources are patchy in space and time, and they vary independently of each other. Most theories on the functional evolution of plants include a major consideration of the stresses associated with these patchy resource environments.

Synergistic Interactions Between Stressors

1. The sensitivity of plants to light-induced reduction in photosynthesis is dependent on tissue temperature. The colder the tissue temperature, the more sensitive it is to high-light photoinhibition.
2. Three important processes are related to the sensitivity of cold tissue to photochemical damage:
   a. Under cold temperature the photochemical reactions are predisposed to absorb excess photon energy.
   b. Biochemical reactions are slower under cold temperature.
   c. Alternative energy dissipation mechanisms are inhibited.
3. Metabolic processes induced by cold temperature may enhance protection against photochemical damage. Photoinhibition reduces the amount of electron transport and the likelihood of photo-oxidation. Induction of free-radical-scavanging proteins will help repair damage due to high light.

4. In temperate forest subcanopy situations, cold-temperature and high-light stresses limit the number of evergreen species.
5. One species (*Rhododendron maximum*) utilizes thermonastic leaf movements to avoid high-light-induced damage to leaves under cold conditions.
6. Another mechanism for reducing photochemical damage under chilling conditions is to increase the amount of xanthophylls and the proportion of total xanthophylls in the A and Z form.
7. The ability to accumulate nutrients is inhibited by drought and low light intensity. Therefore, low nutrient availability is exacerbated by low light intensity or drought.
8. During a drought, nutrients are not available in the soil solution (there is little or no soil solution), and roots lose their ability to accumulate nutrients efficiently.
9. Under low-light conditions, energy resources for root accumulation of nutrients is limited and this slows the rate of nutrient accumulation.
10. There is a trade-off in plants between nitrogen and water use efficiency. Under conditions of low nutrient availability, nutrient use efficiency may be increased, but water use efficiency will decrease. Therefore, both efficiencies cannot be maximized in any given set of conditions.
11. Under low-light conditions, accumulation of ammonium ion instead of nitrate ion will be more efficient because of the low energy cost for absorbing ammonium ion.
12. Nitrogen fixation is not a common attribute of plants under low-light conditions. This is because there is a cost in carbohydrate for the maintanence of nitrogen-fixing nodules. Under high light intensity the cost of nodule maintenance is insignificant, and nitrogen fixation can serve as an excellent way to compensate for nutrient limitation.

Antagonistic Interactions Between Stressors

1. Antagonistic interaction between stressors occurs when one stress enhances the ability of the plant to tolerate another stress.
2. During chilling stress there is often an increase in the concentration of cryoprotectant molecules in tissues. Since the cryoprotectants are osmotically active, they lower the freezing point of the tissue and also increase water stress tolerance.
3. Acclimation to chilling stress will accord a certain amount of water stress tolerance in plants.
4. In the case of *Rhododendron* species, cold-stress tolerance is partly dependent on leaf movement processes. However, high tissue water content is required for leaf movement. Thus, there may be an inverse relationship between cold-stress tolerance and water-stress tolerance in *Rhododendron* species.
5. Plants are more susceptible to the detrimental impact of water stress when under high-light conditions. Under low-light conditions, stomata can close to a greater degree without losing photosynthetic capacity than under high-light conditions.

Anthropogenic Stressors and Natural Resource Limitation

1. There are a multitude of anthropogenic stressors that affect plants as a consequence of the human population, its cultural practices, and its technology.
2. The detrimental impact of those anthropogenic factors will depend on the availability of resources in natural systems.
3. Carbon dioxide concentration is increasing in the atmosphere. In general, plants respond to increased atmospheric carbon dioxide by increasing growth, increasing water use efficiency, and decreasing photorespiration.
4. The impact of elevated CO_2 on growth and metabolism is constrained by nitrogen availability.
5. Elevated CO_2 may have a stronger impact on growth in regions that also receive a relatively large amount of dry deposition from pollution, or in agricultural fields that are fertilized.
6. Elevated CO_2 may also improve nitrogen use efficiency in plants. Under higher atmospheric CO_2, less rubisco is needed to reduce the same amount of carbon. Therefore, the nitrogen that

was formally invested in rubisco can be allocated to other components of photosynthesis, to enhance carbon reduction further.

7. The detrimental impact of atmospheric pollutants on leaf metabolism is also modified by resource availability and climatic conditions.
8. Intercellular dose of pollutants is dependent on atmospheric pollutant concentration and factors that regulate stomatal conductance.
9. High atmospheric pollutant content may have a large impact on leaf tissue metabolism when humidity is high and a small impact in dry air.
10. Vegetation growth form also affects the intensity of pollutant impact on plants. Plants with a slow growth rate and low stomatal conductance (plants of low resource environments) will not suffer as much damage in atmospheres with high pollutant content as will plants that have high growth rates.
11. There is variation in sensitivity among plants to internal pollutant dose. This is due to a different capacity to induce defensive mechanisms among species. Genetic regulation of inducible defense against pollutants may also be restricted by resource availability.
12. A number of different pollutants dissolve in atmospheric water and affect the chemistry of precipitation.
13. Dissolution of some gaseous pollutants cause rainfall acidification and also increase the deposition of nitrogen and sulfur. This wet and dry deposition can be large enough to fertilize a site, particularly when conditions are cool and there is little natural nutrient mineralization.
14. Fertilization can induce new growth on conifers during the autumn, and this new growth is more susceptible to freezing damage. Therefore, pollutant-induced fertilization can increase plant sensitivity to frost.

SELF-STUDY QUESTIONS

1. Compare and contrast the definitions of plant response to resource limitations by Leibig's law of the minimum and compensatory theory.

2. What are three reasons why compensatory allocation of resources maximizes growth in environments with multiple limiting resources?

3. Why is carbon gain considered to be the best indicator of the magnitude of environmental stress on plants?

4. How do you define resource use optimization?

5. In what ways do light availability and tissue nitrogen concentration vary in a canopy?

6. Why is the slow growth rate of plants in resource-limited sites important for theories about investment into herbivory defense or utilization of patchy resources?

7. What factors regulate soil nutrient availability to plants?

8. Why do water limitations and light limitations both decrease nutrient accumulation?

9. What factors might increase nutrient accumulation in regions with frequent drought?

10. Why does there seem to be a trade-off between water use efficiency and nitrogen use efficiency?

11. Under what environmental conditions is nitrogen fixation a good technique for increasing nitrogen availability to plants? Why?

12. What are the reasons why leaves are more susceptible to damage when cold temperatures coincide with relatively high radiation as opposed to cold temperatures in the dark?

13. In what ways can plants that tolerate low-temperature and high-light conditions change metabolic processes to enhance their tolerance of these conditions?

14. What are thermonastic leaf movements, and why are they important to evergreen plants of temperate forest regions?

15. Based on the various processes that regulate nutrient acquisition by plants, why does drought reduce nutrient accumulation in plants?

16. Under low-light conditions, how can plants optimize their nitrogen accumulation and yet conserve carbohydrates for other processes, such as growth and flowering?

17. In what ways are chilling and drought stressors antagonistic?

18. Why is it that plant response to water stress is more dramatic when plants are grown under high-light conditions?

19. What are several important factors in the environment that will regulate photosynthetic response to elevated CO_2?

20. What are the factors that regulate intercellular dose of atmospheric gases in plants?

21. How can increases in atmospheric CO_2 result in increased nutrient use efficiency? What is the most likely ultimate cause for this relationship?

22. What is a potential interaction between acid rain and chilling tolerance in high elevation conifers?

SUPPLEMENTARY READING

Johnson, D. W., and S. E. Lindberg (eds.). 1992. Atmospheric Deposition and Nutrient Cycling in Forested Ecosystems. Springer-Verlag, New York.

Koziol, M. J., and F. R. Whatley (eds.). 1984. Gaseous Air Pollutants and Plant Metabolism. Butterworth, London.

Krahl-Urban, B., H. E. Papke, K. Peters, and C. Schimansky (eds.). 1994. Forest Decline. U.S. Environmental Protection Agency. Corvalis, Oregon.

Mooney, H. A., W. E. Winner, and E. J. Pell (eds.). 1993. Response of Plants to Multiple Stresses. Academic Press, San Diego, CA.

Schlesinger, W. H. 1991. Biogeochemistry: An Analysis of Global Climate Change. Academic Press, New York.

Tilman, D. 1982. Resource Competition and Community Structure. Princeton University Press, Princeton, NJ.

Tilman, D. 1988. Plant Strategies and the Structure and Dynamics of Plant Communities. Princeton University Press, Princeton, NJ.

14 Biotechnology and Environmental Stress

Outline

I. Introduction
II. Methods of Protein Separation and Identification
III. Methods of Gene Isolation
  A. Cutting DNA with Endonucleases
  B. Cloning Genetic Material: Plasmid Vectors
  C. Transforming *E. Coli*
  D. Partial Genomic Library
  E. Screening for Transformations
IV. Methods of Gene Transfer
  A. Indirect Gene Transfer Mechanisms
    1. *Agrobacterium* Vector
    2. Ti Plasmid and Constructs
    3. Coincubation
    4. Agroinfection
    5. Viral Vectors for Transformation
  B. Direct Gene Transfer Mechanisms
    1. Electroporation
    2. Chemically Induced Transformation
    3. Microprojectile Transformation
V. Other Types of Bioengineering Experiments
VI. Bioengineering of Stress-Tolerance Traits in Plants
  A. Thermally Induced Proteins
    1. Heat-Shock Proteins
    2. Cold-Induced Proteins
  B. Dehydration-Induced Proteins
  C. Oxidation-Induced Proteins
  D. Anaerobically Induced Proteins
VII. Summary and Future Directions
  Study-Review Outline
  Self-Study Questions
  Supplementary Reading

Objectives

1. Introduce genetic engineering as it relates to plant stress physiology.
2. Discuss methods of isolating and identifying proteins.
3. Explain several important techniques used to isolate specific genes. Include modes of transforming *E. coli*, creating a genomic and a partial library, and probing the library.

4. Describe the methods used to transfer genes into plants. Include both direct and indirect methods.
5. Explain the types of molecular mechanisms that can be used to transform plant stress tolerance. Include a description of the applied significance of transformation and the significance of transformation procedures to theoretical studies of plant stress physiology.
6. Describe the induction patterns, diversity of structure, and function of stress-induced proteins in the following categories: (a) heat shock, (b) cold shock, (c) dehydration, (d) antioxidant, and (e) anaerobic.
7. Discuss advances in genetic engineering results for agricultural applications for the five categories of stress noted above.

I. INTRODUCTION

Characterization of the structure and function of DNA and RNA over the last 40 years has made genetic engineering possible. The universality of the genetic code among organisms allows for an incredible potential for inserting genes from one species into another species. Such genetic engineering has powerful applied significance and can provide definitive answers about molecular regulation of stress tolerance. On the applied side, it is important to recognize that the genome of species that have evolved in environments with extreme climatic conditions is likely to contain genes that may confer stress tolerance. In contrast, the genome of many crop species may not have genes that confer stress tolerance characters because of their evolutionary heritage or because of the artificial selection process employed in their domestication. Given that one major factor reducing yield in crops is environmental stress (Boyer 1982), it is prudent to search for those genes that confer stress tolerance in a broad array of organisms and attempt to transfer them into the genome of crop species. However, it is particularly difficult to verify that a stress-induced protein actually regulates the stress tolerance trait rather than simply correlating with that trait. Modifying the intensity of gene expression for those genes that produce stress-induced proteins can be used to establish a cause-and-effect relationship between the protein and stress tolerance.

The potential success of genetic engineering in a stress physiology context is dependent on the ability to isolate particular genes that confer stress tolerance. In almost every chapter of this book, we have emphasized the fact that there are a diversity of mechanisms and combinations of those mechanisms that are associated with adaptation to any stress. The complexity inherent in the physiology of plant adaptation to environmental stress makes it difficult to globally identify particular genes that are individually responsible for a particular stress tolerance trait. Some knowledge of basic genetic aspects of stress tolerance mechanisms should be established before discussing the process of, and potentials for, genetic engineering of stress tolerance traits.

In most cases, stress tolerance characteristics appear as quantitative traits. That is, they are controlled by several genes and are potentially affected by other genes through epistasis, pleotropy, and developmental changes in gene regulation processes. These stress-tolerant quantitative traits can also be affected by a number of different environmental characteristics. Quantitative genetic traits are identified by the statistical distribution of trait values in a population (Figure 14.1). Often, the relative stress tolerance of a population of individuals is compared to other populations on the basis of the mean intensity of that trait in each population. However, the variance of the trait in each population may be as

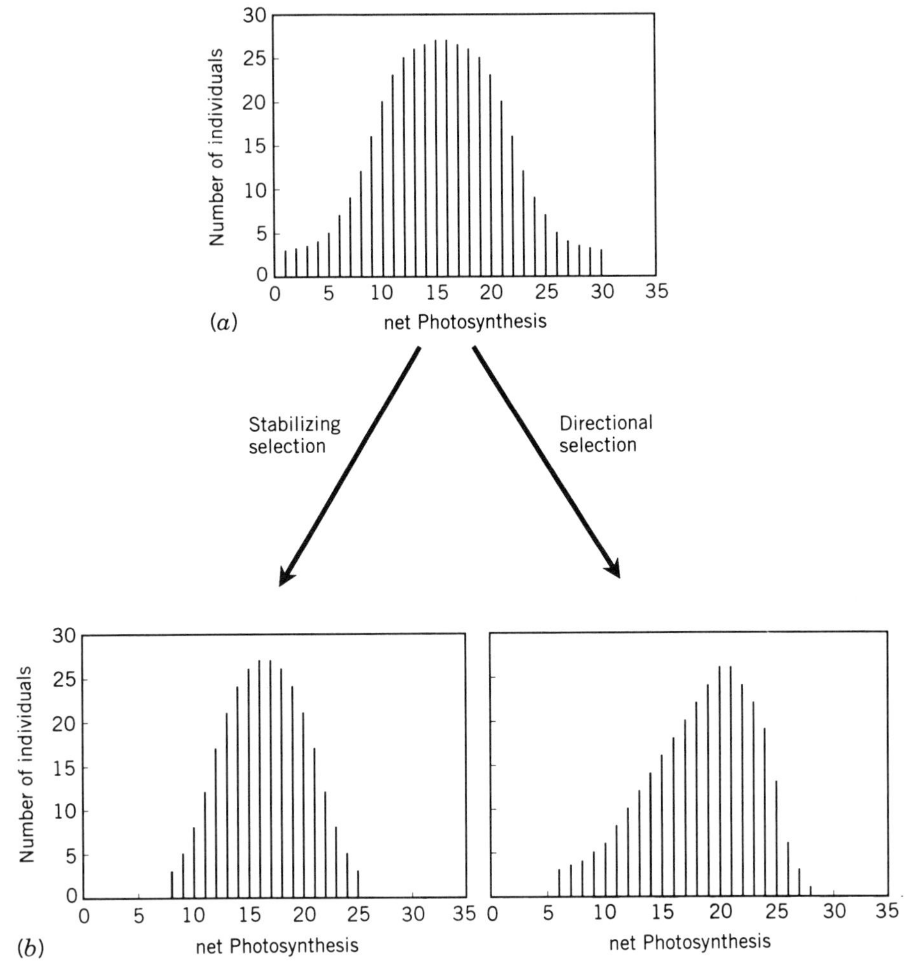

Figure 14.1 (A) Distribution of individuals with various maximum photosynthetic rates in a population. (B) Change in the distribution of photosynthetic rates in a population induced by either stabilizing or directional selection by environmental stress.

or more important for evolutionary processes than the mean. Also, the variance in quantitative stress characters among individuals in a species has provided the resource for artificial selection in crops.

Under stressful environmental conditions, selection pressures affecting the genome of a population can result in a decrease in variance of the stress tolerance trait without changing the mean (stabilizing selection) and/or a skewing of the distribution (directional selection) toward the stress-tolerant genotype (Figure 14.1). Under extreme stress there also may be an excessive restriction in the diversity of alleles for the genes that regulate the stress tolerance character, potentially resulting in fixed alleles (only one type of allele for the gene). If stress conditions limit outcrossing among individuals (e.g., low pollinator abundance or short growth periods), self-pollination will dominate reproduction and the genome of individuals may become extensively homozygous (particularly if the population size is small). Although there is some controversy about the significance of extensive homozygosity to stress resistance in plants, evidence with *Arabidopsis thaliana* indicates that homozygous individuals have decreased fitness compared with heterozygotes.

Asexual reproduction is another characteristic common in plants that have evolved tolerance to stressful environments. Although asexual reproduction limits the number of different genotypes in a population, this model for reproduction can reduce the potential for extensive homozygosity. In small populations gene homozygosity (allele fixing) is promoted by repetitive inbreeding and genetic drift: however, vegetative reproduction reduces the frequency of inbreeding or repetitive self-fertilization, thereby reducing the rate by which alleles get fixed in the population (Silander 1985).

Patterns of temporal and spatial variability in environmental stress may also affect the nature of the genetic system that regulates a stress tolerance trait. Organisms can modify their physiology by a change in alleles (regulation by allele frequency = adaptation) or a change in gene expression (phenotypic regulation = acclimation). The ability to change stress tolerance traits by means of allele frequency requires a diversity of alleles in the population. Consequently, heterozygosity is a prerequisite for adaptation. In contrast, if a species changes its stress tolerance traits by acclimation, phenotypic plasticity is important. Phenotypic plasticity is the phenotypic response of an organism to a change in its environment (Schlichting and Levin 1986). In environments that have a "fine grain" variation in environmental stress (short-term temporal patchiness of resource availability), individuals are likely to experience many different resource availability states in their lifetime, and phenotypic plasticity may be an important mode of adaptation. However, in environments that have "coarse grain" variation in environmental resources (spatial patchiness of resources that are temporally consistent), populations, but not individuals, are likely to experience different resource availability, and genetic diversity may be a more important mode of adaptation than phenotypic plasticity. Any study of potential mechanisms for stress tolerance must identify if stress tolerance traits are a result of differential phenotypic expression or a result of different alleles among individuals or both. Once the basic genetic mechanisms that control the stress tolerance trait are known (phenotypic plasticity versus changes in allele frequencies), careful breeding programs can be established to develop true breeding lines with specific tolerance characteristics (Figure 14.2). These two types of research are not mutually exclusive and either can be used to enhance the other. Plant breeding is an advanced science in which multiple breeding procedures have been developed for optimal artificial selection.

There are several basic strategies for identifying stress tolerance mechanisms at the molecular level after determining the basic mechanism by which stress tolerance traits are genetically regulated. The most important step in identifying a molecular basis for stress tolerance is verifying that there is a cause-and-effect relationship between the presence of particular proteins, or metabolic pathways, and whole-plant stress tolerance. Simply correlating the presence of particular proteins with tolerant genotypes cannot be taken as evidence that those proteins regulate the stress tolerance trait. One strategy to determine cause-and-effect relationships (discovery strategy) would be to discover which molecules (proteins) constitute the basis for tolerance. This could be accomplished by three basic steps. First, differential gene products (proteins) must be identified in the target organisms when grown under stress or nonstress conditions. This is frequently accomplished by examining the protein profile of whole tissue preparations by a protein separation technique such as two-dimensional gel electrophoresis. Once specific gene products associated with

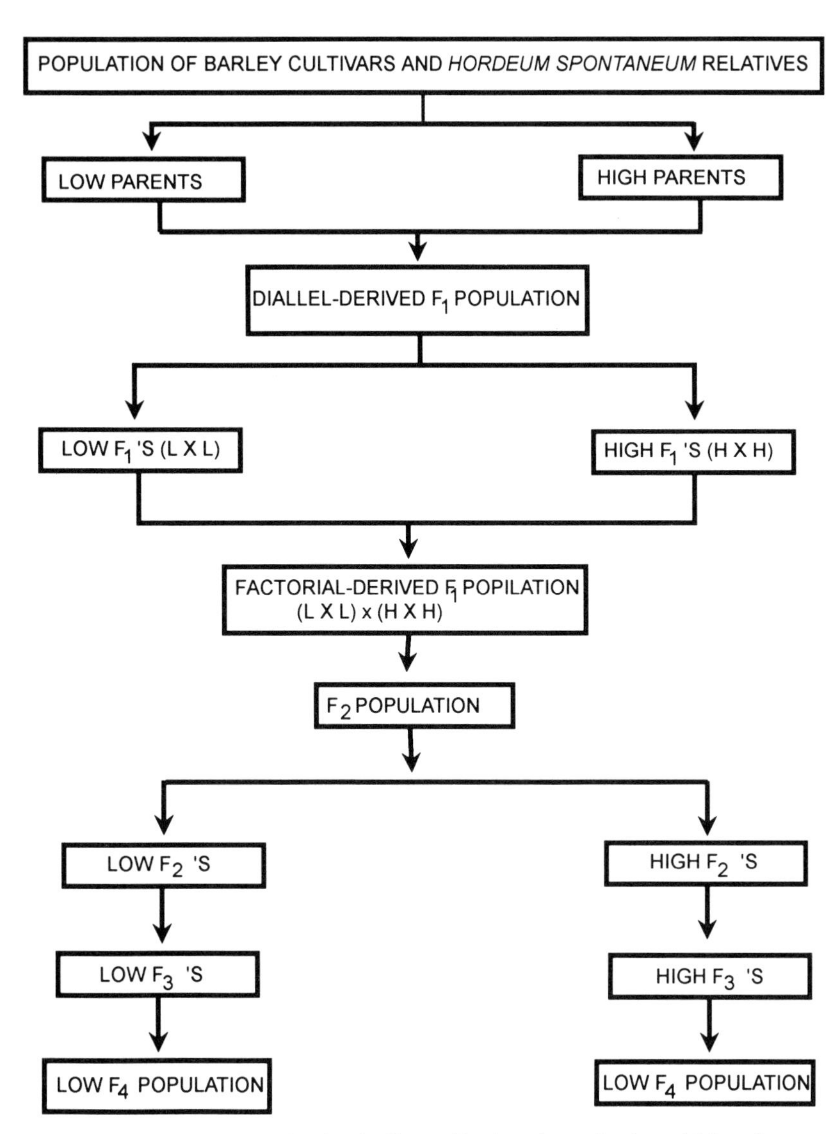

Figure 14.2 Technique used to develop isolines of barley plants that have high or low concentrations of glycine-betain in their tissues. (From Grummet and Hanson 1987.)

stress tolerance have been identified, the second step would be to isolate the genes responsible for those proteins. The techniques used for gene isolation are diverse, and some of them will be discussed below. The third step would be to prove that these specific genes regulate the stress tolerance response. This could be done by transferring the gene into another organism (that does not have the stress tolerance characteristic) and determining if the transgenic organisms have been converted into stress-tolerant genotypes. Many studies have taken the first two steps to identify genes that produce specific cell products which are associated with stress conditions (Bray et al. 1993), but few studies have taken the third step, to show that the isolated genes can confer stress tolerance in transgenic organisms.

A second strategy (inference strategy) would be to transform stress-sensitive plants with various genes from other organisms and determine if any of the transformed lines have increased stress tolerance. This technique presumes that the mechanisms for stress tolerance are known in other species and that the genes have been isolated and their function identified. The crucial characteristic of this strategy is a judicious selection of genes. The genes selected could come from a number of alternative organisms (particularly if they are conserved across wide genetic groups, such as heat-shock proteins), and they could produce gene products not normally found in plants. The key to the inference strategy is to select genes that are likely to be expressed in the target plant and are likely to result in improvement of stress tolerance.

A third strategy (antisense strategy) uses the reverse technique (compared with the first two techniques) to verify that a particular gene confers stress tolerance. In this strategy a DNA sequence is synthesized that will code for an mRNA fragment that is complementary to the mRNA produced by the putative stress tolerance gene. This is termed an *antisense construct.* Stress-tolerant species are then transformed with the antisense construct. The transcription product of the antisense construct will bind to the transcription product of the putative stress tolerance gene, thereby inhibiting translation of the putative stress tolerance mRNA. Therefore, in the presence of the antisense construct, no putative stress tolerance gene product is produced. If the tolerant line loses tolerance when transformed with an antisense construct, the target gene probably confers stress tolerance.

A fourth strategy (mutant strategy) utilizes natural or created mutants to determine if a particular protein confers stress tolerance. Mutant lines of the tolerant species are screened for stress tolerance. Those individuals that do not have the stress-tolerant trait are examined for the presence of the protein that presumably confers stress tolerance. If sensitive mutants consistently do not contain the gene coding for the stress tolerance protein, it is likely that this protein confers stress tolerance. This analysis can be carried further by recomplementation of the mutant. In this experiment, a mutant without the stress-tolerant trait is transformed with a putative stress tolerance gene. If the mutant regains its stress tolerance after transformation, this is good evidence that the gene does regulate stress tolerance.

Molecular techniques for proliferating and cloning plant genes are very well developed, as are the processes for transferring those genes into many crops. However, stable and high expression of transgenes in successive generations of a transgenic line is not always assured. One of the most recalcitrant problems for crop genetic engineers is the establishment of a germ-line transgenic plant that dependably expresses a stress-tolerant trait and is fully fertile, morphologically normal, and high yielding.

In this chapter we cover some of the basic techniques required to understand genetic engineering of crop species. If the reader has a good knowledge of molecular biology and DNA recombination technology, this may be review material. However, the material is important to understand before reading the latter part of this chapter, specific case examples of genetically engineered stress tolerance. The topics covered in this chapter are focused on crop systems because that is where most of the work is being done. Yet it should be noted that there is a great untapped research potential for using the techniques of recombinant DNA technology for studying basic processes of stress tolerance in natural systems, or evolutionary processes of stress tolerance in native plants.

II. METHODS OF PROTEIN SEPARATION AND IDENTIFICATION

Many stress tolerance traits may be regulated by proteins because these molecules constitute the machinery of cell metabolism. When examining a developmental change or varietal difference in a metabolic process, the cause of the change or difference is most commonly a change in the presence, stability, or activity of particular proteins. If differences in character traits among varieties or lines are regulated by genetic factors, this means that there is probably a difference in the presence, functional state, or regulation of a gene. Consequently, the discovery of a particular protein or class of proteins that occur only in stress-tolerant lines or varieties is the first step toward manipulating genes that are responsible for metabolic processes that regulate stress tolerance among varieties. In the case of differential stress tolerance among species, the metabolic differences are likely to be more basic, such as different pathways or different molecular regulation of those pathways.

Protein identification in species that are tolerant or sensitive to a stress starts with a protein preparation from the tissue most likely to be associated with the stress tolerance trait. This may be the organ that is performing the stress-tolerant metabolism (e.g., the leaf for regulation of water use efficiency) or a proximal organ which is the site of synthesis for a metabolic regulator (e.g., roots for ABA production). Many comprehensive texts on cell biology have excellent descriptions of protein extraction procedures (Lodish et al. 1995), so they will not be considered in this chapter.

Proteins can be separated from each other by their chemical reactivities, charge relationships, or molecular weight. Protein separation is normally accomplished by the use of gel electrophoresis or column chromatography. In this chapter we assume that the reader has a familiarity with the general concept of gel electrophoresis. If one characteristic (e.g., pH) is being used to separate proteins, this is called a *one-dimensional protein separation*. One-dimensional separation divides proteins into classes based on the specific chemical character, but this technique does not usually separate specific proteins in each class. *Two-dimensional electrophoresis* is required to separate individual proteins.

Mechanisms for separating proteins by two-dimensional techniques vary in their specificity. General protein separation based on apparent molecular weight is accomplished by one-dimensional SDS polyacrylamide gel electrophoresis and separation based on charge balance by isoelectric focusing. Combining an isoelectric focus in one direction of a gel and SDS electrophoresis in the other direction results in a powerful two-dimensional separation technique for resolving the diversity of proteins in bulk tissue extracts (Figure 14.3). Such two-dimensional gels performed on tissue extracts from stress-tolerant and stress-sensitive species growing under the stress condition can often resolve a series of proteins correlated with stress tolerance.

Several more specific techniques for protein separation are possible if some knowledge of size, charge, and structure of the target proteins is available. When the general size of a protein is known, gel filtration chromatography can separate proteins into packages of defined size. If the charge characteristics of the target protein class is known, ion chro-

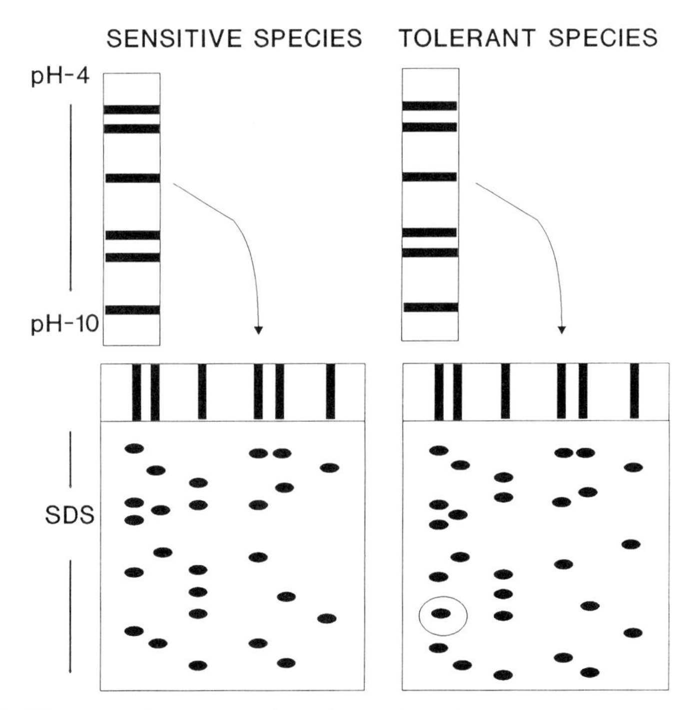

Figure 14.3 Diagrammatic representation of two-dimensional gel electrophoresis of proteins. First, proteins are separated along a gradient of charge by isoelectric focusing. Then the proteins are separated in the other direction on the basis of molecular weight by SDS electrophoresis. The circle on the right hand gel represents the presence of a novel protein in the tolerant species.

matography can be used to separate proteins of specific charge distributions. If an antibody has previously been produced for a similar protein from another organism, affinity chromatography may isolate the target protein in one step (Figure 14.4).

Specific proteins can be identified following two-dimensional gel electrophoretic separation based on their specific function. If the metabolic reaction that the target protein performs is known, chromagenic enzyme reactions can be performed. In this technique, the two-dimensional gel is exposed to a solution containing a specific substrate that binds to the target protein. After incubation with the substrate, the chromagenic reaction is performed, thereby marking the target protein spot by the color change.

Western blotting (Figure 14.5) is a technique that involves first transferring the proteins from an SDS gel into a piece of nitrocellulose. Once in the nitrocellulose, the proteins are held in position and can be exposed to various solutions without being dislodged from their location. After the protein transfer, the nitrocellulose sheet is incubated with an antibody (Ab1) that has specific affinity for the target protein. Ab1 will bind to the target protein. The nonbound antibody is washed off and the nitrocellulose is then exposed to a second antibody (Ab2) that is covalently linked to alkaline phosphatase. This second antibody binds to Ab1, forming an antibody bilayer that has the capacity to form a chromagenic reaction when developed with an appropriate substrate. It is also possible to utilize a three-antibody technique to amplify the signal (make the reaction more prominent).

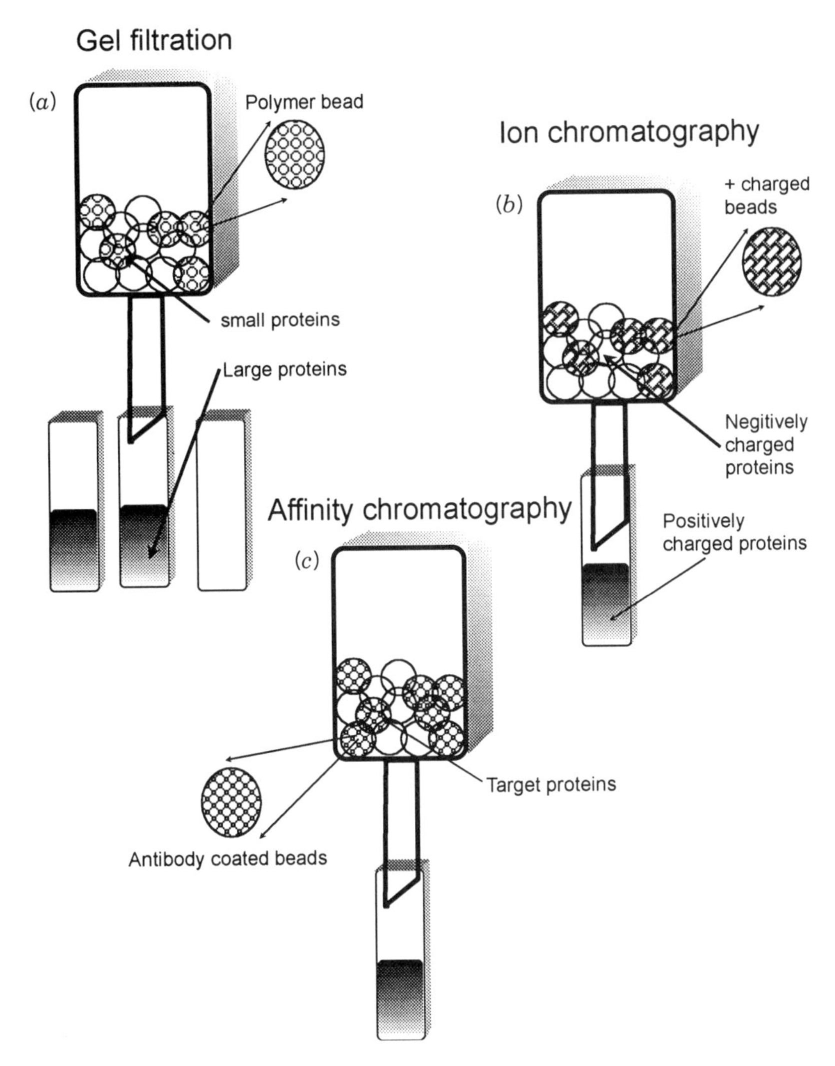

Figure 14.4 Diagrammatic representation of three types of column chromatography. (A) Gel filtration chromotography separates proteins on the basis of molecular size. (B) Ion chromatography separates proteins on the basis of their molecular charge abundances. (C) Affinity chromatography separates those proteins that have affinitiy to a particular antibody from the rest of the proteins in the cell extract.

We have discussed only a few of the techniques available for identifying particular proteins associated with stress-tolerance mechanisms. Other techniques are available to determine a protein's structure, chemical activity domains, and amino acid sequence. Some information about the amino acid sequence (from the N-terminus side) of at least part of the proteins associated with the subject stress-tolerant metabolism may be necessary to identify the specific gene responsible for genetic regulation of that metabolic process (discussed in a later section).

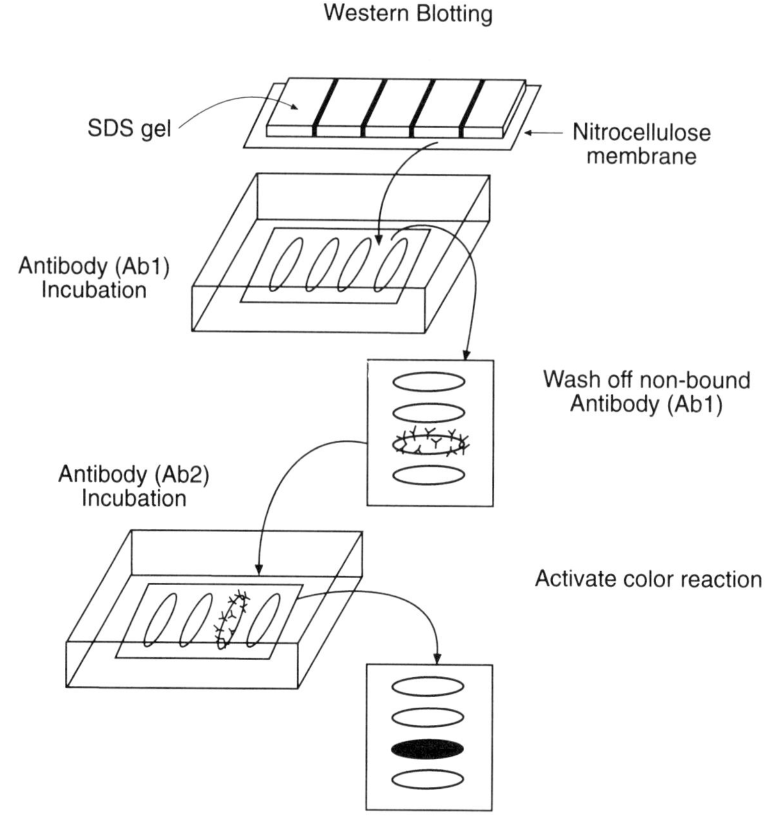

Figure 14.5 Diagrammatic representation of Western blotting. This technique is used to identify particular proteins by using a dual antibody system.

III. METHODS OF GENE ISOLATION

A. Cutting DNA with Endonucleases

The second step in most strategies for genetic engineering is the localization and cloning of a particular gene or genes associated with production of the specific proteins that putatively regulate the stress-tolerance mechanism. In a very general sense, the target gene or genes must be differentiated from the large background of coding DNA in the genome of any organism. Furthermore, the target gene must be cloned into an easily manipulated organism and replicated many thousands of times in order to have enough material for gene transfer into another organism. The following paragraphs cover a "shot in the dark" approach to gene separation (genomic library) and a description of a finer-tuned probing for specific genes (partial genomic library). In either case, the essence of this step is the development of a "library" that contains large numbers of different bacterial or bacteriophage colonies (or plaques in the case of bacteriophages), each of which contains many copies of identical DNA molecules. In this section of the chapter we describe gene cloning procedures that utilize a cellular cloning mechanism. Similar gene multiplication can be accomplished by biochemical reactions called *polymerase chain reactions* (PCRs). Information on the use of PCRs in gene multiplication applications is available in many advanced molecular biology texts (see the Supplementary Readings).

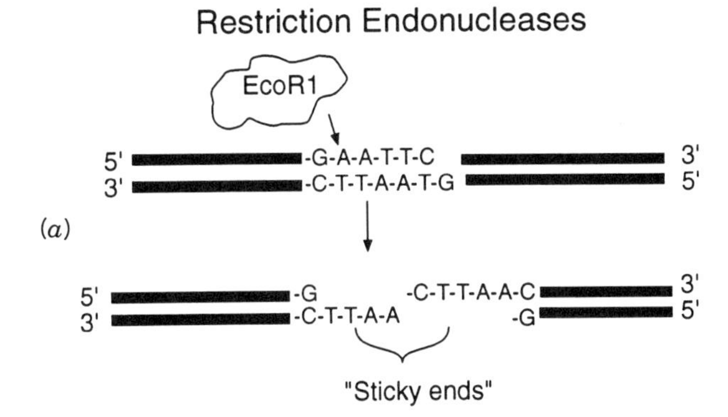

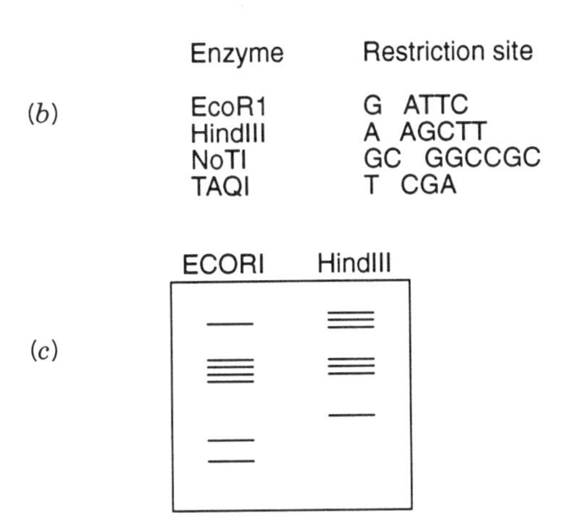

Figure 14.6 Diagrammatic representation of the patterns of restriction endonuclease activity. (A) Enzymatic reaction stimulated by the restriction endonuclease EcoR1. (B) List of several restriction endonucleases and their corresponding restriction site. (C) Diagrammatic representation of a gel in which the same DNA has been separated following being cut with two different restriction endonucleases.

To separate the target gene from a large background of DNA, the DNA must be cut into smaller pieces. The "shot in the dark" approach uses enzymes to cut an extract of whole-tissue DNA into many small pieces. The enzymes used for cutting DNA are restriction endonucleases. Each restriction endonuclease cuts DNA at specific nucleotide sequences called *restriction sites* (Figure 14.6). The specific nucleotide sequence within which an endonuclease cuts DNA is commonly a six- or eight-base-pair-long inverted repeat. Since the restriction endonuclease cuts DNA at specific sequences, this results in a reproducible set of DNA fragments called *restriction fragments.* Endonuclease enzymes evolved in bacteria as a defensive mechanism against infection by bacteriophages (bacterial viruses), and several hundred endonucleases have been isolated from many different bacteria species. A diversity of restriction fragments can be formed by making several DNA extractions from the subject plant and exposing each extraction to a different endonuclease. The endonuclease cuts the DNA in a fashion that results in an extension of six to eight bases of single-stranded DNA at each end of the fragment. These are called "sticky ends" because they can be induced to re-form double-stranded DNA with the use of a DNA repair enzyme called *DNA ligase*. The researcher hopes that one of these fragments of DNA contains the target gene. Once the genomic DNA has been cut by endonucleases, the next step is to isolate the fragments and multiply each fragment individually in order to have enough DNA to search for the particular fragment containing the target gene.

B. Cloning Genetic Material: Plasmid Vectors

Cloning is the process in which each fragment of DNA can be separated and multiplied. The process of cloning is a biotic process that involves either a plasmid/bacterial system or a bacteriophage system. We will consider only the plasmid/bacterial system here, but the reader is encouraged to review the bacteriophage system of cloning, as this system has some advantages and disadvantages compared with the plasmid/bacteria system. A plasmid is a small piece of circular double-stranded bacterial DNA that is not associated with the bacterial chromosome (extrachromosomal). Plasmids range in size from a few thousand base pairs to over 100 kilobases (kb). The plasmid DNA is duplicated before every cell division, so that all daughter cells have at least one copy of the plasmid. These pieces of circular DNA are ideal for study because they are simple in structure and are replicated throughout a functional population of bacteria. If a bacterial population starts with one cell that contains a plasmid, all members of the population will have the same plasmid. The task for genetic engineers is to redesign plasmids so that they contain a fragment of DNA from the genome of the source plant and will be retained in a bacterial population.

Popular plasmids used for DNA cloning reproduce in *Ecshereshia coli* and have been engineered to be small (<3 kb). Most of these plasmids contain only the required genes for their use in genetic engineering (Figure 14.7): (1) a replication origin (sequence required to initiate plasmid replication), (2) one or more drug resistance genes, and (3) a region in which exogenous pieces of DNA can be inserted (multiple restriction site sequence or *polylinker*). To insert DNA fragments into a plasmid, both the source DNA and the plasmid are cut with the same endonuclease (or compatible enzymes). The location

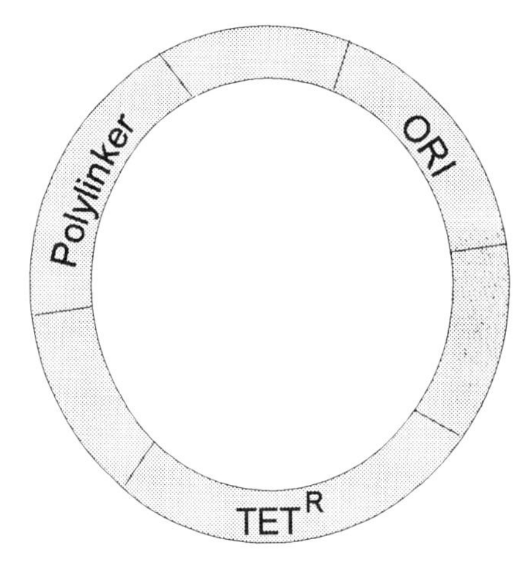

Figure 14.7 Diagram of an *E. coli* plasmid, including several important regions of the plasmid DNA. ORI, genetic material needed to initiate replication of the plasmid; polylinker, a DNA sequence that contains several different nonoverlapping restriction sites.

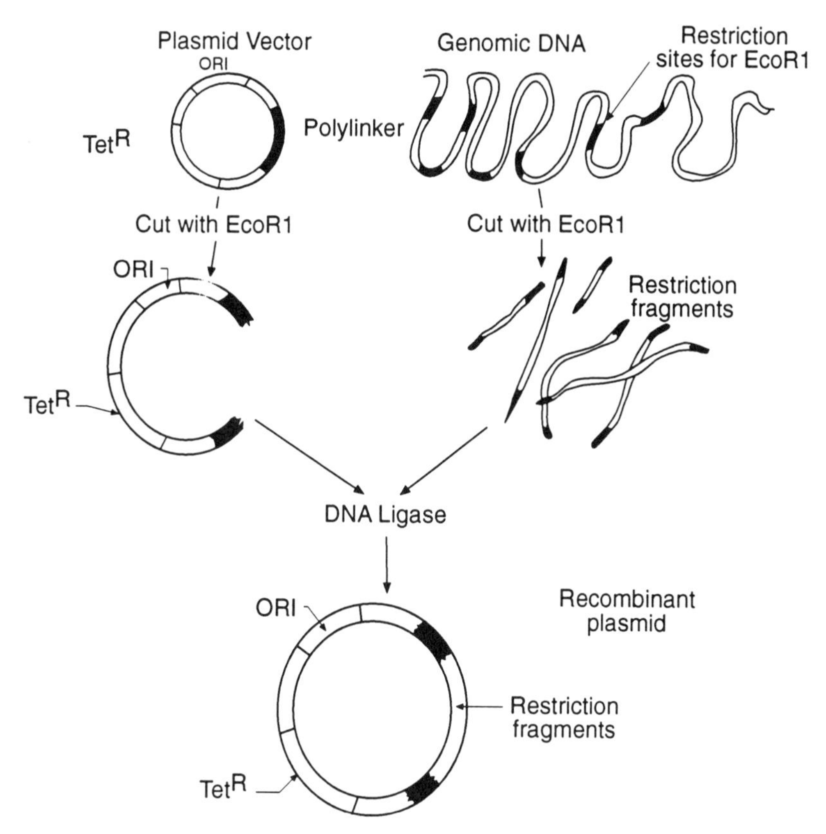

Figure 14.8 Diagram of the technique used to transform a plasmid with a restriction fragment from a eukaryote.

where the DNA is inserted in the plasmid is regulated by adding a polylinker into the specific region of the plasmid genome where exogenous DNA is to be added. This polylinker, also called a *multicloning site,* is a synthesized piece of DNA with 10 or more different sequences corresponding to several endonucleases. Upon mixing the cut plasmid prep with the cut DNA preparation solutions and adding DNA ligase (in vitro ligation reaction), the plasmids may be enlarged by incorporating a fragment of target DNA into the polylinker region (Figure 14.8). Reaction conditions can be chosen to maximize the insertion of one restriction fragment per plasmid cloning vector.

C. Transforming *E. coli*

For the plasmid to be cloned, it must be inside a bacterial cell. In the 1940s it was shown that bacteria could be transformed (genetically altered) by accumulating DNA from the medium when Griffith demonstrated the transformation of nonvirulent to virulent forms of pneumonia. When *E. coli* cells are exposed to high concentrations of certain divalent cations (e.g., $CaCl_2$), a small fraction of the cells are made competent and able to accumulate a plasmid from the surroundings. Therefore, upon exposure of a colony of *E. coli* to the transformed plasmid preparation, some bacterial cells will be transformed with a single plasmid containing a restriction fragment from the source organism. The plasmid that is accumulated from the medium is designed to contain an antibiotic resistance gene; thus the bacterial cells that accumulate a plasmid also gain antibiotic resistance. Transformed bacteria can then be selected by growing them on a medium laced with a specific antibiotic. All transformed bacterial cells will develop individual colonies, while non-transformed cells will die because they do not have antibiotic resistance. The surviving colonies can be transferred to individual plates. The result is a number of different colonies each of which contains a plasmid with a different fragment from the source DNA. This *genomic library* may contain thousands of different clones, each harboring one segment of DNA in the size range of about 25 kb. One of these fragments may contain the gene for the protein in question. The genomic library also includes both coding (transcribed DNA) and noncoding regions of the DNA.

D. Partial Genomic Library

A more discriminating approach to cloning stress-induced genes is the partial genomic library (cDNA or complementary DNA or coding DNA). This technique begins the cloning process at the level of transcribed mRNA rather than at the level of genomic DNA. If a particular targeted protein is expressed in specific cells or at specific times, one can extract the mRNA from cells that are in the process of synthesizing stress-induced proteins. There are several advantages of starting the cloning process with mRNA transcripts instead of DNA. First, there are a smaller number of different mRNA molecules in the cytoplasm at any one time than the total number of genes in the genome. Thus, the final library of clones will be smaller in number and easier to screen. Second, the population of mRNA molecules extracted from the correct cells at the appropriate time (when stress-induced proteins are being synthesized) often contain a high proportion of the transcript from the targeted gene. This makes it easier to identify the stress-induced transcripts for the targeted proteins because they constitute a majority of the mRNA population. Third, mRNA molecules in the cytoplasm have undergone posttranscriptional modification (i.e., intron removal). Therefore, nucleotide sequences of cytoplasmic mRNAs code only for the amino acid sequence of the targeted protein. The main challenge with this technique is to derive double-stranded DNA that is complementary to the stress-induced mRNA sequences in the cytoplasmic extract.

First mRNAs are isolated from the plant tissues of interest, then a single-stranded piece of DNA that is complementary to the mRNA templates can be synthesized by using an enzyme called *reverse transcriptase* (Figure 14.9). This enzyme, isolated from retroviruses, polymerizes deoxyribotriphosphates from the 3′ end of the mRNA (this must be initiated with a multiple thiamin sequence; oligo-dT). After the cDNAs (complementary DNA) are synthesized the solution is treated with alkali, which decomposes the mRNA but not the cDNA. The single-stranded cDNA then needs to be converted into a double-stranded molecule. To do this, the cDNA strands are elongated with a poly sequence of one nucleotide (e.g., dG) by the use of the enzyme terminal transferase. A chemically synthesized polymer of cytosine (oligo-dC) is then added to the solution as a primer for synthesis of the second strand. DNA polymerase is then used to synthesize the double-stranded molecule. At the end of this process each synthesized double-stranded cDNA molecule has a polyG-C sequence at one end and a polyA-T sequence at the other end.

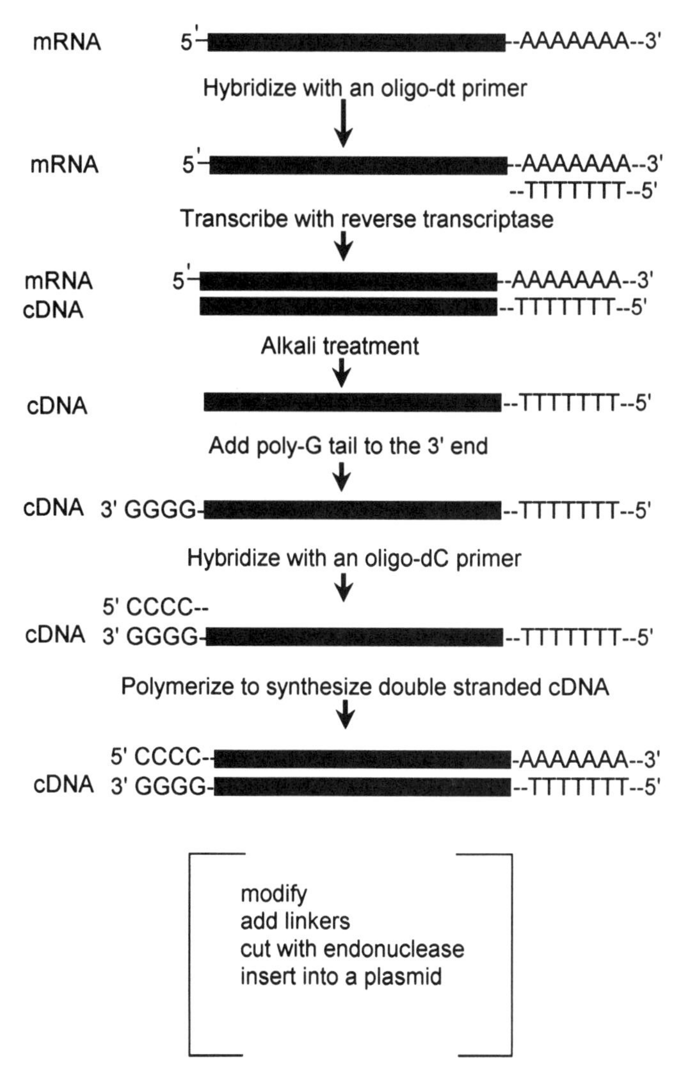

Figure 14.9 Technique used to produce a partial DNA library through the use of cDNA.

Next, specific sequences 10 to 12 oligonucleotides long, with a known restriction site, are attached to each end of the cDNA molecules. This can be done through the use of DNA ligase from the bacteriophage T4 that is able to ligate "blunt end" (no sticky ends) pieces of DNA. The short additional sequences are called *linkers* (they contain a specific restriction site) because they are used to link the cDNA fragments into plasmids. The cDNA fragments can then be cut with a restriction enzyme and joined into a plasmid vector as described previously. Before cutting with the endonuclease (and before adding the linkers) the cDNA must be protected from restriction endonucleases by a modification procedure. This procedure uses an enzyme that methylates the DNA, thereby protecting it from endonucleases. This is the same mechanism that bacteria use to protect their own DNA from the action of restriction enzymes that they produce. A collection of *E. coli* clones that have been transformed with cDNA fragments is termed a *cDNA library.*

E. Screening for Transformations

Once a library of clones has been produced, that library must be screened to select the clone that has the targeted gene of interest. Although there are several techniques for screening libraries, membrane hybridization is the most common. When in the cell, DNA is usually maintained as a double helix by the nature of the double bonds between complementary nucleotides. However, DNA strands can be denatured in vitro by raising the solution pH above 11 or by heating the DNA in a dilute salt solution. The single strands can then be hybridized (alternatively termed annealed or renatured) by lowering the pH or lowering the temperature and increasing the salt solution. During hybridization, only strands that have complementary sequences will join together. Hybridization could occur between complementary strands of DNA or RNA, or between one DNA and a complementary RNA strand. This natural ability of nucleic acids to separate and then anneal only with complementary sequences is used to identify DNA that codes for specific proteins.

At this stage of genetic engineering some information about the proteins produced by the stress-induced target genes is required. The protein must have been isolated, purified, and partially sequenced from the N-terminus. This sequence of six or eight amino acids is used to synthesize a piece of DNA approximately 20 nucleotides long that codes for this sequence of amino acids. The synthesized DNA (termed a *probe*) is labeled with radioactive ^{32}P at the 5′ end. This labeled oligonucleotide will be used as a probe to locate which clone contains the target gene. Since there may be several possible nucleotide sequences that code for the same amino acid sequence (the degenerate aspect of the genetic code), oligonucleotides of all possible sequences are synthesized and combined into one solution (called a *degenerate probe*).

To screen a library of *E. coli* clones by DNA hybridization, the DNA is first extracted from each clone and cut with the restriction endonuclease that was originally used to insert the restriction fragment into the plasmid (Figure 14.10). The DNA is separated by electrophoresis on an agarose gel. Once separated in the gel the DNA is melted (converted to a single-stranded molecule) and transferred by Southern blotting to a piece of nitrocellulose or nylon filter (Figure 14.10). Once in the nitrocellulose the DNA is stable in the single-stranded form. The nitrocellulose is then incubated with a solution containing the labeled probe (synthesized based on the target protein sequence) under conditions that induce hybridization. The labeled probe will hybridize only with a segment of DNA that is complementary to the sequence of the probe. If radioactivity is retained in the nitrocellulose after washing away the unhybridized probe, that particular clone of *E. coli* contains DNA that potentially codes for the target protein. The restriction fragment of source species DNA contained in the plasmid of this particular clone can then be sequenced and the code interpreted to reveal the entire amino acid sequence of the protein coded for by the nucleotide sequence of the cloned DNA. The protein amino acid sequence can then be studied in comparison to sequences of other proteins to determine the likelihood that the targeted gene has been found. In addition, the cloned fragment of DNA can be incorporated into an *E. coli* expression system that will produce large quantities of the protein encoded by the restriction fragment. It is then up to a biochemist to identify whether the protein encoded by the selected restriction fragment is the same as the protein that was originally induced in the stress-tolerant variety.

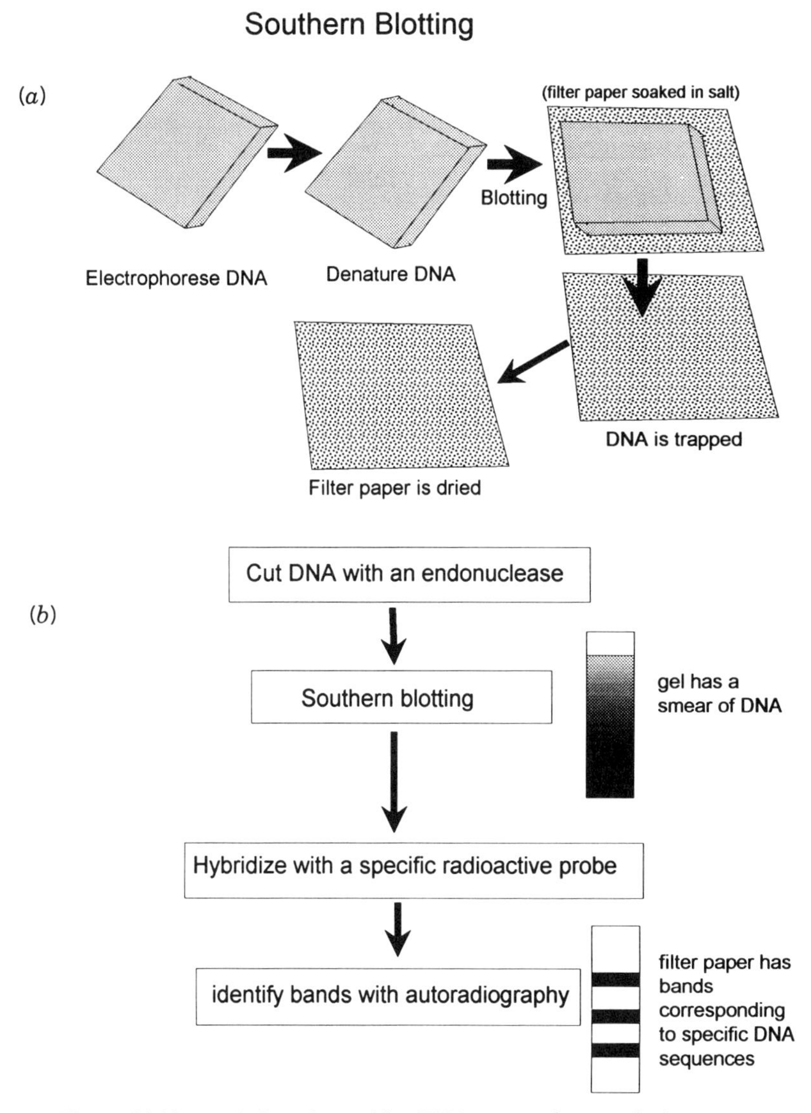

Figure 14.10 Technique for probing DNA sources for a particular gene.

IV. METHODS OF GENE TRANSFER

Once a gene has been identified, isolated, cloned, and verified, the next step toward creating a genetically engineered organism (transgenic organism) is to transfer the gene and a plant promoter into the genetic system of the target organism. There have been a plethora of techniques used to transfer genetic material into target plant species with varying success. Techniques for gene transfer into plants include biological vectors such as viruses and bacteria, and physical processes such as bombardment with gene-coated microprojectiles or electrical shock treatment (electroporation). It is not likely that any one technique will become the dominant mechanism for creating transgenic organisms because of the differential success of various techniques among different species. Some species, such as tobacco (*Nicotiana tabacum*) and *Arabidopsis thaliana,* have proven to be routinely transformed, while other important crop species are particularly recalcitrant for gene transfer mechanisms [soybean (*Glycine max*), rice (*Oryza sativa*), and corn (*Zea mays*). The success of a particular technique can be evaluated by several criteria; however, all techniques should be evaluated on their ability to produce transgenic organisms that are capable of retaining the engineered trait in progeny. This means that the result of gene transfer must be expressed and heritable (Potrykus 1991). Frequently, a less rigorous analysis of success is used, such as indicative results from an associated reporter gene in targeted cells. Accepting indicative evidence as successful gene transfer is potentially dangerous because expression of the reporter gene may not necessarily equate with the successful incorporation of the subject gene into the target species genome.

Genes are transferred into target plants with low frequency when any technique is employed, for the following reasons. First, plant systems are composed of a diversity of cell types, among which only a few are competent for transformation. In many cases it is meristem tissues that have the highest proportion of totipotent cells, and meristems are a small proportion of the total cells of a plant. In addition, of those totipotent cells, only a few will be competent for both transformation and regeneration. A large fraction of totipotent cells are potentially competent if a wound response is elicited mechanically or chemically in these cells. Some species, such as cereals, have only a rudimentary wound response which makes it difficult to transfer genes into these species. In some species it is possible to stimulate the proliferation of cells for regeneration in suspension culture (embryonic suspensions). These cells can be targeted intact for gene transfer, or competence for transformation can be increased by converting the cells into protoplasts (by enzymatically removing the cell wall). In general, the cell wall is a significant barrier for the diffusion of large pieces of DNA. The cell wall needs to be removed, or a technique must be employed to drive the genetic material through the cell wall and into the cytoplasm. Mechanisms of transferring the gene through the cell wall and into the cells include microinjection, microprojectile bombardment, virus infection, and *Agrobacterium tumefaciens* or *A. rhizogens* infection.

A. Indirect Gene Transfer Mechanisms

1.* Agrobacterium *Vector. Among all the potential gene transfer techniques, those using *Agrobacterium tumefaciens* and *A. rhizogens* are the most successful, primarily among dicots. *Agrobacterium*-mediated transformation is so successful that a large number of species have been transformed (Grimsley 1990), and several laboratory manuals have been written on the use of this system (e.g., Draper et al. 1988). *Agrobacterium tumefaciens* is a soil bacterium that infects many angiosperms and gymnosperms that have a wound response mechanism. When a plant is infected with *A. tumefaciens,* tumorous cell growths can be initiated which are commonly called *crown galls.* The tumor is induced by the introduction of DNA into the plant cells at the wound site. The introduced DNA is associated with a large plasmid carried in the *A. tumefaciens* cells (Zaenen et al. 1974). This tumor-inducing plasmid (Ti) contains a small region of T-DNA (transfer DNA) that is transferred into plant wound response cells and covalently incorporated into a plant chromosome (Klee et al. 1987). The T-DNA genes are then expressed in the plant cells and are responsible for tumor induction. Some of the oncogenes (tumor inducing) code for indoleacetic acid and zeatin riboxide, which when overproduced initiate tumorous cell growth in plants. Other genes in the T-DNA produce opines or nopalines that serve as substrate for *A. tumefaciens* growth. When the tumor-inducing genes are removed from the T-DNA of a Ti plasmid, this does not interfere with T-DNA insertion into the plant chromosome. Thus, the basis for gene transfer by *A. tumefaciens* is disarming the Ti plasmid (removal of tumor-inducing capacity) and inserting chimeric genes into the T-DNA. Molecular biologists have shown that the critical aspect of T-DNA for insertion into the plant chromosome is a pair of direct repeats at the flanking portion of the T-DNA. Thus, all other tumor-inducing genes can be removed from T-DNA (disarming) without influencing the ability of T-DNA to insert genes into a plant chromosome. Genetic engineers are able to take advantage of the natural ability of *A. tumefaciens* to transfer DNA into plant chromosomes by cloning foreign DNA into the T-DNA region of disarmed Ti plasmids.

2. Ti Plasmid and Constructs. The Ti plasmid is composed of two regions, the T-DNA region and the vir region (Figure 14.11). The vir region also contains genes that are involved with tumor induction (virulence). The Ti plasmid can be induced to form two

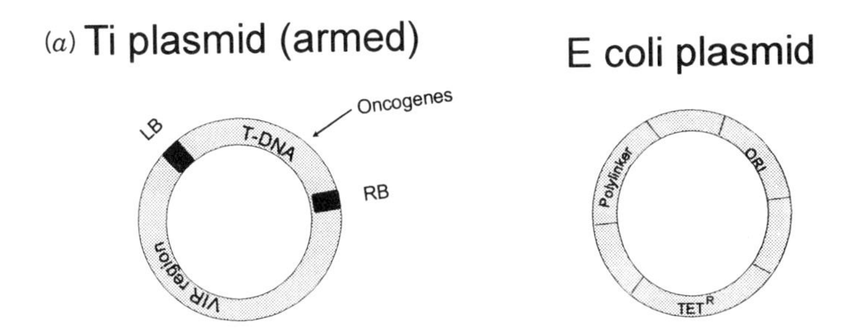

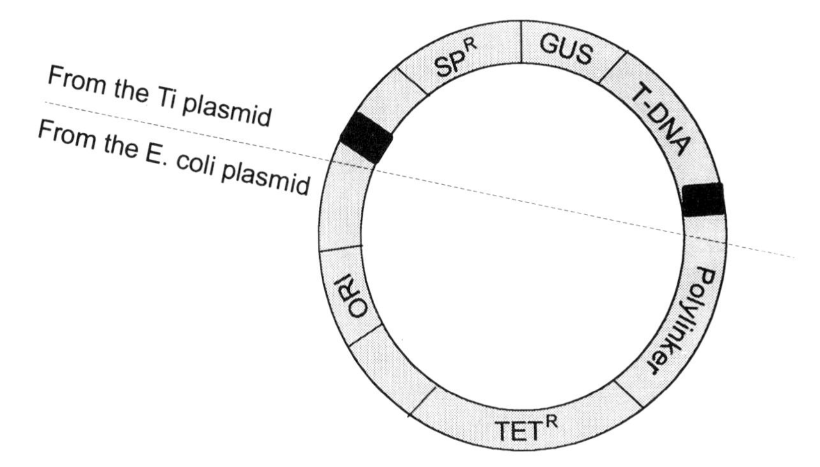

Figure 14.11 (A) Native Ti plasmid before it is dearmed and transformed with foreign DNA. (B) Example of a Ti plasmid construct used to transfere genes into wound cells.

separate plasmids, one containing the vir section and the other containing the T-DNA (Hoekema et al. 1983). Even after the vir region of the Ti plasmid has been removed, the T-DNA does not lose any of its ineffectiveness. Thus, the T-DNA can be incorporated into other plasmids, producing binary plasmids with T-DNA infective potential. The value of creating the binary plasmid is that it can be carried in easily cultured hosts such as *Escherichia coli,* and because new genetic material can easily be incorporated into these vectors. For example, the ColE1 origin of replication is incorporated into Ti plasmid constructs so that increased copies of the plasmid can be maintained in *E. coli,* and genes such as the LacZ′ gene have been incorporated into the T-DNA region to provide several specific restriction enzyme sites. Many other constructs of these binary plasmids have been developed to enhance the ability of maintaining stable, abundant, and selectable clones, to enable easy insertion of foreign DNA into the T-DNA region, and to clearly report the occurrence of transformation. The reporter genes (GUS, bar, CAT, GFP genes) that are included in the T-DNA change the phenotype of the infected cells in a way that is clearly recognizable (e.g., a color change). The expression of a reporter gene indicates those cells that have been transformed. An example construct of a binary plasmid with a reporter gene, selective gene, and restriction site segment is shown in Figure 14.11.

3. Coincubation. Once an appropriate binary construct has been produced, there are several ways that plant tissues can be transformed. In one case, plant cells are co-incubated (co-cultivation) with *A. tumefaciens* containing a modified (engineered) Ti plasmid. After a specified duration of co-incubation, the cultures are washed free of bacteria and the plant cells are incubated in an antibacterial medium. The plant cells that are co-incubated with transformed *A. tumefaciens* could be protoplasts (although this is not often the case), explants of meristematic material (leaf edge disks, lateral meristems, root tips, etc.), or whole plants. Each type of exposure system will have a different plant cell transformation efficiency. In explant systems, plant cell exposure to the transformed bacteria is restricted to outermost cells, while in protoplast systems cell exposure to bacteria is maximized because each protoplast is surrounded by transformed bacteria. In addition, protoplasts are maintained in a highly competent state because they are expressing a wound response and are therefore more likely to be transformed than cells in explant systems. However, the tissue culture protocols for generating whole plants from transformed cells is much easier for explant than protoplast systems. The choice of which explant tissue to utilize in co-incubation is based on totipotency of the tissue (proportion of competent cells), ease of handling in tissue culture, and transformation efficiency. Thus, if an explant tissue can be induced to have many competent cells by wounding, explant systems may be preferable to protoplasts.

Transformation by co-incubation with *A. tumefaciens* can also be accomplished with whole plants or organs rather than protoplasts or explants. In the case of *Arabidopsis thalina,* as well as other small plants, transformation is performed simply by inverting a whole plant into a medium containing *A. tumefaciens* and placing this under a vacuum (Bechtold et al. 1993). Under these conditions it is possible that the germ cells can be infected with the Ti plasmid construct. After the plant has been removed from vacuum and grown normally, some of the seeds will be transgeneic (if the germ cells were transformed). Germinating seeds that are coincubated with *A. tumefaciens* also can be transformed (Chee et al. 1989). Several other whole-plant transformation systems are just being developed and may be the plant transformation system of the future.

4. Agroinfection. A second technique for gene transfer is *agroinfection,* which involves the incorporation of viral DNA into the T-DNA of the Ti plasmid construct. The viral DNA can then be introduced into the plant by the normal *Agrobacterium* transfer process (Gardner et al. 1986). The advantage of incorporating viral DNA into *A. tumefaciens* is that once the Ti plasmid has entered wound cells, the viral DNA releases from the plasmid. The virus can then spread systematically throughout the plant, carrying the viral genes to all cells in the plant. The Ti plasmid (without viral DNA) is restricted to wound response cells, making it impossible for an infection by *A. tumefaciens* alone to incorporate genes systematically through the plant. When a Ti plasmid with viral DNA enters into a wound cell, there is the possibility that the viral DNA (since it is in the T-DNA) will be incorporated into the host cell DNA. This will occur only in the wound cells that are infected with the Ti plasmid. Although the virus travels to other cells in the plant, the viral DNA will not incorporate into chromosomes of any cells other than the wound cells where the T-DNA was present.

As mentioned earlier, it is difficult to transfer genes with *A. tumefaciens* into cereals because the wound response of cereals is rudimentary and wounded cells rapidly die. However, the agroinfection technique has been successful in transferring viral DNA into corn (Grimsley et al. 1987). The viral DNA was transferred through the T-DNA vector, and the viral DNA proliferated systemically through the plant. However, this is not a good mechanism for creating transgenic plants because the viral DNA did not enter the host chromosome and did not enter the generative cells. In fact, the potential for agroinfection to create transgenic cereals is no better than infection by *A. tumefaciens* itself. Although agroinfection may not be an improvement over *A. tumefaciens* infection for creating transgenic individuals, this may be an excellent mechanism for studying the impact of genetically altered viruses on plant processes.

5. Viral Vectors for Transformation. Viral vectors could also be used potentially for creating transgenic plants. Cauliflower mosaic virus is a double-stranded DNA virus that can be used as a viral vector for transferring genes into a target species. By rubbing the virus onto a leaf, the virus infects leaf cells and is transferred systemically throughout the plant. The genes that have been placed into the viral genome are expressed in plant cells, but they are not incorporated into the plant DNA and are not transferred into sexual germ cells. Therefore, viral vectors can stimulate plants to express foreign genes, but they cannot serve as a mechanism for creating transgenic plant lines that maintain a heritable trait. RNA viruses may be the only possibility for using virus vectors to create transgenic plants. Using reverse transcriptase, the viral RNA can be converted into cDNA, which is also infective. If a construct could be made with a transposable element and the cDNA (derived from viral RNA), and the transposable element could be induced to enter the host DNA and be retained, viral vectors may be able to create transgenic organisms.

B. Direct Gene Transfer Mechanisms

Virus vectors and *A. tumefaciens* together constitute only one possible mechanism for transferring genetic material into the genome of plant cells. Several other mechanisms that utilize a physical or chemical technique are also used widely for creating transgenic systems. Some of these techniques are borrowed from microbiology (electroporation), while other techniques were developed on plant systems. Currently, there are a limited number of physical and chemical mechanisms for gene transfer into plants, but the list is likely to grow due to the imagination of scientists and the voracious business of genetic engineering. The main requirement of a physical or chemical gene transfer is first to get the gene through the cell wall into the protoplasm. Therefore, these techniques require either the removal of the cell wall before gene transfer or forceful insertion of the genetic material through the cell wall.

*1. **Electroporation.*** Electroporation is a technique whereby specific pieces of DNA can be induced to enter a cell by exposure to an electric charge. This technique was first used with mammalian cells but is now widely used with plant, yeast, and mammalian cells. In plant systems electroporation is most efficient when protoplasts are used, although electroporation has been used with intact tissues as well (Klöti et al. 1993). There are several reasons why protoplasts are excellent vehicles for gene transfer. First there is no cell wall to interfere with gene transfer. Second, the techniques used to create protoplasts initiate a wound response in the cells and cause a higher proportion of cells to be competent for gene transfer. Third, the gene transfer mechanism does not require a biological vector, so there is no problem with host specificity. Fourth, the foreign DNA comes into contact with a large number of the cells in a solution containing protoplasts, which increases the probability that some of the protoplasts will be transformed. Protoplast transformation by electroporation is generally used only to study transient gene expression in the protoplasts because it is difficult to recover whole plants from protoplasts.

To induce DNA uptake by electroporation, an electric shock of relatively low voltage and high amperage (compared with animal cell electroporation, in which high voltage and low amperage is used) is applied to a co-incubation solution or an explant imbibition system. Theoretically, the electric shock induces small holes in the cell membrane, which allows the foreign DNA to move into the cell. Following the electrical charge the holes reseal, and extracellular DNA may have been transferred into the protoplast. Once inside the protoplast, the foreign DNA may be expressed in the cytoplasm (in the case of an *E. coli* plasmid construct) or inserted into the host chromosome and expressed as a nuclear gene (in the case of a T-DNA construct). Several physical and chemical characteristics of the transfer system may influence the efficiency of transfer by electroporation. For example, the intensity and duration of electrical charge, the size of the plasmid construct containing the DNA to be transferred, the type of plasmid (supercoiled or linear), the concentration of plasmid, and the media conductivity all affect the efficiency of transformation by electroporation.

The effectiveness of transformation depends on the taxonomic relationships of the species and the developmental ages of the cells or tissues (synchronized cells in the M phase have high transformation rates) used to produce protoplasts, as well as the chemical and physical conditions of the coincubation medium. Virtually every species used to make protoplast systems has been transformable, but the efficiency of transformation varies greatly among species (Potrykus 1991). Following successful electroporation, cell or tissue culture is required to initiate whole plants from transformed cells. The difficulties inherent in tissue culture from the protoplast to the whole individual is the major hinderance to successful transformation by electroporation in plant systems. Therefore, many attempts have been made to improve the efficiency of transforming explants or whole plants with electroporation rather than using protoplast systems.

*2. **Chemically Induced Transformation.*** Protoplasts can also be induced to take up DNA from the medium by exposure to a solution with a high osmotic concentration (*chemo-poration*). In combination with a high pH and high Ca^{2+}, both polyethylene glycol (PEG) and polyvinylpyrrollidone (PVP) will induce protoplasts to accumulate DNA from the medium. Commonly, the DNA to be transferred is incorporated into a binary construct containing T-DNA or other plasmid constructs. However, plasmids may not be stable under the harsh chemical treatment required for chemo-poration. Decomposition of plasmids due to high solution osmotic concentration reduces the efficiency of transformation by chemo-poration. Some researchers have protected plasmid constructs from decomposition by using a spheroplast of *A. tumefaciens* (*Agrobacterium* without a cell wall) to house the DNA construct that will be transferred. When a spheroplast solution is combined with a target protoplast solution, cell fusion (or endocytosis) may occur, which transfers the DNA from the spheroplast to the protoplast. In other experiments, plasmid constructs have been protected by encapsulating the DNA to be transferred inside negatively charged liposomes (Gad et al. 1990). When the liposome preparation is combined with the target protoplast suspension, the liposomes fuse with protoplast, transferring their DNA into the protoplast while protecting the DNA from chemical destruction. Protection of the transferred plasmid DNA is also increased by the presence of polyamines such as spermine and spermadine in the coincubation medium. Genes that are induced to enter cells by these chemical means are inheritable in the regenerated transgenic individuals (Paszkowski et al. 1984). Both the electroporation and chemically induced poration techniques may be important for DNA transfer into grasses because *A. tumefaciens* transfer cannot be used in grasses (Potrykus et al. 1985). However, the inherent difficulty of regenerating grasses from protoplasts still limits the applicability of electroporation and chemo-poration protocols for transforming cereals and other grasses.

3. Microprojectile Transformation. In many plant systems protoplast suspensions either are not stable or cannot be induced to generate new individuals. In other species these two poration protocols for creating transgenic individuals are possible but difficult. Therefore, other techniques for transforming cells (without a biotic vector) in callus or in tissue discs have been developed. The most important transformation technique for plant tissues with cell walls is the microprojectile bombardment mechanism. First developed by Klein and Sanford at Cornell University, the DNA to be transferred was coated onto pellets of gold or tungsten. The pellets were loaded into a cartridge and the cartridge was exploded to accelerate the particles down a gun barrel to velocities that were nonlethal and allowed penetration of the pellets through cell walls (Figure 14.12). As a result, this has been termed "gene gun" technology. Most gene guns now utilize a helium mechanism to accelerate the DNA-coated microprojectiles. The commercially available device (originally designed by Dupont, now marketed by Bio Rad) is retrofitted for helium-driven microprojectiles. This is a takeoff of the gun powder device and is called the POS 1000-He. Recently, the particle-in-flow gene gun (PIG) was developed, which uses helium to accelerate microprojectiles that are suspended in droplets of liquid. A third type of mechanism uses electrical charges through water suspensions to accelerate microprojectiles (McCabe and Cristou 1993). Most of the gene gun technology has been developed by industry, and some of this technology is not readily available to researchers at universities.

One of the first studies of microprojectile bombardment techniques involved the transfer and expression of genes associated with anthocyanin production in maize. The general method for transformation by microprojectile bombardment in monocots is to initiate embryonic callus by tissue culture from an immature seed or plant tissue, bombard the callus, select for transformations, and recover transformed individuals. When genes were

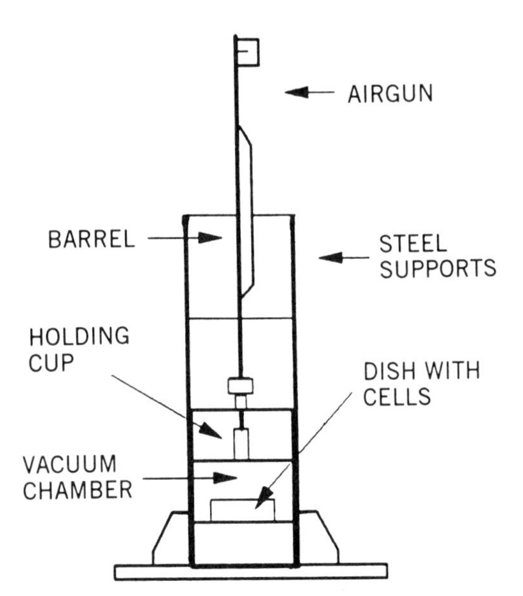

Figure 14.12 Diagram of a particle gun (a particle in-flow device) used to bombard cells with DNA-coated particles in an effort to transform some of the cells. This is one of many different designs for particle bombardment technology. (From Oard 1993.)

transferred into the aleuron layer of maize kernels, the transferred genes were expressed and regulated in thc same way as their native counterparts. The microprojectile transfer of genes has been successful with a number of crop species, such as soybean (McCabe et al. 1988), cotton (Finer and McMullen 1990), and tobacco (Tomes et al. 1990). In fact, the microprojectile technique for gene transfer has become the preferred system for transforming soybean. However, the microprojectile technique has very low efficiency compared with other direct transfer techniques, such as electroporation. The major advantage of microprojectile gene transfer is that differentiated cell tissues (pollen, leaf tissues, meristems) can be used, and these systems have much easier tissue culture protocols than those for protoplast systems.

Several other techniques have been tried to transfer genetic material into plant cells with an intact cell wall. In one case, a UV-microlaser was used to burn holes through cell walls in hopes that after treatment, cells incubated with foreign DNA would be transformed (Weber et al. 1988). Other attempts were to incubate dried seeds with a DNA preparation (Töpfer et al. 1990) or to use a macroinjection technique where the needle is larger than an average cell (De la Pena et al. 1987). In the macroinjection model, extracellular DNA could be transferred into broken cells near the needle because broken cell wall fragments induce a wound response in adjacent cells, making them competent for transformation. Microinjection of foreign DNA into protoplasts or proembryos using microcapillaries or other microscopic devices also has resulted in some success (Neuhaus and Sprangenberg 1990).

V. OTHER TYPES OF BIOENGINEERING EXPERIMENTS

In the sections above we have described one mechanism for modifying the physiology of crops to increase stress resistance. In this case a piece of foreign DNA coding for a gene that confers stress resistance in another species or variety was inserted into the genome of a crop species. Potentially, this can create a transgenic organism that has increased stress tolerance. However, there are other possible ways to adjust plant stress tolerance without adding a novel gene into a plant's genome.

In some cases, plants that are susceptible to stress are incapable of producing enough of a particular protein during the stress to compensate for the environmental perturbation. In such cases, genetic engineering procedures do not need to add foreign genes into the plant's genome because the appropriate gene is already present. Increased stress tolerance may be obtained by increasing the expression of an existing gene. Gene expression could be increased by increasing the number of copies of that gene in the plant's genome. A highly effective promoter could also be linked to the targeted gene to increase the gene expression. In other cases, stress sensitivity may be due to the expression of a particular gene that induces metabolic senescence or otherwise disrupts the normal function of cell metabolism. In such cases, increased stress tolerance may result from decreasing the expression of a particular gene. Gene expression can be reduced by the incorporation of an antisense construct into a plant's genome. Recall that when transcribed, an antisense gene produces mRNA molecules which are complementary to the mRNA transcribed from the targeted gene. The two mRNA molecules hybridize in the cell, decreasing the potential for translating the mRNA transcript for the targeted gene, thereby turning off gene expression.

There are many other ways that genetic engineering could modify the expression of targeted genes and the function of resultant mRNA transcripts or translated proteins. For example, the targeting sequences of stress-induced genes could be altered to induce inappropriate targeting (shuttling the protein to the wrong subcellular location) and thereby inhibit the effectiveness of the induced protein. Moreover, mRNA processing could be circumvented or modified to result in differing and ineffective mRNA sequences. The field of genetic engineering of stress resistance in crops is in its early stages, and many more possibilities will likely arise in the future. A more fledgling application of this technology is to study the existence and evolution of stress-adaptation mechanisms in nondomesticated plants. Although bioengineering of stress tolerance in plants is a young science, some advances have been made in crop systems.

VI. BIOENGINEERING OF STRESS–TOLERANCE TRAITS IN PLANTS

In the following sections, particular stress-tolerance mechanisms that may lend themselves to genetic engineering are considered. In each case the general significance of the stress-tolerance mechanism is reviewed and case studies are presented that relate to specific protein systems that are induced by environmental stress. The material for each section is not meant to be a comprehensive review of all that is known about molecular regulation of each stress-tolerance mechanism. Rather, background information is provided to enhance previous chapters, and examples of how genetic engineering can be used to modify stress tolerance (and elucidate cause-and-effect relationships between proteins and stress tolerance) are considered.

A. Thermally Induced Proteins

1. Heat-Shock Proteins. One of the earliest discoveries of stress-induced protein products were the heat-shock proteins (HSPs). When a wide variety of organisms are exposed to short-duration high-temperature conditions (usually, a 10°C increase over optimal temperatures), several classes of novel proteins can be identified by two-dimensional gel

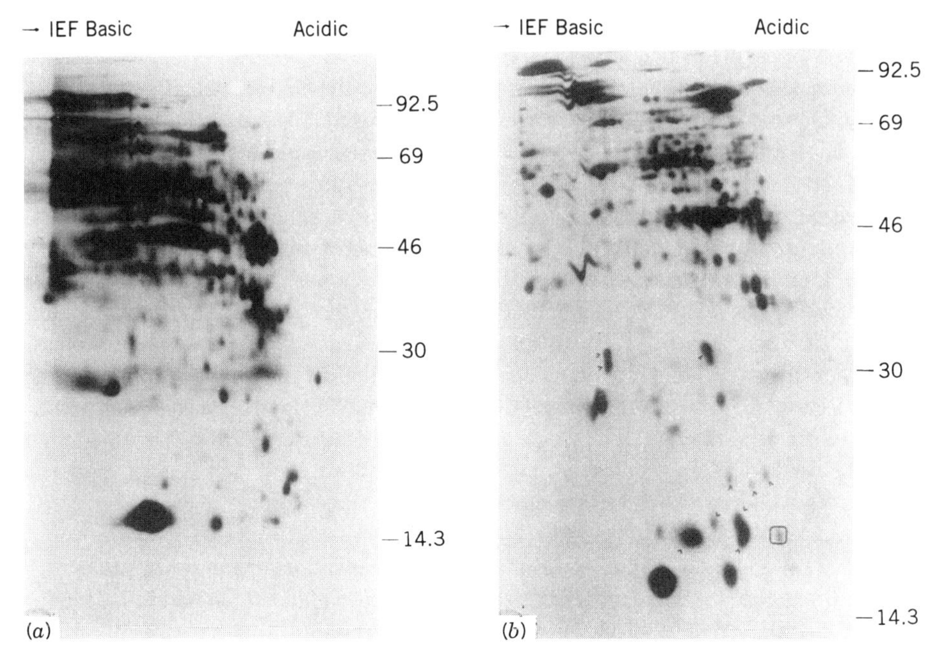

Figure 14.13 Two-dimensional gel electophoresis of total leaf proteins of the wheat cultivar Sturdy before (A) and after (B) heat shock. The square indicates a heat-shock protein unique to the variety Sturdy, and arrows point to low-molecular-weight HSPs. (From Nguyen et al. (1989).

electrophoresis (Figure 14.13). In fact, there is a large change in cell protein composition as pre-heat-shock proteins disappear and heat-shock proteins appear. All HSPs increase in concentration for a limited time period following heat shock (Figure 14.14), and most HSP populations decrease in concentration over time even if temperature remains high.

Induction of proteins following heat shock is dependent on both translation and tran-

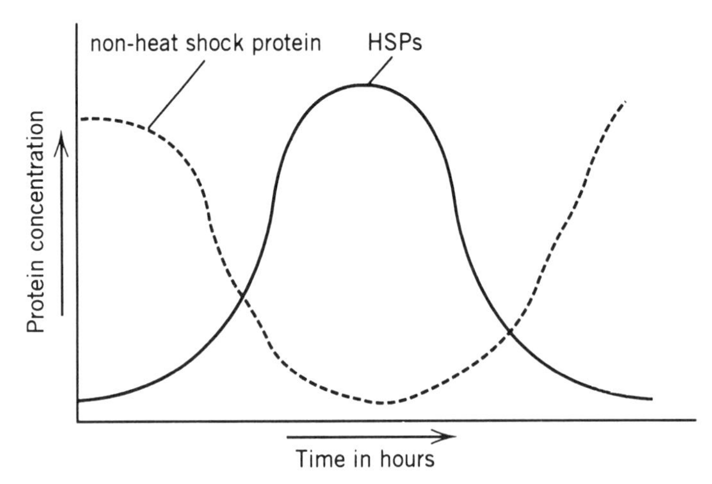

Figure 14.14 Pattern of induction for a group of heat-shock proteins.

scription. Transcription of heat-shock protein genes is regulated by a heat-shock element (HSE). HSEs are found on the DNA upstream from all HSP genes and contain small consensus sequences (sections of nucleotide sequences that are identical among all HSEs from all organisms). Induction of heat-shock gene transcription occurs when a heat-shock factor (HSF, a transcription regulatory protein) binds to the HSE upstream from the HSP gene (Bienz and Pellham 1987). Several HSFs have been identified and sequenced. These factors are rather diverse among organisms, but several DNA binding domains are consistent among all HSFs (Sorger 1991). Heat-shock factors are present constituitively in cells and are activated upon heat shock. Activation of HSFs is a multistep process probably related to phosphorylation of the HSFs and releasing them from binding to a HSP during heat shock. HSP-70 has been found to bind to HSFs and prevent the HSF activation (Abravaya et al. 1992). The up-regulation of HSPs by heat shock induces an increased population of HSP that inhibits HSFs and subsequently results in the down-regulation of HSP genes (Figure 14.15). This interaction between HSP, HSF, and HSE regulates the short-term induction of HSP gene transcription. Thus, the response of HSP gene transcription to heat shock is a fine-tuned process that opens up many possibilities for genetic engineering.

Ubiquitous among kingdoms of organisms, heat-shock proteins can be induced in bacteria, plants, animals, and fungi. There is a diversity of heat-shock proteins falling generally into a high-molecular-weight class (>80 kD), 70-kD HSPs, and a low-molecular-weight class (14 to 30 kD). Upon thermal shock some HSPs are induced in all organisms (HSP 22, 27, 70), while others are inconsistently expressed among organisms (Vierling 1991). Some of the HSPs are constituitivly produced at low levels without being induced by high-temperature shock.

One of the most important questions asked about heat-shock proteins is: Do HSPs confer heat-stress tolerance? Several lines of evidence indicate that HSPs do confer thermotolerance. In several studies with unicellular organisms (both eukaryotes and prokaryotes), mutants that could not produce particular HSPs were found to be heat-shock sensitive in comparison to wild-type cells (McAlister and Finkelstein 1980, Sanchez and Lindquist 1990). The kinetics of the synthesis of HSPs closely match the kinetics of ther-

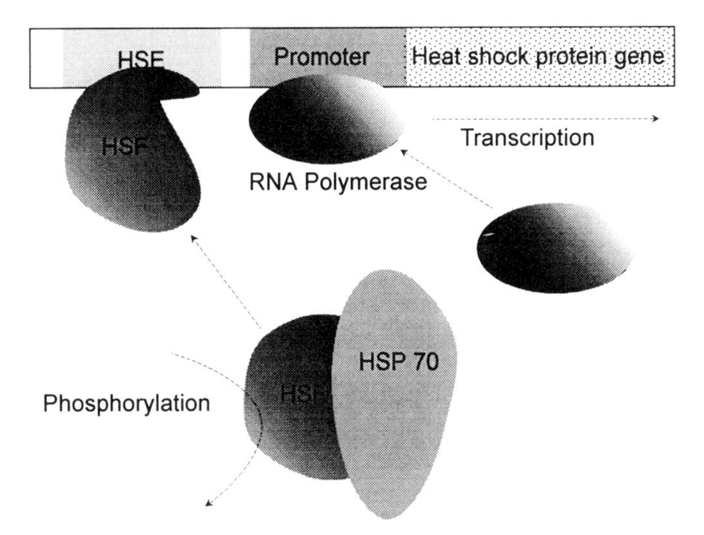

Figure 14.15 Possible molecular regulation mechanism for transcription of heat-shock proteins.

motolerance in many organisms (Landry et al. 1982). When protein synthesis is chemically inhibited, thermotolerance is lost (McAlister and Finkelstein 1980). Conversely, it is possible to induce HSP production at constant temperature by applying arsenite, which also induces thermotolerance (Howarth 1990). In other cases, the induction of HSPs was not associated with thermotolerance (Smith and Yaffe 1991). In some yeast systems, sufficient levels of stress-protective proteins are present at low-temperature growth conditions, so that induced HSPs do not change thermotolerance. Constitutive proteins may be heat activated (thermal-induced change in shape) to become factors that induce thermotolerance without protein synthesis (transcription and translation). Some studies have also shown that HSP production may vary among closely related species in a consistent manner with the habitat in which the various species reside (Coleman et al. 1995). For example, hydra species occupying thermostable pools do not induce HSPs in response to temperature stress. Furthermore, these species are more sensitive to heat stress than hydra species occupying pools with variable temperature (the latter species do have induced HSPs). If all evidence is considered together, induced synthesis of certain (but not all) HSPs is required for heat-shock tolerance in some but not all species. HSPs are not artifacts of heat-induced trauma but are, rather, proteins induced by heat stress, some of which confer heat-stress tolerance. Therefore, engineering of HSP genes in crop plants may result in greater heat tolerance in crops, particularly those crops that are notoriously sensitive to high-heat conditions.

Most classes of HSPs are induced in plant cells following heat stress (Vierling 1991). Many of those same proteins are also induced following other stresses, such as intense water limitation or excessive salinity exposure. Therefore, several of the proteins designated as HSPs in plants are actually general trauma-induced proteins and may have a basic universal metabolic significance such as nonspecific protein recycling or repair. In other organisms (bacteria and animal systems) the function of several different HSPs has been defined. Constitutive HSPs are thought to function as molecular chaperones, which serve to facilitate protein folding and transport of proteins across membranes (Hartl et al. 1994), or proteases that facilitate the recycling of amino acids from heat-damaged proteins into new functional proteins. Other HSPs are likely to have protective function for particularly heat-sensitive proteins.

In plant systems, maintenance of photosynthesis is a critical aspect of thermotolerance, and photosynthesis is particularly sensitive to thermal stress (Weis and Berry 1988). Several HSPs have been shown to be translocated into the chloroplast and correlate with increased thermal tolerance of photosynthesis (Vierling, 1991, Stapel et al. 1993, Clarke and Critchley 1994). In particular, HSP21 and HSP24 are translocated into the chloroplast and may serve to protect PSII activity during high-heat conditions. The modes of action for HSP21 and HSP24 are not known, but they may serve a protective mechanism for thylakoid-bound proteins because HSP24 co-isolates with thylakoid membrane during heat stress (Osteryoung and Vierling 1994).

Heat-shock-protein induction can vary among plant genotypes (Vierling and Nguyen 1990) and among plant cells at different developmental stages (Werner-Washbourne et al. 1989) or plants given different levels of resources (J. Coleman, personal communication). Although the inherent genetic and developmental regulation of HSP induction capacity suggests that HSP induction can be modified in plants by genetic engineering, the control of HSPs by resource levels does suggest that there are resource costs associated with the production of HSPs. Before genetic engineering of HSP production in plants is initiated, the HSPs with the highest likelihood of conferring heat tolerance to a heat-sensitive plant need to be identified (the field of possibilities must be narrowed). Furthermore, a better understanding of the costs (in resources and energy) required for HSP production is necessary before an ecophysiological interpretation of the potential significance of these proteins to stress physiology in nature can be determined.

Which HSPs are the focus of genetic engineering in plants, and what have been the results to thermotolerance of plants by creating transgenic organisms with heat-shock proteins? There are only a few studies in which transgenic organisms have been used to study the consequence of HSP expression on heat tolerance. In one study of mammalian cells, transgenic cells that overproduced low-molecular-weight HSPs had improved short-term thermotolerance (Landry et al. 1989). However, in another study, overexpression of HSP70 in *Drosophila* cells substantially decreased cell growth under normal temperatures (Feder et al. 1992). This suggests that there may be a significant metabolic cost to producing HSPs.

2. Cold-Induced Proteins. In a similar manner to other stress-tolerance traits, low-temperature-induced cold tolerance is a quantitative trait regulated by many genes (Guy 1990a). During the cold acclimation process, changes occur at the physiological, biochemical, and gene level. However, within this matrix of interacting changes in metabolism, two-dimensional electrophoresis has revealed specific changes in polypeptides induced by cold or freezing temperature. In contrast to the results for HSPs, the number of proteins that are induced during cold acclimation are few in number and of relatively low concentration. Furthermore, unlike the heat-shock response, a large majority of pre-cold induction proteins are not dramatically down-regulated by chilling or freezing (Figure 14.16). The proteins that are induced by cold shock are diverse in size, but a high-molecular-weight protein (160 kD) is cold-induced in many plants and animals. Several of the induced proteins are glycoproteins, while other glycoproteins disappear during chilling conditions. Thus, there is a turnover of the glycoprotein pool composition during cold stress, but the abundance of glycoproteins does not necessarily change. Another group of cold-induced proteins are homologous to proteins produced by LEA genes (late-embryo-

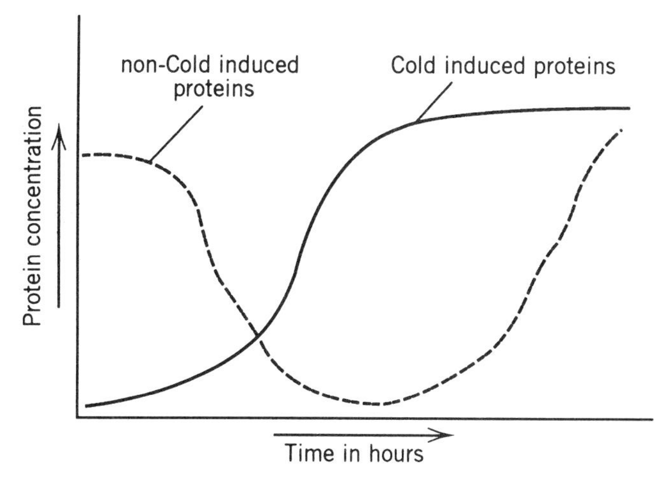

Figure 14.16 Induction kinetics for cold-shock proteins. Note the differences between Figure 14.14 and this figure.

genesis-abundant genes), which encode strongly hydrophobic proteins in seeds that have the unusual property of being stable under boiling conditions.

Although the function of many cold-induced proteins is not known, some specific cases are noteworthy. In maize a 31-kD protein is induced in the chloroplast when plants are exposed to 5°C temperature and high-light conditions. The increasing concentration of the 31-kD protein corresponds to a decrease in the concentration of a 29-kD protein. Since the 29-kD protein is a component of the light-harvesting complex II (LHCII), and the 31-kD protein has high homology to the 29-kD protein, the 31-kD protein has been identified as a precursor to the 29-kD LHCII protein (Hayden et al. 1988). It is most likely that low-temperature-induced photoinhibition (under high-light conditions) inhibits the final steps leading to the production of LHCII proteins, and consequently induces the buildup of the precursor to LHCII protein, which is the 31-kD protein. Thus, this 31-kD protein is a consequence of cold conditions rather that a protein conferring cold tolerance traits to the plant.

Another cold-induced protein (CAP79) has some similarities to HSPs because CAP79 is minimally induced by a heat shock. However, following cold induction the CAP79 proteins increase much more rapidly that they do in response to heat shock, so they are not considered HSPs *sensu stricto.* The CAP79 proteins that are induced in spinach at low temperature correlate with an increase in the corresponding mRNA populations, suggesting cold-induced transcription regulation. The CAP79 proteins are thought to promote renaturation of cold-denatured proteins (Neven et al. 1992).

Although many of the changes in protein activity or composition have been shown to be regulated by post-translational factors, chilling does alter gene expression for a number of different classes of proteins. In several crop species, chilling conditions induce individual translatable messages (novel mRNAs) such as that for the CAP79 proteins. In many of these cases, correlations between increased freezing tolerance and the induction of the novel mRNA have been made (Tseng and Li 1990, Danyluk et al. 1990, Durham et al. 1991). The cold-induced genes (transcription induced by chilling) code for several different classes of proteins, including cryoprotectants, enzymes, and structural proteins.

During cold acclimation the concentrations of many cryoprotectant molecules can build up in plants. Some of these are not related to novel protein synthesis; however, some cold-regulated polypeptides (CORs) have been identified. In *Arabidopsis thaliana,* cold acclimation induced the accumulation of four major mRNAs (Hajela et al. 1990). These mRNA populations code for proteins that are highly hydrophobic and are similar to cryoprotectant proteins produced in spinach. One of the COR proteins, COR15, has been shown to have extreme capacities to protect enzymes from cold denaturation in vitro. COR15 is 1000 times more effective than bovine serum albumin for protecting lactate dehydrogenase from cold denaturation (Lin and Thomashow 1992).

Chilling can result in either up- or down-regulation of enzymes at the transcription and/or translation level. The gene responsible for the synthesis of the small subunit of rubisco is down-regulated following chilling treatment, resulting in a 10-fold decrease in the corresponding mRNA population. In fact, transcript populations of several genes associated with photosynthesis (*Cab* → encoding chla/b protein of PSII, sucrose synthase gene, and several chloroplast-encoded genes) are down-regulated following chilling (Hahn and Walbut 1989). In contrast, several specific gene transcripts for enzymes increase in abundance. The glycerol-3-phosphate acyltransferase gene (GAPT) mRNA populations increase. As discussed earlier, the GAPT enzymes induce a change in the phospholipid composition of the inner chloroplast membrane conferring greater tolerance of cold conditions.

In the case of structural genes, an 80% increase in cell wall hydroxyproline-rich protein occurs in pea when grown at 2°C (Weiser et al. 1990). Extensin is the primary hydroxyproline-rich protein in the cell wall, and northern blots show that three of the seven classes of extensin mRNA increase during this chilling treatment. It is postulated that the increase in extensin increases the rigidity of the cell wall and therefore limits damage to cells during freezing-induced dehydration. Some other structural proteins, such as microtubules, are structurally altered at low temperature, but this is not the result of induced transcription.

Given that cold treatment can induce a diverse array of proteins at the level of transcription in cold-tolerant plants, and that the novel proteins which result from the translation of the induced mRNA result in structural and functional changes in cells, it is highly likely that genetic engineering techniques can alter cold tolerance in sensitive plants. Two of the best cases of the potential for genetic engineering of cold tolerance concern cold induction of GAPT genes (Murata et al. 1992) and the impact of sugars on cell cyroprotection (Sonnewald 1992). As discussed earlier (Chapters 3 and 12), cold tolerance correlates with a high proportion of *cis*-unsaturated fatty acids in the phosphatidylglycerol (PG) molecules in the chloroplast membranes. The degree of unsaturation of PG is dependent on the specific activity of the GAPT. Thus, if plants are transformed to overexpress the GAPT gene, this should result in higher *cis*-unsaturation of the PG in chloroplast membranes and increased cold resistance. Such a result was found for transgenic tobacco (cold sensitive), which were induced to overexpress GAPT genes from *Arabidopsis thaliana,* squash, or *E. coli* (Murata et al. 1992).

One mechanism used by cold-tolerant plants to lower the freezing point of their cytoplasm is an increase in soluble sugars (discussed in Chapter 12). To evaluate the cause-and-effect relationship between soluble sugar concentration and cold tolerance, transgenic tobacco plants were created with increased leaf glucose and sucrose concentrations. In one case the tobacco was transformed with a pyrophosphatase gene from *E. coli* and a yeast invertase gene in the other case. The result was two transgenic tobacco lines with similarly high soluble sugar concentration phenotypes. When plants of these two transgenic lines were exposed to cold stress, only the line transformed with the pyrophosphatase gene gained greater cold tolerance than the wild-type line. Therefore, expression of the pyrophosphatase gene in the transgenic tobacco probably influenced the expression of other genes that conferred cold-stress tolerance.

B. Dehydration-Induced Proteins

Cold, drought, and salinity stress can all cause a reduction in tissue water content. Therefore, it is likely that all of these different stresses induce genes that produce proteins that modify metabolism in a similar manner (Siminovitch and Cloutier 1983). In particular, it is possible that responses to cold shock and drought are linked through responses to ABA. It has long been known that drought can increase cold-stress tolerance in plants (Rosa 1921). As discussed earlier (Chapter 12), freezing of water in intercellular spaces causes the removal of water from the cytoplasm, inducing a desiccation stress, and ABA plays a central role in both cold and drought stress. Many studies have shown an endogenous increase in ABA during cold stress. Furthermore, a cause-and-effect relation between cold tolerance and endogenous ABA has been established with *Arabidopsis thaliana* mutants that are deficient in ABA (Heino et al. 1990). Therefore, it is likely that desiccation stress caused by cold shock or water stress influences the activity of ABA, which regulates genes that code for stress-induced proteins.

Several cold-induced proteins increase in concentration after water stress is initiated at constant temperature (160- and 85-kD proteins in spinach). Severe water stress induces high concentrations of these proteins in spinach, and this causes increased cold tolerance (Guy et al. 1992). Moreover, several membrane-bound proteins and glycoproteins induced by cold stress are also inducible by exogenously applied ABA in alfalfa (Mohapatra et al. 1988a, Robertson and Gusta 1986). One cold-induced protein COR47 has a large amount of sequence homology with the LEA-genes, which may confer dehydration tolerance to seeds just before they become desiccated (Gilmour et al. 1992). It is likely that LEA-like cold-induced proteins like COR47 may confer desiccation protection to tissues that experience freezing temperatures. In contrast, another alfalfa LEA-gene is unregulated by cold, salinity, or drought. Thus, for some cold-regulated proteins there is a linkage between cold, salinity, drought, and ABA induction. Genes induced by cold, salinity, and drought are often not induced by heat shock. There is strong evidence that these genes serve a general desiccation tolerance function.

The linkage between dehydration-induced increases in ABA and cold-inducible proteins is not absolute. There are some cold-induced proteins that are causally related to cold tolerance and are not induced by drought or ABA (Mohapatra et al. 1988b). Furthermore, in ABA-insensitive mutants of *Arabidopsis thaliana,* cold stress up-regulates a series of COR genes, but applying ABA does not affect COR gene transcription. These results were interpreted to mean that cold- and ABA-induction mechanisms for the COR genes were independent. Therefore, cold- and drought-inducible genes are not always linked through the effects of ABA on gene regulation.

Tolerance of low water potential and high salinity is commonly accomplished by changes in tissue osmotic potential. It is widely documented that plants produce compatible solutes (such as amino acids, low-molecular-weight sugars, and sugar alcohols) that accumulate during water limitation. The biochemical pathways producing those solutes are well known, and many genes that encode enzymes in these pathways have been isolated and cloned. Therefore, it is currently feasible to utilize genetic engineering to alter the osmoprotection mechanisms of test species.

One example of an osmoprotectant is glycinebetain, which is synthesized from choline in two steps (Figure 8.9). There is good genetic evidence that glycinebetaine improves salinity and water-stress tolerance in barley and maize (Grumet and Hansen 1986, Rhodes et al. 1989). Isogenic lines of barley or maize that contain relatively high levels of glycinebetain have a greater capacity for osmotic adjustment than isolines with low glycinebetaine concentrations (Figure 14.17). Since only two enzymes are associated with the pathway to glycinebetaine from choline, and glycinebetain is not actively turned over in plant cells, strategies to overexpress these two proteins in transgenic organisms will probably result in higher glycinebetain concentrations in tissues. In a natural system, both low soil water potential and salinity induce both BADH (betain aldahyde dehydrogenase) and CMO (choline monoesterase), the two critical enzymes in the glycinebetain synthesis pathway. All evidence points to a successful increase in osmoprotection in plants by inducing overexpression of two genes associated with glycinebetain synthesis (McCue and Hansen 1990).

Carbohydrates are other compatible osmoprotectants that have implications for desic-

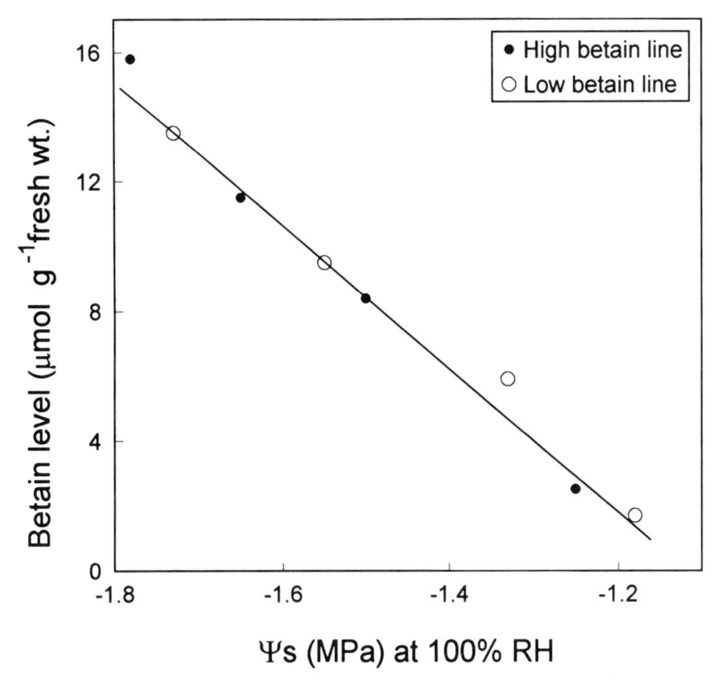

Figure 14.17 Relationship between growth medium, water potential, and the leaf glycine-betain level. (From Grummet and Hanson 1986.)

cation tolerance (Leopold 1990). High concentrations of carbohydrates can suppress temperature-induced membrane phase transitions and prevent membrane leakage (Hoekstra et al. 1989). Furthermore, carbohydrates are responsible for stabilizing proteins and preserving their enzyme activities during water stress. High concentrations of low-molecular-weight sugars such as raffinose can also affect the viscosity of cell water when desiccated. Upon desiccation, several organisms produce a "glass"-like cytoplasm, which is a viscous solution that precludes the activity of all enzymes that require molecular diffusion (Koster 1991). Bruni and Leopold (1991) found that all desiccation-tolerant species enter into a "glass" state under low water potential, which does not occur in sensitive species. Therefore, regulation of carbohydrate synthesis may be an important protective mechanism against the detrimental effects of desiccation (Carpenter and Crow 1988).

In the case of fructans (polyfructose carbohydrates produced by 15% of flowering plants), genetic engineering has been used to evaluate the causal effect of carbohydrate abundance on water stress tolerance. The role of fructans in drought resistance in flowering plants has been demonstrated convincingly (Hendry 1993). A homozygous line of transgenic tobacco (KP12-9) that accumulates fructans was created by inserting the SacB gene from *Bacillus subtilus* with an appropriate promoter into the genome of *Nicotiana tobacum*. Tobacco has no native fructan-producing capacity, so the KP12-9 transgenic organism produces a novel carbohydrate compared to the wild type. Under water stress the KP12-9 line retained 55% higher relative growth rate than the wild type (Pilon-Smits et al. 1995). The fructan present in the transgenic line is of the bacterial levan type and is synthesized by fructosyltransferase expression from the introduced bacterial gene. However, it is difficult to verify a transcript of the SacB gene in the transgenic plants (Van der Meer et al. 1994). Consequently, detecting the molecular regulation of this gene in the transgenic plant may be difficult. In a similar study, enhanced desiccation (salinity) tolerance resulted from introducing a bacterial gene for mannitol 1-phosphate dehydrogenase in tobacco, which resulted in mannitol accumulation (Tarczynski et al. 1993). These results for tobacco are excellent evidence that if carbohydrate concentration (sugar-alcohol in this case) in tissue can be increased by genetic engineering, a more drought-tolerant line of crop is a probable outcome.

Specific proteins induced during water stress, the dehydrins, were first identified in two-dimensional gel electrophoresis in the late 1980s (Close et al. 1989). This specific set of proteins accumulates in plant tissues in response to dehydration stress caused by cold, salinity, or water stress (Chandler and Robertson 1994). The proteins accumulate during the normal process of embryo desiccation and they may be induced by increased ABA content. Other names for the dehydrins are the LEA group 2 proteins or the RAB proteins (Skriver and Mundy 1990, Galau and Close 1992). Dehydrin-encoding cDNA clones have been identified and isolated from a number of species (Vilardell et al. 1990, Robertson and Chandler 1992, Iturriaga et al. 1993, and others). Further study has shown that all the dehydrins have a consensus sequence of 15 amino acids near the carboxy terminus of the proteins. In fact, a core domain of eight amino acids (KIKEKLPG) is repeated one to many times in each protein. Similar proteins have also been found in the cyanobacteria (Close and Lammers 1993), where a 40-kD protein with the dehydrin consensus sequence is induced by osmotic stress in several species. In higher plants there is a repetitive 11-amino acid sequence that is common in dehydrins (Dure et al. 1989, Dure 1993). Furthermore, several dehydrin genes may be coordinately expressed, as they are located in tandem on a single locus (Yamaguchi-Shinozaki et al. 1989). Since the two LEA/RAB/Dehydrin genes are up-regulated during osmotic stress in a diversity of organisms and they are affected by ABA (a water-stress-regulated plant hormone), these genes are highly likely to induce drought tolerance in plants.

C. Oxidation-Induced Proteins

A number of different environmental situations, such as cold temperature combined with high light, or atmospheric ozone, are likely to induce an abundance of free radicals of oxygen (peroxide OOH^-, singlet oxygen $O^\bullet$, superoxide $O_2{}^-$) and other oxidizing compounds, such as hydrogen peroxide (H_2O_2). Free radicals are a normal component of metabolism and are utilized in a number of important metabolic reactions. However, small excesses in active oxygen radicals are not compatible with cytoplasmic function because they can interfere with proper membrane function, denature proteins, and inhibit enzyme function. Thus, plant cells require a regulated balance of free-radical production and degradation to maximize metabolic effectiveness. In previous chapters we discussed the significance of antioxidant systems such as xanthophyll epoxidation, ascorbate, α-tecophorol, and the Mahler reaction pathway for alleviating the stress induced by excessive oxidizing compounds. Therefore, genetic engineering of antioxidant systems is likely to be fruitful for manipulating stress tolerance in plants.

Since many of the same antioxidant systems operate in bacteria and higher plants, many studies have used bacterial genes to evaluate the effectiveness of antioxidant genes for regulating stress tolerance in plants. In *Salmonella typhimurinum* and *E. coli,* two gene products (OxyR and SoxR) positively regulate oxidative gene induction. For example, the oxidized form of the OxyR gene product is a signal to activate RNA polymerase, which will increase the abundance of mRNA coded by oxidative-induced genes (Storz et al. 1990a). In addition, many null mutants for specific antioxidant genes have been studied in bacteria because many mutants can be isolated in bacterial systems fairly quickly. Some of the best studied null mutants are those for manganese and iron forms of superoxide dismutase (MnSOD and FeSOD). These two forms of SOD are coded for by the sodA and sodB genes (Carlioz and Touati 1986). The sodA mutants that lack MnSOD activity are particularly sensitive to methyl viologen (a herbicide that exacerbates O_2 production). The double null mutant for sodA and sodB is killed by exposure to H_2O_2 and has a high frequency of mutation in the presence of oxygen (Farr et al. 1986). In contrast to the null mutant effect, overexpression of SOD genes do not always result in increased tolerance of stress. In fact, it was thought that overproduction of plasmid encoded MnSOD would lead to an increase in H_2O_2 tolerance, but resulted in sensitivity to electron transport disruption (Gruber et al. 1990).

The two different types of SOD in bacteria may have different functions. In a study on strains of *E. coli* that had either MnSOD or FeSOD, those strains with MnSOD were more capable of protecting DNA from oxidative damage, while those strains with FeSOD were better at protecting a cytoplasmic superoxide-sensitive enzyme (Hopkin et al. 1992). Thus, the various forms of SOD may have organ-specific or cytoplasmic-specific activity.

Bacterial genes have been used to transform the antioxidant production in higher plants a number of times. Commonly, one specific gene is induced to overproduce its gene product in transgenic plants. In one case the MnSOD gene from *E. coli* was transferred into tobacco with a soybean-like leader sequence. The overall SOD activity in leaves increased by only 30%, but those plants that expressed the bacterial gene were more tolerant of methyl viologen and low-temperature stress than were the wild-type plants (Van Assche et al. 1989). In another case, the gor gene derived from *E. coli* (codes for glutathione reductase) has been induced to be expressed in tobacco and poplar. Extractable glutathione reductase (GR) activity increased in transgenic plants, but the amount of GR expression was dependent on leaf age and the presence of light (Criessen et al. 1991). Exposure of leaves to light or methyl viologen increased bacterial GR activity in leaves of transformed plants, which may have been the result of an increase in translation rather than transcription of the *gor* gene (Foyer et al. 1991, Aono et al. 1991). When the gor gene was targeted to the chloroplast (Aono et. al. 1993), a high GR activity (three times that in the wild type) was induced in chloroplasts of transgenic plants. Transgenic plants with high chloroplastic GR activity were more tolerant of methyl viologen and sulfur dioxide, but not ozone. Therefore, overexpression of several different bacterial antioxidant genes in plants can increase tolerance to several stresses that induce the buildup of oxygen free radicals in the chloroplast and cytoplasm.

Many plant genes for antioxidant products have been isolated and used in genetic engineering of stress tolerance. One such group, the peroxidases, are widely distributed among plant species and often show little substrate specificity. Among the plant peroxidases, ascorbate peroxidase is best known (Asada 1992). The chloroplast form of ascorbate peroxidase (AsAPOD) is characterized by a narrow pH optimum, high specificity for ascorbate, and lability in the absence of ascorbate (Asada 1992). There is a membrane-bound form of chloroplast AsAPOD, which is structurally different from the soluble chloroplast forms (Miyake et al. 1993). It has been proposed that AsAPOD is a multigene family with at least three genes, two of which code for nearly identical chloroplast enzymes while one codes for a cytosolic AsAPOD (Creissen et al. 1993). Transgenic plants that overexpress particular AsAPOD genes have been created, but little is known of the effects on antioxidant protection.

In the case of plant SOD proteins, studies have been done on the consequences of gene overexpression to antioxidant capacity. When the Cu,Zn SOD (chloroplastic) was induced to be overexpressed by 30 to 50 times, there was no increase in protection against methyl viologen–mediated photoinhibition (Tepperman and Dunsmuir 1990). In a similar study, overexpression of Cu,Zn SOD could not protect against oxidative damage of ozone (Pitcher et al. 1991). Therefore, the overproduction of one enzyme of the Mehler reaction pathway is not likely to provide antioxidant protection because of the H_2O_2 end product and its toxicity. Rather, the enzymes of this antioxidant pathway must be increased in a coordinated manner. In contrast, there is some evidence that simply overproducing this one enzyme can provide protection against photoinhibition. Overexpression of Cu,Zn SOD resulted in a decreased level of cell damage following methyl viologen treatment in tobacco compared with that of the wild type (Gupta et al. 1993).

In summary, plant metabolic processes utilize strong oxidative compounds in several basic reactions, and there are several biochemical mechanisms that remove a small excess of the free radicals of oxygen. However, under several stressful conditions plant antioxidant systems are not regulated enough to compensate for the excessive production of free radicals, particularly in the chloroplast. There is no definitive evidence that oxidative stress per se initiates transcription of antioxidant enzymes. Thus, the bulk of plant cells are responding to some damaged cells rather than responding to the potential for damage. For example, it takes over a day of damaging UV radiation to stimulate the transcription of glutathione reductase significantly (Strid 1993), and the same treatment reduces the SOD mRNA population. In many cases, the introduction of one gene or overexpression of one gene does not result in improved antioxidant protection because one destructive oxygen radical is often replaced with another (see Figure 14.18). Therefore, coordinated upregulation of several enzymes in the antioxidant pathway is required.

D. Anaerobically-Induced Proteins

Anaerobicity is a common characteristic of plants that reside in wetland environments (Chapter 9). However, in nonwetland species anaerobic conditions (due to flooding or oxygen free-gas treatment) induces a moderate set of anaerobic stress proteins (ASPs). For example, a set of 20 ASPs are selectively synthesized in response to low oxygen partial pressure in root tips of maize (Sachs et al. 1980). Similar patterns of anoxic-induced proteins is found in tomato (Tanksley and Jones 1981), pea (Llewellen et al. 1987), and soybeans (Tihanyi et al. 1989), as well as several other species. In general, the kinetics of protein synthesis following anaerobic conditions is first a complete decline in synthesis of aerobic proteins. After about 5 minutes a set of transition proteins (dominated by a 33-kD protein) become abundant for approximately 90 min (Figure 14.18). Anaerobically induced proteins (ASPs) begin to accumulate after approximately 90 min. After 5 h 70% of all protein synthesis is accounted for by ASPs. Most studies on ASPs concern short-term responses because the subject species cannot survive extended periods of anaerobic conditions. The few long-term studies on ASPs have used rice or *Echinocloa* (barnyard grass), because these species can germinate and grow for several weeks in an anaerobic environment. Expression of ASPs is similar among flood-tolerant and flood-intolerant species of *Echinochloa;* however, induc-

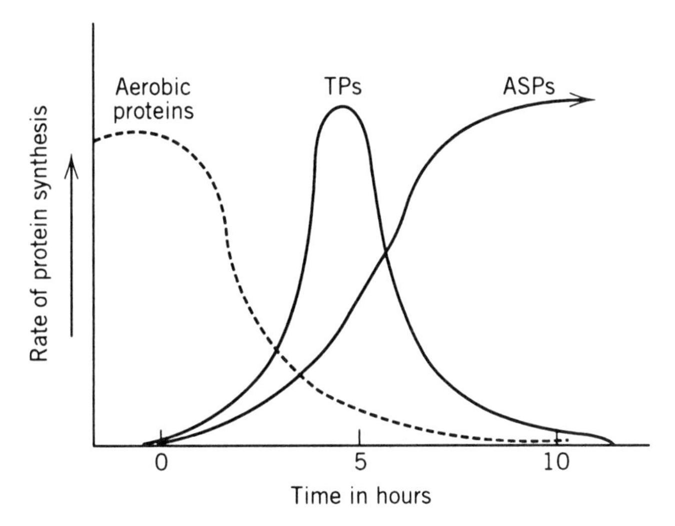

Figure 14.18 Induction kinetics for anaerobic proteins (ASPs) compared with the temporary proteins (TPs), and aerobic proteins. Upon anaerobic shock, aerobic proteins disappear and temporary proteins become abundant for approximately 5 h. After 5 h, anaerobically induced proteins constitute more than 75% of protein synthesis.

tion of ASPs is much faster in tolerant species than in intolerant species (Mujer et al. 1993). Therefore, the speed of up-regulating ASPs may determine if a species is able to survive anoxic conditions.

A large proportion of the ASPs are glycolytic enzymes, such as alcohol dehydrogenase (ADH), fructose-1,6-bisphosphate aldolase, lactate dehydrogenase, and others. Another important ASP is sucrose synthase, which is related to triose-phosphate metabolism (Springer et al. 1986). Among the many ASPs, alcohol dehydrogenases (ADH) and pyruvate decarboxylase are often most abundant.

Alcohol dehydrogenase is encoded by two genes, ADH1 and ADH2. Both genes are up-regulated following anoxia or hypoxia (low but not zero oxygen). In maize root tips exposed to anoxia, ADH1 transcripts become induced after 6 h followed by a rapid decline to minimal at 12 to 18 h (Russell and Sachs 1992). In contrast, under hypoxic conditions (4% O_2) ADH1 activity and mRNA continues to increase 24 h after treatment (Andrews et al. 1993). Therefore, under complete anaerobic conditions, ADH induction is small and fleeting compared to under hypoxic conditions. Pretreatment by hypoxic conditions also increases the anoxic induction of ADH and other glycolytic enzymes in maize (Andrews et al. 1994). Since there seems to be a coordinated simultaneous expression of ASP genes following hypoxic conditions, there may be consensus anaerobic response elements, similar to heat-shock elements, upstream from most glycolytic pathway genes. In the promoter region of maize ADH1, ALD1, and Sh1 (all glycolytic enzymes), a consensus sequence exists that may be an anaerobic induction element (Dennis et al. 1988). Since most ASPs are constitutive genes of a central metabolic pathway (glycolysis), overexpression of these genes in the absence of anaerobic conditions may not be an effective way of increasing anoxic tolerance in plants. However, increasing the copy number of a glycolytic gene construct that has a solely anaerobically regulated promoter may increase plant tolerance of anaerobic conditions.

VII. SUMMARY AND FUTURE DIRECTIONS

It is clear that various environmental stresses induce a diversity of proteins at the transcription and translation levels. Some of the induced proteins have general significance in cellular metabolism, and they are induced by a number of different stress conditions. In other cases, the induced proteins are specific to the particular stress factor. Some of the induced molecules are constitutive proteins that are up-regulated following stress, whereas others are novel proteins that only appear following the stress induction. In all cases, a diversity of proteins are induced for each type or combination of stress conditions.

Many stress-induced proteins are highly conserved among organisms. Consequently, a bacterial gene that is stress induced may be very similar to a gene induced by stress in plants or animals. This conserved character of many stress-induced genes opens the possibility for improving stress tolerance of plants by genetic engineering techniques. Genes from bacteria and animal systems have been transferred into plants and expressed successfully. In many cases these novel genes have enhanced stress tolerance in plants. In other cases it is clear that the expression of a novel gene in plants has not affected stress tolerance.

All of the evidence indicates that physiological mechanisms for stress tolerance involve a suite of genes affecting several possible pathways and regulated by different external factors. Thus, overexpressing one particular gene does not always result in enhanced stress tolerance. Often, a coordinated up-regulation of several points in a pathway is required to increase stress tolerance. Therefore, there has been some dramatic successes in stress tolerance management by genetic engineering, but in many systems a more comprehensive gene adjustment is needed to improve stress tolerance in plants.

STUDY-REVIEW OUTLINE

Introduction

1. Most stress tolerance mechanisms involve several interacting quantitative traits.
2. Genetic heterogeneity in traits may result in greater physiological plasticity, which will result in greater stress acclimation and adaptation.
3. Asexual reproduction has evolved many times in different evolutionary lineages in regions of extreme stress.
4. The advantages of extensive genetic heterogeneity and/or effective plasticity depend on the scale and pattern of temporal and spatial stress variation.
5. One mechanism for identifying proteins that regulate stress tolerance traits is by a discovery technique in which a native protein is discovered in the test species and shown to regulate stress tolerance.
6. A second technique of identifying stress tolerance proteins is by an inference method in which foreign proteins are introduced into the test species and the consequence of that introduction is evaluated.
7. A third strategy of determining if a protein regulates a stress-tolerant trait is to introduce an antisense gene. The antisense gene will produce an mRNA that will bind with the mRNA for the putative stress tolerance protein, precluding the translation of either mRNA.
8. A fourth strategy is to induce mutations in the organisms and search the mutations for one that has specifically lost the putative stress-tolerance protein. The mutants can be screened for stress tolerance capacity.

Methods of Protein Separation and Identification

1. Cellular proteins can be isolated and separated by a number of different techniques, including two-dimensional gel electrophoresis and column chromatography.

2. In two-dimensional gel electrophoresis, commonly used to identify stress-induced proteins, protein extracts are separated by isoelectric focusing in one direction and SDS polyacrylamide electrophoresis in the other direction.
3. Western blotting utilizes a dual-antibody technique to locate specific proteins in a gel.

Methods of Gene Isolation

1. Restriction endonucleases are bacterial enzymes that cut double-stranded DNA at specific nucleotide sequences called restriction sites.
2. Restriction endonucleases are used to separate desired pieces of DNA from the large background of DNA in the cell.
3. Cloning is a mechanism of bulking (multiplying) restriction fragments in a biological vector (usually, a bacteria or virus).
4. Restriction fragments are inserted into bacterial plasmids that contain an antibiotic screening device and a polylinker (multiple restriction site sequence).
5. In a genomic clone, *E. coli* are transformed with plasmids containing restriction fragments. The result is a large number of petri dishes containing one bacterial clone, each with a unique restriction fragment.
6. In partial genomic cloning, mRNA is extracted from the cell and induced to form complementary DNA by reverse transcriptase.
7. The cDNA is incorporated into a bacterial vector in order to multiply each cDNA fragment.
8. The cDNA clone is easier to screen for the desired gene because (a) extraneous introns are removed by the cell during RNA processing before the mRNA was extracted, (b) extracted mRNA from a cell that is actively synthesizing the putative protein is likely to contain a large quantity of the desired mRNA, and (c) there are fewer members of a cDNA clone than of a genomic clone.
9. To screen a genomic or partial genomic library, a probe is needed. The probe can be a small piece of radiolabeled DNA that codes for a six- to eight-amino-acid-long segment of the putative stress tolerance protein.
10. Southern blotting is used to select the clone with the correct DNA or cDNA segment.

Methods of Gene Transfer

Indirect Gene Transfer Mechanisms

1. Due to the low number of competent cells in plants, gene transfer systems are inefficient.
2. The ability to transfer genes into plants is dependent on the evolutionary history of the species. Cereals (grasses) are more difficult than many broadleaf species to transform.
3. The major vector used to transfer genes into plant cells is the plasmid (Ti) from *Agrobacterium tumefaciens* (a soil bacteria that induces cancerous cell growth in plants).
4. The Ti plasmid can be disarmed (made noncancerous) and yet retain its ability to infect cells and incorporate DNA into the host cell genome.
5. T-DNA is the portion of a Ti plasmid that is responsible for incorporating plasmid DNA into the host cells.
6. A binary plasmid construct can be made that contains a restriction fragment from another species and T-DNA from the Ti plasmid. The binary plasmid can be used to transfer genes into host cells in a number of ways.
7. A viral vector can also be used to transfer DNA into a host cell, but the DNA will not be transferred into the genome or the germ cells. Thus, a viral vector cannot be used to develop a true breeding line of transgenic organisms.

Direct Gene Transfer Mechanisms

1. Electroporation is a technique where genes are directly transferred into protoplasts by providing an electrical shock.

2. Electroporation is effective with several species and is becoming an important technique for transforming plants.
3. Chemical shock induction of gene transfer is also possible as long as the genetic material is protected to some degree from the strong chemical treatment.
4. Microprojectile bombardment is a technique in which genetic material is coated on a bead projectile and bombarded into leaf discs, or callus. Genetic material on the beads is incorporated into the genome of some cells with a low frequency.
5. Microprojectile bombardment is the preferred technique of transformation in soybean and many cereals.

Molecular Mechanisms for Transforming Plant Stress Tolerance

1. There are other ways of affecting the stress tolerance traits of a species besides adding new genetic material.
2. An especially effective promoter can be attached to the gene which produces a stress-induced protein, resulting in overexpression.
3. Gene overexpression compared to the wild type can be induced by multiplying the gene copy number.
4. The effectiveness of a particular gene in a stress-tolerance mechanism can be tested by interfering with translation by inserting an antisense gene into the plants genome.

Bioengineering of Stress-Tolerance Traits in Plants

Thermally Induced Proteins

1. Heat-shock proteins were the first stress-induced proteins discovered.
2. A diverse set of heat-shock proteins are induced for a short period of time after heat treatment of 5°C over optimal temperature.
3. Heat-shock protein induction is based on a molecular regulation system involving heat-shock factors and heat-shock elements. Both translation and transcription regulation are involved in heat-shock induction.
4. HSPs are ubiquitous among organisms, including bacteria, animals, plants, and fungi.
5. Some heat-shock proteins are ubiquitous among organisms and others are produced only by certain groups.
6. Several HSPs are constitutive and function as chaperones or proteases.
7. Several HSPs are present only when induced and often serve protective functions for specific heat-labile proteins.
8. In plants, HSP 21 and 24 may be important to protecting PSII activity during high-temperature conditions.
9. In some cases overexpression of HSPs has resulted in thermotolerance, but not in other cases.

Cold-Induced Proteins

1. Cold-induced proteins are not as diverse as heat-shock proteins.
2. Cold-induced proteins remain after cold shock longer than that of heat-shock proteins.
3. Chilling alters the activity of several proteins without inducing transcription.
4. Other cold-induced proteins are associated with an induction of transcription such as CAP79.
5. Several cryoprotectant polypeptides, such as COR15, are induced by cold shock.
6. Chilling stress can result in the down-regulation of enzyme genes, such as that of the low-molecular-weight fragment of rubisco. Other enzymes, such as GAPT, can be up-regulated by cold shock.
7. Overexpression of COR genes have resulted in increased protection from chilling in transgenic plants.

Dehydration-Induced Proteins

1. Several dehydration-induced proteins are also induced by cold shock. This may be due to the impact of cold shock and dehydration on ABA abundance.
2. Some LEA genes (genes induced during early seed development) are induced by both cold and desiccation, while others are only induced by desiccation.
3. Desiccation induction of genes that regulate the synthesis of osmoprotectants much as glycine-betain may be important to desiccation tolerance.
4. Low-molecular-weight carbohydrates such as raphanose and fructans may be important to desiccation tolerance because of their influence on "glass" structure of cytoplasm.
5. Transforming tobacco with fructan accumulation genes caused an increase in desiccation tolerance.
6. Dehydrins are a class of proteins that accumulate in tissues during desiccation stress.
7. Dehydrins may be regulated by lea genes and are up-regulated by osmotic stress in a number of different species.

Oxidation-Induced Proteins

1. Since many different combinations of stresses induce an increase in cellular oxidative compounds, antioxidant systems have general significance to stress physiology.
2. Many null mutants for important antioxidant enzymes such as the SOD enzymes have been isolated. These mutants are very sensitive to stress conditions that increase oxidative compounds in cells.
3. Two different bacterial antioxidant genes have been transferred into sensitive mutants. The result is an increase in antioxidant capacity and an increased tolerance of stress.
4. If only one enzyme of the Mehler pathway is overexpressed, this does not induce greater stress tolerance becausc the rest of the pathway is limiting the effectiveness of this antioxidant system.
5. In plant systems an overexpression of ascorbate peroxidase provides oxidant protection in transgenic plants.

Anaerobically Induced Proteins

1. In general, anaerobic conditions induce an array of enzymes associated with glycolysis.
2. In non-wetland species a group of about 20 anaerobically stimulated proteins (ASPs) are induced.
3. Induction begins after about 90 min of anaerobic treatment, and 70% of all protein synthesis after 5 h are ASPs.
4. Alcohol dehydrogenase (ADH) is an important ASP.
5. ADH induction is maximum under hypoxic conditions, but minimum under anoxic conditions.
6. The promoter region of ADH genes may be an inducible element in much the same manner as heat-shock proteins.
7. Up-regulation of ADH with a solely anaerobic promoter may induce increased tolerance of anaerobic conditions.

SELF-STUDY QUESTIONS

1. Why is it difficult to identify proteins that regulate stress tolerance traits in plants?

2. What are three different types of experiments that can be used to identify proteins and genes that have definite regulation of stress tolerance in plants?

3. What ecological conditions would promote adaptation by phenotypic plasticity versus adaptation by changes in allele frequency?
4. Compare the gel electrophoresis and column chromatography techniques for separating proteins. What are the differences in technique, advantages, and disadvantages of each separation procedure?
5. What is a Western blot, and why is it important for protein identification?
6. What are restriction endonucleases, and how are they related to cloning genetic material?
7. Diagram and describe a plasmid vector that might be used to house restriction fragments.
8. What is the difference between a partial genomic library and a genomic library? Why is it usually more fruitful to probe a cDNA library for a putative restriction fragment than to probe a genomic library?
9. What techniques can be used to screen for genetic transformants?
10. Make a table or chart of the various direct and indirect methods of transforming plant cells. Include the technique name, the transforming vector, and the types of plants most likely to be transformed by each technique.
11. What factors regulate the effectiveness of gene transfer by *A. tumefaciens* in plants?
12. If plant transformation is easiest when protoplasts are used, why is this not the system of choice for most plant genetic engineers?
13. Compare and contrast the advantages and disadvantages of microprojectile bombardment and coincubation with engineered *A. tumefaciens* for plant transformation.
14. What are three other ways of influencing the expression of stress-induced genes in plants other than by adding new genetic material into plants?
15. Compare and contrast the dynamics of heat- or cold-induced proteins.
16. What basic classes of HSPs are induced in plants during heat shock?
17. Describe the potential molecular model of action for HSPs in plants.
18. What is the most likely function of the 31-kD cold-induced protein in plants?
19. What is CAP79, how is it induced, and what is its most probable molecular model of action?
20. What are GAPT genes, and how do they influence cold-stress tolerance in plants?
21. Describe the class of proteins called dehydrins. What other proteins are these related to, and how do they function in cell metabolism?
22. What is a "glass-like" cytoplasm, and how does this relate to drought-induced genes?
23. What are some important antioxidant proteins? When individual genes for antioxidant proteins are up-regulated, what is the consequence to oxidant tolerance in plants?
24. What is the diversity of anaerobically induced proteins in plants?

25. Describe the ADH genes, and explain what happens when they are up-regulated in plants.

SUPPLEMENTARY READING

Binns, A. N., and M. F. Thomashow. 1988. Cell biology of *Agrobacterium* infection and transformation of plants. Annual Review of Microbiology 42:575–606.

Davey, M. R., E. L. Rech, and B. J. Mulligan. 1989. Direct DNA transfer to plant cells. Plant Molecular Biology 13:273–85.

Fraley, R. T., S. G. Rogers, and R. B. Horsch. 1986. Genetic transformation in higher plants. CRC Critical Reviews in Plant Sciences 4:1–46.

Gasser, C. S., and R. T. Fraley. 1989. Genetically engineering plants for crop improvement. Science 244:1293–99.

Hess, D. 1987. Pollen based techniques in genetic manipulation. International Review Cytology 107:169–90.

Lal, R., and S. Lal, 1990. Crop Improvement Utilizing Biotechnology. CRC Press, Boca Raton, FL.

Lindsey, K., and M. G. K. Jones. 1990. Electroporation of cells. Physiologia Plantarum 79: 168–72.

Roest, S., and L. J. W. Gilissen. 1989. Plant regeneration from protoplasts: a literature review. Acta Botanica Neerlandica 38:1–23.

Sanford, J. C. 1990. Biolistic plant transformation. Physiologia Plantarum 79:206–209.

Weising, K., J. Schell, and G. Kahl. 1988. Foreign genes in plants: transfer, structure, expression, and applications. Annual Review of Genetics 22:421–477.

Wray J. L. (ed.). 1992. Inducible Plant Proteins: Their Biochemistry and Molecular Biology. Cambridge University Press, Cambridge.

15 Generalities, Trends, and Future Directions

Outline

Objectives

1. Describe some general patterns of plant response to stress.
2. Explain some important trends in the mechanisms of plant response to stress. Link various responses of plants to stress among many environmental stressors.
3. What are likely to be particularly fruitful future lines of research in plant stress physiology? Stress interdisciplinary and multifactor research.
4. Describe the future significances of research on stress physiology of plants to science and society.

I. INTRODUCTION

The development of research in plant stress physiology has had a long and varied history. Early naturalists in biology were aware of the various responses plants expressed toward environmental factors (Darwin 1865, Schimper 1898). Awareness of the importance of environmental factors as stresses to agricultural systems began during the great dust bowl of the mid-1930s. During this time it became clear that drought and soil salinity could exert a powerful influence on agricultural productivity. A landmark work for plant stress physiology was that of Monsai and Sakai (1953) concerning the light environment within plant communities. Here was the first significant application of biophysics to the function of plants in response to an environmental factor. Important papers on energy balance and the fluxes of gases in and out of leaves followed (Gates 1962). These works provided the basic connection between physical attributes of the environment and plant metabolism. There was now a framework to predict the influences of environmental factors on photosynthesis, water use, and productivity. These initial studies also provided the foundation for the development of plant growth models for crops (Brouwer and deWit 1969) and natural communities (Miller and Tieszen 1972).

In the late 1970s the applied research in plant stress physiology (crop stress physiology) took a slightly different path than that of studies in natural systems (plant ecophysiology). Ecophysiology took a pathway leading to a significant interaction with population and ecosystem biology, while agriculturists studying plant stress physiology developed a closer affinity to molecular biology. This is reflective of a different philosophy about plant responses to environmental stress. Ecophysiologists were stressing the evolutionary processes associated with adaptation in a wide diversity of species. For example, environmental influence on speciation and ecotypes (Clausen et al. 1940), efficiencies of biomass conversion (Penning de Vries 1975), optimization of stomatal behavior (Cowan and Farquhar 1977), the significance of diverse photosynthetic physiology (Björkman et al. 1970), the physiological basis for plant distribution (Lange 1967), and the influence of temperature on photosynthesis (Björkman et al. 1972) all made significant contributions to the development of ecophysiology. In contrast, the philosophy of agricultural research centered on the acclimation of varieties or lines of narrow genetic diversity to variation in agricultural systems. In such systems changes in one gene or one protein can drastically alter the success of a particular variety in the presence of environmental stress.

Another example of the contrast between philosophies of research on stress physiology by agriculturists compared with ecophysiologists is their respective patterns of research on pollutants. When pollutants first became an important component of plant stress physiology, both crop physiologists and ecophysiologists focused their research on toxicology. However, over time, crop physiologist have tended to expand the ideas of toxicology into learning about detoxification mechanisms exhibited by specific lines or mutants. This work is reinforced by research on herbicide mode of action and potential adaptation of tolerance in weeds and crops. In contrast, ecophysiologists have tended to pursue research in atmospheric pollutants at a larger scale. Much effort has been targeted toward modeling whole ecosystem responses to changing atmospheric factors and scaling for the individual physiological response to changes in ecosystem processes (Norman 1993).

Although these two philosophies have tended to drive the two disciplines slightly apart from each other, they are certainly mutually supportive. Today, ecophysiologists occasionally utilize the information developed by crop physiologists to support the metabolic basis of natural adaptation in plants. Some ecophysiologists utilize crop species in studies about basic aspects of plant adaptation to environment. Also, crop physiologists may consider evolutionary aspects of the development of traits in the particular species under consideration. Much has been learned about stress physiology of crops by studying the linkage between physiology and natural environments of wild progenitors. Promoting the sharing of information and techniques between applied physiology at the molecular level, and natural ecophysiology at the organism or population level, is likely to result in significant insight into the intricacies of plant stress physiology.

Agricultural science has also been influenced by ecological and environmental issues. There has been a philosophical trend away from production agriculture toward a sustainable agriculture. The concept of integrated pest management was a progenitor to the philosophy of sustainable agriculture, and the concept of sustainable agriculture is a progenitor of the sustainable biosphere initiative in ecology. Recently, there is interest in the ecological consequence of releasing genetically engineered crops and in the relationship between genetically engineered crops and sustainable agriculture (Giampietro 1994). Interest in this topic is expressed both by ecologists (thus the new field of molecular ecology) and agriculturists. In fact, ecological concepts are becoming a main component of many crop-related texts (Altieri 1994, Carrol et al. 1990).

Linking the expertise of those in different disciplines may result in great advances in the field of plant stress physiology. For example, molecular biologists and ecophysiologists can provide each other with valuable supportive information. The physiological response of whole plants is ultimately dependent on the action and regulation of genes. The action of microevolution on plant fitness (physiological stress) strongly influences the history behind particular molecular regulation schemes. Crop physiologists and evolutionary biologists are likely to benefit from each other's expertise. The selection for mutations and heavily inbred lines of species can yield appropriate systems for the analysis of microevolutionary regulation of metabolism. Understanding metabolic changes that have arisen during natural evolution of stress tolerance can assist plant breeders and genetic engineers target specific metabolic traits for manipulation. Mathematical modelers and plant stress physiologists also can assist each other's research. If mathematical models of plant growth, or cell metabolism, are going to become mechanistic rather than descriptive, they require information on plant stress physiological mechanisms (Figure 15.1).

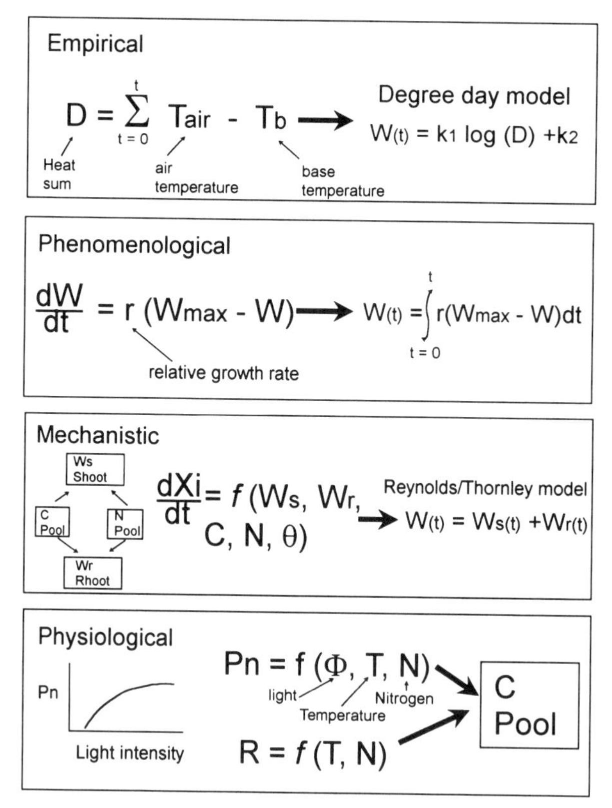

Figure 15.1 Diagrammatic representation of four levels of model complexity. Plant stress physiology will provide the information for models at the physiological level. These physiological models enable a more realistic simulation of the pools and fluxes in mechanistic models. (Adapted from Reynolds et al. 1993.)

Mathematical modeling is an excellent technique for validating the applicability of information learned about plant stress response of one species to that of other species. These are just a few of the possible ways that interdisciplinary research can enhance our comprehension of plant physiological response to stress.

II. TRENDS IN PLANT PHYSIOLOGICAL RESPONSE TO STRESS

Terrestrial plants are a very diverse group with a long evolutionary history compared to many other groups of organisms (except the monera). That wide diversity of species is able to colonize almost all possible habitats on the planet. Plants are also subjected to the wills of the environment once they are established because they are sedentary and lack strong mechanisms for internal homeostasis. As a consequence of habitat diversity and weak homeostasis, plants must have a wide variation in metabolic processes to be able to reside in that diverse continuum of habitats. Although plant metabolism is complex, there are some general aspects of metabolic response to environmental stress that are relatively consistent among most, if not all, species and situations.

A. Acclimation Versus Adaptation

Adaptation and acclimation were defined in the introduction of this book, and these concepts have been consistently referred to throughout the rest of the chapters. Heavy emphasis was placed on these two topics because of their general applicability to all plant responses to environmental stressors. Due to the sedentary nature of plants compared to animals, and the relatively slow generation time of plants compared with microbes, phenotypic plasticity is relatively important in plants compared to that for other kingdoms. Seasonal and daily changes in climatic conditions require that plants adjust physiology. Heterogeneous light availability in subcanopy environments must be utilized efficiently, by rapid acclimation to high light, to gain enough carbon for survival. Flushes of nutrients occur during the development of communities, and these flushes need to be utilized by plants to maintain a competitive edge. In animal systems, acclimation (phenotypic plasticity) is not as important to survival, as is adaptation (changes in allele frequency) because of animal mobility. Furthermore, acclimation is important in plants because of the high abundance of clonal species compared to animal systems. For clonal plants (one genotype) to have a robust lineage over time in a changing environment, they must have phenotypic plasticity, because somatic mutation may not be frequent enough to provide adequate genetic variation for physiological adaptation.

Adaptation is also an important aspect of the evolutionary development of plant stress metabolism. Many cases of ecotypic differentiation have been identified in plants and discussed in previous chapters. Is there a trade-off between phenotypic plasticity and the capacity for genetic change by adaptation? In one sense there is such a relationship because those plants with high phenotypic plasticity are able to cope with a diversity of habitat conditions, which minimizes natural selection for genetic evolution (Figure 15.2). However, phenotypic plasticity is itself genetically regulated. The more interesting question is: What factors of the environment influence the development of gene regulation systems with high or low variance in phenotypic expression?

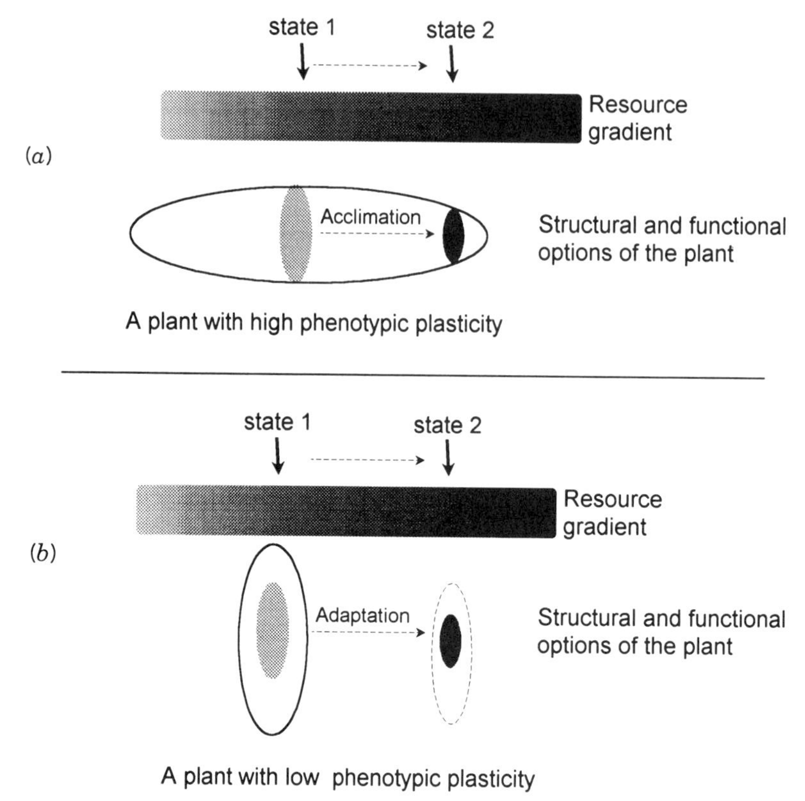

Figure 15.2 Comparison of phenotypic plasticity, adaptation, and acclimation in plants. (A) Condition where a species has wide phenotypic plasticity. When resources change in the environment, metabolism changes by phenotypic plasticity. Acclimation occurs but there is no adaptation. (B) Situation in which a plant species has limited phenotypic plasticity. If environmental resources change outside the species physiological tolerances, adaptation must occur to continue population survival.

This comparison of acclimation and adaptation also relates to the incidence of tolerance and avoidance in plant stress responses. Throughout the previous chapters on every type of environmental stress, it was shown that plants utilize a diverse assemblage of avoidance and tolerance techniques to cope with the stressor. There is no association between particular stressors and the proportion of stress responses that are due to tolerance or avoidance mechanisms. Tolerance mechanisms usually involve a strong metabolic acclimation. Metabolic change of an organism is required to cope with extremes in cytoplasm conditions. Osmotic adjustment, heat- and cold-shock protein synthesis, ACC synthase activity in hypoxic soil, and photosynthetic induction during sun flecks are some examples of tolerance mechanisms established through physiological acclimation. In contrast, avoidance mechanisms often involve adaptation. General attributes of plants, such as deciduousness, the presence of salt glands, the phylotaxi of leaves, timing of phenological events, and inherent allocation patterns, are basically regulated by adaptation, and they are also mechanisms of avoiding stress.

B. Pathway from Environmental Condition to Whole-Plant Response

Most whole-plant responses to environmental stress are dependent on a sensor, a transducer, and a metabolic response (Figure 15.3). The sensors are usually membranes. Membrane composition and function is highly sensitive to heat, water, nutrient status, pH, redox potential, as well as other cellular and environmental factors. Most stresses first influence the structure and function of membranes before they influence cytoplasmic metabolism. The appropriate operation of key metabolic pathways (e.g., photosynthesis and respiration) are dependent on membrane structure and function. Due to the central significance of membrane to plant metabolism, it is logical to expect that membranes act as stress sensors.

Once the membrane has sensed a change in environmental conditions that may affect physiological performance, this information needs to be passed along to the machinery that regulates metabolism. The most common transducer of information from membranes to metabolic regulators are phytohormones. Many phytohormones are inactive when membrane bound (e.g., ABA and phytochrome). Once released from membranes, phytohormones act at many levels in metabolic regulation, including transcription and enzyme activation. Phytohormones also serve to integrate whole-plant response to stress. This is particularly true for plants in comparison to animals, because plant organs experience widely divergent environments (roots in the soil and leaves in the atmosphere).

Metabolic responses of plants are very diverse. Some responses are determined by up- or down-regulation of gene transcription (heat- or cold-shock proteins). In other cases, environmental stress changes the expression of various members of a multigene family

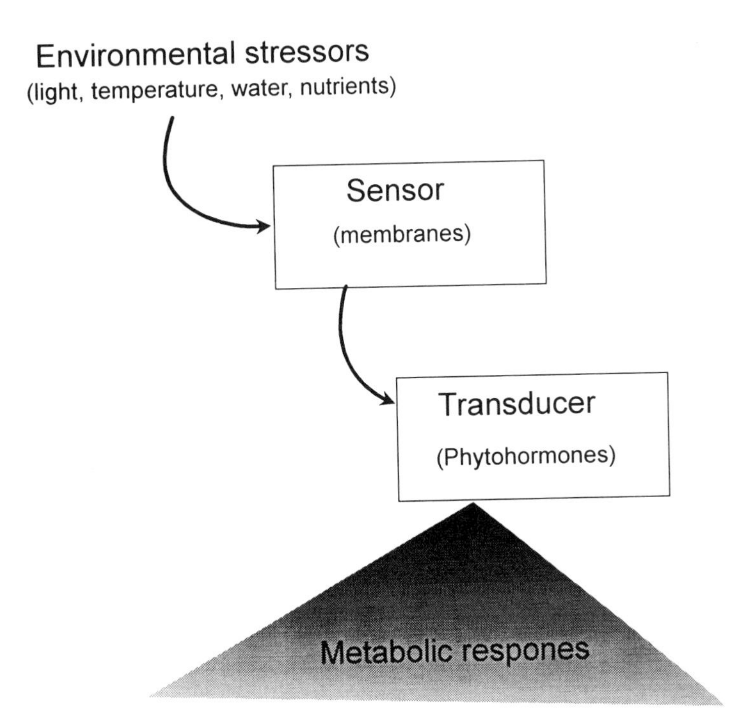

Figure 15.3 Diagrammatic representation of the linkages between stressors, sensors, transducers, and metabolic response.

(antioxidant genes). Some stress responses occur at the level of translation (water stress), while other responses occur by changes in membrane permeability, or subcellular distribution of ions (salinity). However, in almost all cases stress responses occur at multiple levels.

In most reactions of plants to stress, the central metabolic process that integrates responses of most other metabolic processes to stressors is carbon balance. Compensatory theory identified carbon balance as the master integrator of multiple resource limitations in plants. Whenever a stress has a detrimental influence of plant growth, or performance, this is reflected in changes of carbon balance process (Figure 15.4). This is the probable reason why plants of environments with relatively high levels of stress have evolved slow growth rates. Therefore, the measurement of carbon balance may be the best indicator of plant stress because of its central position in metabolic response to stress on an evolutionary level and on an instantaneous level.

C. Diversity of Mechanisms in Plants That Compensate for Stress

Any plant response to stress is not likely to involve only one metabolic change. This is partly because of the weblike interaction between metabolic pathways. Changes in one process is going to influence several other metabolic pathways. For example, changes in the rate of glycolysis influence mitochondria activity, protein synthesis, carbohydrate use, and a multitude of other metabolic processes. Induction of monoterpene synthesis utilizes nitrogen that might have been allocated to other metabolic processes. Acclimation or adaptation to stress also involves parallel or mutually enhancing changes in several different metabolic pathways. Osmotic adjustment in response to water stress by increasing organic acid transport into the vacuole is mirrored by increases in cytoplasmically compatible solutes. In fact, many plant responses to stress (particularly water stress) involve a suite of adaptive changes throughout the plant.

The diversity of metabolic response to stress also is a result of the many ways that plants have evolved to cope with any particular stressor. Plant response to heat is re-

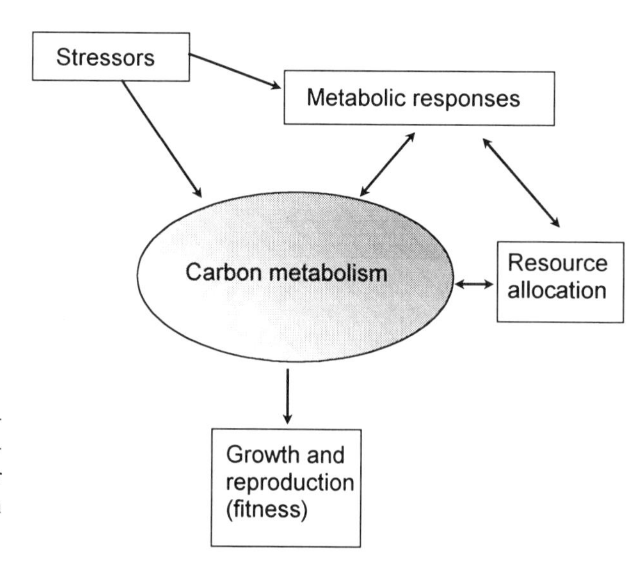

Figure 15.4 Diagrammatic representation of why carbon metabolism is the master integrator of the influences of stressors on plant physiology.

lated to large changes in membrane composition in some species, changes in enzyme thermotolerance in other species, or both in many species. Adaptation to water stress in some species is due to changes in elastic modulus, while others utilize stem succulence, others perform osmotic adjustment, and others change cavitation vulnerability. Species utilizing any or all of these techniques can be found in the same habitat (Nilsen et al. 1984). Adaptation to salinity is accomplished by excluding ions from the roots, compartmentalizing ions in expendable organs, and exuding ions out modified trichomes.

However, there are general attributes of growth form that influence plant response to environmental stress. The classification of physiologically based growth form is relatively easy because suites of traits consistently co-occur, and these suites define overall physiological strategies (Chapin 1980). An important central component of physiological strategies is *relative growth rate* (RGR). To maintain a high RGR value, photosynthetic rates must be high, and resource acquisition from the environment must be high. Such plants are highly responsive to environmental stressors and often utilize proportionally more avoidance than tolerance mechanisms. In contrast, species adapted to high stress environments have low RGR and low responsiveness to environmental stressors. Stress acclimation and adaptation in these species is more frequently accomplished by tolerance than avoidance mechanisms. The RGR among growth forms is a continuum and is highly correlated with net photosynthesis (Figure 15.5); however, this relationship is less clear when comparing RGR and net photosynthesis among species or varieties (Field 1991). By connecting net photosynthesis rate as an indicator of RGR and responsiveness of growth forms to stress, one could argue that the diversity in plant growth form reflects the potential diversity of plant stress response syndromes (particularly, compatible suites of stress response traits).

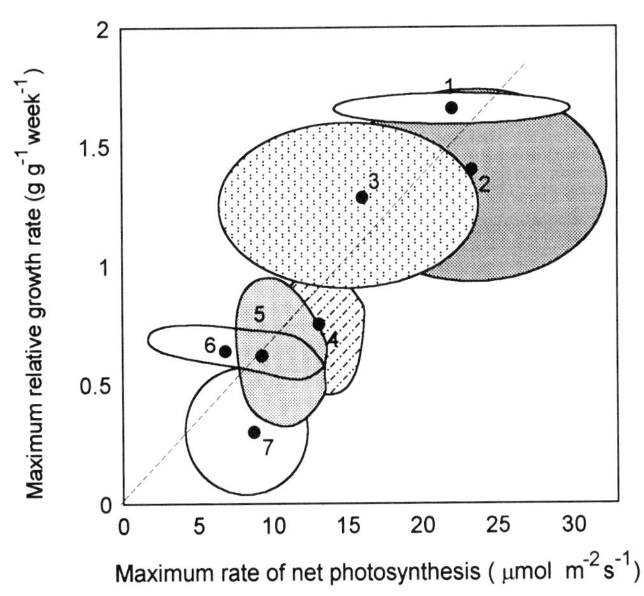

Figure 15.5 Comparison between maximum relative growth rate and maximum net photosynthesis rate among general growth form classes. (Redrawn from Schulze and Chapin 1987.)

D. Significance of Cost–Benefit Analysis

Plants are integrated assemblages of multiple organs. This concept has been used by several researchers by employing population models to evaluate plant growth. In this technique, each individual in the population is replaced by an individual module or growth unit (see Chapter 2 for details). Optimization of performance in a diversity of environments can be attained by changing the allocation of resources among the various organs. However, there is a cost to one process or organ by allocating resources to the other process or organ. Adaptation to environmental stress necessarily includes a balance between the benefit of allocating resources toward one process and the cost of removing those resources from another process. For example, defenses against herbivory damage require an investment of energy, carbon, and nitrogen resources into secondary metabolic pathways. The benefit of resource allocation into secondary metabolism for defensive purposes is balanced against the reduction in growth due to removing those resources from the primary metabolism that regulates growth. Natural selection may favor those genotypes that are most effective at adjusting this balance between the cost/benefit of primary and secondary metabolism in order to optimize growth in environments with significant potential for environmental or biotic stress.

This consideration of the importance of cost/benefit analysis to plant response to stress brings up the concepts of efficiency and magnitude. It is often the case that increased efficiency is considered an adaptation or acclimation to environmental stress. During a drought, increased water use efficiency conserves water. Under low nutrient availability, high nutrient use efficiency enables a higher amount of growth for each amount of nutrient absorbed. However, increased efficiency is usually associated with smaller gains in growth or resource accumulation. Under drought, water use efficiency may increase, but photosynthesis decreases and growth decreases. Under low nutrients, nutrient use efficiency may increase but plants adapted to low nutrients have slow growth rates and are less able to utilize pulses in resource availability compared with plants adapted to high nutrient sites. When considering the importance of efficiencies to plant stress responses, it is important to look past the efficiency ratios and examine the consequence of reduced growth to plant performance.

E. Genetic Engineering and Plant Stress

In this book the topic of genetic engineering has often been connected to studies of plant stress physiology. This is because of the potential for genetic engineering to redesign the physiology of crop plants and make them superior stress tolerators. In addition, genetic engineering can serve as a valuable tool for determining if specific genes or gene products are causally related to stress tolerance mechanisms.

There is an important relationship between genetic engineering and stress tolerance that has been alluded to in many chapters of this book. That is the relationship between the potential for successful genetic engineering and the level of organization in plants (Figure 15.6). At the level of the gene product, genetic engineering has large potential for success. At this level of organization the main type of stress response is shock reaction. The likelihood that genetic engineering will be able to influence whole physiological pathways is less certain. In this case, balanced changes among several genes are required. For example, altering only one gene product of the Mehler reaction did not result in greater oxidant stress tolerance, because the rest of the Mehler reac-

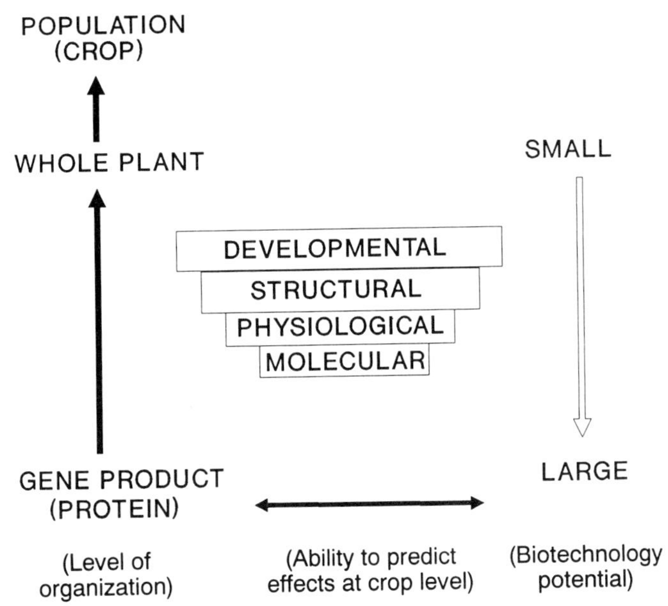

Figure 15.6 Diagrammatic representation of the relationships between the potential for success in genetic engineering and the level of complexity in plant systems. (Redrawn and revised from McCue and Hanson 1990.)

tion enzymes were not up-regulated. Many physiological acclimations and adaptations by tolerance mechanisms occur at the physiological level. As one increase in level of complexity in plant form and function (from the gene product to whole-plant structure and function), the potential for biotechnology decreases, but the connection to plant stress mechanisms increases. Biotechnology will need to increase greatly in sophistication before suites of mechanisms used by plants to cope with environmental stressors can be engineered.

III. FUTURE LINES OF RESEARCH IN PLANT STRESS PHYSIOLOGY

Where is the field of plant stress physiology going in the future? What are the important topics for study in the years to come? These are important questions because the field of plant stress physiology is in a state of flux. The field has matured to include a number of different disciplines, and in the future the field will become increasingly important for a number of related disciplines.

One important line of research is the discovery of linkages between various levels of organization in the organism and their connection to the population and ecosystem. Whether you consider a crop or a natural stand, the processes of the ecosystem are in part determined by the responses of the individual. To model the function of the field or natural stand in a mechanistic way, the linkage between physiological response to resource availability, or stressors, should be known. This "bottom up" approach to systems modeling, achieved by integrating information from plant stress physiology with mechanistic systems models, should increase the ability of those models to predict (forecast) system change in response to environmental perturbation. This is an underdeveloped aspect of plant stress physiology and will be addressed in the years to come.

In many chapters of this book, the environment was considered to be heterogeneous. Patches of stresses occur on various time and spatial scales; consequently, plants experience a mosaic pattern of resources and stresses. One of the future challenges to plant stress physiology is to understand how this heterogeneous environment influences adaptation and acclimation in plants. How important is phenotypic plasticity to plants experiencing various kinds of environmental heterogeneity? What is the cost of exploiting high resource patches, and how does that cost vary with the degree of environmental heterogeneity? Although some of these questions have been addressed with research over the past 10 years (Caldwell and Pearcy 1994), much is yet to be learned about non-steady-state physiology of plants.

Studies in plant stress physiology have focused on two main organs: leaves and fruits. This is because photosynthesis and reproduction determine carbon gain, fruit and seed yield, and fitness; and these are the most important attributes of plants to agriculture and ecology. However, many major responses to environmental stress occur in other organs. Root responses to soil resource limitation and resource heterogeneity and patchiness have strong influences on plant fitness. Although this fact has been often stated, the amount of information on leaf response to stress dwarfs that available on root response to stress. Even more slighted by plant stress research has been the stem. Stems serve as the physical linkage between leaves and roots, and stems are not simply conduits for materials transfer. Stems perform a large number of roles for plants (Gartner 1995). Stems regulate the architecture of canopies, which determines leaf display and carbon gain. Internally, stems transport, store, release, and direct the movement of carbohydrates, nutrients, and water between leaves and roots. These and many other functions of stems have not been studied with the same intensity as that for similar processes in leaves. Research on organs other than leaves should prove to be very fruitful in plant stress physiology in the years to come.

This book has one chapter on multiple stress adaptation in plants. If one compares the material in this chapter to that in the other chapters, it is clear to see that this is a synthetic chapter. The study of multiple stress interactions yields important synthetic information about how plants respond to environmental stress. Since plants experience interactive stressors in all habitats, responses to this amalgam of stressors is likely to be a major influence on plant metabolism. Much more needs to be learned about multiple stresses in the future to be able to understand the consequences of genetic engineering of plants and its interaction with natural systems, and to understand the way that metabolic processes of plants respond to the complex environmental conditions in nature.

There is one particular aspect of research on the association of organisms with their environment that zoologists have emphasized to a greater degree than botanists: the phylogenetic study of metabolism. An important future line of research in plant systems is a phylogenetic view of plant stress physiology. What is the evolutionary history of particular adaptive trait combinations? How has the genetics of particular groups of species interacted with the evolution of plant stress traits? One way of answering some of these questions is first to document the phylogenetic history of particular traits, then evaluate the consequences of altering one or more of those traits to the survival and fitness of species in particular habitats. Phylogenetic trees (phytelic assemblages based on phenetics or cladistics) have become extremely robust in the last 10 years because information at the level of gene structure is used to establish the systematic relationships among taxa. In particular, data on variation in the DNA sequence of chloroplast genes (e.g., the gene for the large subunit of ribulose bisphosphate carboxylase oxygenase: rbsl) and nuclear genes (e.g., the

18S rRNA gene) have been useful for determining evolutionary relationships at the level of family or tribes of plants. Now it is time to use this strong phylogenetic information to help understand the evolution of metabolic responses to stress. Research on crops and progenitor weeds may also be applicable to these questions. Careful consideration of the domestication history of a crop (particularly in relation to the culture that domesticated the crop and the region where the crop was domesticated) can be combined with comparative information between stress-tolerance mechanisms of the wild progenitor and the crop to yield pertinent information on the evolution of stress physiology in plants.

There are several important tools that should be emphasized in future research on plant response to environmental stress. One of those tools is the analysis of tissue isotopic composition. After the influences of metabolic processes on isotopic discrimination are known, a measurement of isotopic composition can be used as an indicator of metabolic response to stress (see Chapter 5). Although great strides have been made in this field over the past 20 years, more information on the validation and applications of isotopic signatures to plant stress physiology is needed. In particular, those studies on multiple isotopic discrimination may be particularly informative. Another important tool for future research in plant stress physiology is chlorophyll fluorescence kinetics. The dynamics of radiation quenching by chlorophyll can yield pertinent information about the status of photosynthesis and its association with environmental stressors in a nondestructive manner. Combining measurements of chlorophyll fluorescence kinetics with other measures of plant stress metabolism may be a powerful mechanism of deciphering plant stress metabolism and its impact on plant fitness in the least invasive manner. Research on plant stress physiology should also make use of genetic engineering technology in future studies. The ability to change the expression of genes and observe the consequences of that change to growth and fitness is an important line of future research. Such studies are going on now extensively with *Arabidopsis thaliana;* however, these studies need to be extended to a wider diversity of species. To completely understand plant adaptation to any particular environmental stressor by genetic engineering studies, combinations of genes need to be manipulated. It may take the manipulation of 10 or 15 genes in one plant to regulate successfully the way this plant copes with environmental stress. Manipulation of multiple genes in one taxon is a challenge for future research.

Plants experience a multitude of stressors, including both abiotic and biotic members. The ability of plants to respond to a biotic stress, such as herbivory, will depend on the availability of resources. There may be a large cost to a plant's ability to acclimate to environmental stress because of simultaneously compensating for a biotic stress. In contrast, some environmental stresses may trigger metabolic attributes that also serve as defense against a biotic stress. It will be important in future studies of plant response to stress in natural systems. Understanding these linkages between biotic and abiotic stress are fundamental to comprehending the ultimate significance of plant physiological response to stress.

IV. FUTURE SIGNIFICANCE OF RESEARCH IN PLANT STRESS PHYSIOLOGY

Why study the physiology of plants under stress? Obviously, the authors of this book think this topic is important. They have spent their careers researching the way plants respond to environmental stress. The applied significance of understanding plant response to stressors has not been ignored by those in biobusiness. The information gathered during studies of plant response to environmental stress has significance to agriculture, forestry, environmental sciences, and basic plant sciences.

The study of plant response to stress will provide important information about how crops cope with less than optimum growing conditions. Crop productivity is significantly inhibited by natural environmental factors and anthropogenic factors such as air pollution. Neither of these environmental stresses are expected to become less intense. As agriculture moves into less arable lands and climatic conditions change, potential environmental stress will increase. As insects evolve adaptation to pesticides, biotic stress on plants may increase. As weeds develop resistance to herbicides, crops may get less of the field resources for growth. As human populations increase, the amount of pollution increases. All of these factors necessitate the development of enhanced stress tolerance in crops. The study of plant stress physiology can identify particularly favorable attributes to strive for in crop breeding programs. Also, in genetic engineering programs, information of physiological adaptation to stress provides the foundation needed before searching for pertinent genes. Once transgenic plants have been created, their physiological responses to stress must be evaluated. Therefore, the study of plant stress physiology is integral to the development of new crop lines for deployment in less than favorable growing conditions.

There are a multitude of studies indicating that global climate change is imminent. The magnitude of that change is a point of contention, but there is little doubt that some changes are likely to occur. How will plants respond to those changes? What might we expect in our natural systems? These are important questions for agriculturists, ecologists, forest scientists, and managers of natural systems. To understand how communities or ecosystems will respond to potential climate change, responses of individual plants to those changes (stresses) need to be understood. Once individual plant responses to the changed environment are understood, linkages to higher levels of complexity are possible. This is the "bottom up" mechanism of interpreting the potential impact of climate change to natural systems. Knowledge of plant physiological response to stress is the foundation of a "bottom up" approach to predicting ecosystem changes in response to the altered environment.

Plant performance is a result of an interacting web of metabolic pathways. As such, it is difficult to determine what aspects of this web are important regulators of plant fitness. For example, we can induce a change in protein complement by heat shock. However, we do not know if those changed proteins influence the fitness of an individual or a population. Studies of metabolic response to stress, and its linkage to plant fitness, is an important mechanism for determining the significance of particular metabolic processes to plants.

There can be little doubt in anyone's mind about the significance of plants to the biosphere. The generation of oxygen, the strong influence on climatic factors, the cycling of nutrients, the scrubbing of the atmosphere, and the provision of almost all reduced carbon for the rest of the biosphere to utilize are just some of the significances of plants. The greater our understanding of plant response to the environment, the greater will be our ability to manage natural and manmade systems to promote a sustainable biosphere.

STUDY-REVIEW OUTLINE

Introduction

1. Although there is a long history of research on plant stress physiology, the dust bowl of the mid-1930s initiated intensive study of this topic in the United States.

2. Landmark research on the application of biophysics to plant physiological process provided the foundation for modern plant stress physiology.
3. A slightly different emphasis developed among scientists studying ecophysiology and those studying agricultural physiology. Ecophysiology tended toward studying physiological response to stress and its impact on the whole organism, while agricultural stress physiology emphasized molecular and biochemical regulation of stress responses in crops.
4. These two fields are coming closer together and utilizing information produced by the other discipline with increasing frequency.
5. Just as ecophysiology has been influenced by molecular studies on crop species, agriculture has been influences by ecophysiology as reflected in the sustainable agriculture initiative.
6. Some of the most significant advances in plant stress physiology of the future may come from linking these two disciplines to a greater degree.

Trends in Plant Physiological Response to Stress

Acclimation Versus Adaptation

1. Adaptation and acclimation are of general importance to plant stress physiology.
2. Phenotypic plasticity (acclimation) may be very important to plants because of their sedentary nature and the temporal heterogeneity of the environment.
3. Adaptation is an important part of plant stress physiology on an evolutionary scale.
4. It is important to learn the genetic constraints to phenotypic plasticity in order to evaluate the relative significance of adaptation and acclimation to plant stress physiology.
5. Acclimation is usually associated with tolerance mechanisms, while adaptation is often associated with avoidance mechanisms.

Pathway from Environmental Condition to Whole-Plant Response

1. Plants respond to environmental stressors by (a) sensing with membranes, (b) transducing with phytohormones, and (c) responding with metabolic acclimation or genetic adaptation.
2. The central integrator of plant stresses is carbon metabolism.

Diversity of Mechanisms in Plants That Compensate for Stress

1. Plant responses to any stress will involve multiple changes at the biochemical, anatomical, morphological, and behavioral levels.
2. In any one environment the same environmental stressors will result in several possible combinations of adaptive traits in plants.
3. There is some association between growth form and suites of stress adaption traits, which are sometimes called *strategies*. Relative growth rate may be an aspect of growth form that is highly related to specific suites of mechanisms used by plants to compensate for general attributes of environmental stress.

Significance of Cost–Benefit Analysis

1. Any adaptive response by plants to stress will have a benefit to growth or fitness. However, there will often be a cost to other metabolic processes associated with that adaptation.
2. Allocating resources to one set of organs may improve growth (the benefit), but it will also result in less resources allocated to the other organs (the cost).
3. Analysis of cost-benefit relationships in plants is an important aspect of plant stress physiology research.
4. One must be very careful with overinterpreting the significance of efficiency indices. There may be a trade-off whereby an increase in one efficiency results in a decrease in another efficiency.

Genetic Engineering and Plant Stress

1. Genetic engineering can be very successful at the level of single gene expression.
2. Adaptation to stress involves multiple genes and entire pathways. Genetic engineering is less developed in the arena of multiple-gene regulation, which is a required area for manipulating stress resistance in plants.

Future Lines of Research in Plant Stress Physiology

1. Important lines of future research are:
   a. The linkage between stress response at the molecular level with whole-plant fitness.
   b. The study of plant response to heterogeneous environments.
   c. Studies on stress physiology of organs other than leaves and fruits.
   d. The study of multiple interaction of stressors and plant form and function.
   e. The phylogenetic study of stress adaptation in plants.
   f. The use of genetic engineering to evaluate the consequences of stress-induced genes to fitness, and the cost associated with those inductions.
   g. The interactions between biotic stresses and abiotic resource limitations may have a profound affect on plant response to stress.

Future Significance of Research in Plant Stress Physiology

1. Some of the future significant focuses of research in plant stress physiology to science and society are:
   a. To provide basic knowledge about how crop and natural plants cope with environmental stressors.
   b. To provide information that enables breeders and genetic engineers to make educated choices in their breeding or engineering programs.
   c. To evaluate plant resistance to environmental stress once transgenic (or genetically bred lines) organisms are created.
   d. To provide information needed to develop "bottom up" integrated models of plant response to global climate change.
   e. To develop experimental protocols for evaluating the significance of various aspects of plant metabolism to species fitness.

SELF-STUDY QUESTIONS

1. What philosophies have the basic field of crop stress physiology and ecophysiology tended to emphasize?
2. What are four general aspects of processes by which plants adapt to stressors?
3. Why is it that the study of multiple stress interaction will be a major part of plant stress physiology in the future?
4. What is so important about phenotypic plasticity to plants?
5. What are the most important significances of research on plant stress physiology?

SUPPLEMENTARY READING

Ehleringer, J. R., and C. B. Field (eds.). 1993. Scaling Processes from Leaf to Globe. Academic Press, San Diego, CA.

Gartner, B. (ed.). 1995. Stems and Trunks in Plant Form and Function. Academic Press, San Diego, CA.

Givnish, T. (ed.). 1986. On the Economy of Plant Form and Function. Cambridge University Press, Cambridge.

Lange, O. L., P. S. Nobel, C. B. Osmond, and H. Zeigler (eds.). 1981. Encyclopedia of Plant Physiology, Vol. 12A. Springer-Verlag, Berlin.

Mooney, H. A., W. E. Winner, and E. J. Pell. 1991. Integrated Responses of Plants to Stress. Academic Press, New York.

APPENDIX A
Symbols and Abbreviations

a	absorbtivity of a tissue
α	variably defined (contact angle in Chapter 7)
Ab	Antibody
ABA	abscisic acid
ACC	1-aminocyclopropane-1-carboxylic acid
ADP	adenosine diphosphate
αj	fraction of reproducing shoots of size j that flowered
$1 - \alpha j$	fraction of reproductive shoots that branched
AM01618	gibberellic acid synthesis inhibitor
AMP	adenosine monophosphate
ANG	aminoethoxyvinylglycine
AOA	aminooxyacetic acid
a^{st}	area of the stomatal aperture
ATP	Adenosine triphosphate
a_w	chemical activity of water
BADH	betain aldehyde dehydrogenase
b_{ij}	proportion of branching shoots of size j that grew to size i
BisP	bis-phosphated molecule
C = C	ethylene
°C	degrees Celsius
C	capacitance
C_3	represents a plant with the photosynthetic physiology designated as C_3
C_4	represents a plant with the photosynthetic physiology designated as C_4
C_a	ambient carbon dioxide concentration
CAM	crassulacean acid metabolism
CAP	cold adaptation proteins
C_c	carbon dioxide concentration in the chloroplast
CCC	chlorocholine chloride (gibberellic acid inhibitor)
CHS	chalcone synthese
C_i	intercellular carbon dioxide concentration
CK	cytokinin
C_l	capacitance of leaves
CL	cardiolipin
cm	centimeters
CMO	choline monooxygenase
CMV	cucumber mosic virus
cos I	cosine of incidence between a leaf and the solar angle
CQT	hyroxycinnamoyl CoA quinate

C_r	root capacitance
c_s	concentration of solutes
C_{st}	capacitance of the stem
c_w	concentration of water
CYT*b*/*f*	cytochrome protein complex of the chloroplast thylakoids
D	dialectric constant
D_1,D_2	proteins in photosystem II core complex
DGDG	digalactacyldiglyceride
DHOAcP	dihydroxyacetone phosphate
δ^{ias}	linear length from mesophyll cell wall to the stomata
DNA	deoxyribonucleic acid
DPA	dihydrophaseic acid
d^{st}	depth of the stomatal aperture
ΔT	difference between tissue temperature and air temperature
DW	tissue dry weight
D_{wv}	diffusion coefficient of water vapor in air
e	object emissivity or atmospheric vapor pressure
E	einsteins (moles of photons)
EER	ethylene evolution rate
EFE	ethylene forming enzyme
Em gene	encodes for most abundant embryo protein in wheat seed
E	transpiration, or modulus of elasticity
E_0	proportionality constant
ϵ_0	electrical potential
E_{IR}	emitted infrared radiation
E_{vat}	slope of the relationship between elastic modulus and turgor pressure
ϕ	proportion of fixed carbon that is lost to respiration or osmotic coefficient or solar constant
$\Phi^{diffuse}$	total diffuse radiation impinging on a leaf surface
$\Phi^{dirrect}$	direct solar radiation impinging on a biological surface
$\Phi_{s(dir)}$	total direct solar radiation
f_{ij}	proportion of flowering shoots of size *j* that grow to size *i*
F	Faraday coefficient
F_0	basal chlorophyll fluorescence
F_{max}	maximum chlorophyll fluorescence
FNR	ferrodoxin NADPH reductase complex of the photochemical reactions in chloroplast thylakoids
FR	radiation in the far-red waveband
FW	tissue fresh weight
g	grams or gravitational force
GA	gibberellic acid
GAP	glyceraldehyde, 3-phosphate
GDR	ratio of the diameter of a gap in a forest canopy to the height of the canopy
g_{ij}	proportion of shoots that grew from size *j* to size *i*
GPP	gross primary production
GSH	glutathione
g_{st}	stomatal conductance

GUS	reporter gene that indicates the presence of transformed cells; GUS is the gene for β-gluronidase
γ_w	activity coefficient of water
h	hours
h	vertical height
HSE	heat-shock element
HSF	heat-shock factor
HSP	heat-shock proteins
H_{sp}	specific heat
HTE	high temperature equilibrated, or high temperature exotherm
H_{vap}	heat of vaporization
H_{wv}	heat of vaporization for water
IAA	indoleacetic acid
IMA pathway	indole-3-acetanide (pathway for IAA synthesis)
IN_s	ice nucleators
IPA pathway	indole-3-pyruvic acid (pathway for IAA synthesis)
IR	infrared radiation
IS_{60}	induction state for leaves adapted to low light after 60 s in high light
J	volumetric flux rate
J^{soil}	flux of water per area of pore space in soil
j^{w}	total flux rate of water per cross-sectional area of vessel
J_{wv}	instantaneous transpiration rate in mmol of water molecules lost per time
KA	2[H]-ent-kaurenoic acid
K	hydraulic conductance
kJ	kilojoules
K_m	Michaelis–Menton coefficient of enzyme efficiency
kPa	kilopascal
K^{soil}	hydraulic conductance in soil
λ	differential of the change of instantaneous transpiration over the change in net photosynthesis
LA	leaf area
LAI	leaf area index
LAR	leaf area ratio
LB	left-hand border sequence for T-DNA genetic information
LD	long day
LD_{50}	time when 50% of a phenomenon has occurred; usually used to refer to leaf abscission, a toxicity response, or a response to high heat
LHCI	light-harvesting complex for photosystem I
LHCII	light-harvesting complex for photosystem II
ln	natural log
LT	low temperature
LTE	low-temperature equilibrated, or low temperature exotherm
LUE	light use efficiency
m	meters
m	mass of a substance
M	mass per mole of a substance
μ	chemical potential

MACC-1	(malonylamino) cyclopropane-1-carboxylic acid
Mb_j	mean number of branches for each branching shoot of size *j*
MDA	malondialdehyde
μE	microeinsteins
Mf_j	mean number of shoots derived from flowering shoots of size *j*
MGDG	monogalactacyldiglyceride
min	minutes
mm	millimeters
mmol	millimoles (10^{-3} moles)
μmol	micromoles (10^{-6} moles)
mol	moles
MPA	megapascals
mV	millivolts (10^{-3} volt)
μ_ω	chemical potential of water
μ^*_ω	reference chemical potential for water
n	number of individual units
NAR	net assimilation ratio
NDVI	normalized difference vegetation index
NIR	near-infrared radiation
nm	nanometers
NPP	net primary production
NUE	nitrogen use effiniency
N_w	mole fraction of water
OAA	oxaloacetic acid
ORI	site of origin for DNA replication in plasmids or bacteria
P	pressure on a system
π	the constant (3.14 . . .)
Π	osmotic pressure
P_{680}	chlorophyll molecule that gets reduced and oxidized to deliver electrons to electron transport in photosystem II
P_{700}	chlorophyll molecule that gets reduced and oxidized to deliver electrons to electron transport in photosystem I
Pa	pascal
PAL	phenylalanine ammonium lyase
PAR	photosynthetically active radiation
PC	plastocyanin or phosphatidylcholine
PE	phosphatidylethanol
PEA	phosphatidylethanolamine
PEG	polyethylene glycol
PEP-Case	phosphoenol pyruvate carboxylase
PEP	phosphoenolpyruvate
PFD	photon flux density
P_{fr}	phytochrome in the far-red configuration
PG	phosphatidylglycerol
PGA	phosphoglyceric acid
PGAL	phosphoglyceraldehyde
PhyI	type I phytochrome
PhyII	type II phytochrome

pH	unit defined as the negative log of hydrogen ion concentration
PI	phosphatidylinositol
PI	plastochron index
pKa	dissociation constant
P_n	net photosynthesis
PPFD	photosynthetic photon flux
PQ	plastoquinone
PR	plastochron rate
P_r	phytochrome in the red configuration
PSI	photosystem I of the chloroplast membranes
PSII	photosystem II of the chloroplast membranes
P^{soil}	hydrostatic pressure in soil
Q_{10}	relationship between reaction rate and temperature
Q_A	quinone that resides in protein D2 of photosystem II
Q_B	quinone that resides in protein D1 of photosystem II
Q_x	ion charge
$\Re$	universal gas constant
r	radius
R	radiation that is in the red waveband
R_a	isotopic ratio in the atmosphere
RAB	genes responsive to abscisic acid
RB	right-hand border sequence for T-DNA genetic information
RGR	relative growth rate
RH	relative humidity
r_j	proportion of shoots of size j that reproduced between time t and $t + 1$
R_j	resistance to water movement across structure or barrier j
R_l	leaf resistance to water flow
RNA	ribonucleic acid
R_p	isotopic ratio in plant tissues
R_r	root resistance to water flow
R_{ref}	isotopic ratio in a reference material
R_s	soil resistance to water flow
r^{st}	radius of the dome of humid air above the stomata
R_{st}	stem resistance to water flow
rubcs	gene for the small subunit of ribulose bisphosphate carboxylase oxygenase
RUBP	ribulose bisphosphate
ρ_ω	density of water
RWC	relative water content of tissues
R^{ias}_{wv}	resistance to water vapor movement in the intercellular spaces of leaves
s	seconds
σ	surface tension
SAM	*S*-adenosylmethionine
SD	short day
SOD	superoxide dismutase
Sp^r	gene that confers resistance to streptomyosin to bacteria
SQDG	sulfoquinovosyldiacyglycerol
SRCM	stimulus/response coupling mechanism

t	time
T	temperature (defined as kelvin or Celsius)
τ	transmissivity
Tc	transition temperature
T-DNA	transfer DNA originally from the TI plasmid
TET^r	tertracycline resistance gene
TIBA	triiodobenzoic acid (auxin transport inhibitor)
TMV	tobacco mosic virus
TP	turgor potential
TR	transpiration of a canopy
triose-P	triose phosphate, a phosphorylated three-carbon sugar
T_{sky}	effective sky temperature based on IR radiation emission from the sky
T_{surr}	temperature of the system surroundings
TW	tissue turgid weight
UV	radiation in the ultraviolet waveband
UV-A	ultraviolet radiation between 320 and 440 nm
UV-B	ultraviolet radiation (280 to 320 nm)
UV-C	ultraviolet radiation (200 to 280 nm)
v	velocity
V_o	original volume of water in a tissue before dehydration
V_e	volume of water exuded from a tissue during a pressure–volume curve
V_{max}	maximum enzyme reaction rate
VIR	genes that code for virulence in the TI plasmid
V_w	partial molar volume of water
ω	efficiency of fertilizer utilization
WD	water deficit of tissues
Ψ	water potential
$1/\Psi$	Inverse of the tissue water potential
Ψ_{crit}	critical water potential required to induce embolism
Ψ_g	gravimetric potential
Ψ_m	matric potential
Ψ_s	solute potential or osmotic potential
Ψ_s^0	osmotic potential at zero turgor pressure
Ψ_s^{100}	osmotic potential at 100% tissue water content
Ψ_τ	pressure potential or turgor potential
ZR	zeatin riboside
Z_w	charge number of water

APPENDIX B
Literature Cited

Abass, M., and C. B. Rajashekar. 1993. Abscisic acid accumulation in leaves and cultured cells during heat acclimation in grapes. HortScience 28:50–52.

Abdul, K. S., and G. P. Harris. 1978. Control of flower number in the first inflorescence of tomato (*Lycopersicon esculentum* Mill.): the role of gibberellins. Annals of Botany 42:1361–1367.

Abdul-Jabbar, A. S., D. G. Lugg, T. W. Sammis, and L. W. Gay. 1984. A field study of plant resistance to water flow in alfalfa. Agronomy Journal 76:765–769.

Abravaya, K., M. P. Myers, S. P. Murphy, and R. Morimoto. 1992. The human heat shock protein HSP70 interacts with HSF, the transcription factor that regulates heat shock gene expression. Genes and Development 6:1153–1164.

Acevedo, E., E. Fereres, T. C. Hsiao, and D. W. Henderson. 1979. Diurnal growth trends, water potential, and osmotic adjustment of maize and sorghum leaves in the field. Plant Physiology 64:476–480.

Acevedo, E. E. 1993. Potential for carbon isotope discrimination as a selection criterion in barley breeding. Pp. 399–417 in: J. R. Ehleringer, A. E. Hall, and G. D. Farquhar (eds.), Stable Isotopes and Plant Carbon–Water Relations. Academic Press, San Diego, CA.

Adams, W. W., III, and B. Demming-Adams. 1994. Carotenoid composition and down regulation of photosystem II in three conifer species during the winter. Physiologia Plantarum 92:451–458.

Aldwinckle, H. S. 1975. Stimulation and inhibition of plant virus infectivity in vivo by 6-benzylaminopurine. Virology 66:341–343.

Allan, A. C., and P. H. Rubery. 1991. Calcium deficiency and auxin transport in *Cucurbita pepo* L. seedlings. Planta 183:604–612.

Altieri, M. A. 1994. Biodiversity and Pest Management in Agroecosystems. Haworth Press, New York.

Amasino, R. M., and C. O. Miller. 1983. Effect of temperature on the morphology and cytokinin levels of tobacco crown gall teratoma tissues. Plant Science Letters 28:245–253.

Ambler, J. R., P. W. Morgan and W. R. Jordan. 1992. Amounts of zeatin and zeatin riboside in xylem sap of senescent and non-senescent sorghum. Crop Science 32:411–419.

Amthor, J. S. 1989. Respiration and crop physiology. Springer-Verlag, Berlin.

Amthor, J. S. 1994. Higher plant respiration and its relationships to photosynthesis. Pp. 71–101 in: E. D. Schulz and M. M. Caldwell (eds.), Ecophysiology of Photosynthesis (Ecological Studies, Vol. 100). Springer-Verlag, Berlin.

Amzallag, G. N., H. R. Lerner, and A. Poljakoff-Mayber. 1992. Interaction between mineral nutrients, cytokinin and gibberellic acid during growth of sorghum at high NaCl salinity. Journal of Experimental Botany 43:8–87.

Anderson, J. M., W. S. Chow, and D. J. Goodchild. 1988. Thylakoid membrane organization in sun/shade acclimation. Pp. 1–26 in: J. R. Evans, S. von Cammerer, and W. W. Adams III (eds.), Ecology of Photosynthesis in Sun and Shade. CSIRO, Australia.

Anderson, M. C. 1964. Studies of the woodland light climate. I. The photographic computation of light conditions. Journal of Ecology 52:2–41.

Andrews, D. L., B. G. Cobb, J. R. Johnson, and M. C. Drew. 1993. Hypoxic and anoxic induction of alcohol dehydrogenase in roots and shoots of seedlings of *Zea mays:* Adh transcripts and enzyme activity. Plant Physiology 101:40–414.

Andrews, D. L., D. M. MacAlpine, J. R. Johnson, P. M. Kelley, B. G. Cobb, and M. C. Drew. 1994. Differential induction of mRNAs for glycolytic and ethanolic fermentative pathways by hypoxia and anoxia in maize seedlings. Plant Physiology 106:1575–1582.

Aono, M., A. Kubo, H. Saji, T. Natori, K. Tanaka, and N. Kondo. 1991. Resistance to active oxygen toxicity of transgenic *Nicotiana tabacum* that expresses the gene for glutathione reductase from *Escherichia coli.* Plant and Cell Physiology 32:691–697.

Aono, M., A. Kubo, H. Saji, K. Tanaka, and N. Kondo. 1993. Enhanced tolerance to photo-oxidative stress of transgenic *Nicotiana tabacum* with high chloroplastic glutathione reductase activity. Plant and Cell Physiology 34:129–135.

Araulo, C. D., A. L. MacKay, K. P. Whittall, and R. J. T. Hailey. 1993. A diffusion model for spin-spin relaxation of compartmentalized water in wood. Journal of Magnetic Resonance 101:248–261.

Armstrong, J., and W. Armstrong. 1994. A physical model involving nucleopore membranes to investigate the mechanism of humidity-induced convection in *Phragmites australis.* Procedings of the Royal Society of Edinburgh 102B:529–540.

Armstrong, W. 1994. Polargraphic oxygen electrodes and their use in plant aeration studies. Proceedings of the Royal Society of Edinburgh 102B:511–528.

Armstrong, W., J. Armstrong, P. M. Beckett, and S. H. F. W. Justin. 1991b. Convective gas flows in wetland plant aeration. Pp. 283–302 in: M. B. Jackson, D. D. Davis, and H. Lambers (eds.), Plant Life Under Oxygen Deprivation. SPB Academic Publishing, The Hague.

Armstrong, W., R. Brandle, and M. B. Jackson. 1994b. Mechanisms of flood tolerance in plants. Acta Botanica Nederlandica 43:307–358.

Armstrong, W., M. Strange. S. Cringle, and P. M. Beckett. 1994a. Microelectrode and modeling study of oxygen distribution in roots. Annals of Botany 74:287–299.

Armstrong, W., E. Wright, S. Lythe, and T. J. Ganard. 1991a. Root adaptation to soil waterlogging. Aquatic Botany. 39:57–73.

Arshad, M., and W. T. Frankenberger, Jr. 1991. Effects of soil properties and trace elements on ethylene production in soils. Soil Science 151:377–386.

Asada, K. 1992. Ascorbate peroxidase: a hydrogen peroxide–scavenging enzyme in plants. Physiologia Plantarum 85:235–241.

Ashworth, E. N. 1993. Deep supercooling in woody plant tissues. Pp. 203–214 in: P. H. Li and L. Christersson (eds.), Advances in Plant Cold Hardiness. CRC Press, Boca Raton, FL.

Ashworth, E. N., and T. L. Kieft. 1995. Ice nucleation activity assocated with plants and fungi. Pp. 137–162 in: R. E. Lee, Jr., G. J. Warren, and L. V. Gusta (eds.), Biological Ice Nucleation and Its Applications. APS Press, St. Paul, MN.

Askenasy, E. 1880. Uber eine neue Methode um die Verteilung der Wachsrumsintensitat in wachsenden teilen zu bestimmen. Verh. Naturh. Medic. Ver. Heidelberg 2:70–153.

Aspinall, D. 1980. Role of abscisic acid and other hormones in adaptation to water stress. Pp. 155–172 in: N. C. Turner and P. J. Kramer (eds.), Adaptation of Plants to Water and High Temperature Stress. John Wiley & Sons, New York.

Asrar, G. 1990. Theory and Applications of Optical Remote Sensing. Wiley-Interscience, New York.

Aubé, C., and W. E. Sachston. 1965. Distribution and prevalence of *Verticillium* species producing substances with gibberellin-like biological properties. Canadian Journal of Botany 43:1335–1342.

Badger, M. R., T. D. Sharkey, and S. von Caemmerer. 1984. The relationship between steady-state gas exchange of bean leaves and the level of carbon-reduction-cycle intermediates. Planta 160:305–313.

Bahman, E., D. Barnhart, and J. G. Waines. 1993. Genetic analysis of transpiration efficiency, carbon isotope discrimination, and growth characters in bread wheat. Pp. 419–450 in: J. R. Ehleringer, A. E. Hall, and G. D. Farquhar (eds.), Stable Isotopes and Plant Carbon–Water Relations. Academic Press, San Diego, CA.

Bailiss, K. W., and I. M. Wilson. 1967. Growth hormones and the creeping thistle rust. Annals of Botany 31:195–211.

Balázs, E., B. Barna, and Z. Király. 1976. Effect of kinetin on lesion development and infection sites in *Xanthi-nc* tobacco infected by TMV: single-cell local lesions. Acta Phytopathologica of Hungary 11:1–9.

Ball, T. 1988. An analysis of stomatal conductance. Ph.D. thesis, Stanford University, Stanford, CA.

Bano, A., H. Hansen, K. Dörffling, and H. Hahn. 1994. Changes in the contents of free and conjugated abscisic acid, phaseic acid and cytokinins in xylem sap of drought stressed sunflower plants. Phytochemistry 17:345–347.

Bao, Y., and E. T. Nilsen. 1988. The ecophysiological significance of leaf movements in *Rhododendron maximum*. Ecology 69:578–587.

Barber, J. (ed.). 1992. The Photosystems: Structure Function and Molecular Biology (Topics in Photosynthesis, Vol. II). Elsevier Press, Amsterdam.

Barker, A. V., and K. A. Corey. 1988. Ethylene evolution by tomato plants under nutrient stress. HortScience 23:202–203.

Barker, A. V., and K. A. Corey. 1990. Ethylene evolution by tomato plants receiving nitrogen nutrition from urea. HortScience 25:420–421.

Barlow, E. W. R. 1986. Water relations of expanding leaves. Australian Journal of Plant Physiology 13:45–58.

Basiouny, F. R., K. Basiouny, and M. Maloney. 1993. Influence of water stress on abscisic acid and ethylene production in tomato under different PAR levels. Journal of Horticultural Science 69:535–541.

Basiouny, F. M., T. K. Van, and R. H. Biggs. 1978. Some morphological and biochemical characteristics of C_3 and C_4 plants irradiated with UV-B. Physiologia Plantarum 42:29–32.

Baskerville, G. L. 1972. The use of logarithmic regression for the estimation of plant biomass. Canadian Journal of Forest Research 2:49–53.

Bates, L. M., and A. E. Hall. 1981. Stomatal closure and soil water depletion not associated with changes in bulk leaf water status. Oecologia 50:62–65.

Bazzaz, F. A. 1990. The response of natural ecosystems to the rising global CO2 levels. Annual Review of Ecology and Systematics 21:167–196.

Bazzaz, F. A., and S. L. Miao. 1993. Successional status, seed size, and response of tree seedlings to CO_2, light, and nutrients. Ecology 74:104–112.

Bechtold, J. K., J. Ellis, and G. Pelletie. 1993. In planta agrobacterium mediated gene transfer by infiltration of adult *Arabidopsis thaliana* plants. Critical Reviews of the Academy of Sciences Paris, Sciences de la Vie 316:1194–1199.

Beckett, P. M., W. Armstrong, S. H. F. W. Justin, and J. Armstrong. 1988. On the relative importance of convective and diffusive gas-flows in plant aeration. New Phytologist 110:463–468.

Beffa, R. H., V. Martin, and P.-E. Pilet. 1990. In vitro oxidation of indoleacetic acid by soluble auxin-oxidases and peroxidases from maize roots. Plant Physiology 94:485–491.

Begum, F., J. L. Karmoher, O. A. Futtah, and A. F. M. Maniruzzaman. 1992. The effect of salinity on germination and its correlation with K^+, Na^+, and Cl^- accumulation in germinating seeds of *Triticum aestivum* L. Plant Cell Physiology 33:1009–1014.

Ben-David, A., Y. Bashan, and Y. Okon. 1986. Ethylene production in pepper (*Capsicum annuum*) leaves infected with *Xanthomonas campestris* pv. *vesicatoria*. Physiology and Molecular Plant Pathology 29:305–316.

Ben-Tal, Y., and S. Marco. 1980. Qualitative changes in cucumber gibberellins following cucumber mosaic virus infection. Physiology and Plant Pathology 16:327–336.

Ben-Yehoshua, S., and B. Aloni. 1974. Effect of water stress on ethylene production by detached leaves of Valencia orange (*Citrus sinensis* Osbeck). Isreal Journal of Botany 22:210.

Bender, M. M. 1971. Variations in the $^{13}C/^{12}C$ ratios of plants in relation to the pathway of carbon dioxide fixation. Phytochemistry 10:1239–1244.

Bensen, R. J., F. D. Beall, J. E. Mullet, and P. W. Morgan. 1990. Detection of endogenous gibberellins and their relationship to hypocotyl elongation in soybean seedlings. Plant Physiology 94:77–84.

Bent, A. F., R. W. Innes, J. R. Ecker, and B. J. Staskawicz. 1992. Disease development in ethylene-insensitive *Arabidopsis thaliana* infected with virulent and avirulent *Pseudomonas* and *Xanthomonas pathogens*. Mololecular Plant-Microbe Interactions 5:372–378.

Bentley, J. C. 1974. Phenological events and their environmental triggers in Mojave desert ecosystems. Ecology 55:856–863.

Berry, J. A., and J. S. Downton. 1982. Environmental regulation of photosynthesis. in: Govindgee (ed.), Photosynthesis, Development, Carbon Metabolism, and Plant Productivity, Vol. II. Academic Press, San Diego, CA.

Berry, J. A., and J. K. Raison. 1981. Responses of macrophytes to temperature. Pp. 277–338 in: O. L. Lange, P. S. Nobel, C. B. Osmond, and H. Zeigler (eds.). Encyclopedia of Plant Physiology, Vol. 12A. Springer-Verlag, Berlin.

Bertrand, A., G. Robitaille, P. Hadeau, and R. Boutin. 1994. Effects of soil freezing and drought stress on abscisic acid content of sugar maple sap and leaves. Tree Physiology 14:413–425.

Bienz, M., and H. R. B. Pellman. 1987. Mechanisms of heat shock gene activation in higher eukaryotes. Advances in Genetics 24:31–72.

Biggs, W. W., A. R. Edison, J. W. Eastin, J. W. Brown, J.W. Maranville, and M. D. Clegg. 1971. Photosynthesis light sensor and meter. Ecology 52:126–131.

Bilger, W., U. Schreiber, and O. L. Lange. 1984. Determination of leaf heat resistance: comparative investigation of chlorophyll fluorescence changes and tissue necrosis methods. Oecologia 63:156–162.

Birch, J. B., and J. L. Cooley. 1982. Production and standing crop patterns of giant cutgrass (*Zizaniopsis miliacea*) in a fresh water tidal march. Oecologia 52:230–235.

Björkman, O., and B. Demming-Adams. 1994. Regulation of photosynthetic light energy capture, conversion, and dissipation in leaves of higher plants. Pp. 17–48 in: E. D. Shulze and M. M. Coldwell (eds.). Ecophysiology of Photosynthesis. Springer-Verlag, Berlin.

Björkman, O., and S. B. Powles. 1984. Inhibition of photosynthetic reactions under water stress: interaction with light level. Planta 161:490–504.

Björkman, O., M. Badger, and P. A. Armond. 1978. Thermal acclimation of photosynthesis: effects of growth temperature on photosynthetic characteristics and components of the photosynthetic apparatus in *Nerium oleander*. Carnegie Institute of Washington Yearbook 77:262–282.

Björkman, O., M. R. Badger, and P. A. Armond. 1980a. Response and adaptations of photosynthesis to high temperatures. Pp. 233–249 in: N. C. Turner and P. J. Kramer (eds.), Adaptation of Plants to Water and High Temperature Stress. John Wiley & Sons, New York.

Björkman, O., B. Demming, and J. T. Andrews. 1988. Mangrove photosynthesis: response to high-irradiance stress. in: J. R. Evans, S. von Cammerer, and W. W. Adams III (eds.), Ecology of Photosynthesis in Sun and Shade. CSIRO, Australia.

Björkman, O., W. J. S. Downton, and H. A. Mooney. 1980b. Response and adaptation to water stress in *Nerium oleander*. Carnegie Institute of Washington Yearbook 79:150–157.

Björkman, O., E. Gauhl, and M. A. Nobs. 1970. Comparative studies of *Atriplex* species with and without β-carboxylation photosynthesis. Carnegie Institute of Washington Yearbook 68:620–623.

Björkman, O., H. A. Mooney, and J. Ehleringer. 1975. Comparison of photosynthetic characteristics of intact plants. Carnegie Institute of Washington Yearbook 74:743–748.

Björkman, O., R. W. Pearcy, A. T. Harrison, and H. A. Mooney. 1972. Photosynthetic adaptation to high temperature: a field study in Death Valley, California. Science 172:786–789.

Björkman, O., S. B. Powles, D. C. Fork, and G. Öquist. 1981. Interaction between high irradiance and water stress on photosynthetic reactions. Carnegie Institute of Washington Yearbook 80:57–59.

Björn, L. O., and Åkerlund, H.-E. 1984. Action spectra for inhibition by ultraviolet radiation of photosystem II activity in spinach chloroplasts. Photochemistry and Photobiology 8:305–313.

Black, V. J., and M. H. Unsworth. 1980. Stomatal responses to sulfur dioxide and water vapor pressure deficit. Journal of Experimental Botany 31:667–677.

Blackman, V. 1919. The compound interest law and plant growth. Annals of Botany 33:353–360.

Bloom, A. J., and J. H. Troughton. 1979. High productivity and photosynthetic flexibility in a CAM plant. Oecologia 38:35–43.

Bloom, A. J., F. S. Chapin III, and H. A. Mooney. 1985. Resource limitation in plants: an economic analogy. Annual Review of Ecology and Systematics 16:363–392.

Boardman, N. K. 1977. Comparative photosynthesis of sun and shade plants. Annual Review of Plant Physiology 28:355–377.

Bohnsack, C. W., and L. S. Albert. 1977. Early effects of boron deficiency on indoleacetic acid oxidase levels of squash root tips. Plant Physiology 59:1047–1050.

Bolduc, R. J., J. H. Cherry, and B. O. Blair. 1970. Increase in indoleacetic acid oxidase activity of winter wheat by cold treatment and gibberellic acid. Plant Physiology 45:461–464.

Boller, T. 1991. Ethylene in pathogenesis and disease resistance. Pp. 293–314 in: A. K. Mattoo and J. C. Suttle (eds.), The Plant Hormone Ethylene. CRC Press, Boca Raton, FL.

Bostock, R. M., and R. S. Quatrano. 1992. Regulation of Em gene expression in rice. Plant Physiology 98:1356–1363.

Bowers, M. C. 1994. Environmental effects of cold on plants. Pp. 391–411 in: R. E. Wilkinson (ed.), Plant–Environment Interactions, Marcel Dekker, New York.

Bown, A. W., D. R. Reeve, and A. Crozier. 1975. The effect of light on the gibberellin metabolism and growth of *Phaseolus coccineus* seedlings. Planta 126:83–91.

Boyer, J. S. 1967. Leaf water potentials measured with a pressure chamber. Plant Physiology 42:133–137.

Boyer, J. S. 1970. Leaf enlargement and metabolic rates in corn, soybeans, and sunflower at various leaf water potentials. Plant Physiology 46:233–235.

Boyer, J. S. 1982. Plant productivity and environment. Science 218:443–448.

Bradford, K. J., and T. C. Hsiao. 1982. Physiological responses to moderate water stress. Pp. 279–286 in: O. L. Lange, P. S. Nobel, C. B. Omond, and H. Ziegler (eds.), Physiological Plant Ecology. II. Water Relations and Carbon Assimilation. Springer-Verlag, New York.

Brailsford, R. W., L. A. C. J. Voesenek, C. W. P. M. Blom, A. R. Smith, M. A. Hall, and M. B. Jackson. 1993. Enhanced ethylene production by primary roots of *Zea mays* L. in response to subambient partial pressures of oxygen. Plant Cell & Environment 16:1071–1080.

Brandle, R. 1991. Flooding resistance of rhizomatous amphibious plants. Pp. 35–46 in: M. B. Jackson, D. D. Davis, and H. Lambers (eds.), Plant Life Under Oxygen Deprivation. SPB Academic Publishing, The Hague.

Bray, E. A. 1991. Regulation of gene expression by endogenous ABA during drought stress. Pp. 81–99 in: W. J. Davies and H. G. Jones (eds.), Abscisic Acid: Physiology and Biochemistry. BIOS Scientific Publishers, Oxford.

Bray, E. A., M. S. Moses, R. Imai, A. Cohen, and A. L. Plant. 1993. Regulation of gene expression by endogenous abscisic acid during drought stress. Pp. 167–176 in: T. J. Close and E. A. Bray, (eds.)., Plant Responses to Cellular Dehydration During Environmental Stress (Current Topics in Plant Physiology: An American Society of Plant Physiologist Series, Vol. 10). American Society of Plant Physiologists, Rockville, MD.

Breeze, V., and J. Elston. 1978. Some effects of temperature and substrate content upon respiration and the carbon balance of field beans (*Vicia faba* L.). Annals of Botany (London) 42:863–876.

Bremner, J. M., and M. A. Tabatabi. 1973. ^{15}N enrichment of soils and derived nitrate. Journal of Environmental Quality 2:363–365.

Briggs, G. M., T. W. Jurik, and D. M. Gates. 1986. Non-stomatal limitation of CO_2 assimilation in three tree species during natural drought conditions. Physiologia Plantarum 66:521–526.

Briggs, P. 1973. Rampage: the story of disastrous floods, broken dams, and human fallibility. McKay, New York.

Brix, H., B. K. Sorrell, and P. T. Orr. 1992. Internal pressurization and convective gas flow in some emergent freshwater macrophytes. Limnological Oceanography 37:1420–1433.

Broadbent, F. E., R. S. Rawschkilb, K. A. Lewis, and G. Y. Chang. 1980. Spatial variability of nitrogen-15 and total nitrogen in some virgin and cultivated soils. Soil Science Society of America Journal 44:524–527.

Brouwer, R., and C. T. deWit. 1969. A simulation model of plant growth with special attention to root growth and its consequences. Pp. 224–244. in: W. A. Whittington (ed.), Root Growth. Plenum Press, New York.

Brown, R. H. 1985. Growth of C3 and C4 grasses under low N levels. Crop Science 25:954–957.

Bruni, F., and Leopold, A. C. 1991. Glass transitions in soybean seed: relevance to anhydrous biology. Plant Physiology 96:660–663.

Brush, R. A., M. Griffith, and A. Mlynarz. 1994. Characterization and quantification of intrinsic ice nucleators in winter rye (*Secale cereale*) leaves. Plant Physiology 104:725–735.

Bunce, J. A. 1982. Effects of water stress on photosynthesis in relation to diuinal accumulation of carbohydrate in source leaves. Canadian Journal of Botany 60:195–200.

Bunce, J. A. 1986. Responses of gas exchange to humidity in populations of three herbs from environments differing in atmospheric water. Oecologia 71:117–120.

Caers, M., P. Rudelsheim, H. Van Onckelen, and S. Horemans. 1985. Effect of heat stress on photosynthetic activity and chloroplast ultrastructure in correlation with endogenous cytokinin concentration in maize seedlings. Plant and Cell Physiology 26:47–52.

Cahill, D. M., and E. W. B. Ward. 1989. Rapid localized changes in abscisic acid concentrations in soybean in interactions with *Phytophthora megasperma* f. sp. *glycinea* or after treatment with elicitors. Physiological and Molecular Plant Pathology 35:483–493.

Cakmak, I., H. Marschner, and F. Bangerth. 1989. Effect of zinc nutritional status on growth, protein metabolism and levels of indole-3-acetic acid and phytohormones in bean (*Phaeseolus vulgaris* L.). Journal of Experimental Botany 40:405–412.

Caldwell, M. M. 1979. Plant life and ultra violet radiation: some perspectives in the history of the Earth's UV climate. Bioscience 29:520–525.

Caldwell, M. M., and L. B. Camp. 1974. Belowground productivity of two cool desert communities. Oecologia 17:123–130.

Caldwell, M. M., and D. M. Eissenstat. 1987. Coping with variability: example of tracer use in root function studies. Pp. 95–106 in: J. D. Tenhunen, F. Catarino, O. L. Lange, and W. C. Oechel (eds.), Plant Response to Stress-Functional Analysis of Mediterranean Systems. Springer-Verlag, Berlin.

Caldwell, M. M., and O. A. Fernandez. 1975. Dynamics of great basin shrub root systems. Pp. 38-51 in: N. F. Haley (ed.), Environmental Physiology of Desert Organisms. Dowden, Hutchinson & Ross, Stroudsburg, PA.

Caldwell, M. M., and R. W. Pearcy. 1994. Exploitation of environmental heterogeneity by plants: Ecophysiological processes above- and belowground. Academic Press, San Diego.

Caldwell, M. M., and J. H. Richards. 1989. Hydraulic lift: water efflux from upper roots improves effectivness of water uptake by deep roots. Oecologia 79:1–5.

Caldwell, M. M., W. B. Sisson, F. M. Fox, and J. R. Brandle. 1975. Growth response to simulated UV radiation under field and greenhouse conditions. Pp. 253–259 in: D. S. Nachtwey, M. M. Caldwell, and R. H. Biggs (eds.), Climatic Impact Assessment Program (CIAP) (Monograph 5 U.S. Department of Transportation, Report No. DOT-TST-75-55). National Technical Information Service, Springfield, VA.

Calkin, H. W., and R. W. Pearcy. 1984. Seasonal progressions of tissue and cell water relations in evergreen and deciduous perennials. Plant Cell & Environment 7:347–352.

Calkin, H. W., A. C. Gibson, and P. S. Nobel. 1985. Xylem water potentials and hydraulic conductance in eight species of ferns. Canadian Journal of Botany 63:632–637.

Campbell. G. 1977. An Introduction to Environmental Biophysics. Springer-Verlag, New York.

Canny, M. J. 1995. Apoplastic water and solute movement: new rules for an old space. Annual Review of Plant Physiology and Plant Molecular Biology 46:215–236.

Cao Y., A. D. M. Glass, and N. M. Crawford. 1993. Ammonium inhibition of *Arabidopsis* root growth can be reversed by potassium and by auxin resistance mutations aux1, axr1, and axr2. Plant Physiology 102:983–989.

Carlioz, A., and Touati, D. 1986. Isolation of superoxide dismutase mutants in E. coli: Is superoxide dismutase necessary for aerobic life? The European Molecular Biology Organization Journal 5:623–630.

Carpenter, J. F., and Crowe, J. H. 1988. Modes of stabilization of a protein by organic solutes during desiccation. Cryobiology 25:459–470.

Carriere, F., P. Chagvardieff, G. Gil, M. Pean, J.-C. Sigoillot, and P. Tapie. 1992. Fatty acid patterns of neutral lipids from seeds, leaves and cell suspension cultures of *Euphorbia characias*. Phytochemistry 31:2351–2353.

Carrol, C. R. J. H. Vandermeer, and P. Rosset. 1990. Agroecology. McGraw-Hill Book Co., New York.

Cary, J. W. 1975. Factors affecting cold injury of sugarbeet seedlings. Agronomy Journal 67:258–262.

Caspar, T., S. C. Huber, and C. Sommerville. 1986. Alterations in growth, photosynthesis, and respiration in a starchless mutant of *Arabidopsis thaliana* deficient in chloroplast phosphoglucomutase activity. Plant Physiology 76:1–7.

Castonguay, Y., P. Nadeau, and R. R. Simard. 1993. Effects of flooding on carbohydrate and ABA levels in roots and shoots of alfalfa. Plant Cell & Environment 16:695–702.

Chalk, P. M. 1985. Estimation of N_2 fixation by isotope dilution: an appraisal of techniques involving ^{15}N enrichment and their application. Soil Biology and Biochemistry 17:389–410.

Chandler, P. M., and Robertson, M. 1994. Gene expression regulated by abscisic acid and its relation to stress tolerance. Annual Review of Plant Physiology and Plant Molecular Biology 45:113–141.

Chapin, F. S., III. 1980. The mineral nutrition of wild plants. Annual Review of Ecology and Systematics 11:233–260.

Chapin, F. S., III. 1989. The cost of tundra plant structures: evaluation of concepts and currencies. American Naturalist 133:1–19.

Chapin, F. S., III. 1991. Effects of multiple environmental stress on nutrient availability and use. Pp. 67–88. in: Mooney, H. A., W. E. Winner, and E. J. Pell (eds.). Response of plants to multiple stresses. Academic Press, San Diego.

Chapin, F. S., III., and G. R. Shaver. 1985. Individualistic growth responses of tundra plant species to environmental manipulation in the field. Ecology 66:564–576.

Chapin, F. S. III, A. J. Bloom, C. B. Field, and R. H. Waring. 1987. Plant response to multiple factors. Bioscience 37:49–57.

Chapin, F. S., III, P. C. Miller, W. D. Billings, and P. I. Coyne. 1980. Carbon and nutrient budgets and their control in coastal tundra in Alaska. Pp. 458–482 in: J. Brown, P. C. Miller, L. L. Tieszen, and F. L. Bunnell (eds.), An Arctic Ecosystem. Dowden, Hutchinson & Ross, Stroudsburg, PA.

Chapin, F. S. III, E. D. Shulze, and H. A. Mooney. 1990. The ecology and economics of storage in plants. Annual Review of Ecology and Systematics 21:423–447.

Chapman, V. J. 1984. Mangrove biogeography. Pp. 15–24 in: F. D. Por and I. Dor (eds.), Hydrobiology of the Mangal: The Ecosystem of the Mangrove Forests (Developments in Hydrobiology, Vol. 20). Kluwer Academic Press, Boston.

Chazdon, R. L. 1988. Sun flecks and their importance to forest understory plants. Advances in Ecological Research 18:1–63.

Chazdon, R., and N. Fetcher. 1984. Photosynthetic light environments in a low land tropical forest in Costa Rica. Journal of Ecology 72:553–564.

Chee, P. P., K. A. Forber, and J. L. Slighton. 1989. Transformation of soybean (*Glycine max*) by infecting germinating seeds with *Agrobacterium tumaciens*. Plant Physiology 91:1212–1218.

Cheeseman, J. M., and J. B. Hanson. 1979. Energy-linked potassium influx as related to cell potential in corn roots. Plant Physiology 64:842–845.

Cheikh, N., and R. J. Jones. 1994. Disruption of maize kernel growth and development by heat stress. Plant Physiology 106:45–51.

Chen, C., J. Ertl, M. Yang, and C. Chang. 1987. Cytokinin-induced changes in the population of translatable mRNA in excised pumpkin cotyledons. Plant Sciences 52:169–174.

Chen, T. H. H., M. J. Burke, and L. V. Gusta. 1995. Freezing tolerance in plants: an overview. Pp. 115–135 in: R. E. Lee, Jr., G. J. Warren, and L. V. Gusta (eds.), Biological Ice Nucleation and Its Applications. APS Press, St. Paul, MN.

Chen, Y.-Z. and B. D. Patterson. 1985. Ethylene and 1-aminocyclopropane-1-carboxylic acid as indicators of chilling sensitivity in various plant species. Australian Journal of Plant Physiology 12:377–385.

Cheung, Y. N. S., M. T. Tyree, and J. Dainty. 1975. Water relations parameters on single leaves obtained in a pressure bomb and some ecological interpretations. Canadian Journal of Botany 53:1342–1346.

Chiarello, N. R., H. A. Mooney, and K. Williams. 1989. Growth, carbon allocation, and cost of plant tissues. Pp. 327–366 in: R. W. Pearcy, J. Ehleringer, H. A. Mooney, and P. W. Rundel (eds.), Plant Physiological Ecology: Field Methods and Instrumentation. Chapman & Hall, New York.

Chrominski, A., S. Halls, D. J. Weber, and B. N. Smith. 1989. Proline affects ACC to ethylene conversion under salt and water stresses in the halophyte, *Allenrolfea occidentalis*. Journal of Environmental and Experimental Botany 29:359–363.

Chumakovskii, N. N. 1986. Stem height and drought resistance in relation to leaf content of abscisic acid and ethylene in wheat plants. Soviet Plant Physiology 33:401–407.

Chung, H.-H., and P. J. Kramer. 1975. Absorption of water and ^{32}P through suberized and unsuberized roots of loblolly pine. Canadian Journal of Forest Research 5:229–235.

Cifuentes, L. A., J. H. Sharp, and M. L. Fogel. 1988. Stable carbon and nitrogen isotopes biogeochemistry in the Delaware estuary. Limnology and Oceanography 33:1102–1115.

Cipollini, M. L., B. G. Drake, and D. Whigham. 1993. Effects of elevated CO_2 on growth and carbon/nutrient balance in the deciduous woody shrub *Lindera benzoin* (L.) Blume (lauraceae). Oecologia 96:339–346.

Clark, J., and R. D. Gibbs. 1957. Studies in tree physiology. IV. Further investigation of seasonal changes in moisture content of certain Canadian forest trees. Canadian Journal of Botany 35:219–253.

Clarke, A. K., and C. Critchley. 1994. Characterization of chloroplast heat shock proteins in young leaves of C_4 monocotyledons. Physiologia Plantarum 92:118–130.

Clausen, J. D. D. Keck, and W. M. Heisey. 1940. Experimental studies of the nature of species. I. Structure of ecological races. Carnegie Institute of Washington No. 520. Washington, D. C.

Cleland, R. 1986. The role of hormones in wall loosening and plant growth. Australian Journal of Plant Physiology 13:93–103.

Clipson, N. J. W., D. R. Lachno, and T. J. Flowers. 1988. Salt tolerance in the halophyte *Suaeda maritima* L. Dum.: abscisic acid concentrations in response to constant and altered salinity. Journal of Experimental Botany 39:1381–1388.

Close, T. J., and P. J. Lammers. 1993. An osmotic stress protein of cyanobacteria is immunologically related to the plant dehydrins. Plant Physiology 101:773–779.

Close, T. J., A. A. Kort, and P. M. Chandler. 1989. A C-DNA based comparison of dehydration-induced proteins (dehydrins) in barley and corn. Plant Molecular Biology 13:95–108.

Cohen, Y., M. Fuchs, and S. Cohen. 1983. Resistance to water uptake in a mature citrus tree. Journal of Experimental Botany 34:451–460.

Colasanti, J., M. Tyres, and V. Sundaresan. 1991. Isolation and characterization of cDNA clones encoding for a functional $p34^{cdc2}$ homologue from *Zea mays*. Proceedings of the National Academy of Science USA 88:3377–3381.

Coleman, B., and G. S. Espie. 1985. CO_2 uptake and transport in leaf mesophyll cells. Plant Cell & Environment 8:449–457.

Coleman, J. S., and K. McConnaughay. 1994. Phenotypic plasticity. Trends in Ecology and Evolution 9:187–191.

Coleman, J. S., S. A. Heckathorn, and R. L. Hallberg. 1995. Heat-shock proteins and thermotolerance: linking molecular and ecological perspectives. TREE 8:305–306.

Coley, P. D., J. P. Bryant, and F. S. Chapin III. 1985. Resource availability and plant antiherbivore defense. Science 230:895–899.

Collatz, G. J., J. T. Ball, C. Grivet, and J. A. Berg. 1991. Physiological and environmental regulation of stomatal conductance, photosynthesis, and transpiration: a model that includes a laminar boundary layer. Agricultural and Forest Meteorology 54:107–136.

Comstock, J. P., and J. R. Ehleringer. 1986. Photoperiod and photosynthetic capacity in *Lotus scoparius*. Plant Cell & Environment 9:609–612.

Comstock, J. P., and J. R. Ehleringer. 1988. Contrasting photosynthetic behavior in leaves and twigs of *Hymenoclea salsola*, a green-twigged warm desert shrub. America Journal of Botany 75:1360–1370.

Comstock, J. P., and J. R. Ehleringer. 1992. Correlating genetic variation in carbon isotope composition with complex climatic gradients. Proceedings of the National Academy of Sciences 89:7747–7751.

Condon, A. G., R. A. Richards, and G. D. Farquhar. 1987. Carbon isotope discrimination is positively correlated with grain yield and dry matter production in field grown wheat. Crop Science 27:996–1001.

Constable, J. V. H., and D. J. Longstreth. 1994. Aerenchyma carbon dioxide can be assimilated in cattail (*Typha latifolia* L.) leaves. Plant Physiology 106:1065–1072.

Corey, K. A., and A. V. Barker. 1989. Ethylene evolution and polyamine accumulation by tomato subjected to interactive stresses of ammonium toxicity and potassium deficiency. Journal of the American Society of Horticultural Science 114:651–655.

Cosgrove, D. J., and D. M. Durachko. 1986. Automated pressure probe for measurement of water transport properties of higher plant cells. Review of Scientific Instrumentation 57:2614–2619.

Costacurta, A., and J. Vanderleyden. 1995. Synthesis of phytohormones by plant-associated bacteria. Critical Reviews of Microbiology 21:1–18.

Cowan, I. R. 1977. Stomatal behavior and environment. Advances in Botanical Research 4:117–227.

Cowan, I. R. 1982. Regulation of water use in relation to carbon gain in higher plants. Pp. 589–614 in: O. L. Lange, P. S. Nobel, C. B. Osmond, and H. Ziegler (eds.). Encyclopedia of Plant Physiology, Vol. II, Physiology Plant Ecology. Springer-Verlag, New York.

Cowan, I. R. 1994. As to the mode of action of the guard cells in dry air. Pp. 205–230 in: E. D. Schultze and M. M. Caldwell (eds.), Ecophysiology of Photosynthesis. Springer-Verlag, Berlin.

Cowan, I., and G. Farquhar. 1977. Stomatal function in relation to leaf metabolism and environment. Symposium of the Society of Experimental Biology 31:471–505.

Cowan, A. K., and P. D. Rose. 1991. Abscisic acid metabolism in salt-stressed cells of *Dunaliella salina*. Plant Physiology 97:798–803.

Craig, H. 1954. Carbon-13 in plants and the relationship between ^{13}C and ^{14}C variations in nature. Journal of Geology 62:115–149.

Crawford, L. A., A. W. Brown, K. E. Breitkreuz, and F. C. Guinel. 1994. The synthesis of α-aminobutyric acid in response to treatments reducing cytosolic pH. Plant Physiology 104:865–871.

Crawford, R. M. M., and M. A. Palin. 1981. Root respiration and temperature limits to the north-south distribution of four perennial maritime plants. Flora 171:338–354.

Creissen, G., E. A. Edwards, C. Enard, A. Wellburn, and P. Mullineaux. 1991. Molecular characterization of glutathione reductase cDNAs from pea (*Pisum sativum* L.). Plant Journal 2:129–131.

Creissen, G., E. A. Edwards, and P. Mullineaux. 1993. Glutathione reductase and ascorbate peroxidase. Pp. 343–364 in: C. H. Foyer and P. M. Mullineaux (eds.). Causes of photooxidative stress and amelioration defense in plants. CRC Press, Boca Raton, FL.

Crick, J. C., and J. P. Grime. 1987. Morphological plasticity and mineral nutrient capture in two herbaceous species of contrasted ecology. New Phytologist 107:403–414.

Crocoll, C., J. Kettner, and K. Dörffling. 1991. Abscisic acid in saprophytic and parasitic species of fungi. Phytochemistry 30:1059–1060.

Crombie, D. S., J. A. Milburn, and M. F. Hipkins. 1985. Maximum sustainable xylem sap tensions in rhododendron and other species. Planta 163:27–33.

Curran, M., M. Cole, and W. G. Allaway. 1986. Root aeration and respiration in young mangrove plants, *Avicennia marina* (Forsk) Vieh. Journal of Experimental Botany 37:1225–1233.

Dacey, J. W. A. 1980. Internal winds in water lilies: an adaptation for life in anaerobic sediments. Science 210:1017–1019.

Dacey, J. W. A. 1981. Pressurized ventilation in the yellow water lily. Ecology 62:1137–1147.

Dacey, J. W. A., and M. J. Klug. 1979. Methane efflux from lake sediments through water lilies. Science 203:1253–1255.

Daie, J., and W. F. Campbell. 1981. Response of tomato plants to stressful temperatures: increase in abscisic acid concentration. Plant Physiology 67:26–29.

Dallaire, S., M. Houde, Y. Gagné, H. S. Saini, S. Boileau, N. Chevrier, and F. Sarhan. 1994. ABA and low temperature induce freezing tolerance via distinct regulatory pathways in wheat. Plant Cell Physiology 35:1–9.

Daly, J. M., and R. E. Inman. 1958. Changes in auxin levels in safflower hypocotyls infected with *Puccinia carthami*. Phytopathology 48:91–97.

Dansgaard, W. 1964. Stable isotopes in precipitation. Tellus 16:436–468.

Danyluk, J., E. Rassart, and F. Sarhan. 1990. Gene expression during cold and heat shock in wheat. Biochemistry and Cell Biology 69:383–391.

Darbyshire, B. 1971. Changes in indoleacetic acid oxidase activity associated with plant water potential. Physiologia Plantarum 25:80–84.

Darwin, C. R. 1865. On the movements and habits of climbing plants. Journal of the Linean Society, London.

Darwin, C. R. 1881. The Power of Movements in Plants. Appleton, New York.

Davenport, T. L., P. W. Morgan, and W. R. Jordan. 1977. Auxin transport as related to leaf abscission during water stress in cotton. Plant Physiology 59:554–557.

Davies, W. J., and J. Zhang. 1991. Root signals and the regulation of growth and development of plants in drying soils. Annual Review of Plant Physiology 42:55–76.

Davies, W. J., J. Metcalf, T. A. Lodge, and A. R. de Costa. 1986. Plant growth substances and the regulation of growth under drought. Australian Journal of Plant Physiology 13:105–125.

Dawson, T. E. 1993. Water sources of plants as determined from xylem-water isotope composition: Perspectives on plant competition, distribution, and water relations. Pp. 465–496 in: J. R. Ehleringer, A. E. Hall, and G. D. Farquhar (eds.). Stable Isotopes and Plant Carbon Water Relations. Academic Press, San Diego.

Dawson, T. E., and J. R. Ehleringer. 1991. Streamside trees that do not use stream water: evidence from hydrogen isotope ratios. Nature 350:335–337.

Decoteau, D. R., and L. E. Craker. 1987. Abscission: ethylene and light control. Plant Physiology 83:970–972.

Dehio, C., K. Grossmann, J. Schell, and T. Schmülling. 1993. Phenotype and hormonal status of transgenic tobacco plants overexpressing the *rol*A gene of *Agrobacterium rhizogenes* T-DNA. Plant Molecular Biology 23:1199–1210.

Deines, P. 1980. The isotopic composition of reduced organic carbon. Pp. 329–406 in: P. Fritz and J. C. Fontes (eds.), Handbook of Environmental Isotope Geochemistry, Vol. 1. Elsevier, Amsterdam.

Dekhuijzen, H. M. 1976. Endogenous cytokinins in healthy and diseased plants. Pp. 527–529 in: R. Heitefuss and P. H. Williams (eds.), Physiological Plant Pathology. Springer-Verlag, Berlin.

De la Pena, A., H. Lorz, and J. Schell. 1987. Transgenic plants obtained by injecting DNA into young floral tillers. Nature 325:274–276.

DeLucia, E. H., W. H. Schlesinger, and W. D. Billings. 1988. Water relations and the maintenance of Sierran conifers on hydrothermally altered rock. Ecology 69:303–311.

Delwiche, C. C., P. J. Zinke, C. M. Johnson, and R. A. Virginia. 1979. Nitrogen isotope distribution as a presumptive indicator of nitrogen fixation. Botanical Gazette 65:69–73.

Demming-Adams, B., and W. W. Adams. 1992. Photoprotection and other responses of plants to high light stress. Annual Review of Plant Physiology and Plant Molecular Biology 43:599–626.

Dennis, E. S., W. L. Gerlach, J. C. Walker, M. Lavin, and W. J. Peacock. 1988. Anaerobically regulated aldolase gene of maize. A chimaeric origin? Journal of Molecular Biology 202:759–767.

Devlin, R. M., and F. H. Witham. 1983. Plant Physiology, 4th ed. Wadsworth Publishing Co., Belmont, CA.

Dileanis, P. D., and D. P. Groeneveld. 1989. Osmotic potential and projected drought tolerance of four phreatophytic shrub species in Owens Valley, California. U.S. Geological Survey Water Supply Paper No. 2370.

Dittmer, H. J. 1937. A quantitative study of the roots and root hairs of a winter rye plant (*Secale cereale*). American Journal of Botany 24:417–420.

Dobert, R. C., S. B. Rood, K. Zanewich, and D. G. Blevins. 1992. Gibberellins and the legume–*Rhizobium* symbiosis. III. Quantification of gibberellins from stems and nodules of lima bean and cowpea. Plant Physiology 100:1994–2001.

Donahue, T. M., J. H. Hoffman, R. R. Hodges, and A. J. Watson. 1982. Venus was wet: a measurement of the ratio of deterium to hydrogen. Science 216:630–633.

Donovan, L. A., N. J. Stumpff, and K. W. McLeod. 1989. Thermal flooding injury of woody swamp seedlings. Journal of Thermal Biology 14:147–154.

Douglas, T. J., and R. R. Walker. 1983. 4-Desmethylsterol composition of citrus rootstocks of different salt exclusion capacity. Physiologia Plantarum 58:69–74.

Downes, R. J., and H. Hellmers. 1975. Environment and the Experimental Control of Plant Growth. Academic Press, London.

Downton, W. J. S., and B. R. Loveys. 1981. Abscisic acid content and osmotic relations of salt-stressed grapevine leaves. Australian Journal of Plant Physiology 8:443–452.

Draper, J., R. Scott, P. Armitage, and R. Walden (eds.). 1988. Plant genetic transformation and gene expression. A laboratory manual. Blackwell Scientific Publishers, Oxford.

Drew, M. C. 1991. Oxygen deficiency in the root environment and plant mineral nutrition. Pp. 303–316. in: M. B. Jackson, D. D. Davis, and H. Lambers (eds.), Plant Life Under Oxygen Deprivation. SPB Academic Publishing, The Hague.

Drew, M. C., and E. J. Sisworo. 1979. The development of waterlogging damage in young barley plants in relation to plant nutrition status and changes in soil properties. New Phytologist 82: 301–314.

Drew, M. C., C.-J. He, and P. W. Morgan. 1989. Decreased ethylene biosynthesis, and induction of aerenchyma, by nitrogen- or phosphate-starvation adventitious roots of *Zea mays* L. Plant Physiology 91:266–271.

Drew, M. C., M. B. Jackson, and S. C. Giffard. 1979. Ethylene-promoted adventitious rooting and development of cortical air spaces (aerenchyma) in roots may be an adaptive response to flooding in *Zea Mays* L. Planta 147:83–88.

Drijver, C. A., and M. Marchland. 1985. Taming the floods. Environmental aspects of floodplain development in Africa. Centre for Environmental Studies, State University of Leiden.

Duke, N. C. 1995. Genetic diversity, distributional barriers, and rafting continents: more thoughts on the evolution of mangroves. Pp. 167–181 in: Y. S. Wong, and N. F. Y. Tam (eds.), Asia-Pacific Symposium on Mangrove Ecosystems (Developments in Hydrobiology No. 106). Kluwer Academic Press, Boston.

Dunn, R. M., P. Hedden, and J. P. Bailey. 1990. A physiologically induced resistance of *Phaseolus vulgaris* to a compatible race of *Colletotrichum lindemuthianum* is associated with an increase in ABA content. Physiological and Molecular Plant Pathology 36:339–349.

Dure, L., III. 1993. A repeating 11-mer amino acid motif and plant desiccation. Plant Journal 3:363–369.

Dure, L., III, M. Crouch, J. Harada, T.-HD. Ho, J. Mundy, R.S. Quatrano, T. Thomas, and Z. R. Sung. 1989. Common amino acid sequence domains among the LEA proteins of higher plants. Plant Molecular Biology 12:475–486.

Durham, R. E., G. A. Moore, D. Haskell, and C. L. Guy. 1991. Cold-acclimation induced changes in freezing tolerance and translatable mRNA content in *Citrus grandis* and *Poncirus trifoliata*. Physiologia Plantarum 82:519–522.

Duzgunes, N., and D. Papahadjopoulos. 1983. Ionotropic effects on phospholipid membranes: calcium–magnesium specificity in binding, fluidity and fusion. Pp. 187–216 in: R. C. Aloia (ed.), Membrane Fluidity in Biology, Vol. 2. Academic Press, New York.

Ehleringer J. 1983. Characterization of a glabrate *Encelia farinosa* mutant: morphology, ecophysiology, and field observations. Oecologia 57:303–310.

Ehleringer, J. R. 1988. Correlations between carbon isotope ratio, water use efficiency and yield. Pp. 165–191 in: J. W. White, G. Hoogenboom, F. Ibara, and S. P. Singh (eds.), Research on Drought Tolerance in Common Bean. CIAT, Cali, Columbia.

Ehleringer, J. R., and O. Björkman. 1977. Quantum yields of CO_2 uptake in C_3 and C_4 plants. Dependence on temperature, CO_2 and O_2 concentration. Plant Physiology 59:86–90.

Ehleringer, J., and O. Björkman. 1978. Pubescence and leaf spectral characteristics of a desert shrub *Encelia farinosa*. Oecologia 36:151–162.

Ehleringer, J. R., and T. A. Cooper. 1988. Correlations between carbon isotope ratio and microhabitat in desert plants. Oecologia 76:562–566.

Ehleringer, J. R., and T. Cooper. 1992. On the role of orientation in reducing photoinhibitory damage in photosynthetic-twig desert shrubs. Plant Cell & Environment 15:301–306.

Ehleringer, J. R., and T. E. Dawson. 1992. Water uptake by plants: perspectives from stable isotope composition. Plant Cell & Environment 15:1073–1082.

Ehleringer, J. R., and C. B. Field. 1993. Scaling physiological processes: Leaf to Globe. Academic Press, San Diego.

Ehleringer, J. R., and I. P. Forseth. 1980. Solar tracking by plants. Science 210:1094–1098.

Ehleringer, J. R., and R. K. Monson. 1993. Evolutionary and ecological aspects of photosynthetic pathway variation. Annual Review of Ecology and Systematics 24:411–440.

Ehleringer, J. R., J. P. Comstock, and T. A. Cooper. 1987b. Leaf-twig carbon isotope ratio differences in photosynthetic-twig desert shrubs. Oecologia 71:318–320.

Ehleringer, J. R., H. A. Mooney, S. L. Gulmon, and P. W. Rundel. 1981. Parallel evolution of leaf pubescencein *Encelia* in coastal deserts of North and South America. Oecologia 49:38–41.

Ehleringer, J. R., S. L. Phillips, W. S. F. Schuster, and D. Sandquist. 1991. Differential utilization of summer rains by desert plants. Oecologia 88:430–434.

Ehleringer, J. R., E.-D. Schulze, H. Ziegler, O. L. Lang, G. D. Farquhar, and I. R. Cowan. 1985. Xylem-tapping mistletoes: water or nutrient parasites? Science 227:1479–1481.

Ehleringer, J. R., Z. F. Lin, C. Field, and C. Y. Kuo. 1987a. Leaf carbon isotope ratios from plants from a tropical monsoon forest. Oecologia 72:109–114.

Eibl, H. 1983. The effect of proton and of monovalent cations on membrane fluidity. Pp. 217–236 in: R. C. Aloia (ed.), Membrane Fluidity in Biology, Vol. 2. Academic Press, New York.

El-D, A. M. S., A. Salama, and P. F. Wareing. 1979. Effects of mineral nutrition on endogenous cytokinins in plants of sunflower (*Helianthus annus* L.). Journal of Experimental Botany 30:971–981.

Ellsworth, D. S., and P. B. Reich. 1993. Canopy structure and vertical patterns of photosynthesis and related leaf traits in a deciduous forest. Oecologia 96:169–178.

Elstner, E. F. 1983. Hormones and metabolic regulation of disease. Pp. 415–434 in: J. A. Callow (ed.), Biochemical Plant Pathology. John Wiley & Sons, New York.

Emanuel, W. R., H. H. Shugart, and M. V. Stevenson. 1985. Climatic change and broad scale distribution of terrestrial ecosystem complexes. Climatic Change 7:29–43.

Erickson, R. 1976. Modeling of plant growth. Annual Review of Plant Physiology 27:407–434.

Erickson, R., and F. Michelini. 1957. The plastochron index. American Journal of Botany 44:297–305.

Ernst, D., W. Schäfer and D. Osterhelt. 1983. Isolation and identification of a new, naturally occurring cytokinin (6-benzylaminopurine-riboside) from an anise cell culture (*Pimpenella anisum* L.). Planta 159:222–225.

Estep, M. F., and T. C. Hoering. 1981. Stable hydrogen isotope fractionation during autotrophic and mixotrophic growth of microalgae. Plant Physiology 67:474–477.

Estruch, J. J., J Schell, and A. Spena. 1991. The protein encoded by the rolB plant oncogene hydrolyses indole glucosides. The European Molecular Biology Organization Journal 10:3125–3128.

Evans, G. C. 1972. The Quantitative Analysis of Plant Growth. Blackwell Scientific Publications, Oxford.

Evans, J. R. 1989. Photosynthesis and nitrogen relationships in leaves of C_3 plants. Oecologia 78:9–19.

Evans, J. R., T. D. Sharkey, J. A. Berry, and G. D. Farquhar. 1986. Carbon isotope discrimination measured concurrently with gas exchange to investigate CO_2 diffusion in leaves of higher plants. Australian Journal of Plant Physiology 13:281–292.

Fanjul, L., and H. G. Jones. 1982. Rapid stomatal responses to humidity. Planta 154:135–138.

Faragher, J. D., and S. Mayak. 1984. Physiological responses of cut flowers to exposure to low temperature: changes in membrane permeability and ethylene production. Journal of Experimental Botany 35:965–974.

Farquhar, G. D. 1989. Models of integrated photosynthesis of cells and leaves. Philosophical Transactions of the Royal Society of London 323:357–367.

Farquhar, G. D., and R. A. Richards. 1984. Isotopic composition of plant carbon correlates with water use efficiency of wheat genotypes. Australian Journal of Plant Physiology 11:539–552.

Farquhar, G. D., M. C. Ball, S. von Caemmerer, and Z. R. Oksandic. 1982a. Effects of salinity and humidity on $\delta^{13}C$ value of halophytes: evidence for diffusional isotope fractionation determined by the ratio of intercellular/atmospheric partial pressure of CO_2 under different environmental conditions. Oecologia 52:121–124.

Farquhar, G. D., J. R. Ehleringer, and K. T. Hubick. 1989. Carbon isotope discrimination and photosynthesis. Annual Review of Plant Physiology 40:503–537.

Farquhar, G. D., M. H. O'Leary, and J. A. Berry. 1982b. On the relationship between carbon isotope discrimination and the intercellular carbon dioxide concentration in leaves. Australian Journal of Plant Physiology 9:121–137.

Farquhar, G., S. von Caemmerer, and J. Berry. 1980. A biochemical model of photosynthetic CO_2 fixation in leaves of C_3 leaves. Planta 149:78–90.

Farr, S. B., R. D'Ari, and D. Touati. 1986. Oxygen-dependent mutagenesis in *Escherichia coli* lacking superoxide dismutase. Proceedings of the National Academy of Sciences 83:8268–8272.

Farrar, J. F., and J. H. Williams. 1991. Control of the rate of respiration in roots: compartmentation, demand and the supply of substrate. Pp. 167–188 in: M. J. Emes (ed.), Compartmentation of Plant Metabolism in Nonphotosynthetic Tissues (Society of Experimental Biology Seminar Series Vol. 42). Cambridge University Press, Cambridge.

Feder, J. H., J. M. Rossi, J. M. Solomon, N. Solomon, and S. Lindquist. 1992. The consequences of expressing HSP 70 in *Drosophila* cells at normal temperatures. Genes and Development 6:1402–1413.

Feeny, P. P. 1976. Plant apparency and chemical defense. Pp. 1–40 in: J. W. Wallace and R. L. Mansell (eds.), Recent Advances in Phytochemistry. Plenum Press, New York.

Fellows, R. J., and J. S. Boyer. 1978. Altered ultrastructure of cells of sunflower leaves having low water potentials. Protoplasma 93:381–395.

Feng, J., and A. V. Barker. 1992. Ethylene evolution and ammonium accumulation by tomato plants under water and salinity stress, II. Journal of Plant Nutrition 15:2471–2490.

Feng, J., and A. V. Barker. 1993. Polyamine concentration and ethylene evolution in tomato plants under nutritional stress. HortScience 28:109–110.

Ferguson, I. B., and R. E. Mitchell. 1985. Stimulation of ethylene production in bean leaf discs by the pseudomonad phytotoxin coronatine. Plant Physiology 77:969–973.

Ferreira, P., A. Hemerley, R. Villarroel, M. Van Montague, and D. Inze. 1991. The *Arabidopsis* functional homologue of the $p34^{cdc2}$ protein kinase. Plant Cell 3: 531–540.

Fetcher, N., and G. R. Shaver. 1982. Growth and tillering patterns within tussocks of *Eriophorum vaginatum*. Holarctic Ecology 5:180–186.

Fichtner, K., and E. D. Shulze. 1992. The effect of nitrogen nutrition on growth and biomass partioning of annual plants originating from habitats of different nitrogen availability. Oecologia 92:236–241.

Fichtner, K., G. W. Koch, and H. A. Mooney. 1994. Photosynthesis, storage, and allocation. Pp. 133–146 in: E. D. Shulze and M. M. Caldwell (eds.), Ecophysiology of Photosynthesis Ecological Studies, Vol. 100. Springer-Verlag, Berlin.

Field, C. B. 1983. Allocating leaf nitrogen for the maximization of carbon gain: leaf age as a control on the allocation program. Oecologia 56:341–347.

Field, C. B. 1991. Ecological scaling of carbon gain to stress and resource availability. Pp. 35–65 in: H. A. Mooney, W. E. Winner, and E. J. Pell (eds.), Response of Plants to Multiple Stresses. Academic Press, San Diego, CA.

Field, C. B., and N. M. Holbrook. 1989. Catastrophic xylem failure: tree life at the brink. Tree 4:124–125.

Field, C. B., and H. A. Mooney. 1983. Leaf age and seasonal effects of light, water, and nitrogen use efficiency in a California shrub. Oecologia 56:348–355.

Field, C. B., and H. A. Mooney. 1986. The photosynthetic–nitrogen relationship in wild plants. Pp. 25–55 in: T. Givnish (ed.), On the Economy of Plant Form and Function. Cambridge University Press, Cambridge.

Field, C. B., N. Chiarello, and W. E. Williams. 1982. Determinants of leaf temperature in California *Mimulus* species at different altitudes. Oecologia 55:414–420.

Field, C. B., J. Merino, and H. A. Mooney. 1983. Compromises between water-use efficiency and nitrogen-use efficiency in five species of California evergreens. Oecologia 60:384–389.

Field, R. J. 1981. A relationship between membrane permeability and ethylene production at high temperature in leaf tissue of *Phaseolus vulgaris* L. Annals of Botany 48:33–39.

Field, R. J. 1984. The role of 1-aminocyclopropane-1-carboxylic acid in the control of low temperature induced ethylene production in leaf tissue of *Phaseolus vulgaris* L. Annals of Botany 54:61–67.

Finer, J. J., and M. D. McMullen. 1990. Transformation of cotton (*Gossypium hirsutum* L.) via particle bombardment. Plant Cell Reports 8:586–589.

Flannagan, L. B., and J. R. Ehleringer. 1991. Stable isotope composition of stem and leaf water: applications to the study of plant water use. Functional Ecology 5:270–277.

Flores, S., and E. Tobin. 1987. Benzyladenine regulation of the expression of two nuclear genes for chloroplast proteins. Pp. 123–132 in: J. Fox and M. Jacobs (eds.), Molecular Biology of Plant Growth Control. Alan R. Liss, New York.

Flowers, T. J. 1975. Halophytes. Pp. 309–344 in: D. A. Baker and J. L. Hall (eds.). Ion Transport in Plant Cells and Tissues. Elsevier, Amsterdam.

Fontes, J.-Ch. 1980. Environmental isotopes in groundwater hydrology. Pp. 411–440 in: P. Fritz and J. Ch. Fontes (eds.) Handbook of environmental geochemistry. Elsevier Publishers, Amsterdam, The Netherlands.

Forseth I. N., and J. R. Ehleringer. 1980. Solar tracking response to drought in a desert annual. Oecologia 44:159–163.

Forseth, I. N., and J. R. Ehleringer. 1982. Ecophysiology of two solar tracking desert winter annuals. II. Leaf movements, water relations and microclimate. Oecologia 54:41–49.

Forseth, I. N., and J. R. Ehleringer. 1983. Ecophysiology of two solar tracking desert winter annuals. III. Gas exchange responses to light, CO_2, and VPD in relation to long-term drought. Oecologia 57:344–351.

Foyer, C. H., M. Lelandais, C. Galap, and K. J. Kunert. 1991. Effects of elevated cytosolic glutathione reductase activity on the cellular glutathione pool and photosynthesis in leaves under normal and stress conditions. Plant Physiology 97:863–872.

Francy, R. J., R. M. Gifford, T. A. Sharkey, and B. Wier. 1985. Physiological influences on carbon isotope discrimination in buon pine (*Lagastrobos franklinnii*). Oecologia 44:241–247.

Fraser, R. S. S. 1991. ABA and plant responses to pathogens. Pp. 189–199 in: W. J. Davies and H. G. Jones (eds.), Abscisic Acid: Physiology and Biochemistry. Bios Scientific Publishers, Oxford.

Fried, M., and H. Broeshart. 1975. An independent measurement of the amount of nitrogen fixed by a legume crop. Plant and Soil 43:707–711.

Fried, M., and V. Middleboe. 1977. Measurement of the amount of nitrogen fixed by a legume crop. Plant and Soil 47:713–715.

Fu, Q. A., and J. R. Ehleringer. 1990. Heliotropic leaf movements in common beans controlled by air temperature. Plant Physiology 91:1162–1167.

Fuente, R. K. D., P. M. Tang, and C. C. DeGuzman. 1985. The requirement for calcium and boron in auxin transport. Pp. 227–230 in: M. Bopp (ed.), Plant Growth Substances. Springer-Verlag, New York.

Furuya, M. 1993. Phytochromes: their molecular species, gene families, and functions. Annual Review of Plant Physiology 44:617–645.

Fusseder, A., A. Wartinger, W. Hartung, E. -D. Schulze, and H. Heilmeier. 1992. Cytokinins in xylem sap of desert-grown almond (*Prunus dulcis*) trees: daily courses and their possible interactions with abscisic acid and leaf conductance. New Phytologist 122:45–52.

Gaastra, P. 1959. Photosynthesis of crop plants as influenced by light, carbon dioxide, temperature, and stomatal diffusion resistance. Laboratory of Plant Physiology Research, Agricultural University of Wageninegen 59:1–68.

Gad, A. E., N. Rosenberg, and A. Altman. 1990. Liposome-mediated gene delivery into plant cells. Physiologia Plantarum 79:177–183.

Gaff, D. F. 1980. Protoplasmic tolerance of extreme water stress. Pp. 207–230 in: N. C. Turner and P. J. Kramer (eds.), Adaptations of Plants to Water and High Temperature Stress. Wiley-Interscience, New York.

Galau, G., and Close, T. J. 1992. Sequences of the cotton group 2 LEA/RAB/dehydrin proteins encoded by *Lea3* cDNAs. Plant Physiology 98:1523–1525.

Gallardo, M., M. delMar Delgado, I. M. Sánchez-Calle, and A. J. Matilla. 1991. Ethylene production and 1-aminocyclopropane-1-carboxylic acid conjugation in thermo-inhibited *Cicer arietinum* L. seeds. Plant Physiology 97:122–127.

Galuslaa, Y. 1984. Heat resistance and energy budget in different Scandinavian plants. Holarctic Ecology 7:1–78.

Gambrell, R. P., R. D. Dulane, and W. H. Patrick Jr. 1991. Redox processes in soils following oxygen depletion. Pp. 101–117 in: M. B. Jackson, D. D. Davis, and H. Lambers (eds.), Plant Life Under Oxygen Deprivation. SPB Academic Publishing, The Hague.

Garcia, F. G., and J. W. Einset. 1983. Ethylene and ethane production in 2,4-D treated and salt treated tobacco tissue cultures. Annals of Botany 51:287–295.

Gardeström, P., and G. E. Edwards. 1983. Isolation of mitochondria from leaf tissues of *Panicum miliaceum,* a NAD-malic enzyme type C_4 plant. Plant Physiology 71:24–29.

Gardner, R., K. Chonoles, and R. Owens. 1986. Potato spindle tuber viroid infections mediated by Ti-plasmid of *Agrobacterium tumefaciens.* Plant Molecular Biology 6:221–228.

Garten, C. T., and G. E. Taylor. 1992. Foliar $\delta^{13}C$ within a temperate forest: spatial, temporal, and species sources of variation. Oecologia 90:1–7.

Gartner, B. L. 1995. Plant Stems: Physiology and Functional Morphology. Academic Press, San Diego.

Gat, J. R. 1980. The isotopes of hydrogen and oxygen in precipitation. Pp. 21–47 in: P. Fritz and J. C. Fortes (eds.), Handbook of Environmental Isotope Geochemistry, Vol. 1, The Terrestrial Environment. Elsevier Publishing Company, Amsterdam.

Gates, D. M. 1962. Energy Exchange in the Biosphere. Harper & Row, New York.

Gates, D. M. 1980. Biophysical Ecology. Springer-Verlag, Berlin.

Gates, D. M. 1990. Climate change and forests. Tree Physiology 7:1–5.

Gates, D. M., and L. E. Papain. 1971. Atlas of Energy Budgets of Plant Leaves. Academic Press, New York.

Gauhl, E. 1976. Photosynthetic response to varying light intensity in ecotypes of *Solanum dulcamara* L. from shaded and exposed habitats. Oecologia 22:275–286.

Geiger, D. R., and S. A. Sovonick. 1975. Effects of temperature, anoxia and other metabolic inhibitors on translocation. Pp. 256–286, in: M. H. Zimmerman and J. A. Milburn (eds.) Transport in plants. I. Phloem transport (Encyclopedia of Plant Physiology, New Series, Vol. 1).

Geyh, M. A., and H. Schleichter. 1990. Absolute Age Determination: Physical and Chemical Dating Methods and Their Application. Springer-Verlag, Berlin.

Gholz, H., C. C. Grier, A. G. Campbell, and A. T. Brown. 1979. Equations for estimating biomass and leaf area of plants in the Pacific northwest (Research Paper 41). Oregon State University, School of Forestry, Corvallis, Oregon.

Giampietro, M. 1994. Sustainability and technological development in agriculture: a critical appraisal of genetic engineering. Bioscience 44:677–689.

Gibson, A. 1983. Anatomy of photosynthetic old stems of nonsucculent dicotyledons from North American deserts. Botanical Gazette 144(3):347–362.

Gibson, A. C., and P. S. Nobel. 1990. The Cactus Primer. Harvard University Press, Cambridge, MA.

Giddings, T. H. Jr., D. L. Brower, and L. A. Staehlin. 1980. Visualization of particle complexes in the plasma membrane of *Micvasterias denticulata* associated with the formation of cellulose fibrils in primary and secondary cell walls. Journal of Cell Biology 84:327–339.

Gildner, B. S., and D. W. Larson. 1992. Photosynthetic response to sunflecks in the desiccation-tolerant fern *Polypodium virginianum*. Oecologia 89:390–396.

Giles, K. L., D. Cohen, and M. F. Beardsell. 1976. Effects of water stress on the ultrastructure of leaf cells in *Sorghum bicolor*. Plant Physiology 57:11–14.

Gill D. S., and B. E. Mahall. 1986. Quantitative phenology and water relations of an evergreen and deciduous chaparral shrub. Ecological Monographs 56:127–143.

Gilmour, S. J., N. N. Artus, and M. F. Thomasho. 1992. Sequence analysis and expression of 2 cold regulated genes of *Arabidopsis thaliana*. Plant Molecular Biology 18:13–21.

Gogala, N. 1971. Growth substances in mycorrhiza of the fungus *Boletus pinicola* Vitt., and the pine tree, *Pinus sylvestris* L. Razprave XIV/5:123–202.

Goldstein, G., J. L. Andrade, and P. S. Nobel. 1991. Differences in water relations parameters for the chlorenchyma and parenchyma of *Opunntia ficus-indica* under wet versus dry conditions. Australian Journal of Plant Physiology 18:95–107.

Goldstein, G., F. Rada, L. Sternberg, J. L. Burguera, M. Burguera, A. Orozco, M. Montilla, O. Zabala, A. Azocar, M. J. Canales, and A. Celis. 1988. Gas exchange and water balance of a mistletoe species and its mangrove hosts. Oecologia 78:176–183.

Gombos, Z., J. Wada, and N. Murata. 1992. Unsaturation of fatty acids in membrane lipids enhances tolerance of the cyanobacterium *Synechocystis* PCC6803 to low-temperature photoinhibition. Proceedings of the National Academy of Sciences 89:9959–9963.

Goodman, R. N., Z. Király, and K. R. Wood. 1986. Pp. 245–286 in: The Biochemistry and Physiology of Plant Disease. University of Missouri Press, Columbia, MO.

Goodwin, P. B. 1978. Phytohormones and growth and development of organs of the vegetative plant. Pp. 31–173 in: D. S. Letham, P. B. Goodwin, and T. J. V. Higgins (eds.), Phytohormones and Related Compounds: A Comprehensive Treatise, Vol. II. Elsevier/North-Holland Biomedical Press, New York.

Goudriaan, J., and H. E. deRuiter. 1983. Plant growth in response to CO_2 enrichment, at two levels of nitrogen and phosphorus supply. 1. Dry matter, leaf area and development. Netherlands Journal of Agricultural Science 31:157–169.

Gould, K., S. Moreno, D. Owen, S. Sazer, and P. Nurse. 1991. Phoshorylation at Thr167 is required for *Schizosaccharomyces pombe* $p34^{cdc2}$ function. The European Molecular Biology Organization Journal 10:3297–3309.

Gowing, D. J. G., W. J. Davies, and H. G. Jones. 1990. A positive root-source signal as an indicator of soil drying in apple, *Malus* × *domestica,* Borkh. Journal of Experimental Botany 41:1535–1540.

Graham, D., and B. D. Patterson. 1982. Responses of plants to low, nonfreezing temperatures: proteins, metabolism, and acclimation. Annual Review of Plant Physiology 33:347–372.

Graham, D., D. G. Hockley, and B. D. Patterson. 1979. Temperature effects on phosphoenol pyruvate carboxylase from chilling sensitive and chilling resistant plants. Pp. 453–461 in: J. M. Lyons, D. Graham, and J. K. Raison (eds.), Low Temperature Stress in Crop Plants: The Role of the Membrane. Academic Press, New York.

Graves, W. R., and R. J. Gladon. 1985. Water stress, endogenous ethylene, and *Ficus benjamina* leaf abscission. HortScience 20:273–275.

Gray, J., and S. J. Song. 1984. Climatic implications of the natural variations of D/H ratios in tree ring cellulose. Earth Planet Science Letters 70:129–138.

Green, J. 1983. The effect of potassium and calcium on cotyledon expansion and ethylene evolution induced by cytokinins. Physiologia Plantarum 57:57–61.

Greenway, H., and R. Munns. 1980. Mechanisms of salt tolerance in non-halophytes. Annual Review of Plant Physiology 31:149–190.

Greitner, C. S., and W. E. Winner. 1988. Increases in $\delta^{13}C$ values of radish and soybean plants caused by ozone. New Phytologist 108:489–494.

Grime, J. P. 1977. Evidence for the existence of three primary strategies in plants and its relevance to ecological and evolutionary theory. American Naturalist 111:1169–1194.

Grimsley, N. H. 1990. Agroinfection. Physiologia Plantarum 79:147–153.

Grimsley N. H., T. Hohn, J. W. Davis, and B. Hohn. 1987. Agrobacterium-mediated delivery of infectious maize-streak virus into maize plants. Nature 325:177–179.

Grodzinski, B., and I. Boesel. 1982. The effect of light intensity on the release of ethylene from leaves. Journal of Experimental Botany 33:1185–1193.

Grodzinski, B., I. Boesel, and R. F. Horton. 1983. Light stimulation of ethylene release from leaves of *Gompherena globsoa* L. Plant Physiology 71:588–593.

Grosse, W., and C. Bauch. 1991. Gas transfer in floating-leaved plants. Vegetatio 97:185–192.

Grout, B. W. W. 1987. Higher plants at freezing temperatures. Pp. 293–314 in: B. W. W. Grout and G. J. Morris (eds.), The Effects of Low Temperatures on Biological Systems. Edward Arnold, London.

Gruber, M. Y., B. R. Glick, and J. Thompson. 1990. Cloned manganese superoxide dismutase reduces oxidative stress in *Escherichia coli* and *Anacystis nidulans.* Proceedings of the National Academy of Sciences USA 87:2608–2612.

Grumet, R., and A. D. Hanson. 1986. Genetic evidence for an osmoregulatory function of glycinebetaine accumulation in barley. Australian Journal of Plant Physiology 13:353–364.

Grunwald, C. 1971. Effects of free sterols, steryl ester, and steryl glycoside on membrane permeability. Plant Physiology 48:653–655.

Grunwald, C. 1974. Sterol molecular modifications influencing membrane permeability. Plant Physiology 54:624–628.

Grunwald, C. 1975. Plant sterols. Annual Review of Plant Physiology 26:209–236.

Grunwald, C. 1978. Shading influence on the sterol balance of *Nicotiana tobacum* L. Plant Physiology 61:76–79.

Guenther, A. B., R. K. Monson, and R. Fall. 1991. Isoprene and monoterpene emission rate variability: observations with eucalyptus and emission rate algorithm development. Journal of Geophysical Research 96:10799–10808.

Guinn, G., and D. L. Brummett. 1988. Changes in free and conjugated indole-3-acetic acid and abscisic acid in young cotton fruits and their abscission zones in relation to fruit retention during and after moisture stress. Plant Physiology 86:28–31.

Gupta, A. S., J. L. Heinen, A. S. Holaday, J. J. Burke, and R. D. Allen. 1993. Increased resistance to oxidative stress in transgenic plants that over-express chloroplastic Cu/Zn superoxide dismutase. Proceedings of the National Academy of Sciences USA 90:1629–1633.

Gutschick, V. B., M. H. Barron, D. A. Waechter, and M. A. Wolf. 1985. A portable monitor for solar radiation that accumulates irradiance histograms for 32 leaf-mounted sensors. Agricultural and Forestry Meteorology 33:281–290.

Guy, C. L. 1990. Cold acclimation and freezing stress tolerance: role of protein metabolism. Annual Review of Plant Physiology and Plant Molecular Biology 41:187–223.

Guy, C. L., and J. V. Carter. 1984. Characterization of partially purified glutathione reductase from cold-hardened and nonhardened spinach leaf tissue. Cryobiology 21:454–464.

Guy, R. D., M. F. Fogel, J. A. Berry, and T. C. Hoering. 1987. Isotope fractionation during oxygen production and consumption by plants. Pp. 597–600 in: J. Biggins (ed.), Photosynthetic Research, Vol. III. Martinus Nijhoff, Dordrecht, The Netherlands.

Guy, C. L., D. Haskell, L. Neven, P. Klein, and C. Smelser. 1992. Hydration-state responsive proteins link cold and drought stress in spinach. Planta 188:265–270.

Guy, R. D., D. M. Reid, and H. R. Krause. 1986. Factors affecting $^{13}C/^{12}C$ ratios of inland halophytes. II. Ecophysiological interpretation of patterns in the field. Canadian Journal of Botany 64:2400–2407.

Guye, M. G., L. Vigh, and J. M. Wilson. 1987. Chilling-induced ethylene production in relation to chill-sensitivity in *Phaseolus* spp. Journal of Experimental Botany 38:680–690.

Hadid, A. F. A., A. R. Smith, A. S. El-Beltagy, and M. A. Hall. 1986. Ethylene production by leaves of various plant species in response to stress imposition. Acta Horticulturae. 190:415–422.

Haeder, H. E., and H. Beringer. 1981. Influence of potassium nutrition and water stress on the content of abscisic acid in grains and flag leaves of wheat during grain development. Journal of the Science of Food and Agriculture 32:552–556.

Hahn, M., and V. Walbot. 1989. Effects of cold-treatment on protein synthesis and mRNA levels in rice leaves. Plant Physiology 91:930–938.

Hajela, R. K., D. P. Horvath, S. J. Gilmour, and M. F. Thomashow. 1990. Molecular cloning and expression of *cor* (cold-regulated) genes in *Arabidopsis thaliana.* Plant Physiology 93:1246–1252.

Hall, F., K. Huemmrich, and S. Goward. 1990. Use of narrow-band spectra to estimate the fraction of absorbed photosynthetically active radiation. Remote Sensing of the Environment 32(1):47–54.

Hall, M. A., J. A. Kapuya, S. Sivakumaran, and A. John. 1977. The role of ethylene in the response of plants to stress. Pesticide Science 8:217–223.

Halliwell, B., and J. M. C. Gutteridge. 1985. Free Radicals in Biology and Medicine. Claredon Press, Oxford.

Handley, L. L., and J. A. Raven. 1992. The use of natural abundance of nitrogen isotopes in plant physiology and ecology: commissioned review. Plant Cell & Environment 15:965–986.

Hanscom, Z., III, and I. P. Ting. 1978. Irrigation mannifies CAM-photosynthesis in *Opurtia basilgris* (Cactaceae). Oecologia 33:1–15.

Hanson, A. D., and W. D. Hitz. 1982. Metabolic responses of mesophytes to plant water deficits. Annual Review of Plant Physiology 33:163–203.

Hardisky, M. A., F. A. Daiber, C. T. Roman, and V. Klemas. 1984. Remote sensing of biomass and annual net aerial primary productivity of a salt marsh. Remote Sensing of the Environment 16:91–106.

Hartl, F. U., R. Hlodan, and T. Langer. 1994. Molecular chaperones in protein folding, the art of avoiding sticky situations. Trends in Biochemical Sciences 19:20–25.

Harwood, J. L. 1983. Adaptive changes in the lipids of higher-plant membranes. Biochemical Society Transactions 11:343–346.

Harwood, J. L., and A. L. Jones. 1989. Lipid metabolism in algae. Advances in Botanical Research. 16:1–53.

Harwood, J. L., A. L. Jones, H. J. Perry, A. J. Ruter, K. L. Smith, and M. Williams. 1994. Changes in plant lipids during temperature adaptation. Pp. 107–118 in: A. R. Cossins (ed.), Temperature Adaptation of Biological Membranes. Portland Press, London.

Hata, S., H. Kouchi, I. Suzaka, and T. Ishii. 1991. Isolation and characterization of cDNA clones for plant cyclins. The European Molecular Biology Organization Journal 10:2681–2688.

Hatcher, E. S. J. 1959. Auxin relations of the woody shoot. Annals of Botany 23:409–423.

Hattersley, P. W. 1982. $\delta^{13}C$ values of C_4 type in grasses. Australian Journal of Plant Physiology 9:139–154.

Hattersley, P. W. 1983. The distribution of C_3 and C_4 grasses in Australia in relation to climate. Oecologia 57:113–128.

Hayden, D. B., P. S. Covello, and N. R. Baker. 1988. Characterization of a 31-kDa polypeptide that accumulates in the light-harvesting apparatus of maize leaves during chilling. Photosynthesis Research 15:257–270.

Hazebroek, J. P., and J. D. Metzger. 1990. Thermoinductive regulation of gibberellin metabolism in *Thlaspi arvense* L. I. Metabolism of [2H]-ent-kaurenoic acid and [^{14}C] gibberellin A12-aldehyde. Plant Physiology 94:157–165.

Hazebroek, J. P., J. D. Metzger, and E. R. Maansager. 1993. Thermoinductive regulation of gibberellin metabolism in *Thlaspi arvense* L. II. Cold induction of enzymes in gibberellin biosynthesis. Plant Physiology 102:547–552.

He, C.-J., M. C. Drew, and P. W. Morgan. 1994. Induction of enzymes associated with lysigenous aerenchyma formation in roots of *Zea mays* during hypoxia or nitrogen starvation. Plant Physiology 105:861–865.

He, C.-J., P. W. Morgan, and M. C. Drew. 1992. Enhanced sensitivity to ethylene in nitrogen- or phosphate-starved roots of *Zea mays* L. during aerenchyma formation. Plant Physiology 98:137–142.

Heichman, K. A., and J. Roberts. 1994. Rules to replicate by. Cell 79:557–562.

Heino, P., G. Sandman, V. Long, K. Nordin, and E. T. Palva. 1990. Abscisic acid deficiency prevents development of freezing tolerance in *Arabidopsis thaliana* (L.) Heynh. Theoretical and Applied Genetics 79:801–806.

Hendry, G. A. F. 1993. Evolutionary origins and natural functions of fructans: a climatological, biogeographic and mechanistic appraisal. New Phytologist 123:3–14.

Henson, I. E. 1983. Effects of light on water stress-induced accumulation of abscisic acid in leaves and seedling shoots of pearl millet (*Pennisetum americanum* [L.] Leeke). Zeitschrift fuer Pflanzenphysiologic 112:257–268.

Herbert, T. J. 1984. Axial rotation of *Erythrina herbacea* leaflets. American Journal of Botany 71:76–79.

Herbert, T. J. 1991. Variation in interception of the direct solar beam by top canopy layers. Ecology 72:17–22.

Herms, D. A., and A. J. Mattson. 1992. The dilemma of plants: to grow or defend. Quarterly Review of Biology 67:283–335.

Hester, P., and W. Stillwell. 1984. Effect of plant growth substances on membrane permeability of urea and erythritol. Biochimica et Biophysica Acta 770:105–107.

Hewett, A. M., and P. F. Wareing. 1973. Cytokinins in *Populus* × *robusta* (Schneid): light effects on endogenous levels. Planta 114:119–129.

Hill-Braunchweig, S. 1993. The acclimation ability of the shale baren endemic *Eriogonum alleni* to light and heat. Ph.D. dissertation, Biology Department, Virginia Polytechnic Institute and State University, Blacksburg, VA.

Hinkley, T. M., F. Duhme, A. R. Hinckley, and H. Richter. 1980. Water relations of drought hardy shrubs: osmotic potential and stomatal reactivity. Plant Cell & Environment 3:131–140.

Hirose, T., M. J. A. Werger, and J. W. A. vanRheenen. 1989. Canopy development and leaf nitrogen distribution in a stand *of Carex acutiformis*. Ecology 70:1610–1618.

Hirsch, A. M., and J. G. Torrey. 1980. Ultrastructural changes in sunflower root cells in relation to boron deficiency and added auxin. Canadian Journal of Botany 58:856–866.

Hoekema, A., P. R. Hirsch, P. J. Hooykaas, and R. A. Schiperoort. 1983. A binary plant vector strategy based on separation of vir and T-region of the *Agrobacterium*. Nature 303:179–181.

Hoekstra, F. A., J. H. Crowe, and L. M. Crowe. 1989. Membrane behavior in drought and its physiological significance. Pp.71–88 in: R. B. Taylorson (ed.), Recent Advances in the Development and Germination of Seeds. Plenum Press, New York.

Hoffman-Benning, S., and H. Kende. 1992. On the role of abscisic acid and gibberellin in the regulation of growth of rice. Plant Physiology 99:1156–1161.

Holbrook, N. M., and T. R. Sinclair. 1992. Water balance in the arborescent palm, *Sabal palmetto*. II. Transpiration and stem water storage. Plant Cell & Environment 15:401–409.

Holbrook, N. M. 1995. Stem water storage. Pp. 151–169 in: B. Gartner (ed.), Stems and Trunks in Plant Form and Function. Academic Press, San Diego, CA.

Holden, M. J., and G. W. Patterson. 1982. Taxonomic implication of sterol composition in the genus *Chlorella*. Lipids 17:215–219.

Hollinger, D. Y. 1989. Canopy organization and foliage photosynthetic capacity in a broad leaved evergreen montane forest. Functional Ecology 3:53–62.

Hook, D. D. 1984. Adaptations to flooding with fresh water. Pp. 265–294 in: T. T. Kozlowski (ed.), Flooding and Plant Growth. Academic Press, New York.

Hopkin, K. A., M. A. Papazian, and H. M. Steinman. 1992. Functional differences between manganese and iron superoxide dismutases in *Escherichia coli* K-12. Journal of Biological Chemistry 267:24253–24258.

Hopkins, W. G. 1995. Introduction to Plant Physiology. John Wiley & Sons, New York.

Horgan, J. M., and P. F. Wareing. 1980. Cytokinins and the growth responses of seedlings of *Betula pendulatha*., and *Acer pseudoplatanus* L. to nitrogen and phosphorus deficiency. Journal of Experimental Botany 31:525–532.

Houssa, C., A. Jacqmard, and G. Bernier. 1990. Activation of replicon origins as a possible target for cytokinins in shoot meristems of *Sinapis*. Planta 181:324–326.

Howard, R. V., and D. M. Orcutt. 1976. Investigation of the lipid composition of three freshwater *Rhodophyta*. Pp. 35–49 in: B. C. Parker and M. K. Roane (eds.), The Distributional History of the Biota of the Southern Appalachians. IV. Algae and Fungi, Biogeography, Systematics, and Ecology. University Press of Virginia, Charlottesville, VA.

Howarth, C. J. 1990. Heat shock proteins in sorghum and pearl millet; ethanol, sodium arsenite, sodium malonate and the development of thermotolerance. Journal of Experimental Botany 41:877–883.

Hsiao, T. C. 1973. Plant responses to water stress. Annual Review of Plant Physiology 24:519–570.

Hsiao, T. C., E. Acevedo, E. Fereres, and D. W. Henderson. 1976. Water stress growth, and osmotic adjustment. Philosophical Transactions of the Royal Society of London, Series B, 278:479–500.

Huberman, M., E. Pressman, and M. J. Jaffe. 1993. Pith autolysis in plants. IV. The activity of polygalacturonase and cellulase during drought stress induced pith autolysis. Plant Cell Physiology 34:795–801.

Hubick, K. T., and G. D. Farquhar. 1987. Carbon isotope discrimination: selecting for water use efficiency. Australian Cotton Grower 8:66–68.

Hubick, K. T., and G. D. Farquhar. 1989. Genetic variation of transpiration efficiency among barley genotypes is negatively correlated with carbon isotope discrimination. Plant Cell & Environment. 12:795–804.

Hubick, K. T., G. D. Farquhar, and R. Shorter. 1986a. Correlation between water-use-efficiency and carbon isotope discrimination in diverse peanut (*Arachis*) germplasm. Australian Journal of Plant Physiology 13:803–816.

Hubick, K. T., R. Shorter, and G. D. Farquhar. 1988. Heritability and genotype × environment interactions of carbon isotope discrimination and transpiration efficiency in peanut. Australian Journal of Plant Physiology 15:799–813.

Hubick, K. T., J. S. Taylor, and D. M. Reid. 1986b. The effect of drought on levels of abscisic acid, cytokinins, gibberellins and ethylene in aeroponically-grown sunflower plants. Plant Growth Regulation 4:139–151.

Hunt, P. G., R. B. Campbell, R. E. Sojka, and J. E. Parsons. 1981. Flooding-induced soil and plant ethylene accumulation and water status response of field-grown tobacco. Plant and Soil 59:427–439.

Hunt, R. 1978. Plant Growth Analysis (Studies in Biology, No. 96). Edward Arnold, London.

Hunt, R. 1979. Plant growth analysis: the rational behind the use of the fitted mathematical function. Annals of Botany 43:245–249.

Hüsken, D., E. Steudle, and O. Zimmerman. 1978. Pressure probe techniques for measuring water relations in cells of higher plants. Plant Physiology 61:158–163.

Huskies, H. L., and J. L. Harper. 1979. The demography of leaves and tillers of *Ammophila arenaria* in a sand dune sere. Oecologia Plantarum 14:435–446.

Hutchison, B. A., D. R. Matt, and R. T. McMillen. 1980. Effects of sky brightness distribution upon penetration of diffuse radiation through canopy gaps in a deciduous forest. Agricultural and Forestry Meteorology 22:137–147.

Hwang, S-Y., and T. T. Van Toai. 1991. Abscisic acid induces anaerobiosis tolerance in corn. Plant Physiology 97:593–597.

Ilker, R., R. W. Breidenbach, and J. M. Lyons. 1979. Sequence of ultrastructural changes in tomato cotyledons during short periods of chilling. Pp. 97–113 in: J. M. Lyons, D. Graham, and J. K. Raison (eds.), Low Temperature Stress in Crop Plants: The Role of the Membrane. Academic Press, New York.

Ingemarsson, B. S. M., E. Lundqvist, and L. Eliasson. 1991. Seasonal variation in ethylene concentration in the wood of *Pinus sylvestris* L. Tree Physiology 8:273–279.

Iturriaga, G., K. Schneider, F. Salamini, and D. Bartels. 1992. Expression of desiccation-related proteins from the resurrection plant *Craterostigma plantagineum* in transgenic tobacco. Plant Molecular Biology 20:555–558.

Jackman, R. L., R. Y. Yada, A. Marangoni, K. L. Parkin, and D. W. Stanley. 1988. Chilling injury: a review of quality aspects. Journal of Food Quality 11:253–278.

Jackson, M. B., S. F. Young, and K. C. Hall. 1988. Are roots a source of abscisic acid for the shoots of flooded pea plants? Journal of Experimental Botany 39:1631–1637.

Jackson, W. T. 1956. Flooding injury studied by approach-graft and split root systems. American Journal of Botany 43:496–502.

Jacobs, T. 1992. Control of the cell cycle. Developmental Biology 153:1–15.

Jacobs, T. W. 1995. Cell cycle control. Annual Review of Plant Physiology and Plant Molecular Biology 46:317–339.

Johnson, A. H. 1988. Red spruce decline. Pp.100–103 in: B. Krahl-Urban, H. E. Papke, K. Peters, and C. Schimansky (eds.). Forest Decline. U.S. Environmental Protection Agency, Corvalis, Oregon.

Johnson, D. B. 1976. A survey of crop-weather models with data on application of selected models to Centre County, Pennsylvania. M.S. Paper, Deptartment of Mereorology, Pennsylvania State. University, University Park, PA.

Johnson, D. G., Schwartz, J. K., Cress, W. D., and J. R. Nevens. 1993. Expression of the transcription factor E2F1 induces quiescent cells to enter S-phase. Nature 365:349–352.

Johnson, P. L., and D. M. Atwood. 1970. Aerial sensing and photographic study of the El Verde rain forest. Pp. B63–B78 in: H. T. Odum and P. F. Pigeon (eds.), A Tropic Rain Forest: A Study of Irradiation and Ecology at El Verde, Puerto Rico, Book 1. Office of Information Services, U.S. Atomic Energy Commission, Oak Ridge, TN.

Jones, A. M., D. S. Cochran, P. M. Lamerson, M. L. Evans, and J. D. Cohen. 1991. Red light-regulated growth. I. Changes in the abundance of indoleacetic acid and a 220 kilodalton auxin-binding protein in the maize mesocotyl. Plant Physiology 97:352–358.

Jones, C. S. 1984. The effect of axis splitting on xylem pressure potentials and water movement in the desert shrub *Ambrosia dumosa* (Gray) Payne (Asteraceae). Botanical Gazette 145:125–131.

Jones, H. G. 1992. Plants and Microclimate, 2nd ed. Cambridge University Press, Cambridge.

Jones, R. A., and S. O. El-Abd. 1989. Prevention of salt-induced epinasty by α-aminooxyacetic acid and cobalt. Plant Growth Regulation 8:315–323.

Joslin, J. D., and G. S. Henderson. 1984. The determination of percentages of living tissue in woody fine root samples using triphenyltetrazolium chloride. Forest Science 30:965–970.

Junk, G., and H. V. Suek. 1958. The absolute abundance of the nitrogen isotopes in the atmosphere and compressed gas from various sources. Geochemica Cosmochimica Acta 14:234–243.

Jurik, T. W., Z. Hanzhong, and J. Pleasants. 1990. Ecophysiological consequences of non-random leaf orientation in the prairie compass plant, *Silphium laciniatum*. Oecologia 82:180–186.

Justin, S. H. F. W., and W. Armstrong. 1987. The anatomical characteristics of roots and plant response to soil flooding. New Phytologist 105:465–495.

Kaber, K., and S. Baltepe. 1990. Effects of kinetin and gibberellic acid in overcoming high temperature and salinity (NaCl) stresses on the germination of barley and lettuce seeds. Phyton 30:65–74.

Kaku, S., and M. Iwaya-Inoue. 1987. Estimation of chilling sensitivity and injury in *Gloxinia* leaves by the thermal historesis of NMR relaxation times of water protons. Plant Cell Physiology 28:509–516.

Kaku, S., and M. Iwaga-Inove. 1990. Factors affecting the prolongation of NMR relaxation times of water protons in leaves of woody plants affected by formation of insect galls. Plant Cell Physiology 31:627–637.

Kannangara, T., R. C. Durley, and G. M. Simpson. 1982. Diurnal changes of leaf water potential, abscisic acid, phaseic acid and indole-3-acetic acid in field grown *Sorghum bicolor* L. Moench. Zeitzgrift für Pflanzenphysiology 106:55–61.

Kannangara, T., N. Seetharama, R. C. Durley, and G. M. Simpson. 1983. Drought resistance of *Sorghum bicolor*. 6. Changes in endogenous growth regulators of plants grown across an irrigation gradient. Canadian Journal of Plant Sciences 63:147–155.

Kao, C. H., and F. S. Yang. 1982. Light inhibition of the conversion aminocylcopropane-1-carboxylic acid to ethylene in leaves is mediated through carbon dioxide. Planta 155:262–266.

Kappen, L., G. Schultz, and R. Vanselow. 1994. Direct observations of stomatal movements.

Pp. 231–246 in: E. D. Shulze and M. M. Caldwell (eds)., Ecophysiology of Photosynthesis. Springer-Verlag, Berlin.

Karamanos, R. E., R. P. Voroney, and D. A. Rennie. 1981. Variation in natural ^{15}N abundance of central Saskatchewan soils. Soil Science Society of America Journal 45:826–828.

Kasamo, K., and T. Shimomura. 1977. The role of the epidermis in local lesion formation and the multiplication of tobacco mosaic virus and its relation to kinetin. Virology 76:12–18.

Kasperbauer, M. J. 1994. Light and plant development. Pp. 83–124 in: R. E. Wilkinson (ed.), Plant–Environment Interactions. Marcel Dekker, New York.

Kaufmann, K. W. 1981. Fitting and using growth curves. Oecologia 49:293–299.

Kawanabe, Y., H. Yamane, T. Murayama, N. Takahashi, and T. Nakamura. 1988. Identification of gibberellin A3 in mycelia of *Neurospora crassa*. Agricultural and Biological Chemistry 49:2447–2450.

Keeley, J. E. 1991. Interactive role of stresses on structure and function in aquatic plants. Pp. 329–343 in: H. A. Mooney, W. W. Winner, and E. Pell (eds.), Response of plants to multiple stresses. Academic Press, San Diego.

Keeley, J. E., and B. A. Morton. 1982. Distribution of diurnal acid metabolism in submerged plants outside the genus *Isoetes*. Photosynthetica 16:546–553.

Keeling, C. D., W. M. Mook, and P. P. Tans. 1979. Recent trends in the $^{13}C/^{12}C$ ratio of atmospheric carbon dioxide. Nature 277:121–123.

Kemp, P. L. 1983. Phenological patterns of Chihuahuan desert plants in relation to the timing of water availability. Journal of Ecology 71:427–436.

Kennedy, R. A., T. C. Fox, J. D. Everard, and M. Rumpho. 1991. Biochemical adaptations to anoxia: potential role of mitochondrial metabolism to flood tolerance in *Echinolcloa phyllopogon* (barnyard grass). Pp. 217–227 in: M. B. Jackson, D. D. Davis, and H. Lambers (eds.), Plant Life Under Oxygen Deprivation. SPB Academic Publishing, The Hague.

Ketchie, D. O., and R. Kammereck. 1987. Seasonal variation of cold resistance in *Malus* woody tissue as determined by differential thermal analysis and viability tests. Canadian Journal of Botany 65:2640–2645.

Khan, A. A., and X.-L. Huang. 1988. Synergistic enhancement of ethylene production and germination with kinetin and 1-aminocyclopropane-1-carboxylic acid in lettuce seeds exposed to salinity stress. Plant Physiology 87:847–852.

Khan, A. A., R. Thakur, M. Akbar, D. HilleRisLambers, and D. V. Seshu. 1987. Relationship of ethylene production to elongation in deepwater rice. Crop Science 27:1188–1196.

Kim, K.-Y., and H. Craig. 1990. Two-isotope characterizations of N_2O in the Pacific Ocean and constraints on its origin in deep water. Nature 347:58–61.

King, R. W., P. K. Jackson, and M. W. Kirshner. 1994. Mitosis in transition. Cell 79:563–571.

Klee, H. J., M. B. Hayford, K. A. Kretzmer, G. F. Barry, and G. M. Kishore. 1991. Control of ethylene synthesis by expression of a bacterial enzyme in transgenic tomato plants. The Plant Cell 3:1187–1193.

Klee, H., R. Horsche, and S. Rogers. 1987. *Agrobacterium*-mediated plant transformation and its further applications to plant biology. Annual Review of Plant Physiology 38:467–486.

Klöti, A. V., A. Iglesias, J. Wünn, P. K. Burkhardt, S. K. Datta, and I. Potrykus. 1993. Gene transfer by electroporation into intact scutellum cells of wheat embryos. Plant Cell Reproduction 12:671–675.

Knapp, A. K. 1992. Leaf gas exchange of *Quercus macrocarpa* (Fagaceae) rapid stomatal responses to variability in sunlight in a tree growth form. American Journal of Botany 79:599–604.

Knapp, A. K., and W. K. Smith. 1987. Stomatal and photosynthetic responses during sun and shade transitions in subalpine plants: influence on water use efficiency. Oecologia 74:62–67.

Knapp, A. K., and W. K. Smith. 1989. Influence of growth form on ecophysiological response to variable sunlight in subalpine plants. Ecology 70:1069–1082.

Knapp, A. K., and W. K. Smith. 1990. Contrasting stomatal response to variable sunlight in two subalpine herbs. American Journal of Botany 77:226–231.

Kohl, D. H., and G. Shearer. 1980. Isotopic fractionation associated with symbiotic N_2-fixation and uptake of NO_2^- by plants. Plant Physiology 66:52–56.

Kolb, K. J., and S. D. Davis. 1993. Drought-induced xylem embolism in co-occurring species of coastal sage and chaparral of California. Ecology: 75:648–659.

Konings, H., and H. Lambers. 1991. Respiratory metabolism, oxygen transport, and the induction of aerenchyma in roots. Pp. 247–265 in: M. B. Jackson, D. D. Davis, and H. Lambers (eds.), Plant Life Under Oxygen Deprivation. SPB Academic Publishing, The Hague.

Körner, C., and W. Larcher. 1988. Plant life in cold climates. Pp. 25–57 in: S. F. Long and F. I. Woodward (eds.), Plants and Temperature. Company of Biologists Limited, Cambridge.

Koster, K. L. 1991. Glass formation and desiccation tolerance in seeds. Plant Physiology 96:302–304.

Kozlowski, T. T. 1984. Flooding and Plant Growth. Academic Press, New York.

Kraepiel, Y., P. Rousselin, B. Sotta, L. Kerhoas, J. Einhorn, M. Caboche, and E. Miginiac. 1994. Analysis of phytochrome-and ABA-deficient mutants suggest that ABA degradation is controlled by light in *Nicotiana plumbaginifolia*. The Plant Journal 6:665–672.

Kramer, P. J. 1951. Causes of injury to plants resulting from flooding of the soil. Plant Physiology 26:722–736.

Kubik, M. P., J. G. Buta, and C. Y. Wang. 1992. Changes in the levels of abscisic acid and its metabolites resulting from chilling of tomato fruits. Plant Growth Regulation 11:429–434.

Kuiper, D. 1983. Genetic differentiation in *Plantago major:* growth and root respiration and their role in phenotypic adaptation. Physiologia Planturum 57:222–230.

Kuiper, D. 1988. Growth responses of *Plantago major* L. ssp. *pleiosperma* (Pilger) to changes in mineral supply. Plant Physiology 87:555–557.

Kuiper, D., J. Schuit, and P. J. C. Kuiper. 1989. Effects of internal and external cytokinin concentrations on root growth and shoot to root ratio of *Plantago major* ssp. *pleiosperma* at different nutrient conditions. Pp. 183–188 in: B. C. Loughman, O. Gaspariková, and J. Kolek (eds.), Structural and Functional Aspects of Transport in Roots. Kluwer Academic Publishers, London.

Kull, O., and U. Niinemets. 1993. Variation in leaf morphometry and nitrogen concentration in Betula pendula Roth., *Corylus avellana* L., and *Lonicera xylosteum* L. Tree Physiology 12:311–318.

Kumar, P. K. R., and B. K. Lonsane. 1990. Solid state fermentation: physical and nutritional factors influencing gibberellic acid production. Applied Microbiology and Biotechnology 34:145–148.

Kummerow, J., D. Krause, and W. Jow. 1977. Root systems of chapparal shrubs. Oecologia 29:163–177.

Kuo, C. G., and C. T. Tsai. 1984. Alternation by high temperature of auxin and gibberellin concentrations in the floral buds, flowers, and young fruit of tomato. HortScience. 19:870–872.

Laan, P., M. Tosserams, C. W. P. M. Bloom, and B. W. Veen. 1990. Internal oxygen transport in *Rumex* species and its significance for respiration under hypoxic conditions. Plant and Soil 122:39–46.

Lacheene, Z., A. El-S., and A. S. El-Beltagy. 1986. Tomato fruit growth pattern and endogenous ethylene, indoleacetic acid and abscisic acid under normal and stress conditions. Acta Horticultrae. 190:325–338.

Lamb, B., D. Gay, H. Westberg, and T. Pierce. 1993. A biogenic hydrocarbon emission inventory for the USA using a simple forest canopy model. Atmospheric Environment 27A:1673–1690.

Lambers, H. 1980. The physiological significance of cyanide-resistant respiration. Plant Cell & Environment 3:293–302.

Lambers, H. 1982. Cyanide resistant respiration: a non-phosphorylating electron transport pathway acting as an energy overflow. Physiologia Plantarum 55:478–485.

Lambers, H. 1985. Respiration in intact plants and tissues. Pp. 418–473 in: R. Douce and A. Day (eds.) Higher Plant Cell Respiration. Springer-Verlag, Berlin.

Lambers, H., and E. Steingröver. 1978. Efficiency of root respiration of a flood-tolerant and flood-intolerant *Senecio* species as affected by low oxygen tension. Physiologia Plantarum 42:179–184.

Lamoreaux, R. J., W. R. Chaney, and K. M. Brown. 1978. The plastochron index: a review after two decades of use. American Journal of Botany 65:586–593.

Landry, J., P. Chretien, H. Lambert, E. Hickley, and L. A. Weber. 1989. Heat shock resistance conferred by expression of the human HSP27 gene in rodent cells. Journal of Cell Biology 109:7–15.

Landry, J., D. Bernier, P. Chretien, L. M. Nicole, R. M. Tanguay, and N. Marceau. 1982. Synthesis and degradation of heat shock proteins during development and decay of thermotolerance. Cancer Research 42:2457–2461.

Lång, V., and E. T. Palva. 1992. The expression of a rab-related gene, *rab*18, is induced by abscisic acid during the cold acclimation process of *Arabidopsis thaliana* (L.) Heynh. Plant Molecular Biology 20:951–962.

Lange, O. L. 1967. Investigations on the variability of heat resistance in plants. Pp. 131–141 in: A. S. Troshin (ed.) The Cell and Environmental Temperature. Pergamon Press, Oxford.

Lange, O. L., R. Lösch, E. D. Schulze, and L. Kappen. 1971. Responses of stomata to changes in humidity. Planta 100:76–86.

Larcher, W. 1980. Physiological Plant Ecology. Springer-Verlag, New York.

Larigauderie, A. D., W. Hilbert, and W. C. Oechel. 1988. Interaction between high CO_2 concentrations and multiple environmental stresses in *Bromus molis*. Oecologia 77:544–549.

Larson, K. D., B. Schaffer, and F. S. Davies. 1993. Floodwater oxygen content, ethylene production and lenticel hypertrophy in flooded mango (*Mangifera indica* L.) trees. Journal of Experimental Botany 44:665–671.

Larson, P. R. 1963. The indirect effect of drought on tracheid diameter in red pine. Forest Science 9:52–62.

Lauenroth, W. K., J. L. Dodd, and P. L. Simms. 1978. The effects of water and nitrogen induced stresses on plant community structure in a semiarid grassland. Oecologia 36:211–222.

Leaney, F. W., C. B. Osmond, G. B. Allison, and H. Ziegler. 1985. Hydrogen isotope composition of leaf water in C_3 and C_4 plants: its relationship to the hydrogen isotope composition of dry matter. Planta 164:215–220.

Leavitt, S. W., and A. L. Long. 1982. Evidence for $^{13}C/^{12}C$ fractionation between tree leaves and wood. Nature 298:742–744.

Ledgard, S. F., J. R. Freney, and J. R. Simpson. 1984. Variations in natural enrichment of ^{15}N in the profiles of some Australian pasture soils. Australian Journal of Soil Research 22:155–164.

Ledgard, S. F., R. Mortan, J. R. Freney, and P. J. Bergersen. 1985. Assessment of the relative uptake of added and indigenous soil nitrogen by nodulated legumes and reference plants in the ^{15}N dilution measurement of N_2 fixation: derivation of the method. Soil Biology and Biochemistry 17:317–321.

Lee, A. H. 1983. Lipidphase transitions and mixtures. Pp. 43–88 in: R. C. Aloia (ed.) Membrane Fluidity in Biology, Vol. 2. Academic Press, New York, NY.

Lee, K. H., and T. A. LaRue. 1992. Ethylene as a possible mediator of light- and nitrate-induced inhibition of nodulation of *Pisum sativum* L. cv. *sparkle*. Plant Physiology 100:1334–1338.

Lefkovitch, L. 1965. The study of population growth in organisms grouped by stages. Biometrics 21:1–18.

Lenaz,G., and G. P. Castelli. 1985. Membrane fluidity: molecular basis and physiological significance. Pp. 93–136 in: G. Benga (ed.), Structure and Properties of Cell Membranes: A Survey of Molecular Aspects of Membrane Structure and Function, Vol. I. CRC Press, Boca Raton, FL.

Leopold, A. C. 1990. Coping with desiccation. Pp. 37–56 in: R. G. Alscher and J. R. Cummings (eds.), Stress Responses in Plants: Adaptation and Acclimation Mechanisms. Wiley-Liss, New York.

Leopold, A. C., and M. E. Musgrave. 1980. Respiratory pathways in aged soybean leaves. Physiologia Plantarum 49:49–54.

Lerdau, M. 1991. Plant function and monoterpene emission. Pp. 121–134 in: T. Sharkey, E. Holland, and H. A. Mooney (eds.), Trace Gas Emissions from Plants. Academic Press, San Diego, CA.

Leshem, Y. Y. 1992. Plant Membranes: a biophysical approach to structure, development and senescence, Kluwer Academic Publisher, Boston, p. 266.

Leslie, P. 1945. The use of matrices in certain population mathematics. Biometrika 33:183–212.

Letham, D. 1973. Cytokinins from *Zea maize*. Phytochemistry 12:2445–2455.

Levitt, J. 1980. Responses of Plant to Environmental Stresses, Vol. 1. Academic Press, New York.

Levitt, J. 1982. Stress Terminology. Pp. 437–439 in: N. C. Turner and P. J. Kramer (eds.), Adaptations of Plants to Water and High Temperature Stress. Wiley-Interscience, New York.

Lieth, H. 1973. Primary production: terrestrial ecosystems. Human Ecology 1:303–332.

Lin, C., and Thomashow, M. F. 1992. A cold-regulated *Arabidopsis* gene encodes a polypeptide having potent cryoprotective activity. Biochemical and Biophysical Research Communications 183:1103–1108.

Lin, C. H., and C. H. Lin. 1992. Physiological adaptation of waxapple to waterlogging. Plant Cell & Environment 15:321–328.

Lindeman, W. 1979. Inhibition of photosynthesis in *Lemna minor* by illumination during chilling in the presence of oxygen. Photosynthetica 13:175–185.

Lindow, S. E. 1995. Control of epiphytic ice nucleation-active bacteria for management of plant frost injury. Pp. 239–256 in: R. E. Lee, Jr., G. J. Warren, and L. V. Gusta (eds.), Biological Ice Nucleation and Its Applications. APS Press, St. Paul, MN.

Ling, G. N., and M. Tucker. 1980. Nuclear magnetic resonance relaxation and water contents in normal mouse and rat tissues and cancer cells. Journal of the National Cancer Institute 64:1199–1207.

Lipscomb, M. V. 1991. The response of four ecicaceous shrubs to multiple environmental resource variation. Ph.D. dissertation, Biology Department, Virginia Polytechnic Institute and State University, Blacksburg, VA.

List, R. J. 1968. Smithsonian Meteorological Tables, 6th rev. ed. Smithsonian Institution Press, Washington, DC.

Little, C. H. A. 1975. Inhibition of cambial activity in *Abies balsamea* by internal water stress: role of abscisic acid. Canadian Journal of Botany 53:3041–3050.

Little, C. H. A., and P. F. Wareing. 1981. Control of cambial activity and dormancy in *Picea sitchensis* by indole-3-acetic and abscisic acids. Canadian Journal of Botany 59:1480–1493.

Liu, P., and J. Loy. 1976. Action of gibberellic acid on cell proliferation in the subapical shoot meristem of watermelon seedlings. American Journal of Botany 63:700–704.

Liu, Z., and D. I. Dickmann. 1992. Abscisic acid accumulation in leaves of contrasting hybrid poplar clones affected by nitrogen fertilization plus cyclic flooding and soil drying. Tree Physiology. 11:109–122.

Llewellyn, D. J., E. J. Finnegan, J. G. Ellis, E. S. Dennis, and W. J. Peacock. 1987. Structure and expression of an alcohol dehydrogenase 1 gene from *Pisum sativum* (cv. 'Greenfast'). Journal of Molecular Biology 195:115–123.

Lo Gullo, M. A., and S. Salleo. 1992. Water storage in the wood and xylem cavitation in 1-year-old twigs of *Populus deltoides*. Plant Cell & Environment 15:431–438.

Lobréaux, S., T. Hardy, and J.-F. Briat. 1993. Abscisic acid is involved in the iron-induced synthesis of maize ferritin. The European Molecular Biology Organization Journal 12:651–657.

Lockhart, J. 1965. An analysis of irreversible plant cell elongation. Journal of Theoretical Biology 8:264–275.

Lodish, H., D. Baltimore, A. Berk, S. L. Zipursky, P. Matsudaira, and J. Darnell. 1995. Molecular Cell Biology, 3rd edition. W. H. Freeman, NY.

Loik, M. E., and P. S. Nobel. 1991. Water relations and mucopolysaccharide increases for a winter hardy cactus during acclimation to subzero temperatures. Oecologia 88:340–346.

Loik, M. E., and P. S. Nobel. 1993. Exogenous abscisic acid mimics cold acclimation for cacti differing in freezing tolerance. Plant Physiology 103:871–876.

Loomis, R. S., and G. E. Worker, Jr. 1963. Responses of sugar beet to low moisture at two levels of nitrogen nutrition. Agronomy Journal 55:509–516.

Loomis, W. D., and R. Croteau. 1973. Biochemistry and physiology of lower terpenoids. Recent Advances in Phytochemistry 6:147–185.

Loreth, F., and T. D. Sharkey. 1990. A gas exchange study of photosynthesis and isoprene emission in red oak (*Quacur rubia* L.). Planta 182:523–531.

Lösch, R. 1993. Plant Water Relations. Progress in Botany 54:102–133.

Lovett, G. M. 1994. Atmospheric deposition of nutrients and pollutants in North America: an ecological perspective. Ecological Applications 4:629–650.

Loveys, B. R. 1979. The influence of light quality on levels of abscisic acid in tomato plants, and evidence for a novel abscisic acid metabolite. Physiologia Plantarum 46:79–84.

Lu, D. B., R. G. Sears, and G. M. Paulsen. 1989. Increasing stress resistance by in vitro selection for abscisic acid insensitivity in wheat. Crop Science 29:939–943.

Lundgarth, H. 1960. Anion respiration. Pp. 185–233 in: W. Rutland (ed.), Encyclopedia of Plant Physiology, Vol. XII/2. Springer-Verlag, Berlin.

Lundgren, K., N. Walworth, R. Booher, M. Dembski, M. Dirschner, and D. Beach. 1991. *mik*1 and *wee*1 cooperate in the inhibitory tyrosine phosphorylation of *cdc*2. Cell 64:1111–1122.

Luo, M., J.-H. Liu, S. Mohapatra, R. D. Hill, and S. S. Mohapatra. 1992. Characterization of a gene family encoding abscisic acid– and environmental stress–inducible proteins of alfalfa. Journal of Biological Chemistry 267:15367–15374.

Luo, Y., and L. S. L. Sternberg. 1991. Deuterium heterogeneity in starch and cellulose nitrate of CAM and C_3 plants. Phytochemistry 30:1095–1098.

Lurie, S., and J. D. Klein. 1991. Acquisition of low-temperature tolerance in tomatoes by exposure to high-temperature stress. Journal of the American Society of Horticultural Science 116:1007–1012.

Lyons, J. M., and J. K. Raison. 1970. Oxidative activity of mitochonria isolated from plant tissues sensitive and resistant to chilling injury. Plant Physiology 45:386–389.

MacFall, J. S., P. Spaine, R. Doudrick, and G. A. Johnson. 1994. Alterations in growth and water transport processes in fusiform rust galls of pine as determined by magnetic resonance microscopy. Phytopathology 84:288–293.

MacFall, J. S., F. W. Wehrli, R. K. Greger, and G. A. Johnson. 1987. Methodology for the measurement and analysis of relaxation times in proton imaging. Magnetic Resonance Imaging 5:209–220.

Macintosh, D. J. 1983. Riches lie in tropical swamps. Geographical Magazine 55:184–188.

Maillette, L. 1982. Structural dynamics of silver birch. II. A matrix model of the bud population. Journal of Applied Ecology. 19:219–238.

Mair-Maerker, U. 1979. "Peristomatal transpiration" and stomatal movement: a controversial view. I. Additional proof of peristomatal transpiration by hydrophotography and a comprehensive discussion in the light of recent results. Zeitzgrift fur Pflanzen Physiology 91:25–43.

Maltby, E. 1991. Wetlands: their status and role in the biosphere. Pp. 3–21 in: M. B. Jackson, D. D. Davis, and H. Lambers (eds.), Plant Life Under Oxygen Deprivation. SPB Academic Publishing, The Hague.

Maltby, E., and R. E. Turner. 1983. Wetlands of the world. Geographical Magazine 55:12–17.

Mansfield, T. A., and P. H. Freer-Smith. 1984. The role of stomata in resistance mechanisms. Pp. 131–146 in: M. J. Koziol and F. R. Whatley (eds.), Gaseous Air Pollutants and Plant Metabolism. Butterworth, London.

Mao, Z., L. E. Craker, and D. R. Decoteau. 1989. Abscission in coleus: light and phytochrome control. Journal of Experimental Botany 220:1273–1277.

Margaris, N. S. 1975. Effect of photoperiod on seasonal dimorphisim of some Mediterranean plants. Berlin Schweiz Botanic Gessette 96–102.

Margaris, N. S. 1977. Physiological and biochemical observations in seasonal dimorphic leaves of *Sarcopoterium spinorum* and *Phlomis fruticosa*. Oecological Plant Physiology 12:343–350.

Markhart III, A. H. 1986. Chilling injury: A review of possible causes. HortScience 21:1329–1333.

Mariotti, A. 1983. Atmospheric nitrogen is a reliable standard for natural ^{15}N abundance measurements. Nature 303:685–687.

Mariotti, A., F. Mariotti, N. Amarger, G. Pizelle, J. M. Ngambi, M. L. Champigney, and A. Moyse. 1980. Fractionnements isotopiques de l'azote lors des processus d'absorption des nitrates et de fixation de l'azote atmosphérique par les plants. Physiology Vegetal 18:163–181.

Martin, B., and Y. R. Thorstenson. 1988. Stable carbon isotope composition (^{13}C), water use efficiency, and biomass productivity of *Lycopersicon esculentum, Lycopersicon pennelli,* and the F_1 hybrid. Plant Physiology 88:213–217.

Martin, G. J., M. L. Martin, and B.-L. Zhang. 1992. Site specific natural isotope fractionation of hydrogen in plant products studied by nuclear magnetic resonance. Plant Cell & Environment 15:1037–1050.

Martin, M. L., G. L. Martin, and C. Guillou. 1991. A site specific and multielement isotopic approach to origin inference of sugars in foods and beverages. Mikrochemica Acta II 81–91.

Masle, J., and G. D. Farquhar. 1988. Effects of soil strength on the relation of water-use efficiency and growth to carbon isotope discrimination in wheat seedlings. Plant Physiology 86:32–38.

Matheussen, A.-M., P. W. Morgan, and R. A. Frederiksen. 1991. Implication of gibberellins in head smut (*Sporisorium reilianum*) of sorghum bicolor. Plant Physiology 96:537–544.

Mathur-DeVre, R. 1984. Biochemical implications of the relaxation behaviour of water related to NMR imaging. British Journal of Radiology 57:955–976.

Matsubara, S. 1990. Structure–activity relationships of cytokinins. Plant Sciences 9:17–57.

Mattoo, A. K., and W. B. White. 1991. Regulation of ethylene biosynthesis. Pp. 21–42 in: A. K. Mattoo and J. C. Suttle (eds.), The Plant Hormone Ethylene. CRC Press, Boca Raton, FL.

Mayoral, M. L., D. A. Atsmon, D. A. Shimshi, and Z. Gromet-Elhanan. 1981. Effect of water stress on enzyme activities in wheat and related wild species: carboxylase activity, electron transport and photophosphorylation in isolated chloroplasts. Australian Journal of Plant Physiology 8:143–146.

McAlister, L., and Finkelstein, D. B. 1980. Heat shock proteins and thermal resistance in yeast. Biochemical and Biophysical Research Communications 93:819–824.

McCabe, D. E., and P. Cristou. 1993. Direct DNA transfer using electric discharge particle acceleration (ACCELL™ technology). Plant Cell Tissue and Organ Culture 33:227–236.

McCabe, D. E., W. E. Swain, B. J. Martinell, and P. Cristou. 1988. Stable transformation of soybean (*Glycine max*) by particle acceleration. Biotechnology 6:923–926.

McCue, K. F., and A. D. Hanson. 1990. Drought and salt tolerance: towards understanding and application. Tibtech 8:358–362.

McCue, K. F., and Hanson, A. D. 1992. Salt-inducible betaine aldehyde dehydrogenase from sugar beet: cDNA cloning and expression. Plant Molecular Biology 18:1–11.

McGowan, M., M. J. Armstrong, and J. A. Corrie. 1983. A rapid fluorescent-dye technique for measuring root length. Experimental Agriculture. 19:209–216.

McGraw, J., and J. Antonovics. 1983. Experimental ecology of *Dryas octopetala* ecotypes. II. A demographic model of growth, branching and fecundity. Journal of Ecology 71:899–912.

McLaughlin, S. B., and G. E. Taylor. 1981. Relative humidity: important modifier of pollutant uptake by plants. Science 211:167–168.

McMichael, B. L., W. R. Jordan, and R. D. Powell. 1972. An effect of water stress on ethylene production by intact cotton petioles. Plant Physiology 49:658–660.

McNamara, S. T., and C. A. Mitchell. 1991. Roles of auxin and ethylene in adventitious root formation by a flood-resistant tomato genotype. HortScience 26:57–58.

McWilliam, J. R., W. Manokaran, and T. Kipnis. 1979. Adaptation to chilling stress in sorghum. Pp. 491–505 in: J. M. Lyons, D. Graham, and J. K. Raison (eds.), Low Temperature Stress in Crop Plants: The Role of the Membrane. Academic Press, New York.

Meinder, H. 1976. Vapor loss through stomatal pores with the mesophyll tissue excluded. Journal of Experimental Botany 27:172–174.

Meinzer, F. C., G. Goldstein, and D. Grantz. 1990. Carbon isotope discrimination in coffee genotypes grown under limited water supply. Plant Physiology 92:130–135.

Meinzer, F. C., G. H. Goldstein, and P. W. Rundel. 1985. Morphological changes along an altitudinal gradient and their consequences for an Andean giant rosette plant. Oecologia 65:278–283.

Meinzer, F. C., P. W. Rundel, M. R. Sharifi, and E. T. Nilsen. 1986. Turgor and osmotic relations of the desert shrub. *Larrea tridentata*. Plant Cell & Environment 9:467–475.

Mengel, K., and M. Viro. 1978. The significance of plant energy status for the uptake and incorporation of NH_4-nitrogen by young rice plants. Soil Science and Plant Nutrition 24:407–416.

Merino, J., C. Field, and H. A. Mooney. 1984. Construction and maintenance costs of Mediterranean-climate evergreen and deciduous leaves. II. Biochemical pathway analysis. Oecologia Plantarum 5:211–223.

Michalczuk, B., and R. M. Rudnicki. 1993. The effect of monochromatic red light on ethylene production in leaves of *Impatiens balsamina* L., and other species. Plant Growth Regulation 13:125–131.

Michelozzi, M., J. D. Johnson, and E. I. Warrag. 1995. Response of ethylene and chlorophyll in two eucalyptus clones during drought. New Forests 9:197–204.

Milchunas, D. G., W. K. Laurenroth, J. S. Singh, C. V. Cole, and H. W. Hunt. 1985. Root turnover and production by ^{14}C dilution: implications of carbon partitioning in plants. Plant and Soil 88:353–365.

Millar, J., and P. Russell. 1992. The cdc25 m-phase inducer: an unconventional protein phosphatase. Cell 68:407–410.

Miller, C., F. Skoog, M. von Saltza, and F. Strong. 1955. Kinetin, a cell division factor from deoxyribonucleic acid. Journal of the American Chemical Society 77:1392.

Miller, P. C., and L. Tieszen. 1972. A preliminary model of processes affecting primary production in arctic tundra. Arctic and Alpine Research 4:1–18.

Minchin, P. E. H., and M. R. Thorpe. 1982. Evidence for the flow of water into sieve tubes associated with phloem loading. Journal of Experimental Botany 33:233–240.

Mineyuki, Y., M. Yashamita, and Y. Nagahama. 1991. p34^{cdc2} kinase homologue in the proprophase band. Protoplasma 162:182–186.

Mishra, R. K., and G. S. Singhal. 1992. Function of photosynthetic apparatus of intact wheat leaves under high light and heat stress and its relationship with peroxidation of thylakoid lipids. Plant Physiology 98:1–6.

Miyake, C., W. H. Cao, and K. Asada. 1993. Purification and molecular properties of the thylakoid-bound ascorbate peroxidase in spinach chloroplasts. Plant Cell Physiology 34:881–889.

Mizutani, H., and E. Wada. 1988. Nitrogen and carbon isotope ratios in sea bird rookeries and their ecological implications. Ecology 69:340–349.

Mohandass, S., and N. Natarajaratnam. 1988. Variations in endogenous growth hormones in rice under two light intensities. Journal of Agronomy and Crop Science 161:178–180.

Mohapatra, S. S., R. J. Poole, and R. S. Dhindsa. 1988a. Detection of two membrane polypeptides induced by abscisic acid and cold acclimation: possible role in freezing tolerance. Plant Cell Physiology 29:727–730.

Mohapatra, S. S., L. Wolfraim, R. J. Poole, and R. S. Dhindsa. 1988b. Molecular cloning and relationship to freezing tolerance of cold-acclimation-specific genes of alfalfa. Plant Physiology 89:375–380.

Mohr, H., and P. Schopfer. 1995. Plant Physiology. Springer-Verlag, Berlin.

Moncur, M. W., and O. Hasan. 1994. Floral induction in *Eucalyptus nitens.* Tree Physiology 14:1303–1312.

Monsai, M., and T. Sakai. 1953. Uber den Lichtfactor in den Pflantzengesellschaften und seine Bedeutung für die Stroffproduktion. Japanese Journal of Botany 14:22–52.

Monson, R. K., A. B. Guenther, and R. Fall. 1991a. Physiological reality in relation to ecosystem- and global-level estimates of isoprene emission. Pp. 185–207 in: T. D. Sharkey, E. A. Holland, and H. A. Mooney (eds.). Trace Gas Emissions from Plants. Academic Press, San Diego, CA.

Monson, R. K., A. J. Hills, P. R. Zimmerman, and R. Fall. 1991b. Studies of the relationship between photosynthesis and isoprene emissions using an on-line isoprene analyzer. Plant Cell & Environment 14:517–523.

Monson, R. K., M. T. Lerdau, T. D. Sharkey, D. S. Schimel, and R. Fall. 1995. Biological aspects of constructing volatile organic compound emission inventories. Atmospheric Environment: 29:2989–3002.

Monteith, J. L. 1972. Solar radiation and productivity in tropical ecosystems. Journal of Applied Ecology 9:747–766.

Monteith, J. L. 1981. Coupling of plants to the atmosphere. Pp. 1–29 in: J. Grace, E. D. Ford, and P. G. Jarvis (eds.), Plants and Their Atmospheric Environment. Blackwell Press, Oxford.

Montero, E., J. Sibole, and C. Cabot. 1994. Abscisic acid content of salt-stressed *Phaseolus vulgaris* L. Journal of Chromatography 658:83–90.

Mook, W. G., M. Koopmans, A. F. Carter, and C. D. Keeling. 1983. Seasonal, latitudinal, and secular variation in the abundance of isotopic ratios of atmospheric carbon dioxide. I. Results from land stations. Journal of Geophysics Research 88:10915–10933.

Mooney, H. A., and C. Chu. 1983. Stomatal responses to humidity of coastal and interior populations of a California shrub. Oecologia 57:148–150.

Mooney, H. A., and S. L. Gulmon. 1982. Constraints on leaf structure and function in relation to herbivory. Bioscience 32:198–206.

Mooney, H. A., O. Björkman, and C. J. Collatz. 1978. Photosynthetic acclimation to temperature in the desert shrub *Larrea divaricata.* I. Carbon dioxide exchange characteristics of intact leaves. Plant Physiology 61:406–410.

Mooney, H. A., B. G. Drake, R. J. Luxmore, W. C. Oechel, and L. F. Pitelka. 1991. Predicting ecosystem responses to elevated CO_2 concentrations. Bioscience 41:96–104.

Mooney, H. A., J. Ehleringer, and O. Bjorkman. 1977. The energy balance of leaves of the evergreen shrub (*Atriplex hymenolytra*). Oecologia (Berlin) 29:301–310.

Mooney, H. A., S. L. Gulmon, J. Ehleringer, and P. W. Rundel. 1980. Atmospheric water uptake by an Atacama desert shrub. Science 209:693–694.

Morgan, D. C., and H. Smith. 1979. Simulated sunflecks have large, rapid effects on plant stem extension. Nature 273:534–536.

Morgan, J. M. 1984. Osmoregulation and water stress in higher plants. Annual Review of Plant Physiology 35:299–319.

Morgan, P. W., D. M. Taylor, and H. E. Joham. 1976. Manipulation of IAA-oxidase activity and auxin-deficiency symptoms in intact cotton plants' manganese nutrition. Physiologia Plantarum 37:149–156.

Moritz, T., J. J. Philipson, and P. C. Odén. 1990. Quantitation of gibberellins A1, A3, A4, A9 and a putative A9-conjugate in grafts of Sitka spruce (*Picea sitchensis*) during the period of shoot elongation. Plant Physiology 93:1476–1481.

Morré, J. D., and C. E. Bracker. 1976. Ultrastructural alteration of plant plasma membranes induced by auxin and calcium ions. Plant Physiology 58:544–547.

Mudd, J. B., C. H. Moeller, and R. E. Garcia. 1984. Biosynthesis and function of sterol derivatives in higher plants. Pp. 349–366 in: W. D. Nes, G. Fuller, and L-S Tsai (eds.), Isopentenoids in Plants: Biochemistry and Function. Marcel Dekker, New York.

Mujer, C. V., M. E. Rumpho, J.-J. Lin, and R. A. Kennedy. 1993. Constitutive and inducible aerobic and anaerobic stress proteins in the *Echinochloa* complex and rice. Plant Physiology 101:217–226.

Muller, W. H., and C. H. Muller. 1964. Volatile growth inhibitors produced by *Salvia* species. Bulletin of the Torrey Botanical Club 91:327–330.

Mulligan, D. R. 1989. Leaf phosphorus and nitrogen concentrations and net photosynthesis in eucalyptus seedlings. Tree Physiology 5:149–157.

Mulroy, T. W., and P. W. Rundel. 1977. Annual plants: Adaptations to desert environments. Bioscience 27:109–114.

Munoz, G. A., and E. Agosin. 1993. Glutamine involvement in nitrogen control of gibberellic acid production in *Gibberella fujikuroi*. Applied Environmental Microbiology 59:4317–4322.

Murali, N. S., A. H. Teramura, and S. K. Randall. 1988. Response differences between two soybean cultivars with contrasting UV-B radiation sensitivities. Photochemistry and Photobiology 48:653–657.

Murata, N., O. Ishizaki-Nishizawa, S. Higashi, H. Hayashi, Y. Tasaka, and I. Nishida. 1992. Genetically engineered alteration in the chilling sensitivity of plants. Nature 356:710–713.

Murray, R. B., and M. Q. Jacobson. 1982. An evaluation of dimension analysis for predicting shrub biomass. Journal of Range Management. 35:451–454.

Nandi, S., L. Palni, D. Letham, and O. C. Wong. 1989. Identification of cytokinins in primary crown gall tumors of tomato. Plant Cell & Environment 12:273–283.

Narayana, I., S. Lalonde, and H. S. Saini. 1991. Water-stress-induced ethylene production in wheat. Plant Physiology 96:406–410.

Navari-Izzo, F., M. F. Quartacci, D. Melfi, and R. Izzo. 1993. Lipid composition of plasma membranes isolated from sunflower seedlings grown under water-stress. Physiologia Plantarum 87:508–514.

Naylor, A. W. 1983. The many faceted problem of chilling injury. Pp. 55–74 in: A. Purvis (ed.), Molecular and Physiological Aspects of Stress in Plants (Proceedings of the Southern Section of the American Society of Plant Physiology). University of Tennessee, Knoxville, TN.

Nemecek-Marshall, M., R. C. MacDonald, J. F. Franzen, C. L. Wojciechowski, and R. Fall. 1995. Methanol emission from leaves: enzymatic detection of gas-phase methanol and relation of methanol fluxes to stomatal conductance and leaf development. Plant Physiology 108:1359–1368.

Nes, W. R. 1984. Uniformity vs. diversity in the structure, biosynthesis, and functions of sterols. Pp. 349–366 in: W. D. Nes, G. Fuller, and L.-S. Tsai (eds.), Isopentenoids in Plants: Biochemistry and Function. Marcel Dekker, New York.

Neskovic, M., and T. Sjaus. 1974. The role of endogenous gibberellin-like substances and inhibitiors in the growth of pea internodes. Biologia Plantarum 16:57–66.

Neuhaus, G., and G. Sprangenberg. 1990. Plant transformation by microinjection techniques. Physiologia Plantarum 79:213–217.

Neuman, D. S., S. B. Rood, and B. A. Smit. 1990. Does cytokinin transport from root-to-shoot in the xylem sap regulate leaf responses to root hypoxia? Journal of Experimental Botany 41:1325–1333.

Neuman, H. H., G. W. Thurtell, and K. R. Stevenson. 1974. In situ measurements of leaf water potential and resistance to water flow in corn, soybean, and sunflower at several transpiration rates. Canadian Journal of Plant Science 54:175–184.

Neven, L. G., D.W. Haskell, C. L. Guy, N. Denslow, P. A. Klein, L. G. Green, and L. Silverman. 1992. Association of 70-kilodalton heat-shock cognate proteins with acclimation to cold. Plant Physiology 99:1362–1369.

Nevin, J. M., I. L. Eaks, C. J. Lovatt, and H. D. Ohr. 1990. Separation of the effects of drought and infection by *Phytophthora cinnamomi* on 'Hass' avocado. Acta Horticulturae 275:729–736.

Nguyen, H. T., M. Krishnan, J. J. Burke, D. R. Porter, and R. A. Vrieling. 1989. Genetic diversity of heat shock protein synthesis in cereal plants. Pp. 319–342 in: J. H. Cherry (ed.), Environmental Stress in Plants, Biochemical and Physiological Mechanisms. Springer-Verlag, New York.

Nightingale, G. T. 1935. Effects of temperature on growth, anatomy, and metabolism of apple and peach roots. Botanical Gazette 96:581–639.

Niklas, K. J. 1994. Plant Allometry: The Scaling of Form and Process. The University of Chicago Press. Chicago.

Nilsen, E. T. 1985. Seasonal and diurnal leaf movements of *Rhododendron maximum* L. in contrasting irradiance environments. Oecologia: 65:296–302.

Nilsen, E. T. 1986. Quantitative phenology and leaf survivorship of *Rhododendron maximum* L. in contrasting irradiance environments of the southern Appalachian Mountains. American Journal of Botany 73:822–831.

Nilsen, E. T. 1991. The relationship between freezing tolerance and thermotropic leaf movement in five *Rhododendron* species. Oecologia 87:63–71.

Nilsen, E. T. 1992a. The influence of water stress on leaf and stem photosynthesis in *Spartium junceum* L. Plant Cell & Environment 15:455–461.

Nilsen, E. T. 1992b. Thermonastic leaf movements: a synthesis of research with *Rhododendron*. Botanical Journal of the Linnean Society 110:205–233.

Nilsen, E. T. 1995. Stem photosynthesis: extent, patterns, and role in plant carbon economy. Pp. 223–238 in: B. Gartner (ed.), Stem Form and Function. Academic Press, San Diego, CA.

Nilsen, E. T., and Y. Bao. 1987. The influence of age, season, and microclimate on the photochemistry of *Rhododendron maximum*. I. Chlorophylls. Photosynthetica. 21(4):535–542.

Nilsen, E. T., and W. H. Muller. 1981a. The influence of low plant water potential on the growth and nitrogen metabolism of the native California shrub *Lotus scoparius* (Nutt. in T&G) Ottley. American Journal of Botany 68:402–407.

Nilsen, E. T., and W. H. Muller. 1981b. Phenology of the drought-deciduous shrub *Lotus scoparius:* climatic controls and adaptive significance. Ecological Monographs 51:323–341.

Nilsen, E. T., and W. H. Muller. 1982. The influence of photoperiod on drought induction of dormancy in *Lotus scoparius* (Nutt. in T&G). Oecologia 53:79–83.

Nilsen, E. T., and M. R. Sharifi. 1994. Seasonal acclimation of stem photosynthesis in woody legume species from the Mojave and Sonoran deserts of California. Plant Physiology 105:1385–1391.

Nilsen, E. T., and M. R. Sharifi. 1996. Carbon isotope composition of legumes with photosynthetic stems from Mediterranean and desert habitats. Bulletin of the Ecological Society of America. 77(3):326.

Nilsen, E. T., and A. Tolbert. 1993. Does winter leaf curling confer cold stress tolerance in *Rhododendrons*. Journal of the American Rhododendron Society 47(2):98–104.

Nilsen, E. T., D. Karpa, H. A. Mooney, and C. Field. 1993. Patterns of stem assimilation in two species of invasive legumes in coastal California. American Journal of Botany 80:1126–1136.

Nilsen, E. T., M. R. Sharifi, P. W. Rundel, W. M. Jarrell, and R. A. Virginia. 1983 Diurnal and seasonal water relations of the phreatophyte *Prosopis glandulosa* (honey mesquite) in the Sonoran desert of California. Ecology 64:1381–1393.

Nilsen, E. T., M. R. Sharifi, and P. W. Rundel. 1984. Comparative water relations of phreatophytes in the Sonoran desert of California. Ecology 65:767–778.

Nilsen, E. T., M. R. Sharifi, P. W. Rundel, R. A. Virginia. 1986. Influences of microclimatic conditions and water relations on seasonal leaf dimorphism of the desert phreatophyte *P. glandulosa*. Oecologia 69:95–100.

Nilsen, E. T., M. R. Sharifi, and P. W. Rundel. 1987b. Leaf dynamics in an evergreen and a deciduous species with even-aged leaf cohorts from two different environments. American Midland Naturalist 118:46–55.

Nilsen, E. T., M. R. Sharifi, P. W. Rundel, I. N. Forseth, and J. R. Ehleringer. 1990. Water relations of stem succulent trees in north-central Baja California. Oecologia 82:299–303.

Nilsen, E. T., M. R. Sharifi, and P. W. Rundel. 1996. Diurnal gas exchange characteristics of two stem photosythesizing legumes in relation to the climate at two contrasting sites in the California desert. Flora 191:105–116.

Nilsen, E. T., M. R. Sharifi, and P. W. Rundel. 1991. Quantitative phenology of warm desert legumes: seasonal growth of six *Prosopsis* species at the same site. Journal of Arid Environments 20:299–311.

Nilsen, E. T., M. R. Sharifi, R. A. Virginia, and P. W. Rundel. 1987a. Phenology of warm desert phreatophytes: seasonal growth and herbivory in *Prosopis glandulose* var. *torreyana* (honey mesquite). Journal of Arid Environments 13:217–219.

Nilsen, E. T., D. A. Stetler, and C. A. Gassman. 1988. The influence of age and microclimate on the photochemistry of *Rhododendron maximum* leaves. II. Chloroplast structure and photosynthetic light response. American Journal of Botany 75(10):1526–1534.

Nobel, P. S. 1978. Surface temperatures of Cacti: influences of environmental and morphological factors. Ecology 59:986–996.

Nobel, P. S. 1991. Physiochemical and Environmental Plant Physiology. Academic Press, New York.

Nobel, P. S., and P. W. Jordon. 1983. Transpiration stream of desert species: resistance and capacitances for a C_3, a C_4, and a CAM plant. Journal of Experimental Botany 34:1379–1391.

Nobel, P. S., J. Cavelier, and J. L. Andrade. 1992. Mucilage in cacti: its apoplastic capacitance, associated solutes, and influence on tissue water relations. Journal of Experimental Botany 43:250.

Norman, J. 1993. Scaling processes between leaf and canopy levels. Pp. 41–76 in: J. R. Ehleringer and C. B. Field (eds.), Scaling Physiological Processes: Leaf to Globe. Academic Press, San Diego, CA.

Norman, J. M., and T. J. Arkebauer. 1991. Predicting canopy light use efficiency from leaf characteristics. In: J. T. Ritchie and J. Hanks. (eds.), Modeling Plant and Soil Systems (Agronomy Monographs 31). American Society of Agronomists, Madison, WI.

Northcote, D. 1972. Chemistry of the plant cell wall. Annual Review of Plant Physiology 23:113–132.

Oard, J. 1993. Development of an airgun device for particle bombardment. Plant Cell Tissue and Organ Culture 33:247–250.

O'Leary, M. H. 1981. Carbon isotope fractionation in plants. Phytochemistry 20:553–567.

O'Leary, H. H. 1984. Measurement of the isotopic fractionation associated with diffusion of carbon dioxide in aqueous solutions. Journal of Physical Chemistry 88:823–825.

O'Leary, M. H. 1993. Biochemical basis of Carbon Isotope fractionation. Pp. 19–28 in: J. R. Ehleringer, A. E. Hall, and G. D. Farquhar (eds.), Stable Isotopes and Plant Carbon-Water Relations. Academic Press, San Diego.

O'Leary, M. H., I. Treichel, and M. Rooney. 1986. Short term measurement of carbon isotope fractionation in plants. Plant Physiology 80:578–582.

O'Toole, J. C., and T. T. Chang. 1979. Drought resistance in cereals—rice: a case study. Pp. 374–405 in: H. Mussell and R. C. Staples (eds.), Stress Physiology of Crop Plants. John Wiley & Sons, New York.

O'Toole, J. C., R. K, Crookston, K. J. Treharne, and J. L. Ozburn. 1976. Mesophyll resistance and carboxylase activity: a comparison under water stress conditions. Plant Physiology 57:465–468.

Ogren, E., and G. Öquist. 1985. Effects of drought on photosynthesis, chlorophyll fluorescence and photoinhibition susceptibility in intact willow leaves. Planta 166:380–388.

Ogren, W. L. 1984. Photorespiration: pathways, regulation and modification. Annual Review of Plant Physiology 35:415–422.

Omran, R. G. 1980. Peroxide levels and the activities of catalase, peroxidase, and indoleacetic acid oxidase during and after chilling cucumber seedlings. Plant Physiology 65:407–408.

Oquist, G., and N. P. A. Hunter. 1989. Effects of cold acclimation on the susceptibility of photosynthesis to photoinhibition. Pp. 471–474 in: M. Baltscheffsky (ed.), Current Research in Photosynthesis, Vol. II. Kluwer Academic Publishers, Dordrecht, The Netherlands.

Orcutt, D. M., and G. W. Patterson. 1975. Sterol, fatty acid, and elemental composition of diatoms grown in chemically-defined media. Comparative Biochemistry and Physiology 50:579–583.

Orshan, G. 1972. Morphological and physiological plasticity in relation to drought. Pp. 245–254 in: C. M. McNeil, J. P. Blaidsel, and J. R. Goodia (eds.), Wildland Shrubs: Their Biology and Utilization. U.S. Department of Agriculture Forest Service, Ogden, Utah.

Osmond, C. B., O. Björkman, and D. I. Anderson. 1980. Physiological Processes in Plant Ecology. Toward a Synthesis with Atriplex. Springer-Verlag, Berlin.

Osmond, C. B., M. P. Austin, J. A. Berry, W. D. Billings, J. S. Boyer, J. W. H. Dacey, P. S. Nobel, S. D. Smith, and W. E. Winner. 1987. Stress physiology and the distribution of plants. Bioscience 37:38–48.

Osteryoung, K. W., and E. Vierling. 1994. Conserved cell and organelle division. Nature 376:473–474.

Otte, M. L., M. J. Dekkers, J. Rozema, and R. A. Broekman. 1991. Uptake of arsenic by *Aster trifolium* in relation to rhizosphere oxidation. Canadian Journal of Botany 69:2670–2677.

Ozga, J. A., and F. G. Dennis, Jr. 1991. The role of abscisic acid in heat stress–induced secondary dormancy in apple seeds. HortScience 26:175–177.

Palta, J. P., and L. S. Weiss. 1993. Ice formation and freezing injury: an overview on the survival mechanisms and molecular aspects of injury and cold acclimation in herbaceous plants. Pp. 143–176 in: P. H. Li and L. Christersson (eds.), Advances in Cold Hardiness. CRC Press, Boca Raton, FL.

Parks, L. W., C. D. K. Bottema, and R. J. Rodriguez. 1984. Physical and enzymic studies on the function of sterols in fungal membranes. Pp. 433–452 in: W. D. Nes, G. Fuller, and L.-S. Tsai (eds.), Isopentenoids in Plants: Biochemistry and Function. Marcel Dekker, New York.

Parups, E. V. 1984. Effects of ethylene, polyamines and membrane stabilizing compounds on plant cell membrane permeability. Phyton 44:9–16.

Passioura, J. B. 1980. The meaning of matric potential. Journal of Experimental Botany 31:1161–1169.

Passioura, J. B. 1982. Water in the soil–plant–atmosphere continuum. Pp. 5–31 in: O. L. Lange, P. S. Nobel, C. B. Osmond, and H. Ziegler (eds.), Encyclopedia of Plant Physiology, New Series, Vol. 12B. Springer-Verlag, Berlin.

Passioura, J. B. 1988. Water transport in and to roots. Annual Review of Plant Physiology 39:245–265.

Paszkowski, J., R. D. Shillito, M. W. Saul, V. Mandak, and T. Hohn. 1984. Direct gene transfer to plants. European Molecular Biology Organization Journal 3:2712–2722.

Patrick, W. H. Jr. 1982. Nitrogen transformations in submerged soils. Pp. 449–465 in: Nitrogen in Agricultural soils (Agronomy Monograph No. 22). American Association of Agronomists.

Pavlic, B. M. 1984. Seasonal changes in osmotic pressure, symplastic water content, and tissue elasticity in the blades of dune grasses growing in situ along the coast of Oregon. Plant Cell & Environment 7:531–539.

Peace, W. J. H., and P. J. Grubb. 1982 Interaction of light and mineral nutrient supply in the growth of *Impatiens parviflora*. New Phytologist 90:127–150.

Pearce, F. 1994. Not warming, but cooling. New Scientist 143:37–44.

Pearcy, R. W. 1978. Effect of growth temperature on the fatty acid composition of leaf lipids in *Atriplex lentiformis*. Plant Physiology 61:484–486.

Pearcy, R. W. 1987. Photosynthetic gas exchange responses of Australian tropical forest trees in canopy, gap, and understory microenvironments. Functional Ecology 1:169–178.

Pearcy, R. W., and H. Calkin. 1983. Carbon dioxide exchange of C_3 and C_4 tree species in an understory of a Hawaiian forest. Oecologia 58:26–32.

Pearcy, R. W., and J. R. Seemann. 1990. Photosynthetic induction state of leaves in a soybean canopy in relation to light-regulation of ribulose-1, 6-bisphosphate carboxylase and stomatal conductance. Plant Physiology 94:628–633.

Pearcy, R. W., J. A. Berry, and D. C. Fork. 1977. Effects of growth temperature on the thermal stability of the photosynthetic apparatus of *Atriples lentiformis* (Torr.) Wats. Plant Physiology 59:873–878.

Pearcy, R. W., R. L. Chazdon, L. J. Gross, and K. A. Mott. 1994. Photosynthetic utilization of sunflecks: A temporally patchy resource on a time scale of seconds to minutes. Pp. 175–208 in: Caldwell, M. M. and R. W. Pearcy (eds.). Exploitation of environmental heterogeneity by plants: Ecophysiological processes above- and belowground. Academic Press, San Diego.

Pearcy, R. W., K. Osteryoung, and D. Randall. 1982. Carbon dioxide exchange characteristics of C_4 Hawaiian *Euphorbia* species native to diverse habitats. Oecologia 55:333–341.

Pearcy, R. W., E. D. Schulze, and R. Zimmerman. 1989. Measurement of transpiration and leaf conductance. Pp. 137–160 in: R. W. Pearcy, J. R. Ehleringer, H. A. Mooney, and P. W. Rundel (eds.), Plant Physiological Ecology: Field Methods and Instrumentation. Chapman & Hall, New York.

Pearson, R. L., L. D. Miller, and C. J. Tucker. 1976. Hand held spectral radiometer to estimate graminaceous biomass. Applied Optics 15:416–418.

Pegg, G. F. 1973. Occurrence of gibberellin-like growth substances in *Basidiomycete* sporophores. Transactions of the British Mycological Society 61:277–286.

Pegg, G. F. 1985. Pathogenic and non-pathogenic microorganisms and insects. Pp. 599–624 in:

R. P. Pharis and D. M. Reid (eds.), Encyclopedia of Plant Physiology: Hormonal Regulation of Development, Vol. III. Springer-Verlag, Berlin.

Pence, V. C., and J. L. Caruso. 1986. Auxin and cytokinin levels in selected and temperature-induced morphologically distinct tissue lines of tobacco crown gall tumors. Plant Science 46:233–237.

Penning de Vries, F. W. T. 1975. The cost of maintenance processes in plant cells. Annals of Botany 39:77–92.

Penning de Vries, F. W. T., A. H. M. Bunsting, and Van Laar 1974. Products, requirements and efficiency of biosynthetic processes: a quantitative approach. Journal of Theoretical Biology 45:339–377.

Peuke, A. D., W. D. Jeschke, and W. Hartung. 1994. The uptake and flow of C, N and ions between roots and shoots in *Ricinus communis* L. III. Distance transport of abscisic acid depending on nitrogen nutrition and salt stress. Journal of Experimental Botany 45:741–747.

Pezeshki, S. R. 1994. Plant Response to Flooding. Pp. 289–321 in: R. E. Wilkinson (ed.), Plant–Environment Interactions. Marcel Decker, New York.

Pezeshki, S. R., and J. L. Chambers. 1985. Responses of cherrybark oak seedlings to short-term flooding. Forest Science 31:760–771.

Pfitsch, W. A., and R. W. Pearcy. 1989. Steady-state and dynamic photosynthetic response of *Adenocaulon bicolor* in its redwood forest habitat. Oecologia 80:471–476.

Phillips, W. S. 1963. The depth of roots in soil. Ecology 44:424.

Phillips, I. D. J., J. Miners, and J. G. Goddick. 1980. Effects of light and photoperiodic conditions on abscisic acid in leaves and roots of *Acer pseudoplatanus* L. Planta 149:118–122.

Phinney, B. O. 1957. Growth response of single-gene dwarf mutants in maize to gibberellic acid. Proceedings of the National Academy of Sciences 42:185–189.

Pillay, I., and C. Beyl. 1990. Early responses of drought-resistant and -susceptible tomato plants subjected to water stress. Journal of Plant Growth Regulation 9:213–219.

Pilon-Smits, E. A. H., M. J. M. Ebskamp, M. J. Paul, M. J. W. Jeuken, P. J. Weisbeck, and S. C. Smeekens. 1995. Improved performance of transgenic fructan-accumulating tobacco under drought stress. Plant Physiology 107:125–130.

Pitcher, L. H., E. Brennan, A. Hurley, P. Dunsmuir, J. M. Tepperman, and B. A. Zilinskas. 1991. Overproduction of petunia copper/zinc superoxide dismutase does not confer ozone tolerance in transgenic tobacco. Plant Physiology 97:452–455.

Poljenkoff-Mayber, A. 1981. Ultrastructural consequences for drought. Pp. 389–401 in: L. G. Paleg and D. A. Aspinal (eds.), The Physiology and Biochemistry of Drought Resistance in Plants. Academic Press, New York.

Poorter, H., and C. Remkes. 1990. Leaf area ratio and net assimilation rate of 24 wild species differing in relative growth rate. Oecologia 83:553–559.

Poorter, H., C. Ramkes, and H. Lambers. 1990. Carbon and nitrogen economy of 24 wild species differing in relative growth rate. Plant Physiology 94:621–627.

Potrykus I. 1991. Gene transfer to plants: assessment of published approaches and results. Annual Review of Plant Physiology and Plant Molecular Biology 42:205–225.

Portykus, I., J. Paszkowski, M. W. Saul, J. Petruska, and R. D. Shillito. 1985. Molecular and general genetics of a hybrid foreign gene introduced into tobacco by direct gene transfer. Molecules, genes, and genetics 199:167–177.

Potter, T. I., and S. B. Rood. 1993. Light intensity, gibberellins and growth in *Brassica*. Plant Physiology Supplement 102:9.

Powles, S. B. 1984. Photoinhibition of photosynthesis induced by visible light. Annual Review of Plant Physiology 35:15–44.

Pradet, A., and P. Raymond. 1983. Adenine nucleotide ratios and adenylate energy charge in energy metabolism. Annual Review of Plant Physiology 34:199–224.

Prakash, L., and G. Prathapasenan. 1990. NaCl- and gibberellic acid-induced changes in the content of auxin and the activities of cellulase and pectin lyase during leaf growth in rice (*Orzya sativa*). Annals of Botany 65:251–257.

Prince, S. D. 1991. Satellite remote sensing of primary production: comparison of results for sehal grasslands, 1981–1988. International Journal of Remote Sensing 12:1301–1330.

Qamaruddin, M., and E. Tillberg. 1989. Rapid effects of red light on the iospentenyladenosine content in Scots pine seeds. Plant Physiology 91:5–8.

Quetier, F. B., L. Lejeune, S. Delorme, and D. Falconet. 1985. Molecular organization and expression of the mitochondrial genome of higher plants. Pp. 25–36 in: R. Douce and D. A. Day (eds.), Higher Plant Cell Respiration. Springer-Verlag, Berlin.

Quinby, J. R., J. D. Hesketh, and R. L. Voight. 1973. Influence of temperature and photoperiod on floral initiation and leaf number in sorghum. Crop Science 13:243–246.

Rada F., G. Goldstein, A. Azocar, and F. Meinzer. 1985. Freezing avoidance in Andean giant rosette plants. Plant Cell & Environment 8:501–507.

Rademacher, W., and J. E. Graebe. 1979. Gibberellin A4 produced by *Sphaceloma manihoticola*, the cause of super-elongation disease of cassava (*Manihot esulenta*). Biochemical and Biophysical Research Communications 91:35–40.

Radford, P. 1967. Growth analysis formulae: their use and abuse. Crop Science 7(3):171–175.

Radin, J. W. 1984. Stomatal responses to water stress and to abscisic acid in phosphorus-deficient cotton plants. Plant Physiology 76:392–394.

Raison J. K., and G. R. Orr. 1990. Proposals for a better understanding of the molecular basis of chilling injury. Pp. 145–164 in: C. Y. Wang (ed.), Chilling Injury of Horticultural Crops. CRC Press, Boca Raton, FL.

Rajagopal, V., and A. S. Andersen. 1980. Water stress and root formation in pea cuttings. III. Changes in the endogenous level of abscisic acid and ethylene production in the stock plants under two levels of irradiance. Physiologia Plantarum 48:155–160.

Rashke, K. 1975. Stomatal action. Annual Review of Plant Physiology 26:309–340.

Rashke, K. 1987. Action of abcissic acid on guard cells. Pp. 253–279 in: E. Zeiger, G. D. Farquhar, and I. R. Cowan (eds.), Stomatal Function. Stanford University Press, Stanford, CA.

Raskin, I., and H. Kende. 1984. Regulation of growth in stem sections of deep-water rice. Planta 160:66–72.

Raven, J. A., and G. D. Farquhar. 1990. The influence of N metabolism and organic acid synthesis on the natural abundance of isotopes of carbon in plants. New Phytologist 116:505–529.

Regehr, D. L., F. A. Bazzaz, and W. R. Boggess. 1975. Photosynthesis, transpiration, and leaf conductance of *Populus deltoides* in relation to flooding and drought. Photosynthetica 9:52–61.

Reich, P. B., and A. W. Schoettle. 1988. Role of phosphorus and nitrogen in photosynthetic and whole plant carbon gain and nutrient use efficiency in eastern white pine. Oecologia 77:25–33.

Reinero, A., G. Shearer, B. A. Bryan, J. L. Skeeters, and D. H. Kohl. 1983. Site of natural ^{15}N enrichment of soybean nodules. Plant Physiology 72:256–258.

Reynolds, J. F., D. W. Hilbert, and P. R. Kemp. 1993. Scaling ecophysiology from the plant to the ecosystem: a conceptual framework. Pp. 127–141 in: J. R. Ehleringer and C. B. Field (eds.), Scaling Processes from Leaf to Globe. Academic Press, San Diego, CA.

Rhodes, D., P. J. Rich, D. G. Brunk, G. C. Ju, J. C. Rhodes, M. H. Pauly, and L. A. Hansen. 1989. Development of two isogenic sweet corn hybrids differing for glycinebetaine content. Plant Physiology 91:1112–1121.

Rice, E. L. 1979. Allelopathy: an update. Botanical Review 45:15–109.

Rice, E. L. 1984. Allelopathy. Academic Press, San Diego.

Rich, P. M. 1989. A Manual for the Analysis of Hemispherical Canopy Photography. Los Alamos National Laboratory, New Mexico.

Richards, L. A., and G. Ogata. 1958. Thermocouples for vapor pressure measurement in biological and soil systems at high humidity. Science 128:1089–1090.

Ristic, Z., and E. N. Ashworth. 1993. Ultrastructural evidence that intracellular ice formation and possibly cavitation are the sources of freezing injury in supercooling wood tissue of *Cornus florida* L. Plant Physiology 103:753–761.

Ritchie, G. A., and J. R. Rodin. 1985. Comparison between two methods of generating pressure–volume curves. Plant, Cell & Environment 8:49–53.

Ritchie, G. A., and R. G. Shula. 1984. Seasonal changes of tissue-water relations in shoots and root systems of Douglas-fir seedlings. Forest Science 30:538–548.

Roberts, S. W., and K. R. Knoerr. 1978. In situ estimates of variable plant resistance to water flux in *Ilex opaca* Ait. Plant Physiology 61:311–313.

Roberts, S. W., B. R. Strain, and K. R. Knoerr. 1980. Seasonal patterns of leaf water relations in four co-occurring forest tree species: parameters from pressure–volume curves. Oecologia 46:330–337.

Roberts, S. W., B. R. Strain, and K. R. Knoerr. 1981. Seasonal variation of leaf tissue elasticity in four forest tree species. Physiologia Plantarum 52:245–250.

Robertson, A. J., and Gusta, L. V. 1986. Abscisic acid and low temperature induced polypeptide changes in alfalfa (*Medicago sativa*) cell suspension cultures. Canadian Journal of Botany 64:2758–2763.

Robertson, M., and Chandler, P. M. 1992. Pea dehydrins: identification, characterization and expression. Plant Molecular Biology 19:1031–1044.

Robichaux, R. H. 1984. Variation in tissue water relations of two sympatric Hawaiian *Daubautia* species and their natural hybrid. Oecologia 65:75–81.

Robichaux, R. H., K. E. Holsinger, and S. R. Morse. 1986. Turgor maintenance in Hawaiian *Debautia* species: the role of variation in tissue osmotic and elastic properties. Pp. 353–380 in: T. J. Givnish (ed.), On the Economy of Plant Form and Function. Cambridge University Press, Cambridge.

Rodecap, K. D., and D. T. Tingey. 1983. The influence of light on ozone-induced 1-aminocyclopropane-1-carboxylic acid and ethylene production from intact plants. Zeitgrift fur Pflanzenphysioly 110:419–427.

Roeske, C. A., and M. H. O'Leary. 1984. Carbon isotope effects on the enzyme catalyzed carboxylation of ribulose-bisphosphate. Biochemistry 23:6275–6284.

Roessler, P. G., and R. K. Monson. 1985. Midday depression in net photosynthesis and stomatal conductance in *Yucca glauca:* relative contributions of leaf temperature and leaf-to-air water vapor concentration difference. Oecologia 67:380–387.

Romera, F. J., and E. Alcántara. 1994. Iron-deficiency stress responses in cucumber (*Cucumis sativus* L.) roots. Plant Physiology 105:1133–1138.

Rosa, J. T. 1921. Investigations on the hardening process in vegetable plants. Missouri Agriculture Station Research Bulletin 48:1–97.

Rosenberg, N, J., B. L. Blad, and S. B. Verma. 1983. Microclimate: The Biological Environment, 2nd ed. John Wiley & Sons, New York.

Rosenzweig, M. L. 1968. Net primary productivity of terrestrial communities: prediction from climatalogical data. American Naturalist 102:67–74.

Ross, J. J., C. L. Willis, P. Gaskin, and J. B. Reid. 1992. Shoot elongation in *Lathyrus odoratus* L.: gibberellin levels in light- and dark-grown tall and dwarf seedlings. Planta 187:10–13.

Rudnicki, R. M., T. Fjeld, and R. Moe. 1993. Effect of light quality on formation in leaf and petal discs of *Begonia × hiemalis Fotsch* cv. Schwabenland red. Plant Growth Regulation 13:281–286.

Rumbaugh, M. D., D. H. Clark, and B. M. Pendry. 1988. Determination of root mass ratios in alfalfa–grass mixtures using near infrared reflectance spectroscopy. Journal of Range Management 41:488–490.

Rundel, P. W. 1980. The ecological distribution of C_4 and C_3 grasses in the Hawaiian Islands. Oecologia 45:354–359.

Rundel, P. W. 1982. Water uptake by organs other than roots. Pp. 111–134 in: O. L. Larrge, P. S. Nobel, C. B. Osmond, and H. Ziegler (eds.), Physiological Plant Ecology, Vol. 2. Springer-Verlag, Berlin.

Rundel, P. W., M. O. Dillon, B. Palma, H. A. Mooney, S. L. Gulmon, and J. R. Ehleringer. 1991. The phytogeography and ecology of the Atacama and Peruvian deserts. Aliso 13:1–49.

Russell, D. A., and Sachs, M. M. 1992. Protein synthesis in maize during anaerobic and heat stress. Plant Physiology 99:615–620.

Ryerson, E., A. Li, J. P. Young, and M. C. Heath. 1993. Changes in abscisic acid levels in bean leaves during the initial stages of host and nonhost reactions to rust fungi. Physiological and Molecular Plant Pathology 43:265–273.

Sachs, M. M., M. Freeling, and R. Okimoto. 1980. The anaerobic proteins of maize. Cell 20:761–767.

Sage, R. F., and R. W. Pearcy. 1987. The nitrogen use efficiency of C_3 and C_4 plants. II. Leaf nitrogen effects on the gas exchange characteristics of *Chenopodium album* (L.) and *Amaranthus retroflexus* (L.). Plant Physiology 84:959–963.

Sage, R. F., T. D. Sharkey, and J. R. Seeman. 1989. Acclimation of photosynthesis to elevated CO_2 in five C_3 species. Plant Physiology 89:590–596.

Saglio, P. H., M. Rancilliac, F. Bruzan, and A. Pradet. 1984. Critical oxygen pressure for growth and respiration of excised and intact roots. Plant Physiology 76:151–154.

Saini, H. S., E. D. Consolacion, P. K. Bassi, and M. S. Spencer. 1989. Control processes in the induction and relief of thermoinhibition of lettuce seed germination. Plant Physiology 90:311–315.

Sakai, A., and W. Larcher. 1987. Frost Survival of Plants: Responses and adaptation to freezing stress. Springer-Verlag, New York, p. 321.

Sakurai, N., M. Akiyama, and S. Kuraishi. 1985. Roles of abscisic acid and indoleacetic acid in the stunted growth of water-stressed, etiolated squash hypocotyls. Plant Cell Physiology 26:15–24.

Sala, O., A. Deregibus, T. Schlichter, and H. Alippe. 1981. Productivity dynamics of a native temperate grassland in Argentina. Journal of Range Management 34:48–51.

Saliendra, N. Z., and F. C. Meinzer. 1989. Relationship between root/soil hydraulic properties and stomatal behavior in sugarcane. Australian Journal of Plant Physiology 16:241–250.

Salisbury, F. B., and C. W. Ross. 1992. Plant Physiology, 4th ed. Wadsworth Publishing Co., Belmont, CA.

Saltveit, M. E., and L. L. Morris. 1990. Overview on chilling injury of horticultural crops. Pp. 3–15 in: C. Y. Wang (ed.), Chilling Injury of Horticultural Crops. CRC Press, Boca Raton, FL.

Sanchez, Y., and S. L. Lindquist. 1990. HSP104 required for the induction of thermotolerance. Science 248:1112–1115.

Saugy, M., G. Mayor, and P.-E. Pilet. 1989. Endogenous ABA in growing maize roots: light effects. Plant Physiology 89:622–627.

Sauter, K. J., D. W. Davis, P. H. Li, and I. S. Wallerstein. 1990. Leaf ethylene evolution level following high-temperature stress in common bean. HortScience 25:1282–1284.

Sawada, K. 1912. Disease of agricultural products in Japan. Formosan Agricultural Review 36:1015.

Sawada, K., and E. Kurosawa. 1924. On the prevention of the bakanae disease of rice. Experimental Station Bulletin of Formosa 21:1–9.

Schäfer, E., K. Apel, A. Batschauer, and E. Mösinger. 1986. The molecular biology of action. Pp. 83–98 in: R. E. Kendrick and G. H. M. Kronenberg (eds.). Photomorphogenesis in Plants. Martinus Nijhoff, Dordrecht, The Netherlands.

Scherer, G. F. E., and B. André. 1993. Stimulation of phospholipase A2 by auxin in microsomes from suspension-cultured soybean cells is receptor-mediated and influenced by nucleotides. Planta. 191:515–523.

Schimper, A. F. W. 1898. Pflantzengeographie auf physiologisch Grundlage. Verlag Gustav Fisher, Jena, Germany.

Schindler, T., and D. Kotziab. 1989. Comparison of monoterpene volatilization and leaf-oil composition of conifers. Naturwissenschaften 76:475–476.

Schleser, G. H., and R. Jayasekera. 1985. $\delta^{13}C$-variation of leaves in forests as an indication of reassimilated CO_2 from the soil. Oecologia 65:536–542.

Schlesinger, W. H. 1984. Soil organic matter: a source of atmospheric CO_2. Pp. 111–127 in: G. M. Woodwell (ed.), The Role of Terrestrial Vegetation in the Global Carbon Cycle. John Wiley & Sons, New York.

Schlesinger, 1991. Biogeochemistry: An Analysis of Global Climate Change. Academic Press, New York.

Schlicting, C. D., and D. A. Levin. 1986. Phenotypic plasticity: an evolving plant character. Biological Journal of the Linnean Society 29:37–47.

Scholander, P. F., L. vanDam, and S. I. Scholander. 1955. Gas exchange in the roots of mangroves. American Journal of Botany 42:92–98.

Scholander, P. F., H. T. Hammel, E. A. Hemmingsen, and E. D. Bradstreet. 1965. Sap pressure in vascular plants. Science 148:339–346.

Schulze, E.-D. 1986. Carbon dioxide and water vapor exchange in response to drought in the atmosphere and in the soil. Annual Review of Plant Physiology 37:247–274.

Schulze, E.-D., and F. S. Chapin III. 1987. Plant specialization to environments of different resource availability. Pp. 120–148 in: E.-D. Schulze and H. Zwolfer (eds.), Potentials and Limitations in Ecosystem Analysis. Springer-Verlag, Berlin.

Schulze, E.-D., J. Cermak, R. Matyssek, M. Penka, R. Zimmerman, F. Vasieck, W. Gries, and J. Kucera. 1985. Canopy transpiration and water fluxes in the xylem of the trunk of *Larix* and *Picea* trees: a comparison of xylem flow, parameter, and curvette measurements. Oecologia 66:475–483.

Schulze, E.-D., G. Grebauer, H. Ziegler, and O. L. Lange. 1991. Estimates of nitrogen fixation by trees on an aridity gradient in Namibia. Oecologia 88:451–455.

Schulze, W., and E.-D. Schulze. 1994. The significance of assimilatory starch for growth in *Arabidopsis thalian* wild-type and starch less mutants. Pp. 123–131 in: E. D. Shulze and M. M. Caldwell (eds.), Ecophysiology of Photosynthesis. Springer-Verlag, Berlin.

Schuch, U. K., and L. H. Fuchigami. 1992. Flowering, ethylene production, and ion leakage of coffee in response to water stress and gibberellic acid. Journal of the American Society of Horticultural Science 117:158–163.

Schwartz, A., and E. Zeiger. 1984. Metabolic energy for stomatal opening: roles for photophosphorylation and oxidative phosphorylation. Planta 161:129–136.

Seeley, S. 1990. Hormonal transduction of environmental stresses. Hort Science 25:1369–1376.

Seeman, J. R., and C. Critchley. 1985. Effects of salt stress on growth, ion content, stomatal behaviour, and photosynthetic capacity of a salt sensitive species, *Phaseolus vulgaris* L. Planta 164:151–162.

Seemann, J. R., T. D. Sharkey, J. L. Wang, and C. B. Osmond. 1987. Environmental effects on photosynthesis, nitrogen use efficiency and metabolic pools in leaves of sun and shade plants. Plant Physiology 84:796–802.

Sellers, P. 1987. Canopy reflectance, photosynthesis, and transpiration, II. The role of biophysics in the linearity of their interdependence. Remote Sensing of the Environment 21:143–183.

Sellers, P., J. W. Shuttleworth, J. L. Dorman, A. Dalcher, and J. M. Roberts. 1989. Calibrating the simple biosphere model (SiB) for Amazonian tropical forest using field and remote sensing data. 1. Average calibrating with field data. Journal of Applied Meteorology 28:727–759.

Setter, T. L., I. Waters, H. Greenway, B. J. Atwell, and T. Kupkanchanakul. 1987. Carbohydrate status of terrestrial plants during flooding. Pp. 411–433 in: R. M. M. Crawford (ed.), Plant Life in Aquatic and Amphibious Habitats. Blackwell Scientific Publications, Oxford.

Sfakiotakis, E. M., and D. R. Dilley. 1974. Induction of ethylene production in 'Bosc' pears by postharvest cold stress. HortScience 9:336–338.

Shah, C. B., and R. S. Loomis. 1965. Ribonucleic acid and protein metabolism in sugar beet during drought. Physiologia Plantarum 18:240–254.

Sharifi, M. R., and P. W. Rundel. 1993. The effect of vapor pressure deficit on the carbon isotope discrimination in the desert shrub *Larrea tridentata* (creosote bush). Journal of Experimental Botany 44:481–487.

Sharifi, M. R., E. T. Nilsen, and P. W. Rundel. 1982. Biomass and net primary production of *Prosopis glandulosa* (Fabaceae) in the Sonoran desert of California. American Journal of Botany 69:760–767.

Sharifi, M. R., F. C. Meinzer, E. T. Nilsen, P. W. Rundel, R. A. Virginia, W. M. Jarrell, D. J. Herman, and P. C. Clark. 1988. Effect of manipulation of water and nitrogen supplies on the quantitative phenology of *Larrea tridentata* (creosote bush) in the Sonoran desert of California. American Journal of Botany 75:1163–1174.

Sharkey, T. D. 1985. Photosynthesis in intact leaves of C_3 plants: physics, physiology, and rate limitations. Botanical Review 51:53–105.

Sharkey, T. D., and E. Singsaas. 1995. Why plants emit isoprene. Nature 374:769.

Shearer, G., and D. H. Kohl. 1978. 15N abundance in N-fixing and non-N-fixing plants. Pp. 605–622 in: A. Frigerio (ed.), Mass Spectrometry in Biochemistry and Medicine. Plenum Press, New York.

Shearer, G., and D. H. Kohl. 1988. Natural ^{15}N abundance as a method for estimating the contribution of biologically fixed nitrogen to N_2-fixing systems: potential for non legumes. Plant and Soil 110:317–327.

Shearer, G., and D. H. Kohl. 1989. Estimates of N_2 fixation in ecosystems: the need for and basis of the ^{15}N natural abundance method. Pp. 342–374 in: P. W. Rundel, J. R. Ehleringer, and K. A. Nagy (eds.), Stable Isotopes in Ecological Research. Springer-Verlag, Berlin.

Shearer, G., L. Feldman, B. A. Bryan, J. Skeeters, D. H. Kohl, N. Amarger, F. Mariotti, and A. Mariotti. 1982. ^{15}N abundance of nodules as an indication of N metabolism in N_2-fixing plants. Plant Physiology 70:465–468.

Shearer, G., D. H. Kohl, and S. H. Chien. 1978. The nitrogen-15 abundance in a wide variety of soils. Soil Science Society of America Journal 42:899–902.

Shearer, G., D. H. Kohl, R. A. Virginia, B. A. Bryan, J. L. Skeeters, E. T. Nilsen, M. R. Sharifi, and P. W. Rundel. 1983. Estimates of N_2-fixation from variation in the natural abundance of ^{15}N in Sonoran Desert ecosystems. Oecologia 56:365–373.

Sherr, C. J. 1994. G1 phase progression: cycling on cue. Cell 79:551–555.

Shipley, B., and R. H. Peters. 1990. A test of the Tilman model of plant strategies: relative growth rate and biomass partitioning. American Naturalist 136:139–153.

Shirazi, A. M., L. H. Fuchigami, and T. H. H. Chen. 1993. Ethylene production of heat-treated stem tissues of Red-osier dogwood at several growth stages. HortScience 28:1117–1119.

Shkol'nik, M. J., T. A. Krupnikova, and N. N. Dmitrieva. 1964. Influence of boron deficiency on some aspects of auxin metabolism in the sunflower and corn. Soviet Plant Physiology 11:164–169.

Silander, J. A., Jr. 1985. Microevolution in clonal plants. Pp. 107–152 in: J. B. C. Jackson, L. W. Buss, and R. E. Cook (eds.), Population Biology and Evolution of Clonal Plants. Yale University Press, New Haven, CN.

Silver, G. M., and R. Fall. 1991. Enzymatic synthesis of ioprene from dimethylallyl diphosphate in aspen leaf extracts. Plant Physiology 97:1588–1591.

Silvester, W. B. 1983. Analysis of N_2-fixation. Pp. 172–212 in: J. C. Gordan and C. T. Wheeler (eds.), Biological Nitrogen Fixation in Forest Ecosystems: Foundations and Applications. Martinus Nijhoff/Dr. W. Junk, Boston.

Siminovitch, D., and Y. Cloutier. 1983. Drought and freezing tolerance and adaptation in plants: some evidence of near equivalences. Cryobiology 20:487–503.

Singh, J. S., and D. C. Coleman. 1983. A technique for evaluating functional root biomass in grassland ecosystems. Canadian Journal of Botany 51:1867–1870.

Sivagami, S., K. P. Vijayan, and N. Natarajaratnam. 1988. Effect of nutrients and growth regulating chemicals on biochemical aspects and hormonal balance with reference to apical dominance in mango. Acta Horticulture 231:476–482.

Skoog, F., and D. Armstrong. 1970. Cytokinins. Annual Review of Plant Physiology 21:359–384.

Skriver, K., and J. Mundy. 1990. Gene expression in response to abscisic acid and osmotic stress. Plant Cell 2:503–512.

Slatyer, R. O. 1967. Plant Water Relationships. Academic Press, New York.

Smakman, G., and R. Hofstra. 1982. Energy metabolism of *Plantago lanceolata* L. as affected by a change in root temperature. Physiologia Plantarum 56:33–37.

Smith, B. J., and M. P. Yaffe. 1991. Uncoupling thermotolerance from the induction of heat shock proteins. Proceedings of the National Academy of Sciences 88:11091–11094.

Smith, B. N., and S. Epstein. 1971. Two categories of $^{13}C/^{12}C$ ratios for higher plants. Plant Physiology 47:380–384.

Smith, H. 1981. Light Quality as an ecological factor. Pp. 93–110 in J. Grace, E. D. Ford, and P. G. Jarvis (eds.), Plants and their atmospheric environment. Blackwell Scientific Publishers, Oxford, UK.

Smith, H. 1982. Light quality, photoperception, and plant strategy. Annual Review of Plant Physiology 33:481–518.

Smolenska, G., and P. J. C. Kuiper. 1977. Effect of low temperature upon lipid and fatty acid composition of roots and leaves of winter rape plant. Physiologia Plantarum 41:29–35.

Smucker, A. J., M. McBurney, and S. L. Srivastava. 1985. Quantitative separation of roots from intact profiles by hydrodynamic elutriation. Agronomy Journal 74:500–503.

Snyder, F. W., and J. A. Bunce. 1983. Use of the plastochron index to evaluate effects of light, temperature, and nitrogen on growth of soy bean (*Glycine max* L. Merr.). Annals of Botany 52:895–903.

Sobrado, M. A. 1991. Cost-benefit relationships in deciduous and evergreen leaves of tropical dry forest species. Functional Ecology 5:608–616.

Sonnewald, U. 1992. Expression of *E. coli* inorganic pyrophosphatase in transgenic plants alters photoassimilate partitioning. Plant Journal 2:571–581.

Sorger, P. K. 1991. Heat shock factor and the heat shock response. Cell 65:363–366.

Sperry, J. S. 1986. Relationship of xylem embolism to xylem pressure potential, stomotal closure, and shoot morphology in the palm *Rhapis excelsa*. Plant Physiology 80:110–116.

Sperry, J. S. 1995. Limitations on stem water transport and their consequences. Pp. 105–124 in: B. Gartner (ed.), Stem Form and Function. Academic Press, San Diego, CA.

Sperry, J. S., and J. E. M. Sullivan. 1992. Xylem embolism in response to freeze–thaw cycles and water stress in ring-porous, diffuse-porous, and conifer species. Plant Physiology 100:605–613.

Sperry, J. S., and M. T. Tyree. 1988. Mechanisms of water stress induced embolism. Plant Physiology 88:582–587.

Sperry, J. S., M. T. Tyree, and J. R. Donnelly. 1988. Vulnerability of xylem to embolism in a mangrove vs. an inland species of Rhizophoraceae. Physiologia Plantarum 74:276–283.

Spikman, G. 1986. The effect of water stress on ethylene production and ethylene sensitivity of *Freesia* inflorescences. Acta Horticulturae 181:135–140.

Sponsel, V. M. 1986. Gibberellins in dark- and red-light grown shoots of dwarf and tall cultivars of *Pisum sativum:* the quantification, metabolism, and biological activity of gibberellins in Progress No. 9 and Alaska. Planta 168:119–129.

Springer, B., W. Werr, P. Starlinger, D. C. Bennett, M. Zokolica, and M. Freeling. 1986. The shrunken gene on chromosome 9 of *Zea mays* L. is expressed in various plant tissues and encodes an anaerobic protein. Molecular Genes and Genetics 205:461–468.

Sprugel, D. G. 1983. Corrections for bias in log-transformed allometric equations. Ecology 64:209–210.

Stall, R. E., and C. B. Hall. 1984. Chlorosis and ethylene production in pepper leaves infected by *Xanthomonas campestris* pv. *vesicatoria*. Phytopathology 3:373–375.

Stapel, D., E. Kruse, K. Kloppstech. 1993. The protective effect of heat shock proteins against photoinhibition under heat shock in Barley (*Hordeum vulgare*). Journal of Photochemistry and Photobiology, B 21:211–218.

Steadman, J. R., and L. Sequeira. 1969. A growth inhibitor from tobacco and its possible involvement in pathogenesis. Phytopathology 59:499–503.

Steele, K. W., B. M. Borish, R. M. Daniel, and G. W. Ottara. 1983. Effects of rhizobial strains and host plant on nitrogen isotope fractionation in legumes. Plant Physiology 72:1001–1004.

Steinmüller, D., and M. Tevini. 1985. Composition and function of plastoglobuli. I. Isolation and purification from chloroplasts and chromoplasts. Planta 163:201–207.

Sternberg, L. S. L. 1988. D/H ratio of environmental water recorded by D/H ratios of plant lipids. Nature 333:59–61.

Sternberg, L. S. L., N. Ish-Shalom-Gordon, M. Ross, and J. Obrien. 1991. Water relations of coastal plant communities near the ocean/freshwater boundary. Oecologia 88:305–310.

Stewart, G. R., J. S. Pate, M. Unkovich. 1993. Characteristics of inorganic nitrogen assimilation of plants in fire-prone Mediterranean-type vegetation. Plant Cell and Environment 16:351–363.

Stewart, C. N., Jr., and E. T. Nilsen. 1995. Phenotypic plasticity and genetic heterogeneity of *Vaccinium macrocarpon* (Ericaceae), the American cranberry. I. Reaction norms of central and marginal clones in a common garden. International Journal of Plant Science 156:687–697.

Stillwell, W., and P. Hester. 1984. Abscisic acid increases membrane permeability by interacting with phophatidylethanolamine. Phytochemistry 23:2187–2192.

Stillwell, W., and S. R. Wassall. 1993. Binding of IAA and ABA to phospholipid bilayers. Phytochemistry 34:367–373.

Stillwell, W., P. Hester, and B. Brengle. 1985. Effect of cytokinins on erythritol permeability to phosphatidylcholine bilayers. Journal of Plant Physiology 118:105–110.

Stitt, M. 1993. Flux control at the level of the pathway: illustrated in studies with mutants and trangenic plants having a decreased activity of enzymes involved in photosynthesis partitioning. Pp. 13–36 in: E. D. Shulze (ed.), Flux Control in Biological Systems. Academic Press, San Diego, CA.

Stokes, M. A., and T. L. Smiley. 1968. An Introduction to Tree Ring Dating. University of Chicago Press, Chicago.

Storz, G., L. A. Tartaglia, and B. N. Ames. 1990a. Transcriptional regulator of oxidative stress-inducible genes: direct activation by oxidation. Science 248:189–194.

Storz, G., L. A. Tartaglia, S. B. Farr, and B. N. Ames. 1990b. Bacterial defences against oxidative stress. Trends in Genetics 6:363–368.

Strain, B., and F. Bazzaz. 1983. Terrestrial plant communities. Pp. 177–222 in: E. Lemon (ed.), CO_2 and plants: the response of plants to rising levels of carbon dioxide. (AAAS Selected Symposium No. 84). American Society for the Advancement of Science, Washington, DC.

Strid, A. 1993. Alteration in expression of defense genes in *Pisum sativum* after exposure to supplementary ultraviolet-B radiation. Plant Cell Physiology 34:949–953.

Sung, F. J. M., and D. R. Krieg. 1979. Relative sensitivity of photosynthetic assimilation and translocation of ^{14}carbon to water stress. Plant Physiology 64:852–856.

Suri, R. A., and C. L. Mandahar. 1985. Involvement of cytokinin-like substances in the pathogenesis of *Alternaria brassicae.* (Berk.) Sacc. Plant Science 41:105–109.

Svoboda J., and L. C. Bliss. 1974. The use of autoradiography in determining active and inactive roots in plant population studies. Arctic Applied Research 6:257–260.

Svejcar, R. T. J., and T. W. Boutton. 1985. The use of stable carbon isotope analysis in rooting studies. Oecologia 67:205–208.

Sziráki, I., E. Balázs, and Z. Király. 1980. Role of different stresses in inducing systemic acquired resistance to TMV and increasing cytokinin level in tobacco. Physiological Plant Pathology 16:277–284.

Taiz, L., and E. Zeiger. 1991. Plant Physiology. Benjamin/Cummings Publishing Company Redwood City, CA.

Takahashi, N., B. Phinney, and J. MacMillan. 1990. Gibberellins. Springer-Verlag, Berlin.

Takemoto, B. K., and R. D. Noble. 1986. Differential sensitivity of duckweeds (Lemnaceae) to sulfite. I. carbon assimilation and frond replication rate as factors influencing sulfite phytotoxicity under low and high irradiance. New Phytologist 103:525–539.

Talbot, A. J. B., M. T. Tyree, and J. Dainty. 1975. Some notes concerning the measurement of water potential of leaf tissue with specific reference to *Tsuga canadensis* and *Picea abies.* Canadian Journal of Botany 53:784–788.

Talon, M., and J. A. D. Zeevaart. 1990. Gibberellins and stem growth as related to photoperiod in *Silene armeria* L. Plant Physiology 92:1094–1100.

Talon, M., J. A. D. Zeevaart, and D. A. Gage. 1991. Identification of gibberellins in spinach and effects of light and darkness on their levels. Plant Physiology 97:1521–1526.

Tang, Z. C., and T. T. Kozlowski. 1984a. Ethylene production and morphological adaptation of woody plants to flooding. Canadian Journal of Botany 62:1659–1664.

Tang, Z. C., and T. T. Kozlowski. 1984b. Water relations, ethylene production, and morphological adaptation of *Fraxinus pennsylvanica* seedlings to flooding. Plant and Soil 77:183–192.

Tanksley, S. D., and R. A. Jones. 1981. Effects of O_2 stress on tomato alcohol dehydrogenase activity: description of a second ADH coding gene. Biochemical Genetics. 19:397–409.

Tarczynski, M. C., R. G. Jensen, and H. J. Bohnert. 1993. Stress protection of transgenic tobacco by production of the osmolyte mannitol. Science 259:508–510.

Taylor, I. B. 1991. Genetics of ABA synthesis. Pp. 23–38 in: W. J. Davies and H. G. Jones (eds.), Abscisic Acid: Physiology and Biochemistry. BIOS Scientific Publishers, Oxford.

Taylor, J. S., M. K. Bhalla, J. M. Robertson, and L. J. Piening. 1990. Cytokinins and abscisic acid in hardening winter wheat. Canadian Journal of Botany 68:1597–1601.

Teeri, J. A., and L. G. Stowe. 1976. Climatic patterns and the distribution of C_4 grasses in North America. Oecologia 23:1–12.

Teeri, J. A., L. G. Stowe, and D. A. Murawski. 1978. The climatology of succulent plant families: Cactaceae and Crassulaceae. Canadian Journal of Botany 56:1750–1758.

Telfer, E. S. 1969. Weight–diameter relationships for 22 woody plant species. Canadian Journal of Botany 47:1851–1855.

Tepperman, J. M., and Dunsmuir, P. 1990. Transformed plants with elevated levels of chloroplastic SOD are not more resistant to superoxide toxicity. Plant Molecular Biology 14:501–511.

Teramura, A. H. 1983. Effects of ultraviolet-B radiation on the growth and yield of crop plants. Physiologia Plantarum 58:415–427.

Tevini, M., and A. H. Teramura. 1989. UVB effects on terrestrial plants. Photochemistry and Photobiology 50:479–487.

Thayer, S. S., and O. Björkman. 1992. Carotenoid distribution and deep oxidation in thylakoid pigment-protein complexes from cotton leaves and bundle sheath cells of maize. Photosynthesis Research 33:213–225.

Thimann, K. 1980. The development of plant hormone research in the past 60 years. Pp. 15–33 in: F. Skoog (ed.), Plant Growth Substances. Springer-Verlag, Berlin.

Thomas, J. C., E. F. Mcelwain, and J. J. Bohnert. 1992. Covergent induction of osmotic stress-responses. Plant Physiology 100:416–423.

Thomas, J. C., A. C. Smigocki, and H. J. Bohnert. 1995. Light-induced expression of ipt from *Agrobacterium tumefaciens* results in cytokinin accumulation and osmotic stress symptoms in transgenic tobacco. Plant Molecular Biology 27:225–235.

Thomas, R. B., J. D. Lewis, and B. R. Strain. 1994. Effects of leaf nutrient status on photosynthetic capacity of lobloly pine (*Pinus taeda* L.) seedlings grown in elevated atmospheric CO_2. Tree Physiology 14:947–960.

Thomas, T. L., J. Vivekananda, and M. A. Bogue. 1991. ABA regulation of gene expression in embryos and mature plants. Pp. 23–38 in: W. J. Davies and H. G. Jones (eds.), Abscisic Acid: Physiology and Biochemistry. BIOS Scientific Publishers, Oxford.

Thornburn, P. J., and G. R. Walker. 1993. The source of water transpired by *Eucalyptus camaldulensis:* Soil, groundwater or streams. Pp. 511–528 in: J. R. Ehleringer, A. E. Hall, and G. D. Farquhar (eds.) Stable Isotopes and Plant Carbon Water Relations. Academic Press, San Diego.

Thornburn, P., G. Walker, and T. Hatton. 1992. Are river red gums taking water from soil, groundwater, or streams? Pp. 37–42 in: Catchments of Green. National Conference on Vegetation and Water Management, Adelaide, Australia.

Thornley, J. H. M. 1976. Mathematical Models in Plant Physiology. Academic Press, New York.

Thorsteinsson, B., and L. Eliasson. 1990. Growth retardation induced by nutritional deficiency or abscisic acid in *Lemna gibba:* the relationship of growth rate and endogenous cytokinin content. Plant Growth Regulation 9:171–181.

Tiessen, H., R. E. Karamonos, J. W. B. Stewart, and F. Selles. 1984. Natural nitrogen 15 abundance as an indicator of soil organic matter transformations in native and cultivated soils. Soil Science Society of America Journal 48:312–315.

Tihanyi, K., A. Fontanell, and J.-P. Thirion. 1989. Gene regulation during anaerobiosis in soya roots. Biochemical Genetics 27:719–730.

Tilman, D. 1988. Plant Stratagies and the Structure and Dynamics of Plant Communities. Princeton University Press, Princeton, NJ.

Ting, I. P. (ed.). 1982. Crassulacean Acid Metabolism (Proceedings of the Fifth Annual Symposium in Botany). American Society of Plant Physiologists, Rockville, Md.

Tingley, D. T., M. Manning, L. C. Grothaus, and W. F. Burns. 1980. Influence of light and temperature on monoterpene emission from slash pine. Plant Physiology 65:797–801.

Tingley, D. T., D. P. Turner, and J. A. Weber. 1991. Factors controlling the emissions of monoterpenes and other volatile organic compounds. Pp. 93–119 in: T. D. Sharkey, E. A. Holland, and H. A. Mooney (eds.), Trace Gas Emissions from Plants. Academic Press, San Diego, CA.

Tinoco-Ojanguren, C., and R. W. Pearcy. 1992. Dynamic stomatal behavior and its role in carbon gain during light flecks of a gap phase and understory piper species acclimated to high and low light. Oecologia 92:222–228.

Tinoco-Ojanguren, C., and R. W. Pearcy. 1993. Stomatal versus bichemical limitations to CO_2 assimilation during transient light in sun and shade adapted *Piper* species. Oecologia 94:395–402.

Tobiessen, P. 1982. Dark opening of stomata in successional trees. Oecologia 52:356–359.

Todd, G. W. 1972. Water deficits and enzyme activity. Pp. 177–216 in: T. T. Kozlowski (ed.), Water Deficits and Plant Growth. Vol. 3. Academic Press, New York.

Tokutomi, S., M. Nakasako, J. Sakai, M. Kataoka, K. T. Yamamoto, M. Wada, F. Tokunaga, and M. Furuya. 1989. A model for the diametric molecular structure of phytochrome based on small-angle x-ray scattering. FEBS Letters 247:139–142.

Tomes, D., A. K. Weissinger, M. Ross, R. Higgins, B. J. Drimmond, S. Schaaf, J. Malone-Schomesberg, M. Straebell, P. Flynn, J. Anderson, and J. Howard. 1990. Transgenic tobacco plants and their progeny created by microprojectile bombardment of tobacco leaves. Plant Molecular Biology 14:261–267.

Tong, C. B. S., and S. F. Yang. 1987. Chilling-induced ethylene production by beans and peas. Journal of Plant Growth Regulation 6:201–208.

Töpfer, R., B. Gronenburn, S. Schafer, J. Schell, and H. H. Steinbiss. 1990. Expression of engineered wheat dwarf virus in seed derived embryos. Physiologia Plantarum 79:158–162.

Toyomasu, T., K. Yamane, I. Yamaguchi, N. Murofushi, N. Takahashi, and Y. Inoue. 1992. Control by light of hypocotyl elongation and levels of endogenous gibberellins in seedlings of *Lactuca sativa* L. Plant Cell Physiology 33:695–701.

Trewavas, A. 1986. Understanding the control of plant development and the role of growth substances. Australian Journal of Plant Physiology 13:447–457.

Trewavas, A. J., and H. G. Jones. 1991. An assessment of the role of ABA in plant development. Pp. 169–188 in: W. J. Davies and H. G. Jones (eds.), Abscisic Acid: Physiology and Biochemistry. BIOS Scientific Publishers, Oxford.

Troughton, J. H., K. Card, and O. Bjorkman. 1974. Temperature effects on the carbon isotope ratio of C_3, C_4, and CAM plants. Carnegie Institute Washington Yearbook 73:780–784.

Tseng, M. J., and P. H. Li. 1990. Alternations of gene expression in potato (*Solanum commersonii*) during cold acclimation. Physiologia Plantarum 78:538–547.

Tucker, D. J. 1977. The effects of far-red light on lateral bud outgrowth in decapitated tomato plants and the associated changes in the levels of auxin and abscisic acid. Plant Science Letters 8:339–344.

Tucker, D. J., and T. A. Mansfield. 1972. Effects of light quality on apical dominance in *Xanthium strumarium* and the associated changes in endogenous levels of abscisic acid and cytokinins. Planta 102:140–151.

Tudela, D., and E. Primo-Millo. 1992. 1-Aminocyclopropane-1-carboxylic acid transported from roots to shoots promotes leaf abscission in Cleopatra Mandarin (*Citrus reshni* Hort. ex Tan.) seedlings rehydrated after water stress. Plant Physiology 100:131–137.

Turian, G. 1961. Accumulation of 3-indole acetic acid and activation of enzymes in the *Ustilago*-induced tumor of maize. Phytopathology 1:35–39.

Turner, N. C. 1981. Techniques and experimental approaches for the measurement of plant water stress. Plant and Soil 58:339–366.

Turner, N. C. 1986. Adaptation to water deficits: a changing perspective. Australian Journal of Plant Physiology 13:175–189.

Turner, N. C., E.-D. Schulze, and T. Gollam. 1984. Response of stomata and leaf gas exchange to vapor pressure deficits and soil water content. I. Species comparisons at high soil water contents. Oecologia 63:338–342.

Tyree, M. T., and M. A. Dixon. 1986. Water stress induced cavitation and embolism in some woody plants. Physiologia Plantarium 66:397–405.

Tyree, M. T., and H. T. Hammel. 1972. The measurement of turgor pressure and the water relations of plants by the pressure bomb technique. Journal of Experimental Botany 23:267–282.

Tyree, M. T., and P. G. Jarvis. 1982. Water in tissues and cells. Pp. 35–78 in: O. Larrge, P. S. Nobel, C. B. Osmond, and H. Ziegler (eds.), Encyclopedia of Plant Physiology, New Series, Vol. 12B. Springer-Verlag, Berlin.

Tyree, M. T., and J. S. Sperry. 1989. Vulnerability of xylem to cavitation and embolism. Annual Review of Plant Physiology and Molecular Biology 40:19–38.

Tyree, M. H., and S. Yang. 1990. Water capacity of *Thuja, Tsuga* and *Acer* stems measured by dehydration isotherms: the contribution of capillary water and cavitation. Planta 182:420–426.

Ueckert, J., T. Hurek, I. Fendrik, and E. -G. Niemann. 1990. Radial gas diffusion from roots of rice (*Oryza sativa* L.) and kallar grass (*Leptochloa fusca* L. Kunth) and the effects of innoculation with *Azospirillum brasilense* Cd. Plant and Soil 122:59–65.

Van Assche, C. J., H. M. Davies, and J. K. O'Neal. 1989. Superoxide dismutase expression in plants. European patent application, publication number EP 0356061 A2.

Van Bel, A. J. E. 1995. The low profile directors of carbon and nitrogen economy in plants: parenchyma cells associated with translocation channels. Pp. 205–222 in: B. L. Gartner (ed.), Plant Stems, Physiology and Functional Morphology. Academic Press, San Diego, CA.

Van der Krieken, W., A. Croes, M. Smulders, and G. Wullums. 1990. Cytokinins and flower bud formation in vitro in tobacco. Plant Physiology 92:565–569.

Van der Meer, I. M., M. J. M. Ebskamp, R. G. F. Visser, P. J. Weisbeek, and S. C. M. Smeekens. 1994. Fructan as a new carbohydrate sink in transgenic potato plants. Plant Cell 6:561–570.

van Loon, L. C. 1982. Regulation of changes in proteins and enzymes associated with active defense against virus infection. Pp. 247–273 in: R. K. S. Wood (ed.), Active Defense Mechanisms in Plants (NATO Advanced Study Institute Series, No. 37). Plenum Press, New York.

van Meeteren, U., and M. de Proft. 1982. Inhibition of flower bud abscission and ethylene evolution by light and silver thiosulphate in *Lilium*. Physiologia Plantarum 56:236–240.

Vanicková-Zemlová, D., H. W. Liebisch, and G. Sembdner. 1981. Role of gibberellins and endogenous inhibitiors in the pathogenesis of dwarf smut infected winter wheat (*Triticum aestivum* L. /*Tilletia controversa* Kuhn). Biochemie und Physiologie der Pflanzen 176:291–304.

Vantoai, T. T. 1993. Field performance of abscisic acid-induced flood-tolerant corn. Crop Science 33:344–346.

Vartepetian, B. B. 1991. Flood-sensitive plants under primary and secondary anoxia: ultrastructural and metabolic responses. Pp. 201–216 in: M. B. Jackson, D. D. Davis, and H. Lambers (eds.), Plant Life Under Oxygen Deprivation. SPB Academic Publishing, The Hague.

Vernieri, P., A. Pardossi, G. Serra, and F. Tognoni. 1994. Changes in abscisic acid and its glucose ester in *Phaseolus vulgaris* L. during chilling and water stress. Plant Growth Regulation 15:157–163.

Vernieri, P., A. Pardossi, and F. Tognoni. 1991. Influence of chilling and drought on water relations and abscisic acid accumulation in bean. Australian Journal of Plant Physiology 18:25–35.

Vickery, P. J. 1976. Grazing and net primary production of a temperate grassland. Journal of Applied Ecology 9:307–314.

Vieira da Silva, J., A. W. Naylor, and P. J. Kramer. 1974. Some ultrastructural and enzymatic effects of water stress in cotton (*Gossypium hirsutum* L.) leaves. Proceedings of the National Academy of Sciences USA 71:3243–3247.

Vierling, E. 1991. The roles of heat shock proteins in plants. Annual Review of Plant Physiology and Plant Molecular Biology 42:579–620.

Vierling, R. A., and H. T. Nguyen. 1990. Heat-shock proein synthesis and accumulation in diploid wheat. Crop Science 30:1337–1342.

Vigh, L., Z. Gombos, I. Horváth, and F. Joó. 1989. Saturation of membrane lipids by hydrogenation induces thermal stability in chloroplast inhibiting the heat-dependent stimulation of photosystem. I. Mediated electron transport. Biochimica Biophysica Acta 979:361–364.

Vilardell, J, A. Goday, M. A. Freire, M. Torrent, M. C. Martínez, J. M. Torné, and M. Pagès. 1990. Gene sequence, developmental expression, and protein phosphorylation of RAB-17 in maize. Plant Molecular Biology 14:423–432.

Virginia, R. A, and C. C. Delwiche. 1982. Natural ^{15}N abundance of presumed N_2-fixing and non-N_2-fixing plants from selected ecosystems. Oecologia 54:317–325.

Virginia, R. A., L. M. Baird, J. S. LaFavie, W. M. Jarrell, B. A. Gryan, and G. Shearer. 1984. Nitrogen fixing efficiency, natural ^{15}N abundance, and morphology of mesquite (*Prosopis glandulosa*) root nodules. Plant and Soil 79:273–284.

Virginia, R. A., W. M. Jarrell, P. W. Rundel, G. Shearer, and D. H. Kohl. 1989. The use of variation in the natural abundance of ^{15}N to assess the symbiotic nitrogen fixation by woody plants. Pp. 375–394 in: P. W. Rundel, J. R. Ehleringer, and K. A. Nagy (eds.), Stable Isotopes in ecological research. Ecological Studies 68. Springer-Verlag, Berlin.

Voesenek, L. A. C. J., and R. Van Der Veen. 1994. The role of phytohormones in plant stress: too much or too little water. Acta Botanica Nederlandica 43:91–127.

Voesenek, L. A. C. J., and C. W. P. M. Blom. 1989. Growth responses of *Rumex* species in relation to submergence and ethylene. Plant Cell & Environment 12:433–439.

Voesenek, L. A. C. J., M. Banga, R. H. Thier, C. M. Mudde, F. J. M. Harren, G. W. M. Barendse, and C. W. P. M. Blom. 1993. Submergence-induced ethylene synthesis, entrapment, and growth in two plant species with contrasting flooding resistances. Plant Physiology 103:783–791.

Voesenek, L. A. C. J., F. J. M. Harren, G. M. Bögemann, C. W. P. M. Blom, and J. Reuss. 1990. Ethylene production and petiole growth in *Rumex* plants induced by soil waterlogging. Plant Physiology 94:1071–1077.

Vogel, J. C. 1978. Recycling of carbon in a forest environment. Oecologia Plantarum 13:89–94.

Vogel, J. C. 1980. Fractionation of Carbon Isotopes During Photosynthesis. Springer-Verlag, Berlin.

Vogel, S. 1981. Life in Moving Fluids: The Physical Biology of Flow. Willard Grant Press, Boston.

Von Cammerrer, S., and G. D. Farquhar. 1981. Some relationships between the biochemistry of photosynthesis and the gas exchange of leaves. Planta 153:376–387.

Von Schaeven, A., M. Stitt, R. Schmidt, U. Sonnewald, and L. Willmitzer. 1990. Expression of a yeast-derived invertase in the cell wall of tobacco and *Arabidopsis* plants leads to accumulation of carbohydrate and inhibition of photosynthesis and strongly influences growth and phenotype of transgenic tobacco plants. The European Molecular Biology Organization Journal 9:3033–3044.

Vu, C. V., L. H. Allen, and L. A. Garrard. 1982a. Effects of supplemental UV-B radiation on primary photosynthetic carboxylating enzymes and soluble proteins in leaves of C_3 and C_4 crop plants. Physiologia Plantarum 55:11–16.

Vu, C. V., L. H. Allen, and L. A. Gerrard. 1982b. Effects of UV-B radiation (280–320nm) on photosynthetic constituents and processes in expanding leaves of soybean [*Glycine max* (L.) Mell.]. Environmental and Experimental Botany 22:465–473.

Wada, E., and A. Hattori. 1978. Nitrogen isotope effects in the assimilation of inorganic nitrogen compounds by marine diatoms. Geomicrobiology Journal 1:85–110.

Wada, H., and N. Murata. 1990. Temperature-induced changes in the fatty acid composition of the Cyanobacterium, *Synechocystis* PCC6803. Plant physiol. 92:1062–1069.

Wada, H., Z. Gombos, and N. Murata. 1990. Enhancement of chilling tolerance of a cyanobacterium by genetic manipulation of fatty acid desaturation. Nature 347:200–203.

Wadman-van Schravendijk, H., and O. M. van Andel. 1986. The role of ethylene during flooding of *Phaseolus vulgaris*. Physiologia Plantarum 66:257–264.

Wagner, B. M., and E. Beck. 1993. Cytokinins in the perennial herb *Urtica dioica* L. as influenced by its nitrogen status. Planta. 190:511–518.

Walker, M. A., and E. B. Dumbroff. 1981. Effects of salt stress on abscisic acid and cytokinin levels in tomato. Zeitschrift für Pflanzenphysiologie 101:461–470.

Wallace, H. A. 1920. Mathematical inquiry into the effect of weather on crop yield in the eight corn belt states. Monthly Weather Review 48:439–446.

Walters, M. B., and P. B. Reich. 1989. Responses of *Ulmus americana* seedlings to varying nitrogen and water status. I. Photosynthesis and growth. Tree Physiology 5:159–172.

Walton, J. D., and P. M. Ray. 1981. Evidence for receptor function of auxin binding sites in maize. Plant Physiology 68:1334–1338.

Wample, R. L., and D. M. Reid. 1979. The role of endogenous auxin and ethylene in the formation of adeventitious roots and hypocotyl hypertrophy in flooded sunflower plants (*Helianthus annuus*). Physiologia Plantarum 45:219–226.

Ward, D. A., and D. W. Lawlor. 1990. Abscisic acid may mediate the rapid thermal acclimatization of photosynthesis in wheat. Journal of Experimental Botany 41:309–314.

Waring, R. H., and S. W. Running. 1978. Sapwood water storage: its contribution to transpiration and effect upon water conductance through the stems of old-growth Douglas-fir. Plant Cell & Environment 1:131–140.

Watson, D. J., T. Motomatsu, K. Loach, and G. F. J. Milford. 1972. Effects of shading and of seasonal differences in weather on the growth, sugar content and sugar yield of sugar-beet crops. Annals of Applied Biology 71:159–185.

Weber, G., S. Monajembashi, K. D. Greulich, and J. Wolfrum. 1988. Injection of DNA into plant cells with a UV laser microbeam. Naturwissenschaften 75:35–36.

Weber, J. A., I. N. Jurik, J. D. Tenhunen, and D. M. Gates. 1985. Analysis of gas exchange in seedlings of *Acer saccrarum:* integration of field and laboratory studies. Oecologia 65:338–347.

Weis, E., and J. A. Berry. 1988. Plants and high temperature stress. Pp. 329–346 in: S. P. Long and F. I. Woodward (eds.), Plants and Temperature. The Company of Biologists Ltd., Cambridge, UK.

Weiser, R. L., S. J. Wallner, and J. W. Waddell. 1990. Cell wall and extension mRNA changes during cold acclimation of pea seedlings. Plant Physiology 93:1021–1026.

Welbaum, G. E., and K. J. Bradford. 1988. Water relations of seed development and germination in muskmelon (*Cucumis melo* L.) Plant Physiology 86:406–411.

Weller, J. L., J. J. Ross, and J. B. Reid. 1994. Gibberellins and phytochrome regulation of stem elongation in pea. Planta. 192:489–496.

Werner-Washbourne, M., J. Becker, J. Kosic-Smithers, and E. A. Craig. 1989. Yeast HSP70 levels vary in response to the physiological status of the cell. Journal of Bacteriology 171:2680–2688.

Westman, W. E. 1981. Seasonal dimorphism of foliage in California coastal sage shrub. Oecologia 51:385–388.

Whenham, R. J. 1982. Abscisic acid metabolism in tobacco (*Nicotiana tabacum* L.) infected with tobacco mosaic virus. Ph. D. thesis, University of Birmingham, Birmingham, England.

Whenham, R. J., and F. S. S. Fraser. 1990. Plant growth regulators, viruses and plant growth. Pp. 287–310 in: R. S. S. Fraser (ed.), Recognition and Response in Plant–Virus Interactions. Springer-Verlag, Berlin.

Whenham, R. J., R. S. S. Fraser, and A. Snow. 1985. Tobacco mosaic virus–induced increases in abscisic acid concentration in tobacco leaves: intracellular location and relationship to symptom severity and to extent of virus multiplication. Physiological Plant Pathology 26:379–387.

Whenham, R. J., I. G. Burns, D. A. Stone, and R. S. S. Fraser. 1989. Effect of nitrogen nutrition and water regime on abscisic, phaseic and dihydrophaseic acid metabolism in leaves of field-grown kale (*Brassica oleracea*): consequences for plant growth and crop yield. Journal of the Science of Food and Agriculture 49:143–155.

White, J. W. C., E. R. Cook, J. R. Lawrence, and W. S. Broecker. 1985. The D/H ratio of sap in trees: implications for water sources and tree ring D/H ratios. Geochemica Cosmochemica Acta 49:237–246.

White, J. 1979. The plant as a metapopulation. Annual Review of Ecology and Systematics 10:109–145.

White, L. M. 1973. Carbohydrate reserves of grasses: a review. Journal of Range Management 26:13–18.

Whittaker, R. H., and P. L. Marks. 1975. Measurement of net primary productivity on land. Pp. 55–118 in: H. Lieth and R. H. Whittaker (eds.). Primary Productivity of the Biosphere. Springer-Verlag, New York.

Whittaker, R. 1975. Communities and Ecosystems. Macmillan, New York.

Whittaker, R., and G. M. Woodwell. 1968. Dimension and production relations of trees and shrubs in the Brookhaven forest, New York. Journal of Ecology 56:1–25.

Wien, H. C., and A. D. Turner. 1989. Hormonal basis for low light intensity–induced flower bud abscission of pepper. Journal of the American Society of Horticultural Science 114:981–985.

Wiese, M. V., and J. E. DeVay. 1970. Growth regulator changes in cotton associated with defoliation caused by *Verticillium albo-atrum*. Plant Physiology 45:304–309.

Wilkinson, R. E. 1994. Isoprenoid biosynthesis. Pp. 149–183 in: R. E. Wilkinson (ed.), Plant–Environment Interactions. Marcel Dekker, New York.

Willert, D. J. von, E. Brinckmann, B. Scheitler, and B. M. Eller. 1985. Availability of water controls cranulacean acid metabolism in succulents of the Richtersveld (Namib desert, South Africa). Planta 164:44–55.

Williams, K., C. B. Field, and H. A. Mooney. 1989. Relationships among leaf construction cost, leaf longevity and light environment in rain-forest plants of the genus *Piper.* American Naturalist 133:198–211.

Williams, S. A., S. C. Weatherwax, E. A. Bray, and E. M. Tobin. 1994. NPR genes which are negatively regulated by phytochrome action in *Lemna gibba* L. G-3, can also be positively regulated by abscisic acid. Plant Physiology 105:949–954.

Wilson, D. 1982. Response to selection for dark respiration rate of mature leaves in *Lolium perenne* and its effects on growth of young plants and simulated swards. Annals of Botany 49:303–312.

Wilson, J. M. 1987. Chilling injury in plants. Pp. 272–292 in: B. W. W. Grout and G. J. Morris (eds.), The Effects of Low Temperatures on Biological Systems. Edward Arnold, London.

Wilson, J. M., and A. C. McMurdo. 1981. Chilling injury in plants. Pp. 145–172 in: G. J. Morris and A. Clarke (eds.), Effects of Low Temperatures on Biological Membranes. Academic Press, New York.

Winer, A. M., J. Arey, R. Atkinson, S. M. Aschman, W. D. Long, C. L. Morrison, and D. M. Olszky. 1992. Emission rates of organics from vegetation in California's Central Valley. Atmospheric Environment 262:2647–2659.

Winner, W. E., and H. A. Mooney. 1980a. Ecology of SO_2 resistance. I. Effects of fumigations on the gas exchange of deciduous and evergreen shrubs. Oecologia 44:290–295.

Winner, W. E., and H. A. Mooney. 1980b. Ecology of SO_2 resistance. II. Photosynthetic changes in shrubs in relation to SO_2 absorption and stomatal behaviour. Oecologia 44:296–302.

Winter, K., V. Lüttge, E. Winter, and J. H. Troughton. 1978. Seasonal shift from C_3 photosynthesis to crassulacean acid metabolism in *Mesembryanthemum crystallinum* in its native environment. Oecologia 34:225–237.

Witty, J. F., and K. E. Giller. 1991. Evaluation of errors in the measurement of biological nitrogen fixation using ^{15}N fertilizer. Pp. 59–72 in: Stable Isotopes in Plant Nutrition, Soil Fertility, and Environmental Studies. IAEA, Vienna, Austria.

Wolf, O., W. D. Jeschke, and W. Hartung. 1990. Long distance transport of abscisic acid in NaCl-treated intact plants of *Lupinus albus*. Journal of Experimental Botany 41:593–600.

Wolter, F. P., R. Schmidt, and E. Heinz. 1992. Chilling sensitivity of *Arabidopsis thaliana* with genetically engineered membrane lipids. The European Molecular Biology Organization Journal 11:4685–4692.

Wright, M. 1974. The effect of chilling on ethylene production, membrane permeability and water loss of leaves of *Phaseolus vulgaris*. Planta 20:63–69.

Wright, S. T. C. 1981. The effect of light and dark periods on the production of ethylene from water-stressed wheat leaves. Planta 153:172–180.

Yabuta, T., and T. Hayashi. 1939. Biochemical studies on "bakanae" fungus of the rice. II. Isolation of "gibberellin," the active principle which makes the rice seedlings grow slenderly. Journal of the Agricultural Chemical Society (Japan) 15:257–263.

Yakir, D., and M. J. DeNiro. 1990. Oxygen and hydrogen isotope fractionation during cellulose metabolism in *Lemna gibba* L. Plant Physiology 93:325–332.

Yakir, D., M. J. DeNiro, and P. W. Rundel. 1989. Isotopic inhomogeneity of leaf water: evidence and implications for the use of isotope signals transduced by plants. Geochemica et Cosmochemica Acta 53:2769–2773.

Yamada, T. 1993. The role of auxin in plant-disease development. Annual Review of Phytopathology 31:253–273.

Yamaguchi-Shinozaki, K., J. Mundy, and N.-H. Chua. 1989. Four tightly linked *rab* genes are differentially expressed in rice. Plant Molecular Biology 14:29–39.

Yamamoto, F., and T. T. Kozlowski. 1987a. Effect of flooding of soil on growth, stem anatomy, and ethylene production of *Cryptomeria japonica* seedlings. Scandinavian Journal of Forest Research 2:45–58.

Yamamoto, F., and T. T. Kozlowski. 1987b. Regulation by auxin and ethylene of responses of *Acer negundo* seedlings to flooding of soil. Environmental and Experimental Botany 27:329–340.

Zaenen, I., N. Van Larcbeke, M. Van Montagu, and J. Schell. 1974. Supercoiled circular DNA in crown gall inducing *Agrobacterium* strains. Journal of Molecular Biology 86:109–127.

Zarambinski, T. I., and A. Theologis. 1993. Anaerobiosis and plant growth hormones induce two genes encoding 1-aminocyclopropane-1-carboxylate synthase in rice (*Oryza sativa* L.). Molecular Biology of the Cell 4:363–373.

Zeigler, H. 1989. Hydrogen isotope fractionation in plant tissues. Pp. 105–123 in: P. W. Rundel, J. R. Ehleringer, and K. A. Nagey (eds.), Stable Isotopes in Ecological Research (Ecological Studies. No. 68). Springer-Verlag, Berlin.

Zhang, J., and W. J. Davies. 1987. ABA in roots and leaves of flooded pea plants. Journal of Experimental Botany 38:649–659.

Zhang, J., V. Schurr, and W. J. Davies. 1987. Control of stomatal behavior by abcissic acid which apparently originates in the roots. Journal of Experimental Botany 38:1174–1181.

Zimmerman, J. K., and J. R. Ehleringer. 1990. Carbon isotope ratios are correlated with irradiance in leaves of the Panamanian orchid *Catasetum viriflavum*. Oecologia 83:247–249.

Zimmerman, U., and E. Steudle. 1975. The hydraulic conductivity and volumetric elastic modulus of cells and isolated cell walls of *Nitella,* and *Chara* spp.: pressure and volume effects. Australian Journal of Plant Physiology 2:1–12.

Zimmerman U., A. Haase, D. Langbein, and F. Meinzer. 1993. Mechanisms of long distance water transport in plants: a re-examination of some paradigms in the light of new evidence. Philosphical Transactions of the Royal Society of London 341:19–31.

APPENDIX C
Glossary

Abiotic. Not involving or produced by organisms; used loosely for factors of the environment and their effects on plants, producing stress and stress symptoms; physiogenic.

Abscisic acid. A phytohormone that generally acts to inhibit metabolic function and to promote dormancy.

Acclimation. Nonheritable modification caused by exposure to new environmental conditions; organisms undergoing changes in physiological processes rendering them tolerant or resistant to stress; nonheritable alterations of phenotype.

Activation energy. The initial investment of energy in a reaction that is required before the reaction will occur on a thermodynamic basis. Only those molecules with energy above the activation energy can react at a given time.

Active site. The specific aspect of an enzyme that is involved in the chemical reaction that the enzyme facilitates.

Acropetal. Movement or translocation of solutes and hormones from the base to the tip of the plant.

Acute. An injurious event that usually involves necrosis, such as fleck, scorch, or bifacial necrosis. In air pollution usage, injury caused by an air pollutant over a short time period; not chronic.

Adaptation. Processes of evolution causing a plant to be fit or adapted for the environment in which it thrives; if not successful, the species perishes.

Air pollution. Gaseous or particulate substances released into the atmosphere in sufficient quantities or concentrations to cause injury to plant life.

Age structure. The proportional distribution of individuals in a population among age classes. An age structure may refer to a population of individuals or a population of organs such as leaves on a plant.

Aldehyde. An organic molecule that has a carbonyl group located at the end of the molecule.

Alleles. Alternative forms of genes. There are normally two alleles for each gene in each individual.

Allelochemicals. Chemical compounds released from a living organism that affect other organisms in the vicinity.

Allelopathy. Biochemical interactions between living organisms; any direct or indirect effect by one plant on another through production of chemical compounds that escape into the environment.

Allocation. The proportional distribution of a particular chemical or resource among the various different functional organs in plants.

Amino acids. Organic molecules possesing both amino and carboxylic acid groups. Between these two groups is a small variable organic constituent.

Amino group. A combination of atoms that contains a nitrogen atom bonded to two hydrogen atoms. This functional group can act as a base in a solution by accepting a hydrogen to form a positively charged moiety.

Amphiphatic. Phospholipids exhibiting polar charges on one end and nonpolar characteristics on the other end.

Ampipoteric. Phospholipid molecules having a polar hydrophilic head and a nonpolar hydrophobic tail.

Anion. A negatively charged ion.

Antiport. Permeases that transport two different substances in opposite directions simultaneously across membranes.

Apical meristem. An embryonic region at the apex of plant shoots in which cells are undergoing rapid cell division. This is the region that regulates shoot growth in length.

Apoplastic. Free space; upon entering the root on its way to the xylem, water moves through intercellular spaces and along cell walls following the pathway of least resistance.

Avoidance. Stress avoidance caused by the plant not coming to thermodynamic equilibrium with the stress; prevention or decrease in the penetration of a stress into tissues.

Basipetal. Movement or translocation of solutes and hormones from the tip to the base of the plant.

Biosphere. The entire portion of the earth that is inhabited by living organisms.

Biotechnology. The industrial use of living organisms to improve human health, environmental contamination, and food production.

Biotic. Induced or caused by action of living organisms: a combining form used to complete words (e.g., antibiotic, endobiotic, exobiotic, of or relating to life).

Boundary layer. A layer of undisturbed air or solvent next to a surface; commonly the layer of air at the surface of a leaf.

Buffer. A substance that consists of acid and base forms that are interchangeable. When this substance is in solution it is able to minimize the change in pH when an acid or base is added to the solution.

C_3 cycle. Carbon assimilation that results in formation of three-carbon compounds such as phosphoglycerate; the Calvin cycle.

C_4 cycle. Carbon assimilation that results in the formation of four-carbon compounds such as malate or aspartate.

Calcicole. A plant that is able to grow and develop on calcareous soils.

Calcifuge. A plant that is injured, killed, or inhibited on calcareous soils.

CAM. *See* Crassulacean acid metabolism.

Capillary water. That quantity of water that is held in small spaces between soil particles or in the apoplastic space of plant tissues. This water is held in place against gravity because of the adhesive and cohesive attractive forces of water.

Carbonyl group. A functional group of atoms consisting of a carbon atom double bonded to an oxygen atom. These groups are found in many molecules, including ketones and aldehydes.

Carboxylic group. A functional group of atoms consisting of a carbon double bonded to an oxygen atom and also single bonded to a hydroxyl (OH) group. This group is found in organic acids, amino acids, and a number of other molecules.

Casparian strip. The heavily suberized transverse and radial walls of the endodermis; water and solutes are unable to leak past the endodermis through intercellular spaces and must pass through the cytoplasm.

Chelate. A metal ion plus a ligand; two important metal chelates found in nature are chlorophyll and heme; chelates play an important role in metabolism and in detoxification of heavy metals.

Chlorosis. Disappearance of chlorophyll as a result of altered metabolism brought on by disease or stress.

Chronic. An injurious event that occurs over time; exposure to an unfavorable environmental factor for a prolonged time, resulting in injury; as opposed to acute, in air pollution injury.

Common garden. A site in which species that naturally grow in different regions or climatic conditions are grown together. The site may be an agricultural field, greenhouse, or growth chamber.

Coordination number. The number of positions within each central atom's (metal) coordination sphere that can be occupied by ligands, irrespective of whether the ligands are neutral or charged. The maximal number of functional groups with which a metal will complex.

Crassulacean acid metabolism. CAM metabolism occurs in species of plants that have open stomates at night and fix CO_2 into organic acids such as malic acid, which is stored until daylight occurs; decarboxylation occurs in daylight, releasing CO_2; CAM plants are able to photosynthesize effectively without open stomatas during periods favoring high transpiration rates.

Critical concentration. That concentration of a nutrient in plant tissue just at the level giving optimal growth.

Cryoprotectant. A chemical compound whose presence prevents injury during freezing; may occur naturally or be applied.

Cuticle. The waxy deposit on the surface of the plant outside the epidermis.

Cuticular resistance. A measure of the reduction in diffusion through the plant tissues as a result of the presence of the cuticle; the reciprocal of conductance; cuticle may vary in thickness, porosity, and composition, rendering more or less resistance to diffusion of gases such as water vapor and carbon dioxide.

Cuticular transpiration. Outward diffusion of water vapor through the cuticle.

Darcy's laws. The first law is for isothermal, steady-state liquid water flow in a porous medium; the second law states that water will flow across an air–water interface only when the pressure potential is sufficiently greater than zero to overcome the surface tension of the fluid.

Dehydration avoidance. Prevention of loss of water by closing of stomata, resulting in the maintenance of turgor during periods favoring high rates of transpiration.

Dehydration postponement. Reduction of transpiration or increase of water absorption because of morphological characteristics of drought-stressed plants.

Dehydration tolerance. The ability of plants to withstand injury at the protoplasmic level when plants are under drought stress.

Desiccation tolerance. Protoplasmic resistance to injury under extreme water-stress conditions; upon rehydration, respiration and photosynthesis begin within a relatively short time.

Disease. A condition of the living animal or plant body or one of its parts that impairs the performance of a vital function; impairment or modification of the performance of a vital function or functions; response to environmental factors or inherent genetic defects.

Dormancy. A state of inactive growth; typically used in reference to buds and seeds; important in resistance to stresses.

Dose. Exposure to a measured concentration of a toxicant for a known duration of time.

Drought. A meteorological term for prolonged periods of reduced precipitation.

Drought resistance. Innate mechanisms by which injury is prevented or minimized during drought stress; some scientists prefer to use the term *drought tolerance.*

Drought tolerance. Result of postponement of dehydration or because of innate dehydration tolerance characteristics. *See also* Dehydration tolerance.

Electrochemical potential. A potential across a membrane composed of two components, a concentration component and an electrical component; final direction of diffusion of a solute is determined by the component that has the steepest gradient across the membrane.

Entropy. A measure of the degree to which a system is removed from equilibrium; as the degree of randomness increases in a system, entropy increases.

Escape. Because of the timing sequence in development and growth, some plants are not subjected to a stress; the life cycle may be completed during a period or season when a particular stress does not occur.

Euhalophytes. Halophytes that grow in conditions of extreme salinity.

Euxerophytes. Plants that grow under dry conditions, such as in deserts.

Facultative halophytes. Plants that acclimate to saline conditions.

Flux. Volume flow across a unit area per unit time; usually given in mol cm^{-2} s^{-1}, for example.

Freezing avoidance. Effects of freezing are excluded from protoplasm.

Frost hardiness. Frost tolerance; may be induced by gradual exposure to lowered temperatures.

Frost resistance. Frost tolerance.

Frost tolerance. Injury does not occur at temperatures of 1 to 3°C.

Gene. A sequence of nucleotides in the DNA of an organism that codes for a particular polypetide and contains other sequences that regulate the transcription of the sequence coding for the polypeptide.

Gibberellins. Growth regulators that cause cell enlargement, promote shoot growth.

Gibbs free energy. Expressed as $G = E + PV - TS$, where E is the internal energy, PV the pressure–volume constant, T the absolute temperature, and S the entropy; the energy isothermally (constant temperature) available for conversion to work.

Glycophytes. Plants that are sensitive to relatively high soil salt concentrations.

Growth inhibitors. Organic chemicals that slow growth; may be naturally occurring or applied.

Growth regulators. Organic chemicals that promote growth when present in relatively low concentrations and may inhibit growth at higher concentrations.

Halophytes. Plants that are able to grow in the presence of high soil salt concentration. *See also* Euhalophytes; Facultative halophytes; Obligate halophytes.

Hardening. Acclimation; increased tolerance of stress; may be caused by gradual exposure to stress; changes may be phenotypic or occur at the protoplasmic level.

Hardiness. Ability to withstand or tolerate stress conditions.

Health. The condition of being sound in body; freedom from disease.

Heat tolerance. High-temperature tolerance; means of dissipating high-temperature effects are present, or physiological changes occur that prevent injury.

Homeostasis. A relatively stable state of equilibrium or tendency toward such a state between the different but interdependent elements or groups of elements of an organism or group.

Homogeneous nucleation. Supercooled water freezes instantaneously when a nucleator is introduced, but at a temperature of $-38.1°C$, water in small droplets freezes automatically without a nucleator.

Hormone. A specific organic product produced in one part of a plant or animal and transported to another part, where it is effective in small amounts in controlling or stimulating processes.

Hydrogen bonding. The elements N, O, and F are the three most electronegative elements; because of this, they apparently attract the electrons of the covalent bond to such an extent as to leave the H nucleus almost completely exposed. H is the smallest of all atoms; therefore, it is possible for it to come very close to an electronegative atom of another molecule. The resulting electrostatic attraction, while weak compared to a covalent bond, is much stronger than the attraction between dipoles or ordinary van der Waals forces and for this reason is called a hydrogen bond.

Hydrophytes. Plants that live in water or in virtually water-saturated soils.

Ice nucleators. Materials that cause freezing of water when it is supercooled; ice crystals may serve this purpose when introduced into supercooled water.

Injury. A change in metabolism that results in altered functions and/or appearance with deleterious effects.

Ionic bond. The strong electrostatic attraction between two opposite charged ions.

Ionization constant. Written as K or K_b; an indication of the constant strength of an acid or base; the greater the degree to which the acid or base ionizes at a given concentration, the stronger it is and the larger is the value of its ionization constant.

Lateral phase separation. Separation of phospholipids in a heterogeneous mixture into more homogeneous groups in membranes when the temperature is reduced.

Lewis acid. Any substance capable of acting as an electron pair acceptor; yields a hydrogen ion in solution.

Lewis base. Any substance possessing a pair of electrons available for sharing with a proton to form a covalent bond; yields a hydroxyl ion in solution.

Ligand. Can be either neutral molecules or negative ions, both of which possess unshared electron pairs and are capable of functioning as electron donors, that is, as Lewis bases. Each molecule or ion possesses, therefore, at least one donor center capable of occupying a position in the coordination sphere of the metal atom. One donor center is described as *unidentate*; a ligand with two or more donor centers is said to be *multidentate*. Multidentate ligands, such as EDTA, are commonly called *chelating agents* because of the manner in which they enclose the central atom in a pincerlike manner reminiscent of the action of a crab's claw (Greek: chele); the complexes are called *chelates*.

Light-harvesting complex. Complex of pigments and proteins that intercept light impinging on leaves; chlorophyll plus other pigments.

Light saturation. The light intensity at which no further increase in intensity can cause an increase in photosynthesis.

Liquid crystalline. Fluid state of pure phospholipids; related to phase transitions in membranes; caused by temperature.

Luxury consumption. Absorption of mineral nutrients above and beyond those that cause increases in growth or yield.

Matric potential. (Ψ_m). The term is used to account for all the forces of capillary, adsorption, and hydration that cause imbibition or holding of water in a matrix of any sort.

Membrane carriers. Molecules or compounds that assist materials to cross membranes; associated with transport of charged particles such as ions; many are proteins.

Membrane fluidity. Changes with temperature and amount of cross linkage.

Membrane transition. Change in the physical state of membrane components as temperature is lowered; change from liquid-crystalline state to gel structure.

Mesophytes. Plants that thrive best with a moderate water supply.

Metalloenzymes. Enzymes that contain a transition metal as the prosthetic group.

Mycorrhizae. A symbiotic association between a fungus and usually the root of a higher plant; the fungus normally makes available to the root forms of elements (especially N and P) that would otherwise be unavailable; mycorrhizal associations occur throughout the plant kingdom.

Natrophilic. "Sodium loving"; refers to plants that grow in sodium-rich soils.

Necrosis. Death.

Niche. A theoretic space defined by the physiological tolerances of a species. The physiological tolerances usually refer to resource availability but they may also refer to biotic or abiotic stresses of the environment.

Noninfectious disease. A disease that is caused by an environmental factor, not by a pathogen; caused by a physiogen; a physiogenic disease.

Nutrient balance. The total number, on an equivalent basis, of cations and anions absorbed by a plant. Plants under uniform environmental conditions tend to take in a constant number of cations and anions. Since the total number of cations and anions absorbed remains constant, if K^+ in the plant were increased, Ca^{2+} and Mg^{2+} would tend to decrease, and vice versa.

Obligate halophytes. Halophytes that are able to grow only in saline soils.

Obligate shade plants. Plants that grow well in reduced light intensities but are injured when placed in relatively high light intensity.

Oligohalophytes. Halophytes that grow in moderately saline soils.

Osmoprotectants. Organic solutes produced as a result of salt stress.

Osmoregulation. The control of plant functions by the regulation of water potentials across differentially permeable barriers or membranes. Water diffusing down a potential gradient is losing energy and can be made to do work; an example is stomatal control by the regulation of the turgidity of guard cells.

Osmotic adaptation. See osmotic adjustment.

Osmotic adjustment. Increase in organic solute content as a result of stress; lowers the osmotic potential of cell sap.

Osmotic potential. The component of water potential that becomes increasingly negative with the addition of solute.

Pathogen. A disease-initiating biotic entity; usually restricted to organismic causes (e.g., bacteria, fungi, viruses).

Pathogenesis. The origination and development of a pathogenic disease.

Permanent wilting percentage (PWP). A moisture content at which the soil is incapable of maintaining succulent plants in an unwilted condition even though transpirational loss is negligibly low; approximately equivalent to 1.5 MPa.

Permeability. A property of a membrane describing material movement across the membrane.

Phase separation. As water in tissues freezes, the solid phase (ice) is separated from liquid, unfrozen water.

Phase transition. Change from liquid to solid as from liquid water to solid ice.

Phenotypic plasticity. An alteration in a trait expression for a given genotype. The change in trait expression is not due to a change in genetic composition; rather, the change is due to a change in gene expression.

Photoinhibition. Light-reduced activity such as growth or an enzymatic reaction.

Photomorphogenetic. Ability to change morphology as a result of exposure to light.

Photophosphorylation. The synthesis of ATP in the illuminated chloroplasts.

Phylogenic. Having to do with the evolutionary lineage of a particular process or organism.

Physiogen. A disease-initiating abiotic entity or event; caused by physical or chemical factors in the environment.

Physiogenesis. The origination and development of a physiogenic disease.

Plasmolysis. The condition of a cell in which the turgor has been reduced by removal of water by osmosis; the cytoplasm may separate from the wall of the cell.

Poikilotherms. Organisms that attain the temperature of the environment while still alive; plants and some animals.

Polymerase Chain Reaction. A reaction that utilizes a thermally stable DNA polymerase to multiply particular fragments of a DNA molecule. The reaction involves thermal cycling between high and low temperatures to induce DNA replication and melt DNA for the next replication cycle.

Population. All the individuals of one species that are located in a specific region. The delineation of the population is often based on environmental attributes, but the actual criterion defining the individuals in a population is the likelihood of interbreeding.

Reciprocal transplant experiment. An experiment in which species that naturally grow in different habitats are both transplanted into the other species habitat. This experiment also requires that both species be transplanted into their own habitat as a control.

Recombinant DNA. Inserting or replacing a fragment of a DNA molecule, thus changing the genome of an organism or plasmid.

Resistance. Tolerance or avoidance; the ability of an organism to withstand or overcome partially or completely the injury caused by a physiogenic or a pathogenic factor; the means of resisting, fighting against, or counteracting the injurious effects of environment.

Rhizosphere. A volume of soil near the root surface that is affected by the presence of the root and the chemical and physical interactions that occur between the root and its immediate environment; the root's sphere of influence.

Relative humidity (RH). The percentage of water vapor present in the atmosphere relative to saturation (100%) at the existing temperature.

Saturated flow. Water flow in a soil in which all pores are filled with water; the hydraulic conductance (K) in a specific saturated soil is at its maximum and is constant.

Scale. The magnitude in volume, area, or time of the particular process under consideration. The study of whole leaf photosynthesis occurs at a greater scale than the study of rubisco activation.

Solar constant. Intensity of unfiltered sunlight entering the ionosphere; 1.39 kW/m^{-2}.

Stability. As applied to chelates, thermodynamically evaluates the feasibility and predictability of resistance or destruction of complexes under various conditions.

Static. A condition when a resource, compound, or process is stable. This is not a dynamic condition because positive gains and negative losses are both equal to zero.

Steady state. A condition under which a resource, compound, or physiological process has reached stability. This is a dynamic condition when positive gains equal negative losses, resulting in a stable process or quantity.

Stem. The central cylinder inside the cortex of roots and stems of vascular plants; composed of tissues inside the endodermis.

Stomatal conductance. The reciprocal of stomatal resistance.

Stomatal resistance. The impedence to gas movement through stomata.

Strain. An injury of a body part or organ resulting from excessive tension, force, influence, or factors causing excessive physical tension.

Stress. A physical or chemical factor that causes bodily tension and may be a factor in disease causation; a state or condition caused by factors that tend to alter an equilibrium.

Stressor. A factor of the environment (biotic or abiotic) that has the capability of causing bodily injury, disease, or aberrant physiology.

Stress proteins. Proteins induced to form as a result of the stress; for example, high-temperature-induced proteins.

Sun fleck. A short period of high radiation that interrupts a general background of low diffuse radiation. This is a common characteristic of the subcanopy environments of deciduous, conifer, and tropical forests.

Symplastic. Refers to the pathway that water and solutes may follow upon entering the root if they cross the plasmalemma and move through the interconnecting plasmodesmata from cell to cell.

Symport. Permeases (proteins) that transport two different molecules in the same direction simultaneously (usually across a membrane).

Symptom. Evidence of disease or physical disturbance; something that indicates the presence of a bodily disorder; an evident reaction of a plant to a pathogen or physiogen.

Syndrome. A group of factors, signs, and symptoms that collectively indicate a disease or disorder.

Synergistic. Two substances separately produce an effect but when used together cause a greater effect than the sum of the two acting separately.

Tolerance. Ability of a plant to survive a stress with little or no injury; the relative capacity of an organism to grow, thrive, or survive when subjected to an unfavorable environmental factor; the capacity to resist or sustain the effects of a disease or stress without dying or suffering irreparable injury.

Trait. A definable characteristic of a species commonly regulated by genetic factors.

Turgor (Ψ_p). The pressure exerted on the cell contents by the walls of a turgid cell; or conversely, the pressure exerted by the water in the vacuole against the cytoplasm and the cell wall.

Uniport. Permease that transports one substance in one direction across a membrane.

Unsatuation. The number of double bonds of a molecule.

van der Waals forces. Forces that arise from attractive and repulsive forces between molecules; the repulsive forces diminish with distance more rapidly than the attractive ones, leading to a net attraction at short distances; forces arise from the attraction of the positively charged nuclei of one molecule and the negatively charged electrons of another; the forces are relatively weak.

van'T Hoff relationship ($\Pi = -C_5 RT$). Relates osmotic potential to universal gas laws.

Water potential (Ψ). The difference between the activity of water molecules in pure distilled water at atmospheric pressure and 30°C and activity of water molecules in any other system; the addition of solutes to water decreases the water potential.

Water stress. Usually reserved for short-term situations in which transpiration rate exceeds the absorption rate of water; may be caused by a number of a different combination of soil and atmospheric factors.

Wilting. Reduction in cell turgor that reduces the rigidity of a plant; result of water stress.

Xerophytes. Plants that are able to resist drought and live in very dry habitats; often, the plants have thick cuticles, water storage tissue, and other morphological features that conserve water.

Zero stress. A level of exposure to an environmental factor that leads neither to injury nor to reduction in growth, yield, or value.

INDEX